Progress in Theoretical Chemistry and Physics

VOLUME 10

Honorary Editors:

W.N. Lipscomb *(Harvard University, Cambridge, MA, U.S.A.)*
I. Prigogine *(Université Libre de Bruxelles, Belgium)*

Editors-in-Chief:

J. Maruani *(Laboratoire de Chimie Physique, Paris, France)*
S. Wilson *(Rutherford Appleton Laboratory, Oxfordshire, United Kingdom)*

Editorial Board:

H. Ågren *(Royal Institute of Technology, Stockholm, Sweden)*
D. Avnir *(Hebrew University of Jerusalem, Israel)*
J. Cioslowski *(Florida State University, Tallahassee, FL, U.S.A.)*
R. Daudel *(European Academy of Sciences, Paris, France)*
E.K.U. Gross *(Universität Würzburg Am Hubland, Germany)*
W.F. van Gunsteren *(ETH-Zentrum, Zürich, Switzerland)*
K. Hirao *(University of Tokyo, Japan)*
I. Hubač *(Komensky University, Bratislava, Slovakia)*
M.P. Levy *(Tulane University, New Orleans, LA, U.S.A.)*
G.L. Malli *(Simon Fraser University, Burnaby, BC, Canada)*
R. McWeeny *(Università di Pisa, Italy)*
P.G. Mezey *(University of Saskatchewan, Saskatoon, SK, Canada)*
M.A.C. Nascimento *(Instituto de Quimica, Rio de Janeiro, Brazil)*
J. Rychlewski *(Polish Academy of Sciences, Poznan, Poland)*
S.D. Schwartz *(Yeshiva University, Bronx, NY, U.S.A.)*
Y.G. Smeyers *(Instituto de Estructura de la Materia, Madrid, Spain)*
S. Suhai *(Cancer Research Center, Heidelberg, Germany)*
O. Tapia *(Uppsala University, Sweden)*
P.R. Taylor *(University of California, La Jolla, CA, U.S.A.)*
R.G. Woolley *(Nottingham Trent University, United Kingdom)*

The titles published in this series are listed at the end of this volume.

EPR OF FREE RADICALS IN SOLIDS

EPR of Free Radicals in Solids

Trends in Methods and Applications

Edited by

Anders Lund
*Department of Physics and Measurement Technology,
Linköping University, Sweden*

and

Masaru Shiotani
*Department of Applied Chemistry,
Graduate School of Engineering, Hiroshima University, Japan*

KLUWER ACADEMIC PUBLISHERS
DORDRECHT / BOSTON / LONDON

Library of Congress Cataloging-in-Publication Data

ISBN 1-4020-1249-7

Published by Kluwer Academic Publishers,
P.O. Box 17, 3300 AA Dordrecht, The Netherlands.

Sold and distributed in North, Central and South America
by Kluwer Academic Publishers,
101 Philip Drive, Norwell, MA 02061, U.S.A.

In all other countries, sold and distributed
by Kluwer Academic Publishers,
P.O. Box 322, 3300 AH Dordrecht, The Netherlands.

Printed on acid-free paper

All Rights Reserved
© 2003 Kluwer Academic Publishers
No part of this work may be reproduced, stored in a retrieval system, or transmitted
in any form or by any means, electronic, mechanical, photocopying, microfilming, recording
or otherwise, without written permission from the Publisher, with the exception
of any material supplied specifically for the purpose of being entered
and executed on a computer system, for exclusive use by the purchaser of the work.

Printed in the Netherlands.

Contents

Preface   xi

Acknowledgments   xiii

Contributing Authors   xv

Part I: Trends in Methods   1

Chapter 1.
Continuous wave EPR of radicals in solids   3
Anders Lund and Wei Liu

    1. INTRODUCTION
    2. RADICAL STRUCTURE
    3. SATURATION PROPERTIES OF RADICALS
    4. INTERNAL MOTION OF RADICALS
    5. DIFFUSION OF RADICALS
    6. REFERENCES

Chapter 2.
Pulsed EPR of paramagnectic centers in solid phases   39
Marina Brustolon and Antonio Barbon

    1. INTRODUCTION
    2. THEORY

3. THE DETERMINATION OF ELECTRON SPIN LONGITUDINAL RELAXATION
4. THE DETERMINATION OF ELECTRON SPIN TRANSVERSE RELAXATION
5. THE DETERMINATION OF HYPERFINE INTERACTIONS
6. REFERENCES

Chapter 3.
Dynamical effects in CW and pulsed EPR   95
Nikolas P. Benetis

1. INTRODUCTION
2. TYPES OF DYNAMICAL PROCESSES ACCESSIBLE BY EPR
3. SELECTED EXAMPLES OF CONFORMATIONAL REORGANIZATION AND LIBRATION
4. FUNDAMENTAL DYNAMICAL PARAMETERS
5. MAGNETIC RELAXATION
6. CHEMICAL EXCHANGE
7. INERTIAL EFFECTS OF ROTATION
8. DYNAMICS BY CW-ENDOR SPECTROSCOPY
9. PULSED-EPR TECHNIQUES
10. LIBRATIONAL MOTION STUDIED BY ED-EPR
11. REFERENCES

Chapter 4.
Quantum effects in deuterium labelled radicals at low temperature   153
Masaru Shiotani and Kenji Komaguchi

1. INTRODUCTION
2. HIGH RESOLUTION ESR AND NUCLEAR SPIN-ROTATION COUPLING OF METHYL RADICALS
3. HYDROGEN ATOM - HYDROGEN MOLECULE PAIR FORMATION IN ARGON
4. PARA-HYDROGEN MOLECULE AND HIGH RESOLUTION ESR SPECTRA
5. JAHN-TELLER DISTORTION OF T_d AND D_{3h} MOLECULES AND H/D ISOTOPE EFFECTS
6. DEUTERIUM ISOTOPE EFFECTS ON METHYL HYDROGEN CONFORMATION
7. STATIC AND DYNAMIC STRUCTURES OF CYCLOHEXANE AND RELATED RADICAL CATIONS
8. HYDROGEN ATOM ABSTRACTION VIA TUNNELLING

9. REFERENCES

Chapter 5.
Xsophe - Sophe – XeprView: A computer simulation software suite for the analysis of continuous wave EPR spectra 197
Graeme R. Hanson, Kevin E. Gates, Christopher J. Noble, Anthony Mitchell, Simon Benson, Mark Griffin, Kevin Burrage

1. INTRODUCTION
2. THE XSOPHE X-WINDOW GRAPHICAL USER INTERFACE
3. SOPHE
4. ROLE OF FREQUENCY (AND TEMPERATURE) IN EXTRACTING SPIN HAMILTONIAN PARAMETERS
5. FUTURE DIRECTIONS FOR XSOPHE
6. CONCLUSIONS
7. REFERENCES

Chapter 6.
The calculation of the hyperfine coupling tensors of biological radicals 239
Fuqiang Ban, James W. Gauld, Stacey D. Wetmore and Russell J. Boyd

1. INTRODUCTION
2. THEORETICAL BACKGROUND
3. THEORETICAL STUDIES OF AMINO ACID RADICALS
4. CONCLUDING REMARKS
5. REFERENCES

Chapter 7.
Ab initio and density functional calculations of electronic g-tensors for organic radicals 267
Martin Kaupp

1. INTRODUCTION
2. THE HAMILTONIAN
3. QUANTUM CHEMICAL APPROACHES
4. PERFORMANCE OF AB INITIO AND DFT METHODS, VALIDATION STUDIES
5. APPLICATIONS TO BIOLOGICALLY RELEVANT RADICALS
6. CONCLUSIONS AND OUTLOOK

7. REFERENCES

Chapter 8.
Radiolabelled radicals derived from volatile organic compounds (VOCs) sorbed on reactive surfaces: implications for atmospheric chemistry and pollution control 303
Christopher J. Rhodes

1. INTRODUCTION
2. METHODS
3. ZEOLITES AND CLAY SURFACES
4. CARBON PARTICLES
5. ONGOING FURTHER STUDIES
6. CONCLUSIONS
7. REFERENCES

Part II: Trends in Applications 335

Chapter 9.
EPR studies of atomic impurities in rare gas matrices 337
Henrik Kunttu and Jussi Eloranta

1. INTRODUCTION
2. EXPERIMENTAL TECHNIQUES
3. ATOMIC IMPURITIES IN RARE GAS MATRICES
4. THEORETICAL TREATMENTS FOR ATOMIC IMPURITIES
5. REFERENCES

Chapter 10.
Organic radical cations and neutral radicals produced by radiation in low-temperature matrices 363
Vladimir Feldman

1. INTRODUCTION
2. EXPERIMENTAL APPROACHES AND OVERVIEW OF RESULTS
3. POSITIVE HOLE MIGRATION AND TRAPPING
4. MATRIX EFFECTS ON TRAPPING AND REACTIONS OF RADICAL CATIONS

5. SELECTIVITY OF THE PRIMARY RADIATION-INDUCED CHEMICAL EVENTS
6. CONCLUSIONS AND OUTLOOK
7. REFERENCES

Chapter 11.
Molecule-based exchange-coupled high-spin clusters 407
Takeji Takui, Hideto Matsuoka, Kou Furukawa,
Shigeaki Nakazawa, Kazunobu Sato, Daisuke Shiomi

1. INTRODUCTION
2. THEORETICAL BACKGROUND
3. SPECTRAL SIMULATION BASED ON A HYBRID EIGENFIELD METHOD AND PERTURBATION TREATMENTS
4. SOLUTION ESR SPECTROSCOPY FOR MOLECULAR HIGH-SPIN SYSTEMS WITH EXCHANGE INTERACTION COMPARABLE TO HYPERFINE INTERACTIONS
5. HIGH SPIN CHEMISTRY OF VARIOUS MOLECULAR CLUSTERS; UTILIZATION OF HIGH-FIELD/HIGH-FREQUENCY ESR AND PULSED ESR SPECTROSCOPY
6. METAL HIGH-SPIN CLUSTERS OF BIOLOGICAL IMPORTANCE; MANGANESE CLUSTERS IN PHOTOSYSTEM II
7. CONCLUSIONS
8. REFERENCES

Chapter 12.
High spin molecules directed towards molecular magnets 491
Martin Baumgarten

1. INTRODUCTION
2. BIRADICALS – THE TRIPLET STATE
3. TRIRADICALS - THE QUARTET STATE
4. TETRARADICALS- THE QUINTET STATE
5. HIGHER SPIN STATES, $S \geq 5/2$
6. CONCLUSION AND OUTLOOK
7. REFERENCES

Chapter 13.
Electron transfer and structure of plant photosystem II 529
Asako Kawamori

1. INTRODUCTION
2. METHODS APPLIED TO PHOTOSYNTHESIS
3. SAMPLE PREPARATIONS
4. STUDIED COMPONENTS
5. STRUCTURE OF PS II; COMPARISON WITH X-RAY ANALYSIS
6. REFERENCES

Chapter 14.
Recent development of EPR dosimetry 565
Nicola D. Yordanov and Veselka Gancheva

1. INTRODUCTION
2. PRINCIPLES OF EPR DOSIMETRY AS A METHOD FOR ESTIMATION OF THE ABSORBED DOSE AND FOR POST RADIATION PROCESSING DETECTION
3. SOLID STATE/EPR DOSIMETERS
4. TRENDS IN THE FUTURE STUDIES ON SS/EPR DOSIMETERS
5. IDENTIFICATION OF RADIATION PROCESSING OF FOODSTUFFS BY EPR
6. DETECTION OF PHARMACEUTICALS STERILIZED BY HIGH ENERGY RADIATION
7. EMERGENCY DOSIMETRY
8. CONCLUSIONS
9. REFERENCES

Chapter 15.
Optically detected magnetic resonance of defects in semiconductors 601
Weimin M. Chen

1. INTRODUCTION
2. BACKGROUND OF THE ODMR TECHNIQUE
3. VARIETIES OF THE ODMR TECHNIQUE
4. APPLICATIONS OF THE ODMR TECHNIQUE
5. RECENT DEVELOPMENTS AND TRENDS
6. REFERENCE

Index 627

Preface

The purposes of this book are to present *methods* and *applications* of modern EPR for the study of free radical processes in solids. The first part is concerned with trends in experimental and theoretical *methods*. In the first chapter continuous wave (CW) EPR and ENDOR methods for studies of radical structure in single crystals and powders are reviewed. Most of the following seven chapters give accounts of novel developments that so far are only available in the journal literature. The chapter by Brustolon and Barbon describes the different pulsed techniques as applied to radicals and spin probes in solid matrices. Methods to extract dynamical parameters from CW and pulsed EPR are summarised in the chapter by Benetis, which also contains an account of relaxation phenomena. One chapter deals with quantum effects in isotopically labelled radicals, which are especially manifested in high resolution EPR at low temperature. The theoretical interpretation of measured parameters, *i.e.* the g- and hyperfine coupling tensors in terms of electronic properties is treated in two chapters by Boyd and his coworkers, and by Kaupp. The trend is towards *ab initio* and density functional methods even for the biological systems. The new experimental methods to determine accurate g-tensors with high field EPR have been reviewed elsewhere but the modern theoretical treatment described here is probably less well known. Single crystal measurement is the most straightforward but not always applicable method for complex systems, where often only powder spectra can be obtained. For these systems analysis by simulation techniques based on exact diagonalisation are beginning to replace the previously used perturbation methods, as described in the chapter by Hanson and coworkers. Furthermore, myon spin resonance provides new means to considerably lower the detection limit in heterogeneous systems as described in a chapter by Rhodes.

In terms of *applications* there is a trend both towards simplification by using matrix isolation in frozen noble gas matrices with accompanying increase of resolution as illustrated in two chapters and towards studies of complex systems treated in the chapters by Takui and his coworkers, Baumgarten and Kawamori. In the former case Kuntu and Eloranta overview the matrix isolation technique for studies of atoms embedded in solid rare gases and Feldman presents recent development in the EPR studies of reactive intermediates from irradiation of moderately large organic molecules using matrix isolation. In the latter case two chapters address the issue of high-spin systems, the one by Baumgarten in organic systems, the other by Takui also in metal-based molecular clusters, which are fields that have strongly developed toward molecular magnets recently. The chapter by Kawamori describes studies of plant photosystem II by pulsed EPR and dual mode CW EPR, and pulsed electron-electron double resonance, the latter to obtain distances between radical pairs trapped after illumination. For EPR dosimetry and other kinds of quantitative EPR, the problem of calibration is an important issue that is addressed in the chapter by Yordanov and Gancheva. Recent developments of optical detection to lower the detection limit and to obtain time-resolution in the characterisation of defect centres in semiconductor materials are presented in the final chapter.

Anders Lund   Masaru Shiotani

December 2002

Acknowledgments

We were pleasured and privileged to work with a distinguished collection of authors in putting together this volume and would like to express our sincere appreciation for their cooperation. Dr Wei Liu and Dr Kenji Komaguchi have greatly assisted in the final editing of the chapters. We are indebted to several colleagues for reading and commenting the chapters. We also acknowledge the help of Dr R. Bjorklund and Ms Mari Löfkvist in the language corrections of several chapters, and the nice co-operation with the editorial staff of Kluwer Academic publishers.

Contributing Authors

Ban Fuqiang - Department of Chemistry, Dalhousie University, Halifax, Nova Scotia, Canada B3H 4J3

Barbon Antonio - Department of Physical Chemistry, University of Padova, Via Loredan 2 35131 Padova, Italy

Baumgarten Martin - Max Planck Institute for Polymer Research, Ackermannweg 10, D-55128 Mainz or PO Box 3148, D-55 021 Mainz, Germany

Benetis Nikolas P. - Department of Biological Chemistry, School of Medicine, University of Ioannina, 451 10 Greece

Benson Simon - Center for Magnetic Resonance, The University of Queensland, St. Lucia, Queensland, Australia, 4072.

Boyd Russell J. - Department of Chemistry, Dalhousie University, Halifax, Nova Scotia, Canada B3H 4J3

Brustolon Marina - Department of Physical Chemistry, University of Padova, Via Loredan 2 35131 Padova, Italy

Burrage Kevin - Department of Mathematics, The University of Queensland, St. Lucia, Queensland, Australia, 4072.

Chen Weimin M. - Department of Physics and Measurement Technology, Linkoping University, S-581 83 Linkoping, Sweden

Eloranta Jussi - Department of Chemistry, University of Jyväskylä, P.O. Box 35,FIN-40351 Jyväskylä, Finland

Feldman Vladimir - Karpov Institute of Physical Chemistry, 10 Vorontsovo Pole Str., Moscow 105064,Russia; Institute of Synthetic Polymeric Materials of RAS, 70 Profsoyuznaya Str., Moscow 117393, Russia; Department of Chemistry, Moscow State University, Moscow 119992, Russia

Furukawa Kou - Departments of Chemistry and Materials Science, Graduate School of Science, Osaka City University, Osaka 558-8585, Japan

Gancheva Veselka - EPR Laboratory, Institute of Catalysis,Bulgarian Academy of Sciences, 1113 Sofia Bulgaria

Gates Kevin E. - Department of Mathematics, The University of Queensland, St. Lucia, Queensland, Australia, 4072.

Gauld James W. - Department of Chemistry & Biochemistry, University of Windsor, Windsor, Ontario, Canada N9B 3P4

Griffin Mark - Center for Magnetic Resonance and the Department of Mathematics, The University of Queensland, St. Lucia, Queensland, Australia, 4072.

Hanson Graeme R. - Center for Magnetic Resonance, The University of Queensland, St. Lucia, Queensland, Australia, 4072.

Kaupp Martin - Institut für Anorganische Chemie, Universität Würzburg, Am Hubland, D-97074 Würzburg, Germany

Kawamori Asako - School of Science and Technology, Kwansei Gakuin University, Sanda, Japan

Komaguchi Kenji - Department of Applied Chemistry, Graduate School of Engineering, Hiroshima University, Higashi-Hiroshima 739-8527, Japan

Kunttu Henrik - Department of Chemistry, University of Jyväskylä, P.O. Box 35,FIN-40351 Jyväskylä, Finland

Liu Wei - Department of Physics and Measurement Technology, Linköping University, S-581 83 Linköping, Sweden

Lund Anders - Department of Physics and Measurement Technology, Linköping University, S-581 83 Linköping, Sweden

Matsuoka Hideto - Departments of Chemistry and Materials Science, Graduate School of Science, Osaka City University, Osaka 558-8585, Japan

Mitchell Anthony - Center for Magnetic Resonance, The University of Queensland, St. Lucia, Queensland, Australia, 4072.

Nakazawa Shigeaki - Departments of Chemistry and Materials Science, Graduate School of Science, Osaka City University, Osaka 558-8585, Japan

Noble Christopher J. - Center for Magnetic Resonance, The University of Queensland, St. Lucia, Queensland, Australia, 4072.

Rhodes Christopher J. - School of Pharmacy and Chemistry, Liverpool John Moores University, Byrom St., Liverpool L3 3AF, U.K.

Sato Kazunobu - Departments of Chemistry and Materials Science, Graduate School of Science, Osaka City University, Osaka 558-8585, Japan

Shiomi Daisuke - Departments of Chemistry and Materials Science, Graduate School of Science, Osaka City University, Osaka 558-8585, Japan

Shiotani Masaru - Department of Applied Chemistry, Graduate School of Engineering, Hiroshima University, Higashi-Hiroshima 739-8527, Japan

Takui Takeji - Departments of Chemistry and Materials Science, Graduate School of Science, Osaka City University, Osaka 558-8585, Japan

Wetmore Stacey D. - Department of Chemistry, Mount Allison University, Sackville, New Brunswick, Canada E4L 1G8

Yordanov Nicola D. - EPR Laboratory, Institute of Catalysis, Bulgarian Academy of Sciences, 1113 Sofia Bulgaria

PART I: TRENDS IN METHODS

Chapter 1

CONTINUOUS WAVE EPR OF RADICALS IN SOLIDS

Anders Lund and Wei Liu
Department of Physics and Measurement Technology, Linköping University, S-581 83 Linköping, Sweden

Key words: solid state, free radicals, EPR, ENDOR, structure, dynamics, adsorption, diffusion, simulations.

Abstract: Continuous wave (CW) EPR and ENDOR methods for studies of radical structure in single crystals and powders are reviewed. Improvements of the standard Schonland procedure to obtain hyperfine- and for I>½ nuclei also quadrupolar tensors from single crystal ENDOR measurements are described. A recently developed method to simulate powder ENDOR spectra of radicals is suggested as an alternative method of analysis when single crystal data are not available. Good agreement with experimental spectra for the l-alanine powder system was obtained. Control of power saturation has become essential in quantitative EPR, and a recently devised numerical procedure has proved useful to obtain relaxation times by the classical CW saturation method. Spectrum simulation methods are described that take into account the different influence of the microwave power on normal and spin flip EPR lines often appearing in the solid state. Internal motion of radicals, frequently occurring also in solid matrices, has been analysed with a slight modification of the established procedure for isotropic exchange in cases when the anisotropy of the hyperfine couplings is not affected by the dynamics. Results contained in thesis studies that are not easily available elsewhere concerning dynamics of neutral and ionic radicals in disordered matrices and in zeolites are presented. Recent results concerning the adsorption/desorption and diffusion of odd-electron NO and NO_2 are also presented. During the course of the work several computer programmes have been developed or modified to help the analysis.

1. INTRODUCTION

Continuous wave electron paramagnetic resonance, CW-EPR, has been in use since 1945 to study paramagnetic species. Other abbreviations are ESR and EMR, the latter standing for electron magnetic resonance. Studies by EPR of free radicals in solids have been made for nearly fifty years. Several treatises have appeared both on the general technique[1-9] and on applications[10-16]. In previous general treatises most emphasis is put on electronic structure, and how this structure is deduced from measurements of g, hyperfine and for nuclei with I > ½ also nuclear quadrupole coupling tensors. This subject is extensively treated also in works about radiation effects in solids, in the theory of magnetic resonance,[16] and in other chapters of this work. Specialised treatises of free radicals in solids involve studies of inorganic systems,[10] radiation chemistry,[11] radiation biophysics,[12] disordered systems,[13] radical ionic systems[14] and radicals on surfaces[15]. The book by Rånby and Rabek[16] about ESR spectroscopy in polymer research contains more than 2500 references.

Studies by EPR of radicals formed in radiation chemistry processes are ubiquitous, especially in early works.[10-12] The treatise[12] dealing with studies of primary radiation effects and damage mechanisms in molecules of biological interest is also valuable as a source of information of ENDOR spectroscopy of primary paramagnetic components formed after irradiation in liquid helium.

More recent literature reviews are contained in the series of books issued annually by the Royal Society of Chemistry.[17] Finally, the tabulated data in the Landolt Börnstein series[18] provide useful structural information. Progress in CW-EPR methodology is mainly in the field of high field EPR. The primary advantage in studies of free radicals is the possibility to resolve g-value anisotropy of a radical or to separate spectra of radicals with different g-factors. The technique is discussed in the chapter of optical detection of this volume and in more detail in a recent review by Smith and Riedi [Ref. 17, vol. 17 pp 164-204].

Although the measurement of parameters used to determine the electronic structure is considered in great detail in previous works some problems remain. In this chapter, the analysis of single crystal data by variants of the Schonland procedure is reviewed, hopefully in sufficient detail to enable the non-specialist to understand the principles of extracting g- hyperfine- and nuclear quadrupolar tensors from single crystal EPR and ENDOR data. Most emphasis is, however, on methods of analysis when either the nuclear quadrupolar interaction (nqi) is of comparable magnitude to the hyperfine energy, or when the hyperfine interaction (hfi) is exceptionally large. The first case with large nqi is of concern *e.g.* in ENDOR studies of radicals interacting with ^{14}N, ^{23}Na, ^{27}Al and other nuclei with nuclear spin I > ½. In this case there is no obvious experimental means to simplify the analysis, and an improved treatment of the data is required. The last case is less likely to cause problems with modern high field spectroscopy, but is still of relevance for X-band EPR or in the analysis of high-precision ENDOR data. Methods to treat this case have been developed several years ago,[19] but have not been generally applied in subsequent EPR studies of free radicals.

The measurement and analysis of dynamic parameters like relaxation times, rates of internal motion within radicals and external motion of radicals, are treated in varying detail in the literature.[1-9] The measurement of relaxation times by saturation is only briefly mentioned in modern literature, because of the development of pulsed EPR. However, in quantitative EPR, control of power saturation is of interest to ensure that spin concentration is correctly measured. The degree of saturation is governed by the relaxation times, and since almost all quantitative measurements are made with CW-EPR, relaxation data obtained with the same technique are relevant. Moreover, CW-EPR is standard equipment in laboratories specialising in *e.g.* radiation dosimetry, control of irradiated food or geological dating, while pulsed EPR may not be available. The methods of analysis of saturation curves developed several years ago have therefore been reviewed and an up-dated technique involving computer-based fitting to all the data, rather than the use of a few special data points used earlier is described.

The analysis of dynamics involving chemical exchange *i.e.* processes involving jumps of a nucleus from one environment to another one with a different environment in liquids is well treated in the general literature.[1,4,5] The analysis of chemical exchange in the general case when the nuclei feature anisotropic hfi is more complex. Fortunately, the dynamics of chemical exchange in solids frequently involves only nuclei with approximate isotropic hyperfine couplings, *e.g.* from hindered rotation of methyl groups in β-position to the radical centre and from ring puckering and other internal motion found in many cation radical systems. In this case the simulation method developed for exchange in liquids can be applied. Nuclei with anisotropic hyperfine couplings can be included in the simulations, provided that they do not participate in the exchange process. This simple approach termed "modified Heinzer" method described in the chapter has also proved useful to analyse cage effects on stability and dynamics of amino radicals generated in zeolites. The method fails, however, when nuclei with anisotropic hfi undergo exchange.

Dynamics in condensed phase, studied by introducing spin probes or labels in the material under study is a vast subject that has been treated in detail elsewhere, particularly in the biological field[20] and is omitted here. In this chapter only the results with NO_x probe molecules undergoing diffusion on surfaces are summarised.

2. RADICAL STRUCTURE

Radicals trapped in solid materials usually possess anisotropic g- and hyperfine tensors. For nuclear spin I > ½ quadrupole interaction also has to be taken into account according to the following spin-Hamiltonian

$$H = \mu_B BgS + IAS - g_N \mu_N BI + IQI \tag{1}$$

The directions of the g, A and Q principal axes do as a rule not coincide with each other or with the crystallographic axes.

2.1 Analysis of single crystal spectra

The most informative method to determine the tensor data is by single crystal measurements. The Schonland method[21] originally developed for the determination of the principal g-values in electron spin resonance is applicable also for the determination of hyperfine coupling tensors from ESR and ENDOR data and in principle also for nuclear quadrupole tensors from ENDOR of nuclei with I > ½. Non-linear least squares procedures are better suited for this case, especially when the nqi is of comparable magnitude to the hfi.

2.1.1 Schonland method

The g-factor with the magnetic field B along the unit vector **l** is given by

$$g^2 = \mathbf{l}\,g^2\mathbf{l} \tag{2}$$

1. Continuous wave EPR of radicals in solids

In the Schonland procedure the g-factors are measured by ESR as functions of the angle of rotation of the crystal with respect to the magnetic field in three different planes. It is usual, but not necessary, to make the measurements in mutually orthogonal planes xy, yz and zx. For rotation in the xy-plane one has

$$g^2 = T^g_{xx}\cos^2\theta + T^g_{yy}\sin^2\theta + 2T^g_{xy}\sin\theta\cos\theta \qquad (3)$$

where θ is measured from the x-axis. The expressions for the orientation dependence of g^2 in the yz and zx planes are obtained by cyclic permutations of the subscripts in (3). In Schonland's original treatment a fitting of the equations to the data was performed individually for each plane. The principal values and directions were obtained by diagonalisation of the $\mathbf{g}^2$ tensor. The principal g-values are obtained as the square root of the corresponding $\mathbf{g}^2$ values. The principal directions for $\mathbf{g}^2$ and $\mathbf{g}$ coincide. Computer programs available today make a simultaneous fit to all data.[22] It is then not necessary to employ orthogonal planes for the measurements. Moreover, error estimates of the principal values and directions are provided.

In the case of dominant electron Zeeman energy the hyperfine energy to first order for a state (M,m) is

$$E(M,m) = g\mu_B BM + G_M m \qquad (4)$$

M and m denote the electronic and nuclear quantum numbers, respectively. The quantity G_M, given by[23]

$$G_M^2 = |(\frac{M}{g}\mathbf{gA} - v_N\mathbf{1})(\frac{M}{g}\mathbf{Ag} - v_N\mathbf{1})| \qquad (5)$$

contains contributions from the hyperfine tensor **A** and from the nuclear Zeeman term $v_N = g_N \mu_N B/h$, both given in frequency units. **1** is the unit tensor.

Depending on the relative magnitudes of the hyperfine and nuclear Zeeman terms, modifications of the procedure are adopted depending also if EPR or ENDOR is employed.

EPR: when $v_N \ll$ hyperfine coupling K, the energy for the transition (M,m→M-1,m) is[24]

$$\Delta E = g\mu_B B - K \cdot m \tag{6}$$

with

$$K^2 g^2 = |\mathbf{T^h}| \tag{7}$$

$$\mathbf{T}^h = \mathbf{g}\mathbf{A}^2\mathbf{g} \tag{8}$$

The measurement of the hyperfine coupling K, as a function of the angle of rotation of the crystal, makes it possible to obtain the tensor $\mathbf{T}^h$ by the Schonland procedure. The $\mathbf{A}^2$ tensor is then obtained by multiplying (8) from the left and right with $\mathbf{g}^{-1}$. The tensor $\mathbf{A}^2$ is then diagonalised to obtain the principal values and directions. This procedure (neglect of v_N) is that usually adopted to obtain hyperfine coupling tensors from ESR measurements.

The opposite case with $v_N \gg$ hyperfine coupling is rare at X-band and lower frequencies, but is of interest for measurements at the higher frequency bands (Q,W...) which have become commercially available.

In the case g is isotropic one has

$$K = |A| \tag{9}$$

The relative signs of the principal values of the tensor can then be obtained. In the case of proton hyperfine coupling splittings for example two lines with m=±½ appear. If they cross over in a plane of rotation one can conclude that there are principal values of opposite signs.

When none of the two extreme cases is applicable as for example in X-band measurements on carbon-centred radicals with hyperfine structure from a hydrogen atom in the α-position a non-linear fit employing the general equation (5) must be employed.[25] The analysis is difficult, however, because of the admixture of nuclear-spin states, resulting in the breakdown in the $\Delta m = 0$ selection rule.[9] Q-band measurements may then be attempted to reduce the intensity of the forbidden transitions.[26]

ENDOR: In this case it is not necessary to make any assumption regarding the relative magnitudes of the nuclear Zeeman and hyperfine energies. For an I=½ nucleus two lines with frequencies G_M appear corresponding to the two values of M=±½. In case of isotropic g Eq. (5) simplifies to

1. Continuous wave EPR of radicals in solids

$$G_M^2 = |\mathbf{T}^2| \tag{10}$$

with $\mathbf{T} = \mathbf{M}\cdot\mathbf{A} - v_N\cdot\mathbf{1}$

In case v_N is constant during the measurements, *i.e.* the static magnetic field is the same, the tensor $\mathbf{T}^2$ can be obtained by a Schonland type fit. After diagonalization, the absolute values of the principal values are obtained as

$$|T_i| = |MA_i - v_N| \tag{11}$$

The ambiguity in evaluating the principal values A_i from (11) can be resolved if both of the two ENDOR transitions corresponding to $M=\pm\frac{1}{2}$ can be observed. In this case one has

$$T_i^2(M = -\tfrac{1}{2}) - T_i^2(M = \tfrac{1}{2}) = 2v_N \cdot A_i \tag{12}$$

The absolute signs are not obtained experimentally, since one cannot assign specific values of M to the observed transitions. For many systems of practical interest the signs are known from theoretical considerations, for instance of α-H($A_i < 0$) and β-H($A_i > 0$) couplings in carbon-centred π-electron radicals, and the sign test (12) is rarely applied.

More complicated equations result in the presence of nqi from $I>\frac{1}{2}$ nuclei.[23] When the ENDOR frequency contributions from the hyperfine and nuclear Zeeman interactions, dominate over that caused by the nqi the ENDOR transition (M,m→m-1) occurs at a frequency given by[12,23,27]

$$hv = G_M + 3P_M\left(m - \frac{1}{2}\right) \tag{13}$$

$$G_M^2 P_M = \left|\left(\frac{M}{g}\mathbf{g}\mathbf{A} - v_N\mathbf{1}\right)\mathbf{Q}\left(\frac{M}{g}\mathbf{A}\mathbf{g} - v_N\mathbf{1}\right)\right| \tag{14}$$

A simple method involves separate measurements of G_M and P_M as illustrated for the case of I=1 corresponding to the energy diagram in Fig 1.

The magnitudes of G_M and P_M are obtained from v_1 and v_2 or v_3 and v_4 (Fig. 1) as a function of orientation in three crystal planes. The quantities can therefore be determined separately as described for the I=½ case. When g is isotropic eq. (14) reads

$$G_M^2 P_M = |TQT| \tag{15}$$

where **T** is given by eq. (10)

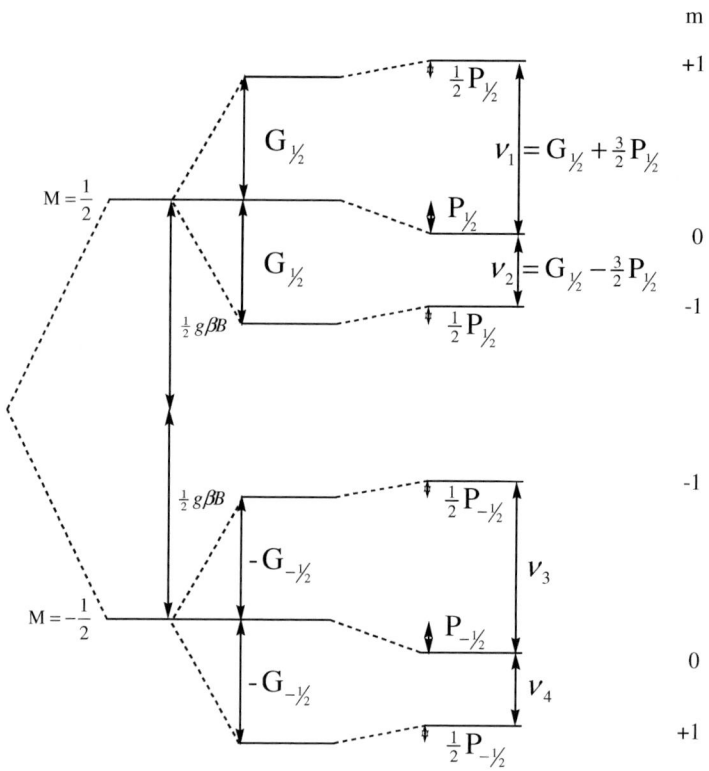

Fig. 1 Energy level diagram for a radical with S=½ and I=1. The ENDOR frequencies v_1 and v_2 for M=½ are given.

The ENDOR tensor **T** is first obtained from the Schonland fit of G_M to the experimental data. Next the tensor $T^q = T \cdot QT$ is calculated from a similar Schonland fit to the experimental $G_M^2 P_M$ data. Finally $Q = T^{-1} T^q T^{-1}$ is calculated and diagonalised to obtain the principal nuclear quadrupole coupling values and the directions of the principal axes.

When g-factor anisotropy occurs this simple approach fails. In the favourable situation when the M=±½ pairs of transitions are observable one may employ

$$G^2\left(M=\frac{1}{2}\right)+G^2\left(M=-\frac{1}{2}\right)-2v_N^2 = \mathbf{l}\frac{\mathbf{gA^2g}}{2g^2}\mathbf{l} \qquad (16)$$

This equation shows that a Schonland fitting procedure will give the elements of $\mathbf{T^a=gA^2g}$ from which $\mathbf{A^2=g^{-1}T^ag^{-1}}$ is obtained as discussed above. This method, suggested several years ago,[27] has become outdated by the more general, non-linear least squares methods described below.

2.1.2 Non-linear least squares method

Using equations (5) and (14) a direct fit of the theoretical ENDOR frequencies (13) to the experimental data by a non-linear least squares procedure can be made. This procedure has been adopted by Sørnes to determine the ^{14}N quadrupole coupling in a nitrogen-containing radical.[28]

In this case the derivatives of eq. (13) with respect to the hyperfine coupling tensor elements are required. Analytical formulae are given in,[25,28] and the latter also contains equations for error estimates of the principal values and direction cosines. This method is probably more satisfactory from an accuracy point of view than the Schonland fitting procedure. But, when the quadrupolar interaction is of the same magnitude as the hyperfine and/or nuclear Zeeman interactions, the perturbation treatment leading to equation (13) cannot be applied. This case occurs frequently *e.g.* for halogenated inorganic radicals, but the accurate method of analysis developed by Byberg *et al*[19] does not seem to have been adopted by others. Equivalent treatments employed by us are therefore summarised.

First order analysis: In case that the nuclear and hyperfine interactions are much weaker than the electronic Zeeman energy a perturbation treatment is still possible with the perturbation operator

$$H' = \mathbf{I A u}S_u + \mathbf{IQI} - g_N\beta_N\mathbf{B}\cdot\mathbf{I} \qquad (17)$$

Here S_u is the component of S along the effective Zeeman field direction $\mathbf{u=gl}/g$.

This method has been used, mostly, to analyse EPR and ENDOR powder spectra, in which case the tensor data are either known from other measurements or obtained by trial and error simulations of the spectra.[29,30]

The analysis of single crystal ENDOR data under those conditions has been only briefly described[31] and is therefore presented in more detail below.

The ENDOR frequencies are given by

$$\nu_{theory}(M,i-k) = (E(M,i) - E(M,k))/h \qquad (18)$$

Here i and k denote states belonging to the same M-manifold, *i.e.* either $M=+\frac{1}{2}$ or $M=-\frac{1}{2}$. In general the states are mixtures of nuclear spin states. The experimental frequencies ν_{exp} are obtained as in the Schonland method with the magnetic field at different angles in three different planes. The error sum

$$\sum (\nu_{exp} - \nu_{theory})^2$$

is then minimised as a function of the elements of **A** and **Q**. For this purpose the derivatives of the frequencies with respect to the hyperfine and nuclear quadrupole tensor elements are required. These can be obtained as

$$\frac{\partial E_i}{\partial A_{\alpha\beta}} = \sum_{m,n} \frac{\partial E_i}{\partial H'_{mn}} \frac{\partial H'_{mn}}{\partial A_{\alpha\beta}} \qquad (19)$$

E_i denotes the nuclear energy state for a given electronic quantum number M. The derivatives of the matrix elements of H' are obtained in analytical form from expressions given elsewhere.[27]

From the equations given in [32] it follows (* denotes complex conjugate)

$$\frac{\partial E_k}{\partial H'_{mn}} = c^*_{mk} c_{nk} \qquad (20)$$

where

$$|M,k\rangle = \sum_{m=-I}^{I} c_{mk} |m\rangle \qquad (21)$$

is the eigenstate vector corresponding to the energy E_k expressed as a combination of nuclear spin states m>. The derivatives of the frequencies with respect to the tensor elements are thus obtained analytically. An analysis of the ^{23}Na hyperfine and quadrupolar interactions with CO_2^- in a $CO_2^- \cdots Na^+$ complex trapped in sodium hydrogen oxalate single crystal was performed by this method,[31] with the exception that the derivatives were calculated numerically, rather than with the analytical method outlined here.

In this particular case it was sufficient to consider the nuclear transitions between adjacent nuclear levels, *i.e.* "$\Delta m = 1$", where the quotes signify that m is not a good quantum number. A possible complication in other cases could be the occurrence of "forbidden" transitions between nonadjacent states due to the state mixing.

Higher order and exact analysis: Higher order perturbation theory can be applied to the case when the first order analysis is inadequate, *e.g.* when the hfi is large. The second order corrections given in [33-35] are in a suitable form to be applied in single crystal analysis *e.g.* in ENDOR measurements of ^{14}N hyperfine and quadrupole couplings. Energy cross terms have to be taken into account in the case of several interacting nuclei.[36,37]

An iterative fitting procedure has been described by Byberg *et al.*[19] The fitting procedure developed by Fox, Holuj and Baylis[32] can also be applied in an analysis involving exact diagonalisation of the spin-Hamiltonian H at each crystal orientation.

They find for the derivatives of the energies with respect to the parameters P_i

$$\frac{\partial E_k}{\partial P_i} = \left\langle k \left| \frac{\partial H}{\partial P_i} \right| k \right\rangle \tag{22}$$

where $|k\rangle$ is the eigenvector corresponding to the eigenvalue E_k. Insertion of (20) in (22) gives for the transition $k \to l$

$$\frac{\partial \nu_{kl}}{\partial P_i} = \left(\mathbf{C}_k^\dagger \frac{\partial \mathbf{H}}{\partial P_i} \mathbf{C}_k - \mathbf{C}_l^\dagger \frac{\partial \mathbf{H}}{\partial P_i} \mathbf{C}_l \right) / h \tag{23}$$

$\mathbf{C}_k$ and $\mathbf{C}_l$ are the eigenvectors of the form (21) and † denotes Hermitean conjugate. The Hamiltonian matrix elements are linear functions of the parameters P_i. The derivatives $\partial \mathbf{H}/\partial P_i$ are therefore simply computed by putting $P_i = 1$, $P_j = 0$ $j \neq i$ in the $\mathbf{H}$ matrix.

Until now the fitting procedure has not been applied to the free radical case – the system under study by Fox *et al*[32] was a rare earth metal ion complex. The method is, however, a straightforward extension of that used in[31] and should be useful particularly in ENDOR where the line positions are accurately measured. The analysis of experimental data collected as with the approximate methods, involves the construction and diagonalisation of the Hamiltonian matrix, followed by the calculation of the derivatives used in the least squares fitting. The procedure is iterated until convergence of the parameters P_i is achieved – *i.e.* of the hyperfine and quadrupolar coupling elements. This method can also handle the interaction with several nuclei. An advantage of this method is that a statistical error analysis can be easily implemented.[32]

2.1.3 Computer programs

The computer programs that the authors are aware of contain several sections, (1) to provide the experimental data (g-factors, hyperfine couplings, ENDOR frequencies) as function of crystal orientation, (2) to perform a least squares fitting of a theoretical model to the data, and (3) to make an error analysis of the parameter values, *i.e.* the principal values and direction cosines of the principal axes of the coupling tensors. The last section is model independent and similar code can be applied in all cases. The first section requires minor modification depending on the model. In general the programs differ most in the corresponding second section. For the Schonland type analysis of the g and hyperfine tensors obtained by EPR and ENDOR, the Fortran program MAGRES[22] is the best documented. A similar program written in Quickbasic used in this laboratory is a modified version of an older Fortran code written by Claesson.[25] Input data consists of crystal orientation and g-factors, hyperfine splittings or ENDOR frequencies, measured in at least three planes. The planes need not be mutually orthogonal. The MAGRES program provides error estimates of the principal values and direction cosines. In an experimental version of the Quickbasic program constraints like axial symmetry can be imposed. The advantage with the programs for the Schonland method is the simple input data. They are particularly suited for analysis of g and hyperfine data, although the nuclear quadrupole tensor of I>½ nuclei can, in principle also be obtained by these programs if at least two ENDOR frequencies are measured at each orientation (*e.g.* those given in Fig. 1 for I=1). ENDOR data of radicals containing I>½ nuclei can, however, be analysed more conveniently by the program ENDPAQ, written by Sørnes.[28] The input is slightly more complex and consists of the crystal orientation, the ENDOR frequency, the magnetic field strength, and the quantum numbers for the transition $|M,m\rangle \rightarrow |M,m-1\rangle$. All observed transitions for different M and m quantum numbers can be included simultaneously in the fit. This is an advantage, since it helps to avoid the ambiguity that may occur in the analysis with the Schonland procedure (eq. 11). It is not necessary to measure pairs of frequencies at each orientation, contrary to the Schonland method. In a version used by the authors, the influence of an anisotropic g according to eq. (5) and (14) is taken into account. Error estimates of the parameters are computed. The program is thus more flexible than the programs based on Schonland´s method. The limitation is that the perturbation theory on which it is based is invalid when the quadrupolar and hyperfine interactions are of similar magnitude. Programs that overcome this limitation have been developed previously by Byberg *et al* for the analysis of ESR data[19] and recently by us for ENDOR analysis.

The first program (AQFIT) uses eq. (19) to obtain the necessary derivatives for a non-linear least squares fit. The input data is similar to that employed in the ENDPAQ program with the difference that the nuclear levels between which the ENDOR transition occurs are specified by integer numbers in ascending energy order, rather than by the nuclear quantum number. Virtually identical tensors are computed with the AQFIT and ENDPAQ programs when the theoretical model for the latter is valid. This is not always the case, however. ^{14}N hyperfine couplings for example, vary much in magnitude, while the principal values of the ^{14}N quadrupolar tensor are typically of the order 1 MHz. In a case under study of a radical with ^{14}N ENDOR frequencies in the range 1.5-5.5 MHz the principal values of the hyperfine coupling tensor differed by up to 0.3 MHz between the methods. The difference was larger than the standard deviations of the tensor elements by a factor 4-5. The accuracy of the ENDPAQ fit can in this case be increased by including second order corrections to the nuclear energy terms, but this treatment is not always sufficient.[31]

The second method, based on the theory by Fox, Holuj and Baylis[32] has been implemented in a preliminary programme using available Fortran libraries for matrix diagonalisation and non-linear fitting.

2.2 Analysis of powder ENDOR spectra

The possibility to apply ENDOR for studies on radical systems in glassy, polycrystalline or amorphous samples has been considered in several studies, including radical ions on catalytic surfaces,[38-41] in frozen matrices,[42-49] in molecular crystals[50] and in biological systems[51,52].

In the ENDOR spectra of transition metal ions and other systems with pronounced g-factor anisotropy one can take advantage of orientational selectivity in the analysis.[53] In this case the ENDOR spectrum obtained at a specific magnetic field setting contains contributions from a limited range of orientations, e.g. along a line or in a plane, for an axially symmetric system with the field set at $g_\parallel$ and $g_\perp$. Methods to analyse ENDOR spectra for this case have been described.[53-59] In many free radical systems the g-anisotropy is quite small and it is then impossible to obtain single crystal like ENDOR spectra, which instead are made up of a large number of orientations. Computer simulations are then essential for the analysis. A programme developed by Erickson has been used in the simulations for this chapter.[60]

The theoretical method has recently been described in detail.[61] Several nuclei with any value of nuclear spin I can be handled. In brief a spin Hamiltonian of the type (17) is assumed for each nucleus. As mentioned this Hamiltonian arises when the electronic Zeeman term is the dominating one, while the hyperfine, quadrupolar and nuclear Zeeman interactions are treated, simultaneously, as a joint perturbation. Thus, no assumption need to be done regarding their relative magnitude or the relative orientation of the principal axes of the tensors. The ENDOR frequencies for the M=±½ electron states are obtained by diagonalisation of the perturbation matrices of each nucleus as described previously.[60,61] The ENDOR intensity to first order for the transition between the states |M, a> ⇔ |M, b> in eq. (21) is

$$\overline{W}_{ab}^2(\theta,\varphi) = \frac{1}{2}\left(\frac{B_2}{B}\right)^2 (\alpha^* T^2 \alpha - \alpha^* Tl \cdot \alpha Tl), \tag{24}$$

where the bar indicates it applies to the powder. **l** is the unit vector of the static magnetic field (<u>not</u> the radiofrequency field) **T** is defined in eq. (10) and α is a complex vector with the components <a |I_x | b>, <a |I_y | b> and <a |I_z | b>.

The powder line shape at the ENDOR frequency ν and static magnetic field B given previously[60,61] is rewritten here in a slightly different form as:

$$Y'(B,\nu) = \int_\theta \sin\theta \int \sum_{ij} s(B - B_{ij}) V_{ij}^2 \left\{ \sum_k t(\nu - \nu_{ik})\overline{W}_{ik}^2 + \sum_l t(\nu - \nu_{jl})\overline{W}_{jl}^2 \right\} \tag{25}$$

Indices *i* and *k* denote nuclear states within the M=-½ manifold, while *j* and *l* refer to levels with M=½. The EPR line shape function (in absorption) s is a weighting function to select the transitions that contribute to the ENDOR signal, and thus gives rise to the angular selection at the magnetic field B. Only those nuclear transitions that have an energy level in common with the EPR transition i → j will contribute to the signal. The line shape function, t is in first derivative to obtain the first derivative ENDOR spectrum Y'(B, ν).

In Fig. 2b and c are shown powder ENDOR spectra obtained with B at the centre (b) and the outermost high field (c) lines of the EPR spectrum in Fig. 2a of irradiated alanine together with their simulations.[62] At the former position ENDOR signals of the two radicals R1 = $CH_3\dot{C}HCOOH$ and R2 = $H_3N^+\dot{C}(CH_3)COO^-$ contribute to the spectrum. One reason for the lack of agreement between the relative intensities of the experimental and simulated curves is that the influence of relaxation is not taken into account in the simulations. The line positions are, however, in good agreement, suggesting that ENDOR powder spectroscopy in conjunction with simulations is a usable alternative in cases where single crystal analysis is not feasible. On request, the Fortran 77 programme written by R. Erickson is available.[60]

3. SATURATION PROPERTIES OF RADICALS

Quantitative measurements of spin concentrations are performed by CW-EPR, see chapter by Yordanov *et al.*, often employing high microwave powers. Microwave saturation may occur because of slow relaxation. It is therefore of interest to briefly review the methods used in the past to determine relaxation times and to propose a variant that is convenient to apply in quantitative ESR. It is also of interest to compare the relaxation times obtained in this manner by those measured directly by pulsed EPR and to be able to predict spectral line shapes at saturation.

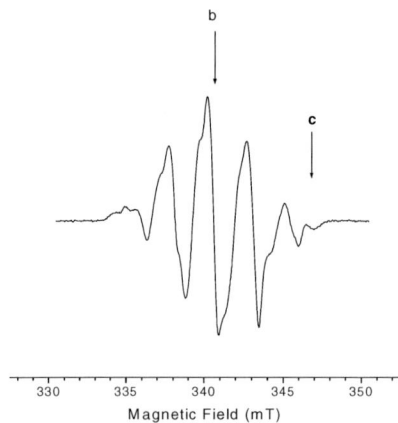

Fig. 2a: First-derivative X-band EPR spectrum from a polycrystalline sample of alanine. x-irradiated (dose 63 kGy) at 295 K and measured at 221 K. The arrows indicate field positions of ENDOR spectra in Fig. 2b and Fig. 2c.

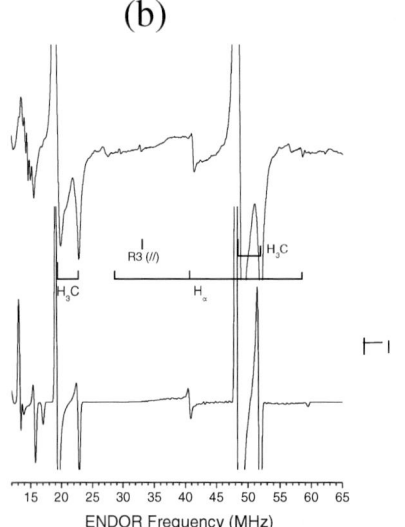
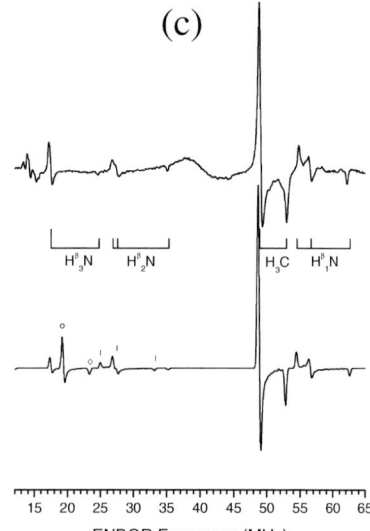

Fig. 2b: Experimental (top) and simulated powder ENDOR spectrum due to radical R1 at 221 K. The experimental spectrum was obtained by saturating the central EPR resonance line (at 340.70 mT, fig 2a). The stick spectra represent the assigned hyperfine coupling tensor principal values as deduced from the previous single crystal study for radicals R1 and R3 (parallel feature, only). Other weak resonance features are due to radical R2, as shown in Fig. 2c. ***2c***: Experimental (top) and simulated powder ENDOR spectrum due to radical R2 at 221 K. The experimental spectrum was obtained by saturating the very high-field EPR resonance line (at 346.79 mT, fig 2a). The stick spectra represent the assigned hyperfine coupling tensor principal values as deduced from the previous single crystal study. In the simulated spectrum (bottom) the lines marked with ○ and | are the low-frequency branches of CH_3 and NH^β_1 ENDOR transitions, respectively. They are not observable in the experimental spectrum due to relaxation effect.

1. Continuous wave EPR of radicals in solids

3.1 Saturation parameters

The saturation properties of a radical depend on the values of the spin-lattice and spin-spin relaxation times T_1 and T_2. To obtain these values a measurement of the signal intensity as a function of the amplitude B_1 of the oscillating microwave magnetic field is made. Two limiting cases termed homogeneous and inhomogeneous broadening give rise to different shapes of the saturation curves.[63,64] In the first case the resulting ESR absorption is a Lorentzian function.

$$g(B - B_0) = \frac{1}{1 + s^2 + \left(\frac{B - B_0}{\Delta B_L}\right)^2} \quad (26)$$

where $s^2 = \gamma^2 B_1^2 T_1 T_2$, $T_2 = (\gamma \Delta B_L)^{-1}$, and $\Delta B_{pp} = (2\sqrt{3}) \Delta B_L$

Experimentally it is convenient to use the linewidth ΔB_L rather than T_2 as parameter. In the usual case that the first derivative is recorded, it is customary to employ the peak-peak linewidth ΔB_{pp}. The microwave power P is measured, and the value of B_1 is calculated as

$$B_1 = K\sqrt{P}$$

where the constant K depends on the type of instrument used. From an experimental point of view it is therefore easier to employ microwave power P, rather than B_1 as a variable and P_0 as saturation parameter using $s^2 = P/P_0$ in (26).

One detail that is occasionally overlooked is that the power dependence is different for the absorption and the first derivative. The first derivative amplitude as function of P is

$$A = \frac{C\sqrt{P}}{(1 + \frac{P}{P_0})^{3/2}} \quad (27)$$

C is a constant, the maximum amplitude occurs at $P = P_0/2$, while for the absorption signal maximum occurs at $P = P_0$. General equations for the power dependence of higher derivatives have been obtained.[64]

The relaxation broadening determined by T_2 in the homogeneous case is often superimposed on other broadening effects, caused *e.g.* by unresolved hyperfine structure. In this inhomogeneously broadened case, the observed lineshape is composed of individual lines at different resonance fields. The lines at these fields are usually assumed to have intensities corresponding to a Gaussian distribution, while the individual lines (the spin packets) have a Lorentzian shape. When the spin packet linewidth is much smaller than the Gaussian width, the amplitude is

$$A = \frac{C\sqrt{P}}{\sqrt{1+\frac{P}{P_0}}} \qquad (28)$$

The amplitude does not depend on if the spectrum is in absorption or derivative form. The saturation curve grows monotonously towards a limiting value.

It is often observed that a fitting of the data cannot be made to any of the two cases. Two methods are proposed to handle this case. In the first the amplitude is assumed to be

$$A = \frac{C\sqrt{P}}{\left(1+\frac{P}{P_o}\right)^\alpha} \qquad (29)$$

where a reasonable range for the empirical parameter α is $\frac{1}{2} \leq \alpha \leq \frac{3}{2}$, when the first derivative is recorded. The parameters are readily obtained by employing a least squares fit of eq. (29) to the saturation curve data as shown in Fig. 3a for the γ-irradiated glycyl-glycine powder with the experimental data shown as filled squares. The estimated value of $\alpha = 0.83 \pm 0.01$ for the data in Fig. 3 indicates that an intermediate situation between homogeneous and inhomogeneous saturation occurs. This probably implies that the assumption that the spin packet linewidth is much smaller than the Gaussian width is not valid.

The limitation with this phenomenological analysis is that it mainly serves to characterise saturation curves with a few parameters. When values of T_1 and T_2 are required the analysis must be modified.

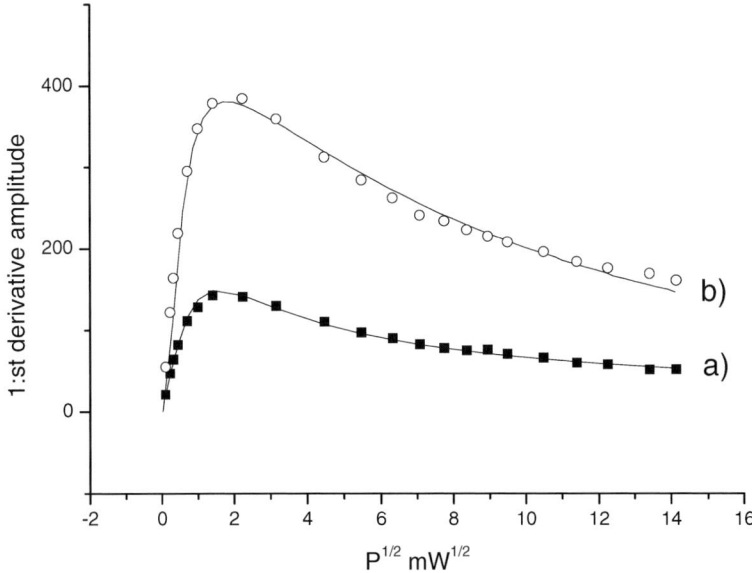

Fig. 3 EPR microwave saturation curves at X-band obtained from γ–irradiated glycylglycine recorded under slow passage conditions (ν_{mod}= 35 or 270 Hz) at room temperature. a) fitting to experimental data using eq (29), α=0.83, P_0=0.5 mW. b) fitting using eq (30), ΔB_G=3.0, ΔB_L=0.3G, P_0=0.3 mW.

Several variants of similar methods have been developed for this case. Castner calculated the amplitude of the absorption signal as a function of the microwave field strength.[65] The lineshape is then a convolution of a Gaussian with a Lorentzian function.

Maruani has extended Castner's method to multiple-level systems (S>½) and applied the analysis to polycrystalline systems using graphical methods.[66] These methods were designed for the case when absorption curves could be obtained but were also applied to the analysis of derivative spectra, in cases where the difference between the saturation behaviour of the absorption and the derivative was small.[67]

Zhidkov et al have taken into account the effect of field modulation to obtain saturation curves for the derivative signal of Lorentzian spin packets with a Gaussian or Lorentzian distribution.[68] Formulae for the first derivative amplitude were developed. Bowman et al have considered the case of second derivative spectra.[69] The following numerical procedure devised by Maruani (unpublished) makes use of all measured data, rather than the few used in earlier treatments.

The line shape function g is assumed to be

$$g(B-B_0) \propto \int_0^\infty \frac{e^{-(\frac{B'-B_0}{\Delta B_G})^2}}{1+\beta_0 s^2 + (\frac{B-B'}{\Delta B_L})^2} dB' \tag{30}$$

in which the nominator is the Gaussian distribution of width ΔB_G centred at B_0 of spin packets and the denominator is the Lorentzian absorption spin packet line centred at B'. Here β_0 is the transition probability, and $s^2 = P/P_0$ as in the homogeneous case. In the case of a single EPR line, $\beta_0 = 1$ is assumed. The function is evaluated, numerically, as a function of the field B, for a fixed value of the microwave power P, using a published algoritm.[70] The derivative shape function is then obtained, numerically, whereafter the maximum amplitude is obtained, also numerically. The procedure is repeated for different values of P to obtain the saturation curve. Finally, a curve-fitting to the experimental data is carried out to find the parameters $\Delta B_L, \Delta B_G$ and P_0. In the provisional programme written by Maruani the fitting of the model to the experimental data is made by manually adjusting the parameters, Fig. 3b.

The analysis is based upon the assumption of slow passage,[71] i.e. $2\pi \nu_m \cdot B_m \ll \Delta B_L / \sqrt{T_1 T_2}$, where ν_m is the frequency and B_m the amplitude of the field modulation. CW-EPR spectra are usually recorded with $\nu_m = 100$ kHz, but to obtain reliable values of relaxation times it is preferable to use the lowest modulation frequency available on the instrument. The data in Fig. 3 were obtained at $\nu_m = 35$ and 270 Hz.

Relaxation parameters obtained by CW and pulsed EPR of systems of interest as ESR dosimeters[72,73,74] are compared in Table 1. In the pulsed measurements biexponential decay occurred frequently. The shortest relaxation times agree reasonably well with the CW-data. In the CW measurements the slow passage condition was not satisfied in all cases. Nevertheless, the comparison indicates that the CW-method is usable in the context of quantitative ESR measurements to estimate the relaxation properties.

3.2 Microwave power effects on ESR spectral shape

To achieve high sensitivity in CW-ESR it is advantageous to apply high microwave power. Under microwave saturation conditions the intensity of spin flip satellite lines, particularly from distant protons, can increase considerably. The spectral shape is affected, which introduces uncertainties in the read-out of the intensity. It is therefore an advantage if the effects of power saturation can be taken into account by simulation. The theory for microwave saturation of inhomogeneously broadened lines mentioned above has proved useful also to simulate spectra of radicals under saturation conditions. Two approaches have been attempted.

In the first method,[75] except for a microwave saturation factor the simulation procedure is identical to that for simulation in the unsaturated case, with a line shape function that is unaffected by saturation. The influence of microwave saturation is taken into account with a saturation factor as in eq. (31).

$$S(B,P) = \sqrt{P} \sum_{k=1}^{L} \frac{\beta_k (B - B_k)}{(1 + \beta_k P / P_0)^\alpha} \qquad (31)$$

Table 1. Comparison of longitudinal (T_1) and transverse (T_M, T_2) relaxation times for X-irradiated single crystal samples determined with pulsed and CW-EPR at room temperature. The changes in T_1 and T_M with magnetic field orientation in the pulsed experiments are indicated. Two different relaxation times τ_a and τ_b were often observed in the pulsed experiments. The shortest times agree reasonably well with the CW-data.

Sample	Pulsed EPR T_1 μs	Pulsed EPR T_M μs	CW-EPR T_1 μs	CW-EPR T_2 μs
L-alanine[72,74]	τ_a= 2.1 τ_b= 21.1	—	2.8	—
2-methylalanine[73]	τ_a= 2 τ_b= 10-35	τ= 0.3-0.4 (estimated)	1.5	0.16
partially deuterated malonic acid[76]	τ= 35 (H and D lines)	τ_a= 0.5-1 τ_b= 5-8 (H and D lines)	40	1.0

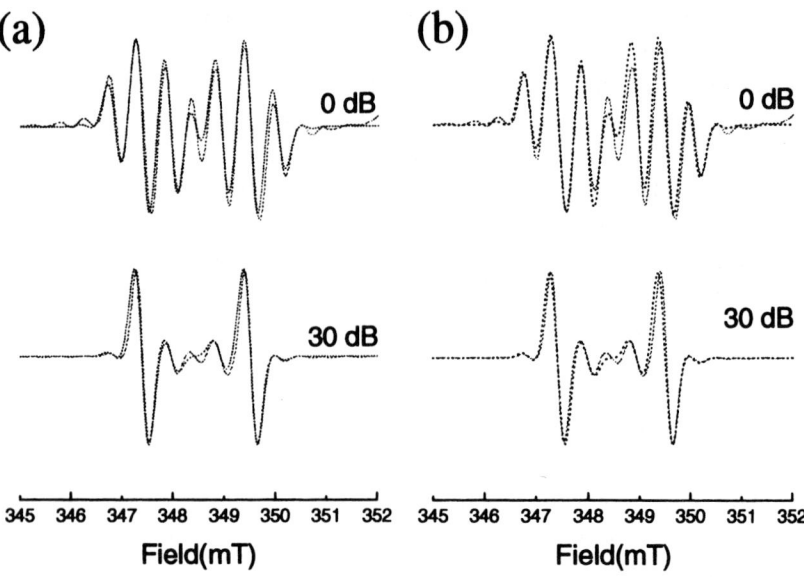

Fig. 4 X–band EPR spectra of X-irradiated malonic acid single crystal recorded at room temperature with magnetic field at $\theta= 49.9°$, $\phi= -39.8°$ in the axes system of ref 73. a) with simulations (dashed curves) using eq (31), $\alpha=0.68$, $P_0=0.1$ mW, $\Delta B_G=3.0$ G. b) using eq (30), $\Delta B_G=2.0$ G, $\Delta B_L=0.018$ G, $P_0=0.04$ mW.

where β_k is the calculated transition probability for each line k. Experimental and simulated spectra of the malonic acid radical, $\dot{C}H(COOH)_2$ are seen to be in agreement over a wide microwave power range, Fig. 4a. The saturation factor accounts for the difference in saturation between allowed and spin flip lines, the latter being caused by weak anisotropic hyperfine coupling with protons at neighbour malonic acid molecules. In initial work different saturation factors were employed for the allowed and forbidden transitions in the alanine radical $CH_3\dot{C}HCOOH$, resulting in very good agreement between experimental and simulated spectra. The physical background for this difference is, however, unclear and the saturation factor of eq. (31) has been used in this chapter.

In the second method a theory utilising a line shape function that takes into account the different degrees of saturation between transitions with different transition probabilities has been employed as a physically more satisfactory simulation method.[76] The simulations with this method are also in good agreement with the experimental data, Fig. 4b. In this method the integral (30) is used as the lineshape function.

Single crystal and powder spectra can be simulated with both methods employing hfi, nuclear g and nqi (I>½) data as input. These data are used in the calculation of the resonance fields B_k and intensities β_k. Both methods also require the power parameter (P_0). The effect of overmodulation is included in the first, but not in the second method. In the first method a value of the parameter α must be provided. This empirical constant is not needed in the second method, where the lineshape is adjusted by the linewidth parameters ΔB_G and ΔB_L for the Gaussian envelope and the Lorentzian spin packet functions in eq. (30). The two programs, written in Fortran 77/90, are available on request.

4. INTERNAL MOTION OF RADICALS

Internal motion such as methyl group rotation, puckering motion (cyclic molecules) and similar types of molecular rearrangement occur frequently for free radicals trapped in solids. The structure and presumably also the dynamics are affected by the matrix. This influence is best understood for matrices with an ordered structure, like the zeolites. In general, the analysis requires advanced theoretical methods, further described by Benetis.[77] The purpose of this section is to present some experimental cases in which relatively simple analysis can provide information about the activation barriers for different types of internal motion. The analysis is possible when the hyperfine coupling constants of the nuclei involved in the dynamics are isotropic. This applies, approximately, to radicals with protons in β-position with respect to the radical centre, and to cyclic radicals undergoing ring puckering. Results[78] that are not easily available of previous studies of radical dynamics in solid matrices are summarised. Most emphasis is, however, on recent studies of dynamics in zeolite matrices. In this case a simple analysis proved possible even though the substituents are attached to a radical centre containing nuclei with anisotropic hyperfine structure, because these nuclei were not involved in the dynamics.

Cyclic radicals: Five-membered ring radicals may have a planar, twisted or envelope conformation, as shown in Fig. 5 for the cyclopentane cation. The ESR spectra recorded in a halocarbon matrix (CF_3CCl_3) showed a temperature dependent lineshape in the region 4-110 K, originally attributed to a puckering motion with E_a = 15 kJ/mol and an exchange between two equivalent envelope structures with E_a = 5 kJ/mol.[79] It was found that the latter process could also be attributed to the exchange between two

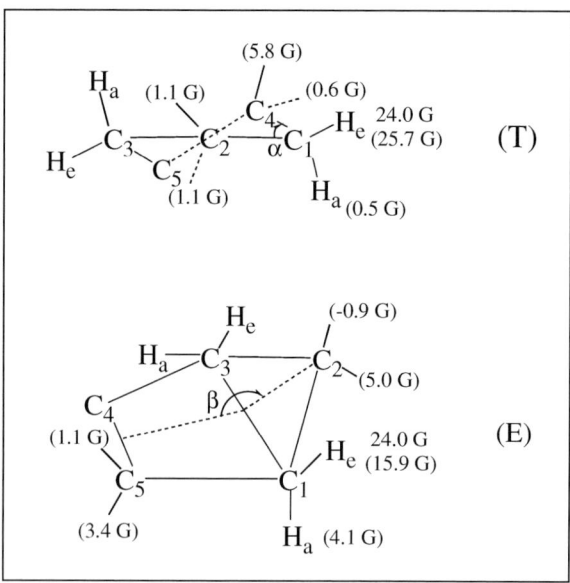

Fig. 5 Geometrical models for the cyclopentane radical cations, (T)= twisted, (E)= envelope structure, in a CF_3CCl_3 matrix. The experimental and calculated hyperfine coupling constants are indicated. Dynamics involving exchange between two equivalent (E) or (T) structure (E_A= 5 kJ/mol) and puckering (E_A = 15 kJ/mol) occurs in the temperature range 4-110 K.

equivalent twisted structures. This model yields better agreement between experimental and calculated hyperfine coupling constants than the envelope model.[78] However, the theoretical methods have improved considerably since the original analysis was carried out, and it would be very interesting to reanalyse the dynamics and structure employing the methods in the chapter by Boyd et al. In the case of cyclopentyl neutral radical trapped in a $CF_2ClCFCl_2$ matrix the dynamics could be analysed by assuming an interconversion between two twisted conformations in the temperature range 110-135 K with E_a = 12 kJ/mol.[80] In agreement with this scheme, the magnitude of the α-H hfi is unaffected by the motion, and the Heinzer model[81] is applicable to account for the spectral changes due to dynamics. The observed activation energy is comparable to that found in other matrices[82] except for the value of 5.4 kJ/mol found in adamantane at a higher temperature range, 134-198 K. This has been explained as a matrix effect, although onset of other kinds of dynamics at higher temperatures cannot be excluded.[78]

For the 5-membered heterocyclic radical species pyrrolidine cation and 1-pyrrolidinyl the intramolecular motion was interpreted in terms of ring inversion between two energetically equivalent twisted structures.[83] The activation energies are 6.7 and 8.0 kJ/mol, for the cation and the neutral radical. It was concluded that the activation energies of the studied five membered ring species show similar dynamics, governed by an interconversion between twisted C_2 structure, and activation energies in the range 5-12 kJ/mol. A significant difference was observed for cyclopentane$^+$, in which case it was necessary to include ring-puckering throughout the ring in the dynamical model in order to reproduce the temperature changes of the experimental ESR spectra. It is probable that this motion is too slow in the heterocyclic species to be visible on the ESR time scale.

The dynamics of the cyclohexane cation is affected by Jahn-Teller distortion.[84] The dynamics can be explained by the conversion between three non-equivalent energy minima with an activation energy as low as 0.71 and 1.00 kJ/mol obtained through a modified Bloch analysis. When the symmetry was lowered by alkyl substitution it was possible to analyse the dynamics by a two-site model, with a proposed interconversion between two structures with elongated carbon-carbon bonds.[85] The activation energy is low, 0.8 and 1.3 kJ/mol for the dynamics of methylcyclohexane$^+$ and 1,1-dimethylcyclohexane$^+$. A possible explanation for the low activation energy is that site-jumping occurs on the potential surface with shallow energy barriers. By contrast, in the five-membered rings reorganisation of the molecular structure is required, particularly for the proposed puckering motion.

In the case of decalin two different electronic cationic states are trapped depending on the matrix.[86] The energy difference between the two states is estimated to be of the order 40 kJ/mol, and an interconversion between them may occur in certain matrices, giving rise to temperature dependent lineshape.

Heterocyclic radicals: The 1-azetidinyl radical observed in halocarbon matrices features dynamical effects above 100 K ascribed to ring inversion between two equivalent bent structures.[78] To analyse the dynamics it was necessary to take into account the reorientation of the axes of the ^{14}N hyperfine structure tensor between the two bent structures. A similar situation applies for the morpholin-1-yl radical where the change of the ESR lineshape was governed by the averaging of the ^{14}N hyperfine anisotropy.[87] In the latter case it was shown that the modified Heinzer method described below is not applicable.

Modified Heinzer method: In contrast to the Heinzer model for isotropic spectra, a decomposition of the Liouville matrix in blocks of size nxn where n is the number of sites has not proved possible when the hfi is anisotropic.

In view of the associated higher demands on computer power an approach has been attempted, in which the small block-size could be retained.[78,88,89] In this model the anisotropic hfi are treated as effective couplings

$$A_{eff}^{[k]}(\theta, \varphi) = \left| \mathbf{A}^{[k]} \cdot \mathbf{u} \right| \qquad (32)$$

for each site k, where **u** is the effective Zeeman field direction as in eq. (17).

As mentioned, the treatment is insufficient for cases with different orientations of the principal axes between different sites, but is applicable to systems in which the anisotropic couplings are not affected by the dynamics. Several cases of dynamics in zeolites mentioned below have been studied recently by this so called modified Heinzer method.

Dynamics of radicals in zeolites: The structure and reactivity of radicals on surfaces have been considered in other works, see[15] for a review. Until recently, detailed studies of the dynamics of radicals trapped in the cages and channels of zeolites have not been made, however. Liu *et al* found that the confined space affected the stability and motion of radicals trapped at those cavities and channels.[90-93] The radicals are generated by γ-irradiation of the zeolites, containing amine molecules used as templates during synthesis in the zeolite.

$[(CH_3)_3N]^{+\bullet}$ and $[(CH_3)_3NCH_2]^{+\bullet}$ radical cations in γ-irradiated Al-offretite, SAPO-37 and SAPO-42 show different stability and dynamics. The strongly temperature dependent EPR spectra observed in the temperature range 140 to 270 K for the $[(CH_3)_3NCH_2]^{+\bullet}$ species in Al-offretite and SAPO-42 were attributed to methyl-group rotation about the C-N bond, Fig. 6. The modified Heinzer program[78] could be used in the analysis, since the two anisotropic α-H in the radical did not take part in the exchange process. Quite good fits could be obtained in simulations, as shown in Fig. 7. The dynamics involves a three-site jump model for each methyl group, combined with free rotation of the methylene group about the N-CH$_2$ bond. The exchange rates are in the order SAPO-37 < Al-offretite < SAPO-42 in the temperature range 110-300 K. In the SAPO-37 the size of the sodalite cage (~6Å) is too small to permit exchange rates that are observable by CW-EPR (c:a 10^7 s^{-1}), whereas in Al-offretite with wider cages (~6×7.4Å) and main channels (~6.5Å) and in SAPO-42 with large (~11Å) channels internal motion is feasible because of less interaction with the matrix. The activation energy in Al-offretite is E_a = 9 kJ/mol, and in SAPO-42 E_a = 11 kJ/mol. In the case of triethylamine (Et$_3$N) and tripropylamine (Pr$_3$N) radical cations generated in γ-irradiated AlPO$_4$-5 from the corresponding templates, strongly temperature dependent EPR spectral line-shapes were apparent in the range 77 to 300 K. The anisotropic ^{14}N hyperfine couplings $A_\parallel$ = 4.4 mT, $A_\perp$ = 0 mT in Et$_3$N$^+$ and $A_\parallel$ = 4.0 mT, $A_\perp$ = 0 mT in Pr$_3$N$^+$ are constant, *i.e.* not averaged by the motion in the temperature range 77 to 300 K. A two-site model with exchanging methylene protons at the C$_\beta$ position accounts for the temperature dependence of the EPR spectra in both cases, when analysed with the modified Heinzer program. The activation energies E_a = 9.1 kJ/mol for Et$_3$N$^+$ and E_a = 11.4 kJ/mol for Pr$_3$N$^+$ correlate well with the potential energy curves for the hindered rotation of the methylene groups although the calculated value at the UHF/3-21G* level for Pr$_3$N$^+$ is slightly lower than the experimental one.[90,92] A possible explanation is that the Et$_3$N$^+$ diameter is smaller than the ring channel size while Pr$_3$N$^+$ has the same diameter as the channel, and therefore may interact with the channel wall.

In conclusion the CW EPR method allows studies of internal motion not only in the liquid phase as studied before but also of radicals trapped in the solids. In the case that the motion involves nuclei with isotropic hfi the Heinzer method is still applicable, even when anisotropic interaction occurs with other nuclei, that do not take part in the exchange process. The modified version of the Heinzer program also allows to use different linewidths for lines with different values of the nuclear quantum number. The size of the channels or cages where the radicals are trapped in the zeolite matrices affects the dynamics. Similar effects may also occur in other matrices, although the results are less clear.

Internal motion studied by ENDOR: Brustolon and coworkers have developed a method to study the rotation of methyl groups by ENDOR spectroscopy.[94-96] The method makes use of the enhancement of the ENDOR signal depending on the methyl rotation rate. A maximum signal is obtained when the methyl rotation rate is equal to the electron Zeeman frequency. The heights of the potential barriers in CH$_3$ĊH – COOH (I) trapped in l-alanine and phenoxyl radical (II) trapped in 4-methyl-2,6-di-tert-butylphenol were quite different, E_I = 17.3 kJ/mol and E_{II} = 3.2 kJ/mol.

Another ENDOR method also developed by Brustolon, which however does not give the activation energy consists in observing the line positions as a function of temperature.[97,98] This method is applicable in the case when the internal motion is due to libration of bonds *e.g.* at β-position to the radical

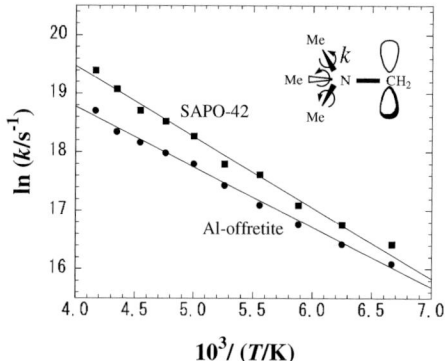

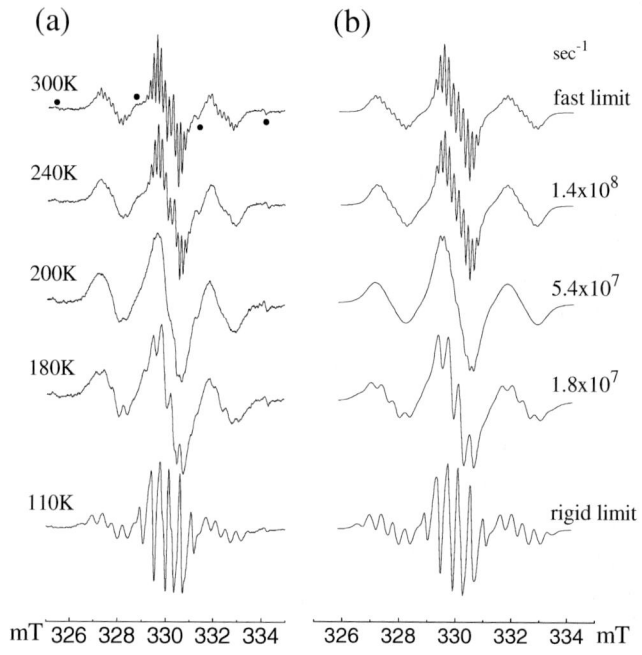

Fig. 6 Arrhenius plots of the exchange rates, ln $k(s^{-1})$ vs. T^{-1} (K^{-1}) for $[(CH_3)_3NCH_2]^{+\bullet}$ in SAPO-42 (■) and Al-offretite (●).

Fig. 7 (a) Temperature dependent EPR spectra of $[(CH_3)_3NCH_2]^{+\bullet}$ generated and stabilized in γ-irradiated Al-offreite. The lines marked (●) are attributed to $[(CH_3)_3N]^{+\bullet}$ (b) EPR spectra simulated using the three-site exchange model for the methyl hydrogen rotation of $[(CH_3)_3NCH_2]^{+\bullet}$. The best-fit M_I and temperature dependent linewidths employed were: ΔB_{PP}= 0.12 mT (M_I=±1) and 0.08 mT (M_I=0) for the 300 K spectrum; ΔB_{PP}= 0.20 mT (M_I=±1) and 0.10 mT (M_I=0) for the spectra above 200 K; ΔB_{PP}= 0.20 mT (M_I=±1) and 0.20 mT (M_I=0) below 200 K. The other EPR parameters used are the same as those of $[(CH_3)_3NCH_2]^{+\bullet}$ in the SAPO-42 system.[93] (1 mT=10 G)

centre as found for radicals trapped in urea and other molecular crystals. The analysis gives the equilibrium values of the dihedral angle and the temperature variation of the amplitude of the angular libration. The numerical values depend on the model assumed for the libration motion, however. The method has been applied also to other radical systems, *e.g.* in glutarimide single crystals after room temperature X-irradiation.[99,100] In this case, the β-proton hyperfine couplings changed with temperature, whereas the α-proton coupling was unaffected, for the heterocyclic type of radicals formed by H atom loss from a methylene group.

5. DIFFUSION OF RADICALS

Spin labels *i.e.* stable free radicals containing a nitroxide (-NO) group covalently linked to the object under study are frequently employed *e.g.* to study the dynamics of biological systems. Spin probes as defined in [101] are paramagnetic molecules that add noncovalently and may diffuse in the medium. In the context of diffusion dynamics in heterogeneous systems nitrogen oxides have been employed in studies by Shiotani and coworkers.[102-107] ESR spectra of NO_2 adsorbed on zeolites were obtained from 4 K up to the temperature at which excessive broadening sets in. The spectra were analyzed based on the slow-motion ESR theory developed by Freed *et al.*[20] The type and degree of motion depended on the properties of the zeolite.[103-107] Rotational diffusion of NO_2 predominates in X- and Y-zeolites. In other cases like in Na-ZSM5, Na-mordenite and K-and L-type zeolites, however, the simulations did not agree well with the experimental spectra, and the possibility of translational diffusion along the zeolite channels was considered. Heisenberg spin exchange can then occur between NO_2 molecules. In Na-ZSM-5 zeolites with different SiO_2/Al_2O_3 ratios for instance, the corresponding ESR spectra of NO_2 at 10 K differed considerably, Fig. 8. This was attributed to slower motion when the Al_2O_3 content is high. At lower Al_2O_3 content the concentration of Na^+ ions is higher, and it was assumed that the mobility of NO_2 is hindered by the presence of the exchangeable cations. The spectral broadening increased with temperature, depended on the structure and the size of the channels, being more pronounced in multiple-channel zeolites than in single-channel ones, and also increased with increasing channel size. These observations are consistent with the assumption that the exchange rate increases with increasing diffusion speed. But, the simulated slow-motional EPR spectra differed significantly from the experimental ones in several other cases as *e.g.* in Fig.8. A possible explanation is that diffusion occurs with a distribution of rates. A procedure to analyse this case has been described by

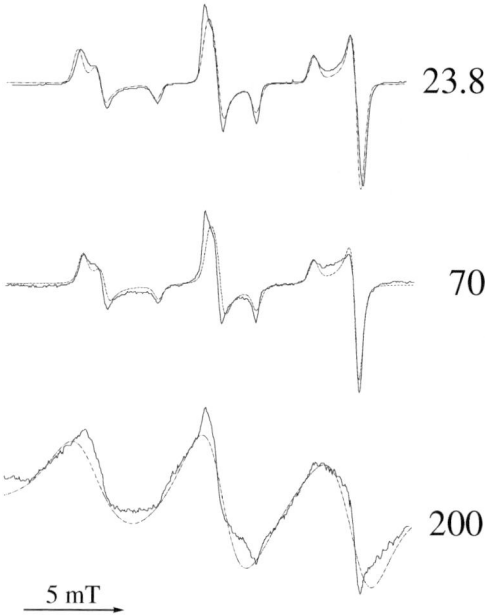

Fig. 8 ESR spectra at 10 K of NO_2 adsorbed on Na-ZSM5 zeolites with SiO_2/Al_2O_3 ratio of 23.8, 70 and 200. Computed spectra with Heisenberg exchange rates of 5.0×10^6, 2.6×10^7 and 3.5×10^8 s^{-1} are shown in dashed lines.

Li.[103] First a series of theoretical slow-motional spectra with different exchange rates was calculated. Next, a principal component analysis of the spectra was made to obtain a collection of eigenspectra and corresponding eigenvalues.[108] The eigenspectra with the 4-5 largest eigenvalues were then used as a basis to approximate the experimental spectra. Finally, the weight of each rate in the experimental spectrum was calculated. The fit to the experimental spectra obtained at different temperatures was much improved, indicating that the assumption of a distribution of rates at each temperature is reasonable. The width of the distribution could also be estimated. In the NO_2/mordenite system for example the distribution is relatively narrow at temperatures below 110 K and broadens above 160 K. One advantage with this technique over least squares fit is that the number of eigenspectra is an order of magnitude fewer than the number of simulated spectra. Similar ESR methods have been applied to the distribution of radical species in irradiated DNA.[109] The number of studies in the EPR field with this and related techniques is, however, small, and it may be too early to conclude if the method is of general applicability in analysing motional effects by EPR.

Bonding at surfaces: Studies of ligand bonding of molecules like water, ammonia, and alcohols to transition metal ions dispersed in zeolites require high resolution and have mainly been carried out with pulsed EPR. Adsorption of paramagnetic probe molecules like nitric oxide has, however, been studied by CW-EPR and ENDOR. The studies by Pöppl and coworkers contain several EPR applications of general interest for the characterisation of adsorption and desorption behaviour.[110-112] Nitric oxide exhibits orbital degeneracy in the gaseous state. Upon adsorption the degeneracy is lifted and it is possible to observe a conventional EPR spectrum, with anisotropic g-factors that depend on the strength of the adsorption. Experimentally, the difference between the g-factors is small, resulting in low resolution at X-band. Measurements at Q and W band have therefore been attempted[110] to obtain the g-tensor for NO in Na-A and Na-ZSM-5 zeolites. Especially, for the W-band measurements the sample preparation in vacuum was difficult and the g-tensor for NO in Na-A and Na-ZSM-5 zeolites has been determined only recently at this band. Only one component of the ^{14}N hyperfine coupling tensor could be resolved by EPR, but the smaller couplings perpendicular to the N-O axis have been determined with ENDOR. It was also possible to observe interaction with Na^+ ions by ENDOR, and from this a geometric model of the NO-Na^+ complex was constructed. An EPR method to study the desorption behaviour of NO on ZSM-5 and Na-A was developed in which desorbed NO was detected in the gas phase at above a certain temperature.[111] The desorption activation energies determined from the temperature dependent linewidth of the adsorbed NO followed the relation E_A (Na-ZSM-5) < E_A (Na-A) << E_A (H-ZSM-5) reflecting the difference between the adsorption strength at an aluminum center and the weak interaction with the sodium cation. In all cases the magnitude of the activation energies and the low desorption temperatures showed that the NO molecules are physically adsorbed. The detailed information that has been collected for the adsorption-desorption of NO on zeolites using EPR is remarkable.

A peculiar property of the Na-A zeolite/NO system is the formation of NO-NO radical pairs by about 4.5 Å.[113-115] The two NO molecules are probably aligned along their N-O bond axes. A confirmation of this model is complicated by the difficulty of simulating the EPR spectra when the axes are not aligned. This difficulty has recently been overcome (see chapter by Hanson) which should open the possibility to deduce the structure of radical pairs in this and similar systems by simulations.

Acknowledgements *We are indebted to Prof. E. Sagstuen for valuable comments on the manuscript and for providing the powder ENDOR spectra of irradiated alanine.*

6. REFERENCES

1. A. Carrington and A.D. McLachlan, *Introduction to magnetic resonance with applications to chemistry and chemical physics*, Harper & Row, New York, 1967.
2. J.E. Wertz and J.R. Bolton, *Electron spin resonance. Elementary theory and practical applications*, McGraw-Hill Book Company, New York, 1972.
3. J.A. Weil, J.R. Bolton and J.E. Wertz, *Electron paramagnetic resonance. Elementary theory and practical applications*, John Wiley & Sons, Inc. New York, 1994.
4. N.M. Atherton, *Electron spin resonance: Principles and applications*, Ellis Horwood, Chichester, 1973.
5. N.M. Atherton, *Principles of electron spin resonance*, Ellis Horwood, New York, 1993.
6. C.P. Poole, Jr. and H.A. Farach, *Theory of magnetic resonance*, 2nd ed. John Wiley & Sons, New York, 1987.
7. P.B. Ayscough, *Electron spin resonance in chemistry*, Methuen & Co. Ltd. London, 1967.
8. L.A. Bljumenfeld, W.W. Wojewodski, A.G. Semjonov, *Anwendung der paramagnetischen Elektronenresonanz in der Chemie*, Akademische Verlagsgesellschaft, Guest & Portig K.-G., Leipzig, 1966.
9. W. Gordy, *Theory and applications of ESR*, Wiley, New York, 1980.
10. P.W. Atkins and M.C.R. Symons, *The structure of inorganic radicals. An application of ESR to the study of molecular structure*, Elsevier publishing company, Amsterdam, 1967.
11. S. Ya. Pshezhetskii, A.G. Kotov, V.K. Milinchuk, V.A. Roginski and V.I. Tupilov, *EPR of free radicals in radiation chemistry*. John Wiley & Sons, New York, 1974.
12. H.C. Box, *Radiation effects, ESR and ENDOR analysis*, Academic Press, New York, 1977.
13. N.D. Yordanov, editor, *Electron magnetic resonance in disordered systems*, World Scientific, Singapore, 1992.
14. A. Lund and M. Shiotani, editors, *Radical Ionic systems*, Kluwer academic publishers, Dordrecht, 1991.
15. A. Lund and C. Rhodes, editors, *Radicals on surfaces*, Kluwer academic publishers, Dordrecht 1995.
16. B. Rånby and J.F. Rabek, *ESR spectroscopy in polymer research*, Springer-Verlag, Berlin 1977.
17. *Electron paramagnetic resonance*, Royal Society of Chemistry, Vol. **1** (1973) to 17 (2000) Thomas Graham House, Cambridge UK.
18. Landolt-Börnstein, *Numerical data and functional relationships in science and technology: Magnetic properties of free radicals*, H. Fischer, editor Springer-Verlag, Berlin 1965-89.
19. J.R. Byberg, S.J.K. Jensen and L.T. Muus, *J. Chem. Phys.* **46** (1967) 131.

1. Continuous wave EPR of radicals in solids

20. D. Schneider and J. Freed, *Biological Magnetic Resonance, Vol. 8, Spin labeling: theory and applications*, ed. L. Berliner and J. Reuben, Plenum Press New York, 1989.
21. D.S. Schonland, *Proc. Phys. Soc.* (London) **73** (1959) 788.
22. W.H. Nelson, *J. Magn. Reson.* **38** (1980) 71.
23. J.A. Weil and Y.H. Anderson, *J. Chem. Phys.* **35** (1961) 1410.
24. A. Lund and T. Vänngård, *J. Chem. Phys.* **42** (1965) 2979.
25. O. Claesson, A. Lund, J.-P. Jørgensen and E. Sagstuen, *J. Magn. Reson.* **41** (1980) 229.
26. M. Lindgren, T. Gustafsson, J. Westerling and A. Lund, *Chem. Phys.* **106** (1986) 441.
27. K.-Å. Thuomas and A. Lund, *J. Magn. Reson.* **18** (1975) 12.
28. A.R. Sørnes, E. Sagstuen and A. Lund, *J. Phys. Chem.* **99** (1995) 16867.
29. K.-Å. Thuomas and A. Lund, *J. Magn. Reson.* **22** (1976) 315.
30. R. Erickson, *Chem. Phys.* **202** (1996) 263.
31. J. Westerling and A. Lund, *Chem. Phys. Lett.* **147** (1988) 111.
32. B. Fox, F. Holuj and W.E. Baylis, *J. Magn. Reson.* **10** (1973) 347.
33. A. Rockenbauer and P. Simon, *J. Magn. Reson.* **11** (1973) 217.
34. A. Rockenbauer and P. Simon, *Mol. Phys.* **28** (1974) 113.
35. M. Iwasaki, *J. Magn. Reon.* **16** (1974) 417.
36. J.A. Weil, *J. Magn. Reson.* **18** (1975) 113.
37. A. Hosseini, E. Sagstuen and A. Lund, *PCCP.* (2002) in press.
38. R.B. Clarkson, R.L. Belford, K.S. Rothenberger and H.C. Crookham, *J. Catal.* **106** (1987) 500.
39. Y. Wu, L. Piekara-Sady and L.D. Kispert, *Chem. Phys. Lett.* **180** (1991) 573.
40. R. Erickson, M. Lindgren, A. Lund and L. Sjöqvist, *Colloids Surf. A.* **72** (1993) 207.
41. R. Erickson, A. Lund and M. Lindgren, *Chem. Phys.* **192** (1995) 89.
42. F. Gerson and X.-Z. Qin, *Chem. Phys. Lett.* **153** (1988) 546.
43. F. Gerson, *Acc. Chem. Res.* **27** (1994) 63.
44. M. Lindgren, R. Erickson, N.P. Benetis and O.N. Antzutkin, *J. Chem. Soc. Perkin Trans.* II. (1993) 2009.
45. P.J. O'Malley and G.T. Babcock, *J. Am. Chem. Soc.*, **108** (1986) 3995.
46. R.M. Kadam, R. Erickson, K. Komaguchi and A. Lund, *Chem. Phys. Lett.* **290** (1998) 371.
47. R.M. Kadam, Y. Itagaki, R. Erickson and A. Lund, *J. Phys. Chem. A.* **103** (1999) 1480.
48. R.M. Kadam, Y. Itagaki, N.P. Benetis, A. Lund, R. Erickson and W. Hilczer, *PCCP.* **1** (1999) 4967.
49. Y. Itagaki, R.M. Kadam, A. Lund and N.P. Benetis, *PCCP.* **2** (2000) 2683.
50. L.R. Dalton and A.L Kwiram, *J. Chem. Phys.* **57** (1972) 1132.
51. C.J. Bender, M. Sahlin, G.T. Babcock, B.A. Barry, T.K. Chandrasekar, S.P. Salowe, J. Stubbe, B. Lindström, L. Pettersson, A. Ehrenberg and B.-M. Sjöberg, *J. Am. Chem. Soc.* **111** (1989) 8076.
52. R. LoBrutto, Y.-H. Wei, R. Mascarenhas, C.P. Scholes and T.E. King, *J. Biol. Chem.* **258** (1983) 7437.
53. G.H. Rist and J.S. Hyde, *J. Chem. Phys.* **52** (1970) 4633.
54. B.M. Hoffmann, J. Martinsen and R.A Venters, *J. Magn. Reson.* **59** (1984) 110; B.M. Hoffman and R.J. Gurbiel, *J. Magn. Reson.* **82** (1989) 309.
55. B.M. Hoffman, R.J. Gurbiel, M.M. Werst and M. Sivaraja, in: *Advanced EPR: Applications in Biology and Biochemistry*, ed. A.J. Hoff (Elsevier, Amsterdam, 1989), pp 541-591; R.J. Gurbiel, C.J. Batie, M. Sivaraja, A.E. True, J.A. Fee, B.M. Hoffman and D. P. Ballou, *Biochemistry.* **28** (1989) 4861.

56. G.C. Hurst, T.A. Henderson and R.W. Kreilick, *J. Am. Chem. Soc.* **107** (1985) 7294; T.A. Henderson, G.C. Hurst and R.W. Kreilick, *J. Am. Chem. Soc.* **107** (1985) 7299.
57. G.P. Gochev and N.D. Yordanov, *J. Magn. Reson.* **102** (1993) 180.
58. A. Kreiter and J. Hüttermann, *J. Magn. Reson.* **93** (1991) 12.
59. C.P. Keijzers, E.J. Reijerse, P. Stam, M.F. Dumont and M.C.N. Gribnau, *J. Chem. Soc. Faraday Trans.* 1, **83** (1987) 3493.
60. R. Erickson, *Electron magnetic resonance of free radicals. Theoretical and experimental EPR, ENDOR and ESEEM studies of radicals in single crystal and disordered solids.* Ph. D thesis, Linköping studies in science and technology. Dissertation No. 391, 1995.
61. R. Erickson, *Chem. Phys.* **202** (1996) 263.
62. M. Heydari, E. Malinen, E.O. Hole and E. Sagstuen, *J. Phys. Chem.* A. (2002) in press.
63. C.P. Poole, Jr, *Electron spin resonance. A comprehensive treatise on experimental techniques.* Dover Publications, Inc. Mineola, 1996.
64. F. Schneider and M. Plato, *Experimental techniques in electron spin resonance.* Verlag Karl Thiemig München, 1971.
65. T.G. Castner, Jr, *Phys. Rev.* **115** (1959) 1506.
66. J. Maruani, *J. Magn. Reson.*, **7** (1972) 207.
67. T. Gillbro and A. Lund, *Chem. Phys.* **5** (1974) 283.
68. O.P. Zhidkov, Ya.S. Lebedev, A.I. Mikhailov and B.N. Provotorov, *Theor. Exp. Chem.* **3** (1967) 135.
69. M..K. Bowman, H. Hase and L. Kevan, *J. Magn. Reson.* **22** (1976) 23.
70. W. Gautschi, *Comm. ACM.* **12** (1969) 635; *SIAM J. Numer. Anal.* **7** (1970) 187.
71. S. Schlick and L. Kevan, *J. Magn. Reson.* **22** (1976) 171.
72. M. Brustolon and U. Segre, *Appl. Magn. Reson.* **7** (1994) 405.
73. S. Olsson, E. Sagstuen, M. Bonora and A. Lund, *Radiation Research.*, **157** (2002) 113.
74. J.R. Harbridge, S.S. Eaton and G.R. Eaton, *J. Phys. Chem.* A. submitted.
75. E. Sagstuen, E.O. Hole, S.R. Haugedal, A. Lund, O.I. Eid and R. Erickson, *Nucleonica* **42** (1997) 353.
76. E. Sagstuen, A. Lund and J. Maruani, *J. Phys. Chem.* A. **104** (2001) 6362.
77. N.P. Benetis, M. Lindgren, H.S. Lee and A. Lund, *J. Appl. Magn. Reson.* **1** (1990) 267.
78. L. Sjöqvist, *The Electron Structure and Dynamics of Organic Radical Cations Studied by ESR Spectroscopy.* Linköping Studies in Science and Technology, Dissertations. No. 247, Linköping, 1991.
79. L. Sjöqvist, A. Lund and J. Maruani, *Chem. Phys.* **125** (1988) 293.
80. L. Sjöqvist, M. Lindgren and A. Lund, *Chem. Phys. Lett.* **156** (1989) 323.
81. J. Heinzer, *Mol. Phys.* **22** (1971) 167.
82. R.V. Lloyd and D.E. Wood, *J. Am. Chem. Soc.* **99** (1977) 8269.
83. L. Sjöqvist, A. Lund, L.A. Eriksson, S. Lunell and M. Shiotani, *J. Phys. Chem.* **94** (1990) 8081.
84. K. Toriyama, *Radical Ionic Systems*, Ch I4, A. Lund and M. Shiotani, Editors, Kluwer Academic Publishers, Dordrecht, 1990.
85. L. Sjöqvist, M. Lindgren, M. Shiotani and A. Lund, *J. Chem. Soc., Faraday Trans* **86** (1990) 3372.
86. V.I. Melekhov, O.A. Anisimov, L. Sjöqvist and A. Lund, *Chem. Phys. Lett.* **174** (1990) 95.
87. O.N. Antzutkin, N.P. Benetis, M. Lindgren and A. Lund, *Chem. Phys.* **169** (1993) 195.
88. L. Sjöqvist, N.P Benetis, A. Lund and J. Maruani, *Chem. Phys.* **156** (1991) 457.
89. N.P Benetis, L. Sjöqvist, A. Lund and J. Maruani, *J. Magn. Reson.* **95** (1991) 523.

90. W. Liu, *Structure and molecular dynamics of radicals trapped in organic-inorganic composites*. Ph.D. Thesis, Department of Applied Chemistry, Faculty of Engineering, Hiroshima University, 2001.
91. W. Liu, P. Wang, K. Komaguchi, M. Shiotani, J. Michalik and A. Lund, *PCCP.* **2** (2000) 2515.
92. W. Liu, S. Yamanaka, M. Shiotani, J. Michalik and A. Lund, *PCCP.* **3** (2001) 1611.
93. W. Liu, M. Shiotani, J. Michalik and A. Lund, *PCCP.* **3** (2001) 3532.
94. M. Brustolon, T. Cassol, L. Micheletti, U. Segre, *Mol. Phys.* **57** (1986) 1005.
95. M. Brustolon, T. Cassol, L. Micheletti, U. Sagre, *Mol. Phys.* **61** (1987) 249.
96. M. Brustolon, A.L. Maniero, U. Sagre, *Mol. Phys.* **65** (1988) 447.
97. M. Brustolon, U. Sagre, *J. Chim. Phys.-Chim. Biol.* **91** (1994) 1820.
98. F. Bonon, M. Brustolon, A.L. Maniero and U. Sagre, *Appl. Magn. Reson.* **3** (1992) 779.
99. N.A. Salih, O.I. Eid, N.P. Benetis, M. Lindgren, A. Lund, E. Sagstuen, *Chem. Phys.* **212** (1996) 409.
100. O.I. Eid, Ph.D. Thesis, Department of Physics, Faculty of Science, University of Khartoum, 1999.
101. G.I. Likhtenshtein, *Biophysical labeling methods in molecular biology.* Cambridge University Press, New York, 1993.
102. M. Shiotani and J. H. Freed, *J. Phys. Chem.* **85** (1981) 3873.
103. H. Li, *Bonding and diffusion of molecules in zeolites by ESR and ENDOR*. Linköping studies in science and technology. Dissertation No. 544, 1998.
104. H. Li, H. Yahiro, M. Shiotani and A. Lund, *J. Phys. Chem.* **102** (1998) 5641.
105. H. Yahiro, M. Nagata, M. Shiotani, M. Lindgren, Haitao Li and A. Lund, *Nukleonika*, **42** (1997) 557.
106. D. Biglino, H. Li, R. Erickson, A. Lund, H. Yahiro, M. Shiotani, *PCCP.* **1** (1999) 2887.
107. H. Yahiro, M. Shiotani, J.H. Freed, M. Lindgren and A. Lund, *Study Surf. Sci. Catal.* **94** (1995), 673 (Catalysis by Microporous Materials).
108. H. Wadsworth, editor, *Handbook of Statistical Methods for Engineers and Scientists*, McGraw-Hill, NewYork, 1990.
109. J. Barnes and W. Bernhard, *Radiat. Research.* **143** (1995) 85.
110. T. Rudolf, A. Pöppl, W. Hofbauer and D. Michel, *PCCP.* **3** (2001) 2167.
111. T. Rudolf, W. Böhlmann, A. Pöppl, *J. Magn. Res.*, **155** (2002) 45.
112. A. Pöppl, T. Rudolf, P. Monikanolen, D. Goldfarb, *J. Am. Chem. Soc.* **122** (2000) 10194.
113. P. H. Kasai, R.J. Bishop Jr, *ESR studies of zeolite chemistry and catalysis*, ed. J.A. Rabo, ACS Monograph 171, pp 350-391. American Chemical Society, Washington DC, 1976.
114. H. Yahiro, A. Lund, R. Aasa, N.P. Benetis and M. Shiotani, *J. Phys. Chem. A.* **104** (2000) 7950.
115. D. Biglino, *Employment of EPR techniques in the study of DeNOx heterogeneous catalysis: interactions between adsorbed molecules and catalytic surface*. Ph. D thesis, Linköping studies in science and technology. Dissertation No. 702, 2001.

Chapter 2

PULSED EPR OF PARAMAGNETIC CENTERS IN SOLID PHASES

Marina Brustolon and Antonio Barbon
Department of Physical Chemistry, University of Padova,Via Loredan 2 35131 Padova, Italy

Key words: pulsed EPR, electron spin echo, echo decay, Hahn echo, stimulated echo, ESEEM, HYSCORE, pulsed ENDOR.

Abstract: We present an overview of the most used Electron Spin Echo techniques and their applications to the study of structure and dynamics of paramagnetic centers in solid phases. A short theoretical section presents the tools necessary to understand the experiments. Three sections describe the experiments that are used for measuring longitudinal relaxation, transverse relaxation and hyperfine interactions. Many examples of applications to different research fields are given.

1. INTRODUCTION

Pulsed EPR spectroscopy has been developed later than pulsed NMR. The pioneering work started in the 60s, and a further development bringing pulsed EPR to become a quite common spectroscopic tool, with pulsed spectrometers also available commercially, had to wait to the 80s. Pulsed EPR techniques are available nowadays in many EPR laboratories.

The basic principles of pulsed EPR and pulsed NMR are the same, as is the physics of the nuclear and electron spins. In fact both types of pulsed magnetic spectroscopies use the same theoretical approaches and models.[1] However, much more stringent conditions for EPR, as faster relaxation times (three orders of magnitude) and larger spectral ranges (three-four orders of magnitude), has required further technological developments. Today, also with the best available technology it is in general impossible to achieve a non selective excitation of all the EPR spectrum, and an important part of the initial time signal cannot be recorded due to the presence of the tail of the excitation pulse (dead time). These limitations make impossible at the moment for pulsed EPR to have a fate similar to that of pulsed NMR, that of completely substituting cw-EPR.

Nevertheless, pulsed EPR gives information on the structure and dynamics of the spin system not reachable with other techniques. For paramagnetic species in solid state the most used EPR pulsed technique is the Electron Spin Echo (ESE) spectroscopy. This technique is used for paramagnetic metals, paramagnetic defects, free and transient radicals and photoexcited triplets. An important feature of echo spectroscopy is that it allows to eliminate of the inhomogeneous broadening of the spectral lines, therefore achieving information on the transverse spin relaxation times of the homogeneous components. Another important feature is the ESEEM effect (Electron Spin Echo Envelope Modulation), a time modulation of the echo decay that gives detailed information on the hyperfine and superhyperfine couplings. These types of applications, together with the measurement of spin-lattice relaxation, are the most frequently found in current research. However, as for NMR, a wealth of different experiments in one, two and also three time dimensions have been described in literature in recent years.

In this contribution we will limit ourselves to describe the simplest experiments and give the basic principles for understanding them. Only general outlines of the formal treatments will be given, providing the references to the papers and books where the theory is fully developed. We tried on the other hand to give a good descriptive physical insight in treating the experiments. We choose to exclude in general from the subjects of this chapter a description of the instrumentation and also detailed descriptions of the more technical aspects of the experiments, as for example phase cycling or treatments of data. On the other hand we will describe a number of applications to different fields, as examples of the many types of information that can be obtained by ESE spectroscopy.

Recently all the methodological tools necessary to understand pulsed EPR experiments, and a large number of examples of applications, have appeared in the book by Arthur Schweiger and Gunnar Jeschke.[2] This was strongly needed by the EPR community, since up to now very few books had appeared some years ago on the subject of pulsed EPR, treating specific fields,[3-12] or very recently analytical aspects.[13]

2. THEORY

The aim of this section is to give the basic concepts and definitions in pulse EPR techniques. General theoretical treatments are reported in references.[2-4,6,13-17]

Pulsed experiments are normally performed in three steps: preparation of the system, evolution (mixing) and detection. The preparation is given by a sequence of pulses applied to the sample to create non-equilibrium magnetization. During the evolution time different processes make the system to evolve, then modifying its non-equilibrium state. In some experiments a mixing period can be present at this stage. A net non-zero transverse magnetization is eventually detected. In pulse techniques applied to solid state samples an echo is formed along the detection axis in the xy plane, perpendicular to the static field B_0. The variation of the echo intensity depending on the preparation and evolution steps can be related to the properties of the system.

A classical approach can be used to treat the dynamics of a single $S=1/2$ spin, or an ensemble of non-interacting $S=1/2$ spins. This approach (which is often referred to as *vector model*) is equivalent to the quantum mechanical one for this case (since an $S=1/2$ spin in magnetic field has two energy levels only), and it can be fruitfully used to have an insight of the events during a pulse experiment for more complex systems.

2.1 Classical treatment: free induction decay and spin echo

The classical equation of the evolution of the magnetization vector $\mathbf{M}$ given by the sum of identical electron spins $\mathbf{\mu}$ in a magnetic field $\mathbf{B}$ is:

$$\frac{d\mathbf{M}}{dt} = -\gamma_a \mathbf{M} \times \mathbf{B} \tag{1}$$

where γ_a is the absolute value of the electron magnetogyric ratio $\gamma_a = |\gamma| = g_e \mu_B / \hbar$.

In a magnetic resonance experiment the field has a static component, B_0, assumed to be along the **k** director of the z axis, and a field $B_1(t)$ oscillating in the plane perpendicular to z at a microwave (mw) frequency ω.

The equation of motion of the single components of the magnetization can be conveniently described in a right-hand rotating frame at frequency ω with the x axis parallel to B_1, so that B_1 is time independent in this axes system.

In the rotating frame eq. (1) is written as

$$\frac{dM_x}{dt} = -(\gamma_a B_0 - \omega)M_y = -\Delta\omega M_y$$

$$\frac{dM_y}{dt} = (\gamma_a B_0 - \omega)M_x - \gamma_a B_1 M_z = \Delta\omega M_x - \omega_1 M_z \qquad (2)$$

$$\frac{dM_z}{dt} = \gamma_a B_1 M_y = \omega_1 M_y$$

where $\Delta\omega = \gamma_a B_0 - \omega = \omega_0 - \omega$ and $\omega_1 = \gamma_a B_1$ is called nutation or Rabi frequency. For $\Delta\omega = 0$ (resonance condition) the motion of the magnetization is therefore a precession around the B_1 direction at frequency ω_1. After a time t_p the magnetization is tilted by an angle $\theta_p = \gamma B_1 t_p = \omega_1 t_p$.

The motion in presence of B_1 is more complex if the pulse is applied with an offset resonance $\Delta\omega \neq 0$. In this case the nutation is around an effective field tilted from B_1 by an angle $\zeta = \arctan(\omega_1 / \Delta\omega)$ with a nutation frequency $\omega_{eff} = (\Delta\omega^2 + \omega_1^2)^{1/2}$.

So far the model does not take into account the spin relaxation. Bloch introduced in a phenomenological way two relaxation terms, relative to two separated relaxation processes: a transverse and a longitudinal relaxation processes, with times T_2 and T_1.

2. Pulsed EPR of paramagnetic centers in solid phases

$$\frac{dM_x}{dt} = -\Delta\omega M_y - \frac{M_x}{T_2}$$

$$\frac{dM_y}{dt} = \Delta\omega M_x - \omega_1 M_z - \frac{M_y}{T_2} \qquad (3)$$

$$\frac{dM_z}{dt} = \omega_1 M_y - \frac{M_z - M_0}{T_1}$$

These equations give the basis to describe a simple real experiment: the Free Induction Decay (FID) of an ensemble of spins with the same frequency $\Delta\omega$ (isochromats).

Longitudinal relaxation of the magnetization requires that energy is exchanged between the spin system and some degree of freedom of the surrounding (*e.g.*, molecular rotations or vibrations). During the transverse relaxation process, instead, the total amount of magnetization energy accumulated in the spin system is kept fixed. Therefore, the relaxation of the transverse components of the magnetization is faster than the longitudinal one. In solids, the difference is usually large and $T_2 \ll T_1$.

Let us apply a mw pulse of frequency ω to a system with equilibrium magnetization $\mathbf{M}=M_0\mathbf{k}$ for a time $t_{\pi/2} = \pi/2\omega_1 \ll T_1, T_2$. This is called a $\pi/2$ pulse. The effect of the pulse is to bring the magnetization from the z to the $-y$ direction.

The evolution of the system in the rotating frame is given by the solution of the eq. (3) with initial conditions $\mathbf{M}(t=0) = -M_0\mathbf{j}$. The solution is

$$M_x = M_0 \sin(\Delta\omega t)\exp(-t/T_2)$$

$$M_y = -M_0 \cos(\Delta\omega t)\exp(-t/T_2) \qquad (4)$$

$$M_z = M_0[1-\exp(-t/T_1)]$$

In resonance condition ($\Delta\omega = 0$) during the free evolution time following the end of the pulse the magnetization lies along $-y$ and decays due to the transverse and longitudinal relaxations. The Fourier Transform (FT) of the FID gives a lorentzian line of half width at half height $1/T_2$ centered at zero frequency. If the resonance condition is not met, the magnetization decays and rotates in the xy plane at a frequency $\Delta\omega = \omega_0 - \omega$. The FT of the FID is then a lorentzian line of the same width centered at $\Delta\omega$.

In a real solid sample the magnetic anisotropic interactions are not averaged by the molecular reorientation as in solution, and therefore it is almost impossible to have an isochromatic spin system (apart from conditions of very strong spin exchange in concentrated spin systems). Due to the large number of hyperfine and superhyperfine interactions of each electron spin many EPR lines are overlapping, and this gives rise to the inhomogeneous line broadening in the cw-EPR spectrum.

It should be noted that the FT of a rectangular pulse of width t_p at frequency ω_0 contains frequency components in the excitation bandwidth $\omega_0 \pm \delta\omega$, with $\delta\omega \approx 1/\tau_p$.[2] Therefore spins with resonances in this range can be excited by the pulse. When the excitation bandwidth is larger than the overall frequency extension of the EPR spectrum, the pulse excites all the electron spins, and it is called *non selective*, whereas if only a part of the spectrum is excited the pulse is *selective*. In general in the case of pulsed EPR the pulses are selective, due to the instrumental limitations in producing short and powerful mw pulses for excitation bandwidths comparable to the usual widths of EPR spectra. Therefore, for an accurate analysis of the experiment, in general the shape of the FT of the pulse should be taken into account, since not all the spins affected by the pulse will be tilted by the same angle.[2]

Groups of spins with the same resonance frequency are called *spin packets*. Each spin packet i would give rise to an homogeneously broadened component with half width at half height $1/T_{2i}$ in the cw spectrum. A number of spin packets with resonance frequencies within the excitation bandwidth will be rotated in the xy plane, all along the $-y$ direction, where they will then start to precess, each with its frequency ω_{0i} (see figure 1). Due to their different resonance frequencies the spin packets are dephasing very rapidly, with a characteristic time $T_2^* \ll T_2$. This dephasing is reversible, and the spin packets can be rephased at time 2τ by applying a π pulse after a time τ. The $\pi/2 - \tau - \pi - \tau$ -echo sequence is called Hahn Echo or 2p-ESE (see figure 1 and 2).[18-19]

2. Pulsed EPR of paramagnetic centers in solid phases

On the other hand, the *xy* magnetization of each spin packet decays irreversibly during τ with characteristic time T_{2i} due to stochastic interactions. As a consequence the total echo intensity decays on increasing τ with a characteristic time T_M (phase memory time, see section 4).

The classical magnetization vector picture given above is not any more adequate for $S>1/2$ or when interactions between spins are taken into account. A quantum-mechanical treatment is therefore necessary. We will give here only the results of this treatment, referring to the specialized literature for the derivation.

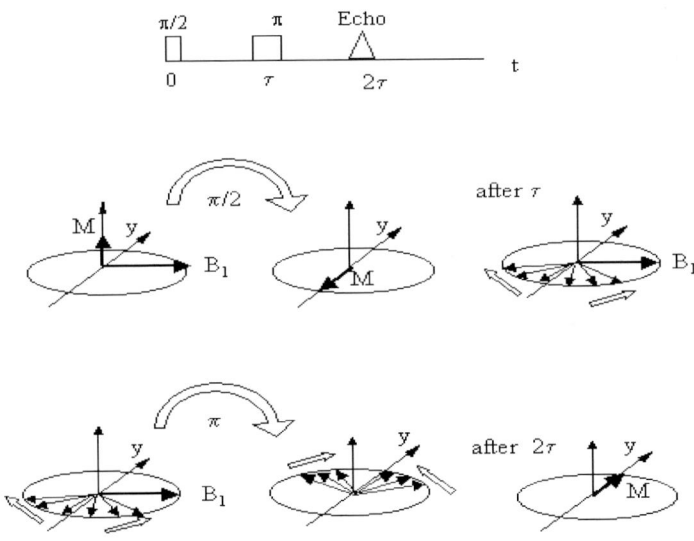

Figure 1. Vector model of the Hahn echo formation.

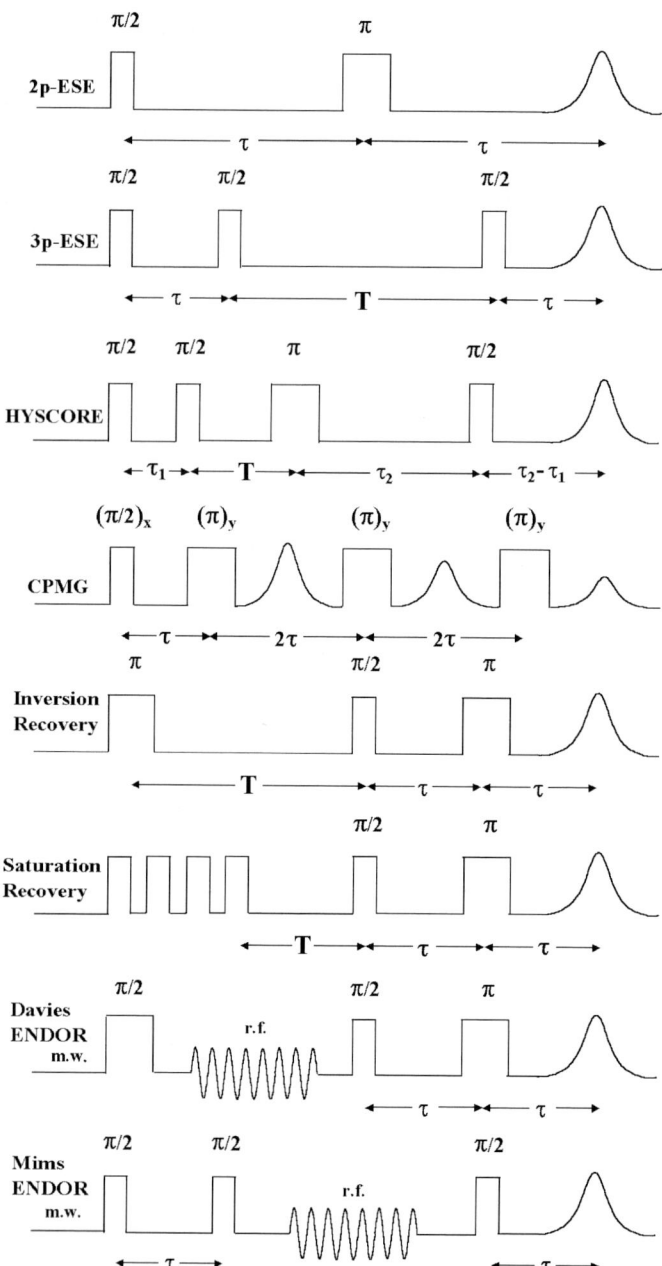

Figure 2. Pulse sequences used in the described experiments.

2.2 Quantum mechanical treatment of pulse experiments

We introduce now the concept of *density matrix*. A *reduced* density matrix relative to the spin system $\sigma(t)$ is generally used for treating of spin dynamics.[16] A reduced matrix is based on wavefunctions of the spin coordinates only, and not on the complete functions of spin and spatial coordinates. In the following we will restrict to spin coordinates.

For a single spin $S=1/2$ the wavefunction can be expressed in the basis of the eigenfunctions of the spin Hamiltonian, $|\alpha>$ and $|\beta>$:

$$|\psi> = c_\alpha|\alpha\rangle + c_\beta|\beta\rangle \tag{5}$$

where c_α and c_β are complex coefficients:

$$|\psi> = e^{-i\varphi_\alpha}|c_\alpha\||\alpha\rangle + e^{-i\varphi_\beta}|c_\beta\||\beta\rangle \tag{6}$$

The state of an ensemble of spins all with the same wavefunction is called a *pure state*. All the spins have the same phase in the pure state, and the state is said to be *coherent*.

For this system the density matrix σ is defined as

$$\sigma = \begin{vmatrix} |c_\alpha|^2 & c_\alpha c_\beta^* \\ c_\alpha^* c_\beta & |c_\beta|^2 \end{vmatrix} \tag{7}$$

The expectation value of any operator A in terms of the density matrix is given by:

$$\langle A \rangle = tr\{\sigma \cdot A\}. \tag{8}$$

Then the expectation values of the components $<S_x>$, $<S_y>$, $<S_z>$, of the spin operator S are given by

$$<S_x> = \frac{1}{2}(c_\alpha^* c_\beta + c_\alpha c_\beta^*) = |c_\alpha| |c_\beta| \cos\Delta\varphi$$

$$<S_y> = \frac{1}{2}(c_\alpha^* c_\beta - c_\alpha c_\beta^*) = |c_\alpha| |c_\beta| \sin\Delta\varphi \quad .\tag{9}$$

$$<S_z> = \frac{1}{2}(|c_\alpha|^2 - |c_\beta|^2)$$

An ensemble of spins with different wave functions $|\psi_k>$ is said to be in a *mixed state*. The elements of the density matrix for a mixed state are given by averages of the elements in matrix (7):[20]

$$\sigma = \begin{vmatrix} \overline{|c_\alpha|^2} & \overline{c_\alpha c_\beta^*} \\ \overline{c_\alpha^* c_\beta} & \overline{|c_\beta|^2} \end{vmatrix} \tag{10}$$

For a system at thermal equilibrium the diagonal elements of the density matrix are the equilibrium *populations* the off diagonal terms are called *coherences* and they are equal to zero:

$$\overline{c_i^* c_j} = \overline{c_i c_j \exp(i\Delta\varphi_{ij})} = 0 \tag{11}$$

as the phase differences are random.

The components of the macroscopic magnetization are proportional to the ensemble averages of eq. (9). Therefore the diagonal elements of the density matrix are related to the longitudinal magnetization, whereas the off diagonal elements are related to transverse one. In a system at thermal equilibrium in an external magnetic field the transverse magnetization is zero and only the longitudinal magnetization is present.

During the preparation period the spin system is prepared by a sequence of pulses and coherences or polarized non-equilibrium states are created. In the evolution period the system evolves under the influence of the spin Hamiltonian, and different types of magnetization transfers may take place. In the detection period coherences are detected as transverse magnetization.

2. Pulsed EPR of paramagnetic centers in solid phases

By using the time-dependent Schrödinger equation, the Liouville-von Neumann equation is obtained[2,16]

$$\frac{d\sigma(t)}{dt} = -\frac{i}{\hbar}[H(t), \sigma(t)], \tag{12}$$

where $H(t)$ is the spin Hamiltonian, which has formal solution:

$$\sigma(t) = \exp\left[-\frac{i}{\hbar}\int_0^t H(t')dt'\right]\sigma(0)\exp\left[\frac{i}{\hbar}\int_0^t H(t')dt'\right] \tag{13}$$

By selecting a suitable rotating frame system, the Hamiltonian can often be made time-independent within each evolution period. Then the evolution of the density matrix can be obtained in successive steps in which a static Hamiltonian acts on the system. In each step, from t to $t+t_1$, the evolution of the density operator is obtained as

$$\sigma(t+t_1) = \exp\left[-\frac{i}{\hbar}Ht_1\right]\sigma(t)\exp\left[\frac{i}{\hbar}Ht_1\right] \tag{14}$$

The echo intensity at the end of an experiment is obtained as

$$\langle M(t_1 + t_2 + t_3 + ...)\rangle = tr\{\sigma(t_1 + t_2 + t_3 + ..) \cdot M\} \tag{15}$$

Accordingly, with this procedure the magnetization arising from any pulsed EPR experiment could be obtained.

2.2.1   2p- and 3p-ESE experiments. A qualitative description

In the following we will use the concept of density matrix to describe the 2p-ESE and 3p-ESE experiments on the basis of the modifications of its elements during the different steps of the experiment. Complete calculations can be found for example in refs. 1, 6. For a spin system $S=1/2$, $I=1/2$, coupled by a weak hyperfine interaction either isotropic or anisotropic, the spin Hamiltonian H is

$$\begin{aligned}H &= v_e S_z - v_N I_z + STI \\ &\approx v_e S_z - v_N I_z + T_{xz}I_x S_z + T_{yz}I_y S_z + T_{zz}I_z S_z\end{aligned} \tag{16}$$

where ν_e and ν_N are the electron and nuclear resonance frequency and **T** is the hyperfine tensor; for weak couplings the terms in S_x and S_x can be neglected.[14]

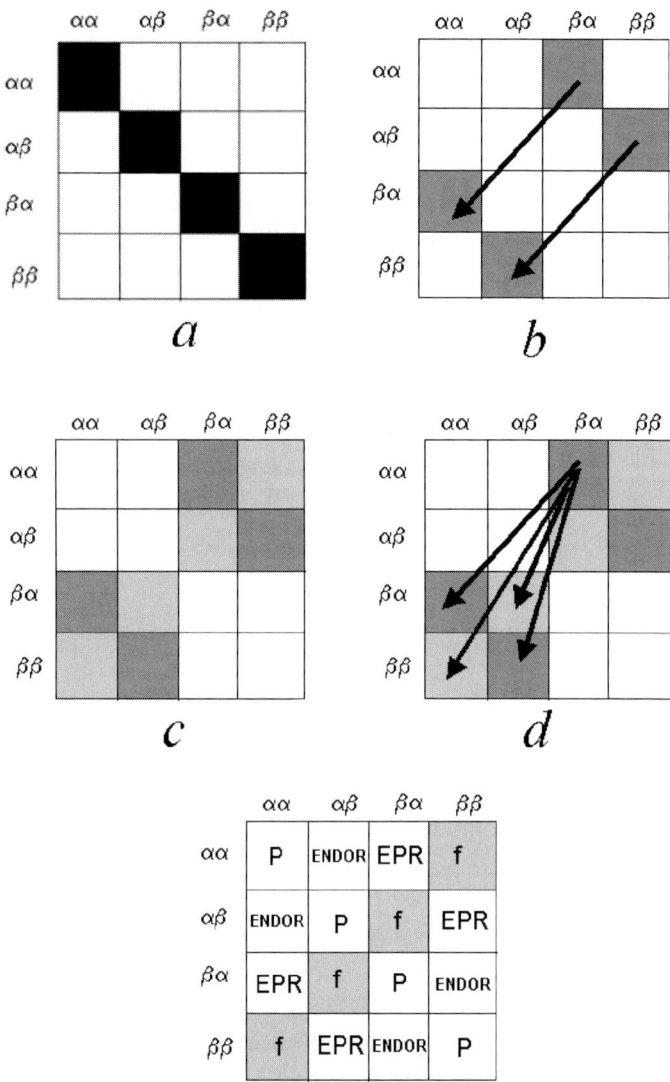

Figure 3. Evolution of the density matrix during a 2p-ESE experiment for a $S=1/2$, $I=1/2$ spin system, coupled either with isotropic (*a* and *b*) or anisotropic (*c* and *d*) interaction. Bottom: definition of the density matrix elements; P is the polarization of the single spin state; ENDOR, EPR are coherences corresponding to allowed transitions; f correspond to prohibited transitions.

2. Pulsed EPR of paramagnetic centers in solid phases

It is convenient to use a basis of four coupled spin functions $\alpha\alpha$, $\alpha\beta$, $\beta\alpha$, $\beta\beta$, product of the electronic and nuclear eigenfunctions of the Zeeman terms in eq. (16). In figure 3 (bottom) one can see the definitions of the elements of spin density matrix σ for this spin system. Spin populations (P) along the trace and coherences off diagonal are indicated. The coherences corresponding to allowed ($\Delta m_s = \pm 1, \Delta m_I = 0$) and forbidden ($\Delta m_s = \pm 1, \Delta m_I = \pm 1$) EPR transitions are indicated respectively with EPR and f, those corresponding to ENDOR transitions ($\Delta m_s = 0, \Delta m_I = \pm 1$) with $ENDOR$.

Let us first describe the 2p-ESE experiment for a $S=1/2$, $I=1/2$ system with an isotropic hyperfine interaction. In figure 3 a and b, the effect on σ of the $\pi/2$ and π pulses is shown. The populations (a) are transferred to EPR coherences (b) by a $\pi/2$ pulse. After a time τ the π pulse changes the phases of the coherences (arrows in b). The complete treatment shows that the rephasing of the coherences is taking place after another interval τ, then the rephased coherence is detected.

When an anisotropic hyperfine coupling is present, allowed and forbidden EPR transitions can both be excited. In fact the quantization axis for the nucleus is determined by the resultant of the applied field and of the anisotropic hyperfine field, the two fields having different directions, then the eigenfunctions of the Hamiltonian (16) are mixing of α and β nuclear states. The effect of a $\pi/2$ pulse for such a system is to create coherences as shown in c. In d one can see the effect of the π pulse on just one of the coherences created previously. It has the effect of redistributing (branching) the coherence considered on four allowed and forbidden EPR coherences, with probabilities depending on the angle between applied and hyperfine fields. This effect is due to the fact that during the flipping of the hyperfine field consequent to the change in m_s produced by the π pulse there is a probability of flipping also for the nuclear spin. This branching prevents a complete rephasing of the spin packets, and it is the basis for observation of ESEEM, as we will discuss in section 5.1.1.

The effect on the density matrix of the pulse sequence in 3p-ESE or stimulated echo is shown schematically in figure 4. An anisotropic coupling for the $S=1/2$, $I=1/2$ system is considered. The first $\pi/2$ pulse creates the same electron coherence pattern as for the 2p-ESE (b). Then the magnetization evolves freely during τ. In the vector model the spin packets rotate in the xy plane and they will distribute in the plane with different phases depending on their Larmor frequencies. Let us take into account those spin packets which after τ are again aligned along the axis perpendicular to B_1 (therefore along y or $-y$). They will be rotated respectively along $-z$ or $+z$ by the second pulse, therefore giving a contribution to the magnetization along the z axis. As one can see in figure 4-c, diagonal elements in σ are in fact created by the second $\pi/2$ pulse. On the other hand the spin packets which after τ are aligned along x will be unaffected by the $\pi/2$ pulse. Therefore the z polarization after the second pulse has a modulation at frequency $\pi/2\tau$.

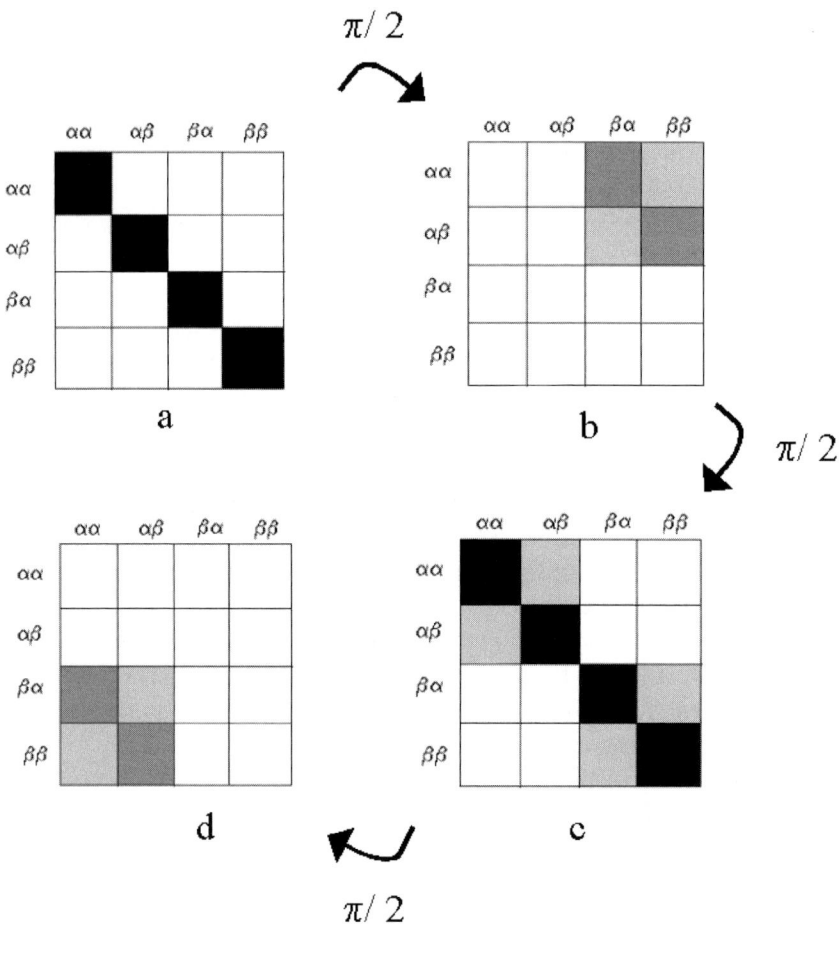

$\pi/2 - \tau - \pi/2 - T - \pi/2 - \tau - $ echo

Figure 4. Evolution of the density matrix during a 3p-ESE experiment for a coupled $S=1/2$, $I=1/2$ spin system.

2. Pulsed EPR of paramagnetic centers in solid phases

It should be noted that the electron spin polarization is decaying with rates depending on the spin-lattice relaxation and in general on *spectral diffusion* as explained later. These relaxations are slower than the rate of dephasing in the *xy* plane typical of the 2p-ESE, and therefore the decaying of the stimulated echo with time T is slower than the Hahn echo decay on increasing τ. Besides the electron spin polarization, the second $\pi/2$ pulse creates also nuclear coherences, as can be seen in *c* in the figure. Finally, the third $\pi/2$ pulse transfers the polarization back to *xy* plane, where it starts to precess. We should expect to detect an FID, that is in fact formed, but due to the modulation of the polarization it refocuses after a time τ. Therefore the *stimulated echo* is in reality an FID.[2]

2.2.2 Spin relaxation and echo decay

Information on the system can be obtained from echo experiments by observing the profile of the echo decays on increasing one of the interpulse intervals. In the case of 2p-ESE the decay as a function of τ is observed, whereas in the case of 3p-ESE the dependence of the decay on both τ and T can be obtained.

The two properties of echo decay are given by the effect of spin relaxation and by the nuclear envelope modulation (ESEEM) effect. ESEEM can be calculated for any spin system by using the theory treated above, in terms of the reduced spin density matrix. On the other hand, to calculate the relaxation behavior one should take into account the interactions of the spin system with other degrees of freedom of the system.

The general evolution of the density operator under a time-independent Hamiltonian H in the presence of relaxation processes is given by the quantum mechanical master equation[2,14,16]

$$\frac{d\sigma(t)}{dt} = -\frac{i}{\hbar}[H,\sigma(t)] - \Gamma\{\sigma(t) - \sigma_0\} \qquad (17)$$

where σ_0 is the equilibrium density operator and Γ the relaxationsuperoperator that accounts for the dissipative interactions between the spin system and the lattice.

The matrix elements of the relaxation superoperator Γ_{klkl} describe the decay of the element σ_{kl} of the density matrix, while elements Γ_{klmn} describe relaxation between elements σ_{kl} and σ_{mn}. The correspondence to Bloch equations for the $S=1/2$ system is recovered by setting $\Gamma_{kkll} = \pm 1/2T_1$ and $\Gamma_{mnmn} = 1/T_2$ (with m≠n).

Although the correct time evolution of the system is fully described by eq. (17), generally for spin systems in solids the relaxation is treated separately with respect to the spin dynamics treated previously. As a consequence, the time evolution of the echo intensity $V(t)$, proportional to the magnetization, is calculated as[21]

$$V(t) = V_m(t)V_d(t) \qquad (18)$$

where $V_m(t)$ describes the modulation of echo intensity due to the electronspin echo envelope modulation (ESEEM). The function $V_d(t)$ introduces phenomenologically the decay function of the signal due to any relaxation process.

The spin relaxation processes affecting the longitudinal and transverse relaxation rates are discussed respectively in sections 3.1 and 4.1. The expressions for $V_d(t)$ obtained for the 2p-ESE are given in the Appendix. Some expressions for $V_m(t)$ are discussed in section 5.1.1.

An example of the application of eq. (18) is shown in figure 5.[22] In the figure one can see two ESE spectra at different temperatures for a nitroxide radical doping a single crystal. Both $V_d(t)$ and $V_m(t)$ change with the temperature. $V_d(t)$ can be fitted with one linear exponential for T = 300 K, as the relaxation process is dominated by the motions of the radical in the matrix. On the other hand, the decay is fitted with a gaussian for T = 50 K, since at this temperature the dominant relaxation mechanism is the fast flip-flop of neighbor proton spins. The modulation is simulated as due to the four freely rotating methyl groups of the radical at 300 K, and to the eight methyl groups of the host neighbor molecules at 50 K.

2.2.3 Product operator method

A convenient method for treating the spin dynamics in pulse magnetic resonance experiments is that of decomposing the density operator in a set of product operators.[1,2,16]

2. Pulsed EPR of paramagnetic centers in solid phases

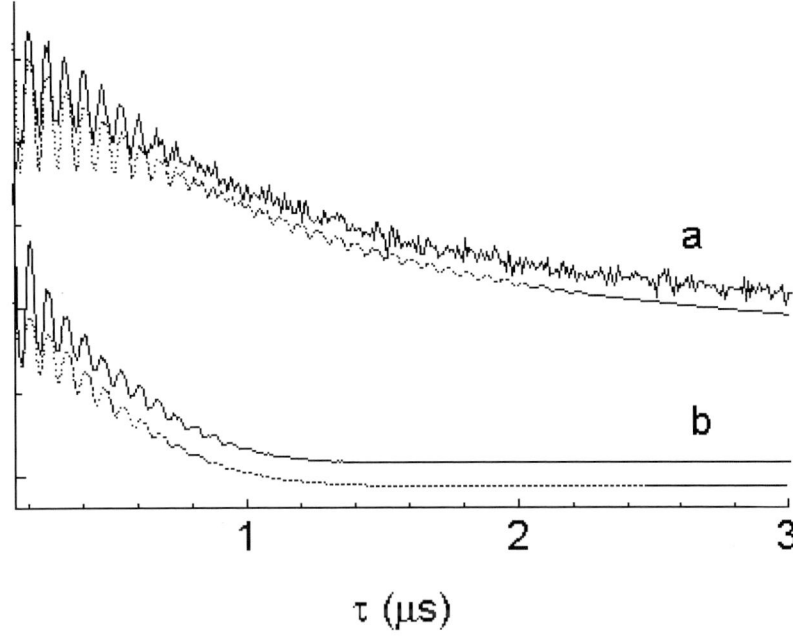

Figure 5. Echo decays of 4-oxo-2,2,6,6,-tetramethylpiperidin-1yloxyl (tempone) doping substitutionally a single crystal of the diketone 2,2,4,4-tetramethyl-cyclobutan-1,3-dione (upper solid lines) for *a*: T=300 K; *b*: T=50 K and the relative simulations (lower dotted lines) for T=300 K and T= 50 K (adapted from ref. 22). The decay at T = 300 K is obtained from a crystal deuterated in the methylene positions. The simulation of the modulation is obtained by assuming dipolar couplings respectively (*a*) with the four methyl groups of the radical; (*b*) with the eight methyl groups of the first shell of host molecules.

This method combines elements of both vector model and density matrix approach. It makes the physical picture easier and is advantageous when one is interested to put in evidence only some properties of the evolution of the system relative to some operators. The method is based on the expansion of the density operator on a suitable base of operators B_s,

$$\sigma(t) = \sum_{s=1}^{n^2} b_s(t) B_s \tag{19}$$

where the time dependence is given by the coefficients $b_s(t)$. B_s elementsbelong to the Liouville space obtained by direct product of the basis set of the single spins^2 that, for spin systems, are normally Cartesian spin operators.

The theory is commonly used to calculate the spin physics of pulse sequence in NMR.[23] For pulse EPR experiments, the theory can be found in ref. 2, section 4.2.3.

3. THE DETERMINATION OF ELECTRON SPIN LONGITUDINAL RELAXATION

The measurement of electron spin-lattice relaxation rates with cw-EPR techniques can be done with the *progressive saturation* technique. This method is illustrated in this book in chapter 1.

Pulsed EPR methods for measuring electron spin-lattice relaxation times T_1 are *inversion recovery*, *saturation recovery* with long saturating pulse, and saturation recovery with a train of short saturating pulses. It should be noted that a systematic comparison between the applicability and results obtainable with these different methods could help in the development of a reliable experimental protocol for the measurement of spin-lattice relaxation times for systems with different properties. In fact a drawback of these techniques is the selectivity of the preparation pulses, as usually the width of the spectrum is larger than the bandwidth of the pulses. Thus only a part of the spin packets belonging to the EPR spectrum is affected by the pulse. As a consequence, the dynamics of the electron spins after the preparation procedure is determined by more than one process, and the recovery is given in general by a multiexponential expression:[24]

$$V(t) = \sum_j V_{0j}[1 - n\exp(-t/T_{1j})] \tag{20}$$

(n=1 for the saturation recovery, n=2 for inversion recovery)

Very often two main processes are active, *i.e.* the exchange of energy between the spins and the *lattice*, and the exchange of energy between the spins affected by the pulses and those not affected. The first process is the true spin-lattice relaxation, the second one is the so called *spectral diffusion*.[25] The saturation recovery experiments should be designed in

2. Pulsed EPR of paramagnetic centers in solid phases

particular to exclude as much as possible the interference of spectral diffusion.

The resulting traces are described usually by a sum of two exponentials, and a least-squares fitting yields the rates of spectral diffusion, generally faster, and of the spin-lattice relaxation.

When a distribution of relaxation times due to intrinsic disorder in the sample is present, the recovery curves are given by stretched exponentials

$$V(t) = V_0[1 - n\exp(-(t/T_1)^\beta)] \qquad (21)$$

where β depends on the distribution of relaxation times.

3.1 Spin-lattice relaxation processes

Phonons and spin-lattice relaxation:
The exchange of energy between the spin system and the lattice takes place by exchange of phonons. This latter can occur in different ways.[24,26]

The *direct process* is the emission of a phonon with frequency equal to the electron Larmor frequency into a lattice mode of the same frequency. By emission of the phonon the spin state changes. This process therefore involves only low frequency phonons, and it is important only at very low temperatures. It should be noted that the importance of this mechanism is related to the working frequency of the spectrometer.[27] For X-band EPR the direct process exhibits a temperature dependence $1/T_1 = aT$ above 1-2 K.

The *Raman process* involves a simultaneous interaction with two latticemodes whose difference frequency is the Larmor frequency ω_0. This process can be considered as an inelastic scattering between the paramagnetic species and a phonon of frequency ω_A higher than the Larmor frequency ω_0, which is absorbed taking the paramagnetic system to a virtual state, and then emitted at a minor frequency ω_B, where $\omega_A - \omega_B = \omega_0$. The temperature dependence at low temperature is given by $1/T_1 \propto T^9$ for half-integer S systems (Kramer systems), and $1/T_1 \propto T^7$ for integer S (non-Kramer systems). In the high temperature limit, $1/T \propto T^2$.[28]

The *Orbach-Aminov process* involves an absorption of a phonon with transition from the ground state to a real excited state, and then transition back to the ground state in a different spin state. To be effective, it requires the presence of low lying excited states. The temperature dependence is given by $T_1 \propto 1/[\exp(-\Delta/k_B T) - 1]$.

Motions and spin-lattice relaxation:
Intramolecular and intermolecular motions with frequencies near the electron Larmor frequency can affect the spin-lattice relaxation due to the modulation of some spin Hamiltonian parameter. Motional frequencies of the order of 10 GHz corresponding to X-band Larmor frequency are not very common in solid phases. Guest molecules in inclusion compounds can have intermolecular motion with high frequencies. Intramolecular motions with such a frequency can be ring interconversions or methyl groups rotation. In the latter case the time modulation of the beta protons hyperfine coupling constant (h.c.c.) has been shown to be a powerful mechanism for non secular relaxations.[29-30]

The effect of tunneling modes for protons in glassy solutions of radicals has also been considered a possible spin-lattice relaxation mechanism.[31]

Magnetic interactions between different paramagnetic species:
The spin-lattice relaxation of a species A can be affected by the presence of a faster relaxing species B. The influence on the T_1 values of species A by species B depends on the spectra of the two species, on their spin exchange and dipole-dipole interaction. In the cases in which it is possible to distinguish the effect on the relaxation of species A due to the presence of species B, information on the distance of the two species can be be achieved.[2,32]

3.2   Methods

The different pulsed methods used for measuring the spin-lattice relaxation rates all consist of two parts. The first part is the preparation of the spin system to a non equilibrium population distribution, and the second one is the observation of the equilibrium recovery. Experiments taking advantage also of magnetic field sweeps or jumps have been described, and a review can be found in ref. 2.

3.2.1 Saturation recovery with cw-EPR detection.

This method has been used for many years.[24] A long pulse of microwaves partially saturate the EPR signal, that is then followed in its evolution towards equilibrium recovery by the monitoring of the cw-EPR signal.

3.2.2 Saturation recovery with pulsed detection of recovery

The EPR signal is saturated with a long cw pulse, or with a train of shorter pulses (see figure 2). Then the recovery of the magnetization can be followed either by a detection pulse giving rise to a FID, or by a two pulses sequence giving rise to an echo. This second method is normally used for solid state samples, due to the fast decay time T_2^* of the FID for paramagnetic species in solids.

3.2.3 Inversion recovery

The spin state populations are inverted by a convenient mw pulse, and then the recovery of the signal is followed by observing the echo (see figure 2). Due to the limited spectral range affected by the inverting pulse, this method is not usually convenient for systems giving rise to a spectrum wider than the latter spectral range, *i.e.* wider than ca. 10 G. In fact a single pulse can in the latter case "burn a hole" in the EPR spectrum, and therefore the recovery would be due to a competition between spin-lattice relaxation and spectral diffusion. To ascertain the importance of spectral diffusion in the recovery it is convenient to compare the form of the inversion recovery and saturation recovery curves, where in the second case the use of trains of saturating pulses allows to extend the saturated range and therefore to reduce the extent of spectral diffusion. Usually the cross checking of the results of the two techniques is useful to have a better insight into the physical meaning of the measured relaxation times.

3.2.4　Stimulated or 3p-ESE

In 3p-ESE (see figure 2) during time T a non equilibrium magnetization is stored along z, see figure 4, and therefore it decays in principle with time T_1. Therefore it should be possible to measure T_1 by following the intensity of the stimulated echo by varying the time delay T. The drawback is that the effects of spectral and spin diffusion are more important in this case than in the other described experiments.[2] Therefore a comparison between the rate of stimulated echo decay with the rates of magnetization recoveries gives a criterion for assessing the importance of diffusion of non equilibrium magnetization in the system.

3.2.5　Decay of photoexcited species

Photoexcitation with a laser pulse can create paramagnetic species with non equilibrium populations, and their kinetics to the equilibrium can be followed by probing the magnetization with a two pulses echo on varying the time after the laser pulse.[33] A contribution of both chemical and spin lattice kinetics can be present.

3.3　Examples

Radicals in glassy phases:
A series of nitroxide radicals in different glassy phases have been studied inthe temperature range between 20 K and 150 K by Du, Eaton and Eaton.[34] The T_1 values have been found to depend on the temperature and on the resonance field. The temperature dependence was fitted to $\log(1/T_1) = n\log(T) + C$, with $n \approx 2$. The average values are $n=2.34\pm0.05$ and $C=7.0\pm0.3$. The relaxation is explained by a two phonons Raman process, but the relevant vibrations have been considered as belonging to the molecule itself and not to the lattice modes. At higher temperatures effects due to the increasing mobility of the radicals in the matrices are relevant. The dependence of T_1 on the orientation of the probe in the magnetic field is attributed to the symmetries of the vibrational modes and to the orientation dependence of the spin-orbit coupling.[34]

The T_1 values have been determined for the frozen solutions of the radical anion of fullerene C_{60}^- [35] and for a series of fulleropyrrolidines mono and bisadducts[36]. Values of spin-lattice relaxation times much shorter than the average ones for organic radicals have been obtained. The T_1 values approach the normal values as the stiffness of the fullerene sphere increases

2. Pulsed EPR of paramagnetic centers in solid phases   61

on going from C_{60}^- to the radical anions of N-methylfulleropyrrolidine monoadducts to bisadducts. The short T_1 has been attributed to the near degeneracy of the π-electrons energy levels of fullerenes, allowing fast radiationless transitions between the different electronic states.

Radicals in molecular crystals:
The determination of the spin concentration from the EPR signal of theradical generated by high energy irradiation of crystalline L-alanine is used for EPR dosimetry of irradiation dose. Therefore the radical formed in the latter substance, and in particular its spin-lattice relaxation properties, have been studied with several EPR techniques. Due to rapid spectral diffusion in the irradiated alanine samples at room temperature, the estimated relaxation times are strongly dependent upon the measurement technique.[37,38] In the latter paper it was shown that a biexponential recovery curve was obtained with a saturation recovery experiment, giving two times attributed respectively to the spin-lattice relaxation process and to the spectral diffusion. Two sets of pulses with different lengths have been tested, in the first one relatively short saturating pulses (16 ns) and longer detection pulses (48-96 ns), in the second one the saturating and detecting pulses were 80 ns and 16 ns long respectively. The second experiment allowed to obtain the saturation recovery for a small range of spin packets inside a wider hole, and as a consequence longer times were measured with the second type of experiment. T_1 determined in this way was very similar to the value measured by an independent cw electron and nuclear double resonance experiment, Longitudinally Modulated ENDOR technique.[39] It can be noted that also the stimulated echo decay has been studied, and its biexponential fit shows that the equilibrium magnetization in this experiment is achieved only by spectral diffusion.

Ghim et al.[40] measured the saturation and inversion recovery of irradiated alanine at different EPR frequencies at room temperature. They found that in any case the recovery curves were given by the sum of two exponentials, and the contribution of the faster relaxing component, *i.e.* the spectral diffusion contribution, decreases with increasing microwave frequency. In addition to the changing weights of the two components, the values of relaxation times for both components increased with increasing microwave frequency. The much shorter characteristic times measured in any case with the inversion recovery experiment are attributed to the rapid spectral diffusion in the L-alanine sample at room temperature.

The effect of methyl groups rotational motions on spin-lattice relaxation has been observed in irradiated alanine at different temperatures.[30] Previous

ENDOR amplitude studies had shown that the cross electron-nuclear flip-flop relaxation was strongly affected by the modulation of the hyperfine coupling of beta protons due to the methyl group rotation in the radical.[29] The biexponential fit of the inversion recovery curves at different temperatures have shown that the fast component of the recovery is completely dominated by the same relaxation mechanism. Clear cut results are obtained by comparing the different values of the latter fast component for recoveries obtained on different methyl protons hyperfine lines, respectively affected and non affected by the exchange. The slow component of the recovery is shown on the other hand to be dominated by the relaxation process due to the rotation of neighbor -CH_3 and -NH_3 groups in the undamaged molecules.

Triplet states:
Spin lattice relaxation has been studied in naphtalene doped withquinoxaline at very low temperature (T<2 K) at variable magnetic field. It was shown that the direct process is taking place, dominated by the coupling with the $\Delta m_S = 2$ transition of the triplet states.[27]

Transition metals complexes:
In Cu(II) complexes in glassy frozen solution the EPR spectrum isdominated by the g anisotropy, and a striking dependence of T_1 on the resonance field has been obtained, attributed to local intramolecular vibrations modulating anisotropically the spin – orbit coupling.[2,41]

$[Fe_3S_4]^+$ clusters ($S = 1/2$) in four proteins have been studied by pulsed X- and Q-band EPR. The T_1 values have been determined by saturation recovery technique in the range 2-4.2 K. The T_1 values vary sharply across the EPR envelope, due to a distribution in cluster properties. The temperature dependence of $1/T_1$ is analyzed in terms of the Orbach mechanism involving a doublet excited state at 20 cm^{-1} above the ground state.[42]

The T_1 values at different temperatures have been measured by inversion recovery for the S_2 state multiline signal of the Mn cluster in Photosystem II oxygen evolving complex in different states.[43] The temperature dependence of T_1 shows that the signal relaxes via an Orbach relaxation pathway, indicative of a low lying excited state at an energy varying in the range 33.8-39.7 cm^{-1} above the ground state.

High-spin Fe(III) octaethylporphyrin in single crystal of the Ni analogue shows T_1 values obtained by inversion recovery that are originated by an

2. Pulsed EPR of paramagnetic centers in solid phases

Orbach process under 6 K, through the intermediate Kramers states. A D value of 7 cm^{-1} was obtained from the temperature dependence. A spin-lattice relaxation time $T_1 = 4.15$ μs was obtained for the magnetic field perpendicular to the porphyrin plane, twenty times longer than with the field in the plane. This allows to obtain information on the states mixing due to spin orbit interaction.[44]

Other systems:
The spin lattice relaxation has been measured by saturation recovery with atrain of saturating pulses in metallic silver nanoparticles embedded in amorphous SiO$_2$ and crystalline TiO$_2$ matrices in the range 4-300 K. T_1 of the order of 10^{-2}-10^{-1} s have been obtained, the temperature dependence being different in the two matrices, with a Raman-type temperature dependence in the TiO$_2$ matrix.[45]

Hydrogenated amorphous silicon (a-Si:H) spin-lattice relaxation in undoped[28] and n- and p- doped[46] a-Si:H has been measured. The recovery traces are given by stretched exponentials. Dangling bond signals recover with times of the order of 10^{-2}-10^{-4} s, whereas conduction electrons signals recover with times of the order of 10^{-5}-10^{-7} s.

The kinetics of spin lattice relaxation of C$_{60}^+$ radicals in C$_{60}$ powder at room temperature shows a stretched exponential behavior, eq. (21), which is attributed to a distribution of relaxation times. A strong effect of the presence of oxygen on the spin lattice relaxation suggests that the distribution of T_1 values might be due to the different interaction of the cation with the adsorbed oxygen.[47]

4. THE DETERMINATION OF ELECTRON SPIN TRANSVERSE RELAXATION

The measurement of the electron spin transverse relaxation in the solid phase is performed by using echo experiments. The relaxation is characterized by the rate $1/T_M$ of the irreversible dephasing of the *xy* magnetization, where T_M is the so called *phase memory time*. This parameter is related but not coincident with the T_2 transverse relaxation time of a single spin packets, as explained later.

Two methods are currently used to measure the spin dephasing in the *xy* plane: the 2p-ESE decay (or Hahn echo decay) and the Carr-Purcell-

Meiboom-Gill (CPMG) experiment (see figure 2). The experiments are described in detail below. The two experiments are different, as in 2p-ESE decay echoes obtained at different delay times between the two pulses are observed, whereas in the CPMG experiment the same echo is refocused periodically for several times. The refocusing of the initial spin packets minimizes the effect of the different magnetic energy transfer processes.[48] As a consequence, the decay times obtained from CPMG experiments are normally longer than those obtained from 2p-ESE decay.[49]

This latter method is much more popular than the CPMG experiment. In fact the 2p-ESE decay is an easier experiment, suitable also for systems with short phase memory times. Moreover a considerable amount of theoretical work has been done in order to get information on the relaxation mechanisms acting in the system from the shape of the decay curves.[25,50,54] As a consequence a clear relation between the decay functions and the properties of the system can sometimes be found. Nevertheless a systematic comparison of the results obtained from both experiments on different systems would be very useful to give a better insight into the very complicated competition of different relaxation processes in determining the echo dephasing kinetics.

Large differences in phase memory times can be found along the spectrum; this effect can be due to the residual mobility of the paramagnetic species, or to their interactions with the environment (for example for different distances from fast relaxing paramagnetic metals, or to different symmetries of the coordinating sites) that induce anisotropic effects of the spin relaxation.

In field-swept echo detected EPR (EchoEPR), the intensity of the Hahn echo is recorded as a function of the magnetic field B_0 at a fixed time delay τ between the two pulses. This spectroscopy is interesting for samples where the phase memory time T_M varies substantially for the different spin packets. In fact in this case the relative intensities of the EPR lines in cw-EPR and EchoEPR spectra can be very different, since the profile of the EchoEPR spectra depend on the echo decay rate.

4.1 Processes of spin dephasing

The loss of the phase memory is due to: i. processes affecting the individual spin packets, similar to those giving rise to the homogeneous EPR linewidth in solution, originating from the residual intramolecular and intermolecular motions in the matrix modulating in time the hyperfine and g tensor parameters; ii. collective relaxation processes, due to magnetic interactions between different spin packets, and to the interaction with the nuclear spins bath.

The relative importance of the different relaxation mechanisms at a given temperature depends on the selectivity of the pulses, on the concentration of the radicals, on their residual mobility in the solid matrix, on the presence of intramolecular motions such as methyl group rotation and on the concentration and type of nuclear spins present in the diamagnetic matrix.

In asystem with high dilution of the electron spins not interacting with the nuclear spins bath the collective relaxation processes could be neglected. In this case the echo decay would give $1/T_2$, the homogeneous linewidths of the spin packets (in the hypothesis of the same homogeneous linewidth for all of them). The residual motions of the paramagnetic centers in the solid matrix, as librational, intramolecular, etc. give rise to a relaxation behaviour that can be modelled by using the same theory developed for relaxation in liquids. In particular, for "fast" motions the relaxation effects can be analyzed in the framework of the well known Redfield-Freed theory[51-52] by taking into account the orienting effects of the anisotropic environment.[53]

The different processes bringing about the spin dephasing due to spin-spin interactions have been treated in a series of theoretical and experimental works.[25,50,54-56] In the following we give a very short account of them.

Let us consider the relaxation mechanisms due to the electron-electron interactions. The pulses are in general selective, and they excite only a part of the spin packets in the sample (usually called spins A, whereas spins B are the ones not excited). Time dependent dipolar interactions A-A and A-B both contribute to the dephasing of spins A. Very often the time dependence of A-A interaction is dominated by the flipping of the spins produced by the pulses themselves (*instantaneous diffusion*). In this case a clear dependence of $1/T_M$ on the microwave power is found. On the other hand both interactions A-A and A-B are modulated in time by the intrinsic processes of the spin system, *i.e.* spatial and spin dynamics.

The dephasing of spins A can be produced by a relaxation of the first kind, *i.e.* a time fluctuation of the dipolar magnetic interaction parameters due to spatial dynamics. A relaxation of the second kind is due to random modulation of the electron-electron dipolar interaction via spin flips of the coupled spins. These spin flips can be due mainly to spin-lattice relaxation (the so called T_1 samples) or to spin flip-flops (T_2 samples).[20,25,56] The intrinsic relaxation processes and instantaneous diffusion contribute to the rate of spin dephasing in an additive way if the former are not much faster than the latter one.[25]

Also nuclear-electron dipolar interactions contribute to the spin dephasing. The most important contribution comes from the bath of the matrix protons, which fluctuate because of the nuclear flip-flop transitions. When this relaxation mechanism is important, the concentration and type of matrix protons is determining the rate of the echo decay.[57]

All these relaxation mechanisms give rise to an echo decay (measured by 2p-ESE or CPMG) that can be represented by a stretched exponential. The effective decay function, and the temperature dependence of T_M depend on the dominating relaxation mechanism, and can give information on the dynamics and concentration of the spins.

Theoretical treatments take into account the evolution of the density matrix under the time dependent dipolar interactions between the observed spin packets (spins A) and all the spins present in the system (spins A, spins B and nuclear spins). A detailed treatment can be found in ref. 50.

In the Appendix the 2p-ESE decay functions for relaxation processes described above are reported.

As for longitudinal relaxation, transverse relaxation can be induced by a residual motion (intra- or inter-molecular) that affects the resonance field value of the species. This effect is well known for paramagnetic species in solution.[14]

4.2 Methods

4.2.1 2p-ESE decay

The 2p-ESE experiment has been described in section 2 and an example is also given. The choice of the pulses length and power depends on the system and on the desired information. Short pulses have a wider bandwidth than long pulses, therefore the number of excited spin packets is higher for short pulses, and an echo of higher intensity is expected. On the other hand, as discussed above, different pulse lengths determine different relative concentration of A- and B-spins, therefore affecting also the dephasing rate.

2. Pulsed EPR of paramagnetic centers in solid phases

Moreover shorter pulses give rise to a more complex ESEEM modulation, since they can excite a larger range of allowed and forbidden EPR transitions.

The echo decay functions are always stretched exponentials of type:

$$V_d(2\tau) = V_0 \exp[(-2\tau/T_M)^x] \qquad (22)$$

The explicit forms are given in the Appendix. From the value of the x exponent one can get information on the mechanism that determines the transverse relaxation.

If instantaneous diffusion is present[58] the relaxation rate is dependent on the microwave power and this effect is used as a clear cut indication of the latter contribution to spin dephasing.[59-62]

In this case for non selective pulses the echo intensity depends on the tilting angles θ_i of the i-th pulse; if $2\theta_1 = \theta_2$ the echo intensity is given by the expression:

$$V_d(2\tau) \propto \sin^3\theta_1 \exp(-2\tau/T_M) \qquad (23)$$

the phase loss rate has the form:

$$1/T_M = A + B \cdot \sin^2(\theta_1) \qquad (24)$$

where A is the contribution due to any other process, and B is due to instantaneous diffusion and given by:

$$B = (4\pi^2 \gamma^2 \hbar C / 9\sqrt{3}) = b \cdot C \qquad (25)$$

where $b = 8.2 \; 10^{-13}$ cm^{3}s^{-1}, and C is the concentration of the spins affected by the pulses in spins/cm^3.[55] The selectivity of the pulses can be taken into account.[59]

4.2.2 CPMG

This pulse sequence has been introduced for NMR by Carr and Purcell, and modified by Meiboom and Gill.[1] A $\pi/2$-τ-π sequence is used to refocuse the magnetization at $t=2\tau$, where an echo is formed. If a π pulse is applied after a time 3τ, another echo will be formed at time 4τ. In this manner successive π-pulses at $(2n+1)\tau$ $(n=0,1,2,..)$ form progressively decaying echoes at $(2n+2)\tau$, that are acquired. The advantage is that with a single experiment several points are obtained, reducing the acquisition time. On the other hand the time resolution is limited by the spectrometer dead time t_d, since the condition must hold $\tau \geq t_d$.

It has been shown that the relative contribution from the spectral and spin diffusion to the dephasing is reduced in this experiment with respect to the 2p-ESE decay due to the shorter τ values for each recorded echo. In fact the relative contribution to the dephasing of the diffusion terms compared to the T_2 term depends on τ, and it can be decreased on decreasing τ.[1,48,63]

The intensities of unwanted echoes overlapping regular echoes can be strongly reduced by using a hard $\pi/2$ pulse, followed by more selective long pulses.[63] To overcome the problem of the cumulative effects due to the deviation of the pulse rotation from π a 90° phase shift between subsequent π pulses is used.[1]

A detailed full analysis of CPMG experiments has been carried by Schwartz et al. for the case of EPR of nitroxides in viscous solution both in fast and slow motional regime.[48] This method is applicable to situations in which slow reorientations of the molecules are possible, like, for example, in soft glasses close to the melting point. The method is based upon the evaluation of the eigenvectors and the eigenvalues of the stochastic Liouville operator. The primary echo is calculated to be a sum of exponential decays with characteristic times related to the differences of eigenvalues.

4.2.3 EchoEPR

EchoEPR spectra are obtained by recording the echo intensity, for a given τ, as function of the magnetic field B_0. In general the latter for a given B_0 and τ can be calculated by:[64]

$$E(2\tau, B_0) = E_{CW}(B_0) V_m(2\tau, B_0) V_d(2\tau, B_0) \tag{26}$$

where E_{cw} is the intensity of the cw-EPR spectrum, V_m and V_d the nuclear modulation and decay functions. The dependence of the echo intensity on the nuclear modulation can help in disentangling EchoEPR spectra due to species with different modulations.[65] However in general the effect of the modulation is strongly reduced by using long selective pulses. To get a better spectral resolution $\Delta\omega$ the echo is integrated with a window $\Delta t_{obs} > 2\pi/\Delta\omega$.[2,49]

For an oriented paramagnetic center the terms in eq. (26) depend on the orientation of the magnetic field with respect to the molecular axes. For a disordered distribution the angular dependence has to be integrated over all the possible orientations of the magnetic field. For $S = 1/2$ coupled with a nuclear moment I, by neglecting the effect of the nuclear modulation, the echoEPR spectrum can be computed by the expression:[66]

$$E(2\tau, B_0) = \sum_{M=-I}^{I} \frac{1}{4\pi} \iint \sin\theta d\theta d\varphi f\left[(B_0 - B_M(\theta,\varphi))/\delta\right] R_M(2\tau,\theta,\varphi) \quad (27)$$

where the sum runs into the M values, $f\left[(B_0 - B_M(\theta,\varphi))/\delta\right]$ is the residual lineshape function, and $R_M(2\tau,\theta,\varphi)$ the relaxation decay function for the particular spin packet.

Distortion of the lineshape with respect to the cw spectrum can be due to any anisotropic decay of the magnetization, and to the effect of instantaneous diffusion that depends on the concentration of the spin packets at any field value.

4.3   Systems studied and some applications

Echo decay of nitroxide spin probes
By means of the analysis of the electron spin echo decay of nitroxideradicals, a wealth of information has been obtained on the relaxation mechanisms inducing spin dephasing for the spin probes at different temperatures and in different phases. These studies therefore allow to exploit the echo decay results and provide an insight into dynamical and structural properties of condensed phases of different types (see for example[67-68]).

At very low temperatures in diluted nitroxide glassy solutions in proton-containing solvents the fluctuations of nuclear spins dominate the echo dephasing.[22,25,54,69-70] The protons of the matrix are known to produce dephasing of the electron spins thanks to the random modulation of the local magnetic field due to their nuclear flip-flop transitions. The echo decay depends therefore on the proton concentration, but also on the type of protons. Particularly effective in producing dephasing are methyl groups, suggesting an important contribution from tunneling processes. Eaton et al. determined a dependence of the rate of dephasing on the barrier hindering the methyl groups rotations.[57,70] At higher concentration of radicals and in deuterated solvents the instantaneous diffusion dominates.[70]

On increasing the temperature the echo decay becomes more and more dominated by relaxation processes due to different motions of the probes. Contributions from the softening of the glassy phase and from the intramolecular motions of the probe itself can be present.

In ref. 22 a detailed study of the echo decay at different temperatures for the tempone radical hosted in a single crystal allows to study the effect on the dephasing of the intrinsic motions of the probe. Methyl groups rotation, conformational interconversion and a libration in the matrix contribute to the echo decay for T>80 K, giving rise to a minimum in phase memory time T_M < 150 ns in the range 130-210 K.

Echo decay and motions
The temperature dependence of phase memory time can be used to studythe motion of the paramagnetic species in a given matrix.

Echo decays of hydrazinium ions in lithium hydrazinium sulphate (LHS) show two local minima for T_M, at 115 K and at 175 K.[71] The minimum at 115 K is attributed to hydrazinium ion librations, whereas the one at 175 K is produced by the hindered rotation of the $-NH_3^+$ group. The minima are explained as due to the transitions from slow to fast limit motions as compared with the splitting between the hyperfine proton lines.

An application of the CPMG method to a study of rotational motions in *t*-butyl groups of the phenoxy radical formed in γ-irradiated single crystal of 4-methyl-2,6-di-*t*-butyl phenol in the temperature range 130-290 K has allowed to detect three different kinds of motions. The resulting values of the dynamical parameters are in good agreement with those obtained for the undamaged precursor molecule by previous NMR studies.[72]

Instantaneous diffusion and concentration
When instantaneous diffusion is a relevant relaxation mechanism, a study of the microwave power dependence of the echo decay can give the microscopic average concentration of radicals in the sample. This determination has been done in a number of cases.[59-61]

2. Pulsed EPR of paramagnetic centers in solid phases

In most of them it is difficult to extract the instantaneous diffusion rate from the simple 2p-ESE decay, since other relaxation mechanisms dominate in causing the loss of phase memory time.

Applications of EchoEPR spectroscopy
Dzuba studied the librational motions of nitroxide spin probes in organicglassy media at different temperatures by using the X-band EchoEPR spectroscopy.[73]

The anisotropic libration motion of semiquinones in photosynthetic reaction centers of Rhodobacter sphaeroides R26 and in frozen isopropanol solution have been studied by Rohrer et al.[74] at W-band which allows a sensitive detection of the orientation-dependent echo-decay functions, since the high Zeeman field leads to a spectral resolution of the anisotropic g-tensor.

The symmetries of different coordinating sites of Fe(III) incorporated into aluminosilicate and aluminophosphate sodalite affect the relaxation times of the paramagnetic ions, the ones with a more symmetric environment relaxing more slowly. X- and W-band EchoEPR spectroscopies allow to distinguish between them.[75]

Extensive use of the capabilities of the EchoEPR technique has been done in a study on Cu(II)-doped inorganic glasses. X- and S-band echoEPR yield information on the local symmetry of the Cu(II) coordination polyhedra, the chemical nature of the atoms in the second and higher coordination spheres, the distribution of the parameters of the static spin Hamiltonian and the low-temperature motions of the dopant-containing structural units.[76]

A clear cut example of the effect of instantaneous diffusion on EchoEPR spectra has been shown in partially deuterated irradiated ammonium tartrate (see figure 6).[64] In the figure one can see two echoEPR spectra obtained with different mw power levels. The spectrum with low power (upper trace) is similar to the cw one, showing the main spectrum due to the radicals with ^{12}C and weak sidebands due to radicals with a ^{13}C. On the other hand in the echoEPR spectrum with higher power level the instantaneous diffusion affects strongly the phase memory times of the spin packets corresponding to

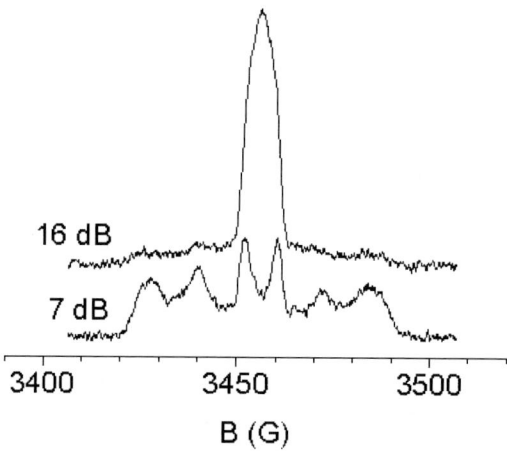

Figure 6. EchoEPR of partially deuterated irradiated ammonium tartrate at room temperature. The main cw-EPR spectrum is given by a single line for this orientation of the single crystal in the magnetic field. Satellite lines are due to molecules with ^{13}C in natural abundance. The two EchoEPR spectra are obtained at different microwave attenuations by means of a 2p-ESE sequence with a $\pi/2$ pulse of 24 ns, a π pulse of 48 ns and a τ of 1216 ns. More details are given in ref. 64.

the more concentrated radical (with ^{12}C), whereas the instantaneous diffusion contribution to the total relaxation is much smaller for the satellite bands, due to the much lower concentration of the relative species, thus giving a stronger signal for these lines.

EchoEPR has also been used to study the motion of molecules in their triplet state. Different spectral lineshapes, with different temperature dependence, have been obtained for a 4-4' disubstituted dithiophene in two different solid matrices, glassy toluene and a spirocyclophosphazene inclusion compound[77] using continous-wave or spin echo detection. This behaviour has been attributed to spin relaxation due to modulation of the Zero Field Splitting tensor induced by fast librational motion of the dithiophene triplet. The echoEPR lineshapes in the two matrices were reproduced considering a librational motion occurring preferentially around different molecular axes.[77]

4.4 Electron spin-spin distance determination

The fine interaction parameter D between two localized electron spins at a distance r~2 nm is D~3.4 G [78] therefore contributing only to the linewidth in the cw-EPR spectrum. On the other hand, in this case it is sometimes possible to obtain information on the electron-electron distance by pulsed EPR. A very concise summary of the experiments follows. Reviews of the methods can be found in refs. 2 and 79.

4.4.1 2+1 ESE

The insertion of an extra pulse between the $\pi/2$ and π pulses of a 2p-ESE sequence produces an increase of instantaneous diffusion. This one, as discussed in section 4.1, depends on the dipolar interactions between the spins giving rise to the echo. The dependence of the echo intensity on the position of the extra pulse is then modulated by the electron-electron interaction.[80-81]

4.4.2 DEER (Double Electron-Electron Resonance) or PELDOR (pulsed ELDOR)

This technique is called either DEER (Double Electron-Electron Resonance) or PELDOR (Pulsed Electron Electron Double Resonance). It allows to measure distances from 1.5 up to 13 nm.[82] The technique is similar to the previous one, the main difference being that the extra pulse (pumping pulse) is at a different mw frequency with respect to the two pulses giving rise to the echo. In this way the spins observed (spins A) can be different from those affected by the extra pulse (spins B). The pumping pulse gives rise to flips of spins B changing the dipolar interactions with spins A. The decay of the PELDOR signal occurs with a time depending on $1/\omega_{dip}$, where ω_{dip} is the dipolar spin coupling.[83] In the case of equal spin-spin distances as in biradicals the echo envelope is modulated. The modulation depends on the fine interaction between the two electrons.[83]

4.4.3 Out-of-phase ESEEM

This technique has been applied to the measurement of electron-electron distance in spin-correlated radical pairs. The modulation of the out-of-phase echo is dominated by spin-spin couplings between the two radicals.[78,84] Exchange and dipolar interactions can be both measured.

Important information on light-induced structural changes as the distance between acceptor and donor cofactors in the reaction centers has been obtained from the study of the radical pairs formed following photoexcitation in photosynthetic systems (see refs. 33 and 85 and references therein).

5. THE DETERMINATION OF HYPERFINE INTERACTIONS

The determination of hyperfine interactions is important both for assessing the structure of the paramagnetic species and for obtaining information on the 3D local geometry surrounding the electron spin probe. In solid state, due to the large number of hyperfine dipolar couplings with surrounding nuclei and to different types of disorder, the cw-EPR spectrum exhibit inhomogeneously broadened lines. The poor resolution hinders the determination of the hyperfine couplings, except those large with respect to the inhomogeneous linewidths.

Two well established methods are mainly used to measure the hyperfine tensors of weakly coupled nuclei, ENDOR (cw and pulsed) and ESEEM. In principle the same information can be obtained with both spectroscopies.

For a doublet, ENDOR frequencies $v_{\alpha,\beta}$ of weakly coupled nuclei are given for a particular orientation (l,m,n) of the magnetic field in an arbitrary reference system by

$$v^2_{\alpha,\beta} = \frac{1}{4} \left\{ [(T_{xx} \mp 2v_N)l + T_{xy}m + T_{xz}n]^2 + [T_{xy}l + (T_{yy} \mp 2v_N)m + T_{zy}n]^2 + [T_{xz}l + T_{yz}m + (T_{zz} \mp 2v_N)n]^2 \right\}$$

(28)

2. Pulsed EPR of paramagnetic centers in solid phases

where v_N is the nucleus Larmor frequency, T_{ij} are the components of the hyperfine coupling tensor in the reference system. If $|v_\alpha - v_\beta|$ is of the order of magnitude of the ENDOR linewidths the nuclear spin contributes to the so-called matrix ENDOR band. In this case only the ESEEM technique can give information on the number and hyperfine tensors of weakly coupled nuclei. On the other hand, the condition for observing ESEEM is the simultaneous excitation of allowed and forbidden EPR transitions. As a consequence, due to the limited bandwidth of microwave pulses, only small couplings can be studied with ESEEM. This condition is not stringent in the case of ENDOR techniques that therefore allow the determination also of large couplings. The two types of techniques are in conclusion complementary. This is true also by considering the relative intensities in different frequency ranges, since ESEEM is particularly suited to determine low frequencies, as the modulation depths are more pronounced at low frequency, whereas the sensitivity of pulse ENDOR is higher for high frequency transitions.

The ESEEM effect was interpreted by Mims as a modulation of the 2p-ESE decay as related to the presence of coupled nuclei.[86] Now several different pulses sequences are used giving rise to the ESEEM effect.[87]

Among these techniques, it is worth to mention the hyperfine-sublevel correlation (HYSCORE) technique, that is a 2D experiment particularly useful for the assignment of ESEEM frequencies to different spin systems and for the detection of broad signals.

5.1 ESEEM spectroscopy

The most popular experiments are based on the modulation of the 2p-ESE and the 3p-ESE, thus obtaining the 2p- and 3p- ESEEM, but many other experiments have been described.[2,6,21,88-89]

5.1.1 2p-ESEEM

The ESEEM effect due to the hyperfine interaction with a nuclear spin can be detected when the mw pulse is creating coherences corresponding to the relative allowed and forbidden EPR transitions, see figure 3. After the $\pi/2$ pulse the four coherences (the "allowed" σ_{13} and σ_{24}, and the "forbidden" σ_{14} and σ_{23}) precess in the xy plane each with its frequency. After precessing for a time τ at frequency ω_{ij} each coherence ij has acquired a phase $\Delta\varphi_{ij} = \omega_{ij}\tau$.

For each coherence, the π pulse has the effect of reversing the phase, but also it can induce a nuclear spin flip, as discussed in the theory section. As a consequence, only a fraction of each initial coherence σ_{ij} is refocused after time τ, i.e. the fraction which has conserved the same frequency after the π pulse. The other fractions are now precessing at a different frequency $\omega_{ij} + \Delta\omega$, and therefore they will refocus at different times with respect to τ, unless their frequency jump $\Delta\omega$ is such that $\Delta\omega\tau = 2n\pi$. The echo intensity recorded for different τ is therefore showing a "modulation", with frequencies that are the differences between those of the allowed and forbidden EPR transitions, i.e. the ENDOR frequencies and their combination.[21,90-91]

A straightforward analysis for this system can be done with a density matrix approach to obtain the function $V_m(t)$ in eq. (18) for a 2p-ESEEM experiment.

The result for the $S=1/2$, $I=1/2$ system is[6]

$$V_m(2\tau) = 1 - 2k[\sin^2(\pi v_\alpha \tau)\sin^2(\pi v_\beta \tau)] =$$
$$= 1 - \frac{k}{2}\left[1 - \cos(2\pi v_\alpha \tau) - \cos(2\pi v_\beta \tau) + \frac{1}{2}\cos(2\pi v_+ \tau) + \frac{1}{2}\cos(2\pi v_- \tau)\right] \quad (29)$$

where v_α and v_β are respectively the two ENDOR frequencies, $v_+ = v_\alpha + v_\beta$ and $v_- = v_\alpha - v_\beta$ and k the modulation depth

$$k = \left(\frac{v_N B}{v_\alpha v_\beta}\right)^2. \quad (30)$$

The B parameter, given in a generic reference system xyz in which the external magnetic field is in the plane xy with direction cosines $(l,m,0)$, is:[92]

$$B = \sqrt{(T_{xz}l + T_{yz}m)^2 + [(T_{xx} - T_{yy})lm - T_{xy}(l^2 - m^2)]^2} \quad (31)$$

In the expression above we can note that combination frequencies appear in the spectrum with opposite sign and with half intensity with respect to the ENDOR frequencies. When the applied field is along the principal directions of the hyperfine tensor $B = 0$, then the modulation depth is also equal to zero.

Analytical expressions have been obtained also for a nucleus with $I=1$ as

$$V_m(2\tau) = 1 - \frac{16k}{3}\sin^2(\pi v_\alpha \tau)\sin^2(\pi v_\beta \tau) + \frac{16k^2}{3}\sin^4(\pi v_\alpha \tau)\sin^4(\pi v_\beta \tau) \quad (32)$$

For weak interactions the term in k^2 can be neglected and an expression similar to the $S = 1/2, I = 1/2$ case is obtained.

For $I>1/2$ systems, non-negligible nuclear quadrupolar interactions have to be taken into account. For the solution of these problems we send the reader to selected publications,[6,13,21] where details for analytical as well as numerical solutions are given.

It can be seen that if n interacting nuclei are present, the modulation function is given by the product of the modulation functions of each single nucleus as[21]

$$V_m^{tot} = \prod_{i=1}^{n} V_m^i \quad (33)$$

The effect of the power and length of the pulses on the modulation has been analyzed in ref. 93-94. Pulses of 10 ns can excite allowed and forbidden transitions in a range of several gauss (typical values are around 10 G ca).

Modulation of the echo intensity can also be induced by electron-electron interactions. This effect is observed in photoproduced radical pairs.[78,84] In these systems the normal in-phase echo is zero, whereas a modulated out-of-phase echo is observed; the modulation frequency depends on the electron-electron distance, as discussed shortly in section 4.3.3.

5.1.2 3p-ESEEM

In 3p-ESEEM the modulation of the echo envelope is observed by increasing the T delay between the second and third $\pi/2$ pulses, see figure 2. This technique offers the advantage that the echo modulation usually occurs on longer time scales, thus obtaining a better frequency resolution with respect to 2p-ESEEM. For the latter experiment the echo decay is determined by transverse relaxation processes. On the other hand in 3p-ESEEM during T the decay is due to slower processes like electron spin-lattice relaxation, spectral diffusion and transverse nuclear relaxation, as a consequence of the polarization and coherences present in this evolution period (see figure 4).

The analysis done for the 2p-ESEEM can be extended to 3p-ESEEM. The modulation of the stimulated echo depends on the nuclear coherences present during time T, see figure 4). During T the matrix elements corresponding to nuclear coherences precess at the ENDOR frequencies v_α and v_β. Therefore these frequencies only are seen in the modulation of the stimulated echo.

Exact solutions can be obtained for the $S = 1/2$, $I = 1/2$ case, for which the expression is:[4]

$$\begin{aligned}V_m(2\tau+T) = &\ 1 - k\left[\sin^2(\pi v_\alpha \tau)\sin^2(\pi v_\beta(\tau+T))\right.\\ &\ \left. + \sin^2(\pi v_\beta \tau)\sin^2(\pi v_\alpha(\tau+T))\right]\\ = &\ 1 - \frac{k}{2} + \frac{k}{4}\{\cos(2\pi v_\alpha \tau) + \cos(2\pi v_\beta \tau)\\ &\ + [1 - \cos(2\pi v_\beta \tau)]\cos(2\pi v_\alpha(\tau+T))\\ &\ + [1 - \cos(2\pi v_\alpha \tau)]\cos(2\pi v_\beta(\tau+T))\}\end{aligned} \quad (34)$$

It is worth to note that the combination frequencies v_+ and v_- are absent when T is varied, therefore the spectrum is simpler with respect to 2p-ESEEM.

Another important point is that modulations due to the two frequencies v_α and v_β are weighted by factors that depend on τ, that are equal to zero

2. Pulsed EPR of paramagnetic centers in solid phases

when the condition $v\tau = n$ is fulfilled. The latter condition gives rise to the so called *blind spots*: for particular values of τ, some frequencies are not appearing in the spectrum. One can use this condition to eliminate, for example, modulation due to one type of nuclei to put in evidence the modulations due to other nuclei. On the other hand, to be sure to detect all the modulation frequencies, several decays corresponding to different τ values must be recorded. A systematic variation of both T and τ is also possible, giving rise to a 2D-experiment. However, the same information given by such experiment is given with better resolution by HYSCORE, described below.

Similarly, for the 3p-ESEEM of a $S=1/2$, $I=1$ system with negligible quadrupolar interaction, one obtains

$$V_m(2\tau+T) = 1 - \frac{4k}{3}\{\sin^2(\pi v_\alpha \tau)[1-\cos(2\pi v_\beta(\tau+T))] + \\ + \sin^2(\pi v_\beta \tau)[1-\cos(2\pi v_\alpha(\tau+T))]\} + \\ + \frac{4k^2}{3}\{\sin^4(\pi v_\alpha \tau)[1-\cos(2\pi v_\beta(\tau+T))] + \\ + \sin^4(\pi v_\beta \tau)[1-\cos(2\pi v_\alpha(\tau+T))]\} \tag{35}$$

As for the 2p-ESEEM, other cases require particular treatments to obtain analytical or numerical solutions.[6,13,21]

5.1.3 HYSCORE

HYSCORE[95] is a two-dimensional experiment derived from the 3p-ESE where a π pulse has been introduced in between the second and third $\pi/2$ pulses (v. figure 2). The effect of the introduction of π pulse is a transfer of population from one m_s manifold to the other, and a mixing of nuclear coherences in one manifold with those of the other manifold. In the following evolution time these new created coherences evolve with new frequencies corresponding to correlations between sublevels of the two m_s manifolds. These correlations show up as cross peaks in the 2D-FT of the modulation of the echo.

The echo intensity is recorded by varying τ_1 and τ_2. After a double FT with respect to these times, a 2D spectrum with axes ω_1 and ω_2 is obtained. The correlation of two ENDOR frequencies $\omega_\alpha, \omega_\beta$, relative to the same nucleus, is indicated by a cross peak in correspondence of the two frequencies. Since only pairs of frequencies belonging to the same paramagnetic center are correlated, this technique allows to disentangle the spectra due to different species. As an example, in figure 7 one can see the HYSCORE absolute value spectrum of an irradiated succinic acid single crystal at room temperature.

The two pairs of peaks on the diagonal (C, D, E and F) are the regular ESEEM peaks originating from two α-proton couplings. They are attributed to two different sites in the crystal.

Cross-peaks are expected between nuclear transitions that belong to the same manifold. Cross peaks between C and E (peak A) and between D and F (peak B), and the relative symmetric peaks with respect to the diagonal (not labeled in the figure) are present, then indicating that C and E belong to one spin site and that D and F belong to a different site.

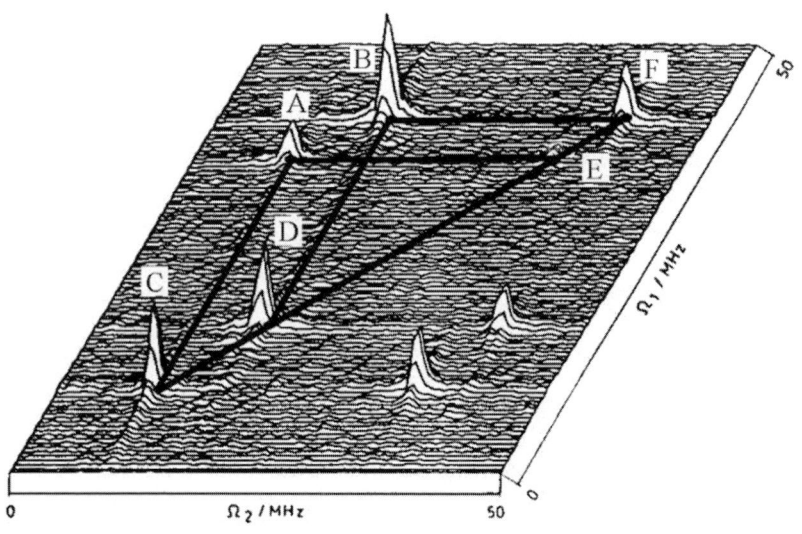

Figure 7. HYSCORE absolute value spectrum of an irradiated succinic acid single crystal at room temperature. Pulse separation $\tau_1=200$ ns, pulse width 10 ns for $\pi/2$ and 14 ns for π. Recorded data matrix (130,130) with steps of 8 ns; zero filled to (256,256) and apodized in τ_1 and τ_2 prior to FT (Adapted from Figure 8 in ref. 96.).

Moreover, if two types of nuclei are detected, with frequencies $\omega_{\alpha 1}, \omega_{\beta 1}, \omega_{\alpha 2}, \omega_{\beta 2}$, cross peaks appear also at $\omega_{\alpha 1}, \omega_{\beta 2}$ and $\omega_{\beta 1}, \omega_{\alpha 2}$. If the two h.c.c.'s of nuclei 1 and 2 have the same sign, $\omega_{\alpha 1}, \omega_{\beta 2}$ and $\omega_{\alpha 2}, \omega_{\beta 1}$ are pairs of high/low (or low/high) frequencies, whereas if the two h.c.c.'s of nuclei 1 and 2 have opposite sign $\omega_{\alpha 1}, \omega_{\beta 2}$ and $\omega_{\alpha 2}, \omega_{\beta 1}$ are pairs of high/high (or low/low) frequencies. Therefore the relative signs of the h.c.c.'s are obtained.

With respect to 1D techniques, the resolution is therefore improved. Moreover, it overcomes the problem of instrumental deadtime that introduces lineshape distortion and loss of broad lines in the 1D spectra. Thus this technique is particularly useful in the study of disordered samples.

Details on the theory of HYSCORE with ideal and non-ideal pulses can be found in refs. 97-98.

5.1.4 Signal analysis

The time trace of ESEEM is normally very little informative without proper analysis. One way to proceed is to perform a spectral analysis in order to extract the frequency components out of the ESEEM signal, that are related to hyperfine interactions. Normally this is achieved by subtracting the unmodulated part of the signal, very often well approximated by an exponential decay, and then by performing the Fourier Transform of the modulated signal. Standard procedures are available to reach this goal.[99]

ESEEM signals suffer for strong limitations due either to the instrumental dead time or to the limited extent of the signal. To improve the signal-to-noise ratio and to avoid introduction of artifacts on the spectrum, some advanced mathematical methods are currently available. A discussion about the comparison among the methods can be found in refs. 2, 100. An efficient approach for powder samples, within the point-dipole approximation, is to use the spherical approximation model[101] to fit directly the time profile from structural information. This has the advantage that time profiles are calculated by assuming a given geometry (distances of paramagnetic nuclei from the paramagnetic center, number of equivalent nuclei).

5.2 Pulsed ENDOR spectroscopy

ENDOR spectroscopy takes advantage of the use of a second radiating field in the radiofrequency (rf) band to induce nuclear transitions. cw-ENDOR is a very well known and powerful spectroscopy. However the interplay of the relaxation rates often limits its applicability. Pulsed ENDOR spectroscopy in general overcomes the problem of unfavorable relaxation properties.

By application of microwave and radiofrequency pulses, selective transfer of either polarization or coherence can be obtained. Several methods are available for this spectroscopy. We limit ourselves to introduce the most popular pulse sequences based on polarization transfer: the Davies[102] and the Mims[103] sequences.

During the preparation step a spin polarization inversion is obtained by application of a proper microwave sequence. This part of the experiment is followed by application of a selective rf pulse that induce a nuclear spin transitions. The inverted polarization of the spin packets with nuclear transitions on-resonance with the rf field is transferred to microwave off-resonance spin packets in the Davies experiment. In the Mims experiment, the inverted spin packets simply exchange polarization between nuclear sublevels. The transfer is complete if the pulse is a nuclear π-pulse.

The magnetization unaffected by the rf pulse is monitored in the detection step by means of an electron spin echo. ENDOR transitions are then determined in these techniques as a reduction of the echo intensity by sweeping the frequency of the rf pulse.

Radiofrequency is applied to the sample in the EPR cavity by means of a properly positioned broadband radiofrequency coil.[104,105] Typically, nuclear transitions are obtained with 10 μs ca pulses. Low temperatures are normally used to have longer relaxation times for the inverted magnetization.

It should be noted that pulsed ENDOR at W-band (95 GHz) has many advantages with respect to X-band, in particular because of the much larger separation in frequency of the lines due to different types of nuclei. Therefore pulsed ENDOR at W-band is one of the more powerful spectroscopies in particular for paramagnetic centers in disordered and biological systems.[106-107] A further advantage of this technique in the latter systems is the better orientation selection which is possible for the increased

2. Pulsed EPR of paramagnetic centers in solid phases

separation of the EPR peaks corresponding to the g-tensor principal directions due to the higher field.[74]

5.2.1 Davies ENDOR

Davies ENDOR pulse sequence is presented in figure 2.

The magnetization is rotated by a selective microwave π-pulse (generation of polarization). Then, during the mixing period, a selective rf pulse is sent to the sample (polarization transfer) and, finally, the magnetization is monitored by a 2p-ESE ($\pi/2$-τ-π-τ) sequence (detection of the polarization). If the rf pulse is resonant with a nuclear transition also the electron polarization is affected and, as a consequence, also the intensity of the echo. The theoretical maximum effect is the complete quenching of the echo intensity.[2]

As several microwave pulses are involved, attention has to be paid to the bandwidth of the different pulses in order to have good selectivity and sensitivity.[2] The bandwidth of the first pulse determines the bandwidth of the inverted packets when broad lines are present. As the pulse must be selective, then the bandwidth of the first pulse is a lower limit for the hyperfine interactions that can be seen with pulse ENDOR. Moreover, the bandwidth of the echo sequence has to be matched to the bandwidth of the rotated spin packets in order not to generate echo signals from unrotated magnetization, and to have the maximum intensity effect. Spectral diffusion processes can induce a modification of the initial bandwidth.

5.2.2 Mims ENDOR

Mims ENDOR (see figure 2) is based on a stimulated echo sequence obtained with unselective microwave pulses, to which a rf pulse is added between the second and the third $\pi/2$ pulses. As discussed in section 2, the z polarization after the second pulse has a modulation at frequency $\pi/2\tau$. It can be shown that the effect of a resonant rf pulse is to reduce this periodic polarization and therefore the amplitude of the stimulated echo.

For an $S=1/2$, $I=1/2$ system, it can be shown[2] that by averaging over all resonance offsets of an infinitely broad EPR line, the echo intensity after nuclear inversion depends on τ as $V=[1+\cos(a_{iso}\tau)]/4$. In Mims ENDOR therefore blind spots occur as in stimulated echo ESEEM.

This technique is convenient for measuring hyperfine couplings smaller than c.a. 5 MHz, whereas for larger couplings Davies ENDOR is a better choice.[2]

In figure 8 the cw-ENDOR and the Mims ENDOR spectra of the [2Fe-2S]$^+$ cluster of *Anabaena* ferredoxin in D$_2$O solvent are shown (from ref. 108).

The cw-ENDOR spectrum (figure 8-A) is dominated by the broad *distant ENDOR* signal from deuterons. On the other hand, a resolved ^{2}H ENDOR spectrum is obtained with the Mims sequence (figure 8-B) at the high-field edge of the EPR spectrum from which a splitting of about 0.60 MHz can be obtained. In fact in Mims ENDOR the *distant ENDOR* signal is not visible because the rf pulse length is shorter than the time for the spin diffusion processes, a necessary condition to obtain the signal.[108]

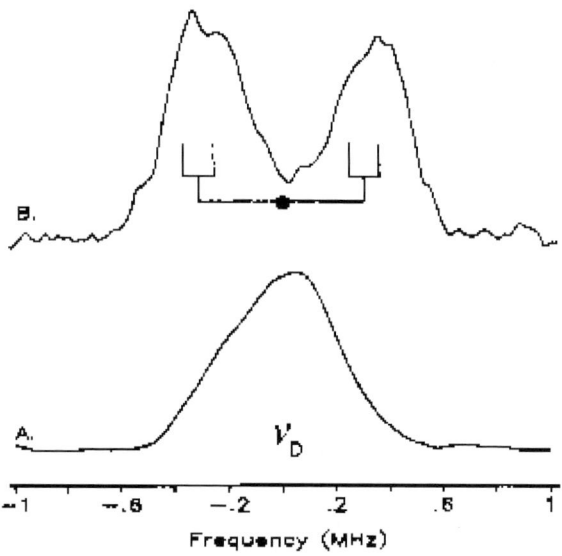

Figure 8. Comparison between the cw-ENDOR (A-bottom) and the Mims ENDOR (B-top) spectra of the [2Fe-2S]$^+$ cluster of Anabaena ferredoxin in frozen D$_2$O solvent (from ref. 108). The two spectra are plotted as function of $\delta\upsilon=\upsilon-\upsilon_D$. Experimental conditions: A) B$_0$=12670 G, υ_e=34.54 GHz, scan rate 0.25 MHz/s, 200 scans. B) B$_0$=3357 G, υ_e=9.15 GHz, mw pulse 16 ns, rf width 40 μs, τ=420 ns, 64 scans.

5.3 Applications

There are many reports in literature that use the above described 1D-, 2D-ESEEM and pulse ENDOR techniques to determine structures of the paramagnetic probe and its surrounding. Many other possible experiments have been also described and suggested in particular by A. Schweiger and co-workers[2] and this wealth of methods will be certainly more and more exploited by the EPR community. Moreover, the advent of High Frequency EPR, both cw and pulsed, increases greatly the power of the described spectroscopic tools.

For the reasons previously mentioned, very often it is the combination of many different EPR methods that allows to have a real insight into the problems. It should be noted that for many systems these techniques represent the only possibility of obtaining a clear cut structural information in particular on complex and disordered systems.

In this section we give an overview of the fields in which the described techniques have been applied, giving references to specific papers as examples of applications, without any pretence to give a complete account of the very abundant research activity in this field.

Applications can be found for different classes of compounds: they spread from inorganic glasses to *in-vivo* biological samples, and in these classes the studied paramagnetic centers are radicals, defects and transition metal centers.

Paramagnetic metals:
The pulsed techniques described here have made possible the detailed study of many metal complexes containing paramagnetic ions, and of the use of paramagnetic metals as probes in many systems. In fact these spectroscopies allow the determination of the superhyperfine interactions of the weakly coupled nuclei surrounding the probe.

Metal complexes: some of the most important applications to metallo centers are reviewed in ref 6,109-110, although for many aspects this is a field very close to metalloproteins[111] discussed below.

Zeolites and inorganic catalysts: pulsed spectroscopies have been used to distinguish and characterize paramagnetic centers of vanadia supported on silica, titania, and magnesia[112]; to interpret the EPR spectra of VO^{2+}-exchanged zeolites[113]; to characterize monocyclopentadienyl Ti(III) catalysts.[114] Inclusion of NO in Na-A zeolites and the interaction of NO with $^{23}Na^+$ has been studied with ESEEM spectroscopy.[115] ESEEM technique has been used in the determination of the coordination geometry of zeolite-encapsulated copper(II)-histidine (CuHis) complexes.[106]

Metalloproteins and biological systems: a recent review by Prisner, Rohrer and MacMillan is devoted to biological application of pulsed EPR.[116] Cammack *et al.* reviewed the studies of hyperfine interactions in [2Fe-2S] proteins by EPR and ENDOR.[117] The information obtained by EPR methods on oxygen-evolving complex of photosystem II has been reviewed by Britt *et al.*[118] and for photosystem I a review can be found in ref. 119.

Pulsed techniques have been applied to the study of blue copper proteins,[120-121] to copper binding sites in prion proteins[122], to proteins and enzymes with Mn(II) or VO^{2+} paramagnetic probes substituting Mg(II), as H^+-ATPase.[123]

Chlorophyll and carotenoid radicals in photosystem II have been studied by pulsed ENDOR.[124] Anisotropic hyperfine components of chemically prepared carotenoid radical cations have been determined by 1D and 2D ESEEM and pulsed ENDOR study.[125]

Electronic structure of antiferromagnetically coupled dinuclear manganese (Mn(IV)-Mn-(III)) complexes has been studied.[126]

High-field (95 GHz) pulsed EPR and electron-nuclear double resonance (ENDOR) techniques have been used for the first time to determine coordinates of ligand protons in the Mn(II) binding site of concanavalin.[127]

The coordination structure and oxidation state of the VO^{2+} ion in vivo in bone samples has been determined with ESEEM spectroscopy.[128]

Inorganic systems and glasses:
Paramagnetic species in $6Li_2O$ $3P_2O_5$ $6SiO_2$ and $6Na_2O$ $3P_2O_5$ $6SiO_2$ glassesproduced under γ-ray irradiation, assigned to the $Q^{n=2}$-type center (POHC-defect), have been studied by ESEEM.[129]

Cu(II)-doped inorganic glasses[76] and borate glasses[130] have been studied. Fe(III) incorporated into $AlPO_4$-20 by has been studied by X- and W-band pulsed EPR spectroscopies.[75] Structure of Cu(II)-histidine complexes in frozen aqueous solutions has been determined from high-field pulsed electron nuclear double resonance.[131]

2. Pulsed EPR of paramagnetic centers in solid phases

Fullerene systems:
Pulsed EPR and ENDOR techniques have been used to study several endohedral fullerenes. Pulse ENDOR experiments on ^{14}N atoms encapsulated in C_{60} (S=3/2) allowed to detect a symmetry lowering induced by a phase transition in polycrystalline C_{60} at T_c=258 K.[132] Hyperfine interaction of the ^{13}C nuclei in the librating C_{60} molecules was detected using pulse ENDOR in both phases.

Pulsed ESR investigations of anisotropic interactions in M@C_{82} (M=Sc,Y,La) in frozen solutions allowed to obtain the principal values of the hyperfine tensor A of the ^{89}Y nuclear spin (I=1/2) with the electron spin on the C_{82} cage in Y@C_{82}. The relative orientation of g and A tensors were determined by applying three- and four-pulse electron spin echo envelope modulation techniques (ESEEM).[133]

Organic semiconductors:
Pulse ENDOR applied to unoriented and oriented Feast-type poliacetylene[134] allowed the determination of the spin density distribution of neutral solitons (radical electrons) present in the system. By fitting the ENDOR lineshape, a distribution over ca 60 carbon units has been found, with alternant even/odd distribution for the nearest-neighbor units.

APPENDIX

The 2p echo decays are all given by a stretched exponential function

$$V_d(2\tau) = V_0 \exp(-2\tau/T_M)^x \tag{A.1}$$

1. In the case of relaxation due to time modulation of anisotropic terms of the spin Hamiltonian, solution of eq. (17) in fast motion régime (Redfield limit)[14], x=1,

$$1/T_M = \Delta^2 \frac{\tau_c}{1+\omega^2\tau_c^2} \tag{A.2}$$

where Δ is the magnetic anisotropy (in frequency units) averaged by the motion and τ_c is the correlation time of the motion.

2. In the case of instantaneous diffusion, x=1 and with $2\theta_1 = \theta_2$ (see section 4.1.1)

$$1/T_M = A + B \cdot \sin^2(\theta_2/2) \tag{A.3}$$

where A is the contribution due to any other process, and B is due to instantaneous diffusion and it is proportional to the concentration see eq. (25), and θ_2 is the tilting angle of the π pulse (assuming non selective pulses).

In the case of spin-spin interactions (spectral diffusion) the values of the exponent and of $1/T_M$ depend on the particular process and model, see Table A.1.

The general expression found for the echo intensity is obtained from solution of eq. (13) as:[50]

$$V_Y(2\tau + T) = n(\omega_k)\hbar^2 \gamma \omega_k/(4kT)\mathrm{Re}\left[\exp\left[i\int_0^{2\tau} s(t')\delta\omega_k(t')dt'\right]\right] \tag{A.4}$$

where ω_k is the resonance frequency of the k-th spin packet, $\delta\omega_k$ its frequency shift due to coupled-spin flips, and s(t') is a step function (equal to +1,0 or –1 during the different free evolutions).

Evaluation of the integral can be performed on the basis of the type of spin-flip mechanism with different models. The expression is given for the three pulses (stimulated) echo, for the 2p-echo the time T = 0.

Table A.1. Analytical expressions for echo decay in the presence of spectral diffusion for given models and approximations. $\Delta\omega_B = 4\pi^2 \gamma_A \gamma_B \hbar C_B / 9\sqrt{3}$, τ_c is the correlation time of the electronic flip-flop process and $m = 32\pi(3 a \gamma_e \gamma_N \hbar)^{3/4} nW 0.01840$ where W is the nuclear flip-flop frequency.

Relaxation mechanism	x	$(1/T_M)^x$	Model	Ref.
Nuclear spin flip-flop	7/4	m		69
B-electronic spin flip-flop	1/2	$2\Delta\omega_B\sqrt{\tau_c/\pi}$	Gauss-Markov ($\tau/\tau_c \gg 1$)	50
B-electronic spin flip-flop	3/2	$2\Delta\omega_B\sqrt{1/6\pi\tau_c}$	Gauss-Markov ($\tau/\tau_c \ll 1$)	50
B-electronic spin flip-flop	1/2	$2\Delta\omega_B\sqrt{\tau_c/2\pi}$	Sudden-jump ($\tau/\tau_c \gg 1$)	50

Relaxation mechanism	x	$(1/T_M)^x$	Model	Ref.
B-electronic spin flip-flop	2	$\Delta\omega_B / 2\tau_c$	Sudden-jump ($\tau/\tau_c \ll 1$)	50

Acknowledgements *We thank Prof. U. Segre for critical discussions on theoretical aspects of this chapter and Mr M. Bellinazzi for help in the preparation of the manuscript.*

6. REFERENCES

1. C.P. Slichter, *Principles of Magnetic Resonance*, Springer-Verlag, Berlin, 1990.
2. A. Schweiger, G. Jeschke, *Principle of pulse electron paramagnetic resonance*, Oxford University Press, New York, 2001.
3. L. Kevan, R.N. Schwartz Ed., *Time domain Electron Spin Rersonance*, John Wiley & sons, New York, 1979.
4. L. Kevan, M.K. Bowman Ed., *Modern Pulsed and continuous-wave electron spin Resonance*, John Wiley & sons, New York, 1990.
5. C. P. Keijzers, E. J. Reijerse, J. Schmidt Ed., *Pulsed EPR: A new field of applications*, North Holland, Amsterdam, 1989.
6. S.A. Dikanov, Y.D. Tsvetkov, *Electron Spin Echo Envelope Modulation (ESEEM) Spectroscopy*, CRC Press, Boca Raton, 1992.
7. N.D. Chasteen, P.A. Snetsinger, in *Physical Methods in Bioinorganic Chemistry, Spectroscopy and Magnetism* Ed. by L. Que, University Science Books, Sausalito, 2000.
8. A. J. Hoff Ed., *Advanced EPR Applications in Biology and Biochemistry*, Elsevier, Amsterdam, 1989.
9. D.J. Kosman, in *Structural and resonance techniques in biological research* Ed. by D.L. Rousseau, Academic Press, New York, 1984.
10. W.B. Mims, J. Peisach in *Biological magnetic resonance* Vol 3, Ed. by L.J. Berliner, J. Rubens, Plenum Press, New York, 1981.
11. J. Schmidt, D.J. Singel, *Ann. Rev. Phys. Chem.* **38** (1987) 141.
12. H. Thoman, L.R. Dalton, L.A. Dalton in *Biological magnetic resonance*, Vol. **6**, Ed. by L.J. Berliner, J. Reuben, Plenum, New York (1984) 143.
13. M. Mehring, V.A. Weberuβ, *Object oriented EPR Classes of objects, Calculations and Computations*, Academic Press, New York, 2001.
14. N.M. Atherton, *Principles of ESR*, Ellis Horwood and Prentice Hall, London, 1993.

15. J.A. Weil, J.R. Bolton, J.E. Wertz, *Electron Paramagnetic Resonance Elementary Theory and Practical Application*, John Wiley & sons, New York, 1994.
16. R.R. Ernst, G. Bodenhausen, A. Wokaun, *Principles of Nuclear Magnetic Resonance in one and two dimensions*, Oxford University Press, Oxford, 1987.
17. J.H. Freed Ch. 3 in *Spin Labeling. Theory and applications,* Ed by L.J. Berliner, Academic Press Inc., New York, 1976.
18. E.L. Hahn, *Phys. Rev.* **80** (1950) 580.
19. H.Y. Carr, E.M. Purcell, *Phys. Rev.* **94** (1954) 630.
20. A. Abragam, *Principles of Nuclear Magnetism*, Clarendon Press, Oxford, 1961.
21. L. Kevan Ch. 5 in *Modern Pulsed and continuous-wave electron spin Resonance*, Ed. by L. Kevan, M. K. Bowman, John Wiley & sons, New York, 1990.
22. A. Barbon, M. Brustolon, A.L. Maniero, M. Romanelli, L. C. Brunel, *Phys. Chem. Chem. Phys.* **1** (1999) 4015.
23. P.B. Kingsley, *Con. Magn. Res.* **7** (1995) 29; ibid. **7** (1995) 115.
24. M.K. Bowman, L.Kevan Ch. 3 in *Modern Pulsed and continuous-wave electron spin Resonance*, Ed. by L. Kevan, M. K. Bowman, John Wiley & sons, New York, 1990.
25. K.M. Salikhov, Y.D. Tsvetkov Ch. 7 in *Time domain Electron Spin Rersonance*, Ed. by L. Kevan, R.N. Schwartz, John Wiley & sons, New York, 1979.
26. W. Gordy *Theory and Applications of Electron Spin Resonance* vol XV of *Techniques of Chemistry* Ed. by A. Weissberger, John Wiley & sons, New York, 1980.
27. K.F. Renk, H. Sixl, H. Wolfrum, *Chem. Phys. Lett.* **52** (1977) 98.
28. R. Durny, S. Yamasaki, J. Isoya, A. Matsuda, K. Tanaka, *J.non Cryst. Solids* **164-166** (1993) 223.
29. M. Brustolon, T. Cassol, L. Micheletti, U. Segre, *Mol. Phys.* **61** (1987) 249.
30. B. Rakvin, N. Maltar-Strmecki, P. Cevc, D. Arcon, *J. Magn. Reson.* **152** (2001) 149.
31. M.K. Bowman, L. Kevan, *J. Phys. Chem.* **81** (1977) 456.
32. M. Seiter, V. Budker, J.-L. Du, G.R. Eaton, and S.S. Eaton, *Inorg. Chim. Acta* **273** (1998) 354.
33. D. Stehlik, K. Moebius, *Annu. Rev. Phys. Chem.* **48** (1997) 745.
34. J-L. Du, G.R. Eaton, S.S. Eaton, *J. Magn. Reson.* A **115** (1995) 213.
35. A.J. Shell-Sorokin, F. Mehran, G.R. Eaton, S.S. Eaton, A. Viehbeck, T.R. O'Toole, A.A. Brown, *Chem. Phys. Lett.* **195** (1992) 225.
36. M. Brustolon, A. Zoleo, G. Agostini, M. Maggini, *J. Phys. Chem. A.* **102** (1998) 6331.
37. M. Brustolon, U. Segre, *Appl. Magn. Reson.* **7** (1994) 405.
38. K. Nagakawa, S.S. Eaton, G.R. Eaton, *Appl. Radiat. and Isot.* **44** (1993) 73.
39. R. Angelone, C. Forte, C. Pinzino, *J. Magn. Reson.* A**101** (1993) 16.
40. B.T. Ghim, J.L. Du, S. Pfenninger, G.A. Rinard, R.W. Quine, S.S. Eaton, G.R. Eaton *Appl. Radiat. Isot.* **47** (1996) 1235.
41. J-L. Du, G.R. Eaton, S.S. Eaton, *J. Magn. Reson.* A **117** (1995) 67.
42. J. Telser, H. Lee, B.M. Hoffman, *J. Biol. Inorg. Chem.* **5** (2000) 369.
43. G.A. Lorigan, R.D. Britt, *Photosyntesis Research.* **66** (2000) 189.
44. T. Nishio, S. Yokoyama, K. Sato, D. Shiomi, A.S. Ichimura, W.C. Lin, D. Dolphin, C.A. McDowell, T. Takui, *Synthetic Metals.* **121** (2001) 1820.
45. G. Mitrikas, C.C. Trapalis, G. Kordas, *J. Chem. Phys.* **111** (1999) 17.
46. C. Malten, J. Müller, F. Finger, *Phys. Stat. Sol. B* **201** (1997) R15.
47. G.G. Fedoruk, *Phys. of Solid State.* **42** (2000) 1147.
48. L.J. Schwartz, A.E. Stillman, J.H. Freed, *J. Chem. Phys.* **77** (1982) 5410.
49. K. Holczer, D. Schmalbein, P. Baker, *Bruker application note* 113.
50. M. Romanelli, L. Kevan, *Concepts Magn. Res.* **9** (1997) 403; ibid. **10** (1998) 1.

2. Pulsed EPR of paramagnetic centers in solid phases 91

51. A.G. Redfield, *Adv. Magn. Reson.* **1** (1966) 1.
52. J.H. Freed, G.K. Fraenkel, *J. Chem. Phys.* **39** (1963) 326.
53. P.L. Nordio Ch. 2 in *Spin Labeling. Theory and applications*. Ed. by L. J. Berliner, Academic Press Inc., New York, 1976.
54. I.M. Brown Ch. 6 in *Time domain Electron Spin Rersonance*, Ed. by L. Kevan, R. N. Schwartz, John Wiley & sons, New York, 1979.
55. A.M. Raitsimiring, K.M. Salikhov, B.A. Umanskii, Y. D. Tsvetkov, *Sov. Phys. Solid State* **16** (1974) 492; A.M. Raitsimring, K.M. Salikhov, Y.D. Tsvetkov, *Fiz. Tverd. Tela* **16** (1974) 756.
56. D.C. Doetschmen, G.D. Thomas, *Chem. Phys. Lett.* **232** (1995) 242.
57. A. Zecevic, G.R. Eaton, S.S. Eaton, M. Lindgren, *Mol. Phys.* **95** (1998) 1255.
58. J.R. Klauder, P.W. Anderson, *Phys. Rev.* **125** (1962) 912.
59. M. Brustolon, A. Zoleo, A. Lund, *J. Magn. Reson.* **137** (1999) 389.
60. I.M. Brown, *J. Chem. Phys.* **58** (1973) 4242.
61. S.S. Eaton, G.R. Eaton, *J. Magn. Reson.A* **102** (1993) 354.
62. R. Boscaino, M. Gelardi, *Phys. Rev. B* **46** (1992) 14550.
63. V.V. Kurshev, A.M. Raitsimring, *J. Magn. Reson.* **88** (1990) 126.
64. M. Brustolon, M. Romanelli, M. Bonora, A. Barbon, A. Lund, *Appl. Magn. Reson.* **20** (2001) 171.
65. D. Goldfarb, L. Kevan, *J. Magn. Reson.* **76** (1998) 276.
66. S.A. Dzuba, *Spectrochim. Acta A* **56** (2000) 227.
67. I. Hiromitsu, L. Kevan, *J. Am. Chem. Soc.* **109** (1987) 4501; T. Hiff, L. Kevan, *J. Phys. Chem.* **93** (1989) 1572; G. Martini, S. Ristori, M. Romanelli, L. Kevan, *J. Phys. Chem.* **94** (1990) 7607.
68. S. Saxena, J. Freed, *J. Phys. Chem. A* **101** (1997) 7998.
69. A.D. Milov, K.M. Salikhov, Y. D. Tsvetkov, *Sov. Phys. Solid State* **15** (1973) 802.
70. M. Lindgren, G.R. Eaton, S.S. Eaton, B.-H. Jonsson, P. Hammarström, M. Svensson, U. Carlsson, *J.Chem.Soc. Perkin Trans2*, (1997) 2549.
71. J. Goslar, W. Hilczer, P. Morawsky, *Solid state ionics* **127** (2000) 67.
72. M. Brustolon, A.L. Maniero, M. Bonora, U. Segre, *Appl. Magn. Reson.* **11** (1996) 99.
73. S.A. Dzuba, *Phys. Lett. A*, **213** (1996) 77.
74. M. Rohrer, P. Gast, K. Moebius, T.F. Prisner, *Chem. Phys. Lett.* **259** (1996) 523.
75. D. Arieli, D.E. W. Vaughan, K.G. Strohmaier, H. Thomann, M. Bernardo, D. Goldfarb, *Magn. Reson. In Chem.* **37** (1999) S43.
76. R. Stosser, S. Sebastian, G. Scholz, M. Willer, G. Jeschke, A. Schweiger, M. Nofz, *Appl. Magn. Reson.* **16** (1999) 507.
77. A. Barbon, M. Bortolus, M. Brustolon, A. Comotti, A.L. Maniero, U. Segre, P. Sozzani submitted to *J. Phys. Chem. B*.
78. A.J. Hoff, P. Gast, S.A. Dzuba, C.R. Timmel, C.E. Fursman, P.J. Hore, *Spectrochim. Acta A* **54** (1998) 2283.
79. K.V. Lakshmi, G.W. Brudvig, *Curr. Opin. Struct. Biol.* **11** (2001) 523.
80. V.V. Kurshev, A.M. Raitsimring, Yu.D. Tsvetkov, *J. Magn. Res.* **81** (1989) 441.
81. A.V. Astashkin, H. Hara, A. Kawamori, *J. Chem Phys.* **108** (1998) 3805.
82. A.V. Milov, A.G. Maryasov, Yu.D. Tsvetkov, J. Raap *Chem. Phys. Lett.* **303** (1999) 135.
83. A.V. Milov, Yu.D. Tsvetkov, F. Formaggio, M. Crisma, C. Toniolo, J. Raap, *J. Am. Chem. Soc.* **123** (2001) 3784.
84. J. Tang, M.C. Thurnauer, J.R. Norris, *Chem. Phys. Lett.* **219** (1994) 283.
85. M. Iwaki, S. Itoh, H. Hara, A. Kawamori *J. Phys. Chem. B* **102** (1998) 10440.
86. W.B. Mims, *Phys. Rev. B* **5** (1972) 2409.

87. A. Schweiger, *Appl. Magn. Reson.* **5** (1993) 229.
88. L. Kevan, Ch. 8 in *Time domain Electron Spin Rersonance*, Ed. by L. Kevan, R. N. Schwartz, John Wiley & sons, New York, 1979.
89. W.B. Mims, J. Peisach, Ch. 1 in *Advanced EPR Applications in Biology and Biochemistry*, Ed by A. J. Hoff, Elsevier, Amsterdam, 1989.
90. N.M. Atherton, *Chem. Soc. Rev.* 293 (1993).
91. A. Schweiger, *Angew. Chem. Int. Ed. Engl.* **30** (1991) 265.
92. M. Brustolon, A.L. Maniero, S. Jovine, U. Segre, *Res. Chem. Intermed.* **22** (1996) 359.
93. L. Braunschweiler, A. Schweiger, J.M. Fauth, R.R Ernst, *J. Magn. Reson.* **64** (1985) 160.
94. H. Barkhuijsen, R. de Beer, B. J. Pronk, D. van Ormondt, *J. Magn. Reson.* **61** (1985) 284.
95. P. Höfer, A. Grupp, H. Nebenfuehr, M. Mehring, *Chem. Phys. Lett.* **132** (1986) 279.
96. P. Höfer *Bruker Application notes* n. 118.
97. R. Szosenfogel, D. Goldfarb, *Mol. Phys.* **95** (1998) 1295.
98. C. Gemperle, G. Aebli, A. Schweiger, R.R. Ernst, *J. Magn. Res.* **88** (1990) 241.
99. W.H. Press, B.P. Flannery, S.A. Teukolsky, W.T. Vetterling, *Numerical Recipes, The Art of Scientific Computing*, Cambridge University Press, Cambridge, 1986.
100. D. van Ormond, Ch. 4 in *Pulsed EPR: A new field of applications*, Ed. by C. P. Keijzers, E. J. Reijerse, J. Schmidt, Ed. North Holland, Amsterdam, 1989.
101. L. Kevan, Sec. 6 in *Pulsed EPR: A new field of applications*, Ed. by C. P. Keijzers, E. J. Reijerse, J. Schmidt, North Holland, Amsterdam, 1989.
102. E.R. Davies, *Phys. Lett. A* **47**A (1974) 1.
103. W.B. Mims, *Proc. R. Soc. London* **283** (1965) 452.
104. J. Forrer, S. Pfenninger, J. Eisenegger, A. Schweiger, *Rev. Sci. Instrum.* **61** (1990) 3360.
105. W. Bietsch, J.U. vin Schuetz, *Bruker Report* **139** (1993) 12.
106. R. Grommen, P. Manikandan, Y. Gao, T. Shane, J.J. Shane, R.A Schoonheydt, B.M. Weckhuysen, D. Goldfarb, *J. Am. Chem. Soc.* **122** (2000) 122.
107. B. Epel, C.S. Slutter, F. Neese, P.M.H. Kroneck, W.G. Zumft, I. Pecht, O. Farver, Y. Lu, D. Goldfarb, *J. Am. Chem. Soc.* **124** (2002) 8152.
108. C. Fan, M.C. Kennedy, H. Beinert, B.M. Hoffman, *J. Am. Chem. Soc.* **114** (1992) 374.
109. C. Calle, R.-A. Eichel, C. Finazzo, J. Forrer, J. Granwehr, I. Gromov, W. Groth, J. Harmer, M. Kälin, W. Lämmler, L. Liesum, Z. Mádi, S. Stoll, S. Van Doorslaer A. Schweiger, *Chimica* **55** (2001) 763.
110. J.R. Pilbrow *Transition ion Electron Paramagnetic Resonance*, Oxford University Press, New York, 1990.
111. Y. Deligiannakis, M. Louloudi, N. Hadjiliadis, *Coo. Chem. Rev.* **204** (2000) 1.
112. V. Luca, D.J. MacLachlan, R. Bramley, *Phys. Chem. Chem. Phys.* **1** (1999) 2597.
113. P.J. Carl, S.L. Isley, S.C. Larsen, *J. Phys. Chem. A* **105** (2001) 4563.
114. S. Van Doorslaer, J.J. Shane, S. Stoll, A. Schweiger, M. Kranenburg R.J. Meier, *J. Organomet. Chem.* **634** (2001) 185.
115. D. Biglino, M. Bonora, A. volodin, A. Lund, *Chem. Phys. Lett.* **349** (2001) 511.
116. T. Prisner, M. Rohrer, F. MacMillan, *Ann. Rev. Phys. Chem.* **52** (2001) 279.
117. J.K. Shergill, R. Cammack, J.H. Weiner, *J. Chem. Soc. Faraday Trans.* **89** (1993) 3685.
118. J.M. Peloquin, R.D. Britt, *Biochim. Biophys. Acta-Bioenerg.* **1503** (2001) 96.
119. Y. Deligiannakis, A.W. Rutherford, *Biochim. Biophys. Acta genetics* (2001) 1507
120. M. van-Gastel, J.W.A. Coremans, J. Mol, L.J.C. Jeuken, G.W. Canters, E.J.J. Groenen, *J. Biol. Inorg. Chem.* **4** (1999) 257.
121. C.E. Slutter, I. Gromov, B. Epel, I. Pecht, J.H. Richards, D. Goldfarb, *J. Am. Chem. Soc.* **123** (2001) 5325.

2. Pulsed EPR of paramagnetic centers in solid phases

122. C.S. Burns, E. Aronoff-Spencer, C.M. Dunham, P. Lario, N.I. Avdievich, W.E. Antholine, M.M. Olmstead, A. Vrielink, G.J. Gerfen, J. Peisach, W.G. Scott, G.L. Millhauser, *Biochemistry* **41** (2002) 3991.
123. B. Schneider, C. Sigalat, T. Amano, J.L. Zimmermann, *Biochemistry* **39** (2000) 15500.
124. P. Faller, T. Maly, A.W. Rutherford, F. MacMillan, *Biochemistry* **40** (2001) 320.
125. T.A. Konovalova, S.A. Dikanov, M.K. Bowman, L.D. Kispert, *J. Phys. Chem. B* **105** (2001) 8361.
126. K.O. Schafer, R. Bittl, W. Zweygart, F. Lendzian, G. Haselhorst, T. Weyhermuller, K. Wieghardt, W. Lubitz, *J. Am. Chem Soc.* **120** (1998) 13104.
127. R. Carmieli, P. Manikandan, A.J. Kalb, D. Goldfarb, *J. Am. Chem. Soc.* **123** (2001) 8378.
128. S.A. Dikanov, B.D. Liboiron, K.H. Thompson, E.Vera, V.G. Yuen, J.H. McNeill, C. Orvig, *J. Am. Chem. Soc.* **121** (1999) 11004.
129. G. Kordas, *J.non Cryst. Solids* **133** (2001) 281.
130. G.Kordas, *Phys. And Chem. of glasses* **42** (2001) 226.
131. P. Manikandan, B. Epel, D. Goldfarb, *Inorg. Chem* **40** (2001) 781.
132. N. Weiden, H. Kass, K.P. Dinse, *J. Phys. Chem. B* **103** (1999) 9826.
133. S. Knorr, A. Grupp, M. Mehring, U. Kirbach, A. Bartl, L. Dunsch, *Appl. Phys. A* **66** (1998) 257.
134. H. Kaess, P. Höfer, A. Grupp, P.K. Kahol, R. Weizenhoefer, G. Wegner, M. Mehring, *Europhys. Lett.* **4**, 947 (1987).

Chapter 3

DYNAMICAL EFFECTS IN CW AND PULSED EPR

Nikolas P. Benetis
Department of biological Chemistry, School of Medicine, University of Ioannina,451 10 Greece

Key words: Libration, site exchange, step-free rotation, Field-Sweep Echo- Detected EPR, local fields, adiabatic broadening, lifetime broadening, methyl tunnelling, free planar-methyl rotor, spin-rotation coupling, torsional oscillation, quenched (stopped) rotor, stretched exponential decay, spin-diffusion, spectral diffusion, instantaneous diffusion

Abstract: Typical molecular processes with EPR accessible dynamical parameters are catalogued and evaluation of their timescales according to the different EPR methods used for this purpose is described. The detection and description of the dynamics of small cyclic radicals and related nitroxide labels in solids and the connection to the structural parameters are given. Both fast-motion averaging and the method of lineshape modification due to chemical exchange are outlined. Some usually overlooked anomalies concerning the activation parameters of the thermally activated rotary motion and their relation to the microscopic variables of the spin-motion system are discussed. The definition of the thermodynamic limits differentiating diffusional motion from quantum motion and the particular ways of the couplings of these motions to the spin system are exemplified. Several experiments manifesting their difference, such as comparison of EPR spectra for classical and for tunnelling rotors, as well as severely distorted EPR spectra including totally quenched (stopped) methyl-type rotors are reviewed and explained. Spin-lattice relaxation and broadening are discussed for fast and slow motions in solids and the characterization of the dynamics according to the effects of motion on the ESE decay are considered in the framework of pulsed EPR. Finally, some standard biological applications in determining the timescales and the motional pattern of disordered matter at the molecular level are described with emphasis to the modern pulsed EPR techniques and some relevant recent developments.

1. INTRODUCTION

As opposed to the *statics* in physics that is applicable to bodies in rest, and the *kinematics* that describes the motions of the bodies, *dynamics* has the objective to determine which motion will take place under the influence of given forces.[1]

In solids, several types of paramagnetic centers can be found, such as stabilized electrons, atoms, paramagnetic ions, ion pairs, radicals, and biradicals. The *motional dynamics* and *spin dynamics* phenomena to be discussed in this chapter concern the paramagnetic species as they interact with their closest environment in solid matrices, leading to modification of the EPR measurables. In particular the measurables of the CW EPR (Continuous Wave Electron Paramagnetic Resonance) and transient (pulsed) EPR spectroscopy for free radicals are determined by the motional degrees of freedom which are allowed for these radicals.

One cannot avoid considering some properties of *fluid motional dynamics* as detected by EPR, either because of their similarities or of their contrasting characteristics compared to dynamics in solids. For instance even though the environment of the paramagnetic probes in composite phases such as *zeolites* and *liquid crystals*, particularly *membranes*, is highly anisotropic, the very slow and restricted overall motion of some large units can coexist with much faster *local motions* and internal rotary motions.

In most cases, approximate simulation of the macroscopic magnetic properties can be obtained from first principles. The spins are usually treated *quantum mechanically* while the rest of the degrees of freedom, the *motional part* of the environment with which the spins interact, are treated as *classical systems* with a continuum of energy levels, or at least as *random processes*. Interestingly, some of the motional degrees of freedom have to be treated quantum mechanically,[2] and vice versa, some fast relaxing electron spins must be included in the "lattice",[3] as it will be seen further in this text.

Often, very crude dynamical models, such as the Arrhenius law of exponential temperature dependence of the rate of thermally activated processes, work nicely at the macroscopic level.[4,1] As it was seen however, the activation model of Arrhenius was not always consistent with the microscopic model for hindered rotational motion.[5,6] In some cases neither *free diffusion* nor a *jump model* were particularly successful to describe a classical rotor in presence of a periodic *restoring force*[5] and more general models such as a Smoluchowski drift diffusion model was necessary in order

3. Dynamical effects in CW and pulsed EPR

to account for the motional subsystem. The rotation of methyl-type rotors at the lowest temperatures had to be described quantum mechanically.[2,7] Both a quantum and a classical description of the motion were needed simultaneously for the *transition regime* between *inertial* and *diffusional limits* at intermediate temperatures.[6] For many classical rotors it appears that the motion does not really stop even for temperatures lower than the melting point, but at least eventually the molecule (radical) becomes (partially) immobilized.

Methyl-type radicals, for which the methyl radical $^\bullet CH_3$ with D_3 symmetry is the prototype, or similar small fragments with a different central atom than carbon and/or protons exchanged by deuteron, can be isolated in inert gas matrices at cryogenic temperatures. Their EPR spectra display quantum effects such as *extremely sharp* and easily saturated EPR signals and in general they behave rather differently than strongly hindered rotors. The spectral properties of these "free" rotors indicate that there is not good thermal contact to the lattice, a fact also verified by the apparent free rotation of these radicals for very low temperatures. Interestingly, those rotors stop at temperatures close and under 5 K "in spite" that they are not hindered![7]

Almost all biological ESEEM applications of paramagnetic metal ions are possible only under cryogenic conditions.[8] and the same situation is found also in some *triplets*.[9,10] This is predominantly due to the *homogeneous broadening* that arises when the anisotropic interactions of the electron spin are modulated by *thermal motion* at higher temperatures. The ESE (Electron Spin Echo) methods eliminate the inhomogeneous broadening, but the homogeneous broadening is also usually large and it is still always there to tell about the dynamics. Some direct FID (Free Induction Decay) methods are possible in EPR, however. In particular motional narrowed spectra of *e.g.* nitroxides can routinely be obtained by *one-shot registration*[11] after a single pulse, and subsequent FFT (Fast Fourier Transformation) of the time resolved signal. Also, special soft pulse sequences involving Extended Time Excitation[12] (ETE), allow for the whole ESEEM signal to be registered by using a single two-pulse echo experiment.[13]

Relaxation in relatively spin-dense systems of radicals and echo decay starts to become substantial along with the realization of effects such as *spin-diffusion, spectral-diffusion,* and *instantaneous diffusion*. The relaxation phenomena that are important in solids have to be considered in this case and if the concentration of the paramagnetic centers is still rather low, even the pure *nuclear spin diffusion* can also affect the electron spins by modulations of the local electron-nuclear *S-I* dipole-dipole interaction.[14,15,16] These phenomena will be described in this text along with the pulsing of inhomogeneous lines, and will be used for the interpretation of the Echo Detected EPR spectra of nitroxides.

Several important concepts such as *spin-diffusion* are also known from the nuclear magnetism (NMR) and the book of Abragam,[17] which is a good source of additional reading, also for other general concepts such as the *spin temperature*. One of the important early successes of spin diffusion was the explanation of the strong enhancement of the relaxation rates and the broadening of solid diamagnetic samples by introduction of very small amount of *paramagnetic impurities*.[17,18]

2. TYPES OF DYNAMICAL PROCESSES ACCESSIBLE BY EPR

In the following we list several kinds of dynamics that have been typically investigated by EPR (and NMR). We have tried to collect also the timescales of some interesting dynamical processes that have been studied by the EPR methods in the following Table 1. The classification used is certainly not unique, but the collected cases constitute an almost complete overview of the most common motions studied by magnetic resonance in solids and fluids.

The term *libration* refers to small-amplitude overall *restricted-reorientation* motion.[19] This motion is very important in modulating the magnetic interactions for partially immobilized anisotropic systems, such as nitroxide radicals, used in studies of polycrystalline phase or glassy materials. Libration can induce very similar reorientation effects as internal motion in cyclic nitroxides as it was observed by the group of Brustolon.[14] For the lower temperatures, $T < 200$ K, libration was the cause of the overall restricted reorientation of the tempone around the N-O axis. At higher temperatures, $T > 230$ K, the motion of a characteristic direction of the (hyperfine) hf tensor of the nitrogen atom was induced by *intramolecular* conformation change of the tempone molecule between two possible twisted-crossover conformations of the 6-membered ring, see Fig. 1. A more sophisticated motional model involving libration was investigated by the group of Dzuba in order to describe aggregates with long-range order[20] with simultaneous local motions.

3. Dynamical effects in CW and pulsed EPR

A different type of *restricted motion* was the overall molecular rotation occurring around a certain axis defined with respect to the molecular frame found for temperatures lower than 144 K in the morpholine radical,[21] with a

Table 1. The dynamic range of the most usual motions. The correlation times τ_C, given here in experimental situation and give the corresponding reference

Intramolecular processes
o *Conformation changes*, i.e. motions involving internal molecular degrees of freedom
÷ *Hindered rotation* and *torsional oscillations* of small fragments such as methyl with respect to the rest of the molecule[22-25]
÷ *Methyl rotation*: fast[26] $10^{-10} < \tau_C < 10^{-7}$, slow[27] $10^{-5} < \tau_C < 10^{-2}$
÷ *Twist-crossover interconversion*[14] in 6-membered-rings, as in cyclic nitroxides $\tau_C \sim 10^{-7}$
÷ *Puckering motion* and generally large amplitude[28] *pseudorotation* type of motion in saturated 4-, 5-, and 6-membered rings: $\tau_C < 10^{-8}$, see also Chapter 1 in this book
o *Exchange* of *protons* or any other small group between different sites
o *Exchange* of *electrons* $\tau_C \sim 10^{-10} - 10^{-12}$
o *Isomerization* of some kind e.g. *valence isomerization*
o *Pseudorotation* due to the Jahn-Teller effect
o *Spin-Rotation* interaction[2] in methyl radical

Intermolecular processes
o Overall *tumbling-rotation*.[11] Slow motion: $\tau_R \geq 10^{-6}$, fast motion: $\tau_R \leq 10^{-11}$
o Overall *restricted rotation* of molecules[21] and molecular fragments[7]
o Overall small-amplitude *librational motion*,[29] slow motion at ca 77 K, $\tau_C > 3 \cdot 10^{-9}$
o Libration of ring protons[30]
o *Diffusion of defects* and *radicals* in crystals[44]
o *Exchange* of small fragments between a molecule and the surrounding bulk[18]
o *Electron transfer*, ref. [31] pag. 207, $\tau_C \sim 2 \cdot 10^{-8}$
o *Proton transfer*, ref. [31] pag. 204, $\tau_C \sim 6 \cdot 10^{-9}$
o *Translational motion* (*Heisenberg spin-exchange*[1]). Liquids $10^{-11} < \tau_C < 10^{-9}$
o *Ion-pair interconversions* (association- dissociation) $\tau_C \sim 10^{-9}$
o Electron-spin exchange, ref. [31] pag. 204, $\tau_C \sim 7 \cdot 10^{-11}$
o Tempone in perdeuterated toluene[11] at 21° C, $\tau_C \sim 2 \cdot 10^{-7}$
o *Doppler effect* and *collisions in gases*, Abragam[17] pags. 427, 322

---

[2] Spin-rotation coupling is used in two different senses: a) In relaxation theory it means the coupling of the electron spin with the orbital angular momentum. b) In treating quantum rotors it means the mixing of the rotation of the molecule with the nuclear-spin degrees of freedom.

ring skeleton similar to that of tempone. The relevant correlation time was ca 10^{-7} sec, in the temperature range where this motion was accessible by CW EPR. The interconversion of this molecule between *e.g.* two stable chair conformations was not possible at that temperature range due to the high barrier, a fact verified also by the absence of the exchange of the protons of the ring.[21] On the other hand, the morpholine skeleton was flat enough to rotate inside the matrix cavities around an axis with a restricted direction.

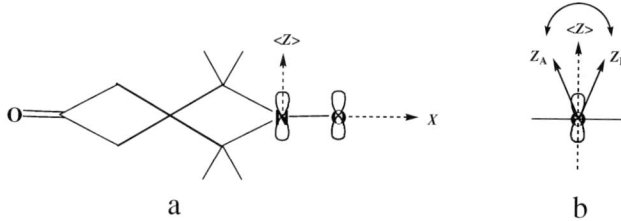

Figure 1. For temperatures lower than 200 K the tempone molecule is assumed to perform restricted reorientation in the host matrix around the molecular *x*-axis. Exchange between twist crossover conformers of the same molecule can be involved for temperatures higher than 260 K, modulating the direction of the hfi tensor. (Fig. 1 ref. [14])

The probability density of the orientation of the libration axis of elongated molecules involving a restoring potential was used in that work in order to relax the condition of perfect order. At the lowest experimental temperatures this kind of model is a reasonable step toward a *wobbling* motion that is assumed for higher temperatures, meaning a slow drift of the direction of the libration axis simultaneously active to libration.

Except from the above pure cases, some exchange processes can act simultaneously, as *e.g.* in small saturated cyclic hydrocarbon radical ions under matrix isolation, where several kinds of ring conformers are possible.[33]

Another interesting case is a correlation between the "vastly different" processes of the *overall reorientation* (tumbling) and some *intramolecular* chemical exchange reactions[34] that can occur in parallel in the coordination sphere of complexes with bulky ligands in viscous solutions.

3. SELECTED EXAMPLES OF CONFORMATIONAL REORGANIZATION AND LIBRATION

Many detailed conformation studies of *saturated hydrocarbon rings* by EPR, involving temperature dependent lineshape simulations for the determination of the experimental *activation parameters*, exist in literature.[33] More systematic coverage of the subject with examples of several studied systems, are given in this book by Lund and Liu. Some conformation studies will be discussed further together with the EPR-lineshapes in exchanging systems, where also some questions about the equivalence between the *microscopic barriers* and the experimental *activation energies* are raised.[5,6]

Averaging of the tensorial magnetic parameters with respect to (intermolecular) overall motion is usually performed directly over the orientation dependent parameters of the spin-Hamiltonian. The set of the magnetic parameters used in simulations that include dynamics is assumed to be obtained in as low temperature as possible, and it is implicitly assumed that the rotational invariants (trace and determinant) of these parameters do not change significantly with temperature.

The contribution of each ring proton of the "rigid" unsaturated rings with conjugated π-bond systems to the hf interaction can be estimated by using the semi-empirical relations of Heller-McConnell type. It is based on the proportionality of the hf coupling to the *electron density* of the π-orbitals of the adjacent carbon atoms; see *e.g.* ref. [26] Chapt. 6-7. On the other hand, in the versatile saturated 5- and 6-heterocyclic rings containing nitrogen, such as the case of pyrrolidine and piperidine skeletons (involved in cyclic nitroxides), thermally induced deformations occurring by interconversion between the stable conformations of the ring carbons, modulate the hf couplings at relatively high temperatures. Each intermediate stable configuration determines automatically the location of the axial and equatorial protons, as well as the configuration of the C-NO-C fragment of the ring in cyclic nitroxides, and by this way their hf parameters.

The effect of averaging of the magnetic parameters for conditions of rapid motion in cyclic nitroxides is given by Rockenbauer *et al.*[35] for several internal motions and can be used in high resolution EPR studies.

3.1 Saturated rings and cyclic nitroxides

Here are some examples of the involved processes, which are responsible for the most important intramolecular *large-amplitude* motions[28,35] affecting the averaging of the magnetic parameters of small cyclic radical ions and nitroxides.

a) *Ring puckering* can be seen as part of the conformational changes giving finally the *chair* and *boat* transformations of the 6-rings and the *"envelope"* structure of the 5-membered rings.

b1) 5-membered rings conformations: i) *Envelope* ii) *Twist*

b2) 6-membered rings conformations: i) *Chair* and ii) *Boat* iii) *Twist-crossover*, see Fig. 1.

c) *Pyramidal* (out of plane) *inversion/distortion* of the C-NO-C fragment vs. planar sp^2 hybridization.

d) *Pseudorotation*. The above three types of distortions can "rotate" around the ring, exchanging position with the neighbouring corners of the polygon. This rotation does not involve any actual displacement of mass in the plane of the ring, therefore the adjective pseudo (= false in Greek)

e) *Internal rotation* of the methyl groups.

To give some orders of magnitude for the experimentally obtained activation energies involved in the above processes we reproduce the values for different nitroxide molecules as they were reported in ref. [35], where they were compared also to *ab-initio* computations. To gain some feeling about the temperatures that are required for the thermal activation of these motions the equivalent of the energies in Kelvin is given in parenthesis. The conversion was made according to the relations: 1.0 kcal/mole ↔ 4.182 k Joule/mole ↔ 503.2 K.

Chair-chair interconversion activation energy: 7-8 kJ/mole (ca 902 K).

Methyl rotation activation energy: 20-30 kJ/mole (ca 3000 K!) in 4-methyl and 4-propyl substituted piperidine-1-oxide.[35]

The pyramidal inversion barrier of the C_2NO fragment at the nitrogen atom is referred to as a small one compared to the above two processes.

The energy difference between the axial and equatorial configurations of the NO-fragment was 2-3 kJ/mole (ca 300 K).

Comparing also to the regular methyl rotation with a potential barrier of ca 1000 K, *e.g.* 9 kJ/mole (1082 K) in tempone,[14] and the inversion of ammonia which needs ca 3047 K to be activated, it is obvious that a) The activation energies for methyl rotation in nitroxides are relatively high while the pyramidal inversion at the Nitrogen atom has a relatively small value b)

They are widely varying among the activation energies of the above processes as seen by the presented values.

3.2 The hyperfine interaction of beta protons

As an example about how the averaging can be performed one can take the usual dihedral-angle dependence of the scalar hyperfine spin-spin coupling for the internal rotation of the beta-protons with respect to the carbon with the unpaired p_z electron, ruled to a good approximation by the following relation[36]

$$a = a_0 + a_2 \cos^2 \vartheta \qquad (1)$$

The mechanisms of *spin polarization* and the *hyperconjugation* for the contact term of the hf coupling correspond to the above coefficients a_0 and a_2, respectively. The first term in the above equation, with a value of $a_0 \sim 10$ MHz, is sometimes ignored compared to the major contribution from the second term, which is proportional to $a_2 \sim 140$ MHz.

The dihedral angle ϑ between the two planes defined by the p_z axis and the C–H bond, having in common the C–C bond, varies in time with rotation or by thermally activated hindered rotary motion of the methyl fragment. In the latter case it can be considered as a stochastic process $\vartheta(t)$ and through the cosine functional dependence the contact coupling is also modulated randomly by motion. The average of the above relation for a very "flat" potential, *i.e.* a freely rotating methyl, is $<a> = a_0 + a_2/2$. For high potential barriers the hf coupling depends on the *twist angle* ϑ_0, which is an important parameter in the configuration of the quenched (stopped) rotor, determining the positions of the wells of the potential energy function of the methyl fragment compared to the direction of the p_z orbital.[5,6] Some of the above motional effects on the EPR lineshape will be discussed further for the different conditions of *diffusional* and *inertial* rotation.

It can be generally shown that for several motional models, in particular for libration motion, the average $<\cos^2 \vartheta>$ is not always 1/2 but at least it takes the same functional form, see ref. [37,30] and citations there:

$$<\cos^2 \vartheta> = (1/2)[1 + \cos(2\vartheta_0) F(\alpha)] \qquad (2)$$

Where one can obtain the following functions $F(\alpha)$ of the libration amplitude α.

(i) Uniform distribution[37] $F(\alpha) = \text{sinc } \alpha \equiv (\sin \alpha)/\alpha$
(ii) Two-site jump exchange[37] $F(\alpha) = \cos \alpha$
(iii) Classical harmonic oscillator[38] $F(\alpha) = J_0(2\alpha)$, where J_n are the Bessel functions of the first kind[39]
(iv) Stochastic harmonic oscillator[40] $F(\alpha) = \exp\{-<\alpha^2>\}$

The averages shown for the above models were tested in the following system: A proton detachment from a carbon in the ring of glutarimide in X-ray irradiated single crystal of this substance generated a carbon-centered radical with a pure p_z orbital.[30] The hyperfine couplings of the two protons in beta position with respect to the radical site changed reversibly between 51-38 G and 12-22 G in the temperature range 100-375 K. The correlated motion of the two beta protons of the ring was considered as a possible source of the temperature effects on the EPR lineshapes. Although it was not clear from the EPR data if the motion of the protons was libration or reorientation due to conformation changes, a chemical exchange model considering the temperature-dependent amplitude of the motion was used successfully to analyze the situation.

4.   FUNDAMENTAL DYNAMICAL PARAMETERS

To specify the dynamics of a system in an average way one needs relatively few parameters. As an example we can take the overall reorientation in the thermodynamic limit of diffusional motion. The relevant dynamical parameter could be the *reorientation correlation time* τ_R of a paramagnetic species, *e.g.* a radical or a nitroxide molecule, dissolved in a liquid. It can be defined as the average time during which the paramagnetic species remains at a certain orientation. The range of the reorientation correlation times that give visible effects on the EPR lineshape extends between the pico-second regime, $\tau_R \sim 10^{-12}$ sec (*fast motion*) to $\tau_R \sim 10^{-5}$ sec (*slow motion*), depending also on the strength of the interaction of the spin with the molecular motion. Close to the fastest and the slowest limits within the above range, and often before the limits are reached, a change in the correlation time cannot further alter the CW EPR lineshape.

Beyond these limits, other magnetic resonance methods such as pulsed EPR can still be sensitive. From this point of view the EPR and the NMR complement each other in sensitivity to different *molecular motions* in fluids, such as solutions of paramagnetic substances.[18] Furthermore, other motions such as *internal motions* of molecular fragments in solids can be investigated by selecting the appropriate magnetic resonance method since these motions can exhibit average fluctuation frequencies, or equivalently inverse correlation times $1/\tau_C$, in the order of magnitude of the Zeeman or the hf interaction.

High-frequency/high-field EPR, such as 2-mm-band CW-EPR[41] offers a great advantage in the study of dynamics in regular organic radicals with typically small g-tensor anisotropy. Compared to X-band, which can be considered as a low–resolution spectral method with (resonance) line separation usually smaller than the linewidth, high frequency EPR offers a greater resolution. By estimating the lineshape variation of the nitroxides with temperature, using the characteristic frequencies at high-field conditions of measurement, the dynamical range of the CW-EPR method becomes extended toward faster motions. In a recent work, already the 95 GHz W-band CW lineshape of tempone doped in a single crystal of a host with similar structure and exchanging between two environments with different g-values, was sensitive to motion, while the X-band of the same sample was already fast-motion averaged.[14]

Accessibility of dynamical information to both faster and slower timescales of motion can also be obtained by pulsed-EPR. This is primarily due to two factors, the volatility of the pulsed methods to adjust in each particular case at hand, and the elimination of the inhomogeneous broadening. Dzuba et al.[27] used very early a straightforward and innovative experimental pulse-EPR method to study the internal rotation of the methyl group in the radical anion $CH_3\text{-}{}^{\bullet}CH\text{-}CO_2^-$, obtained in irradiated single crystal of L-alfa-alanine. This method relied on the selective saturation of the different EPR transitions of the methyl spectrum by a nutation of about 90-degrees, followed by a two-pulse sequence with a short and constant pulse delay. The amplitude of the generated echo was registered for variable waiting time after the first saturating pulse and was employed on the top of the T_1 processes to simulate the kinetics of the magnetization signal. By this way correlation times in the range of 10^{-5}-10^{-2} sec were obtained, which are indeed too long to be accessible by regular CW-EPR. The same investigator with his team performed later extensive studies of the effect of restricted reorientation (libration) by using echo-detected EPR spectra of several nitroxide probes.

The detailed kinetics of dephasing of the echo signals contains information about the dynamics of the system within an extensive range of timescales. The two parameters that are necessary to determine the "stretched-exponential" decay form of the echo, shown later in eq. (10), are also often adequate according to theory to discriminate the actual mechanism among several alternatives of the motional and spin dynamics.

Another task of the studies of dynamics is to understand the fundamental microscopic principles and identify the parameters that give rise to the macroscopic measurables. *e.g.* one can connect the reorientation correlation time with more fundamental quantities, such as the dimensions of the molecule and the *viscosity coefficient* η, utilizing the dynamical information from EPR. In such an analysis, the connected quantities depend on the motional model used, as for example Stokes' expression for the *transnational diffusion* constant D for the archetype Brownian particle is given by[17] $D = k\,T/6\pi\,a\,\eta$, where a is the radius of the molecule considered approximately as spherical. For a tumbling molecule one usually assumes that the orientation Ω of a predefined molecular frame at time t (with initial condition Ω_0) satisfies the diffusion equation in the angular space Ω. The *reorientation diffusion constant* and the Stokes' constant for *rotation* are closely related to the *translational diffusion constant*, given in the above relation.[17] The quantity D is identical to the self-diffusion constant measured by the Carr-Purcell method.[42] The experimentally important temperature dependence of the correlations times is given further in this text, first in the next Subsection and later under the Redfield relaxation limit, see eq. (6).

4.1 Parameterization in more accurate models of methyl rotation

Two rather accurate theoretical models of the dynamics of hindered rotation, one classical[5] and one quantum mechanical,[6] were thoroughly tested quite recently in combination with experimental EPR lineshapes of the methyl hyperfine structure in methyl malonic acid radical $CH_3\text{-}\!\dot{C}(COOH)_2$. Some interesting albeit occasionally "peculiar" results of "academic" interest can be summarized as follows:

Model I) The classical model of diffusion in the presence of a periodic potential (Smoluchowski drift diffusion model.)
a) Both the *free diffusion* and the *site jump model* were shown to be inadequate to reproduce the EPR lineshapes in the experimentally important region $0.32 < V_3/k\,T < 5.62$ of many rotors, where V_3 is the *barrier height*. b) The separate-harmonic-oscillators model for the high-barrier limit was not consistent with reasonable barriers but it gave a reasonably good Arrhenius fit with pre-exponential factor $2.08\,D\,V_3/k\,T$, where D is the diffusion constant.

As a byproduct of this study we obtained that c) The widely used relation for the isotropic hf-couplings in eq. (1) must be corrected by a linear cosine term $\sim a_1 \cos\vartheta$ in systems with pyramidal distortion in the alfa-carbon, a fact observed also earlier by other investigators. (This term was also introduced in the pure quantum treatment coming next.) And d) The rigid limit of this purely classical case gave unexpectedly a quartet of equal intensities, which is also the result of the quantum theory for the *stopped methyl rotor*, see Fig. 11(b). This is an indication that the stopped rotor quartet is a consequence of the symmetry of the rotor rather than the quantum effects of motion.

Model II) The quantum rotor in presence of a periodic potential.

a) The Arrhenius barrier for activated methyl rotation was found to be 754 K and was thus significantly different from the used potential barrier used in the Hamiltonian of the quantum rotor $V_3 = 618$ K. This was as unexpected as that b) The deuterated rotor Arrhenius barrier differed also from both the above values being 387 K. c) The deuterated rotor behaved classically even at the lowest temperatures d) The methyl rotor was also behaving classically above 50 K. A more detailed discussion about quantum effects of motion on the EPR spectra is undertaken in a later section.

The above brief account for the extensive treatments given in refs. [5,6], shows that it can be worth it to be careful in describing the dynamics of some quantum systems with high symmetry in terms of the Arrhenius parameters due to the insensitivity of the experimental fits to the well known law. Furthermore, the macroscopically determined activation energies can be different from the microscopic potential barrier.

4.2 Forms and phases of solids subjected to dynamics studied by EPR

The interactions of the electron spin in disordered solids cause broadening of the EPR spectra. The inhomogeneous broadening is due to anisotropic interactions, the Zeeman and the hf interaction for radicals, and eventually the ZFS for triplet electronic states. The solid-state-EPR spectra can be broadened in addition by field inhomogeneities and unresolved hf splitting. It is assumed that the spectrum of a *polyoriented* sample is made up of a broad envelope of homogeneously broadened peaks from spins resonating at different frequencies, each associated with a certain orientation. The different peaks define the *spin packets*. Another definition of a *spin packet* was given recently in the book by Schweiger and Jeschke,[43] as being an ensemble of spin systems that experiences the same time averaged local fields. This definition specifies in a better way the role of dynamics instead of the regular definition that is strictly applicable to the rigid limit of inhomogeneously broadened EPR spectra.

Since it is not possible to crystallize several interesting materials one has to deal with EPR measurements in *disordered* solids, which are categorized as *glasses* or *polycrystalline* matter. The EPR lineshape of such samples are without distinction usually called *powder spectra*. If no bias on the random (uniform) orientational distribution of the molecular units of the spin system is applied, all these materials are treated theoretically in a similar manner, at least with respect to the resonance frequencies. In that case, one of the most appropriate methods to study such systems is the ESEEM, and particularly the 2D-HYSCORE technique. This method is usually combined with pulsed ENDOR, since complementary information is always desirable.

The *microcrystalline* and the *amorphous* solids can be modelled similarly in simulations of EPR spectra at low temperatures where motional dynamics can be disregarded. In the opposite case, *i.e.* when dynamics are important, the magnetic behaviour of the *microcrystalline* state is different than the amorphous *glassy* state and it is possible to discriminate between them by using Field-Sweep Pulse-Detected EPR,[44] further called Echo-Detected EPR ED-EPR. The *pulse delay* dependence of the echo-detected lineshape in glassy materials is namely more prominent. The reason for that is the difference in the *spin-diffusion* conditions that equalizes the *spin temperature* across the whole sample of the above two different kinds of disordered solids. While the spin packets within each microcrystallite are isolated in space and the magnetization of each spin packet develops independently thus minimizing spin diffusion, in the amorphous glassy state the spin packets can interact with each other in the regions of overlap and are more sensitive to cross correlation that increases the spin-spin relaxation rate. A complementary explanation was given in ref. [44] based in the existence of "special type of vibrations in amorphous materials not present in crystalline states."

We can say that *isotropic liquids* exhibit orientation order of *short range*, which can be contrasted to the *crystalline state* where a *long-range order* exists between the perfectly oriented moieties in a crystal. The situation is more complex in mixed phases such as partially oriented macromolecular aggregates. *Mesophase*, *i.e.* intermediates between fluids and crystals, also called *liquid crystals*, is found mainly in three different phases, *nematic*,

3. Dynamical effects in CW and pulsed EPR

cholesteric and *smectic*. The liquid crystals can have long range order up to 10^6 Å!

An important general remark here is that even though the environment of a paramagnetic solute in liquid crystal solvents is highly anisotropic some restricted-reorientation rates of local nature can be relatively fast.[45,20]

A dense fluid which consists of "spherical" molecules is generally characterized by *short-range order* and *long-range disorder*, which can usually be described by a *pair distribution function*.[44,45] The orientational averaging of the second rank harmonic, $S = <P_2(\cos\vartheta)>$, actually the Legendre polynomial, where the angle ϑ is defined between a molecular and a static laboratory direction, is the most important parameter discriminating ordered from disordered solids. The average S is called the *order parameter* and vanishes in the absence of an orientation *restoring potential*. Equivalently, for a constant potential energy the orientation probability density is *uniform* and factors out in the calculation of the average S, which obtains the value zero by symmetry. On the contrary the physics of oriented systems requires that the orientation distribution has necessarily some tensorial components of second rank (or higher) rendering the value of the order parameter finite. *e.g.* for an ensemble of elongated molecules it is required at least a cylindrical symmetry giving $S \neq 0$.

Figure 2. The cholestane nitroxide probe that was inserted in a multibilayer film consisting of the macroscopically oriented lipid dimyristoyl phosphatidyl- choline (L-alfa) during the study of the motional dynamics. (ref. [20], pag. 81)

A feeling for how this works in practice is gained in the following, where in a simple way, orientation restoring potentials are applied on the modelling of small-amplitude *restricted-rotational* motion of nitroxide probes in certain solids. In a particular lamellar system, the libration axis, which is restricted by a potential,[20] was allowed to change direction. That treatment demonstrated both theoretically and experimentally the implications of the structural order of the environment on the dynamics by a combined study of

CW- and ED-EPR lineshapes. One could easily follow the limiting cases that were obtained using a truncated Hamiltonian and only two simple dynamics parameters, a *correlation time* and the average *amplitude* of libration.

4.3 Estimation of the amplitude of libration in nitroxides

The libration amplitude in anisotropic systems can be determined from the field separation of the outer peaks A_{zz} of the CW-EPR spectrum in nitroxides, which is affected by slow overall reorientation motion.[46] This method was originally developed for the restricted-reorientation effects on the CW-EPR lineshapes of nitroxides by Freed and collaborators.[47] The situation is similar for libration, with the difference that libration concerns small-amplitude oscillatory motions instead of rotation.

Commonly the *x*-axis of the principal hf molecular frame in a nitroxide probe is assumed to be directed along the N-O bond, while the *z*-molecular axis is assumed to be along the p_z orbital containing the unpaired electron. Furthermore A_{yy} has a similar value as A_{xx}, while A_{zz} is the largest hyperfine component. Sometimes, however, even in "simple" cases such as the nitroxide 3-carboxy-proxyl, some complications may arise concerning the interpretation of the libration restricted around a certain axis, see ref. [48].

For fast librational motion around the *x*-axis and assuming small amplitude librations, one can derive a simple relation between the time averaged parallel hyperfine component A'_{zz} and the mean square amplitude of the libration angle $<\alpha^2> \approx <\sin^2 \alpha>$, shown in the following equation.

$$A'_{zz} = A_{zz} - (A_{zz} - A_{yy}) <\alpha^2> \qquad (3)$$

The A'_{zz} component is directly related to the field separation ΔH of the outer spectral peaks of the CW spectrum of the nitroxide by the simple relation $\Delta H = 2 A'_{zz}$. The effects of the restricted motion on the nitroxide lineshape are shown in Fig. 3. The averaging of the anisotropic components of the hf tensor was rather mild compared to unrestricted rotation, leaving a large

3. Dynamical effects in CW and pulsed EPR   111

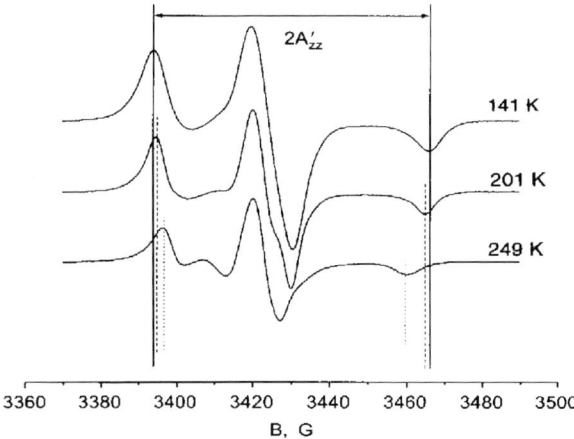

Figure 3. The EPR lineshape of a partially immobilized nitroxide with large residual anisotropy. Libration around the *x*-axis of the molecule decreases the field separation between the outer peaks compared to completely immobilized molecule. (Fig. 1 ref. [19])

amount of *residual anisotropy* to the EPR lineshape even for fast motions. The relatively small decrease of A'_{zz} with temperature was still significant to obtain data for the librational amplitude $<\alpha^2>$.

In order to gain some feeling of the librational amplitude range, we mention that the *standard deviation* of the libration angle $<\alpha^2>^{1/2}$ for tempone in doped glassy alfa-D-glucose[46] varied in the interval 0.20 to 0.38 rad, that is from 11 to 22 degrees, for temperatures between 295 and 340 K. This old method is used to disentangle the libration amplitude and the correlation time of libration that are obtained as a product in echo-detected EPR studies.[46,20]

4.3.1   Relation to the "order parameter"

A variation of the method of the field separation of the outer peaks of a powder spectrum for the estimation of the restricted slow reorientation correlation was given by Freed and collaborators.[46] The reorientation correlation time τ_R for slow isotropic reorientation was directly related to a particular quantity S, defined as the ratio of the field separation of the outer peaks of the EPR lineshape in the presence and in the absence of motion, by the following relation.

$$\tau_R = a \, (1-S)^b \text{ with } S = A'_{zz} / A_{zz} \tag{4}$$

The two involved parameters *a* and *b* depend on the *motional model*, the *intrinsic linewidth*, and the *hyperfine parameters* and could be obtained by fitting the experimental spectra provided that a certain model of simulation of the EPR lineshape including slow motional dynamics was available.

Similarly, a simple relation between τ_R and S was determined from simulated spectra of morpholine radical for which we had experimental data from powder samples.[21] The measurable S was shown to be intimately related to the *order parameter*. The variation of the measured S values (after elimination of the contribution of the isotropic hyperfine proton couplings) for relatively fast motion were computed as in the second expression in the above eq. (4).

The theoretical values of S for different angles of the z-dipolar axis with respect to the rotation axis are shown in Fig. 4. The principal dipolar z-axis of the nitrogen was assumed to perform jumps between three sites, simulating a rotation around an axis perpendicular to the (average) molecular plane, see Fig. 9. The computed of S were almost indistinguishable from the values of the second order harmonic $Y_{2,0}(\beta)$, indicating an intimate relation to the order parameter S. Furthermore, from the intercept of the experimentally obtained curve S, shown as a straight line in the lower part of Fig. 4, with the theoretical curve we could "accurately read" the angle β in the diagram. This angle is the opening of the cone on which the p_z-orbital is confined during the rotation-like process.

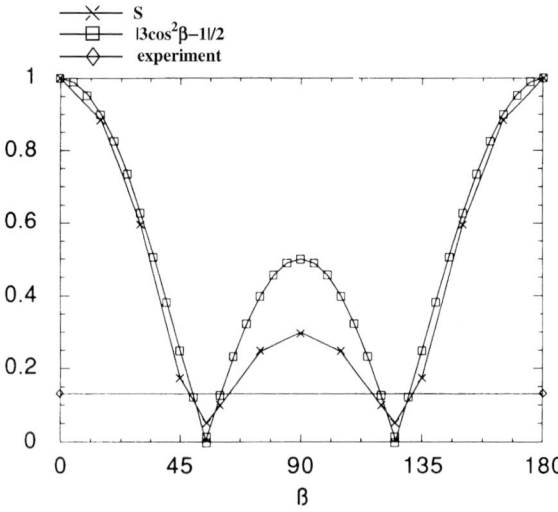

Figure 4. The angular dependence of the quantity $S = A'_{zz} / A_{zz}$ was essentially the same as for the second order harmonic in $Y_{2,0}(\beta)$. The angle β is defined between the rotation axis and the

3. Dynamical effects in CW and pulsed EPR

z-direction of the nitrogen hyperfine tensor **A**. The quantity S defined above was the ratio of the field separation of the outer peaks of the CW EPR spectrum of morpholine for relatively fast restricted rotation/reorientation, divided by the same quantity for immobilized molecules.[21]

5. MAGNETIC RELAXATION

5.1 Spin-lattice relaxation in very cold solids

The most well documented theoretical treatment concerning electronic spin-lattice relaxation times T_1 at liquid He temperatures refers to dilute paramagnetic metal ions hosted in ionic crystals. The important relaxation mechanisms for the corresponding sub-5 K temperatures regime occur through modulation of the ligand field by the crystal-lattice vibrations. The temperature dependence of the T_1 relaxation time can be written by adding together three main contributions to a single expression, see ref. [49] pag. 65.

$$\frac{1}{T_1} = a\coth(\frac{h\nu}{2kT}) + bT^n + \frac{c}{\exp(\Delta/kT)-1} \qquad (5)$$

The symbol T is the *lattice temperature*, assuming that the electron spin is in equilibrium with the lattice. The constant a of the first term is the magnitude of the *direct spin-phonon processes*. The second term is due to *Raman processes* and b determines the magnitude of this source of relaxation. The third term and the value of c depend on the contribution of the *Orbach process*. The variable Δ represents the crystal field splitting. The constants a, b, and c can have widely varying values from system to system depending on which is the dominating part. The exponential n for the Raman two-phonon process can take several unusually large integer values, such as *e.g.* $n = 5$ for multiplets with small splitting, $n = 7$ for non Kramers doublets, and $n = 9$ for Kramers doublets, implying a strong temperature dependence. The strong temperature dependence is also a characteristic of the other two terms in the above equation but the absolute values of the overall coefficients are small.

One can use some of the tendencies given by the above equation as indicative even for the relaxation conditions in our case of covalent crystals and powders as in the more detailed discussion found in ref. [50]. For these systems the spin-phonon mechanisms result to relatively small values of relaxation rates compared to the rates due to thermal motion, which can modulate the *orientation* of the anisotropic interactions and/or the *distances* between interacting spins quantitatively. Since thermal motion is not possible in the limited range of the sub-5 K region, the "crystalline-lattice" *zero-level vibrations* are the only possible motions that are left to modulate the magnetic interactions. As they do not display large amplitudes and they are high-frequent and coherent, they are usually inefficient in inducing fast spin relaxation, if not many-particle (synergetic) and/or particular quantum effects occur. Finally for temperatures higher than ca 40 K, with the onset of *thermally activated* processes and the following "classical" motions that induce spin relaxation rates larger by several orders of magnitude, the spin-phonon mechanisms totally lose importance and can be neglected.

5.2   The Redfield relaxation limit

This is the classical theory of relaxation primarily concerning fluids, but still useful in the examination of the solid state for reasons that will be explained.

The characterization of the spin-motion dynamics as fast, is determined by the relation $<|H_{SL}|> \cdot \tau_C \ll 1$, where $H_{SL}(t)$ is the time dependent interaction between the spin S and the *lattice L* with zero average $<H_{SL}(t)> = 0$, and where the parameter τ_C of the lattice with dimension of time is called the *correlation time*. One of the assumptions of the Redfield-relaxation theory is what is called the *strong narrowing* condition,[51] which implies that $\tau_C \ll T_{1,2}$, where T_1 and T_2 are the *spin-lattice* and *spin-spin relaxation times*, respectively. Equivalently it means that the fluctuations of the lattice are faster than the fluctuations in the decaying macroscopic magnetization.

Even for liquids, the particular conditions of *low temperature* and/or for *high viscosity*, renders the electron spin of relatively bulky molecules or complex ions into the thermodynamic limit of motion called the *slow-motion regime*. In that limit the Redfield theory of narrowing is not valid because the spin fluctuations and the motional degrees of freedom become of the same order of magnitude and can be statistically correlated.[3]

In the Redfield region, both relaxation times are relatively long, in particular the transversal relaxation time T_2 as the term *strong narrowing* refers to linewidths, but T_2 decreases monotonically with increasing correlation time τ_C, *i.e.* for slower motion. This is apparent in Fig. 5 where also the different behavior of the *longitudinal relaxation time* is shown in the

3. Dynamical effects in CW and pulsed EPR

right half of the diagram where the condition $\omega_{0S} \cdot \tau_C \leq 1$ is no longer valid. The variable ω_{0S} here is the electron Larmor frequency. Notice also that always $T_1 \geq T_2$.

The left part of the diagram in Fig. 5, which is called the *extreme narrowing region*, corresponds to faster motions while the right part corresponds to slower motions, as the correlation times become larger from left to right. An interpretation of the *extreme narrowing* condition $\omega_{0S} \cdot \tau_C \leq 1$ is that the fluctuations of the lattice are faster even in comparison to the electron Larmor frequency, *i.e.* $\omega_{0S} \leq 1/\tau_C$. Equivalently, the lattice changes many times during each Larmor period in average, and consequently the precessing spins feel an average lattice interaction, which by definition vanishes.

For temperatures corresponding to the extreme narrowing and higher, the *spin-lattice* relaxation time T_1 becomes equal to, and follows along with, the *transversal relaxation* time T_2 in a monotonically increasing fashion with further increasing temperature. (The primed quantity T_2' designating the *adiabatic part* of the transversal relaxation time T_2 has strictly this property.) In this regime, the relaxation rates are shown to become linear in the correlation time, *i.e.* $1/T_{1,2} \propto \tau_C$, and most importantly the spin dynamics can be characterized as field independent.

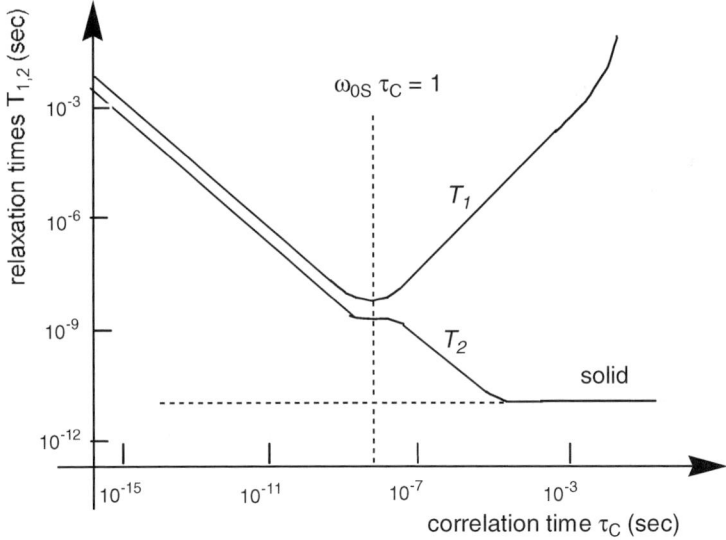

Figure 5. Inverse temperature dependence of the *longitudinal* T_1, and the *transversal* T_2 electron- (or nuclear) spin *relaxation times* in liquids, along the *correlation time* axis. To the left part of the diagram and until the vertical broken line at $\omega_{0S} \cdot \tau_C = 1$, is the *extreme*

narrowing regime. Except from the region far out to the right, designated as "solid" in this figure, the Redfield, also called *strong narrowing* conditions,[51] are tacitly assumed.[26]

The lattice motions are usually considered as *thermally activated* processes and the correlation times exhibit normally a temperature dependence of Arrhenius type, *i.e.*

$$\tau_C = \tau_C^\infty \exp(E_a/kT) \tag{6}$$

This temperature dependence is particularly steep for the lower temperatures until the realm of temperatures equivalent to the activation energy, *i.e.* for $T \cong E_a/k$.

While according to the above discussion *slower modulations* give *faster relaxation* the opposite is also possible in several systems, *i.e.* faster modulation gives faster spin relaxation. Most prominent and interesting example is the longitudinal relaxation at very cold conditions, seen in the previous Subsection.

Another example, where one has a similar behaviour of increasing spin-lattice relaxation rate $1/T_1$ with temperature T, is coming from the region of the Redfield *strong narrowing* but only within the slower motions, the non-extreme narrowing region of the Redfield[51] regime. It implies that the lattice fluctuations are too slow and the *spectral density function* $J^{(n)}$, which has the general Lorentzian shape,

$$J^{(n)}(\omega) \propto \overline{|H_{SL}(t)|^2} \frac{\tau_C}{1 + n^2 \omega^2 \tau_C^2} \tag{7}$$

indicates that there is no strong component in the relaxation power spectrum at as large frequency as the spin resonance frequency ω_{0S}. This does not happen before the motion obtains a large enough modulation rate $1/\tau_C$, *i.e.* when the matching $\tau_C \cdot \omega_{0S} \approx 1$ occurs, as shown in Fig. 6. For the fastest motions on the other hand the resulting "white" relaxation spectrum becomes inefficient for all frequencies. These conclusions are based on the assumption that the total relaxation power, which corresponds to the area under the spectral density functions in Fig. 6, is constant for different correlation times.

3. Dynamical effects in CW and pulsed EPR

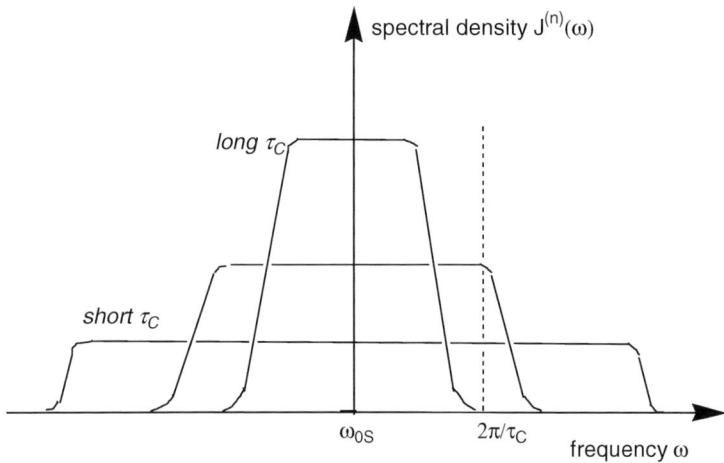

Figure 6. The *spectral density* functions $J^{(n)}(\omega)$ for $n\neq 0$, i.e. Fourier transforms of *decaying correlation functions* of the lattice variables according to the original (Bloembergen-Purcell-Pound) BPP[52] -(Wangness-Bloch) WB[53] -Redfield[51] theory about *magnetic relaxation*. The relaxation for long *correlation time* τ_C (slow motion) is efficient only for transitions inside a narrow region with frequencies $\omega_{0S} \pm 2\pi/\tau_C$, compared to shorter τ_C (faster motion).[54]

The integer n in eq. (7) corresponds to the step of the ladder operators $S_{\pm}^n$ for the electron spin S in the random part of the spin Hamiltonian. The ladder operators induce transitions $|\Delta M_S| = n$, with $n = 0, 1, 2$, between the Zeeman levels M_S. Notice that for the *adiabatic* or *secular* term with $n = 0$, the spectral density is monotonically increasing with the correlation time τ_C, and is frequency/field independent. Finally notice the quadratic dependence of the spectral density in eq. (7) on the interaction strength $|H_{SL}|^2$; the interaction can be *e.g.* the hfi.

5.3 Broadening

The main contribution to the homogeneous broadening in solids is *adiabatic*,[55] merging from the *secular* terms of the Hamiltonian, including *spin-flips* of like spins, and comes from terms of the electron-Zeeman interaction for anisotropic g-tensor which contains the secular S_z operator, as well as the dipolar interaction. When motion becomes faster the *lifetime broadening* due to mutual transitions of the spin system and the lattice can become significant and contribute to the total broadening. The *lifetimes* of the spin states become shortened due to more frequent transitions between different spin states, during which the exchange of energy with the lattice takes place. However, the lifetime broadening is significant only when efficient motional narrowing is already activated and the total broadening is small.

Thus, the higher the temperature the smaller the total linewidth even with the contribution of the lifetime broadening of the states, due to the smallness of the correlation time and the overall proportionality $1/T_2 \propto \tau_C$. Notice that this does not mean that we can neglect the contribution of the lifetime broadening, *i.e.* terms with $n \neq 0$ in eq. (7), in calculating the total broadening, because it becomes significant relative to the adiabatic part for fast motion.

As an outline of the above discussion the following expressions in eqs. (8) and (9) can be given for the total broadening that is strictly valid for *two-level* spin systems.

$$1/T_2 = 1/T_2' + 1/2T_1 \tag{8}$$

T_1 is the spin-lattice relaxation time. This very useful relation separates the two possible sources of broadening, *i.e.* the *adiabatic* and the *lifetime broadening*. Sometimes, also the following relation is given, in particular in NMR, which includes in addition the *inhomogeneous broadening* $|\gamma \Delta H|$.

$$1/T_2^* = 1/T_2 + |\gamma \Delta H| \tag{9}$$

The star signifies that the total broadening of an EPR line is not only due to relaxation, the term designated by $1/T_2$, but also by field inhomogeneity, at least for fast motion. The pattern of these expressions with the different contributions remains the same even in more complex situations.[17]

The point here is that each *spin packet* only contains the first two contributions (homogeneous) shown in eq. (8).

Table 2. Order of magnitude of the electron-spin relaxation times induced by different motions or relaxation mechanisms under various "lattice"-temperature conditions. T_2 values were occasionally deduced from the homogeneous broadening

Temperature-Conditions	MECHANISM Electron-spin-Relaxation	Longitudinal relaxation time T_1 sec	Transversal relaxation time T_2 sec
Cryogenic	o Electromagnetic bath: photons[49,‡]	10^8 ! < T_1	-

‡ All the values spin-lattice relaxation times in this regime are theoretical and they are heavily overestimated due to the failure of the early attempts to assign an appropriate relaxation

3. Dynamical effects in CW and pulsed EPR

(liquid He) 1-5 K	o Lattice vibrations‡: phonons ÷ Direct process ÷ Raman processes ÷ Orbach process	$10^3 < T_1$ $10 < T_1$ $10^{-3} < T_1 < 1$	- - -
Low (liquid N_2) 77 K	o Slow Tumbling o Exchange	10^{-12} -	- 10^{-8}
Room temperature	o Tumbling o Gas collisions	10^{-6} 10^{-4}	10^{-6} 10^{-4}
	Nuclear-spin-relaxation		
Slow motion	o Paramagnetic impurities[56]	$T_1 < 10^{-3}$	$10^{-7} \leq \tau_C \leq T_2$
Other	o Other	sec-min	$T_2 \leq T_1$

For systems dense in unpaired electron spins the most efficient term of the dipolar Hamiltonian and the scalar spin-spin (exchange) interaction is the *flip-flop* term proportional to the bilinear operators $S_{1+} \cdot S_{2-}$ and $S_{1-} \cdot S_{2+}$.

5.4 Electron-spin dynamics through nuclear relaxation

It is a well known fact that even small amounts of paramagnetic impurities enhance the nuclear relaxation rates in both liquid solutions and diamagnetic crystals. On purpose, one can introduce salts of paramagnetic ions in solution or dope the diamagnetic crystals with paramagnetic ions. Concentrations already in the ppm (part per million) range are most of the times enough for a significant enhancement of the nuclear spin relaxation and increased linewidths of the NMR spectra.

The mechanisms of these strongly "disproportional" effects are, however, different in the liquids and in the solids.

The average distance of radicals in a uniform and isotropic solution with concentration C is given by $R \sim C^{-1/3}$. To give an idea about this distance, for particle concentrations of ca $10^{18}/cm^3$ that are considered relatively dilute, e.g. for $C = 1$ mM, one obtains an average distance of about 90 Å. This is a substantial distance for the electron spins to exert any direct interaction to each other, although experimental data about Heisenberg exchange effects start at concentrations of this order of magnitude.[11]

There are thus necessarily processes that bring the electron spins to a "closer contact", at least in a statistical way, to each other and to all the

---

mechanism to solids' spin dynamics, but they have a fundamental interest indicating also the inefficient coupling of the spins with the lattice at low temperatures.

nuclei of the sample, which act as the "lattice". In liquids this is not particularly difficult to understand since the particles in a fluid can rapidly change positions, and furthermore the Brownian motion of all the involved species can contribute to the energy balance during mutual spin flips of the "unlike" nuclear and electron spins.[17] This guarantees the conservation of energy while motion allows also for many reencounters at a high rate. Thus in liquids it is reasonable with a great relaxation effect from impurities of relatively few paramagnetic centers.

In solids, the mechanism must be of a different nature because, excluding diffusion, the particles are static. Thus, the mobility of liquids is replaced partially by the fast (and independent from the mutual interaction) *electron spin relaxation*. Furthermore, nuclear *spin diffusion* replaces the possibility of closer spatial approach of the electron to all nuclei by the effects of the *S-I* spin interaction of a paramagnetic center over the entire sample. The mechanism of spin diffusion involves spin flips between neighboring nuclear spins which transfer polarization disturbances through the sample[17] and was originally introduced by Bloembergen 1949.

The nuclear relaxation enhancement has the additional interest of being a way of determining indirectly the relaxation rates, both $1/T_1$ and $1/T_2$, of the electron spin. These rates are normally many orders of magnitude higher than for nuclear relaxation, a fact that makes the direct measurement of the electron spin relaxation difficult. Several theories of computing the nuclear relaxation in paramagnetic systems of varying levels of accuracy have been developed in the past. One of the most complete theories involving *slow motions for the electron spin* was developed in Stockholm starting with the original reference by Benetis *et al.*[3] In that treatment the electron spin was considered as a part of the lattice for the nuclei along with the regular motional degrees of freedom. Furthermore, all possible correlations between the electron spin and the motional subsystems were retained in this model.

The nuclear relaxation effects of paramagnetic "impurities" are still today extensively used to extract structural and dynamical information about *paramagnetic metal complexes* in solution.[18] There is a wide field of applications in biochemistry, pharmacy, and medicine among other disciplines, and the interest of the field is very large. The apparent similarity of the *active center* in very important for life enzymes and the coordination of paramagnetic complexes is the main reason. The sensitivity of Nuclear Magnetic Relaxation Dispersion (NMRD) curves to the variation of the external field has been found to be an excellent tool for the investigation of the details of the structure of the complexes around the coordination metal

3. Dynamical effects in CW and pulsed EPR

ion. Among the dynamical processes accessible by this method is the *overall reorientation* of the whole complex, *internal motions* such as *vibration-distortion*[57] and internal *rotations*,[58] as well as *exchange* of the ligands from the first coordination sphere to the bulk.[59]

6. CHEMICAL EXCHANGE

The CW EPR has been used from the early days of EPR to study the dynamics in systems with *chemical-molecular reorganization*,[60] or as they are simply called *(chemically) exchanging* systems, see also Chapter 1 about Continuous Wave EPR in this book. Chemical processes can give rise to line broadening and/or to modification of the entire EPR-lineshape pattern if they alter the magnetic environment of the unpaired electron. Suppose that the paramagnetic center can be found in several magnetic environments, or "sites", that are exchanged in time due to some "chemical" process. In particular, the characteristic modification of the shape and the number of the different hyperfine peaks by the "motion" in *isotropic systems* is relatively easy to understand.

If motion exchanges the hf parameters of two lines, their eventually resolved slow motion spectrum acquires extra broadening, in excess to the *intrinsic broadening* of each line, first at the original transition frequencies, cf. Fig. 7(a). The excess broadening is equal to the inverse lifetime of the corresponding site. For faster motion the *lifetime broadening* of the pairs of exchanging transitions, shifts gradually the two lines to an intermediate position, so that they at last overlap and finally coalesce to a single broad signal, cf. Fig. 7(b). The new lines of the spectrum created by this way start to sharpen by motion with further increasing exchange rate, Fig. 7(c). For really fast motion the final spectrum has a totally different appearance than the original slow motion spectrum to conform to a spectrum with two equivalent positions.

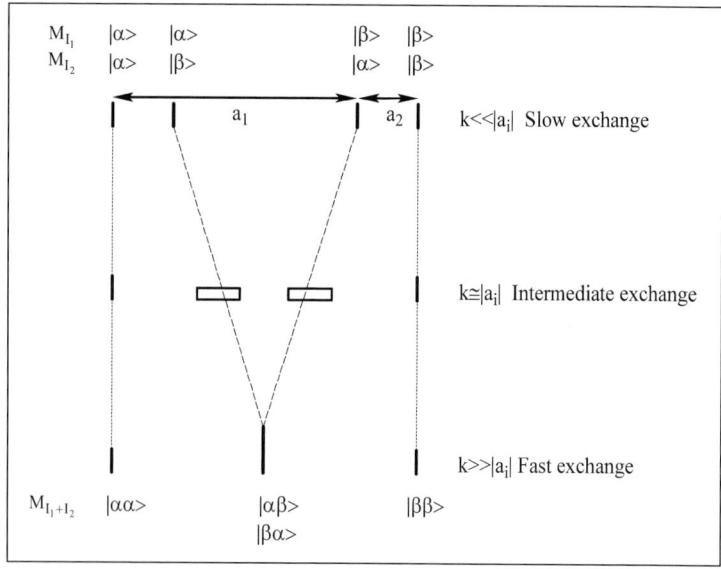

Figure 7. Alteration of the hyperfine lineshape due to chemical exchange in a radical containing two isotropic nuclei with spin $I = 1/2$ for the three characteristic domains of dynamics for the exchange process: (a) *slow*, (b) *intermediate* (coalescence), (c) *fast limit*, of exchange. All possible EPR transitions with the same total nuclear projection $M_I = M_{1I} + M_{2I}$ mix by motion and their resonance position is altered.

Each of the three domains of exchange-dynamics for the above spectral transformations can be followed quantitatively with relatively simple approximate expressions for the *linewidth* and the *resonance positions* of the exchanging peaks. The relevant parameter for the simple two-site exchange is the *lifetime* of the interconverting sites (particles) or the inverse of the lifetime, which is proportional to the *rate of reaction*, see *e.g.* ref. [31] pag. 214. All the above spectral modifications are furthermore occurring within a rather limited temperature interval. The most rapid modification of the lineshapes with the temperature happens close to the temperatures for which the thermal motion modulates the hf interaction with rates (in MHz) that approach the hyperfine interactions, usually 1-35 Gauss for radicals in solids, *i.e.* 3-100 MHz. Thus, the corresponding exchange correlation times are usually longer than 10^{-8} sec, demonstrating that exchange is generally a *slower process* compared *e.g.* to the pico-second (10^{-12} sec) timescales of reorientation correlation in ambient fluids.

3. Dynamical effects in CW and pulsed EPR

After the above discussion it is obvious that for further increase of the temperature over the fast exchange limit, no additional change can be observed in the lineshape rendering faster rate constants inaccessible to plain CW EPR. At the other extreme of low temperatures, slower motions cannot either alter the lineshape, and thus also slower motions are inaccessible for CW EPR. However, in some cases even these "extreme" rates can be studied by using high-field EPR,[41] ENDOR[22,61] or pulsed EPR methods.[27,62,63]

We give further an example of intramolecular exchange concerning the phenyl phosphine ligands in a Cu-complex, see Fig. 8. The triphenyl phosphine ligands contain two magnetically inequivalent ^{31}P nuclei ($I = 1/2$), one in *equatorial* and one in *axial position*, which were found to have isotropic hf couplings.[34] Their exchange was observed in the CW lineshape as alteration of the spin states with nuclear projections $|\alpha\beta\rangle$ and $|\beta\alpha\rangle$ with respect to the two phosphorous. The corresponding peaks were first broadened for slow exchange while for intermediate rates they coalesced to almost disappearance for the higher measurement temperatures. These phenomena were clearly recognizable in spite of the complexity of the system, involving: a) Additional couplings to the two different magnetic isotopes of copper $^{63,65}Cu$ ($I = 3/2$) and b) Anisotropic interactions modulated by the *slow tumbling* of the complex in the viscous medium giving asymmetric broadening.

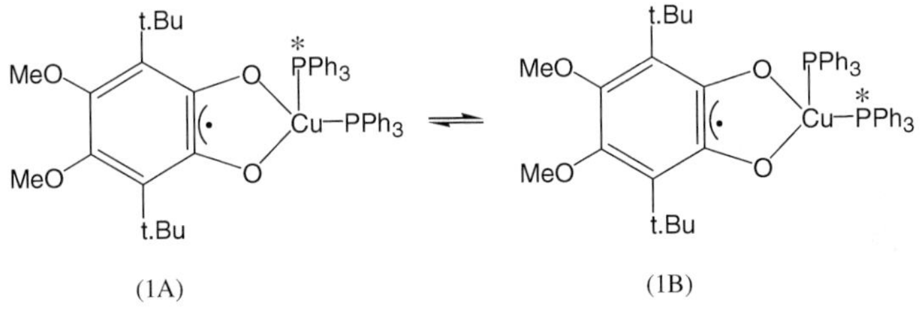

Figure 8. The intramolecular exchange of the ligand triphenyl phosphine between two sites in the o-semiquinone complex of monovalent copper.[34] (The unpaired electron is localized on the quinone ligand in this complex.)

The significant result of this work, studied by the temperature dependent CW- EPR linewidths, was that there is a connection between *chemical kinetics* and *motional dynamics* for bulky complexes with ligands that stick out to the bulk of the solution. Thus, a whole "cage" of particles extended even outside the complex ion has to be included in the model of certain intramolecular reactions[34] in order to explain their kinetics.

6.1 The exchange lineshape of anisotropic systems

In studying anisotropic systems several additional theoretical difficulties appear, *e.g.* the interactions are no longer secular and additional "sites" may be indirectly defined through the orientation change of the g-tensor and/or the hf-tensor. In that case one has to include in the motional part of the problem except from spin populations, corresponding to diagonal elements of the density matrix, even spin-spin interactions that may involve extra couplings in the usually blocked density matrix. If one chooses to work with the density matrix formalism the creation of an exchange (super)operator in the combined space of the EPR transitions and the "sites" is straightforward, leading to a simplified form of the SLE[64] (Stochastic Liouville Equation). The mathematical problem of the kinetics for exchanging particles under chemical reorganization is relatively easy to derive from the chemical equations describing the reorganization at hand. It is also straightforward to implement the population equations in the Liouville type numerical simulation of the EPR lineshapes, or even in simulations of the ESEEM signal.[63] The chemical reactions describing the reorganization of the system lead usually to non-linear kinetics, however, and the equation of motion has to be linearized as a first step. After that, the numerical matrix formalism is straightforward to apply[60] in the simulations with the density matrix theory even for systems with anisotropy,[64] where the meaning of "site" is generalized as discussed above.

Except from simulations based on the *density matrix* formalism the simpler method of a *modified* form of the *Bloch equations* has also often been employed. The Bloch equations, which describe the magnetization classically, are valid for simpler two-level systems without strong spin-spin coupling[65,66] but they give equivalent results as the more general density matrix formalism in the regime of their validity. They can be modified in particular for the purpose of simulating chemically exchanging systems to give the so-called *McConnell equations*,[67] which have been derived for first order reactions.

A general treatment for the EPR lineshapes of isotropic systems within the density matrix formalism was given by Heinzer.[60] Sjöqvist, Maruani and Lund[64] extended this method to anisotropic systems using second order perturbation. The most recent treatment from this laboratory was an accurate non-perturbative solution of the same problem given by Benetis *et al*.[68,69] The latter treatment revealed that one has to be careful when the high field approximation is applied to lineshape simulations of exchanging spin-systems, in particular in systems with *strong anisotropy* and/or *extensive reorganization*.[21] It was found that the nonperturbative exchange theory for

3. Dynamical effects in CW and pulsed EPR

the CW lineshape was necessary to reproduce faithfully the residual anisotropy in the spectra of randomly oriented morpholine molecules in cold Freons, subjected to restricted rotational motion in the matrix, see Fig. 4 and Fig. 9. The high field approximation gave more or less inaccurate simulation of the lineshapes for this system, failing to reproduce the lineshape at intermediate rates, overestimating consistently the field separation of the outer peaks over the whole range of exchange rates.[21] This was attributed to the difficulty of incorporating the varying quantization axes for the nuclear spins of the high-field sublevel Hamiltonians obtained for each site in the computation of the exchange process.

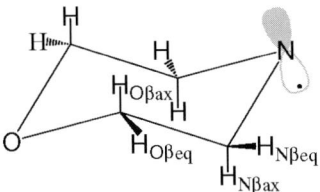

Figure 9. Geometry of the saturated 6-membered ring of morpholine radical, indicating how the chair C_S conformation affects the orientation of the p_z orbital at the nitrogen nucleus.[21]

One has to point out an important fact about the exchange dynamics for internal rotation of methyl-type fragments. For the high temperature *diffusion limit* the isotopic substitution does not change significantly the thermal activation parameters of the Arrhenius type. This is in contrast to the rotational regime that is controlled by *inertial effects* at very low temperatures close to 5 K and sometimes up to 20-30 K. There, the dynamics of deuteron substituted methyl have very different characteristics compared to regular methyl, as it will be analysed further. Thus, each of the following pairs displayed almost identical barriers. a) The methyl fragment $-CH_3$ and the deuterated analogue $-CD_3$ in acetic acid anion radicals[70] and b) The methylene fragment $-{}^{\bullet}CH_2$ and $-{}^{\bullet}CD_2$ in irradiated crystals of Zn-acetate single crystal.[68] The rate of rotation in the first pair was $k_H = 3.3 \; 10^{11}$ Hz $\times$ exp(-2.2 kcal mole^{-1}/R T) for proton and $k_D = 4.8 \; 10^{11}$ Hz $\times$ exp(-2.5 kcal mole^{-1}/R T) for deuteron, in the temperature range 77-170 K. In the second pair the rate of rotation was identical for both the proton and the deuteron fragments, $k = 6.73 \; 10^{12}$ Hz $\times$ exp(-7.17 kcal mole^{-1}/R T), in the temperature range 170-300 K.

7. INERTIAL EFFECTS OF ROTATION

Turning to *quantum effects* of rotation, we notice that *spin-rotation coupling*[2] in the Hamiltonian is obtained through the hyperfine interaction, either isotropic or anisotropic.

The methyl-type rotors XY_3, where the element X belongs to the carbon or the nitrogen column of the periodic table of the elements and Y is proton or deuteron, in inert-gas matrices can be considered as *pure quantum rotors* that occupy only the first few rotational levels at the lowest experimental temperatures. This becomes possible because the *rotational constants* of such fragments are of the order of 7 K, and sometimes even larger, being thus several orders of magnitude larger than the rotational constants of regular "heavy" molecules, with low symmetry. Most importantly, rotors with kinetic energies of this order of magnitude can be considered as free when comparison of the *kinetic energy* to the typical *potential energies* in solid host matrices is undertaken. This is only possible in "ultra light" and "highly symmetric" rotors, where interaction with the matrix is not large enough to quench quantum rotation.[71] The experimental low temperature EPR lineshapes are drastically affected by this property.

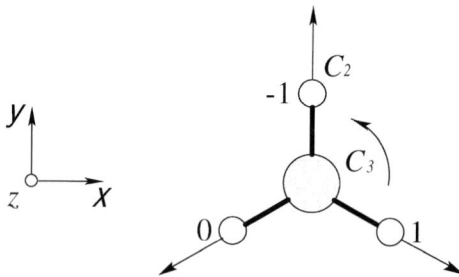

Figure 10. The model of the symmetric-top free-methyl rotor, *i.e.* a planar, methyl fragment of D_3 symmetry[7] considered in three dimensions. The rotational C_3 and C_2 axes are indicated in the figure while the three protons are indexed by -1, 0, and 1.

The citation of the term "ultra light" is utilized since heavy nuclei can be in the position of the central atom X if one of the symmetry axes of the rotor passes through it, resulting to very small *moment of inertia* for the relevant rotation. Most of the studied systems here had D_3 symmetry, see Fig. 10. The citation of the term "highly symmetric" rotors is used since only basic "exchange" symmetry between the outer Y atoms was enough to give quantum effects[7] in the EPR spectra of the rotors, see mixed- isotope proton-deuteron methyls.

7.1 Quantum-rotation of small radicals in inert matrices

Matrix isolation methods of radicals in cold *Freons* (small halogenated paraffins) are developed during a long period of time by among others Lund and coworkers, in cooperation with Shiotani and his laboratory in Hiroshima. The latter group was particularly specialized in experimental CW-EPR methods on radicals under *rare-gas-matrix* (*noble-gas-matrix*) isolation, a specialization rather rare among the EPR laboratories. For more information about *matrix isolation* of free radicals see relevant Chapters in this edition. In particular the rare-gas matrix isolation is very important for testing experimentally the *quantum effects* on the lineshapes of "ultra-light" and "highly symmetric" methyl-type rotor radicals. The quantum mechanical *spin-rotation coupling*[2] in these radicals by the hf-interaction modifies radically the classically expected EPR lineshape of the methyl radical at about 5 K. Considering the interaction of the alfa-proton methyl rotors with the host Ar (Argon) matrix[7] it was shown that, at least for the lowest experimental temperatures $T < 20$ K, a *freely-rotating planar-rotor model* was adequate to account for the experimental findings.[72] In other words, no appreciable hindering potential by the rare-gas matrix was found to apply in these systems.

Still, several significant new *quantum effects* other than *tunnelling* were discovered and explained semiquantitatively[7] by using group theory. It was possible, for example, to observe and identify the EPR signal from the *stopped methyl-rotor* for both the usual proton-methyl radical and the deuteron-substituted methyl radical, at the lowest experimental temperatures close to 5 K. The observed severe distortion of the lineshapes involving exclusion of several EPR transitions, was attributed to the symmetry of the D_3 point group of the radical "figure" and the Pauli principle.

The quantum effects were most dramatic in the deuterated methyl rotor where the entire "classically" expected septet had collapsed to a mere singlet! at the lowest experimental temperatures. The severely distorted EPR "singlet" of a stopped deuterated methyl CD_3 in Ar matrix at 4.1 K is shown in the right uppermost part of Fig. 1 of the Chapter 4 by Shiotani and Komaguchi in this book. It is compared with the classically expected septet which is obtained already from temperatures a little higher than 25 K, seen just bellow the low temperature singlet in the same figure. This was in agreement with a single allowed EPR transition expected theoretically, containing the *antisymmetric* nuclear-spin wave-function (with respect to exchange of any two deuterons) coupled to the lowest rotation state. The expressions for the lowest mixed rotational-and-nuclear-spin part of the quantum states of the system which are allowed by exchange symmetry[7] where derived by Benetis and Sørnes under low temperature conditions and are partially reproduced in Chapter 4.

Surprisingly enough, even mixed CH_2D and CHD_2 rotors, which have solely fundamental C_2 symmetry, were characteristically affected by exclusion of EPR transitions.[7] As it was explained in that work, the fundamental symmetry operation for the observed effects was the exchange of the protons or deuterons of the molecules by the C_2 rotation symmetry, assuming that they are indistinguishable Fermions and Bosons, respectively.

Among the new recently interpreted quantum effects, it was the extreme sharpness of the experimental EPR spectra at the lowest temperatures, which is unexpected for anisotropic systems. Further theoretical investigation, including full simulation of the experimental spectra, remain to be performed, however, in order to obtain information about dynamics for temperatures higher than 20 K where the effects of thermal motion start to appear in the experimental spectra.

Preliminary data show that also other "light" systems such as ammonia-based rotors, *e.g.* the first of each of the following *isoelectronic pairs*: NH_2 with H_2O^+, and NH_3^+ with CH_3, exhibit similar properties. On the other hand, the classical thermal motion increasingly affects "heavier" and non-planar methyl-type rotor radicals, such as the above two as well as SiH_3. In the latter compound of silicon, the central carbon atom in the methyl group was exchanged by a heavier element in the same column of the periodic table of the elements. By this kind of studies the systematically increasing importance of the interaction of the host molecules with the inert-gas matrix could be evaluated. For more recent experimental results see relevant Chapter of this edition by Masaru Shiotani and his team.

Regarding the dynamical information contained in the spectra of the pure quantum rotors one should not expect to obtain the same kind of parameters as in the case of classical motions, *e.g.* in diffusion. An immediate difference is that the rotation is not any more a part of the lattice, and thus we cannot obtain a quantity such as *correlation time* of rotation. We can instead extract the *rotational frequencies* expected for the rotors or equivalently the *rotational quantum number* from the effects on the EPR spectra. Thus, the *rotation barrier* of the methyl type quantum rotors in an inert gas matrix was definitely much smaller than the *rotational constant B*, which is of the order

3. Dynamical effects in CW and pulsed EPR 129

of 5-8 K, since the radical was performing free rotation up to the first two decades of K. As a comparison we report the rotation constant of the H_2 molecule,[17] which is one order of magnitude larger, that is, 86 K.

The motion of methyl rotors has been found to be important in understanding the experimental ESEEM and ENDOR properties in solid state for experiments performed at very low temperatures. In ESE studies of methyl- substituted nitroxides it was obvious that nuclear and electron spin relaxation was affected very distinctly by the quantum properties of methyl-rotation[14,73] in addition to the effects of the classical relaxation theory. The experimental facts about the effect of the hindered internal rotation of methyl fragments in nitroxides, have not been understood in detail so far.

7.2 Hindered rotation of "light" molecular fragments

The quantum effects of rotary motion discussed above are normally important and can be observed at the lowest experimental temperatures close to 5 K in systems that contain the lightest possible atoms, protons and deuterons. They result to relatively *temperature independent* EPR lineshapes, since they are *inertia dependent* as opposed to diffusion. The quantum effects become of minor importance between 40 and 60 K, giving place to classical effects of motion[6] at even higher temperatures. As a *thermal reservoir* here can be considered the vibration motion of the "crystal" lattice to which the rotation degree of freedom is primarily coupled, and indirectly to the spins.[74]

Usually the rotary motion of methyl fragments of larger units is substantially hindered by *intramolecular*, but also by *intermolecular* forces, depending on the phase of the solid (crystal lattice), represented by potential barriers of the order of 1000 K. Cases of much smaller barriers, *e.g.* toluene with barrier 6.8 K(!) with almost free rotation, and much higher potential barriers than 1000 K, up to 2000 – 3000 K with strongly hindered rotation also exist.[75] (The higher values of 3000 K are usually found for the methylene fragment $-{}^{\bullet}CH_2$).

The strongly hindered methyls perform torsional oscillations, where the motion is not possible to stop in the classical meaning, *i.e.* by freezing to very low temperature. Those quantum systems perform zero-level oscillations if the barriers are too high, or otherwise tunnel (coherently) to the adjacent wells[71,23] for lower barriers. In the case of very low barriers compared to the splitting of the pure rotational states, such as for "light", protonated methyl rotors in frozen inert gases, the rotation stops per definition when only the lowest rotational level with zero angular momentum ($\ell = 0$) is populated. The EPR spectra of the *stopped quantum rotors* are characteristic,[7] *i.e.* four equal lines shown in Fig. 11(b), and occur at the lowest liquid He temperatures due the lack of adequate thermal excitation.

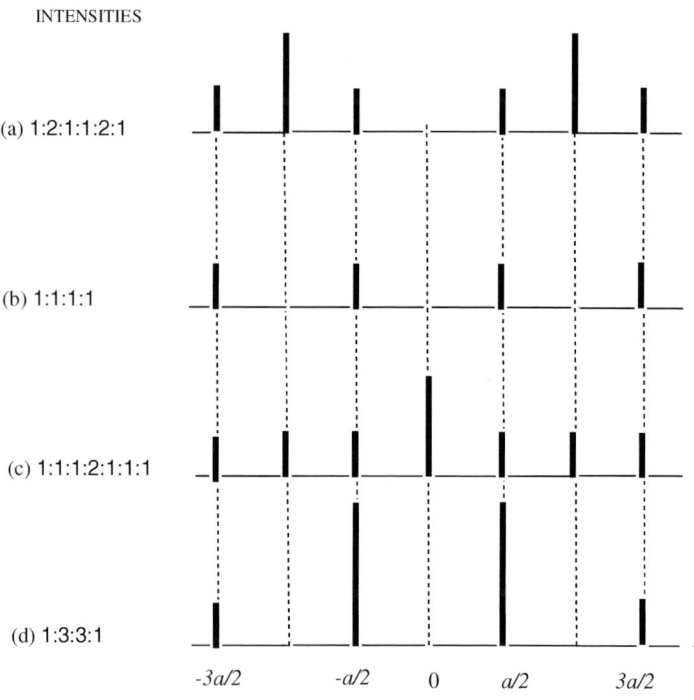

Figure 11. Basic EPR stick-spectra for a *hindered* methyl rotor with isotropic *beta* protons with hf coupling according to eq. (1), along with the appropriate limiting conditions for the motion, see text for details. (a) *Stopped classical rotor* $\tau_{exch} \rightarrow \infty$. (b) *Stopped quantum-rotor* ($\ell = 0$) for negligible barrier or *strongly-hindered diffusional* rotor. (c) *Tunneling quantum rotor* (superposition of $\ell = 0$ and $\ell = \pm 1$ rotational levels). (d) Fast classical- ($\tau_{exch} \rightarrow 0$) and very rapid high energy quantum-rotor (high rotational quantum numbers ℓ). The quantum mechanical *z*-projection quantum number of the rotational angular momentum is denoted by ℓ. Notice that the stopped quantum rotor in (b) requires negligible barrier and that it is also obtained for classical treatment of motion.[5]

Freed[23] seems to be the first who recognized and interpreted the quantum effects of spin-rotation coupling on the CW-EPR spectra in the case of a *hindered methyl rotor* at low temperatures. He derived also the theory of the *tunnelling-rotor septet* seen in the above Fig. 11(c), which is also obtained experimentally in several systems. In the same figure two more characteristic cases of experimentally obtained spectra are shown, in addition to the above

3. Dynamical effects in CW and pulsed EPR 131

two, using as a model the identical methyl fragment for different thermodynamic conditions. The stick spectra of this figure have been approximately computed using identical magnetic parameters. The underlying physics for the differences are briefly explained briefly in the legend of Fig. 11.

8. DYNAMICS BY CW-ENDOR SPECTROSCOPY

ENDOR spectra are free from unresolved hf structure caused by coupling to nearby and distant nuclei that usually broaden the EPR lineshapes. This is one of the major reasons that make ENDOR a preferable method for the determination of small hf splittings. The intensity distribution of the CW-ENDOR lineshapes however is more complicated than the EPR as it is heavily depending on the sensitive balance of the different electron-spin and nuclear relaxation pathways. This was also one of the reasons that the pulsed ENDOR became a preferred choice to the CW technique in many recent studies, as the interpretation of the pulsed spectra are free from complicated spin dynamics. In spite of this, investigation of exchange dynamics is possible by temperature dependence of the CW-ENDOR lineshapes in a parallel fashion to chemical exchange studies by the modification of CW-EPR lineshapes, described elsewhere in this text. This method is developed and exemplified in detail in the book by Harry Kurreck.[61]

As mentioned earlier in this chapter a usually successful combination in the investigation of disordered solids is the pulsed ENDOR with ESEEM and HYSCORE. In a recent work,[76] we presented a simple analytic expression for what is usually called an ideal ENDOR lineshape of disordered systems. In studying the higher multiplets of paramagnetic metal ions a simplified ENDOR pattern is an essential aid in interpreting the usually complex experimental spectra. For $S = 5/2$ in Mn^{2+} complexes for instance, we can ideally have as many as five overlapping pairs of HYSCORE ridges from different allowed EPR transitions and the best way to disentangle this pattern is using the ENDOR lineshapes.

Several theoretical aspects of the spin dynamics in connection to the ENDOR-lineshape intensity distribution were early investigated by Freed.[77] A long-standing subject of investigation by the group of Brustolon[25] from Padova University is the methyl rotation. In particular ENDOR *enhancement* was used in several temperature dependent CW ENDOR-linewidth studies for the investigation of this motion. Relatively simple radicals, such as those from gamma-irradiated L-alanine, CH_3-$^{\bullet}CH$-$COOH$, and simple methyl substituted aromatic molecules, were used as models during the development of this method. In both single crystals and in powders, and it looked like that the values of the activation parameters for the methyl rotation by standard methods could be reproduced rather well. What was in addition very appealing with the temperature ENDOR enhancement studies of this group was the apparent simplicity of the theoretical temperature dependence, expressed by a single spectral density of Lorentzian form.

The expression for the enhancement curve for which an example is shown in Fig. 12, has the same form as the frequency dependence of the spectral density appearing in the discussion about the Redfield theory, *i.e.* eq. (7) with $n = 1$. This is the result of the dominance of a single cross-relaxation pathway determining the EPR *desaturation* and the ENDOR line intensity, as it was deduced by a detailed theoretical consideration by the above investigator.

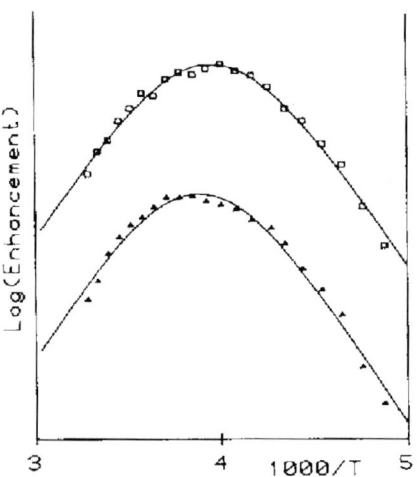

Figure 12. Typical temperature dependence of ENDOR enhancement from the methyl rotation. In this case it includes the maximum of the Lorentz spectral density. (Fig. 1 ref. [25])

9. PULSED-EPR TECHNIQUES

Random reorientation should average out the nuclear modulations of the ESEEM signal,[11] and even restricted motion in crystals would severely broaden the corresponding Fourier-Transforms at intermediate rates.[63] Freed and cooperators in an early spin-echo application investigated the effects of

3. Dynamical effects in CW and pulsed EPR

the slow reorientation on both the nuclear modulations and the ESE *phase memory decay* in viscous fluids.[62,78]

In one particularly interesting study, they calculated the temperature dependence of the *phase memory (decay) time* T_M of the primary echo signals and compared with *spin-echo* experiments for the tempone probe in a glycerol/water solution where slow motion conditions prevailed. Furthermore, some experimental T_2 data from CW spectra of tempone were also compared, as they were plotted in the same diagram together with the theoretical dependence of the transversal relaxation times T_2 and T_M on the temperature, see Fig. 2 ref. [11]. In that Figure the phase memory decay rate $1/T_M$ goes apparently through a maximum at about 0° C and decreases for higher temperatures. Converting the temperature data to corresponding reorientational correlation time τ_R for the three most known models of random reorientation,[78] *i.e. Brownian diffusion, free diffusion,* and *jump model* an interesting property was obtained. The phase memory decay time was given by the simple relations $T_M \propto (\tau_R)^b$, with $b \cong 1/2$, 2/3 and 1 for the above three models, respectively. Freed explained that in the *slow exchange limit* it is reasonable with an exchange broadening equal to τ_R^{-1}, the "jump" frequency. This is analogous to the *lifetime broadening* of the simple site exchange since each jump changes the randomly the orientation-"site" and thus the resonance frequency of the electron spin leading to *spectral diffusion*.

The above dependence of the *relaxation time* T_M on the *correlation time* τ_R in the slow motion region conforms to model simulations of the destruction of the nuclear ESE envelope modulations through faster decay of the time resolved signal in single crystals by motion.[63] In the latter case, the random reorientation of the hfi tensor was modelled by a simple two-site exchange of the principal axes of the hf interaction of the probe molecule between two extreme positions through intramolecular or intermolecular reorganization. In spite of the clear difference in the motion of the two systems, *i.e.* the viscous fluid with unrestricted reorientation vs. the single crystal with a restricted reorientation, respectively, the theoretical dependence of the phase memory decay time T_M on the *exchange / reorientation rate* was astonishingly similar. The "linear" relations $T_M \propto (\tau_{exch})^{\pm 1}$ were found in the single crystal *exchange model*, which is identical to reorientation for the *jump diffusion model* in the slow motion region discussed above.

The relevant diagram in Fig. 13 displays furthermore two distinct branches as in the reorientation case of Freed Fig. 2 ref [11], one for the slow motion and one for fast motion, corresponding to the two different exponents +1 and -1, respectively, in the above relation. In the left, slow motion branch of Fig. 13 the phase memory is lost faster for faster motions, while in the fast motion branch to the right in the same figure the phase memory is lost slower for faster motions. Accordingly, the two branches pass through a maximum phase memory decay rate $1/T_M$ in the same way as the typical longitudinal relaxation rate $1/T_1$ does, and not the typical transversal relaxation rate $1/T_2$ in the Redfield regime. The traditional transversal relaxation rate exhibits instead a monotonically decreasing behaviour with increasing temperature, as it was shown in Fig. 5.

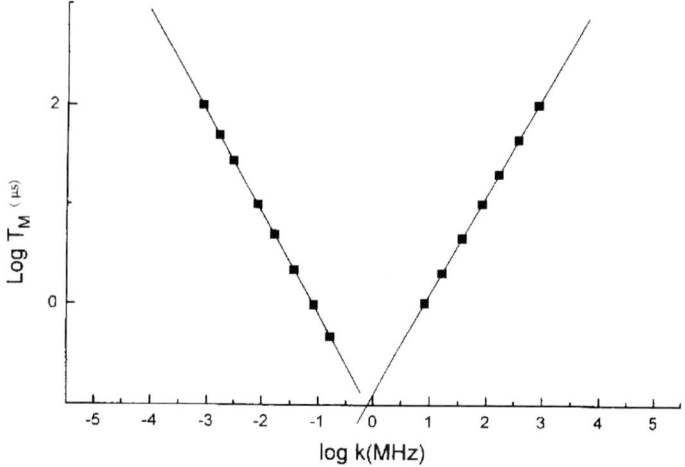

Figure 13. Phase memory decay time T_M dependence on the rate of random jumps between two possible dipolar frames of a radical in a single crystal according to model simulations of the primary ESEEM. (Fig. 3 ref. [63])

The explanation of the single crystal two-"site" exchange case is in accord to the standard CW lineshape results, where the random jumps of the resonance frequency between two discrete values destroys the hyperfine structure due to *lifetime broadening*. For very fast exchange the individual slow motion lines overlap and finally a narrowed average spectrum appears.

Similar results were found for the hindered rotation of the methylene fragment $-^{\bullet}CH_2$ in irradiated zinc acetate dihydrate crystals by other investigators[79] in a work that even the nonsecular nuclear-I-part effects were discussed. It is worth to notify that such terms, which are necessary for regular spin echo modulation, are usually less important when exchange dynamics corresponding to unrestricted overall motion are affecting the lineshape in powders. On the other hand they give significant deviations of

3. Dynamical effects in CW and pulsed EPR 135

the distance between the outer peaks of the CW-EPR lines for restricted motion and contribution of forbidden transitions to the spectra in single crystals.[68]

Finally, the group of Brustolon studied the restricted conformational changes of the saturated tempone ring by ESE and ESEEM, while the *libration* of the same molecule was more appropriately studied by high-field EPR.[14] In that work the above discussed dip in the relaxation time was verified by the total disappearance of the echo signal in the temperature range 130-210 K, and similar observations were made about the relaxation behaviour of several nitroxide labels by the Siberian group.[80]

9.1   ESEEM and tunnelling frequency

Normally, the tunnelling frequency of hindered methyl fragments attached to larger units (radicals) is obtained indirectly, involving complicated experimental setups. According to a relatively recent theoretical prediction, however, nuclear modulations can be observed even in the Electron Spin Echo experiment of methyls with isotropic hyperfine interactions under certain conditions.[81] Furthermore, according to that theory, one can register the tunnelling frequency directly in the ESEEM spectrum, giving this method great potential applicability. Namely, a doublet about the tunnelling frequency split by the hf interaction and two peaks with inverted intensity compared to this doublet, one at the double tunnelling frequency and one close to zero, are expected for a Hahn-echo sequence detection 90 degrees out of phase from the MW field.[81] Due to the high frequency harmonics involved here this method was limited by the contemporary instrumental development. However, choosing appropriate known systems with relatively high potential barrier the prediction of that theory can be checked out, at least in a limited number of samples.

The principle for the expected Echo Modulation for a hindered rotating methyl under quantum conditions of motion is the "slightly" different mixing of the torsional states with the nuclear-spin states in the two electron spin manifolds by the contact term of the hfi. This is a totally different prospect of echo modulation compared to regular ESEEM theory for nuclear modulations, which is based on purely magnetic interactions as the only cause of nuclear mixing. Due to the similar physics,[2] *i.e.* presence of not allowed EPR transitions due to nuclear mixing with the rotational degree of freedom by the hfi, the ESEEM spectra should remind the spin-flip satellites of EPR for tunnelling methyls observed occasionally also experimentally.

A simpler exchange theory for echo modulation in nitroxides induced by the motion of a classically rotating methyl fragment of the probe molecule, was developed just prior the time of writing this Chapter by the group of Dzuba and is presently under publication.[82] The corresponding experimental verification was performed for several methyl substituted nitoxides using a variation of the stimulated echo sequence with variable pulse delay τ between the first and the second pulse instead of the regular T variation.

A newer study involving the primary Quinone A is currently performed for the interpretation of the unexpected electron-electron modulations found in the stimulated echo signal which was observed recently in Novosibirsk. A preliminary theoretical verification of the echo modulations in such systems by using formal spin density theory already exists as it was developed during the summer of 2002. A simple model of a single electron exchanging between two different environments with slightly different Larmor frequencies was used for the paper-and pencil computation. It could be interesting to predict if *e.g.* small orientation changes of the g-tensor in the primary quinine $UQ^{\bullet-}$, which are undetectable by CW EPR are possible to be detected by spin-echo methods. This is an interesting problem connected with the biophysics of photosynthesis.

Such studies can also answer to several questions concerning other recent experimental data, e.g. not explained modulations of high frequency in some irradiated materials[83] and stimulate utilization of transient EPR methods as a complementary tool in the investigation of the quantum effects in methyl-type rotors.

9.2 Two-dimensional experiments

Within the ESEEM (Electron Spin Echo Envelope Modulation) techniques, the *primary Hahn echo* and the *three- pulse stimulated echo* were also used as the basis for the design of two-dimensional methods.[84] Also, a variation of the four-pulse echo, the (two-dimensional) 2D- HYSCORE method, has also a less known one dimensional 1D variant.[76]

A classical example is a series of ED-EPR experiments taken for several values of the pulse delay τ of the Hahn echo. One can finally collect the entire τ series in a two-dimensional spectral representation, where in one dimension is plotted the resonance field and in the other the homogeneous broadening.[11] A brief list of the main classical two-dimensional methods

3. Dynamical effects in CW and pulsed EPR 137

contains three of them: a) EPR-SECSY (Spin Echo Correlated Spectroscopy), b) EPR-COSY (Correlation Spectroscopy) and c) 2D-ELDOR[85] (Electron-Electron Double Resonance). These methods discussed by Gorcester and Freed[11] were based to the transfer of the FT-NMR (Fourier transform NMR) to EPR called FT-EPR.[10] The experimental registration included not only detection of echoes, but also detection of FID's and it is tested mostly for fast motions so far.

Concerning the pulsed ELDOR another series of pulse sequences was proposed (some very recently) with the purpose of a direct evaluation of unpaired electron distances in disordered solids.[43] It is interesting to note that among the first theoretical computations and experimental measurements of the electron-electron two-pulse signal modulations were performed as early as 1969 by a Russian team led by Salikhov and Tsvetkov.[86]

9.3 Spectral diffusion

The echo decay in a *primary* or a *stimulated echo* experiment is due to different intrinsic relaxation processes which for a certain *temperature* depend on the *concentration* of the radicals and their *residual mobility*, as well as on the *interaction* with the nuclear magnetic isotopes, particularly protons, present in the diamagnetic matrix. In the case of broad inhomogeneous EPR lines one can by a pulse excite only a limited part of the spectrum, usually a bandwidth of about 9 Gauss of the CW line for moderate MW power, and thus only a limited group of *spin packets* becomes affected.

In order to organize the discussion of the mechanisms of the echo decay in solids and to distinguish the relaxation mechanisms that are due to the effect of the pulse, one usually labels the spin packets affected directly by the pulse as spins **A**, and the ones not affected by the pulse but interacting with the first, as spins **B**. A simple example of this division of the EPR signal is shown in Fig. 14.

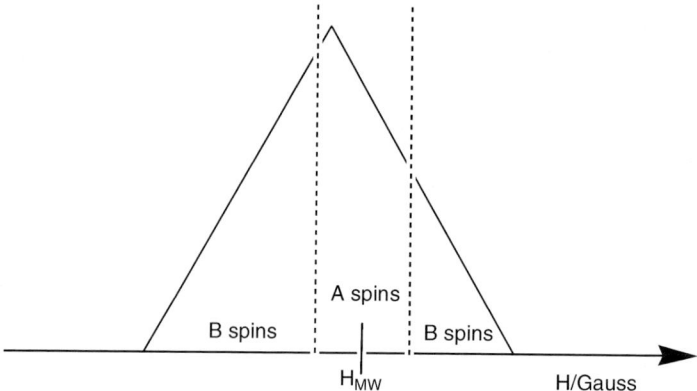

Figure 14. The separation of an inhomogeneously broadened signal into two parts, a region with the **A** spins, actually spin packets which are directly excited by a MW pulse, and the rest of the EPR band with the so called **B** spins that are not directly affected by the pulse.[16] The A-spin and B-spin regions depend in simpler cases on the central MW frequency $\sim H_{MW}$ and the intensity of the MW field $\sim H_1$.

Starting with the intrinsic interactions, we have so far mostly referred to relaxation mechanisms which depend on the thermal fluctuations of the parameters of the spin Hamiltonian. Using the previous assignment, they affect both the directly involved spins **A** as well as the **B** spins. These interactions are thus similar to the *scalar* mechanism of the *first kind* to be distinguished from the mechanisms similar to the scalar mechanism of the *second kind*.[17] In the latter, the modulation of the interaction necessary to give relaxation to the spins **A** is due to the independent fluctuations of the spin orientation of the **B** spins due to their spin-lattice relaxation.[14] We emphasize that this classification is only in analogy to the effect of the *contact term* of the scalar hyperfine interaction and that we have here in addition to consider the intermolecular interactions between the **A** spins and the unobserved spins **B** spins modulated by mutual spin-flips.

The **B** spins generate local fields at the positions of the **A** spins that directly participate to the formation of the ESE signal through the dipolar interaction. Considering a primary, $\pi/2$-τ- π-τ-echo, sequence if the **B** spins do not change orientation during the interpulse periods τ they induce static resonant frequency shifts which contribute to the inhomogeneous broadening of the signal that is unimportant for the echo decay. If however the orientation of the **B**-spins is randomly modulated in some way, *e.g.* by an independent relaxation mechanism, the fluctuations of the *local field* at the positions of the **A** spins give migrations of the resonance frequencies which result to *spectral diffusion*. Spectral diffusion gives *irreversible dephasing* to the electron-spin precession.

3. Dynamical effects in CW and pulsed EPR 139

Summarizing, the random modulation of the **B** spins causing spectral diffusion can be of two kinds: a) In T_1-type systems the random changes of the spin orientation of the **B** spins by spin-lattice relaxation are mediated to the **A** spins by the *S-S* interactions. b) In T_2-type systems the mutual spin flip-flops between **A** and **B** spins are the reason for this kind of spectral diffusion.[15]

The relative importance of the above two mechanisms for the ESE decay varies with *temperature* and radical *concentration*. For systems at relatively high temperature, and thus shorter T_1, as well as dilute matter with respect to paramagnetic centers, the flip-flops are infrequent and random variation of the spin orientation by the spin-lattice interaction dominates in producing spectral diffusion. For lower temperatures and higher concentrations of the paramagnetic centers the flip-flop mechanism is the most important mechanism.

Except from the above adiabatic[55] mechanisms of spin dephasing by spin-resonant frequency shifts there is one more way for the acceleration of the echo decay of the **A** spins which is due to a phenomenon related to spin diffusion called *spin-excitation transfer*. Part of the excitation of the **A** spin by the MW pulse can migrate to **B** spins by mutual flip-flops and cannot participate in the formation of the echo increasing thus the apparent decay of the signal.

In addition to the **B** electron spins in the sample, even the spins of the matrix nuclei are also included in the definition of the **B** spins as they can enhance *spectral diffusion*. A necessary condition is that there is an *S-I* coupling to the **A** spins and that the nuclei have a strong and independent relaxation mechanism. They cause dephasing of the electron spins as *spectral diffusion* can be induced by random modulation of the local fields at the position of the electron by the nuclear spin flips. One should remember that the nuclear spin system in solids is rather dense. It consists usually of the abundant proton nuclei, a fact that compensates for their smaller gyromagnetic ratio in comparison to the electrons. They are interacting with each other, and via this dipole-dipole mechanism, the nuclear spin flips are transferred by nuclear spin diffusion to the rare paramagnetic centers.

9.4   Instantaneous diffusion

Generally speaking, the kinetics of dephasing of the echo signals according to the above mechanisms has a "*stretched exponential*"[87] functional form with a great variation of the parameters x and τ_m.

$$I(t) = A\exp[-(t/\tau_m)^x] \qquad (10)$$

The point here is that these two parameters can specify rather well the dynamics of the system. An example is the *instantaneous diffusion* mechanism which can be shown to display exponential decay, *i.e.* $x = 1$.

The instantaneous diffusion is a source of ESE decay due to the dipolar S-S interaction among only the **A** spins. The decay in this case is brought about by the local-field changes resulting from the spin flips induced by the second MW pulse of the primary echo. Instantaneous diffusion depends on the concentration of the **A** spins and the extend of the perturbation created by the pulse on them, that is the *nutation angle*, and results to a very simple exponential decay. A detailed analysis of the instantaneous diffusion computed under rather general conditions revealed the following relationship:[19]

$$\frac{1}{T_M} = \frac{4\pi^2}{9\sqrt{3}} \gamma^2 \hbar C < \sin^2(\vartheta/2) >_g \qquad (11)$$

C is the concentration and γ is the gyromagnetic ratio of the **A** spins, while the average with respect to the lineshape $g(H)$ indicated in the above equation is an important quantity to be computed. It is a measure the involvement of the variable nutation angles of the different parts of the inhomogeneous EPR lineshape to the decay. The angle $\vartheta \equiv \vartheta(H)$ here is the nutation angle that the second pulse in the primary echo tilts the magnetization resonant at field H. The nutation angle depends on the amplitude of the MW field H_1, giving thus a reliable experimental way to recognize the instantaneous diffusion mechanism by variation of the H_1. We reproduce the suggestive form of this averaging given in refs. [19,87], for pulse duration t_p.

$$<\sin^2(\vartheta/2)> = \frac{\int dH\, g(H) \frac{H_1^2}{(H-H_0)^2 + H_1^2} \sin^2\left(\frac{\gamma t_p}{2}\sqrt{(H-H_0)^2 + H_1^2}\right)}{\int dH\, g(H)} \qquad (12)$$

Application of this type of phase decay in the ED-EPR spectra of spin-labelled peptides[19] is described next.

10. LIBRATIONAL MOTION STUDIED BY ED-EPR

The *Echo-Detected* ED-EPR method was used by the Siberian group for the study of the relaxation due to librational motion in several nitroxides such as the usual tempone, 2,2,6,6- tetramethyl-4- piperidone-1- oxide, and other nitroxides, see Fig. 2.

The experimental approach consists in obtaining the echo-detected ED-EPR spectrum for different *pulse delays* τ, referred to as "the time separation between pulses between the first and the second pulse", in the primary echo.[44] Several structural and dynamical parameters are involved in the amount of the alteration of the ED-EPR spectra by the τ variation as long as the librational motion is not frozen, a condition imposing a minimum required temperature. This phenomenon is due to phase relaxation induced by the libration motion. A temperature of 77 K was thus adequate in most cases to freeze the motion and to forbid a further τ dependence to be observed but not always, since the phenomenon of instantaneous diffusion could be active.

One idea of this group was to use the nitroxides for the structural and dynamical investigation of *glassy* vs. the *crystalline* phases and by this way to discriminate between these components in different frozen biological samples.[44] As it was known previously that the glassy state has a *cryoprotective activity* for cells, while the formation of crystalline ice is considered as the basic cause of cell death at low temperatures, the interest of such a study is great for cryogenic life preservation. An early application of ED-EPR of this kind by the Siberian group that is reviewed here is a study[44] from 1993. More recent work of the group involving also some theoretical developments is found in refs. [20,46,19,48], and most recently new models of dynamics were investigated in a combination of a classical theoretical approach with interesting experimental results.[29] However, the assigned, single type of overall librational motion was not always consistent with the experimental data.[29,80]

10.1 Glasses and polycrystallites

In general, glassy materials displayed a more considerable alteration of the ED-EPR lineshape with an increase of τ than in crystalline matter.

Some of the biological materials tested in this study were: A mixture of soybean phospholipids, defatted embryos and endosperm of wheat kernels doped with tempone. The results were similar to the pulse delay variation effects found in simpler inorganic systems used as models such as, supercooled ethanol, water-glycerin solutions, and also in the supercooled phthalate for temperature variation.

As it was first tested in toluene a stronger sensitivity with respect to τ variation was indeed observed in the amorphous glassy phase in comparison to the crystalline. The sensitivity of the pulse-delay dependent modification of the spectra varied also in the different regions of the inhomogeneously broadened line. Most sensitive to τ variation was the high field component of the ^{14}N hf triplet, which for this reason obtained an additional minimum in the glassy series, see Fig. 1 ref. [44]. It was concluded that this type of dependence could be attributed to the differential magnetic phase relaxation (*phase memory time* T_M) of the nitroxide for the different orientations with respect to the external field.

A quantitative explanation of this behaviour is achieved by considering the spin relaxation due to small-amplitude *librations* of the nitroxide molecular probe, as we discuss further.

We reproduce here the relations for the orientation dependent relaxation factors $R_m(2\tau, \vartheta, \varphi)$ from ref. [46] that have to be multiplied with the primary echo signal for each pair (ϑ, φ) of polar angles and each nuclear projection $m \equiv M_I = 0, \pm 1$, of the hyperfine lines of ^{14}N. They were obtained within the limit of the Redfield relaxation theory:

$$R_{\pm 1}(2\tau, \vartheta, \varphi) = \exp\{-2\tau[r_a(\vartheta,\varphi) - r_{na}(\vartheta,\varphi)]\} \tag{13.a}$$

$$R_0(2\tau, \vartheta, \varphi) = \frac{1}{3}\exp[-2\tau\, r_a(\vartheta,\varphi)] \times \{2\exp[-2\tau\, r_{na}(\vartheta,\varphi)] + 1\} \tag{13.b}$$

Where r_a signifies the *adiabatic* contribution to the decay of the primary two-pulse echo arising from the electron-spin dephasing and r_{na} the *non-adiabatic* contribution to the decay arising from the modulation of the electron spin resonance frequency by the jumps of the nitrogen nuclei projections.[46] We reproduce further only the most important adiabatic part for the simple case of small libration angles α and for libration around the x-axis of the nitroxide. Actually the non-adiabatic (lifetime) part was discarded in the simulations of that work assuming slow motions, that is $\tau_C > 3\ 10^{-9}$ sec, also in agreement of the inefficiency of the non-adiabatic part for slow motions.

3. Dynamical effects in CW and pulsed EPR

$$r_a(\vartheta,\varphi) = \tau_C <\alpha^2> \sin^2\vartheta \cos^2\vartheta \sin^2\varphi \times$$

$$\times \left[\omega(g_{xx}-g_{zz}) + m(A_{xx}^2 - A_{zz}^2)/a_0\right]^2 \qquad (14)$$

Except from the trigonometric factors of the field orientation, the g-tensor components g_{ij}, the spectrometer operating frequency ω, the hyperfine tensor components A_{ij}, and the average $\mathbf{a}_0$ of the vector $\mathbf{a} = (A_{zx}, A_{zy}, A_{zz})$, are involved in this relation. Although the symbolism about m used in that reference is not totally consistent, we note that at least the above cited adiabatic part of the relaxation rate is proportional to the product $<\alpha^2> \cdot \tau_C$. The proportionality to the correlation time agrees with the adiabatic part of the Redfield theory, for $n = 0$ in eq. (7) as it should. Notice, however, that with increasing temperature the product $<\alpha^2> \cdot \tau_C$ can either increase or decrease since $<\alpha^2>$ is an increasing quantity while τ_C a decreasing one, cf. eq. (6).

10.2 Multilayer aggregates

The study of a biologically interesting lipid *multibilayer aggregate*[20] by ED-EPR was undertaken in the experimental and theoretical work of Dzuba and collaborators. The experimental spectra of the dopand, *i.e.* the particularly elongated *cholestane* spin label shown in Fig. 2, depended drastically on the orientation of the multibilayer with respect to the field as shown in the Fig. 15 and Fig. 16. This property shows that the label exhibits long range order in the aggregate as it is oriented with respect to the macroscopic *director* of the multibilayer, *i.e.* the optical axis of the membrane normal to the lamellar surface.

The variation of the lineshapes with the pulse delay τ could be fitted to the product $<\alpha^2> \cdot \tau_C$, used as a single parameter, where α is the angle of the torsional motion and τ_C the correlation time of the responsible stochastic process of libration. The indicated average of the angle is evaluated with respect to time and the above relation is strictly valid for small libration angles. The ED-EPR method alone could not allow for the separate determination of both the variance $<\alpha^2>$ of the libration angle and the correlation time τ_C, and was therefore combined by separate determination of the former by CW-EPR, as shown in Fig. 3.

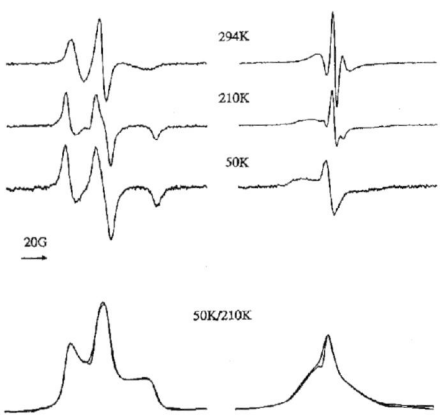

Figure 15. The CW EPR spectra of the elongated nitroxide spin-label shown in Fig. 2, in a *lamellar phase* for several temperatures and for two orthogonal directions of the multibilayer surface with respect to the external magnetic dc field, one parallel (left) and one perpendicular (right). The integrated CW spectra in the lower panel, which are nearly unchanged in the interval 50/210 K, were compared to echo-detected EPR traces (not shown) which exhibited significant temperature dependence. (Fig. 1 ref. [20])

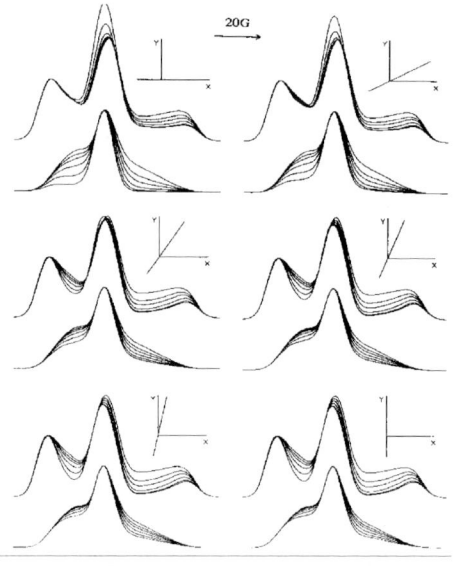

Figure 16. Simulations of the echo-detected EPR spectra of the nitroxide shown in Fig. 2 in a lamellar phase for six different values of the product $\tau \cdot <\alpha^2> \cdot \tau_C$, two characteristic orientations of the static field (parallel and perpendicular) with respect to the *director*, and six different orientations of the libration axis in the molecular frame shown by the XY-axes insets. Each of the six pairs of panels contains bunches of six different spectra with different products $\tau \cdot <\alpha^2> \cdot \tau_C$. Compare the experimental CW spectra in Fig. 15. (Fig. 3 ref. [20])

10.3 Peptide chain mobility

The *local mobility* of the peptides as well as *unfolding* and *denaturation* of larger proteins can be studied by spin labelling. Mutant proteins or polypeptides with predefined residue(s) at certain position(s) of the aminoacid chain by appropriate nitroxides can be used.

Three labelled variants of the small natural peptide peptaibol, containing only 10 aminoacid residues, were prepared by substitution of one aminoisobutyric acid residue by 2,2, 6,6- tetramethyl- piperidine- 1-oxyl- 4-carboxylic acid, at the two terminal and one intermediate positions, and were studied by Dzuba et al.[19] The dependence of the ED-EPR on the *pulse delay*, the *temperature*, and the *nutation angle* variation were used to study the librational dynamics and eventual deviations of the *local concentration* of the label. By observing the lower left part of Fig. 17 we notice that the variation of the pulse delay τ affects most significantly the central hf-component of the nitroxide. In contrast to the *spectral diffusion*, the effect of *instantaneous diffusion* on the spectra of partially immobilized nitroxides can be seen as the selective decrease of the central peak intensity of the nitroxide with the increase of the pulse delay τ.

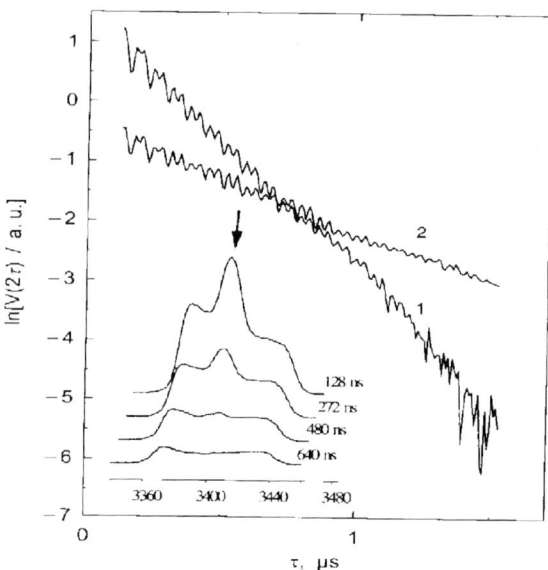

Figure 17. Determination of the (local) radical concentration by using the properties of the instantaneous diffusion, see text. (Fig. 8 ref. [19])

The instantaneous diffusion has furthermore an exponential decay rate proportional to the concentration of the irradiated **A** spin packets, which is largest just at the central hf component $M_I = 0$ of the ^{14}N nucleus. Inspection of the intensity distribution of the immobilized nitroxide spectra, *e.g.* in Fig. 15, to Fig. 17, shows this fact indicating that relatively more spin packets can be affected by irradiating just the middle region of the inhomogeneous spectrum. The central peak attains also for this reason the largest dynamical range concerning pulse-delay variation dependence of the intensity when the decay is due to instantaneous diffusion. This is demonstrated very well in the lower left part of Fig. 17, where the pulse-delay dependent variation of the central peak is by far more impressive compared to the variation of the intensity of the two outer peaks.

An interesting application of the variation of the nutation angle is also displayed in Fig. 17, which with clarity demonstrates that a) The decay of the ESE signal is *exponential*, as the instantaneous diffusion mechanism requires, and also that b) The decay rate varies with the change of the MW intensity. Thus, computing the average $<\sin^2(\vartheta/2)>_g$ according eq. (12) for two different values of the MW field intensity H_1, the authors of that work[19] obtained a larger value of the average for the greater nutation angle and a corresponding greater slope of the time domain signal in the semilogarithmic plot shown in the higher middle of Fig. 17. One interesting issue here was the potential determination of *local* deviations of the concentration of a given radical with respect to the bulk value in the sample. By using the above instantaneous diffusion data and the slopes of the two curves in Fig. 17 the above team was able to compute the concentration C shown in eq. (11). This is not necessarily equal to the average concentration of the sample but to the effective concentration of *paramagnetic centers* within a narrow resonance-frequency band (**A** spins). In that particular case the experimental value was however equal to the average concentration.

10.4 Protein mobility studies

An important application of the nitroxide spin labels is found in a series of studies concerning the mobility of selected regions of certain macromolecules such as peptides[19] and enzymes[73] in particular. Usually the *dephasing* of the Electron Spin Echo signal and not the nuclear modulations, if the signal is modulated at all, is of interest for this method. It is generally accepted that for dilute solutions of organic radicals at low temperature, *electron spin echo dephasing* is dominated by the interaction between the free radical and the nuclear spins (protons) in the surrounding medium.

3. Dynamical effects in CW and pulsed EPR

Herzog and Hahn[88] first recognized the importance of *spin diffusion* mediated by nuclear spin *flip-flops* on the dephasing of the electron spin echo. Secular flip-flop transitions among the nuclear spins, which are induced by the terms $I_{1+}I_{2-} + I_{1-}I_{2+}$ of the nuclear-pair dipolar Hamiltonian modulate the electron-nuclear *S-I* hyperfine coupling providing an efficient mechanism of the second kind for echo dephasing for the electron spin. This is obviously a relaxation mechanism which in many ways resembles the way that electron spins that are not directly affected by the pulse, **B** spins, give fluctuations to the z-component of the resonating **A** spins through their spin-lattice T_1-relaxation. In the case of relatively dilute electron spins both nuclear and electron **B** spins are treated similarly, also including *semisecular* terms of their electronic Hamiltonian, i.e. isolated flips or flops, where the energy balance is assisted by the mediation of the "lattice". The latter processes are heavily depending on the state of the matter and on *cross relaxation*.

A statistical treatment for the fluctuation of the z-component of the magnetization of the **B** spins for different theoretical models can be cast to a "*stretched*"-exponential-decay[14,87] function shown in eq. (10). Extensive theoretical derivations of the *stretched exponential* forms for different models are found in the articles of Romanelli and Kevan ref. [87]. Except for the usual *exponential* and *Gaussian decays* (for exponential power of time x = 1 and 2, respectively) this relation also includes other possible cases such as x = 1/2, 3/2, and 7/4, at least theoretically. The echo decay function of ESEEM can in practice equally well, if not better, be fitted by considering exponential polynomials in the time variable t, which naturally include exponential and Gaussian decays. Third order decays are supported by the self-diffusion mechanism, in connection to the field gradients and the Carr-Purcell sequence,[42] but higher integer powers than that can hardly be given theoretical support.

The stretched-exponential-decay function was used qualitatively in order to simulate the envelope of the electron spin echo decaying signal of spin labels in different solvents as well as spin labels attached to different positions in human carbonic anhydrase II[73c] (HCA II). The solvents were classified tentatively according to the values of the two parameters of this function,[14,73] i.e. τ_m the phase memory decay time, and the exponent x, at a temperatures lower than 60 K. For spin-labeled mutants of HCA II in glassy 1:1 water:glycerol at 11 and 40 K, the shape of a 2-pulse echo decay was different for labels on the surface or in the interior of the protein.[73a] By such a study one could obtain information about the mechanism determining T_2 relaxation by the magnetic nuclei surrounding the spin label. An interesting observation in several studied systems with this method was the great sensitivity of the dephasing rates to the groups of slowly-to-rapidly rotating methyl groups confined in the sphere of interaction of the electron spin.[14,73] A model was developed that allowed to predict the shape of the T_M decay curve at low temperatures. The model used a crystal lattice and/or protein crystal structure of spin-label and surrounding protons and methyl groups in the solvent-protein phase.

Previous results of the same invetigators have shown that the ESR lineshape associated with an introduced spin-label can give very useful information in terms of local structure and mobility at the position of the spin-label, such as the structure of equilibrium folding intermediates. This method was applied in the investigation of the degree of folding of partially *denaturated mutant* protein HCA II containing label molecules in different positions of the primary chain by CW EPR lineshape simulations.[89]

Acknowledgements *I would like to thank the so many persons that gave me the opportunity of working out this text. First of all Professor Anders Lund from Linköping and Professor Dimitrios Galaris in Ioannina University in Greece, as well as Professor Rimma Samoilova from the Institute of Chemical Kinetics and Combustion in Novosibirsk. Professor Samoilova I also thank for showing me how one works with ESEEM in practice. The highly inspiring scientific environment of the interested young students and the seniors in Novosibirsk led by Professor Sergei Dzuba and Professor Yuri Tsvetkov, and particularly the friendliness of my hosts was for me something new. Especially Eugenia Kirillina is highly acknowledged for all practical help and for teaching me some dynamics of nitroxides. I was also always happy to discuss and learn from the theorist Alexandr Maryasov, who also borrowed me his Russian(!) version of Abragam. I also want to thank the unknown reviewer that took away 2/3, or about 80 pages(!) from my original manuscript and Dr. Martina Huber for correcting the manus. Finally Mari Löfkvist is acknowledged for the language corrections.*

11. REFERENCES

1. Keith R. Symon *"Mechanics."* 3rd edt., Addison-Wesley Publ. Comp. Reading, Massachusetts, 1974.
2. A.R. Sørnes and N.P. Benetis, *Chem. Phys.* **226**:1-2 (1998) 151.

3. Dynamical effects in CW and pulsed EPR

3. N. Benetis, J. Kowalewski, L. Nordenskiöld, H. Wennerström and P-O Westlund, *Mol. Phys.* **48**:2 (1983) 329.
4. Donald A. McQuarrie *"Statistical mechanics."* Harper and Row 1976, New York.
5. A.R. Sørnes and N.P. Benetis. *J. Magn. Res.* **125**:1 (1997) 52.
6. A.R. Sørnes, N.P. Benetis, R. Erickson, A.S. Mahgoub, L. Eberson, and A. Lund, *J. Phys. Chem.* A **101**:48 (1997) 8987.
7. T. Yamada, K. Komaguchi, M. Shiotani, N.P. Benetis, and A.R. Sørnes, *J. Phys. Chem.* A **103**:25 (1999) 4823.
8. Y. Deligiannakis, M. Louloudi and N. Hadjiliadis, *Coord. Chem. Rev.* **204** (2000) 1.
9. J. Schmidt and David J. Singel, *Ann. Rev. Phys. Chem.* (1987) 141.
10. M. Bowman *"Modern Pulsed and continuous-wave electron spin resonance."* Chapt. 1, Edts. L. Kevan and M.K. Bowman, John Wiley & Sons, NY 1990, pags. 1-42.
11. a) J. Gorcester, G.L. Millhauser, and J.H. Freed *"Advanced EPR: Applications in Biology and Biochemistry."* Chapt. 5, Edt. A.J. Hoff, Elsevier, Amsterdam, 1989, pag. 177-242; b) Jeff Gorcester, Glenn L. Millhauser, and Jack H. Freed. *"Modern pulsed and continuous-wave electron spin resonance."* Chapt. 3, Edt. Larry Kevan and Michael Bowman. John Willey & Sons, NY 1990, pags. 119-194.
12. Arthur Schweiger *"Modern pulsed and continuous-wave electron spin resonance."* Chapt. 2, Edts. Larry Kevan and Michael Bowman. John Willey & Sons, NY 1990, pags. 43-118.
13. S.A. Dzuba, I.V. Borovykh, and A.J. Hoff, *J. Magn. Res.* **133** (1998) 286.
14. Antonio Barbon, Marina Brustolon, Anna Lisa Maniero, Maurizio Romanelli, and Louis-Claude Brunel, *Phys. Chem. Chem. Phys.* **1** (1999) 4015.
15. K.M. Salikhov and Yu.D. Tsvetkov *"Time Domain Electron Spin Resonance."* Chapt. 7, Edts. L. Kevan and R.N. Schwartz, John Wiley, NY 1979.
16. Ian M. Brown *"Time Domain Electron Spin Resonance."* Chapt. 6, Edts. L. Kevan and R.N. Schwartz, John Wiley, NY 1979.
17. A. Abragam *"The Principles of Nuclear Magnetism."* Clarendon press, Oxford 1961. Page references of specific topics of interest here: *Spin-packet* pag. 397, *spin- diffusion* pags. 378-389 and 103-111, and additional reading pgs.10-38 and 133-144. *Spin temperature* pgs. 133-144. *Relaxation due to paramagnetic impurities* pags. 378-398. Notice that definition of *spin diffusion* in pags. 59 and 61, concerns actually the *self-diffusion* of spin-bearing molecules measured when an external *field gradient* is applied on the top of the static field. An extensive consideration of the broadening in *solid state* is given in Abragam Chapter X, but it concerns nuclear relaxation. Many important aspects and definitions about electron spin relaxation at very low temperatures are found in Abragam and Bleaney, ref. [49], but are specialized to paramagnetic metal ions.
18. J. Kowalewski, L. Nordenskiöld, N. Benetis and P.O. Westlund, *Progr. NMR Spectr.* **17** (1985) 141.
19. Yu.V. Toropov, S.A. Dzuba, Yu.D. Tsvetkov, V. Monaco, F. Formaggio, M. Crisma, C. Toniolo, and J. Raap, *Appl. Magn. Res.* **15** (1998) 237.
20. Sergei A. Dzuba, Hiroshi Watari, Yuhei Shimoyama, Alexandr G. Maryasov, Yoshio Kodera, and Asako Kawamori, *J. Magn. Res.* **115** (1995) 80.
21. O.N. Antzutkin, N.P. Benetis, M. Lindgren and A. Lund, *Chem. Phys.* **169** (1993) 195.
22. F. Bonon, M. Brustolon, A.L. Maniero, and U. Segre, *Appl. Magn. Res.* **3** (1992) 779.
23. J.H. Freed, *J. Chem. Phys.* **43** (1965) 1710.
24. S. Clough and F. Poldy, *J. Chem. Phys.* **51** (1969) 2076.
25. Marina Brustolon, Teresa Cassol, Lauretta Micheletti, and Ulderico Segre, *Mol. Phys.* **57**:5 (1986) 1005.

26. A. Carrington and A.D. McLachlan. *"Introduction to Magnetic Resonance."* Harper, NY 1967.
27. S.A. Dzuba, K.M. Salikhov, and Yu.D. Tsvetkov, *Chem. Phys. Letts.* **79** (1981) 568.
28. D.G. Lister, J.N. Macdonald and N.L. Owen *"Internal Rotation and Inversion. An Introduction to Large Amplitude Motions in Molecules."* Academic press, London 1978.
29. E.P. Kirillina, S.A. Dzuba, A.G. Maryasov, and Yu.D. Tsvetkov, *Appl. Magn. Res.* **21** (2001) 203.
30. N.A. Salih, O.I. Eid, N.P. Benetis, M. Lindgren, A. Lund, and E. Sagstuen, *Chem. Phys.* **212** (1996) 409.
31. J.E. Wertz and J.R. Bolton. *"Electron Spin Resonance. Elementary Theory and Practical Applications."* McGraw-Hill, N.Y. 1972.
32. Frank J. Adrian, *J. Chem. Phys.* **88**:5 (1988) 3216.
33. a) L. Sjöqvist, N.P. Benetis and A. Lund, *Chem. Phys.* **156** (1991) 457; b) M. Lindgren, R. Erickson, N.P. Benetis, and O. Antzutkin, *J. Chem. Soc. Perk. Trans.* 2 (1993) 2009; c) M. Lindgren, N.P. Benetis, M. Matsumoto, and M. Shiotani, *Appl. Magn. Res.* **9** (1995) 45.
34. R.R. Rakhimov, N.P. Benetis, A. Lund, J.S. Hwang, A.I. Prokof'ev, and Y.S. Lebedev, *Chem. Phys. Let.* **255**:1-3 (1996) 156.
35. A. Rockenbauer, M. Györ, H.O. Hankovszky, and K. Hideg. *"Electron Spin Resonance."* Chap. 5, Vol. **11**A, pag. 145-182, Royal Soc. of Chem. 1987.
36. C. Heller and H.M. McConnell, *J. Chem. Phys.* **32** (1960) 1535.
37. F. Bonon, M Bustolon, A.L. Maniero, and U. Segre, *Appl. Magn. Res.* **3** (1993) 779.
38. E.W. Stone and A.H. Maki, *J. Chem. Phys.* **37**:6 (1962) 1326.
39. George Arfken *"Mathematical methods for Physicists."* Second edt. Academic press, NY 1970.
40. Nikolas P. Benetis and Anders R. Sørnes, unpublished.
41. Yakob S. Lebedev *"Time Domain Electron Spin Resonance."* Chapt. 8, Edts. L. Kevan and R.N. Schwartz, John Wiley, NY 1979.
42. H.Y. Carr and E.M. Purcell, *Phys. Rev.* **94** (1954) 630.
43. A. Schweiger and G. Jeschke *"Principles of Pulse Electron Paramagnetic Resonance"*, Oxford University Press 2001.
44. S.A.Dzuba, Ye.A. Golovina, and Yu.D. Tsvetkov, *J. Magn. Res.* B **101** (1993) 134.
45. G.R. Luckhurst *"Electron Spin Relaxation in Liquids."* Chapt. X, Edts. L.T. Muus and P.W. Atkins, Plenum Press, NY 1972.
46. Sergei A. Dzuba, *Phys. Letts.* A **213** (1996) 77.
47. S.A. Goldman, G.V. Bruno, and J.H. Freed, *J. Phys. Chem.* **76**:13 (1972) 1858.
48. J. Buitink, S.A. Dzuba, F.A. Hoekstra, and Yu. D. Tsvetkov, *J. Magn. Res.* **142** (2000) 364.
49. A. Abragam and B. Bleaney *"Electron Paramagnetic Resonance of Transition Ions."* Dover Publications, inc NY 1970.
50. Michael K. Bowman and Larry Kevan *"Time Domain Electron Spin Resonance."* Chapt. 3, Edts. L. Kevan and R.N. Schwartz, John Wiley, NY 1979.
51. a) A.G. Redfield, *IBM J. Res. Dev.* **1** (1957) 19; b) *Adv. Magn. Reson.* **1** (1965) 1.
52. N. Bloembergen, E.M. Purcell, and R.V. Pound, *Phys. Rev.* **73** (1948) 679.
53. a) R.K. Wangsness and F. Bloch, *Phys. Rev.* **89** (1953) 728; b) F. Bloch, *Phys. Rev.* **102** (1956) 104.
54. C.P. Slichter *"Principles of Magnetic Resonance."* Third Edition, Springer Verlag, Berlin 1990.
55. Beware that the term *adiabatic* has been used in several other magnetic resonance contexts concerning both solids and liquids. In the present case we mean the use of the *S*-

3. Dynamical effects in CW and pulsed EPR

secular Hamiltonian. The quantum mechanical or Ehrenfest sense of adiabaticity according to Abragam ref. [17], pags. 135-136, is the one where the variation of the field, or generally any other external parameter of the system, is applied in such a way that *no spin transitions are induced*. Thus, the populations of the levels are maintained during the change, while obviously heat has to flow to or from the ensemble to compensate the parametric change in the energy (Hamiltonian). Such can also be the so called *sudden change* that conserves the *spin temperature*. The *thermodynamic sense* of an *adiabatic transformation* on the other hand, has the profound interpretation of prohibited exchange of energy between the spins and the lattice. Such a *reversible change* of the system close to equilibrium is also *isentropic*, i.e. it is conserving the entropy, Abragam ref. [17] pags. 144-149. Examples belonging in this sense of adiabaticity are the cases of *adiabatic demagnetization* and *adiabatic fast passage*, or simply *fast passage*, discussed in detail also for nuclei by E. Fukushima and S.B.W. Roeder in "Experimental Pulse NMR. A Nuts and Bolts Approach." Addison-Wesley Publishing Company, Inc, London 1981. Adiabatic fast passage means slow enough rate of change of the field through resonance in order to be *reversible* but fast enough so that exchange of energy with the lattice by T_1 processes not to be possible, see also Slichter's book ref. [54] Chapt. 6.

56. N.P. Benetis. *"Nuclear-Spin Relaxation in Paramagnetic Metal Complexes and the Slow-Motion Problem for the Electron Spin."* Ph. D. Thesis, University of Stockholm 1984.
57. P. O. Westlund, N. Benetis and H. Wennerström, *Mol. Phys.* **61**:1 (1987) 177.
58. N. Benetis and J. Kowalewski, *J. Magn. Res.* **65** (1985) 13.
59. N. Benetis, J. Kowalewski, L. Nordenskiöld, H. Wennerström and P.O. Westlund, *J. Magn. Res.* **58** (1984) 261.
60. a) J. Heinzer, *Mol. Phys.* **22**:1 (1971) 167; b) J. Heinzer, *Program 209*, QCPE Indiana University, 1972.
61. Harry Kurreck *"Electron Nuclear Double Resonance Spectroscopy of Radicals in Solution."* Wolfgang Lubitz VCH, 1988.
62. Leslie J. Schwartz, Arthur E. Stillman, and Jack H. Freed, *J. Chem. Phys.* **77** (1982) 5410.
63. U.E. Nordh and N.P. Benetis, *Chem. Phys. Lett.* **244** (1995) 321.
64. a) L. Sjöqvist, A. Lund, and J. Maruani, *Chem. Phys.* **125** (1988) 293; b) N.P. Benetis, L. Sjöqvist, A. Lund, and J. Maruani, *J. Magn. Res.* **95** (1991) 523.
65. J.A. Ladd and H.W. Wardale *"Internal Rotation in Molecules."* Chapt. 5, Edt. W.J. Orville-Thomas, John Wiley & Sons NY 1974.
66. Richard R. Ernst, Geoffrey Bodenhausen, and Alexander Wokaun. *"Principles of Nuclear Magnetic Resonance in One and Two Dimensions."* Clarendon Press, Oxford 1991.
67. H.M. McConnell, *J. Chem. Phys.* **28** (1958) 430.
68. N.P. Benetis, M. Lindgren, H-S. Lee, and A. Lund, *J. Appl. Res.* **1** (1990) 267.
69. N.P. Benetis, A.S. Mahgoub, A. Lund and U.E. Nordh, *Chem. Phys. Lett.* **218** (1994) 551.
70. R. Erickson, U. Nord, N.P. Benetis and A. Lund, *Chem. Phys.* **168**:1 (1992) 91.
71. W. Press *"Tracts in Molecular Physics."* V. 92, Springer-Vela Berlin, 1981.
72. Masaru Shiotani, Tomoya Yamada, Kenji Komaguchi, Nikolas P. Benetis, Anders Lund and Anders R. Sørnes. *"The Meeting on Tunneling Reactions and Low Temperature Chemistry"*, JAERI- Conf. 98-014, (1998) pags. 58-63.
73. a) M. Lindgren, G.R. Eaton, S.S. Eaton, B.-H. Jonsson, P. Hammarström, M. Svensson, and U. Carlsson, *J. Chem. Soc. Perkin Trans.* **2** (1997) 2549; b) A. Zecevic, G.R. Eaton, S.S. Eaton, and M. Lindgren, *Mol. Phys.* **95** (1998) 1255; c) Martina Huber, Mikael

Lindgren, Per Hammarström, Lars-Göran Mårtensson, Uno Carlsson, Gareth R. Eaton, Sandra S. Eaton, *Biophys. Chem.* **94** (2001) 245.
74. S. Clough and F. Poldy, *J. Chem. Phys.* **51** (1969) 2076.
75. N.L. Owen *"Internal Rotation in Molecules."* Chapt. 6, Edt. W.J. Orville-Thomas, John Wiley & Sons NY 1974.
76. Nikolas P. Benetis, Paresh Dave, and Daniella Goldfarb, *J. Magn. Res.* **158** (2002) 126.
77. J.H Freed *"Multiple Electron Resonance Spectroscopy."* Edts. M.M. Dorio and J.H. Freed, Plenum Press 1979.
78. A.E. Stillman, L.J. Schwartz, and J.H Freed, *J. Chem. Phys.* **73** (1980) 3502.
79. L.D. Kispert, M.K. Bowman, J.R. Norris, and M.S. Brown, *J. Chem. Phys.* **76**:1 (1982) 26.
80. S.A. Dzuba, A.G. Maryasov, K.M. Salikhov, and Yu.D. Tsvetkov, *J. Magn. Res.* **58**:1 (1984) 95.
81. A.R. Sørnes and N.P. Benetis, *Chem. Phys. Lett.* **287** (1998) 590.
82. L.V. Kulik, I.A. Grigoryev, E.S. Salnikov, S.A. Dzuba, and Yu.D. Tsvetkov, Submitted, July 2002.
83. Roland Erickson, and Anders Lund, Linköping University, private communication.
84. S.A. Dikanov, Yu.D. Tsvetkov *"Electron Spin Echo Envelope Modulation (ESEEM) Spectroscopy."* Boca Raton: CRC Press, 1992.
85. D. Gamliel and J.H. Freed, *J. Magn. Res.* **89** (1990) 60.
86. V.F. Yudanov, K.M. Salikhov, G.M. Zhidomirov, and Yu.D. Tsvetkov, *Teor. Eksp. Khim.* **5** (1969) 663.
87. M. Romanelli and L. Kevan, *Concepts Magn. Res.* **9** (1997) 403; *ibid.* **10** (1998) 1.
88. B. Herzog, and E.L. Hahn, *Phys. Rev.* **103** (1956) 148.
89. Rikard Owenius *"Studies of Local Interactions between and within Proteins using Site – Directed Labeling Techniques."* PhD thesis, Dept. of Physics and Measurement Technology, Linköping University, Sweden 2001.

Chapter 4

QUANTUM EFFECTS IN DEUTERIUM LABELLED RADICALS AT LOW TEMPERATURE

Masaru Shiotani and Kenji Komaguchi
Department of Applied Chemistry, Graduate School of Engineering, Hiroshima University, Higashi-Hiroshima 739-8527, Japan

Key words: quantum effects, ^{2}D effect, spin-rotation couplings, zero point vibrational energy, Jahn-Teller distortion, low temperature, high resolution ESR, Ar matrix, para-H_2

Abstract: Recent progress in deuterium labelling studies of radicals in the solid state is reviewed. Emphasis is placed on quantum effects at low temperature. The high-resolution ESR spectra of partially ^{2}D labelled methyl radicals were radiolytically generated together with a hydrogen atom - methyl radical pair in an Ar matrix at cryogenic temperatures. The methyl radical spectra are discussed in terms of nuclear spin-rotation couplings using a three-dimensional free quantum-rotor model. A hydrogen atom - hydrogen molecule (H···H_2) pair formation in Ar is discussed in terms of ^{2}D isotope effects on quantum mechanical tunnelling reaction. The H_2 molecule as a "quantum solid" for high-resolution ESR spectroscopy is presented. ^{2}D effects on zero point vibrational energy (ZPVE) were presented in combination with Jahn-Teller distortion of methane and tetramethylsilane radical cations with a high symmetry of T_d structure originally as well as chemically important cyclohexane radical cations. Furthermore, ^{2}D isotope effects on methyl group conformation were exemplified using selectively deuterated dimethylether and monofluoromethane radical cations; the experimental results were also interpreted in terms of ZPVE incorporated with the mass difference of the two hydrogen isotopes.

1. INTRODUCTION

Static and dynamic structure, and the physical-chemical nature of reaction intermediates such as ionic and neutral radicals in chemical reactions have attracted much attention because they play or are expected to play a key role in important chemical processes.[1-11] In the previous book, "Radical Ionic Systems: properties in condensed phases",[2] we reviewed one chapter, "Deuterium labelling studies of cation radicals". In that chapter we noted as follows. Use of selectively ^{2}D labelled compounds is essential for an unequivocal assignment of the ESR spectra of organic radicals because of the difference in magnetic properties, *i.e.* nuclear spin and magnetic moment, between ^{1}H and ^{2}D atoms. In addition the ^{2}D labelling is expected to give rise to important isotope effects on the static and dynamic structures of the radicals due to the mass difference, *i.e.* ^{2}D isotope effects. However, only a limited number of molecules have been subjected to ^{2}D labelling ESR studies because the synthesis of ^{2}D labelled compounds is often difficult and troublesome, time consuming and expensive.

In this decade considerable progress has been made in this field, *i.e.* strong and clear ^{2}D isotope effects have been found on structures, conformations, molecular dynamics and reactions of organic radicals. This chapter reviews "quantum effects in ^{2}D labelled radicals at low temperature". Here, in terms of "quantum effects" we will present topics related to "nuclear spin-rotation couplings"[12-20] connected to "the Pauli principle",[14,16] "quantum tunnelling",[2,11,21] "H_2 molecule as a quantum solid",[23,24] "Jahn-Teller (*J-T*) or *pseudo J-T* distortion",[1,2,5,25-40] etc. In most cases, ^{2}D isotope effects on zero point vibrational energy (ZPVE) may play an important role. We wish to emphasize that low temperature matrix-isolation ESR is useful to study such "quantum effects" in radicals.

The first three sections address quantum effects in high-resolution ESR of ^{2}D labelled radicals in either a solid argon (Ar) or H_2 matrix at low temperature. In section 2 we will deal with the high resolution ESR observation of ^{2}D labelled methyl radicals, CH_3, CD_3, CHD_2 and CH_2D.[12] The experimental spectra are explained in terms of nuclear spin-rotation couplings using a three-dimensional free quantum-rotor model. The application of the Pauli principle in combination with D_3 point group symmetry results in an important exclusion of EPR transitions for the ^{2}D labelled methyl radicals. In addition, we will describe high-resolution ESR spectra and characterization of a hydrogen atom - methyl radical pair (H↑ ↑CH_3) generated radiolytically in Ar.[41] Section 3 presents a hydrogen atom - hydrogen molecule (H⋯H_2) pair formation in Ar.[21,42] The reactions can proceed *via* quantum mechanical tunnelling in which ^{2}D isotope effects play a critically important role. Section 4 will describe usefulness of the H_2 molecule as a "quantum solid" for high-resolution ESR spectroscopy. In addition to the high-resolution ESR spectra of some small organic radicals in

4. Quantum effects in deuterium labelled radicals

the *para*-H_2 matrix, partial orientation and motional dynamics of stable NO_2 as a spin probe in the solid H_2 matrix will be presented.[43]

Further, we will discuss *J-T* or *pseudo J-T* distortion and conformation of some small organic radical cations in which ^2D isotope effects on ZPVE of a C-H bond stretching vibration[22,31,33] play a key role. Section 5 concerns $^1H/^2D$ isotope effects on ZPVE combined with *J-T* distortion of radical cations such as methane[+ 32,33] and tetramethylsilane[+] (TMS^+)[29,34,44,45] whose mother molecules have degenerate highest occupied molecular orbitals (HOMO) in a high symmetry T_d or D_{3h} structure. Section 6 describes ^2D labelling studies on methyl group comformations; ^2D effects are demonstrated by ESR studies of selectively deuterated dimethylether (DME^+)[22] and monofluoromethane $(CFDH_2)$[46,47] radical cations in combination with *ab initio* and density functional theory (DFT)[22,46] calculations. The isotope effects and temperature dependent ^1H *hf* splittings can be interpreted in terms of ZPVE incorporated with the mass difference of the two hydrogen isotopes. In section 7 we will deal again with ^2D effects of ZPVE on static and dynamic *J-T* distortions of chemically important cyclohexane radical cations using selectively deuterated compounds.[5,26,27] Asymmetrically distorted structures of alkane radical cations due to *pseudo J-T* effects are also exemplified by ESR studies using partially deuterated silacyclohexane (*c*SiC5) radical cations.[25,39,40]

In the last section, we will add an ESR study of selectivity and enormous $^1H/^2D$ isotope effects on H atom abstraction from CH_3SiH_3[48] as an example of recently studied quantum mechanical tunnelling reactions in the solid state at cryogenic temperatures.

2. HIGH RESOLUTION ESR AND NUCLEAR SPIN-ROTATION COUPLING OF METHYL RADICALS

This section deals with high-resolution isotropic ESR spectra of isotopically labelled methyl radicals radiolytically generated and isolated in solid argon (Ar) matrix at a low temperature range of 4.2 K to 40 K.[12]

The methyl radical is the most simple and fundamental alkyl radical. The associated ESR studies have attracted much attention; they include methyl radicals isolated in inert gas matrices,[49-52] adsorbed on solid surfaces,[53,54] and trapped in organic media[55-57] at low temperatures. Most studies of quantum effects on ESR lineshapes of methyl-type radicals so far have been concentrated on the tunnelling rotation of methyl hydrogens in *β*-

proton systems (·CRR'-CH$_3$),[13-20] but not the methyl radical itself. On the other hand, among possible inert gases, Ar can provide one of the most appropriate matrices to observe EPR spectra with high resolution, because it has no nuclear spin, including all stable isotopes. However, even in an Ar matrix, ESR spectra of the CH$_3$ radical reported earlier[51,52] did not distinctly reflect its intrinsic high-resolution because of, for example, unresolved super-hyperfine (*hf*) splittings due to ^{127}I ($I = 5/2$) concomitantly generated with the CH$_3$ radical by photolysis of methyl iodide (CH$_3$I) in the matrix. In order to prevent such unnecessary broadening of the ESR spectra due to the interactions between the CH$_3$ radical and other radicals or molecules, we employed the X-ray radiolysis of an Ar matrix containing selectively deuterated methanes at 4.2 K to generate the radical.[12] The observed spectra of CH$_3$, CD$_3$, CHD$_2$ and CH$_2$D are explained by a three-dimensional free quantum-rotor model; *that is*, in contrast to the β-proton systems, the ^{1}H *hf* coupling is anisotropic in its intrinsic nature and no hindering barrier is present. The problem can be treated theoretically in terms of nuclear spin-rotation couplings and the application of the Pauli principle in combination with D_3 point group symmetry and results in an interesting exclusion of EPR transitions for the selectively deuterated methyl radicals.[12]

In addition to the isolated methyl radicals, high resolution ESR spectra attributable to a hydrogen atom-methyl radical pair (H↑ ↑CH$_3$) were observed for X-ray irradiated Ar mixed with methanes at 4.2 K.[41] The ESR spectra of the triplet pair were characterized by a forbidden transition of $\Delta m_s = \pm 2$ and an electron-electron dipole transition at an allowed $\Delta m_s = \pm 1$ band. The latter enabled us to evaluate the separation and orientation of the paired radicals.

2.1 High resolution ESR spectra of methyl radicals in Ar matrix: CH$_3$, CH$_2$D, CHD$_2$ and CD$_3$

Figure 1 shows the high resolution isotropic ESR spectra of the deuterium labelled methyl radicals, CH$_3$, CH$_2$D, CHD$_2$ and CD$_3$, isolated in a solid Ar matrix at 4.2 K and 20 K.[12]

CH$_3$ Radical. A highly resolved quartet ESR spectrum with an equal-intensity was observed below 6 K and was attributed to the isolated CH$_3$ radical with isotropic ^{1}H *hf* splitting of 2.315 mT (after the second-order correction): A_1-lines in D_3 symmetry. Upon increasing the temperature above 12 K, a new doublet, *E*-lines, with the same splitting appeared at $m_F = \pm 1/2$ positions. The line positions were 0.024 mT higher than the inner $m_F = \pm 1/2$

4. Quantum effects in deuterium labelled radicals

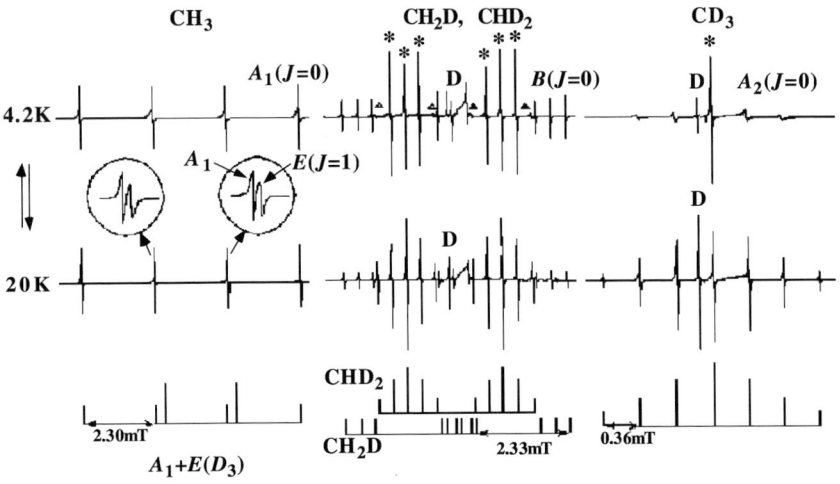

Figure 1. The high-resolution ESR spectra of CH_3 (*left*), CH_2D and CHD_2 (*middle*) and CD_3 (*right*) were observed in an Ar matrix at 4.2 K and 20 K. The radicals were generated by X-ray irradiation of the Ar matrix containing *ca*. 0.1 mol% CH_4, CH_2D_2, or CD_4. At the lower temperatures close to 4.2 K, the ESR lineshapes of these radicals are dominated by *hf* patterns with anomalous intensity, attributed to quantum effects. [taken from ref. 12]

lines of A_1-lines: the difference in the resonance lines originates from the second-order shift between the total nuclear spins of $F = 1/2$ (*E*-lines) and $F = 3/2$ (A_1-lines). The intensity of the *E*-lines increased with increasing temperature and became two times stronger than that of $m_F = \pm 1/2$ (A_1-line) at 40 K. Because of very high spectral resolution with a linewidth of 0.007 mT, the spectrum of CH_2D radical as well as $^{13}CH_3$ was clearly observed with their natural abundance of 0.015% (D) and 1.1% (^{13}C) at 20 K.

CD_3 Radical. Quantum effects of the perdeuterated methyl radical CD_3 on the EPR lineshape were also observed. A strong singlet was superimposed on the central peak of the septet with *hf* splitting of 0.359 mT. The septet has a relative intensity ratio of 1.0:3.8:7.0:105:8.7:4.3:1.5 at 4.1 K; the ratio is close to a binomial intensity of 1:3:6:7:6:3:1 except for the central line. It was concluded that the CD_3 spectrum was observed with superposition of a "classical" high-temperature spectrum due to a freely rotating CD_3 with a very strong transition at the central line. Obviously the CD_3 (the *boson* with $I = 1$) spectra behaved very much differently from the CH_3 spectra. The central line intensity rapidly decreased with increasing temperature and the spectral intensity already reached the binomial one at 10 K.

***CH$_2$D* and *CHD$_2$*.** Partially deuterated methyl radicals, CH$_2$D and CHD$_2$, were generated using CD$_2$H$_2$ as a solute. The 4.2 K spectrum of the CHD$_2$ radical consists of a double quintet due to one hydrogen and two deuterium with *hf* splittings of 2.315 and 0.359 mT. The central triplet of the quintet has a narrow linewidth of 0.01 mT and a very strong transition at m_F = 1, 0, -1 with a 1:1:1 intensity ratio, whereas the outer two lines have about four times broader linewidth. Upon warming the triplet intensity rapidly decreased with temperature and reached a binomial intensity of 1:2:3:2:1 due to a freely rotating CHD$_2$ at 10 K. On the other hand, the 4.2 K CH$_2$D spectrum gave a triple triplet due to two hydrogens and one deuterium. All the *hf* lines were of an equal intensity, but not with a binomial intensity of 1:2:1. Similarly to the case of the CH$_3$ radical, upon warming above 15 K, two new lines became visible at the higher fields by 0.024 mT from the central m_F = 0 transition of the two hydrogens and the relative intensity reached a limiting value of 1:2 at 40 K (figure. 1).

Their relative intensity ratio at 4.2 K is *ca.* 8:1 (CHD$_2$: CH$_2$D), which is quite different from a statistical value of 1:1 expected from the solute methane, CH$_2$D$_2$, used for the experiments. The result indicates that light hydrogen (H) can be de-hydrogeneted much easier than the heavy hydrogen (D). The observed deuterium isotope effect on the reaction is characteristic of quantum mechanical tunneling.[11]

2.2 Possible nuclear spin-rotation couplings

Applying the Pauli principle the observed 1:1:1:1 quartet (A_1-lines) of CH$_3$ (the *fermion* with I = 1/2) was attributed to the four totally symmetric A_1 nuclear spin states coupled with the rotational ground state, J = 0, in D_3 symmetry[12]:

$$\Psi^{A_1}_{n,n,n}(000) = |000\rangle |nnn\rangle$$

$$\Psi^{A_1}_{n,n',n}(000) = \frac{1}{\sqrt{3}}|000\rangle \left[|n'nn\rangle + |nn'n\rangle + |nnn'\rangle \right] \qquad (1)$$

for $n = -n' = \pm 1/2$, for details see refs. 14 and 16 and Fig. 10 in the chapter by Benetis in this book. On the other hand, the *E*-lines were attributed to the nuclear spin states coupled with the J = 1 rotational state[12]:

$$\Psi^{A_1}_{n,n',n}(1qM) = \frac{1}{\sqrt{6}}\left[|1qM\rangle + |1-qM\rangle\right] \sum_j \varepsilon_j^{-q} |\zeta_j\rangle \qquad (2)$$

with $n = -n' = \pm 1/2$ and with $q = \pm 1$. In the earlier study of β-proton rotors with C_{3v} symmetry (·CRR'-CH$_3$ type of radical)[14,16] the isotropic hf splitting of a quadratic cosine form couples the degenerate rotational states with projections $J_z = \pm 1$ and split them, giving the characteristic triplet called E-lines. For the present D_3 rotor the E-lines, however, are almost superimposed on the A_1-lines at $m_F = \pm 1/2$.

The CD$_3$ radical at 4 K shows an abnormally strong singlet at the center of the spectrum. Consistent with this observation, the Pauli principle allows only one spin-rotation state at the lowest rotational level $J = 0$ with the antisymmetric nuclear spin function of A_2 in D_3 symmetry[12]:

$$\Psi^{A_2}_{-1,0,1}(000) =$$

$$= \frac{1}{\sqrt{6}}|000\rangle\{[|-101\rangle+|1-10\rangle+|01-1\rangle]-[|10-1\rangle+|-110\rangle+|0-11\rangle]\} \quad (3)$$

For the same reason as for the CH$_3$ radical, the expected E-lines did not move from their original positions, but were superimposed on the A_2-lines. Here the second order shifts were not of significance because of the 6.5 smaller hf splitting of deuteron than proton. It should be noted that, on the contrary, the CD$_3$ rotor in C_{3v} symmetry (β-CD$_3$) is expected to show the intensity ratio of 1:2:2:3:2:2:1 for A-lines (A_2 for D_3 symmetry) at $J = 0$.

The reader can refer to ref. 12 for the spin-rotation couplings of CH$_2$D and CHD$_2$ radicals

2.3   Radical pair of H↑ ↑CH$_3$ in Ar

A singlet molecule decomposes into a pair of doublet species (radicals) by ionizing radiation (or photolysis), and a pairwise trapping of radicals is inherent in radiation damage processes. When a pair of radicals are trapped with a distance less than ca. 2 nm, singlet or triplet states of radical pairs is formed, the latter being observed by ESR. A number of ESR studies have been reported for the radical pairs generated in irradiated organic crystals, which include pioneer work on a dimethylglyoxime single crystal by Kurita.[58] However, only a few ESR studies have been carried out on the radical pairs of fundamentally important primal organic species such as a hydrogen atom - methyl radical radical pair (H↑ ↑CH$_3$) trapped in organic compounds. Gordy et al. have presented an ESR spectrum attributable to the H↑ ↑CH$_3$ radical pair separated by one methane molecule in irradiated solid methane at 4.2 K.[59] They evaluated the separation distance to be 0.68 nm from the observed axially symmetric electron-electron dipole coupling tensor. Toriyama *et al.* have reported the spatial distribution of the paired H atom and methyl radical in irradiated solid CH_4 and CD_4 at 4.2 K.[60] Here we wish to introduce highly resolved ESR spectra of the H↑ ↑CH_3 radical pairs trapped in solid Ar matrices.[41]

As mentioned in the above section, the high resolution ESR spectra of a series of ^{2}D labelled methyl radicals were observed for X-irradiated Ar mixed with the corresponding partially deuterated methane molecules at 4.2 K.[41] When the irradiated samples were recorded with higher microwave powers and amplitude gains, the ESR spectra attributable to the radical pairs of hydrogen atom - hydrogen atom (H↑ ↑H) and hydrogen atom - methyl radical (H↑ ↑CH_3) were clearly observed at both bands of the allowed $\Delta m_s = \pm 1$ and forbidden $\Delta m_s = \pm 2$ transitions in addition to the isolated H atom and CH_3 radical.

Here we focus on the high-resolution spectra of a H↑ ↑CH_3 radical pair. Three different sets of doublet with 25.9, 25.5, and 25.3 mT splittings, which were further split into a quartet with 1.16 mT, were observed at the $\Delta m_s = \pm 2$ band; see figure 2. The splittings of the double-quartets were very close to one- half of the isotopic ^{1}H *hf* splittings of an isolated H atom (a_{iso} = 51.8 mT) and CH_3 radicals (a_{iso} = 2.3 mT). This enabled us to attribute the double-quartets at $\Delta m_s = \pm 2$ to the triplet state of the H↑ ↑CH_3 radical pair, which lies below the singlet state by 2*J* (a singlet-triplet separation) with |*J*| >> a_{iso}(H atom) = 1.4 GHz.[61] At the allowed $\Delta m_s = 1$ transition, it was observed that the ^{1}H *hf*-lines with one-half of a_{iso}(H) and a_{iso}(CH_3) splittings are further split into anisotropic doublets due to $d_{\parallel}$ and $d_{\perp}$, parallel and perpendicular components of the electron-electron dipole coupling (*i.e.* zero-field splittings). At least three different sets of $d_{\perp}$ splittings were clearly resolved in the spectrum; the most intense lines being attributable to $d_{\perp}$ = 11.5 mT. The radical pair formation was further confirmed by observing the corresponding spectra of D↑ ↑CD_3 radical pairs for irradiated CD_4 in Ar.[41] Note that the observed $d_{\perp}$ value of 11.5 mT is quite different from $d_{\perp}$ = 9.1 mT reported for the H↑ ↑CH_3 radical pair trapped in irradiated pure solid methane at 4.2 K.[60] The separation of the H↑ ↑CH_3 radical pair, *R*, can be evaluated from the experimental $d_{\perp}$ value based on the assumption of a point-dipole interaction model: $R = (3g\beta/2d_{\perp})^{1/3}$, where *g* is the *g*-factor and β is the Bohr magnetron of the electron.[61] The observed value of $d_{\perp}$ = 11.5 mT leads to *R* = 0.62 nm. It is known that there are three different trapping sites in the solid Ar, *i.e.* the interstitial tetrahedral site, interstitial octahedral site, and substitutional site of the *fcc* lattice (see the following section, 3.1).[41] Assuming that the CH_3 radical of the H↑ ↑CH_3 pair occupies the substitutional site as the isolated CH_3 radical,[12] the most probable site of the

4. Quantum effects in deuterium labelled radicals

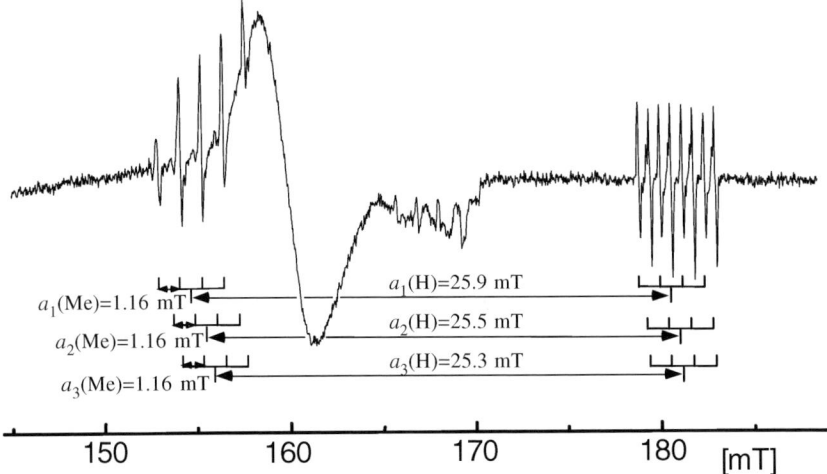

Figure 2. ESR spectrum at Δms = ±2 transition of H↑ ↑CH$_3$ radical pairs trapped in a solid Ar matrix at 4.2 K. The radical pairs were generated by X-ray irradiation of the solid Ar mixed with *ca.* 0.2 mol% CH$_4$. The strong broad singlet at *ca.* 160 mT comes from unidentified triplet species.

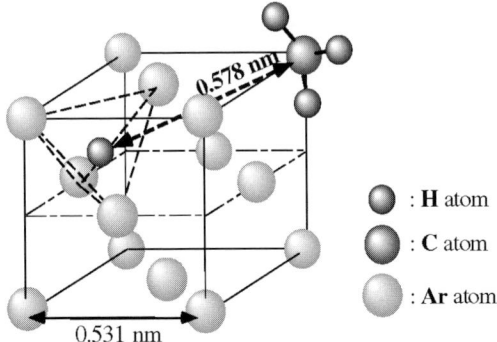

Figure 3. Schematics of trapping sites for H↑ ↑CH$_3$ radical pair in Ar *fcc* lattice. We propose that the CH$_3$ radical and the counter H atom occupy a substitutional site and an interstitial tetrahedral site, respectively. Separation of the two sites is 0.58 nm, which is close to 0.62 nm derived from the observed $d_\perp$ of 11.5 mT.

counter H atom is the interstitial tetrahedral one at $R(theo)$ = 0.58 nm [vs $R(exp)$ = 0.62 nm from the ESR data], see figure 3. The H atom gives a *hf* splittng of 51.8 mT, which is larger than that in the gas phase, 50.8 mT. The H atom, which is trapped in a narrower site, such as the interstitial tetrahedral one, generally gives a larger ^{1}H *hf* splittng than the gas-phase splitting.[21,62,63] This provides additional support for the suggested trapping site.

We will close this section with just noting that Knight and his collaborators have observed successfully ESR spectra of three different triplet radical pairs of H↑ ↑H, H↑ ↑D, and D↑ ↑D generated and trapped in rare gas matrices at 4.2 K and have developed a theoretical model of treating these spin-pairs as weakly interacting atoms.[64,65]

3. HYDROGEN ATOM - HYDROGEN MOLECULE PAIR FORMATION IN ARGON

As mentioned in the foregoing section, Ar can provide a suitable matrix for high resolution ESR because of having no isotopes with nuclear spins and having chemical inertness. In addition, the closest inter-atomic distance of solid Ar with the fcc structure, 3.755 Å, is almost the same as that of a solid hydrogen molecule with hcp structure, 3.79 Å.[62,63] Thus, H_2 molecules are expected to be uniformly isolated in solid Ar so as to occupy its substitutional sites. We have generated hydrogen atoms in a solid Ar matrix at cryogenic temperatures by means of X-ray radiolysis to study the fundamental reaction process between hydrogen atom and hydrogen molecule. As the results we have succeeded in observing the super-*hf* couplings of a hydrogen atom - hydrogen molecule pair that was formed *via* a quantum mechanical tunnelling reaction. Deuterium isotope effects on the pair formation in Ar are introduced in this section.

3.1 Hyperfine splittings and locations of H-atoms

Figure 4 shows low-field components of the doublet ESR spectra of hydrogen atoms generated by the X-ray radiolysis and trapped in a solid Ar matrix at 4.2 K. Just after the radiolysis, the H atoms are trapped in three different sites, *i.e.* interstitial tetrahedral site [$H_{i(t)}$], interstitial octahedral site [$H_{i(o)}$], and substitutional site [H_S], which can be distinguished by the ^1H *hf* couplings, 51.54 mT, 51.42 mT, and 50.73 mT in a solid Ar matrix, respectively; more than 90 % of H atoms generated were predominately occupied in the narrowest $H_{i(t)}$ site. The ^1H *hf* couplings larger than the theoretical value, 50.8 mT, in the gas phase led to hydrogen atoms trapped in the narrower sites in the solid Ar matrix as a result of the Pauli exclusion effect in preference to the van der Waals attraction effect.[62,63]

On annealing the sample to 13 K, the H atoms in $H_{i(t)}$ start to migrate to H_S sites with concomitant decrease in the total amounts due to the recombination reaction, H + H → H_2. The migration is completed below 20

4. Quantum effects in deuterium labelled radicals

K. On the other hand, a new doublet with a ^{1}H *hf* coupling constant of 51.18 mT, indicated by the arrow in figure 4, appeared at 13 K. Figure 5 shows the low-field components of the ESR spectra at 20 K corresponding to the new doublet of H and D atoms in an Ar matrix containing small amount of H_2, D_2, and HD, respectively. The peak with the strongest intensity is due to H atoms in substitutional sites as shown in figure 5(c). In a solid Ar matrix containing HD, the doublet splits further into isotropic nine lines with super-*hf* coupling constants of 0.068 mT and 0.062 mT for the outer two and inner two splittings, respectively, and with peak intensities 1:1:2:1:2:1:2:1:1, in contrast with only the simple doublet observed in Ar matrix containing H_2.

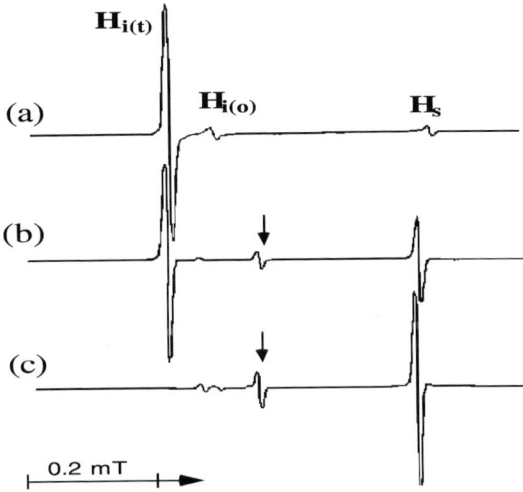

Figure 4. Temperature dependence of the low field components of the ESR spectra observed for hydrogen atoms in a solid Ar matrix containing *ca.* 1.0 mol % H_2; (a) observed at 4.2 K just after X-ray radiolysis at the same temperature, (b) observed at 13 K after annealing the sample from 4.2 K, and (c) observed at 20 K. [taken from ref. 21]

On the other hand, the five isotropic lines with super-*hf* coupling constants of 0.062 mT and peak intensities 1:1:2:1:1 was observed for the new doublet of D atoms in an Ar matrix containing D_2, whereas no ESR signal corresponding to the new doublet can be observed for D atoms in solid Ar containing HD.

3.2   H···H_2 pair formation *via* a tunnelling reaction

During the migration of hydrogen atoms in solid Ar containing HD, the H and D atoms independently encounter an HD molecule isolated in a substitutional site (H_S). The D atoms would react with HD *via* a tunnelling reaction (a) to produce an H atom, which is trapped near the D_2 molecule, while the H atom is merely trapped in the neighbourhood of the HD molecule without producing a D atom because reaction (b) is an endothermic process.[21,42,66,67]

$$HD + D \rightarrow H + D_2 \quad \text{----- (a)}$$
$$HD + H \rightarrow D + H_2 \quad \text{----- (b)}$$

Thus, in an Ar matrix containing HD as additives, the super-*hf* structure with nine isotropic lines for the H atom is attributable to H⋯D_2 and H⋯HD pairs. For the H⋯HD pair the ESR spectrum is expected to be double triplets with super-*hf* couplings of a_1(H) and a_2(D) as shown by stick diagrams in figure 5(c): a_1(H) being expected to be 6.5 times larger than a_2(D).

Now we concentrate on the H⋯D_2 pair. *Para*-D_2 (*p*-D_2) and *ortho*-D_2 (*o*-D_2) must be taken into account as a partner of the H atom. *p*-D_2 and *o*-D_2 are required to take their rotational levels, J = odd and J = even,

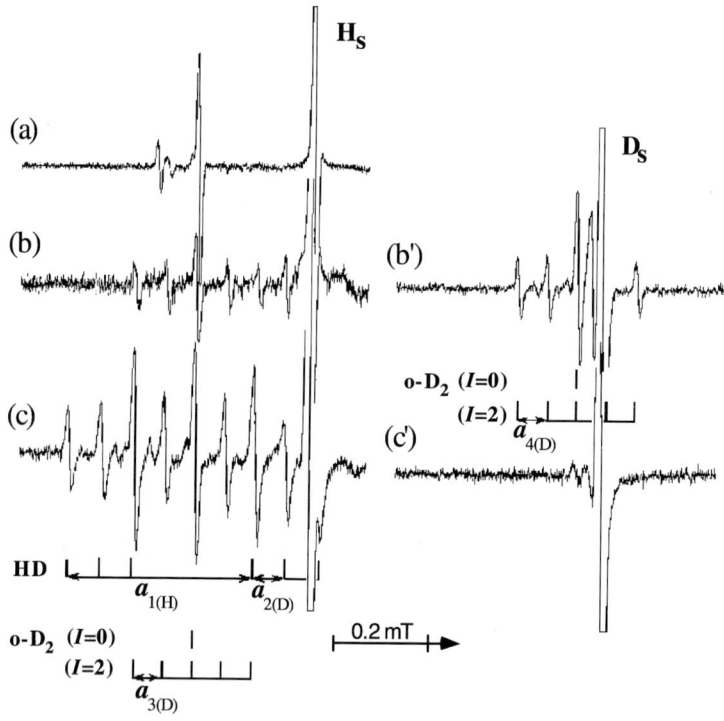

Figure 5. Low-field components of the ESR spectra observed for H atoms (left column) and D atoms (right column) in a solid Ar matrix containing (a) 1.0 mol % H_2, (b) 1.0 mol % D_2, and (c) 3.3 mol % HD at 20 K. The H and D atoms were produced by X-ray radiolysis of the samples at 4.2 K. [taken from ref. 21]

respectively, because of the *Bose* particle of ^{2}D atom, so that o-D_2 ($J = 0$) is stable by *ca.* 85 K than p-D_2 ($J = 1$) in the ground state.[24] The conversion between o-D_2 and p-D_2 is forbidden without any magnetic interaction. The H and D atoms can serve as magnetic perturbations to allow the adjacent p-D_2 to convert into o-D_2 in the lowest rotational level, $J = 0$, at low temperatures. Five isotropic lines of peak intensity 1:1:2:1:1 with a_3(D) are expected for the H$\cdots$$D_2$ pair. Furthermore, the peak intensities of a singlet due to o-D_2 ($I = 0$) and each of five lines with a super-*hf* coupling due to o-D_2 ($I = 2$) are equal because the ratio of the state density, o-D_2 ($I = 2$) : o-D_2 ($I = 0$) is 5:1.

The above assumption was supported by the results that a simple doublet without any super-*hf* structure was observed for the H-H_2 pair where the o-H_2 ($I = 1$) near an H atom is converted into p-H_2 ($I = 0$). A super-*hf* structure similar to that expected for the H atom of the H$\cdots$$D_2$ pair was observed for the D atom of the D$\cdots$$D_2$ pair [figure 5(b')]. Consequently, the ESR results whether the super-*hf* structure can be observed or not, and the structural patterns were successfully explained by a hydrogen atom - hydrogen molecule pair completely isolated in an Ar matrix.

The unpaired electron density of each of the three H atoms, *i.e.* H$\cdots$$H_2$ pair, is theoretically calculated as a function of distance between the H atom and the H_2 molecule. In order to fit between the observed and theoretical super-*hf* coupling constants, the distance was evaluated as 1.86 and 1.64 Å for perpendicular and parallel positions, respectively, thus the mean distance is obtained as 1.73 Å.

4. PARA-HYDROGEN MOLECULE AND HIGH RESOLUTION ESR SPECTRA

The solid hydrogen molecule is a well-known quantum solid[68] because of its large vibrational amplitude at zero-point level. Normal hydrogen (n-H_2) molecules consist of 25% *para*-hydrogen (p-H_2) and 75% *ortho*-hydrogen (o-H_2) molecules. The p-H_2 has a rotational quantum number of $J = 0$ at 4.2 K, whereas the o-H_2 has $J = 1$; as a result, the n-H_2 molecules are in rotational quantum states ($J = 0, 1$) at 4.2 K. Furthermore, the non-magnetic p-H_2 with I (total nuclear spin) $= 0$ is separately obtained from the magnetic o-H_2 with $I = 1$. Thus, the solid p-H_2 is potentially useful as a matrix for high-resolution ESR spectroscopy.[23]

4.1 High resolution ESR spectra of some small organic radicals in p-H$_2$

Miyazaki and his co-workers first demonstrated that the ESR line width (0.011 mT) of H atoms (H$_t$) trapped in solid p-H$_2$ at 4.2 K is ca. three times narrower than that (0.034 mT) of H$_t$ in n-H$_2$.[69] The difference in the line width is attributable to the dipolar hf interactions in the two matrices. As the I value of p-H$_2$ is zero, the dipolar interaction between the electron spin (S) of H$_t$ and the nuclear spins of the surrounding p-H$_2$ molecules vanishes. Thus the p-H$_2$ matrix can give rise to the narrow ESR line width. On the other hand, the o-H$_2$ molecules with $I = 1$ located near H$_t$ in the n-H$_2$ matrix rotate in various orientations and the non-vanishing dipolar interaction results in line width broadening.

The same group has further reported the ESR spectra of some small alkyl radicals, such as methyl and ethyl radicals, trapped in solid H$_2$ matrices.[23,70] Figure 6 shows the spectra of CH$_3$ radicals generated by the UV photolysis of CH$_3$I in n-H$_2$ and p-H$_2$ matrices at 4.2 K. The quartet spectrum of CH$_3$ in p-H$_2$ is much narrower (0.03 mT) than that in n-H$_2$ (0.26 mT) similar to H$_t$ in the matrices. Ethyl radicals were also generated by UV photolysis of C$_2$H$_5$I isolated in solid p-H$_2$.[70] Although the hf anisotropies due to α-^{1}H(-CH$_2$) and β-^{1}H(-CH$_3$) partially remain in the ESR spectra of C$_2$H$_5$ in the temperature range between 3.1-6.7 K, the line width was as narrow as 0.02 mT, which is close to that (0.01 mT) of the isotropic spectrum in liquid ethane reported by Fessenden.[71] Small splittings attributable to the "A" and

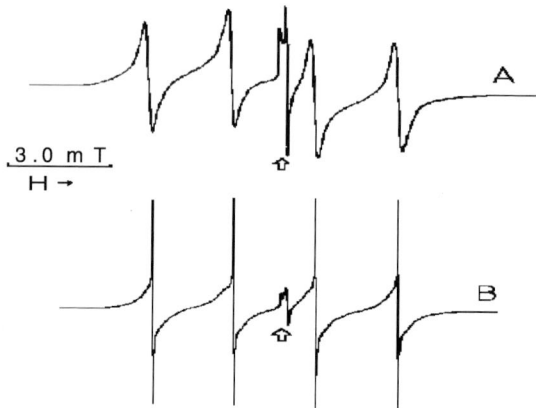

Figure 6. ESR spectra of CH$_3$ radicals produced in the UV-illuminated H$_2$ and CH$_3$I (0.06 mol%) mixtures at 4.2 K. CH$_3$ radicals in the n-H$_2$ (A) and in the p-H$_2$ (B). [taken from ref. 23]

"*E*" transitions were resolved in the spectrum; the former and latter stand for the lowest and second lowest rotational energy levels of the CH_3 group in C_2H_5 radical, respectively. Based on the temperature dependence of the *E/A* intensity ratio, the rotational energy splitting between the *A* and *E* levels was evaluated to be 5.3±0.7 K in *p*-H_2. The narrower line width in *p*-H_2 is consistent with the absence of the dipolar *hf* interactions of the electron spin magnetic moment of a CH_3 radical with nuclear magnetic moments of the *p*-H_2 molecules.

4.2 Partial orientation and dynamics of NO_2 in solid H_2

NO_2 can be potentially useful as a spin probe to study motional dynamics of molecules in a matrix by ESR based on its large anisotropy of both *A* and *g* tensors whose principal axes are coincident with the molecular ones as well as its being a stable gaseous radical.[72,73] High-resolution ESR spectra of NO_2 trapped in solid H_2 permitted us to discuss both on preferential orientation

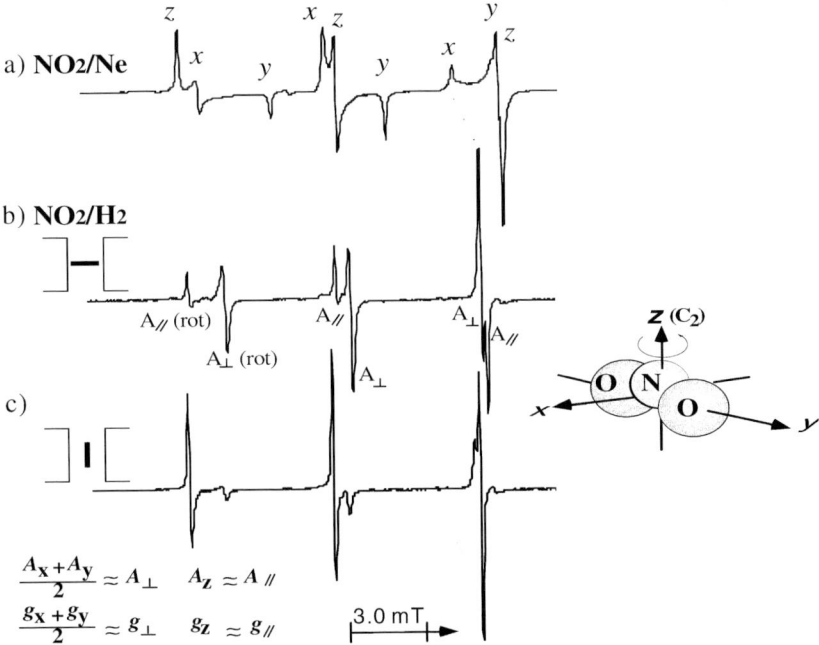

Figure 7. ESR spectra of NO_2 in Ne (a) and *p*-H_2 (b, c) matrices. Spectra (b) and (c) were recorded at the orientation of the flat sample plane parallel and perpendicular to the magnetic field, respectively. Insert below: Schematics of NO_2 whose molecular axes coincide with the principal directions of the *A* and *g* tensors.

and rotational diffusion of NO_2 in the matrix.[43]

Hydrogen gas containing *ca.* 0.1 mol% of NO_2 was prepared in a Suprasil ESR sample tube. The sample tip was then cooled down to 4.2 K to condense the mixture gas for ESR measurements. ESR spectra of NO_2 observed in a solid *p*-H_2 matrix are shown in figure 7. The 4.2 K spectrum was analysed in terms of an axial symmetric A and g tensor components: $g_{//}$ = 2.0022, $g_\perp$ = 1.9920, and $A_{//}$ = 5.85 mT, $A_\perp$ = 5.04 mT. The axial symmetric values are correlated with the non-overall symmetric values of the rigid state NO_2 observed in a Ne matrix as follows: $g_{//} \approx g_z$, $g_\perp \approx (g_x + g_y)/2$, and $A_{//} \approx A_z$, $A_\perp \approx (A_x + A_y)/2$. This suggests that the NO_2 molecule rotates about the molecular z-axis with a frequency high enough to completely average out the x and y components ($\nu \gg |g_x - g_y| \approx 2$ MHz).

The spectral peak intensity was found to depend on the orientation of the sample cell to the external magnetic field (H_0). The orientation dependence was much enhanced when a flat sample tube was used. When the flat plane was parallel to H_0, the maximum and minimum intensities were obtained for the $A_\perp$ and $A_{//}$ components, respectively. When the plane was perpendicular to H_0, the opposite orientation dependence was observed. The results suggest that the NO_2 molecule partially orients in the H_2 matrix to make its z-axis perpendicular to the flat plane of the sample cell even at a low temperature of 4.2 K. On increasing the temperature to 5.5 K, the NO_2 spectra disappeared irreversibly.

We propose the following model for the observed partial orientation and rotation of NO_2 in the solid hydrogen matrix. The NO_2 orients its C_2-axis to the *c*-axis direction of the *h.c.p.* crystalline lattice, to which the crystal preferentially grows during the sample cooling process to 4.2 K from the room temperature. The N atom of NO_2 can be favourably held by three H_2 molecules in the β-plane of the *h.c.p.* crystalline lattice so as to allow NO_2 to rotate freely about the C_2-axis.

5. JAHN-TELLER DISTORTION OF T_D AND $D_{3\square}$ MOLECULES AND H/D ISOTOPE EFFECTS

Tetrahedral molecules in T_d symmetry have a three-fold degenerate t_2 highest occupied molecular orbital (HOMO). When they release one electron, the degeneracy is lifted and the associated radical cations are formed having a lower symmetry of either C_{3v}, C_{2v} or D_{2d} due to the so called static Jahn-Teller (*J-T*) distortion.[1-2,28-31] In this section we deal with a C_{2v} structure distortion of methane[32,33] and tetramethylsilane (TMS)[29,34,44,45] radical cations

4. Quantum effects in deuterium labelled radicals

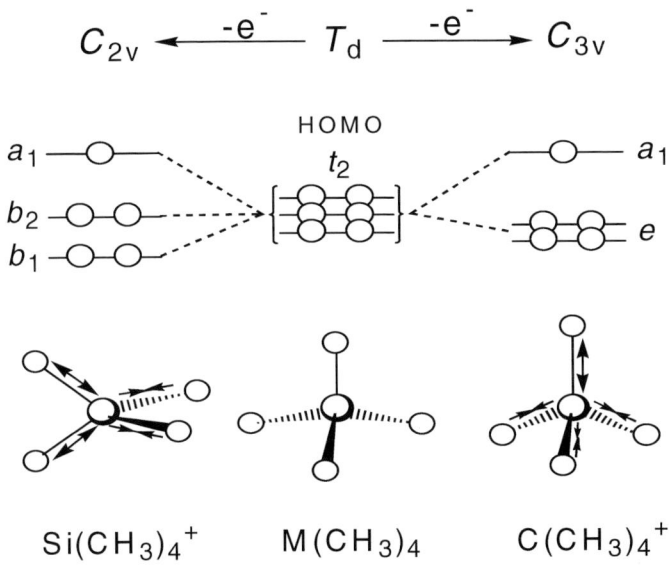

Figure 8. Schematics showing how three-fold degeneracy of the t_2 HOMO is lifted when tetrahedral (T_d) molecules release one electron to yield the radical cations with a lower symmetry of either C_{3v} or C_{2v}. The arrows, ↔ and →←, indicate an elongated and a compressed σ bond, respectively. [taken from ref. 34]

from an original T_d symmetry. As for a C_{3v} distortion from T_d, the neopentane radical cation, $C(CH_3)_4^+$, can be a typical example.[74] Here we just mentioned that an unpaired electron in $C(CH_3)_4^+$ is mainly located in one of the C-C bonds and that the three hydrogens, belonging to each of the three methyl groups with the other C-C bonds and being at the trans position with respect to the bond, gave large couplings of 4.2 mT.

Furthermore, we deal with a structural distortion of the trimethylenemethane radical cation (TMM$^+$). A precursor of TMM$^+$ is the neutral TMM biradical with a D_{3h} symmetry, which is in a triplet state with a doubly degenerate e'' HOMO in the ground state.[75] We conclude that the original D_{3h} symmetry is reduced to a lower one of C_{2v} in TMM$^+$ due to the J-T effect, so as to split the original degenerate e'' orbitals into a_2 and b_1 orbitals in C_{2v} through a structural distortion *via* e' vibration.

5.1 Methane radical cations: CH_4^+, CDH_3^+, $CD_2H_2^+$, CD_3H^+ and CD_4^+

Knight et al. observed ESR spectra of a series of selectively deuterated methane radical cations, CH_4^+, CDH_3^+, $CD_2H_2^+$, CD_3H^+ and CD_4^+, generated and stabilized in a neon matrix at 4 K.[32,33] The observed isotropic hf coupling constants (a_{iso}) are summarized in Table 1. The a_{iso} value of 5.43 mT for ^{1}H in CH_4^+, divided by the ratio of the ^{1}H and ^{2}D nuclear g factors, g_H/g_D = 6.514, yields an equivalent deuterium a_{iso} value of 0.83 mT. This value agrees well with the observed deuterium a_{iso} value of 0.81 mT for CD_4^+, in which all the hydrogens are replaced with ^{2}D atoms. No unusual characteristics are evident. However, for the methane radical cations containing a mixture of ^{1}H and ^{2}D atoms, i.e. CDH_3^+, $CD_2H_2^+$ and CD_3H^+, the simple conversion from the ^{1}H hf scale to the ^{2}D hf scale using the appropriate ratio of g-factors does not account for the observed hf values. For example, in CDH_3^+ the observed a_{iso} value of 7.64 mT would predict a deuterium a_{iso} value of 1.2 mT compared to a directly observed deuterium a_{iso} value of ±0.21 mT. Note that the ^{1}H a_{iso} value for CDH_3^+ (7.64 mT) differs substantially from that observed for CH_4^+ (5.43 mT).

Table 1. Observed isotropic hf values of a_{iso} (mT) for ^{1}H and ^{2}D in methane radical cations in neon at 4 K. [taken from ref. 33]

	CH_4^+	CDH_3^+	$CD_2H_2^+$	CD_3H^+	CD_4^+
a_{iso} :H	5.43	7.64	12.2	12.5	---
a_{iso}^a :H(g_D/g_H)	0.83	1.2	1.9	1.9	---
a_{iso} :D	---	-0.22	-0.22	0.45	0.81
a_{iso}^b :D(g_H/g_D)	---	-1.4	-1.4	2.9	5.3

[a] The experimental $a_{iso}(^1H)$ and $a_{iso}(^2D)$ values have been converted to ^{2}D and ^{1}H hf "scales" by dividing and multiplying by the ratio of the ^{1}H and ^{2}D nuclear g-factor, g_H/g_D=6.514, respectively.

In order to account for the observed a_{iso} values of ^{1}H and ^{2}D atoms the following rules were proposed for the location of ^{1}H and ^{2}D atoms in the radical cation.[33] (a) Methane radical cation possesses a C_{2v} type geometrical structure with two distinctly different electronic sites, i.e. apical "*a*" site and equatorial "*e*" site. (b) Site "*a*" can accommodate two ^{1}H or ^{2}D atoms and is coplanar with the carbon *p*-orbital with an unpaired electron; site "*e*" can accommodate two atoms and lies in the nodal plane of this same carbon *p*-orbital. (c) ^{2}D atoms prefer the nodal plane site "*e*" and ^{1}H atoms prefer the co-planar site "*a*". (d) ^{1}H atoms exchange with other ^{1}H atoms even between the different sites, "a" and "e", but not with ^{2}D atoms, and ^{2}D atoms will likewise exchange with other ^{2}D atoms but not with ^{1}H atoms in a different electronic environment. The orbitals of atoms that occupy the nodal plane site "e" cannot efficiently mix with the p-orbital. Their hf interaction can be dominated by a spin polarization mechanism so as to produce a small negative a_{iso} value. The 1s orbitals of the ^{1}H and ^{2}D atoms in the co-planar site "a" can mix with this carbon p-orbital and yield a large positive a_{iso} value.

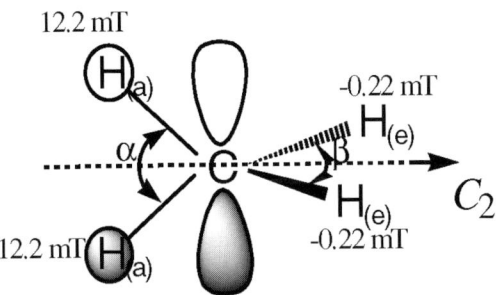

Figure 9. Schematics of the C_{2v} geometrical structure in the CH_4^+ cation with the $^2B_1$ state. Sites marked as "a" are coplanar with the carbon p-orbital with an unpaired electron and sites "e" lies in the nodal plane of the p-orbital. ^{1}H atoms prefer the co-planar "a" site and ^{2}D atoms favor the nodal plane site "e".

Based on the rules outlined above, the a_{iso} values observed for ^{2}D in CDH_3^+ and $CD_2H_2^+$ are almost the same as -0.22 mT. There are two equivalent positions in the nodal plane site "e", which are strongly preferred by the ^{2}D atoms. Likewise for $CD_2H_2^+$ and CD_3H^+, the first two ^{1}H atoms can occupy their preferred co-planar site "a" and exhibit a large hf coupling of 12.2 and 12.5 mT, respectively. Thus we can determine that a_{iso} for site "a" has a value of ca. 12.2 mT and a_{iso} for site "e" has a hydrogen equivalent value of -1.4 mT (-0.22 mT for ^{2}D). A weighted average of these two basic a_{iso} values can account for the observed hf splittings. For CH_4^+, where H atoms exchange with those in different sites, we have 5.39 mT [(2x12.2-2x1.43)/4] vs. the experimental value of 5.43 mT. Furthermore, for CDH_3^+ such H atom exchanges also occur and we have 7.66 mT [(2x12.2-1.43)/3] vs the experimental value of 7.64 mT. However, for $CD_2H_2^+$ and CD_3H^+, the H atoms do not exchange their positions with the ^{2}D atoms in site "e" and the observed hydrogen a_{iso} values are almost the same (12.2 and 12.5 mT).

The C_{2v} structure of the CH_4^+ cation has been theoretically supported by calculations with several different methods.[76-79] For example, the calculations by the LSD(BP)-DFT/DZP method[79] resulted in the two long C-

H bonds (1.208 Å) and the two short C-H bonds (1.104 Å) lying in the mutually orthogonal planes: angle "α" between the long bonds and angle "β" between the short bonds are 58.6° and 124.5°, respectively. The experimental ^{1}H hf splittings were also well reproduced by the theoretical calculations: for 2H at "a" sites, 12.2 (*exp*) vs. 12.11 mT (*cal*), and for 2H at "e" sites, -1.43 (*exp*) vs. -1.63 mT (*cal*).[79]

The deuterium isotope effects can be interpreted in terms of a zero point vibrational energy (ZPVE) of a C-H bond stretching vibration.[22,33] In short, the singly occupied molecular orbital (SOMO) is delocalised from the centre carbon to hydrogens making the corresponding C-H bond weaker and longer. The ZPVE is proportional to $(k/m)^{1/2}$ as the first approximation, where k is the force constant and m is the mass of the hydrogen nucleus. The decrease in ZPVE upon deuteration is greater for bonds having larger force constants. Thus, the deuterium atoms can preferentially occupy positions with a larger force constant, *i.e.* the nodal plane positions with a short bond distance. Readers can see refs. 31 and 33 for more detail theoretical calculations on the ZPVE shifts by the deuterium substitutions for methane radical cations.

5.2 Tetramethylsilane radical cations: $Si(CH_3)_4^+$, $Si(CH_3)_3(CD_3)^+$ and $Si(CH_3)_2(CD_3)_2^+$

Tetramethylsilane (TMS) has a degenerate t_2 HOMO with T_d symmetry similar to CH$_4$. As an extension of the C_{2v} structural distortion and D/H isotope effects of methane radical cations, here we introduce the deuterium substitution effects on static and dynamic structures and conformations of the TMS$^+$ radical cation. The ESR spectra of Si(CH$_3$)$_4^+$ in C_{2v} symmetry were first observed in halocarbon matrices by Williams *et al.*, although incorrect ESR parameters were unfortunately extracted from the insufficiently resolved spectra.[29] Better resolved spectra of Si(CH$_3$)$_4^+$ were obtained together with Ge(CH$_3$)$_4^+$ and Sn(CH$_3$)$_4^+$ by Bonazzola using SiCl$_4$, GeCl$_4$, TeCl$_4$ and SnCl$_4$ as matrices.[44,45]

Well-resolved isotropic ESR spectra were observed for Si(CH$_3$)$_4^+$ (TMS$^+$), Si(CH$_3$)$_3$CD$_3^+$ (TMS-d_3^+), and Si(CH$_3$)$_2$(CD$_3$)$_2^+$ (TMS-d_6^+), which were generated in SnCl$_4$ matrices by ionizing radiation at 77 K.[34] The spectrum of TMS$^+$ was well reproduced by a computer simulation using g = 2.0049 and isotropic hf couplings of $|a_{iso}(6H)|$ = 0.95 mT and $|a_{iso}(6H)|$ = 0.47 mT; the ESR parameters were very close to the reported ones.[44,45] On the other hand, the *ab initio* MO calculations[34] resulted in that TMS$^+$ has a

4. Quantum effects in deuterium labelled radicals

C_{2v} structure having two apical ("*a*") methyl groups with a long Si-C bond distance of 2.014 Å and a small bond angle of 88.6° and two equatorial ("*e*") methyl groups with a short Si-C bond distance of 1.856 Å and a large bond angle of 115.6°, as shown in figure 10. The carbon atoms of the "*e*" methyl groups are in the nodal plane of the Si *p*-orbital containing a relatively large spin density. The ^1H nuclei of the "*a*" methyl groups were calculated to have an average value of (-)0.4 mT, whereas those of the "*e*" methyl groups have an average value of (+)0.21 mT for the theoretical *hf* coupling. Although the absolute values of both theoretical *hf* splittings are about one half of the experimental values, by comparing the calculated couplings with the experimental ones, we can attribute the larger coupling of 0.95 mT to the ^1H nuclei of the "*a*" methyl groups with the longer Si-C bonds and the smaller one of 0.47 G to those of the "*e*" methyl groups with the shorter Si-C bonds.

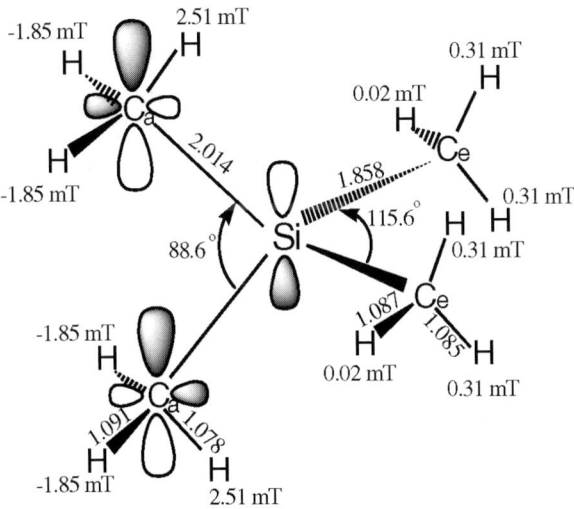

Figure 10. Geometrical structure of TMS$^+$ and schematics of the singly occupied molecular orbital (SOMO) calculated at the UHF/6-31G** level. The ^1H *hf* couplings were evaluated for the structure by the DFT (B3LYP/6-31G**) method. C_a and C_e stand for the carbon atoms at apical ("*a*") and equatorial ("*e*") positions, respectively. [taken from ref. 34]

The ESR spectra of TMS-d_3^+ and TMS-d_6^+ consist of the multiple lines of 0.47 mT, attributable to the ^1H nuclei of the equatorial CH$_3$ groups in the C_{2v} structure of TMS$^+$.[34] Thus, the C_{2v} structure is retained for these deuterated TMS radical cations. If a CD$_3$ group in TMS-d_3^+ were located at random (or statistically), two different conformations having the CD$_3$ group at the apical ("*a*") site and equatorial ("*e*") site would be in a 1:1 ratio. The observed spectra could not be reproduced by simulation for these two mixed conformations, but simulated for a single conformation having the "*e*" CD$_3$ group (model *A* in figure 11). For the spectrum of TMS-d_6^+, computer simulations were conducted using a model for three C_{2v}-like species having two CD_3 groups at the "*a*" and "*e*" positions at random (model A), and using two single-conformation models having two "*e*" CD_3 groups (model B), and having two "*a*" CD_3 groups (model C). Only for model A did the simulation spectrum satisfy the observed one. This is in clear contrast with the results for $CH_2D_2^+$, where the heavier D nuclei are in "*e*" sites, TMS-d_6^+ was concluded to have CD_3 groups at both "*a*" and "*e*" sites in the C_{2v}-like structure.

One can expect asymmetrical displacements in TMS$^+$, which form two pairs of unequal Si-C bonds, *i.e.* elongated and shortened bonds, corresponding to "*a*" and "*e*" carbons, respectively; similar asymmetrically distorted structures have been observed for a series of radical cations of alkyl-substituted cycloalkanes[26,35-38,80,81], silacycloalkanes[5,25,39] and diethylsilanes.[40] Lighter methyl groups, CH_3, would be more advantageous for such simultaneous displacements. For TMS-d_6^+, the first pair of asymmetric bonds may be simultaneously utilized by two lighter CH_3 groups and the last pair by two CD_3 groups. In the case of TMS-d_3, two CH_3 groups take the first pair of asymmetric bonds and a CD_3 and the other CH_3 groups

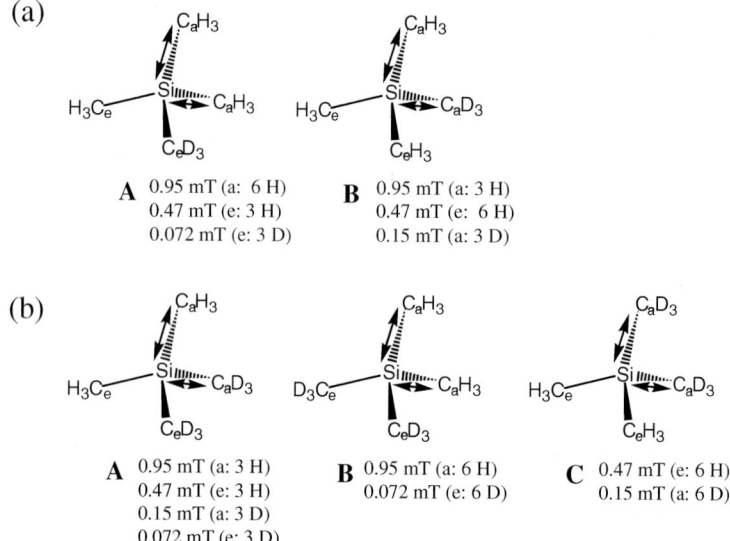

Figure 11. (a) Two possible C_{2v} structures of TMS-d_3^+ with one CD_3 group at equatorial ("*e*") site (model A) and at apical ("*a*") site (model B). (b) Three possible C_{2v} structures of TMS-d_6^+ with one CD_3 group at the "*e*" and one CD_3 group at the "*a*" sites (model A), both CD_3 groups at the "*e*" sites (model B), and both CD_3 groups at the "*a*" sites (model C). C_a and C_e stand for the carbon atoms at the "*a*" and "*e*" sites, respectively, and the arrow denotes the elongated Si-C bond. The numerical values are isotropic ^1H or ^2D *hf* splittings expected for each conformation. [taken from ref. 34]

4. Quantum effects in deuterium labelled radicals

possess the last pair of bonds. The CD_3 group may prefer the "*e*" site in the nodal plane, because this site needs a smaller displacement after releasing one electron. Thus, model A is plausible for TMS-d_3^+, in accord with the conclusion from the observed spectra.

5.3 Trimethylenemethane radical cation

Neutral trimethylenemethane (TMM) biradical is one of the most attractive organic molecules for investigating the electronic structure because of a rather small fundamental molecule with a high symmetry of D_{3h} and a triplet state with a doubly degenerate HOMO in the ground state.[75] Thus, the structural distortion of TMM$^+$, a doublet state with one unpaired electron, was studied by ESR spectroscopy and *ab initio* MO calculations.[82] The

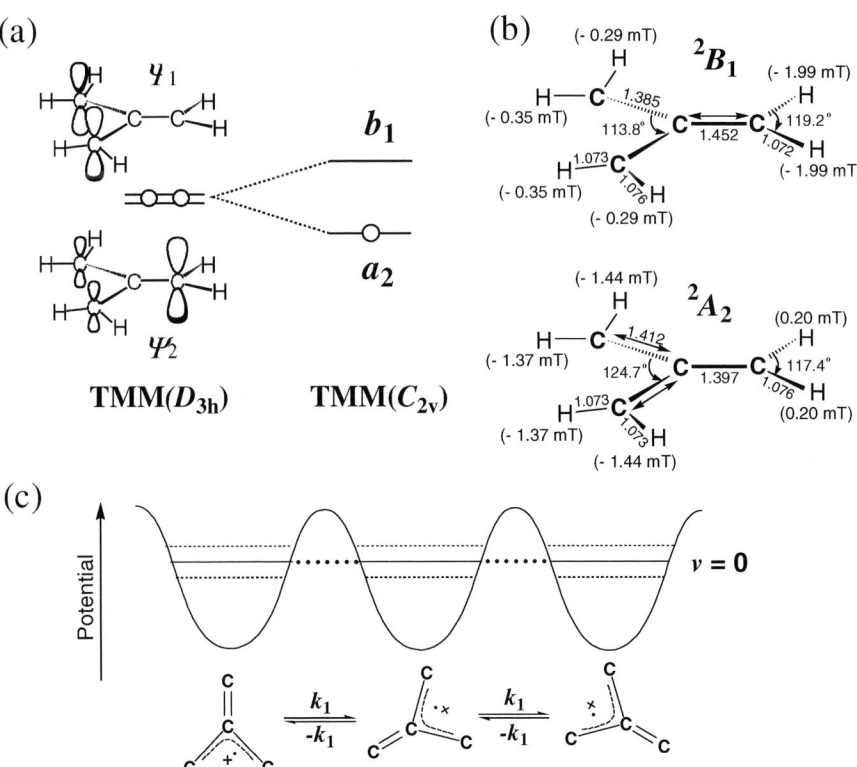

Figure 12. (a) Schematic representations of the degenerate HOMO, *e*", of TMM, which consists of the superposition of two MOs, ψ_1 and ψ_2. (b) Optimised geometrical structures for TMM$^+$ with C_{2v} symmetry at the 6-31G** level. The bond lengths are in Å. The isotropic ^{1}H *hf* splittings in parentheses are evaluated for two states, $^2A_2$ and $^2B_1$, with C_{2v} structure by the INDO MO method. (c) The ESR experimental result can be explained in terms of a dynamical model of interchange among three equivalent distorted C_{2v} structures. (taken from ref. 82).

TMM$^+$ radical cation was generated by γ-ray irradiation of solid solutions containing methylenecyclopropane (MCP) in a CF$_2$ClCF$_2$Cl matrix. The 4.2 K ESR spectrum consists of an isotropic septet with relative intensities close to a binomial one, 1:6:15:20:15:6:1, and is successfully reproduced by employing magnetically equivalent six protons with 0.93 mT *hf* splitting. No appreciable ESR line shape change was observed between 4 K and 125 K. The magnetically equivalent six protons can not be attributed to any radicals with the original ring-closed MCP geometrical structure, but with the ring-opened TMM radical cation.

The original D_{3h} symmetry of TMM is expected to be reduced to a lower one of C_{2v} due to a Jahn-Teller effect so as to split the degenerate SOMO, e'', into a_2 and b_1 orbitals in the C_{2v} symmetrical structure through a structural distortion *via* e' vibration; see figure 12(a). In order to see which state of $^2A_2$ and $^2B_1$ is responsible for the observed TMM$^+$ *ab initio* MO calculations were performed. Figure 12(b) shows two states, $^2A_2$ and $^2B_1$ of the C_{2v} geometrical structure of TMM$^+$, which were optimized at UHF/6-31G** level on the Gaussian 90 program. The $^2A_2$ structure has one shorter C-C bond of 1.397 Å along the C_2 axis, the other two C-C bonds having a longer bond length of 1.412 Å. On the other hand, in the $^2B_1$ structure the unique C-C bond becomes longer (1.452 Å) than the other two C-C bonds (1.385 Å). The ^{1}H *hf* splittings were evaluated for the two optimized structures by means of the INDO MO method (values in parentheses). The total *hf* splittings of 5.26 mT and 5.22 mT calculated for both structures are in fairly good agreement with the experimental value of 5.58 mT. The MO calculations resulted in that the $^2A_2$ structure is more stable by *ca.* 0.3 eV in the total energy than the $^2B_1$ structure. Thus, the former structure seems preferable as the ground state of TMM$^+$ with a distorted C_{2v} structure, though the energy difference is not so large. In the $^2A_2$ structure the unpaired electron mainly resides in the $2p_z$ orbitals of two equivalent terminal carbons, similar to the case of the allyl radical. The spectra were explained by an intramolecular dynamics among three energetically equivalent C_{2v} structures to average out the structural distortion out giving an apparent D_{3h} structure even at a low temperature of 4.2 K; see figure 12(c).

Readers can refer to refs. 2, 3, 31 and 34 for further details of *J-T* distortion of other T_d and D_{3h} molecules.

4. Quantum effects in deuterium labelled radicals

6. DEUTERIUM ISOTOPE EFFECTS ON METHYL HYDROGEN CONFORMATION

In this section, deuterium (^{2}D) isotope effects on the methyl group conformation are demonstrated by ESR studies of selectively deuterated dimethylether (DME) radical cations in combination with *ab initio* and density functional theory (DFT) calculations.[22] The ESR spectra show strong ^{2}D isotope effects on the ^{1}H *hf* splittings for $CD_3OCH_3^+$, $CD_3OCH_2D^+$ and $CD_3OCHD_2^+$ as well as temperature dependence on the *hf* splittings. The ^{2}D isotope effects and temperature dependent *hf* splittings can be interpreted in terms of a zero point vibrational energy (ZPVE) of a C-H bond stretching vibration by incorporating the mass difference of the two hydrogen isotopes in addition to their magnetic properties. A similar deuterium effect on the methyl hydrogen conformation has been reported for the partially deuterated monofluoromethane radical cation of $CFDH_2^+$ by Knight *et al.*,[46] which is also dealt with in this section.

6.1 Dimethylether cations: $CD_3OCH_3^+$, $CD_3OCH_2D^+$ and $CD_3OCHD_2^+$

ESR results. Six different deuterium labelled DME radical cations were generated and stabilized in a halocarbon matrix by ionising radiation at 77 K; they are $CH_3OCH_3^+$, $CH_3OCH_2D^+$, and $CH_3OCHD_2^+$ in addition to the above three radical cations of deuterated dimethylethers.[22] In the spectra of the radical cations of $CD_3OCH_3^+$, $CD_3OCH_2D^+$ and $CD_3OCHD_2^+$, the number of ^{1}H *hf* lines decreases from four to three and two depending on the number of light hydrogen (^{1}H), respectively. The ^{1}H *hf* splitting increases with increasing in the number of ^{2}D atoms substituted from 4.3 mT for CH_3 to 5.2 mT for CH_2D and 6.2 mT for CHD_2 at 77 K. Furthermore, strong temperature dependent ^{1}H *hf* splittings were observed for $CD_3OCH_2D^+$ and $CD_3OCHD_2^+$ in the temperature range from 4 K to 100 K at which the radical cations undergo changes to neutral radicals. For example, the ^{1}H *hf* splitting of $CD_3OCHD_2^+$ decreases with increasing temperature from 8.8 mT at 4 K to 6.2 mT at 80 K; above the temperature the *hf* splitting becomes almost constant up to 100 K (figure 14). On the other hand, no appreciable temperature dependence was observed for the ^{1}H *hf* splitting of the CH_3 group; the splitting of 4.3 mT, which corresponds to the rotationally averaged value of $CH_3OCH_3^+$, remained unchanged. These results lead us to conclude that the ^{1}H atoms of the CH_2D and CHD_2 groups in $CD_3OCH_2D^+$ and $CD_3OCHD_2^+$ take fixed positions at 4 K and that the ^{1}H and ^{2}D atoms cannot be completely averaged out even at a higher temperature of 100 K.

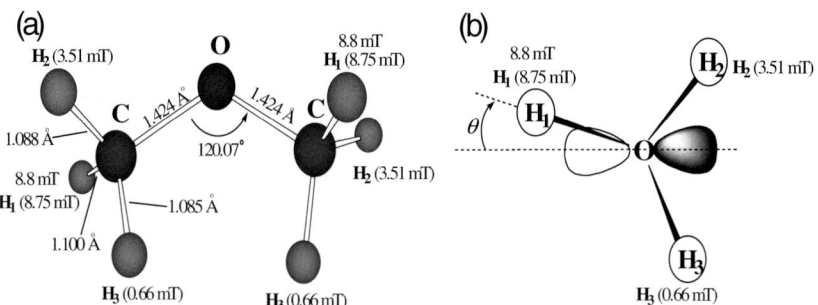

Figure 13. Optimized geometrical structure of DME$^+$ calculated by the *ab initio* MO method at the MP2/6-31G(d,p) level. The values in parentheses are ^{1}H *hf* splittings evaluated for the structure using the DFT method at the BLYP/6-31G(d,p) level. [taken from ref. 22]

^{2}D mass effects on methyl hydrogens conformation. Using a McConnell type of equation for β-protons,[83] $a^\beta_H = B_0 \cos^2\theta$, and the experimental splitting of 4.3 mT for the rotationally averaging, $<a^\beta_H> = 4.3$ mT, we have 8.6 mT for the proportional constant, B_0. Comparing with the experimental value of 8.8 mT for the ^{1}H splitting of CHD$_2$ group, we can conclude that the light hydrogen (^{1}H) preferentially occupies the $\theta = 0°$ position which is parallel to the unpaired electron orbital of oxygen constituting CD$_3$OCHD$_2^+$, figure 13 (b). The preference of the light hydrogen to occupy the $\theta = 0°$ position can be explained as follows. Through the hyperconjugation effect, which is maximum around this position, the singly occupied molecular orbital (SOMO) is delocalised from the oxygen atom to the hydrogen atom so as to making the corresponding C-H bond weaker and longer. Note that the effect is, however, partially countered by the nonbonding, repulsive interaction between the C-H bond(s), which offsets the equilibrium from the perfectly eclipsed conformation.[22] Since the ZPVE of a C-H stretching vibration is proportional to $(k/m)^{1/2}$ in the first approximation (see section 5.1), the decrease in the ZPVE upon deuteration will be larger for bonds having higher force constants.

Temperature dependent ^{1}H hf splittings. The observed temperature dependency of ^{1}H *hf* splittings is successfully analysed using the following model.[22] That is, if a molecular radical can exist in several forms, *i.e.* rotational or geometrical isomers, and the forms are interchanging rapidly enough to average the associated *hf* splittings, a mean total *hf* splitting, $<\Sigma a_i>$, can be obtained:

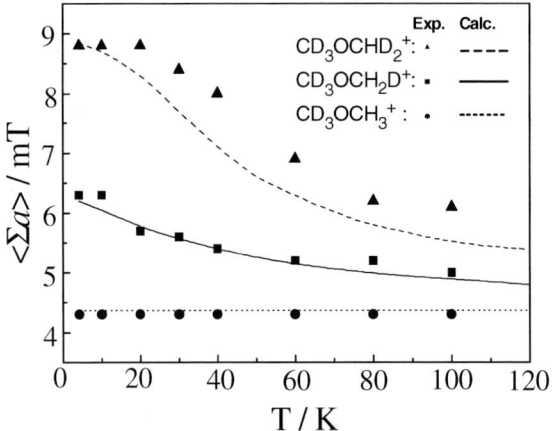

Figure 14. The temperature dependencies of total ^{1}H *hf* splitting, $\langle \Sigma a_i \rangle$, compare well with the theoretical ones for $CD_3OCHD_2^+$, $CD_3OCH_2D^+$ and $CD_3OCH_3^+$. See the details in text. [taken from ref. 22]

$$\langle \Sigma a_i \rangle = p_I \Sigma a_I + p_{II} \Sigma a_{II} + p_{III} \Sigma a_{III} \qquad \cdots (4)$$

where p_i ($i = $ I, II,…) are the probabilities of finding the different forms and Σa_i ($i = $ I, II,…) the total ^{1}H *hf* splittings of the corresponding form.[84] As mentioned above, the ^{2}D atoms in selectively deuterated methyl groups can preferentially occupy sites with larger C-H force constants to minimize the ZPVE. Thus, different rotational isomers of DME$^+$ have different ZPVEs depending both on how many sites and which positions ^{2}D atoms occupy.

Figure 14 shows the mean total ^{1}H *hf* splittings, $\langle \Sigma a_i \rangle$, observed for $CD_3OCHD_2^+$, $CD_3OCH_2D^+$ and $CD_3OCH_3^+$ as a function of temperature.[22] The theoretical values of $\langle \Sigma a_i \rangle$ in the figure were calculated according to equation (4), using a Boltzman distribution to evaluate the probability of finding the different rotational isomers. The ZPVEs were obtained from the frequency calculations at the MP2/6-31G(d,p) level and the *hf* splittings of a_I, a_{II} ⋯ were obtained at the BLYP/6-311+G(2df,p) well level for the MP2 calculated structures.

6.2 Methyl fluoride cations: CH_3F^+ *vs* CH_2DF^+

Similar to methane[32] neutral methyl fluoride (CH_3F) has a doubly degenerate HOMO. The CH_3F^+ cation is expected to show drastic static and dynamic *J-T* effects as has been theoretically discussed.[47] Furthermore, CH_3F^+ is of

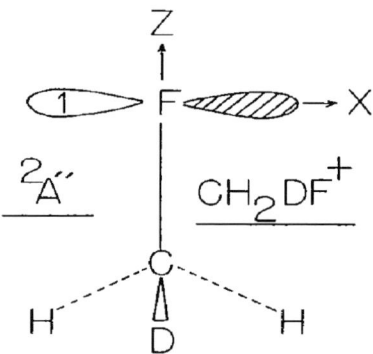

Figure 15. The coordinate system of CH_3F^+ used to define the observed ESR parameters assigned to the X, Y, and Z direction. [taken from ref. 46]

interest since it is isoelectronic with a neutral methoxy radical (CH_3O) and radical cations such as CH_3OH^+, $CH_3NH_2^+$, $CH_3CH_3^+$. Thus, three different isotopic forms of methyl fluoride cation $^{12}CH_3F^+$, $^{13}CH_3F^+$, and $^{12}CH_2DF^+$ have been generated by photoionization at 16.8 eV and other methods in neon matrices at 4 K and subjected to ESR studies.[46]

The experimentally determined g and $hf\ A$ tensors of CH_2DF^+ in neon at 4 K are $g_X = 2.0032$, $g_Y = 2.0106$, and $g_Z = 2.0120$; for ^1H $A_X = 17.2$, $A_Y = 17.0$, and $A_Z = 17.2$ mT; for ^2D $|A_X| = 0.18$, $|A_Y| < 0.11$, and $|A_Z| = 70.25$ mT; for ^{19}F $A_X = 34.4$, $A_Y = -4.64$, and $A_Z = -5.92$ mT. For CH_3F^+, the g tensor and ^{19}F A tensor were similar to those above but the three ^1H atoms were magnetically equivalent with values of $A_X = 11.3$, $A_Y = 11.5$, and $A_Z = 11.1$ mT on the ESR time scale. The results can be reasonably explained in terms of a static C_s conformation with a $^2A''$ state (figure 15), in which a large amount of unpaired electron is in the ^{19}F $2p_x$ orbital, but significant spin density also resides on the hydrogen $1s$ orbitals and the carbon $2p_x$ orbital as suggested by $^{13}CH_3F^+$ experiments; for ^{13}C $A_X = 1.71$, $A_Y = A_Z = -1.93$ mT. In this conformation the two hydrogens connected to carbon by the dashed lines have a significant amount of the spin density, *i.e.* a large isotropic hf value, due to a hyperconjugation mechanism; the conformation similar to that of the planar radicals H_2CO^+,[85] H_2CN,[86] and H_2BO.[86] The unique H atom (replace the D of figure 15 with one H atom) located in the nodal plane of the ^{19}F $2p_x$ orbital should have a very small or negative a_{iso} value. However, these large differences in the H atom environments must be undergoing some type of averaging behaviour given the clear quartet ^1H hf structure for the CH_3 group with $a_{iso}= 11.3$ mT.

The presence of a single ^{2}D atom in CH$_2$DF$^+$ drastically changes the ESR *hf* pattern and confirms the averaging conformation discussed above for CH$_3$F$^+$. That is, the ^{2}D atom acts to prevent the averaging process yielding two hydrogens with an unusually large a_{iso} value of 17.1 mT and one with a ^{2}D a_{iso} value of -0.18 or 1.18 mT on the hydrogen scale. This weighted average of these ^{2}D values for CH$_2$DF$^+$ is 11.1 mT, which is very close to the value of 11.3 mT. One suggested mechanism for this averaging is a combination of dynamic *J-T* distortion and tunnelling interchange of the H atoms.[46] The ZPVE difference between ^{1}H and ^{2}D atoms may be responsible for restricting the averaging or interchange phenomenon as was discussed in the preceding sections of this chapter for the selectively deuterated methane radical cations (CH$_4^+$, CH$_3$D$^+$, CH$_2$D$_2^+$, CHD$_3^+$ and CD$_4^+$)[32] and dimethylether radical cations (CD$_3$OCH$_3^+$, CD$_3$OCH$_2$D$^+$, CD$_3$OCHD$_2^+$, and CD$_3$OCD$_3^+$).[22]

7. STATIC AND DYNAMIC STRUCTURES OF CYCLOHEXANE AND RELATED RADICAL CATIONS

The radical cations of saturated hydrocarbons have been of interest because they are fundamentally important molecules and generally have a degenerate or nearly degenerate HOMO.[1-5,31] Ionization can be followed by a geometrical distortion due to instability of the degenerate orbital (*J-T* or *pseudo J-T distortion*).[35-40,87-100] In this section we deal with three topics on static and dynamic structural distortions of cyclohexane and related radical cations in low temperature halocarbon matrices. The first topic deals with the *J-T* split HOMO of the cyclohexane radical cation in selectively alkyl-substituted cyclohexanes.[37,38] The second topic deals with deuterium isotope effects on static and dynamic structures of the cyclohexane radical cation itself.[5,26,27] The third topic is on the radical cations of silacyclohexane (*c*SiC5), in which the Si atom can be regarded to play a role as a "probe" to detect an asymmetrically distorted geometrical structure (^{2}A in C_1) with one of the two Si-C bonds elongated.[5,25,39]

7.1 Possible Jahn-Teller distorted structures of cyclohexane cation

The radical cation of cyclohexane (*c*C6) is of particular interest because of its high molecular symmetry, D_{3d}.[1,4,35-38,87-90] The introduction of substituents to cyclohexane in a certain symmetrical or asymmetrical manner is a chemically intuitive way to remove orbital degeneracy; therefore electronic

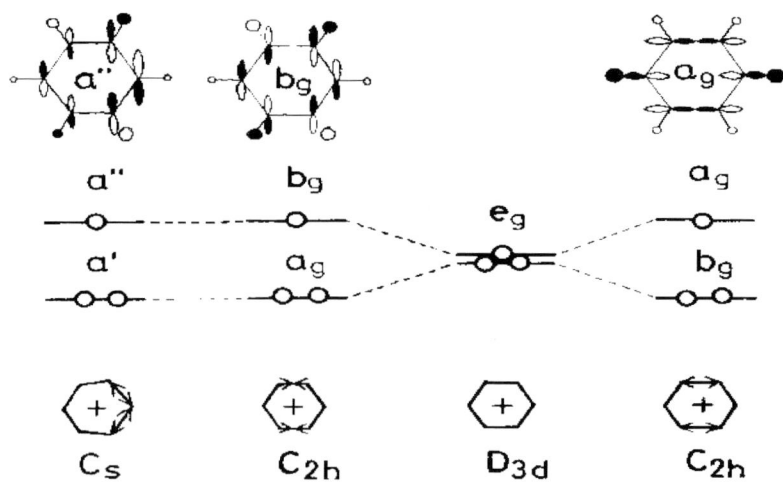

Figure 16. Schematic representation showing how the $4e_g$ degenerate orbital in $cC6^+$ changes by distortion from the original D_{3d} to either C_{2h} or C_s symmetry. The unpaired electron in the cation occupies either the a_g (C_{2h}), b_g (C_{2h}) or $a"$ (C_s) orbital. [taken from Fig. 1 in ref. 37]

states similar to those predicted by the *J-T* theorem could be studied.[37,38] Thus, a series of ESR studies was carried out on structural distortions and dynamics in the radical cations of selectively alkyl-substituted cyclohexanes and related molecules. The radical cations were classified into ^2A_g-like (in an elongated C_{2h} structure), ^2B_g-like (in a compressed C_{2h} structure), and $^2A"$-like (in a C_s structure) with structural resemblance to the cyclohexane cation.[35-38,89,90] For example, the radical cations of methylcyclohexane (Me-cC6) and 1,1-dimethylcyclohexane (1,1-Me$_2$-cC6) were concluded to take an asymmetrically distorted C_1 geometrical structure with one of the two C-C bonds elongated.[35,36] The temperature dependent ESR spectra were analyzed in terms of a selective bond length alternation between the two adjacent C-C bonds and their structures were concluded to be averaged by intramolecular dynamics so as to give apparently a symmetric C_s structure with increasing temperature.

7.2 Deuterium effects on cyclohexane cations

^{2}D-labelling is another way of obtaining information about the static and dynamic *Jahn-Teller* (*J-T*) effects of cyclohexane.[1] Thus a series of selectively deuterated cyclohexane (cC6) radical cations were radiolytically generated in a cC_6F_{12} or CF_3-cC_6F_{11} matrix at low temperatures and subjected to ESR studies combined with *ab initio* and density functional theory calculations. They are $cC6^+$, $cC6$-1,1-d_2^+, $cC6$-1,1,3,3-d_4^+,[26] $cC6$-

4. Quantum effects in deuterium labelled radicals

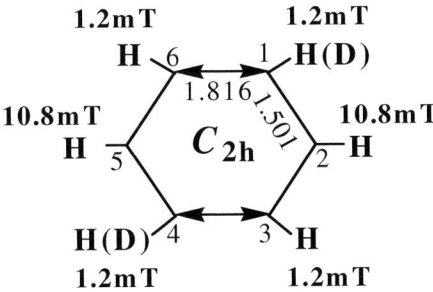

Figure 17. The theoretical isotropic *hf* splittings calculated for the equatorial hydrogens of $cC6^+$ in a C_{2h} structure. The arrows (↔) stand for the elongated C-C bonds: the theoretical C-C bond lengths (Å) are given in the figure. [taken from ref. 27]

$1,1,4,4-d_4^+,$[27] $cC6$-$1,1,2,2,3,3$-d_6^+ and $cC6$-d_{12}^+. Because of a limited number of pages, here our attention is mainly focused on $cC6$-$1,1,4,4$-d_4^+ as a prototype of selectively deuterated cyclohexane radical cations.

ESR spectrum and the C_{2h} distorted structure. The 4.2 K ESR spectrum of $cC6$-$1,1,4,4$-d_4^+ generated radiolytically in CF_3-cC_6F_{11} apparently consists of a triplet of 12.1 mT with a relative intensity of 1:4:1. This relative intensity deviates considerably from a binomial relative intensity of 1:2:1 due to two equivalent protons. This suggests that $cC6$-$1,1,4,4$-d_4^+ is not in the frozen state, but in a dynamic state even at 4.2 K.

The *ab initio* MO calculations of $cC6^+$ at MP2/6-31G** resulted in a C_{2h} structure whose total energy was lower by 1.2 kcal/mol than the C_s structure (next stable structure). The theoretical ^1H *hf* splittings were calculated by the DFT method for the optimized structure and are shown in figure 17. Based on the assumption of the C_{2h} distorted structure with both C(3)-C(4) and C(6)-C(1) bonds elongated, the larger ^1H *hf* splitting of 10.8 mT (a_1) and smaller splitting of 1.2 mT (a_2) were calculated for two equatorial hydrogens at the C(2) and C(5) positions and four equatorial hydrogens at the C(1), C(3), C(4) and C(6) positions, respectively (figure 17). The 12.1 mT splitting of the triplet corresponds very well to the sum of a_1 and a_2. Thus, the 4 K spectrum of $cC6$-$1,1,4,4$-d_4^+ was attributable to the C_{2h} distorted structure with the two equatorial ^2D atoms selectively occupying the C(1) and C(4) positions. The smaller splitting of a_2 was not resolved in the spectrum due to the dynamical state of the cation.

Zero-point vibrational energy (ZPVE). Depending on the sites which the deuterium atoms occupy, three C_{2h} distorted forms, *P*, *Q* and *Q'*, are possible for $cC6$-$1,1,4,4$-d_4^+ (figure 18). The theoretical calculations

(DFT/6-31G** level) resulted in a Q form whose ZPVE was lower by 1.71 kJ·mol^{-1} than that of the P form. The Q and Q' forms are mirror images with the same ZPVE. The probability of finding the different forms (p_P, p_Q, $p_{Q'}$) was assumed to follow a Boltzmann distribution, the same as for the deuterated dimethylether (DME) radical cations in section 6.1 of the present chapter. Using the calculated ZPVE values the following probabilities were evaluated: for example, $p_P : p_Q : p_{Q'} \cong$ 0.000 : 0.500 : 0.500 at 4.2 K and 0.094: 0.453: 0.453 at 130 K. At 4.2 K, the p_P value is essentially zero so that Q and Q' forms determine the ESR spectrum. Based on the assumption of an interchange (or jump) mode (figure 18), the ESR spectrum was successfully reproduced employing the calculated ^{1}H hf splittings of 10.8 mT (2H) and 1.2 mT (2H) as the rigid limit structure and the rate constant of interchanging, k_1, between Q and Q' forms as an adjustable parameter. Thus, we concluded that the original D_{3d} structure of cC6-1,1,4,4-d_4 was distorted into the C_{2h} structure after a one electron oxidation.

Temperature dependent total ^{1}H hf splittings. The ESR spectral line shapes of cC6-1,1,4,4-d_4^+ depended strongly on temperature. The ESR spectrum observed at 4.2 K changes into a quintet with increasing temperature: an averaged hf splitting of the quintet being 5.8 mT (4H) at 80 K. At the same time the total ^{1}H hf splitting decreased from 24.1 mT (4.2 K) to 21.9 mT (130 K). When the three forms are interchanged with rates faster than 2.7 x 10^8 sec^{-1}, which corresponds to the ^{1}H hf difference between 10.8 mT and 1.2 mT, a mean total ^{1}H hf splitting, $<\Sigma a_i>$, can be obtained[27] as mentioned in section 6.1 for the selectively deuterated DME cations[22]:

$$<\Sigma a_i> = p_P \Sigma a_P + p_Q \Sigma a_Q + p_{Q'} \Sigma a_{Q'} \quad \cdots \cdot (5)$$

Using the calculated values of p_P, p_Q (= $p_{Q'}$), Σa_Q (= $\Sigma a_{Q'}$) = 24.0 mT, and Σa_P = 4.8 mT, the temperature dependency of $<\Sigma a_i>$ was evaluated and compared well with the experimental values.[26, 27]

A similar ESR study was carried out for other deuterated cyclohexane radical cations of cC6-1,1-d_2, cC6-1,1,3,3-d_4^+,[26] and cC6-1,1,2,2,3,3-d_6^+. Their temperature dependent ESR spectra were also successfully analyzed by employing the same method as for cC6-1,1,4,4-d_4^+. For example, the theoretical calculations for cC6-1,1,3,3-d_4^+ resulted in a P form whose ZPVE was lower by 836 J·mol^{-1} than that of the Q (or Q') form. Assuming the Boltzmann distribution, the p_P value approaches unity so that the P form dominates the conformation at 4.2 K. In this case, no interchange is possible between P and Q (or Q') forms. With increasing temperature, the three forms (one P form and two Q forms) start to interchange.

4. Quantum effects in deuterium labelled radicals

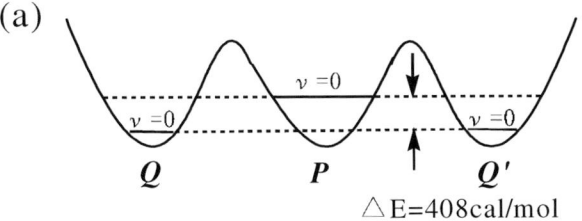

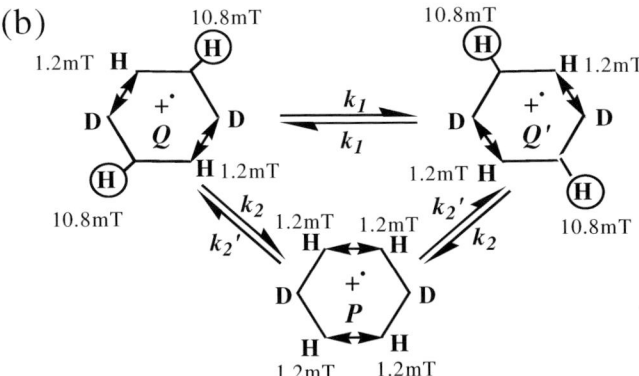

Figure 18. (a) Potential energy curve with zero point vibrational energies (ZPVEs) calculated for three C_{2h} geometrical structures (P, Q and Q' of cC6-1,1,4,4-d_4^+. (b) Schematics showing three site exchange (or jump) among one C_{2h} (P) and two C_{2h} (Q and Q') structures. [taken from ref. 27]

We summarize the ^{2}D isotope effects on the structure of $cC6^+$ as follows. (a) Three C_{2h} structures are possible for $cC6^+$ and their ZPVEs are shifted by varying the number and position of substituted ^{2}D atoms due to the mass effect. (b) By changing temperature the population of each ZPVE level is changed under assumption of a Boltzmann distribution, which is reflected by the temperature dependent ESR spectra. The readers can refer to refs. 26 and 27 for further details.

7.3 Asymmetrically distorted structure and dynamics of silacyclohexane cations

We have first reported that Me-$cC6^+$ and 1,1-Me$_2$-$cC6^+$ take an asymmetrically distorted geometrical structure (^2A in C_1) with one of the two C-C bonds elongated, which are attached to methyl group(s)[35,36,38,81] as mentioned in the beginning of this section, 7.1. Since then, a number of similar asymmetrical structural distortion have been observed for the

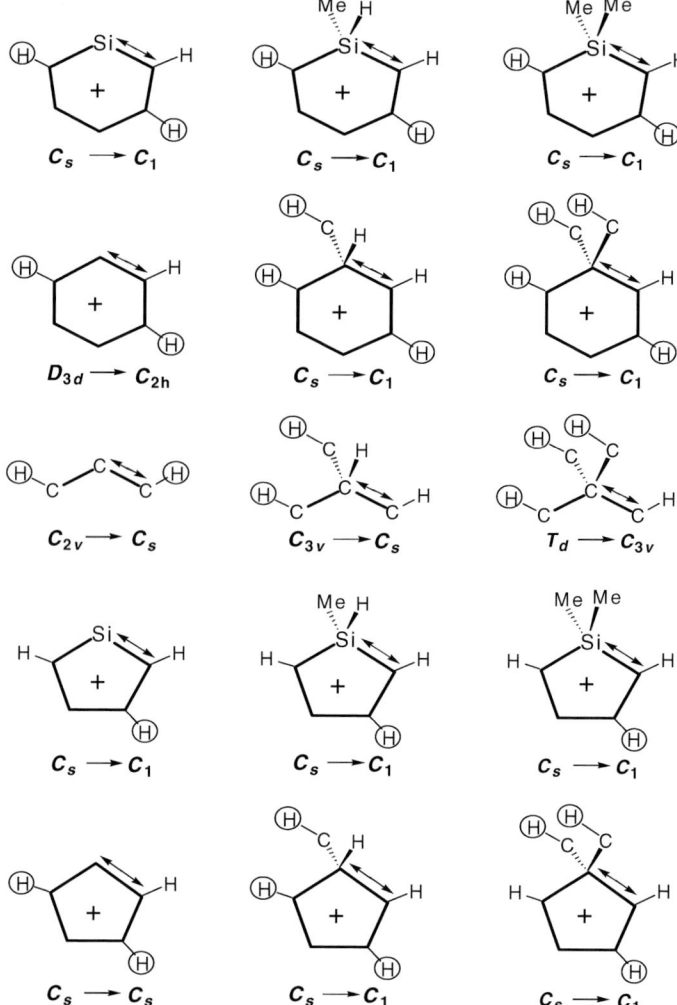

Figure 19. Schematic presentation showing structural distortion in the radical cations of saturated hydrocarbons and those with one Si atom substituent. The arrow (↔) indicates a σ bond in which an unpaired electron mainly resides so as to asymmetrically elongate the corresponding bond length. The hydrogen atoms marked with O are located in the trans position with respect to the elongated σ bond with large unpaired electron density and give rise to a large ^{1}H *hf* splitting detectable by ESR

following cycloalkane and related alkane radical cations in our group and other groups: cSiC3$^+$,[5,91] 1-Me-cSiC3$^+$, 1,1-Me$_2$-cSiC3$^+$,[5] cC5$^+$,[5,93] Me-cC5$^+$, 1,1-Me$_2$-cC5$^+$, cSiC4$^+$,[5] cSiC5$^+$,[5,25,92] Me-cSiC5$^+$, 1,1-Me$_2$-cSiC5$^+$,[25,92] 1,c2,c3-Me$_3$-cC6$^+$, 1,t2,c3-Me$_3$-cC6$^+$,[37] Et$_2$Si$^+$,[40,101] Et$_2$SiMe$^+$, Et$_2$SiMe$_2^+$, and Et$_4$Si$^+$. Based on the experimental results combined with *ab initio* MO calculations, we concluded that the decrease in symmetry with one electron

4. Quantum effects in deuterium labelled radicals

oxidation is an intrinsic nature of alkane radical cations whose parent molecules have certain symmetrical elements such as C_s, C_2, and C_{2v} in the geometrical structure, even though they are not *J-T* active.

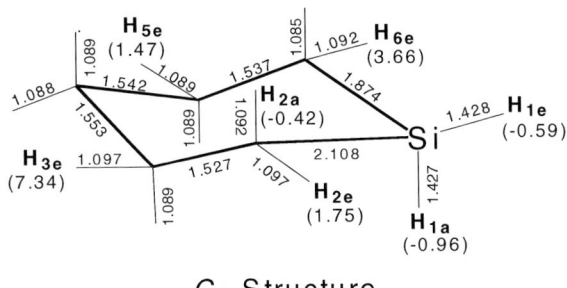

C_1 Structure

Figure 20. Optimised C_1 geometrical structure of $cSiC5^+$ calculated by using *ab initio* MO method at the UHF/STO-3G level of Gaussian 90. The values in parentheses are the isotropic ^{1}H *hf* splittings (in mT) evaluated by the INDO MO method for each structure and the bond lengths in Å.

Here, the asymmetrical structural distortion and dynamics are demonstrated using silacyclohexane (*c*SiC5) radical cation[25,92] as a prototype of saturated hydrocarbons. *c*SiC5 contains one Si atom in the structure; the ionisation potential (I_p) of a Si atom is lower than that of a C atom: I_{p1} = 8.15 eV (Si) and 11.26 (C).[102] Thus the unpaired electron is expected to favourably reside on a particular Si-C bond or on the two Si-C bonds equally, and the Si atom can be regarded as a "probe" to detect whether $cSiC5^+$ has the geometrical structure of an asymmetrically distorted C_1 or a symmetrical C_s.

ESR Spectra. The 4.2 K spectrum of $cSiC5$-2,2,6,6-d_4^+ consists of a doublet of doublets with *hf* splittings of 7.6 mT (1H) and 2.9 mT (1H). The spectrum of 1-Me-$cSiC5$-2,2-d_2^+ consists of a doublet of *ca.* 3 mT in addition to the *hf* pattern of $cSiC5$-2,2,6,6-d_4^+. Based on the results the triplet splitting of *ca.* 3 mT observed for $cSiC5^+$ was attributed to two equatorial hydrogens at C(3) and C(5), *i.e.* H$_{2e}$ and H$_{6e}$. Assuming that the σ(Si-C) bonds have higher spin densities than C-C bonds, the *hf* splittings of 7.6 mT and 2.9 mT were reasonably attributed to the two equatorial hydrogens, H$_{3e}$ and H$_{5e}$. This attribution leads us to an asymmetrically distorted C_1 structure of the *c*SiC5 radical cations in a static structure, in which the unpaired electron is predominately in one of the two Si-C bonds, Si-C(2). The Si-C bond with higher spin density is expected to be weaker and elongated. The asymmetrically distorted C_1 structure derived from the present experimental results was theoretically supported (see figure 20).

Origin of the Structural Distortion. The structural distortion can originate from the second-order Jahn-Teller theory (*pseudo J-T effect*).[103,104] Based on the theory, the energy of the radical cation in the ground electronic state is expressed by the following equation:

$$E = E_0 + Q \langle \psi_0 | \frac{\partial U}{\partial Q} | \psi_0 \rangle + \frac{Q^2}{2} \langle \psi_0 | \frac{\partial^2 U}{\partial Q^2} | \psi_0 \rangle + \sum_k \frac{\left[Q \langle \psi_0 | \frac{\partial U}{\partial Q} | \psi_k \rangle \right]^2}{(E_0 - E_k)}$$

Where E_0 is the energy without distortion, Q is the displacement of the normal coordinate, U is the nuclear-nuclear and nuclear-electronic potential energy, ψ_0 and ψ_k are the wave functions for the ground and excited electronic states. For an asymmetric distortion the second term should be zero. The third and the fourth terms are always positive and negative, respectively. Thus, the asymmetrical structure can be energetically preferable when the sum of the two terms is negative. The fourth term is inversely proportional to the energy difference between the electronic ground and lower lying electronic excited states. Since the energy is smaller than several electron volt (eV) for most of the alkane radical cations, the fourth term predominates over the third term so as to make the asymmetrical C_1 structure.

Temperature Dependent ESR Spectra. Here we are concerned with the temperature dependent ESR spectra of $c\text{SiC5-2,2,6,6-}d_4^+$ as an example, in a temperature range from 4.2 K to 140 K. The 4.2 K spectrum consists of double doublets with *hf* splittings of 7.6 mT and 2.9 mT with the inner doublet less intense than the outer one. With increasing temperature a new singlet appeared at the central position and its intensity increased with temperature. At 140 K the spectrum became a triplet of *ca.* 5.3 mT with the relative intensity close to a binomial one. The ESR spectra can be explained by an interchange of the two hydrogens of H_{3e} and H_{5e}. The $c\text{SiC5}^+$ has two energetically equivalent mirror image structures, one with the Si-C(2) bond elongated and the other with the Si-C(6) bond elongated. It is schematically presented how ^{1}H *hf* splittings of 7.6 mT and 2.9 mT are averaged out by such intramolecular exchange process with the rate k (s^{-1}).

The weaker inner doublet of the 4.2 K spectrum of $c\text{SiC5-2,2,6,6-}d_4^+$ was explained by assuming that the *hf* splittings of H_{3e} and H_{5e} have been partially averaged by the exchange process with a slightly smaller rate constant, k (s^{-1}) than the *hf* splitting difference in the two hydrogens, $a(H_{3e}) - a(H_{5e}) = 1.3 \times 10^7$ (s^{-1}). With a k value nearly equal to the difference in the two *hf* splittings the inner doublet should disappear as observed at ca. 40 K. On further increasing the temperature a new singlet appears at the central position and its intensity grows, suggesting further increase in the rate constant. Arrhenius plots of the rate constants, k (s^{-1}), show a non-linear relationship over the temperature range of 4.2 K to 130 K for all silacyclohexane radical cations studied. Assuming a linear relationship in the higher temperature region between 60 K and 130 K, an activation energy of ca. 0.3 kcal/mol was evaluated. Below 40 K, the rate constants were almost independent of temperature: $k = 3.6 \times 10^7$ s^{-1} for the cSiC5$^+$.

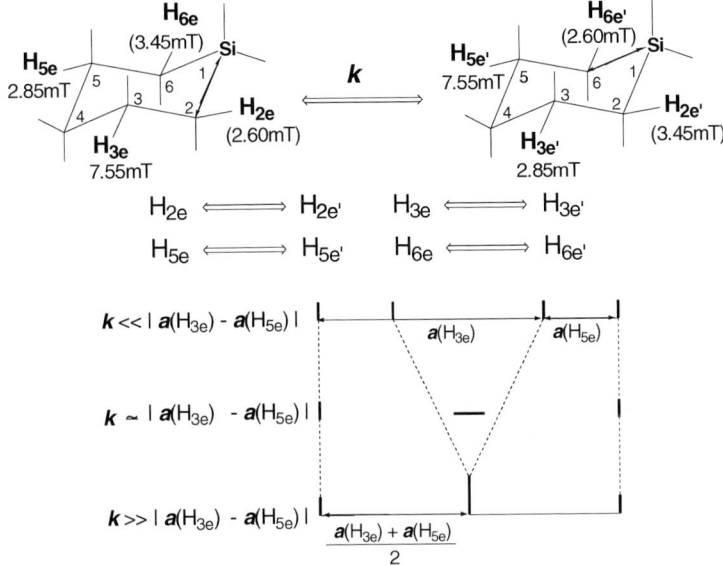

8. HYDROGEN ATOM ABSTRACTION VIA TUNNELLING

The quantum mechanical tunnelling effects are fundamentally important in low temperature chemistry. Chemical reaction *via* tunnelling effects is characterized by a large isotope H/D effect on the rate constant, a non-linear Arrhenius plot, and high selectivity of the reaction.[10,11] Four decades ago a remarkable ^{2}D effect was reported for the decay rate of radicals produced in protiated and perdeuterated organic compounds at 77 K by Sullivan and Koski.[105] Since then a number of studies have been reported on tunnelling reactions in low temperature solids.[21,106-110] For example, in 1971, Williams and collaborators reported the H atom abstraction from CH$_3$CN by the CH$_3$ radical with the low activation energy of 1.4 kcal mol^{-1} and a large "primary" kinetic isotope effect, $k_{(H)}/k_{(D)} > 140$, between 77 K and 87 K.[107] One of the authors (M.S.) has also demonstrated a non-liner Arrhenius plot for the reaction, $CH_3OH + CH_3 \rightarrow CH_2OH + CH_4$, in solid CH_3OH at the low temperatures of 10 K - 90 K.[110] Here we wish to add our recent ESR study on selective hydrogen atom abstraction from the Si-H bond of CH_3SiH_3 in the solid state at cryogenic temperatures to the previous studies on quantum mechanical tunnelling reactions.

8.1 Reaction selectivity and enormous H/D isotope effects on H atom abstraction in solid CH_3SiH_3

Methylsilane (CH_3SiH_3), the simplest alkylsilane, is composed of a CH3 group and a SiH_3 group, which are expected to show remarkably different reactivity from each other. That is, the Si-H bond length of Si with sp^3 hybridization is 0.149 nm, which is 36 % longer than the corresponding C-H bond.[102] In addition, the dissociation energy of the Si-H bond is more than 40 $kJmol^{-1}$ smaller than the C-H bond.[102] These may significantly contribute to either the classical or non-classical (*i.e.* tunnel) reaction kinetics. Thus, the CH_3SiH_3 molecule was chosen as a candidate molecule for studying hydrogen abstraction reactions.

The ESR study was carried out to elucidate a hydrogen atom abstraction reaction (c) from methylsilane (CH_3SiH_3) by the methyl radical,

$$CH_3SiH_3 + CH_3 \rightarrow CH_3SiH_2 + CH_4 \quad \cdots \cdot (c)$$

in a solid solution of CH_3SiH_3 containing 1 mol% of CH_3I in the temperature range of 3 K - 115 K. The ESR spectra observed immediately after UV-photolysis at 77 K were dominated by the methylsilyl radical (CH_3SiH_2), but with a small amount of CH_3 radical. When CH_3SiD_3 was employed, on the contrary, the major radical turned out to be the CH_3 radical, but with a small amount of the CH_3SiD_2 radical. Neither SiH_3CH_2 nor SiD_3CH_2 radicals were observed for the both systems. The results indicate that hydrogen atom abstraction by the CH_3 radical occurs preferentially at the Si-H bond of CH_3SiH_3, but not at the C-H bond. Similar reaction selectivity was observed when the hydrogen atom (H atom) was employed instead of the CH_3 radical.

The CH_3 radicals decayed following first order reaction kinetics with a concomitant formation of the CH_3SiH_2 (CH_3SiD_2) radicals in CH_3SiH_3 (CH_3SiD_3) (figure 21). The associated decay rates of CH_3 showed an anomalously large isotope effect; *i.e.*, for example, $k_{(Si-H)} \approx 4 \times 10^{-2}$ s^{-1} and $k_{(Si-D)} \approx 7 \times 10^{-6}$ s^{-1} ($k_{(Si-H)}/k_{(Si-D)} \approx 5 \times 10^3$) at 77 K. Furthermore, a non-linear

Arrhenius plot was obtained for the rates so as to become almost independent of the temperature below 20 K. Assuming a linear Arrhenius plot above 20 K, an apparent activation energy for the reaction was evaluated to be $E_{a(Si-H)} \approx 1$ kJmol^{-1} and $E_{a(Si-D)} \approx 9$ kJmol^{-1} ($E_{a(Si-H)}/E_{a(Si-D)} \approx 1/10$). The observed large ^{2}D isotope effect on the rates and the non-linear Arrhenius plots strongly suggest that the quantum mechanical tunnelling effects greatly contribute to the H atom abstraction from the Si-H bond of the SiH$_3$ group, especially below 20 K. Hiraoka et al. have recently reported that amorphous silicone thin films are formed by successive H atom abstraction from the silane (SiH$_4$) molecule by H atoms at the low temperature of 10 K.[111] However, they could not provide any direct experimental evidence of paramagnetic species as reaction intermediates, because the reactions were monitored by in situ IR spectroscopy, not by ESR. The present ESR study on the H atom abstraction from the Si-H bond of methylsilane strongly suggests that the quantum mechanical tunnelling reaction can play an important role, too, for the formation of amorphous silicon from silanes.

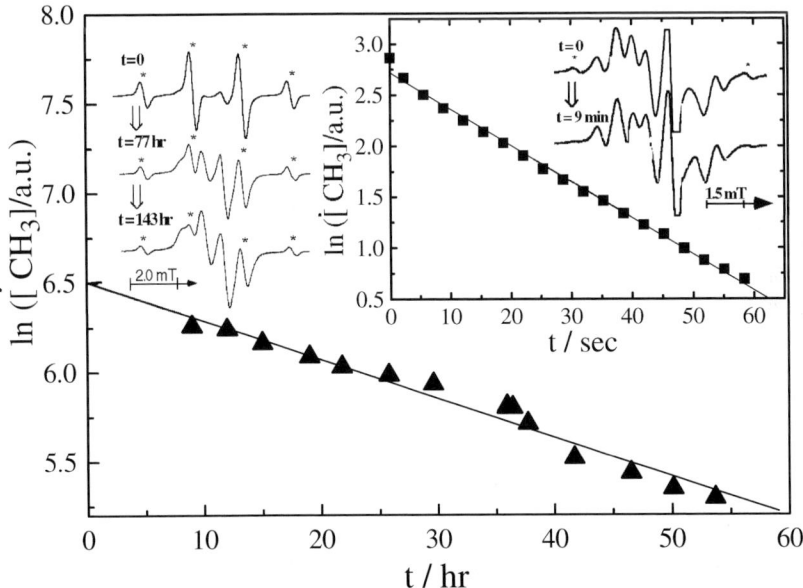

Figure 21. First-order plots of CH$_3$ radical decaying in the CH$_3$SiH$_3$ (■) and CH$_3$SiD$_3$ (▲) systems at 77 K. The CH$_3$ radicals were generated by the UV-photolysis of the two methylsilanes containing 1 mol% of CH$_3$I at the same temperature. [taken from ref. 48]

9. REFERENCES

1. M. Shiotani, *Magn. Reson. Rev.* **12** (1987) 333.
2. A. Lund and M. Shiotani, eds. *Radical Ionic Systems*, Kluwer, Dordrecht, 1991.
3. M. Lindgren and M. Shiotani, in *Radical Ionic Systems*, eds., A. Lund and M. Shiotani, Kluwer, Dordrecht, p. 125, 1991.
2. M. Shiotani and A. Lund, in *Radical Ionic Systems*, eds., A. Lund and M. Shiotani, Kluwer, Dordrecht, p. 151, 1991.
3. M. Shiotani and K. Komaguchi, Houshasen Kagaku (Radiation Chemistry in Japanese) **65** (1998) 2.
6. M. Shiotani, *CRC handbook of Radiation Chemistry*, ed. Y. Tabata, Chap. III.B.7, *CRC Press*, pp.139-144, 1991.
7. M. Shiotani, H. Yoshida, in *CRC handbook of Radiation Chemistry* (ed. Y. Tabata), Chap. VIII.C, *CRC Press*, pp. 440-467, 1991.
8. M. Shiotani, *CRC handbook of Radiation Chemistry*, ed. Y. Tabata, Chap. IX.B, *CRC Press,* pp.544-567, 1991.
9. M. Shiotani and M. Lindgren, *Radicals on Surfaces*, eds A. Lund & C. Rhodes, *Kluwer*, pp. 179-199, 1994.
10. J. E. Willard, in *Chemical Kinetics of Small Organic Radicals*, ed. Z. B. Alfassi, CRC Press, Baca Roton, Fl, Vol. **IV**, pp. 1-29, 1988.
11. V. A. Benderskii, D. E. Makarov, and C. A. Wight, *Chemical Dynamics at Low Temperatures, Advances in Chemical Physics,* Vol. **LXXXVIII**, John Wiley & Sons, New York, 1994.
12. T. Yamada, K. Komaguchi, M. Shiotani, N. P. Benetis, and A. R. Sørnes, *J. Phys. Chem. A* **103** (1999) 4823.
13. H. M. McConnell, *J. Chem. Phys.* **29** (1958) 1422.
14. J. H. Freed. *J. Chem. Phys.* **43** (1965) 1710.
15. S. Clough and F. Poldy, *J. Chem. Phys.* **51** (1969) 2076.
16. R. B. Davidson and I. Miyagawa, *J. Chem. Phys.* **52** (1970) 1727.
17. S. Clough and J.R. Hill, *J. Phys. C: Solid State Phys.* **7** (1974) L20.
18. M. Geoffroy, L. D. Kispert, and J. S. Hwang, *J. Chem. Phys.* **70** (1979) 4238.
19. M. Matsushita, T. Momose, and T. Shida, *J. Chem. Phys.* **92** (1990) 4749.
20. S. Kubota, M. Matsushita, T. Shida, A. Abu-Raqabah, M.C.R. Symons, and J. L. Wyatt, *Bull. Chem. Soc., Jpn.* **68** (1995) 140.
21. K. Komaguchi, T. Kumada, Y. Aratono, and T. Miyazaki, *Chem. Phys. Lett.* **268** (1997) 493.
22. M. Shiotani, N. Isamoto, M. Hayashi, T. Fängström, and S. Lunell, *J. Am. Chem. Soc.* **122** (2000) 12281.
23. T. Miyazaki, K. Yamamoto, and J. Arai, *Chem.Phys. Lett.* **219** (1994) 405.
24. I. F. Silvera, *Rev. Mod. Phys.* **52** (1980) 393
25. K. Komaguchi and M. Shiotani, *J. Phys. Chem.* **101** (1997) 6983.
26. P. Wang, M. Shiotani, and S. Lunell, *Chem. Phys. Lett.* **292** (1998) 110.
27. M. Shiotani, P. Wang, and K. Komaguchi, *Bull. Polish Academy of Chemistry, Physical Chemistry* **47** (1999) 89.
28. C. A. Coulson and H. L. Strauss, *Proc. R. Soc. London, Ser. A* **269** (1962) 443.
29. R. N. Dixon, *Mol. Phys.* **20** (1971) 113.
30. F. A. Grimm and J. Gody, *Chem. Phys. Lett.* **6** (1970) 336.
31. L. N. Shchegoleva and P. V. Schastnev, *Molecular Distortions in Ionic and Excited States*, CRC Press, New York, p. 123, 1995.
32. L. B. Knight, Jr., J. Steadman, D. Feller, and E. R. Davidson, *J. Am. Chem. Soc.* **106**

4. Quantum effects in deuterium labelled radicals

(1984) 3700.
33. L. B. Knight, Jr., G. M. King, J. T. Petty, M. Matsushita, T. Momose, and T. Shida, *J. Chem. Phys.* **103** (1995) 3377.
34. K. Komaguchi, T. Marutani, M. Shiotani, and A. Hasegawa, *Phys. Chem. Chem. Phys.* **3** (2001) 3536.
35. M. Shiotani, N. Ohta, and T. Ichikawa, *Chem. Phys. Lett.* **149** (1988) 185.
36. M. Lindgren, M. Shiotani, N. Ohta, T. Ichikawa, and L. Sjöqvst, *Chem. Phys. Lett.* **161** (1989) 127.
37. M. Shiotani, M. Lindgren, and T. Ichikawa, *J. Am. Chem. Soc.* **112** (1990) 967.
38. M. Shiotani, M. Lindgren, N. Ohta, and T. Ichikawa, *J. Chem. Soc., Perkin Trans. 2* (1991) 711.
39. T. Fängström, S. Lunell, B. Engles, L. Eriksson, M. Shiotani, and K. Komaguchi, *J. Chem. Phys.* **107** (1997) 297.
40. K. Komaguchi, T. Marutani, M. Shiotani, and A. Hasegawa, *Phys. Chem. Chem. Phys.* **1** (1999) 4549.
41. K. Komaguchi, K. Nomura, and M. Shiotani, in manuscript.
42 K. Komaguchi, T. Kumada, T. Takayanagi, Y. Aratono, M. Shiotani, and T. Miyazaki, *Chem. Phys. Lett.* **300** (1999) 257.
43. K. Komaguchi, S. Yamada, M. Shiotani, and P. H. Kasai, *Proceeding of 3rd International Conference on Low. Tem. Chem*, Nagoya, 1999.
44. L. Bonazzola, J. P. Michaut, and J. Roncin, *J. Phys. Chem.* **95** (1991) 3132.
45. L. Bonazzola, J. P. Michaut, and J. Roncin, *New J. Chem.* **16** (1992) 489.
46. L. B. Knight, Jr., B. W. Gregory, D. W. Hill, C. A. Arrington, T. Momose, and T. Shida, **94** (1991) 67.
47. B. F. Yates, W. J. Bouma, and L. Radon, *J. Am. Chem. Soc.* **109** (1987) 2250.
48. K. Komaguchi, Y. Ishiguri, H. Tachikawa, and M. Shiotani, *Phys. Chem. Chem. Phys.* **4** (2002) 5276.
49. C. K. Jen, S. N. Foner, E. L. Cochran, and V. A. Bowers, *Phys. Rev.* **112** (1958) 1169.
50. R. L. Morehouse, J. J. Christiansen, and W. Gordy, *J. Chem. Phys.* **45** (1966) 1751.
51. P. H. Kasai and D. McLeod Jr., *J. Am. Chem. Soc.* **94** (1972) 7975.
52. G. Cirelli, A. Russu, R. Wolf, M. Rudin, A. Schweiger, and H. H. Günthard, *Chem. Phys. Lett.* **92** (1982) 223.
53. M. Fujimoto, H. D. Gesser, B. Garbutt, A. Cohen, *Sience* **154** (1966) 381.
54. M. Shiotani, F. Yuasa, and J. Sohma, *J. Phys. Chem.* **79** (1975) 2669.
55. R.W. Fessenden, *J. Phys. Chem.* **71** (1964) 74.
56. S. Kubota, M. Iwaizumi, Y. Ikegami, K. Shimokoshi, *J. Chem. Phys.* **71** (1979) 4774.
57. A. R. Sørnes, N.P. Benetis, R. Erickson, A.S. Mahgoub, L. Eberson, and A. Lund, *J. Phys. Chem. A* **101** (1997) 8987.
58. Y. Kurita, *J. Chem. Phys.* **41** (1964) 3926.
59. W. Gordy and R. Morehouse, *Phys. Rev.* **151** (1966) 207.
60. K. Toriyama, M. Iwasaki, and K. Nunome, *J. Chem. Phys.* **71** (1979) 1698.
61. N. M. Atherton, *Electron Spin Resonance: Theory and Applications,* John Wiley & Sons, p. 150, New York, 1973.
62. S. N. Foner, and E. L. Cochran, V. A. Bowers, and C. K. Jen, *J. Chem. Phys.* **32** (1960) 963.
63. F. J. Adrian, *J. Chem. Phys.* **32** (1960) 972.
64. L. B. Knight, Jr., W. E. Rice, L. Moore, and E. R. Davidson, *J. Chem. Phys.* **103** (1995) 5275.
65. L. B. Knight, Jr., W. E. Rice, L. Moore, E. R. Davidson, and R. S. Dailey, *J. Chem. Phys.* **109** (1998) 1409.

66. G. C. Hancock, C. A. Mead, D. G. Truhlar, and A. J. C. Varandas, *J. Chem. Phys.* **91** (1989) 3492.
67. T. Takayanagi and S. Sato, *J. Chem. Phys.* **92** (1990) 2862.
68. J. van Kranendonk, *Solid Parahydrogen*, Plenum Press, New York, 1983.
69. T. Miyazaki, T. Hiraku, K. Fueki, and Y. Tsuchihashi, *J. Phys. Chem.* **95** (1991) 26.
70. T. Kumada, J. Kumagai, and T. Miyazaki, *J. Chem. Phys.* **114** (2001) 10024.
71. R. W. Fessenden, *J. Chem. Phys.* **37** (1962) 747.
72. M. Shiotani and J. H. Freed, *J. Phys. Chem.* **85** (1981) 3873.
73. P. H. Kasai, W. Weltner, Jr., and E. B. Whipple, *J. Chem. Phys.* **42** (1965) 1120.
74. M. Iwasaki, K. Toriyama, and K. Nunome, *J. Am. Chem. Soc.* **103** (1981) 3591.
75. O. Claesson, A. Lund, T. Gillbro, T. Ichikawa, O. Edlund, and H. Yoshida, *J. Chem. Phys.* **72** (1980) 1463.
76. W. Meyer, *J. Chem. Phys.* **58** (1973) 1017.
77. M. N. Paddon-Raw, D. J. Fox, J. A. Pople, K. N. Houk, and D. W. Pratt, *J. Am. Chem. Soc.* **107** (1985) 7696.
78. R. F. Frey and E. R. Davidson, *J. Chem. Phys.* **88** (1988) 1775.
79. L. A. Eriksson, S. Lunell, and R. J. Boyd, *J. Am. Chem. Soc.* **115** (1993) 6896.
80. J. G. Aston, R. M. Kennedy, and G.H. Messerly, *J. Am. Chem. Soc.* **63** (1941) 2343.
81. L. Sjöqvst, M. Lindgren, A. Lund, and M. Shiotani, *J. Chem. Soc. Faraday Trans.* **86** (1990) 3377.
82. K. Komaguchi, M. Shiotani, and A. Lund, *Chem. Phys. Lett.* **265** (1997) 217.
83. C. Heller and H. M. McConnell, *J. Chem. Phys.* **32** (1960) 1535.
84. G. Frankel, *J. Chem. Phys.* **71** (1967) 139.
85. L. B. Knight, Jr. and J. Steadman, *J. Chem. Phys.* **80** (1984) 1018.
86. W. Weltner, Jr., *Magnetic Atoms and Molecules*, Van Nostrand, New York, 1983.
87. M. Iwasaki and K. Toriyama, *Faraday Discuss. Chem. Soc.* **78** (1984) 19.
88. S. Lunell, M. B. Huang, O. Claesson, and A. Lund, *J. Chem. Phys.* **82** (1985) 5121.
89. M. Lindgren, M. Matsumoto, and M. Shiotani, M. *J. Chem. Soc. Perkin Trans. 2*, (1992) 1397.
90. M. Shiotani, M. Matsumoto, and M. Lindgren, *J. Chem. Soc. Perkin Trans. 2*, (1993) 1995.
91. K. Komaguchi, M. Shiotani, M. Ishikawa, and K. Sasaki, *Chem. Phys., Lett.* **200** (1992) 580.
92. M. Shiotani, K, Komaguchi, J. Ohshita, M. Ishikawa, and L. Sjöqvist, *Chem. Phys. Lett.* **188** (1992) 93.
93. M. Lindgren, K. Komaguchi, M. Shiotani, and K. Sasaki, *J. Phys. Chem.* **98** (1994) 8331.
94. Y. Ito, H. F. M. Mohamed, and M. Shiotani, *J. Phys. Chem.* **100** (1996) 14161.
95. Y. Itagaki, M. Shiotani and, H. Tachikawa, *Acta Chemica Scandinavia* **51** (1997) 220.
96. A. Hasegawa, Y. Itagaki and, M. Shiotani, *J. Chem. Soc., Perkin Trans. 2* (1997) 1625.
97. R. M. Kadam, R. Erickson, K. Komaguchi, M. Shiotani, and A. Lund, *Chem. Phys. Lett.* **290** (1998) 371.
98. Y. Itagaki, M. Shiotani, A. Hasegawa, and H. Kawazoe, *Bull. Chem. Soc. Japan* **71** (1998) 2547.
99. H. Sakurai, M. Shiotani, and T. Ichikawa, *Radiat. Phys. Chem.* **54** (1999) 235.
100. Y. Itagaki and M. Shiotani, *J. Phys. Chem. A* **103** (1999) 5189.
101. M. Shiotani, unpublished data.
102. *CRC Handbook of Chemistry and Physics* (82nd edition), ed. D. R. Lide, CRC Press, Baca Roton, Fl, 2001.
103. R. G. Pearson, *J. Am. Chem. Soc.* **91** (1969) 1252.

104. R. G. Pearson, *J. Am. Chem. Soc.* **91** (1969) 4947.
105. P. J. Sullivan and W. S. Koski, *J. Am. Chem. Soc.* **85** (1963) 384.
106. W. G. French and J. E. Willard, *J. Phys. Chem.* **72** (1968) 4604.
107. F. Williams and E. D. Sprague, *J. Am. Chem. Soc.* **93** (1971) 787.
108. E. D. Sprague, *J. Phys. Chem.* **77** (1973) 2066.
109. T. Miyazaki, *Radiat. Phys. Chem.* **37** (1991) 635.
110. R. L. Hadson, M. Shiotani, and F. Williams, *Chem. Phys. Lett.* **48** (1977) 193.
111. K. Hiraoka, T. Sato, S. Sato, S. Hishiki, K. Suzuki, Y. Takahashi, T. Yokoyama, and S. Kitagawa, *J. Phys. Chem. B* **105** (2001) 6950.

Chapter 5

XSOPHE - SOPHE – XEPRVIEW

A computer simulation software suite for the analysis of continuous wave EPR spectra

Graeme R. Hanson,[a] Kevin E. Gates,[b] Christopher J. Noble,[a] Anthony Mitchell[a], Simon Benson,[a] Mark Griffin,[a,b] Kevin Burrage[b]
[a]*Center for Magnetic Resonance and the* [b]*Department of Mathematics, The University of Queensland, St. Lucia, Queensland, Australia, 4072.*

Key words: paramagnetic, spin Hamiltonian, electron Zeeman, hyperfine, superhyperfine, nuclear Zeeman, quadrupole, exchange, dipole-dipole, Graphical User Interface, XSophe, Sophe, X windows, XeprView, Xepr, Corba, matrix diagonalisation, transition probability, randomly orientated, frequency swept, Gaussian, Lorentzian, resonant field, lineshape, distribution, single crystal, SOPHE partition scheme, eigenfield, field segmentation, perturbation, eigenvalues, eigenvectors, SOPHE interpolation, looping transitions, mosaic misorientation, computational time, linewidth, Kivelson, parallelisation, D-E strain, g-A strain, Optimisation, non linear least squares, Hooke and Jeeves, Quadratic, Simplex, Monte Carlo, simulated annealing, goodness of fit, energy level diagrams, transition roadmaps, transition surfaces, multifrequency, forbidden transitions, state mixing, variable temperature, energy level.

Abstract: The XSophe-Sophe-XeprView computer simulation software suite provides scientists with an easy-to-use research tool for the analysis of isotropic, randomly orientated and single crystal continuous wave electron paramagnetic resonance (CW EPR) spectra. XSophe provides an X Windows graphical user interface to the Sophe programme allowing; the creation of multiple input files, the local and remote execution of Sophe and display of Sophelog (output from Sophe) and input parameters/files. Sophe is a sophisticated computer simulation software programme with a number of innovative technologies including; the Sophe partition and interpolation schemes, a field segmentation algorithm, the mosaic misorientation line width model, parallelisation (OpenMP - for SGI computers running the Irix operating system) and spectral optimisation. In conjunction with the SOPHE partition scheme and the field segmentation algorithm, the SOPHE interpolation scheme and mosaic misorientation linewidth model greatly increase the speed of simulations for most spin systems. The output of CW EPR spectra (1D and 2D) from the Sophe programme can be visualised in conjunction with the experimental spectrum in XeprView or Xepr. Energy level diagrams, transition roadmaps and transition surfaces aid the interpretation of complicated randomly orientated EPR spectra and can be viewed with a netscape browser and an OpenInventor scene graph viewer.

1. INTRODUCTION

Multifrequency continuous wave electron paramagnetic resonance (CW EPR) spectroscopy[1-7] is a powerful tool for characterising paramagnetic molecules or centres within molecules that contain one or more unpaired electrons. Computer simulation of the experimental randomly orientated or single crystal EPR spectra from isolated or coupled paramagnetic centres is often the only means available for accurately extracting the spin Hamiltonian parameters required for the determination of structural information.[1,2,9-21] EPR spectra are often complex and are interpreted with the aid of a spin Hamiltonian. For an isolated paramagnetic centre (A) a general spin Hamiltonian is:[1,2,8]

$$\mathcal{H}_A = S \cdot D \cdot S + \beta B \cdot g \cdot S + S \cdot A \cdot I$$
$$+ I \cdot Q \cdot I - \gamma I \cdot (1-\sigma) \cdot B \cdot I \qquad [1]$$

where S and I are the electron and nuclear spin operators respectively, D the zero field splitting tensor, g and A are the electron Zeeman and hyperfine coupling matrices respectively, Q the quadrupole tensor, γ the nuclear gyromagnetic ratio, σ the chemical shift tensor, β the Bohr magneton and B the applied magnetic field. Additional hyperfine, quadrupole and nuclear Zeeman interactions will be required when superhyperfine splitting is resolved in the experimental EPR spectrum. When two or more paramagnetic centres (A_i, i = 1, ..., N) interact, the EPR spectrum is described by a total spin Hamiltonian ($\mathcal{H}_{Total}$) which is the sum of the individual spin Hamiltonians ($\mathcal{H}_{Ai}$, Eq. [1]) for the isolated centres (A_i) and the interaction Hamiltonian ($\mathcal{H}_{Aij}$) which accounts for the isotropic exchange, antisymmetric exchange and the anisotropic spin-spin (dipole-dipole coupling) interactions between a pair of paramagnetic centres.[1,9,10]

$$\mathcal{H}_{Total} = \sum_{i=1}^{N} \mathcal{H}_{A_i} + \sum_{i,j=1, i \neq j}^{N} \mathcal{H}_{A_{ij}}$$

$$\mathcal{H}_{A_{ij}} = J_{A_{ij}} S_{A_i} \cdot S_{A_j} + d_{A_{ij}} S_{A_i} \times S_{A_j} + S_{A_i} \cdot D_{A_{ij}} \cdot S_{A_j}$$

[2]

The XSophe-Sophe-XeprView computer simulation software suite[16,17,22,23] consists of:

- the XSophe X-windows interface,
- Sophe authentication and Corba daemons,
- Sophe, a state-of-the-art computational programme for simulating CW EPR spectra and
- XeprView, a programme for comparing experimental and simulated spectra.

The software suite runs on SGI computers running IRIX (6.2, 6.3, 6.4 and 6.5) and personal computers running Linux (RedHat: 6.2, 7.1, 7.2, 7.3, 8.0 and Mandrake: 8.1, 8.2, 9.0). Use of the latest versions of the operating systems is strongly recommended.[24]

2. THE XSOPHE X-WINDOW GRAPHICAL USER INTERFACE

XSophe-Sophe-XeprView computer simulation software suite provides scientists with an easy-to-use research tool for the analysis of isotropic, randomly orientated and single crystal continuous wave (CW) EPR spectra. XSophe provides an X-Windows graphical user interface (Figure 1) to the Sophe program allowing; the creation of multiple input files, the local and remote execution of Sophe and display of sophelog (output from Sophe) and input parameters/files.

XSophe allows transparent transfer of EPR spectra and spectral parameters between XSophe, Sophe and XeprView®, using state-of-the-art platform-independent Corba libraries. This interactivity (Scheme 1) allows the execution and interaction of the XSophe interface with Sophe on the same computer or a remote host through a simple change of the hostname. XSophe contacts the Sophe Corba daemon, which then interacts with the Sophe authentication daemon *via* a socket to validate the username and password which was encrypted with 128 bit encryption and embedded in a

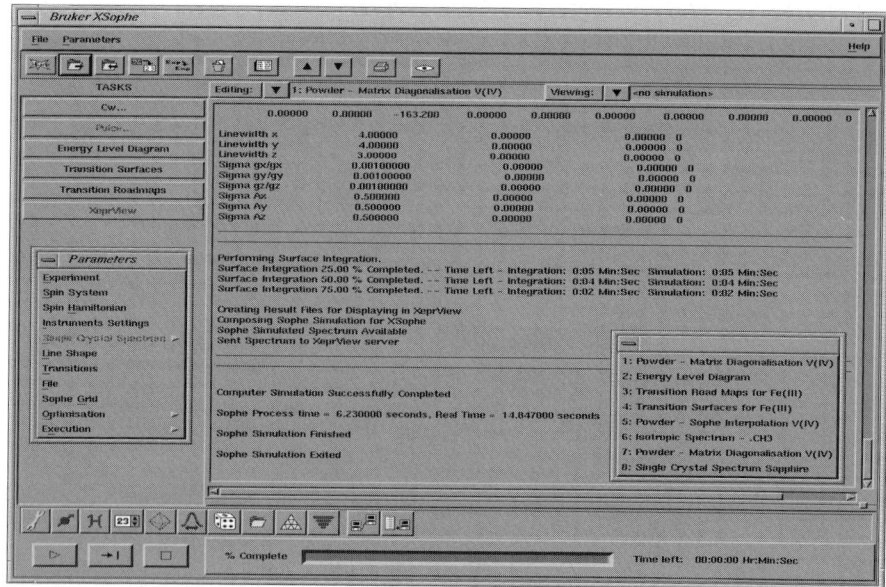

Figure 1. The XSophe (v 1.1) main Window. The interface allows the creation and execution of multiple input files on local or remote hosts. There are macro task buttons to guide the novice through the various menus and two button bars to allow easy access to the menus. For example the bottom bar (left to right), Experimental Parameters, Spin System, Spin Hamiltonian, Instrumental Parameters, Single Crystal Settings, Lineshape Parameters, Transition Labels/Probabilities, File Parameters, Sophe Grid Parameters, Optimisation Parameters, Execution Parameters and Batch Parameters.

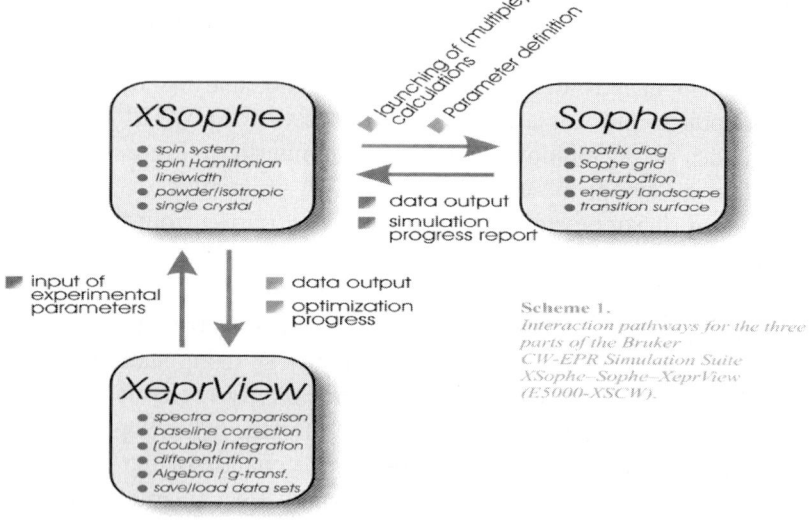

Scheme 1.
Interaction pathways for the three parts of the Bruker CW-EPR Simulation Suite XSophe–Sophe–XeprView (E5000-XSCW).

Scheme 1. A schematic showing the interactivity of XSophe, Sophe and XeprView.

Corba string. Once validated the Sophe authentication daemon forks a Sophe which then performs the simulation.

The output of CW EPR spectra (1D and 2D) from the Sophe program can be visualised in conjunction with the experimental spectrum in XeprView® or Xepr®. Computer simulation of single crystal spectra measured in a plane perpendicular to a rotation axis can be performed by defining the rotation axis and the beginning and end angles in the plane perpendicular to this axis. Energy level diagrams, transition roadmaps and transition surfaces aid interpretation of complicated randomly oriented EPR spectra and can be viewed with a netscape browser and an OpenInventor scene graph viewer.

3. SOPHE

Sophe is a sophisticated computer simulation software programme written in C++ employing a number of innovative technologies including; the SOPHE partition and interpolation schemes, a field segmentation algorithm, the mosaic misorientation line width model, parallelisation (OpenMP - for SGI computers running the Irix operating system) and spectral optimisation. In conjunction with the SOPHE partition scheme and the field segmentation algorithm, the SOPHE interpolation scheme[16,22] and mosaic misorientation linewidth model greatly increase the speed of simulations for most spin systems. For complex spin systems, parallelisation enables the simulation of these systems on a multiprocessor computer and the optimisation algorithms in the suite provide the experimentalist with the possibility of finding the spin Hamiltonian parameters in a systematic manner rather than a trial-and-error process. The functionality of Sophe is described below:[22-24]

Experiments

    Continuous Wave EPR Spectra displayed in XeprView.

    Energy level diagrams, transition surfaces and transition roadmaps displayed in Netscape.

    Homotopy will be available in version 2.x of XSophe.

    Pulsed EPR spectra will be available as an additional component of version 2.x of XSophe.

Spin Systems

    Isolated and magnetically coupled spin systems.

    An unlimited number of electron and nuclear spins is supported with nuclei having multiple isotopes.

Spin Hamiltonian Interactions

    2nd order Fine Structure Interaction, 4th and 6th order corrections [**S.D.S**, B4, B6].[8]

Isotropic and Anisotropic Electron Zeeman [gβ**B.S**, β **B**.g.**S**].
Isotropic and Anisotropic Hyperfine [a**S.I**, **S.A.I**].
Nuclear Zeeman Interaction for nuclei [$g_N \beta_N$ **B.I**].
Quadrupole [**I.P.I**].
Isotropic Exchange [J_{iso} **S**$_i$.**S**$_j$].
Anisotropic Exchange (dipole dipole coupling) [**S**$_i$.**J**.**S**$_j$].

Continuous Wave EPR Spectra

Spectra types:
Solution, randomly orientated and single crystal.
Symmetries:
Isotropic, axial, orthorhombic, monoclinic and triclinic.
Multidimensional spectra:
Variable temperature, multifrequency and the simulation of single crystal spectra in a plane.

Methods

Matrix diagonalization - mosaic misorientation linewidth model.
Sophe Interpolation.

Optimisation (Direct Methods)

Methods:
Hooke and Jeeves.
Quadratic variation of Hooke and Jeeves.
Simplex.
Two Simulated Annealing methods.
Spectral Comparison:
Raw data and Fourier transform.

For nuclear superhyperfine interactions Sophe offers two different approaches; full matrix diagonalization and first order perturbation theory. If all the interactions were to be treated exactly, a Mn(II) (S=5/2, I=5/2) coupled to four ^{14}N nuclei would span an energy matrix of 2,916 by 2,916. To fully diagonalise[25] a Hermitian matrix of this size, it would take some 13 hours on a Silicon Graphics O2 (R5K) workstation, let alone the memory requirement (~68 MB for a single matrix of this size with double precision). In fact, in most systems the electronic spin interacts strongly with one or two nuclei but weakly with other nuclei and the latter approach of first order perturbation may be a satisfactory treatment which will ease the computational burden for large spin systems.

The specification of transition labels is not necessary in Sophe. In the absence of labels a threshold value for the transition probability is required. The program will then perform a search for all transitions which have a transition probability above this threshold value at a range of selected orientations. For a single octant the following orientations (θ,ϕ) are chosen: (0°,0°), (45°,0°), (90°,0°), (45°,90°), (90°,45°), (90°,90°). The transitions found then act as "input" transitions.

The program is designed to simulate CW EPR spectra measured in either the perpendicular ($B_0 \perp B_1$) or parallel ($B_0 \parallel B_1$) modes, where B_0 and B_1 are the steady and oscillating magnetic fields, respectively. It can also easily generate single crystal spectra for any given orientation of B_0 and B_1 with respect to a reference axis system which is normally either the laboratory axis system or the principal axis system of a chosen interaction tensor or matrix in the spin Hamiltonian.

3.1 Theory for the computer simulation of randomly orientated CW EPR spectra

Computer simulation of the experimental randomly orientated or single crystal EPR spectra from isolated or coupled paramagnetic centres is required to accurately determine the spin Hamiltonian parameters [Eqs. 1, 2] and the electronic and geometric structure of the paramagnetic centre. Simulation of randomly orientated EPR spectra is performed in frequency space through the following integration:[1,26]

$$S(B,v_c) = \sum_{i=0}^{N} \sum_{j=i+1}^{N} C \int_{\theta=0}^{\pi} \int_{\phi=0}^{\pi} |\mu_{ij}|^2 f[v_c - v_0(B), \sigma_v] d\cos\theta \, d\phi \quad [3]$$

where $S(B, v_c)$ denotes the spectral intensity, $|\mu_{ij}|^2$ is the transition probability, v_c the microwave frequency, $v_0(B)$ the resonant frequency, σ_v the spectral line width, $f[v_c - v_0(B), \sigma_v]$ a spectral lineshape function which normally takes the form of either Gaussian or Lorentzian, and C a constant which incorporates various experimental parameters. The summation is performed over all the transitions (i, j) contributing to the spectrum and the integrations, performed numerically (Section 3.3), are performed over half of the unit sphere (for ions possessing triclinic symmetry), a consequence of time reversal symmetry.[1,8] For paramagnetic centres exhibiting orthorhombic or monoclinic symmetry, the integrations in Eq. [3] need only be performed over one or two octants respectively. Whilst centres exhibiting axial symmetry only require integration over θ, those possessing cubic symmetry require only a single orientation.

3.2 Field versus frequency swept CW EPR

In practice the CW EPR experiment is a field swept experiment in which the microwave frequency (υ_c) is kept constant and the magnetic varied. Computer simulations performed in field space assume a symmetric lineshape function f in Eq. [3] (f(B-B$_{res}$), σ_B) which must be multiplied by dυ/dB and a constant transition probability across a given resonance.[1,26] Pilbrow has described the limitations of this approach in relation to asymmetric lineshapes observed in high spin Cr(III) spectra and the presence of a distribution of g-values (or g-strain broadening). The following approach has been employed by Pilbrow et al. in implementing Eq. [3] (frequency swept) into computer simulation programmes based on perturbation theory.[1,27] Firstly, at a given orientation of (θ, φ), the resonant field positions (B$_{res}$) are calculated with perturbation theory and then transformed into frequency space (υ_0(B)). Secondly, the lineshape ($f(\upsilon_c-\upsilon_0$(B), σ_υ) and transition probability are calculated in frequency space across a give resonance and the intensity at each frequency stored. Finally, the frequency swept spectrum is transformed back into field space. Performing computer simulations in frequency space produces assymmetric lineshapes (without having to artificially use an asymmetric lineshape function) and secondly, in the presence of large distribution of g-values will correctly reproduce the downfield shifts of resonant field positions.[27]

Unfortunately, the above approach cannot be used in conjunction with matrix diagonalization as an increased number of matrix diagonalizations would be required to calculate f and the transition probability across a particular resonance. However, Homotopy[28,29] which is in general three to five times faster than brute force matrix diagonalization allows the simulations to be performed in frequency space.

3.3 Numerical integration - choice of angular grid

The simulation of a randomly oriented EPR spectrum involves integration over a unit sphere (Eq. [3]) which is perfomed numerically by partitioning a unit sphere and calculating the resonant field positions and transition probabilities at all of the vertex points. The simplest and most popular partition scheme is that of using the geophysical locations on the surface of

5. Computer simulation software XSophe - Sophe - XeprView

the Earth for the presentation of world maps. However, the solid angle subtended by the grid points is uneven and alternative schemes have been invented and used in the simulation of magnetic resonance spectra. For example, in order to reduce computational times involved in numerical integration over the surface of the unit sphere, the igloo,[19] triangular[30] and spiral[31] methods have been invented for numerical investigations of spatial anisotropy. In 1995, we described a new partition scheme, the SOPHE partition scheme[16] in which any portion of the unit sphere ($\theta \in [0, \pi/2]$, $\phi \in [\phi_1, \phi_2]$) or $\theta \in [\pi/2, \pi]$, $\phi \in [\phi_1, \phi_2]$) can be partitioned into triangular convexes. For a single octant ($\theta \in [0, \pi/2]$, $\phi \in [0, \pi/2]$) the triangular convexes can be defined by three sets of curves

$$\theta = \frac{\pi}{2} \frac{i}{N}$$

$$\theta \phi = \frac{\pi}{2} \frac{i-1}{N}(\phi_2 - \phi_1) \quad [4]$$

$$\theta \phi = \theta \; (\phi_2 - \phi_1) - \frac{\pi}{2} \frac{i-1}{N} (\phi_2 - \phi_1), \quad (i = 1,2,......,N)$$

where N is defined as the partition number and gives rise to N+1 values of θ. Similar expressions can be easily obtained for $\theta \in [\pi/2, \pi]$, $\phi \in [\phi_1, \phi_2]$. A three dimensional visualisation of the SOPHE partition scheme is given in Figure 2b.

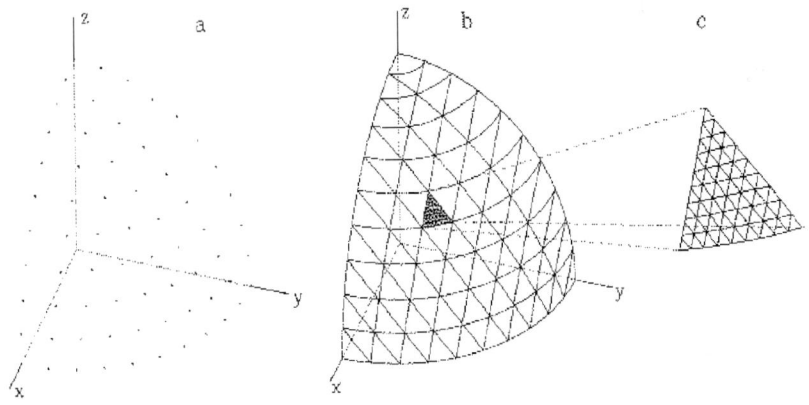

Figure 2. A schematic representation of the SOPHE partition scheme. (a) Vertex points with a SOPHE partition number N = 10; (b) the SOPHE partition grid in which the three sets of curves are described by Eq. [4]. (c) Subpartitioning into smaller triangles can be performed by using either Eq. [4] or alternatively the points along the edge of the triangle are interpolated by the cubic spline interpolation method[30] and each point inside the triangle is linearly interpolated three times and an average is taken.

As can be seen this method partitions the surface of the unit sphere into triangular convexes which resemble the roof of the famous **Sydney Opera House** (SOPHE). In the SOPHE grid there are N curves in each set with the number of grid points varying from 2 to N+1 in steps of 1. In order to produce simulated spectra of high quality, the unit sphere is often required to be finely partitioned, in other words, a large number of vertex points are required to reduce computational noise which is often observed when the spin Hamiltonian parameters are highly anistropic and the linewidths are small. Each triangle in Figure 2b can be easily subpartitioned into smaller triangles, referred to as tiny triangles. In Figure 2c, a selected triangle is further partitioned into 81 tiny triangles with a subpartion number N=10. The grid formed in such a subpartition can still be described by Eq. [4]. In this particular case, θ is stepped in a smaller step of $\pi/(2(N-1)*(M-1))$ from $\theta = 45°$ to $\theta = 54°$, the two corresponding curves which bound the triangle (Figure 2c). A similar process is applied to curves in sets 2 and 3. Alternatively, various interpolation schemes may be used to generate finer grids for simulating randomly orientated EPR spectra.[20,30-32] In 1995 we described a highly efficient interpolation scheme, the SOPHE interpolation scheme (Section 3.4.2).[16]

3.4 Calculation of resonant field positions

The very nature of EPR spectroscopy as a field-swept technique imposes a computational challenge to computer simulation of randomly orientated spectra. In essence, during an EPR experiment, the spin system under investigation is constantly modified through the Zeeman interactions as the magnetic field is swept. In a general situation where two or more interactions have comparable energies, search for resonant field positions is not a trivial task as the dependence of the energies of the spin states on field strength (B_0) can be very complex. The complication involved is best manifested by the presence of multiple transitions between a given pair of energy levels.

3.4.1 Brute force - matrix diagonalization and field segmentation algorithms

A number of search schemes have been used in the full matrix diagonalization approach for locating resonant field positions.[14,20,33-35] Generally, they can be grouped into two categories. In category I, the resonant field position is searched independently for every transition. Among the schemes belonging to this category, the so-called iterative

5. Computer simulation software XSophe - Sophe - XeprView

bisection method is the safest but probably the most inefficient method.[14] Other more efficient methods such as the Newton-Raphson method have also been used.[14] In general, these search schemes are time-consuming as a large number of diagonalizations are normally required. The search schemes belonging to category II may be called segmentation methods. In these schemes, the field sweep range is divided equally into k segments and for each segment, the whole energy matrix is diagonalized once for the centre field value of that segment. Thus only k diagonalizations are performed for each orientation. A perturbation theory is then employed for determining the presence of a transition in each segment. This search scheme is still limited to situations where in each segment there is no more than one possible transition. However, if k is not too small, the chance of having two resonances in a single segment is rare. Reijerse et al.[20] use a first-order perturbation approach for exploring transitions in each segment. However, from our experience, first-order perturbation theory cannot be guaranteed to produce resonant field positions with satisfactory precision. In Sophe we have adopted the second-order eigenfield perturbation theory (described below) originally developed by Belford et al.[36] This method has also been used by other groups.[34]

In Eigenfield perturbation theory[36] the Hamiltonian is first diagonalized at an arbitrary field point X_0. A series of adjustments are then made to the magnetic field strength, the eigenvalues and eigenvectors until the resonant field position can be approximated to sufficient accuracy by B_0.

The initial Hamiltonian $\mathcal{H}_0 = \mathcal{H}_{FI} + X_0 \mathcal{H}_{FD}$ is diagonalized to determine the required pair of eigenvalues ($E_{u,0}$ and $E_{v,0}$) and eigenvectors (U_0 and V_0). The difference between these eigenvalues is called W_0, and the required microwave frequency is given by $W_0 + \delta W$. The magnetic field strength is then adjusted by a factor:

$$X_1 = \frac{\delta W}{A_{vv} - A_{uu}} \quad [5]$$

where $A_{uu} = \langle U_0 | \mathcal{H}_{FD} | U_0 \rangle$ and $A_{vv} = \langle V_0 | \mathcal{H}_{FD} | V_0 \rangle$. The Hamiltonian is adjusted by the term $\mathcal{H}_j = X_j \mathcal{H}_{FD}$, where $\mathcal{H}_{FD}$ is the field dependent Hamiltonian. The equations:

$$\sum_{j=0}^{n} (\mathcal{H}_j - E_{u,j}) U_{n-j} = 0$$

$$(\mathcal{H}_0 - E_{v,0} - W_0)V_n + (\mathcal{H}_1 - E_{v,1} - \mathcal{E}W)V_{n-1}$$
$$+ \sum_{j=2}^{n} (\mathcal{H}_j - E_{v,j})V_{n-j} = 0 \quad [6]$$

are then used to obtain the corresponding change to the eigenvalues and eigenvectors given by $E_{u,n}$ and $E_{v,n}$ and U_n and V_n respectively. The magnetic field strength is then re-adjusted according to:

$$X_n = \sum_{j=1}^{n-1} \frac{X_j A_{vv(n-j)} - A_{uu(n-j)}}{A_{uu} - A_{vv}} \quad [7]$$

where $A_{uu(n-j)} = \langle U_n | \hat{A} | U_{(n-j)} \rangle$. The magnetic field strength, eigenvalues and eigenvectors are iteratively adjusted in this manner until the resonant field position has been determined to sufficient accuracy. The final solutions for the magnetic field strength, eigenvalues (E) and eigenvectors (q) are given by:

$$B_0 = \sum_{k=0}^{n} X_k$$

$$E_i = \sum_{k=0}^{n} E_{u,k} \qquad E_j = \sum_{k=0}^{n} E_{v,k} \quad [8]$$

$$q_i = \sum_{k=0}^{n} U_k \qquad q_j = \sum_{k=0}^{n} V_k$$

This procedure has been incorporated into the Sophe computational programme, where for each orientation over the SOPHE grid the field range is divided into a number of intervals, k. Matrix diagonalization is performed once in each interval, and the above perturbations are then employed to locate the resonant field positions within the interval. The segment number, k, is a user-input parameter. We have found that second-order eigenfield perturbation theory used in conjunction with our segmentation scheme can deal with complicated situations such as multiple transitions and has also proved to be efficient and reliable for locating the resonant field positions in field-swept EPR spectra.

A saving factor in the segmentation method lies in the fact that full matrix diagonalization is only performed k times irrespective of the number of transitions involved. By contrast, in the other schemes, a few diagonalizations are required for each transition and for large spin systems this number can become very large. The precision of the resonant field positions normally depends on the segment number k as well as on the spin

5. *Computer simulation software XSophe - Sophe - XeprView* 209

system. How large the segment number should be depends on the nature of the system under study. However, simulations can be performed with different segmentation numbers providing an easy test of precision.

3.4.2 SOPHE interpolation method

The SOPHE interpolation scheme is divided into two levels of interpolation, a global interpolation using cubic splines[37] and a local interpolation using simple linear interpolation. Given the function values which may represent the resonant field position or the transition probability at the vertex points (Figure 2a), we use the cubic spline interpolation method to interpolate the function values at all other points on the curves described by Eq. [4] (Figure 2b). This is actually carried out in three different sets. In each set, there are N interpolations with the number of knots (vertex points) varying from 2 to N+1. Although in two of the three sets (Eqs. 4b and 4c) both variables θ and ϕ are involved, variable ϕ can be treated as a parameter.[16] First derivative boundary conditions[37] have been employed in our program which has been proved to produce high-quality interpolated data.[16]

After the global interpolation, the integration over the unit sphere can be viewed as integrating through individual triangularly shaped convexes. A second level of interpolation is carried out based on the values globally interpolated and this is schematically shown in Figure 2c. The resonant field position and transition probability are calculated at the vertices (tiny triangles) formed by linear interpolation (up to version 1.0.2 of XSophe) of the points on adjacent sides of the triangular convex. This is repeated for the other two pairs of sides of the triangular convex and the results averaged. Linear interpolation is based on a subpartition scheme and each triangular convex can be subpartitioned differently.[16] Intuitively speaking, the global cubic spline interpolation can be viewed as building up a "skeleton" based on the SOPHE grid and the local linear interpolation can be viewed as a "*tile filling process*". Up until version 1.1 of XSophe we assumed that all of the tiny triangles in a given triangle subtended the same solid angle. In version 1.1 we now calculate the exact areas and use cubic spline interpolation for the *tile filling process*.

The use of the SOPHE interpolation scheme significantly reduces the time-consuming process of locating the resonant field positions and evaluation of the transition probabilities in the brute force matrix diagonalization (Section 3.4.4). The disadvantage of the SOPHE interpolation scheme is that it will fail when there are multiple resonant field positions present at a given orientation (θ, φ) or when looping transitions are present. The mosaic misorientation line width model (Section 3.4.3)[38] or homotopy (Section 3.5)[28,29] are two alternative approaches to solving these problems which have been implemented into Sophe.

3.4.3 Mosaic Misorientation[38]

When the line width is small compared to the anisotropy of the system it is necessary to integrate over a large number of orientations to avoid simulation noise in the simulated spectrum. This increases the computational time considerably. The SOPHE grid and Interpolation schemes were developed as a way to overcome this problem. A new approach based on the mosaic misorientation linewidth model[39] has recently been developed. In the mosaic misorientation model a Gaussian distribution of molecular geometry axes about an average crystal c-axis is assumed. In the current implementation the partial derivatives of the eigenvalues with respect to a rotation about the x and y axes are calculated using first order perturbation theory (Figure 3).

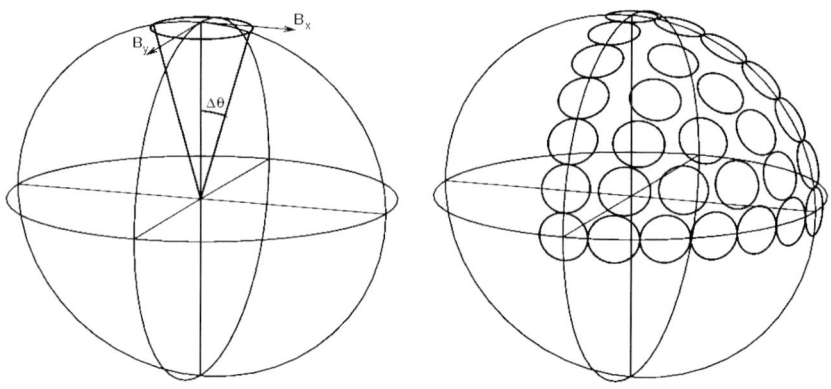

Figure 3. Mosaic misorientation model.

The contribution of a Gaussian distribution, of half-width $\Delta\theta$, to the linewidth can then be calculated with the following equation.

$$\sigma_m^2 = (\Delta\theta \frac{\partial E_{ij}}{\partial \theta_x})^2 + (\Delta\theta \frac{\partial E_{ij}}{\partial \theta_y})^2, \quad E_{ij} = E_i - E_j \quad [9]$$

In the simulation of powder spectra each point in the SOPHE grid is considered to be a microcrystallite with a Gaussian distribution of orientations, $\Delta\theta$ (Figure 3a), such that there is an overlap between adjacent grid points (Figure 3b).

$$\Delta\theta = \frac{\pi}{4(N-1)} \qquad [10]$$

At turning points in the spectrum the partial derivatives of the eigenvalues and hence σ_m are zero and the linewidths are determined by other contributions. At other orientations where the resonant field varies strongly with orientation the line widths will be broadened (smoothed). This model simulates an EPR spectrum where the important features, the turning points, are resolved in a significantly reduced time. Increasing N, the number of bands in the SOPHE grid, will lead to a convergence to the 'true' spectrum. For a large number of spin systems, N can be set to 20. The mosaic misorientation linewidth model can be contrasted to interpolation schemes by considering it an extrapolation method.

3.4.4 A Comparison of Brute Force Matrix Diagonalization, SOPHE Interpolation and Mosaic Misorientation

An example demonstrating the efficiency of the SOPHE partition and interpolation schemes is shown in Figure 4c where we have calculated a randomly orientated spectrum for a high spin rhombically distorted Cr(III) ion for which an appropriate spin Hamiltonian is:

$$\mathcal{H} = g_e B \cdot S + D[S_z^2 - \frac{1}{3}S(S+1)] + E(S_x^2 - S_y^2) \\ + S \cdot A \cdot I - g_n \beta B \cdot I \qquad [11]$$

The spin Hamiltonian parameters employed were $g_e = 1.990$, $D = 0.10$ (cm^{-1}), $E/D = 0.25$, $g_n = 1.50$, $A_x = 120$, $A_y = 120$, $A_z = 240$ (10^{-4} cm^{-1}). A narrow line width was chosen (30 MHz) in order to demonstrate the high efficiency of these schemes. The unit sphere has to be partitioned very finely in order to produce simulated spectra with high signal-to-noise ratios when there is large anisotropy and the spectral linewidths are narrow. The simulated spectra employing brute force matrix diagonalization with N=18 and N=400 are shown in Figures 4a and 4b repectively. Including the SOPHE interpolation scheme with a partition number N=18 (Figure 4c) dramatically improves the signal to noise ratio with a considerable reduction in computational time. Application of the mosaic misorientation line width model (Figure 4d) also dramatically improves the signal to noise ratio and is computationally faster than the SOPHE interpolation method as interpolation has many overheads. Clearly the mosaic misorientation linewidth model is the fastest approach and now replaces the brute force matrix diagonalization approach within the computational programme Sophe and is also preferred over SOPHE interpolation.

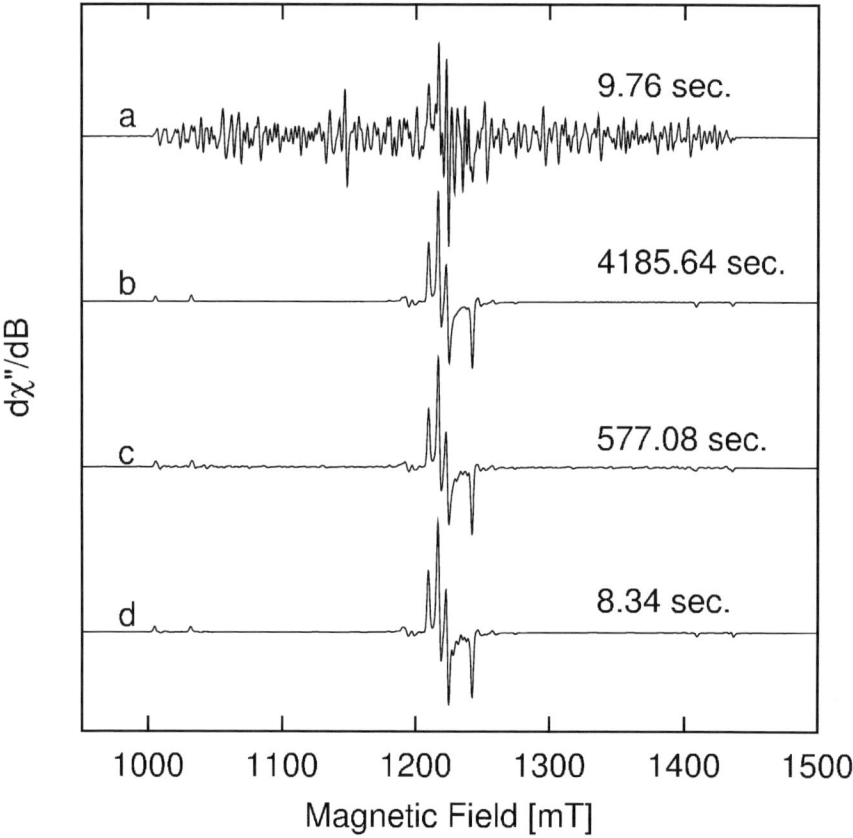

Figure 4. Computer simulations of the powder EPR spectrum from a fictitious spin system (S=3/2; I=3/2) which demonstrates the efficiency of the SOPHE interpolation scheme. (a) Without the SOPHE interpolation scheme, N=18, (b) Without the SOPHE interpolation scheme, N=400, (c) With the SOPHE interpolation scheme, N=18 and (d) With the mosaic misorientation linewidth model, N=18. The computational times were obtained on a SGI O2 R5K (180 MHz). υ=34 GHz; field axis resolution: 4096 points; an isotropic Gaussian lineshape with a half width at half maximum of 30 MHz was used in the simulation.

3.5 Homotopy segmentation algorithm

Methods such as the brute force technique employ full matrix diagonalization to calculate the complete set of eigenvalues and eigenvectors as a function of orientation. However the simulation process generally makes use of only the pair of eigenvalues and eigenvectors directly involved in the current transition. The computational time involved in determining these eigenpairs can be improved through the use of alternative numerical methods such as Homotopy.[28,29] Our group has developed a number of simulation techniques based on the Homotopy algorithm (the Methods of Unresolved Edges[29] and Unmatched Segments,[29] and the Homotopy Segmentation Method[29]).

The Homotopy Segmentation method determines the resonant field positions at each orientation independently (in a similar manner to the Eigenfield method). This is achieved by generating a set of splines which approximate the transition level energies as a function of field (Figure 5, Section 3.5).

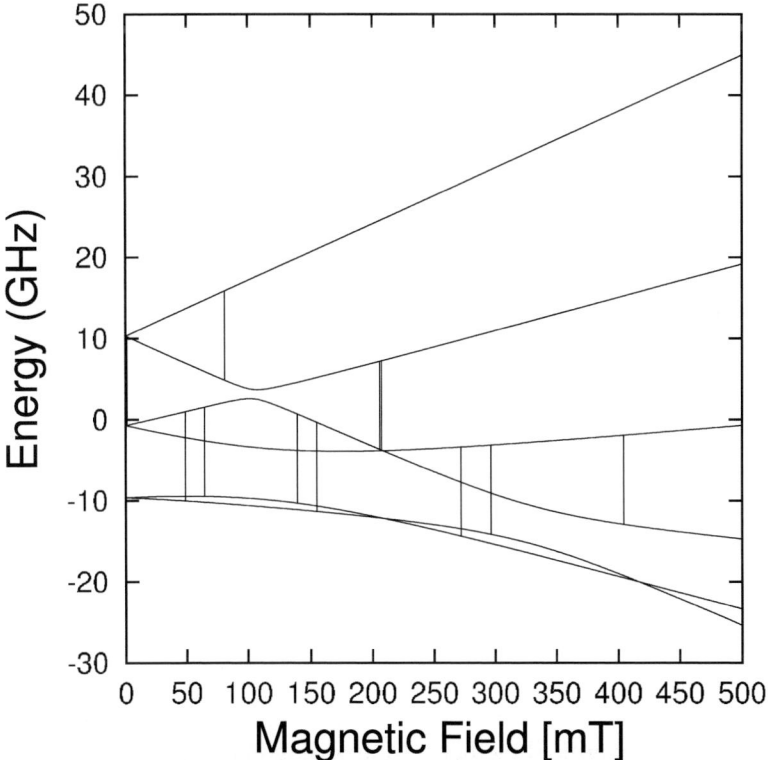

Figure 5. Energy level diagram showing the EPR transtions for a high spin Fe(III) centre with D=0.1 cm^{-1}, E/D = 0.25, g=1.99, θ= 0°, φ=0° and υ= 11GHz.

These energy levels are then used to define the energy difference :

$$F_{ij} = (E_i - E_j) - h\nu \qquad [12]$$

The resonant field positions (*i.e.* where $F_{i,j}$ equals zero) and the turning points of $F_{i,j}$ with respect to B can be determined directly from these splines. EPR spectra can be simulated through either evaluating the resonant field positions at a given set of orientations exactly, or by evaluating the resonant fields at a small number of orientations and then interpolating over the transition surfaces (Figure 6) to produce an approximation for the resonant field positions at the remaining set of orientations. If interpolation is being employed, then it is necessary to connect the resonant field positions from adjacent orientations into a fully-connected surface. This is considered in Section 3.5.2. Each resonant field position irrespective of whether it was determined exactly or is an approximation then contributes a transition lineshape to the final EPR spectrum.

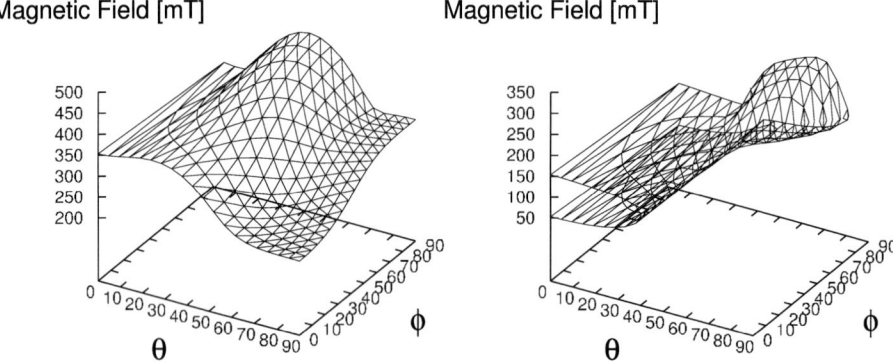

Figure 6. Transition surfaces for the high spin Fe(III) centre described in Figure 5, $\upsilon = 9$ GHz. (a) between levels 3 and 4 and (b) between levels 3 and 5.

3.5.1   Locating the resonant field positions at a single orientation

As with the Eigenfield method, the Homotopy Segmentation method determines the resonant field positions at each orientation independently. The field range over the spectrum is divided into an distribution of points (where the number of points required is significantly smaller than that required by the Eigenfield method). Matrix diagonalization is then employed to compute all of the eigenvalues at each field point. To improve

5. Computer simulation software XSophe - Sophe - XeprView

computational time, a variant of matrix diagonalization is employed at this stage which only computes the complete set of eigenvalues. A set of splines can then be produced which model each eigenvalue as a function of the magnetic field strength. The resonant field positions at the given orientation are predicted by the points where the distance between any two splines matches the microwave frequency.

The initial set of splines were defined from an even distribution of field points. Homotopy is then employed to determine additional eigenvalues in the regions of predicted resonant field positions. These new field points are added to the splines until each of the resonant field positions has been determined to sufficient accuracy.

No eigenvectors have so far been determined at the current orientation. The transition probability and linewidth are a function of the two eigenvectors contributing to each resonant field position. The required eigenvectors, and the resulting transition probability and linewidth, are determined at three points on the spline in the vicinity of each resonant field position. A quadratic function is then produced which describes these values as a function of field, and is used to approximate the transition probability and linewidth at the resonant field position.

This procedure determines all of the resonant field positions present at a single orientation. The resonant field positions are located to a required level of accuracy (as required by the numerical integration of the resonant field position as a function of orientation). If interpolation is employed within the simulation or the surfaces are to be visualised, then adjacent resonant fields need to be matched and the turns traced. These additional procedures employ the turning points of the F_{ij} functions (ie. the local maxima and minima of F_{ij} as a function of magnetic field, see Section 3.5.2). These turning points are not used in the numerical integration, and hence no new points are added to the splines to improve the accuracy of the turning points.

3.5.2 Comparing adjacent orientations

In general, adjacent orientations will contain the same number of resonant field positions and turning points, in which case these points can be matched directly. This however is not the case if:

- a resonant field position or turning point move so they are inside the field range at one orientation, and outside the field range at another,

- two resonant field positions merge, causing a turn in the transition surface,
- or if two turning points merge and cancel each other out.

These phenomena only occur at a few select orientations, and it is unlikely that more than one of these phenomena will occur between any two grid points.

When there are the same number of resonant field positions at two adjacent orientations then these resonant field positions are simply matched up one to one. If the number of resonant fields changes by one between adjacent orientations, then the resonant field positions can still be simply matched one to one. It is assumed that the new resonant field was produced from a field moving from outside the field range at one orientation to inside the range at the other orientation, and so either the first or last resonant field at a given orientation will have no neighbour. The slope of F_{ij} at each resonant field will either be positive or negative, and will be the same at two adjacent points. Hence by matching the sign of the slope at the endpoints, the extra field point can be determined. If there are two extra points at one orientation, the field points might still be matched one to one. The two points might both have moved in from the ends of the field range. This is assumed to have occurred if the sign of the slope does not match at either end of the field range.

When the resonant field positions cannot be matched one to one (such as when there is a turn in the transition surface), the turning points are considered. In most cases there will be the same number of turning points at adjacent orientations, and between any two turning points there will be at most one resonant field position. Hence if two turning points are matched and they both contain a resonant field position between them, then the two resonant field positions can also be matched. If two turning points are matched, but there is a resonant field position at only one orientation, then a turn in the surface has been encountered. In this case the turning point at one orientation will be positive, the turning point at the other orientation will be negative, and there will be a turn in the surface between these two turning points. Turning points might have moved in from the ends of the field range, in a similar way to the resonant field positions. The turning points may still be matched up if there are one or two more turning points between adjacent orientations.

A turn in the surface is encountered if the turning point at one orientation is positive, and the turning point at an adjacent orientation is negative. The surface will turn at an intermediate orientation where the value of F (B,θ) at the turning point is zero. Homotopy is used to trace a line between the two turning points until the turn is resolved to sufficient accuracy.

In highly complex regions of the transition surfaces, the number of resonant field positions and turning points may vary significantly between adjacent orientations in the numerical grid. In these regions it is no longer possible to simply match the orientations point for point, and hence the above procedures are employed to determine the complete set of resonant field positions and turning points at intermediate orientations between these points.

3.6 Linewidth models

A number of linewidth models originally developed for magnetically isolated paramagnetic species have been incorporated into the XSophe computer simulation software suite. For all the linewidth models discussed below the linewidth parameter, σ_v, is given in energy units. In Sophe (field space version), σ_v is converted to a field-domain linewidth parameter σ_B through $\sigma_B = |dB/dE_{ij}| \sigma_v$ (where B is the magnetic field and $E_{ij} = E_i - E_j$).[1,26] $|dB/dE_{ij}|$ is calculated for each transition by using eigenfield perturbation theory.[36] The linewidth models incorporated into Sophe include:

- Kivelson's linewidth model[40] for isotropic spectra

$$\sigma_v = \alpha + \beta M_I + \gamma M_I^2 + \delta M_I^3 \qquad [13]$$

The coefficients α, β, γ and δ can be related to the solvent viscosity, correlation time, molecular hydrodynamics radius and the anisotropy of the spin system under study.[40]

- Angular variation of the g-values.[1]

$$\sigma_v^2 = (\sigma_x^2 g_x^2 l_x^2 + \sigma_y^2 g_y^2 l_y^2 + \sigma_z^2 g_z^2 l_z^2)/g^2 \qquad [14]$$

where $g^2 = g_x^2 l_x^2 + g_y^2 l_y^2 + g_z^2 l_z^2$, σ_i's (i=x,y,z) are the input linewidth parameters and l_i's (i=x,y,z) are the direction cosines of the magnetic field with respect to the principal axes of the g matrix.

- A correlated distribution of g and A values.

A correlated distribution of g and A values also termed g and A strain was originally developed by Froncisz and Hyde[41] and has been used successfully to account for the linewidth variations encountered in spin S=1/2 systems particularly in copper and low spin cobalt (S=1/2) complexes.[1,3,41] When expressed in the frequency-domain,[1,22] the linewidth in this model is based on the formulae

$$\sigma_v^2 = (\sum_{i=x,y,z} \{\sigma_{R_i}^2 + [\frac{c\, g_i}{g_i} v_0(B) + \sigma A_i M_I]^2 \} g_i^2\, l_i^2) / g^2 \quad [15]$$

where the σR_i (i =x, y, z) are the residual linewidths due to unresolved metal and/or ligand hyperfine splitting, homogeneous linewidth broadening, and other sources, σg_i's and σA_i's are the half widths of the Gaussian distributions of the g and A values. The g-A strain model involves nine parameters for a rhombically distorted metal ion site.

- A statistical distribution of D and E.

Wenzel and Kim[42] have described a statistical D-E strain model. In their model, the distributions of D and E are assumed to be Gaussian and independent of each other with the resulting full width at maximum slope due to strain alone given by

$$\sigma_{DE}^2 = \sigma_D^2 \{<\psi_i|S_z^2|\psi_i> - <\psi_j|S_z^2|\psi_j>\}^2$$
$$+ \sigma_E^2 \{<\psi_i|S_x^2 - S_y^2|\psi_i> - <\psi_j|S_x^2 - S_y^2|\psi_j>\}^2 \quad [16]$$

where σ_D and σ_E are the half-widths at maximum slope of the distributions of D and E in energy units, respectively, and ψ_i and ψ_j are the wavefunctions associated with transition i j. A residual linewidth, σ_R, is convoluted with the D-E strain effects

$$\sigma_v^2 = \sigma_{DE}^2 + \sigma_R^2 \quad [17]$$

3.7 Parallelisation

With the advent of multiprocessor computers and the new algorithms described above the simulation of EPR spectra from complex spin systems consisting of multiple electron and or nuclear spins becomes feasible with Sophe. Optimisation of the spin Hamiltonian parameters by the computer will also be possible for these spin systems. Parallelisation of the matrix diagonalization method has been performed at the level of the vertices in the Sophe grid. For example, if a computer has 5 processors then the number of Sophe grid points is divided into groups of five and each group is then processed by one of the processors with the resultant spectra being added to an array shared by the five processors. For the hypothetical Cr(III) spin system shown in Figure 4 a three-fold reduction in computational time is observed. Greater reductions are observed for more complex spin systems.

3.8 Optimisation algorithms

A unique set of spin Hamiltonian parameters for an experimental EPR spectrum is obtained through minimising the goodness of fit parameter (GF)

$$GF = (\sum_{}^{N} (Y_{exp} - S(B,v_c) * \alpha)^2)^{1/2} / (N * \sigma) \quad [18]$$

where the experimental spectrum (Y_{exp}) has been baseline corrected assuming a linear baseline and the simulated spectrum has been scaled by α to Y_{exp}. N is the number of points in common between the experimental and simulated spectra and σ is the magnitude of noise in the spectrum. In the past minimising GF has been performed through a process of trial-and-error by visually comparing the simulated and experimental spectra until a close match was found. Recent progress in reducing computational times for computer simulations (Section 3.2-3.7) and the improved speed of workstations allows the use of computer-based optimisation procedures to find the correct set of spin Hamiltonian parameters from a given EPR spectrum.

The most appropriate technique for optimising a set of spin Hamiltonian parameters is nonlinear least squares.[43] This method has the advantage that the differences (Y_{exp} - $S(B, v_c)$, Eq. [18]) associated with the more extreme positive or negative values are exaggerated, which emphasises genuine peak mis-matching whilst tending to reduce the impact of noise. Unfortunately, evaluation of $S(B, v_c)$ can take a long time and as there is no analytic derivative information available, this method is not really an option for general spin systems. Direct search methods are characterised by evaluating the function at several points within the spin Hamiltonian parameter space (H_p-space)[44] and then using the knowledge gained during the last few evaluations in an attempt to choose a more promising point. These methods have lost popularity over time, and have been largely superseded by methods using derivative information, although they are still used in places where noise is prevalent. They suffer only two drawbacks: the algorithms are particularly susceptible to becoming trapped in local minima; and they tend to be fairly inefficient in their use of function evaluations (at least in comparison with derivative-based methods on functions where derivatives are available). Three of these direct methods were considered particularly promising, the Hooke and Jeeve's,[45] Simplex[46] and a Quadratic method based on the Hooke and Jeeves method. We have also implemented two simulated annealing approaches[47] in Sophe[22].

In addition there are several problems which need to be addressed, including (i) the sensitivity of scaling the various spin Hamiltonian parameters and the method chosen for comparing the experimental simulated spectra.[22] In XSophe we allow the user to control the sensitivity of parameter adjustment throughout the optimisation procedure and secondly the user can compare the spectra directly or the Fourier transformed spectra. The latter method provides increased resolution through separating the high and low frequency components[48] As an aid to optimising the computer simulation XSophe/XeprView has the capability of displaying intermediate spectra (Magnetic Field vs. Intensity vs. Iteration Number) and the corresponding spin Hamiltonian parameters. Ideally, you would like to optimise a set of multifrequency EPR spectra with a single set of spin Hamiltonian parameters. Although this can be achieved in the current version of Xsophe, the methodology is not straight forward. Version 2.0 of XSophe currently under development will allow this and the optimisation of multi-component spectra.

3.8.1 Hooke and Jeeves method

The Hooke and Jeeve's method[45] was first publicised in 1961 and has been widely used for some time. The method starts with an "exploratory" move. Each parameter is considered in turn, and checks are made to see how changes made to the parameter affect the error. A fixed step is made in the positive direction, and it is observed whether the error increases or decreases. If no improvement is made, then a fixed step is made in the negative direction. Again it is observed whether the error is improved. Based upon which fixed steps improved the error (but ignoring how much the error changes), a direction for "pattern" moves is then determined. Fixed steps are then made along the "pattern" direction until no further progress is made, at which stage another exploratory move is made.[44]

3.8.2 Quadratic method

One of the problems observed while using the Hooke and Jeeve's method is that fixed steps are made simply based on whether the error is increasing or decreasing, but ignoring how much each parameter affects the error. During the "exploratory" move, three points are typically evaluated with respect to each parameter. A quadratic can hence be drawn through these three points, and a prediction can be made for the value of each parameter which minimises the error. Hence a quadratic method was designed which is similar to the Hooke and Jeeve's method, but instead of using fixed steps makes steps based upon a quadratic with respect to each parameter.

3.8.3 Simplex method

The Spendley, Hext and Himsworth (or Simplex) method[46] has a very geometrical interpretation. It works by taking an $H_p + 1$ vertex simplex (ie. a regular $H_p + 1$ sided figure in H_p-space) and evaluating GF at each of the vertices. It then takes the vertex point with the largest value of GF (subject to a few restrictions) and replaces it with its reflection in the hyper-plane formed by the other H_p points.

The elegant nature of this algorithm is apparent in considering the case where $H_p=2$. The problem can now be visualized by a three dimensional surface (ie. GF vs. the two H_p parameters) whose lowest point (valley/well) we aim to find. The simplex in this case is an equilateral triangle which 'rests' upon the three dimensional surface. The algorithm, simply dictates that the highest of the three points will lift over the other two, causing the triangle to 'flip' over (so the highest point now points down hill). This process is repeated, and the triangle 'flips' its way down to bottom of a nearby minimum. If the triangle cannot flip to a new minimum, the size of the triangle is contracted and termination takes place when the number of contractions reaches a defined limit.

The simplex method has two advantages over the Hooke and Jeeve's method. The first is its use of H_p+1 evaluations of GF in determining the next point at which to evaluate $S(B, \upsilon_c)$ compared with the Hooke and Jeeve's method which uses at most two points (and then only during the pattern step, it normally uses only one). The second advantage is that the simplex method is somewhat more resilient to local minima. If one of the points is very close and if the simplex is still large enough, the simplex may well 'flip out' of the well. Unfortunately, in some circumstances, the simplex method can fail. To take a two-dimensional case, it is susceptible to becoming trapped in a deep, narrow 'valley' if one of the points lands very close to the bottom and the other two points are on different walls. Having reached this stage, no progress can be made except along the valley (the sides are too steep to allow any climbing). If the point near the bottom is on the down-hill side of the valley, then no movement can be made as the triangle cannot flip over a point, only a line. Consequently, improvement is only possible with contractions occurring about the lowest point which will result in early termination in a local minimum. A possible way to overcome this difficulty, is to combine the standard simplex algorithm described above with a stochastic process (such as simulated annealing). This method would hopefully 'break-out' of such minima. A brief discussion of how this could be implemented is given after the description on simulated annealing.

3.8.4 Simulated Annealing

Simulated annealing is one of a class of methods known as Monte-Carlo methods.[47] The simplest Monte-Carlo method starts at a point $\mathcal{H}_p^0$ and makes a random step of length l in $\mathcal{H}_p$ space to the new point $\mathcal{H}_p^{'0}$. If GF ($\mathcal{H}_p^{'0}$) is less than $\mathcal{H}_p^0$ then the step is accepted and $\mathcal{H}_p^1$ is set to $\mathcal{H}_p^{'0}$ at which point the process is repeated by taking a step from $\mathcal{H}_p^1$. However if GF increases, then the step is rejected and another step is taken from $\mathcal{H}_p^0$. This process gradually forms a sequence of $\mathcal{H}_p^s$ with decreasing values of GF. The algorithm terminates with answer $\mathcal{H}_p^k$ when no more progress is being made. At this point the algorithm may be restarted from this answer with a decreased step size.

The concept of simulated annealing comes (somewhat loosely) from the annealing (solidifying) of liquids (especially molten metals). The molecules in their liquid state have a high kinetic energy which allows (indeed, forces) them to change their location and orientation. By cooling, their energy is depleted and so they gradually 'settle down' into what eventually becomes their fixed location. The structure which results differs significantly depending upon the rate of cooling.

Simulated annealing attempts to emulate the slow cooling process by using one very significant change from the standard Monte-Carlo method.[47] Rather than rejecting all function-increasing steps, it accepts the 'uphill' kth step with probability P($\mathcal{H}_p^k$, T) where P satisfies $\lim_k$ (P=0). The value T is known as the temperature of the system and decreases as k increases in accordance with a cooling schedule. The hope is that the acceptance of a detrimental step will have long-term benefits by allowing the sequence of $\mathcal{H}_p^s$ to 'escape' from a local minimum.

5. Computer simulation software XSophe - Sophe - XeprView 223

Two approaches for choosing a random point have been implemented into Sophe. In the first method a random jump is performed by choosing a random point on a hypershere of radius r about the current point. Both the temperature and the radius are decreased periodically throughout the search. The benefit of this approach is that all $\mathcal{H}_p$ dimensions are searched simultaneously. In the second approach, each of the $\mathcal{H}_p$ axes are searched in turn.

3.8.5 Parameter and Spectral Scaling

An important part of practical multi-variable optimisation is the sensitivity of scaling the various spin Hamiltonian parameters. Since the Hooke and Jeeves and Simplex algorithms take finite steps along various co-ordinate axes, there must be some uniformity in the rate of change in GF in each of these directions. Since a change of δ in one of the g values would have significantly less effect on $S(B, \upsilon_c)$ than a step of δ in one of the A values, then, without scaling, for any given step size, only some of the variables will be usefully optimised. To solve this problem, the $\mathcal{H}_p$-space formed by the parameters to be varied is transformed to one where a given step length will have a more uniform effect on GF in each of the $\mathcal{H}_p$ directions. Given the necessity of avoiding costly evaluations of $S(B, \upsilon_c)$ and derivative information this can be done by performing a linear transformation by multiplying each of the spin Hamiltonian parameters by h_i. The h_i's are manually chosen to keep fluctuations in GF to approximately the same order of magnitude. Parameter scaling is not required in the quadratic and simulated annealing approaches as the step size is dynamically adjusted.

The simulated spectrum is multiplied by α (Eq. [18]) so that it can be directly compared to the experimental spectrum, Y_{exp}. In Sophe we have three methods for scaling, double integration (for derivative EPR spectra), peak to peak extrema and optimisation of the scaling factor.

3.8.6 Spectral Comparison

Many complex randomly orientated EPR spectra contain overlapping resonances with complex lineshapes. Optimisation of such spectra often leads to the computer broadening (increase in linewidth) of one of the resonances to reduce the error. Although we do not profess to have the ultimate solution to this problem, we have employed two approaches to help solve this problem. The first approach involves the ability to control the sensitivity of parameter adjustment throughout the optimisation procedure, whilst the second involves the comparison of the Fourier transformed experimental and simulated spectra. The latter method provides increased resolution through separating the high and low frequency components.[48] As an aid to optimising the computer simulation, XSophe/Xepr has the capability of displaying intermediate spectra (Magnetic Field vs. Intensity vs. Iteration Number) and the corresponding spin Hamiltonian parameters.

3.9 Visual aids for analysing complex CW EPR spectra

XSophe provides three tools (Energy level Diagrams, Transition Roadmaps and Transition Surfaces) for aiding the analysis of complex CW EPR spectra. XSophe outputs a web page (html) describing the spin system and various experimental parameters in conjunction with Portable Network Graphics (png) and postscript (ps) files which can be viewed with a Netscape (4.7x) browser and Ghostview or an another postscript viewer. Both the Portable Network Graphics (png) and postscript (ps) files can be incorporated into OpenOffice.org, StarOffice, Microsoft Office or other wordprocessing and graphics programmes.

3.9.1 Energy level diagrams

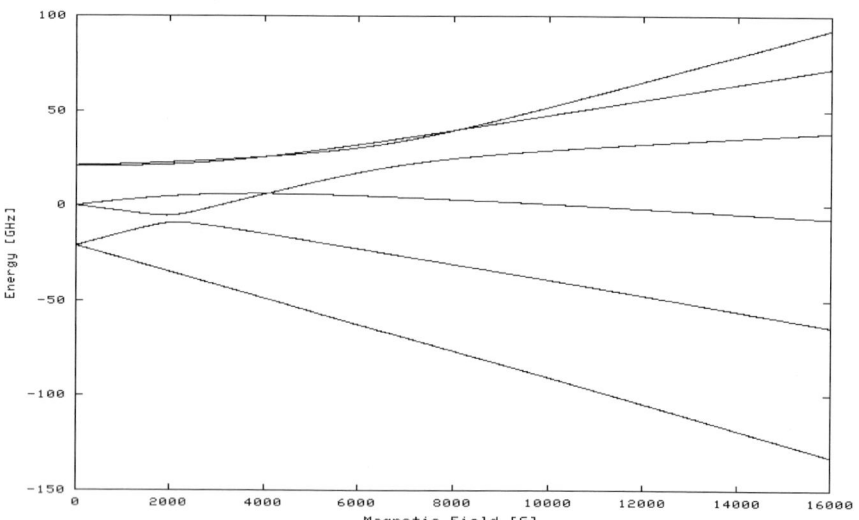

Figure 7. An Energy level Diagram for a hypothetical high spin Fe(III), S=5/2 spin system (D= 6 GHz, E/D=1/3, g=2, ν_c=9.75 GHz, N=1, Field Steps=100, T= 4K) produced by XSophe using the default Eenergy level default.sph data file.

5. Computer simulation software XSophe - Sophe - XeprView

The Sophe programme employs matrix diagonalization to calculate energy level diagrams and EPR transitions along the x, y and z principal directions. An example of such a calculation is shown in Figure 7.

3.9.2 Transition roadmaps

The Sophe programme employs matrix diagonalization to calculate transition roadmaps which can be viewed using a netscape web browser (Figure 8).

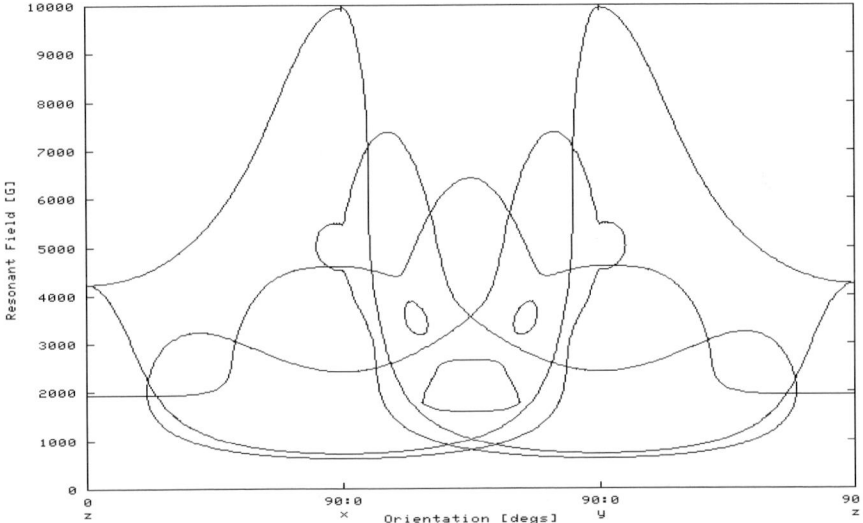

Figure 8. A Transition Roadmap for a hypothetical high spin Fe(III), S=5/2 spin system (D= 4.5 GHz, E/D=1/3, g=2, $\chi, \rho, \tau = 90°$, υ_c=9.75 GHz, N=100, Field Steps=100, T= 4K) produced from XSophe using the default TransitionRoadmap_default.sph data file.

3.9.3 Transition surfaces

The Sophe programme employs matrix diagonalization to calculate transition surfaces producing an output spectrum file (html) and an OpenInventor (iv) file which can be viewed using an OpenInventor viewer on both SGI (Irix) and PC (Linux) computers (Figure 9, 10). On SGI computers running Irix a netscape web browser (v 4.7x) in conjunction with an OpenInventor plugin is used to display the scene graph (Figure 9). Whilst on PC's running Linux a combination of a netscape web browser and ivview (Figure 10) is used to display the scene graph.[24] The scene graphs are plotted using cartesian coordinates, which is extremely useful for highly anisotropic spin

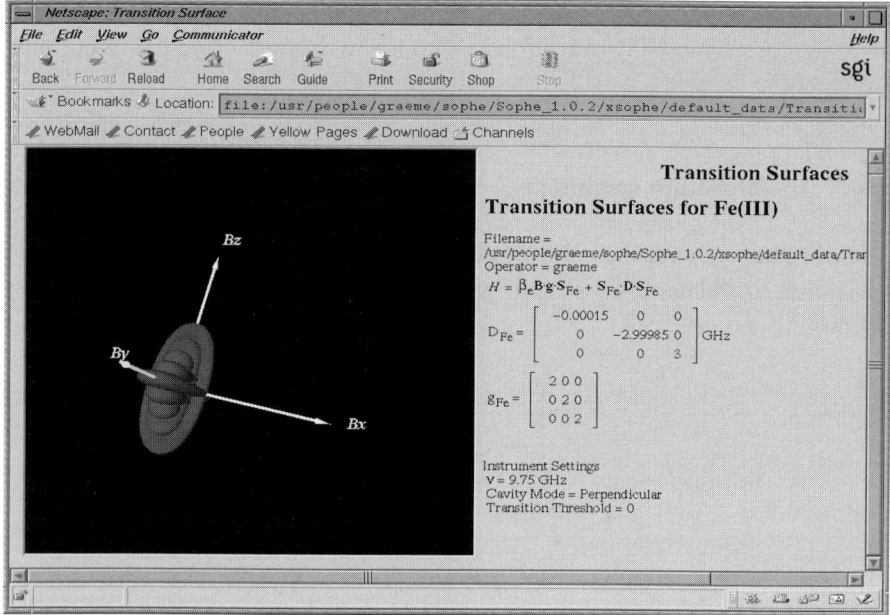

Figure 9. A Transition Surface for a hypothetical high spin Fe(III), S=5/2 spin system (D= 4.5 GHz, E/D=1/3, g=2, υ_c=9.75 GHz, N=1, Field Steps=30, T= 4K) produced from XSophe using the default TransitionSurface_default.sph data file. Results are displayed within a netscape browser (v 4.7x) on an SGI computer running IRIX.

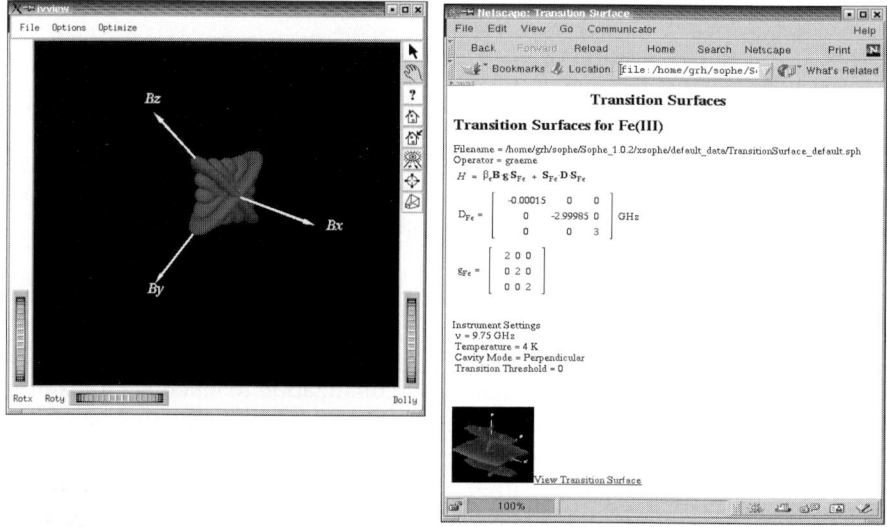

Figure 10. A Transition Surface for a hypothetical high spin Fe(III), S=5/2 spin system (D= 4.5 GHz, E/D=1/3, g=2, υ_c=9.75 GHz, N=1, Field Steps=30, T= 4K) produced from XSophe using the default TransitionSurface_default.sph data file. Results are displayed within a netscape browser and the OpenInventor scene graph viewer ivview on computers running Linux.

systems such as high spin systems. This requires the minimum field to be set to zero which is defined as the origin. In contrast, for nearly isotropic systems, it is important to examine a single transition at a time. The choice of transitions can be made through the Transition Labels/Probabilities window and either setting the transition threshold or defining the transitions.

In high spin systems looping transitions can often be observed and these are seen as bubbles in the surface (Figure 11).

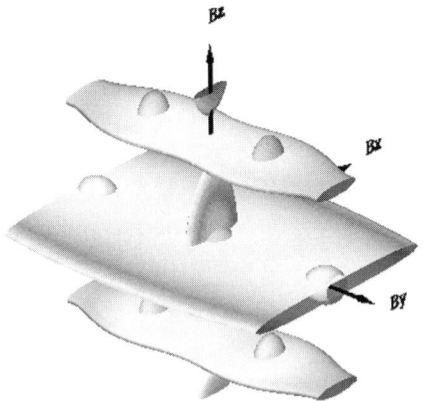

Figure 11. A transition surface for a high spin Fe(III) spin centre found in sweet potato purple acid phosphatase (D= -0.88 cm^{-1}, E/D=0.1925, g=2, υ_c=9.5784 GHz, N=1, Field Steps=100, T= 4K) revealing looping transitions.[49]

4. ROLE OF FREQUENCY (AND TEMPERATURE) IN EXTRACTING SPIN HAMILTONIAN PARAMETERS

Often X-band randomly orientated CW EPR spectra are complex consisting of many overlapping resonances and the only means of extracting a unique set of spin Hamiltonian parameters is to perform a multifrequency CW EPR experiment. The most common microwave frequencies employed for such experiments are listed in Table 1.

The use of microwave frequencies higher than X-band can improve spectral resolution (g-value resolution, Section 4.1) and simplify spectra by eliminating state mixing, energy level crossings and anticrossings and looping transitions (Section 4.4). However, if there is a large distribution of parameters (g, A and D) as is the case in many Jahn Teller distorted copper(II) complexes and low symmetry high spin Fe(III) centres, increasing the microwave frequency can lead to significant broadening of the resonances which can cancel out the gain in spectral resolution (Section 4.5).

Table 1. List of the most common microwave frequency bands for CW EPR experiments.

Microwave Band	Microwave Frequencies	Common Frequency
L	1 - 2 GHz	~ 1 GHz
S	2 - 4 GHz	~ 4 GHz
C	4 - 8 GHz	6 GHz
X	8 - 12 GHz	9.2 - 9.7 GHz
K	18 - 26 GHz	~ 24 GHz
Q	26 - 40 GHz	~ 35 GHz
W	75 - 110 GHz	~ 95 GHz
D	110-170 GHz	

4.1 Electron Zeeman and Hyperfine Interactions

The resonant magnetic field positions for an S=1/2, I=0 axially symmetric spin system are given by:

$$B_{\|} = \frac{h\nu_c}{\beta g_{\|}} \qquad B_{\perp} = \frac{h\nu_c}{\beta g_{\perp}} \qquad [19]$$

The magnetic field separation (resolution) between these two resonances is:

$$B_{\|} - B_{\perp} = \frac{h\nu_c}{\beta(g_{\|} - g_{\perp})} = K\nu_c \qquad [20]$$

Clearly, increasing the microwave frequency increases the g value resolution. High frequencies such as Q- and W-band have been exploited to resolve multiple species present in an equilibrium and g- and A anisotropy (Figure 12).

Another application of high frequency EPR is to separate field independent resonances from those which are field dependent. Often the

5. Computer simulation software XSophe - Sophe - XeprView

preparation and purification of magneticaly coupled species (eg. Binuclear copper(II) complexes) results in the presence of a small amount (<5%) of monomeric (S=1/2) impurity and at X-band frequencies the allowed $\Delta M_S = \pm 1$ resonances overlap with those from the dipole-dipole coupled species. Increasing the microwave frequency to at least 35 GHz allows the resolution of the spectra from both species.

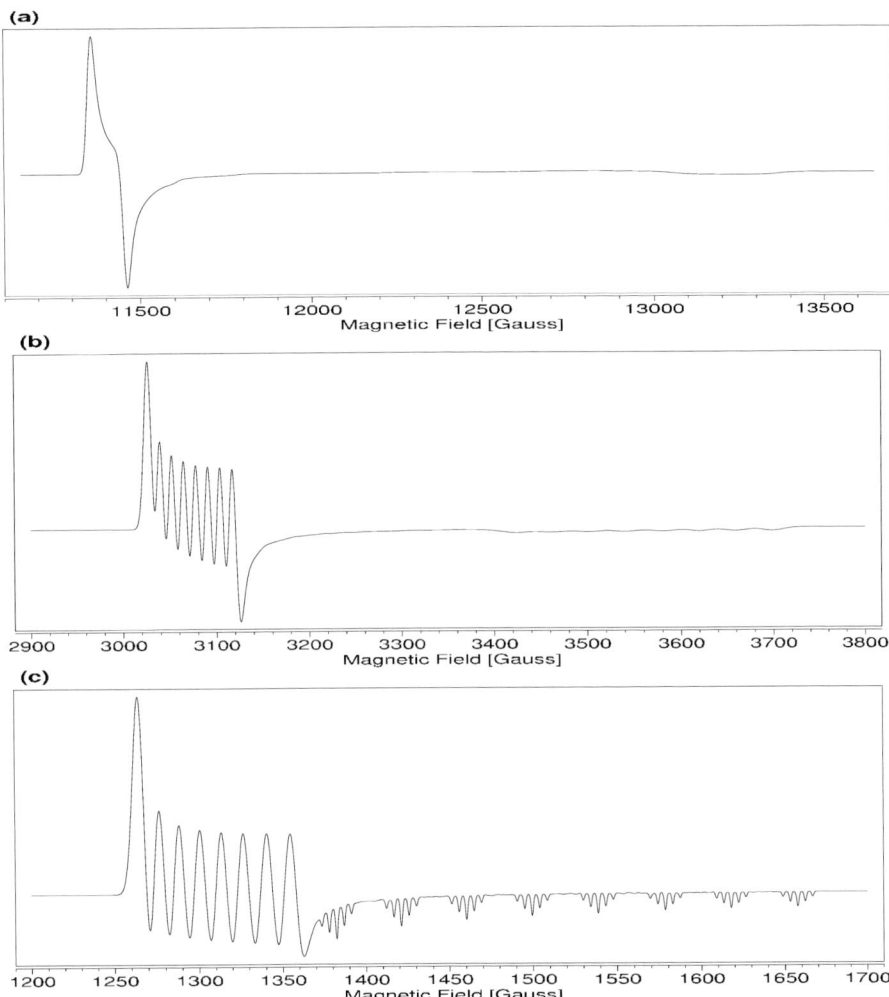

Figure 12. Multifrequency computer simulation of a CW EPR spectra from a low spin Co(II) centre ligated by two magnetically equivalent nitrogen atoms. Spin Hamiltonian parameters are: $g_\parallel$, 1.925; $g_\perp$, 2.230; $A_\parallel$(Co), 106.0 x10^{-4}cm^{-1}; $A_\perp$(Co), 40.0x10^{-4} cm^{-1}; $A_\parallel$(N), x 10^{-4} cm^{-1}; $A_\perp$(N), x 10^{-4}cm^{-1}. (a) υ_c = 35.5962 GHz, (b) υ_c = 9.5962 GHz, (c) υ_c = 4.0962 GHz.

4.2 The fine structure interaction

For many high spin Fe(III) centres found in biological systems, the axial zero field splitting (D) is greater than the X-band microwave quantum (0.3 cm^{-1}) and thus transitions are only observed within Karamers' doublets rather than between them. Consequently, the only method for determining D is to either increase the microwave frequency or perform a variable temperature measurement. In either case computer simulation is required to extract the value of D. An example of a variable tempearture computer simulation for a high spin Fe(III) centre is shown in Figure 13 with the temperauture dependence (slices along the temperature axis at a specific magnetic field) shown in Figures 13b,c for both negative and positive values of D respectively.

Metalloproteins containing high spin Ni(II) (S=1), Mn(III) (S=2) and Fe(II) (S=2) often appear to be EPR silent at X-band frequencies as there is insufficient energy to produce a transition between energy levels. Sometimes, the Fe(II) spin systems contain a significant distribution of zero field splittings which allows the observation of $\Delta M_S=\pm 4$ forbidden transitions whose lineshapes are quite 'strange'.

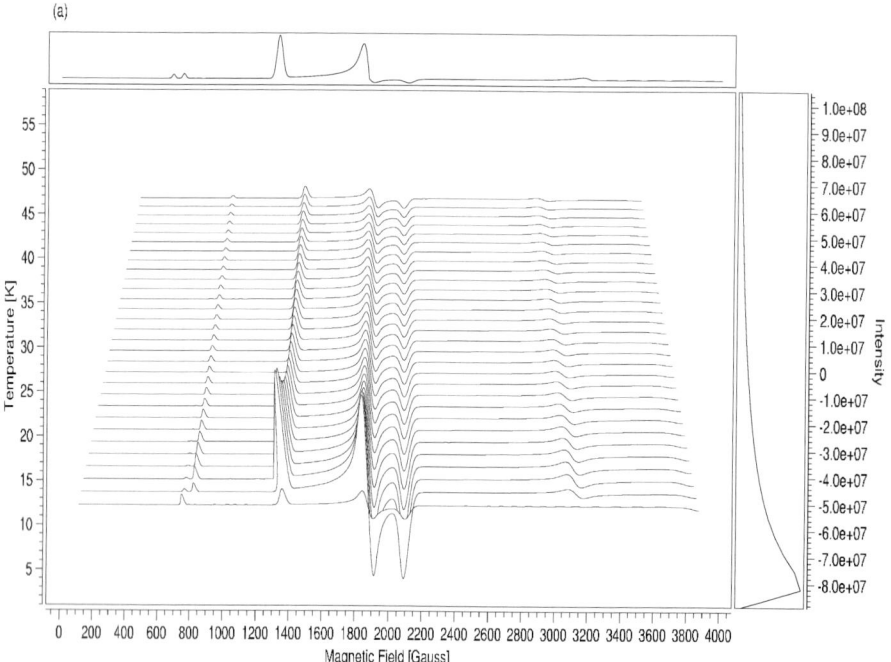

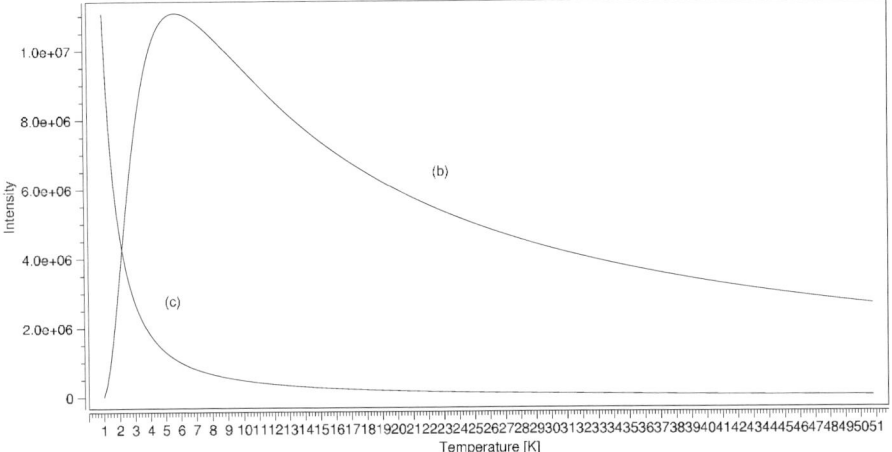

Figure 13. Variable temperature computer simulation of the Fe(III) centre found in sweet potato purple acid phosphatase, ν_c=9.5784 GHz. (a,b) Spin Hamiltonian parameters are D= -0.88 cm^{-1}, E/D=0.1925 and g=2, (c) Spin Hamiltonian parameters are D= 0.88 cm^{-1}, E/D=0.1925 and g=2, (b,c) Temperature dependence of the resonance at B= 748.1 Gauss.

4.3 Isotropic and anisotropic exchange interactions

Variable temperature EPR spectroscopy is often employed to extract the isotropic exchange coupling constant, J_{iso}. An example of an antiferromagnetically coupled spin system is a nitroxide biradical [50] which can be described by the following spin Hamiltonian:

$$\mathcal{H} = J_{iso} S_1 \cdot S_2 + \sum_{i=1}^{2} (g \beta B \cdot S_i + a S_i \cdot I_i[N] - g_n \cdot \beta_n \cdot B \cdot I_i[N]) \quad [21]$$

where g=2.00585, a=13.5 x 10^{-4} cm^{-1} and J_{iso} = 40 cm^{-1}. A variable temperature simulation (Figure 14) of the experimental variable temperature spectrum can be used to extract the isotropic exchange coupling constant. However, if J is large, then a combination of higher frequencies and/or higher temperatures is required to extract the exchange coupling constant.

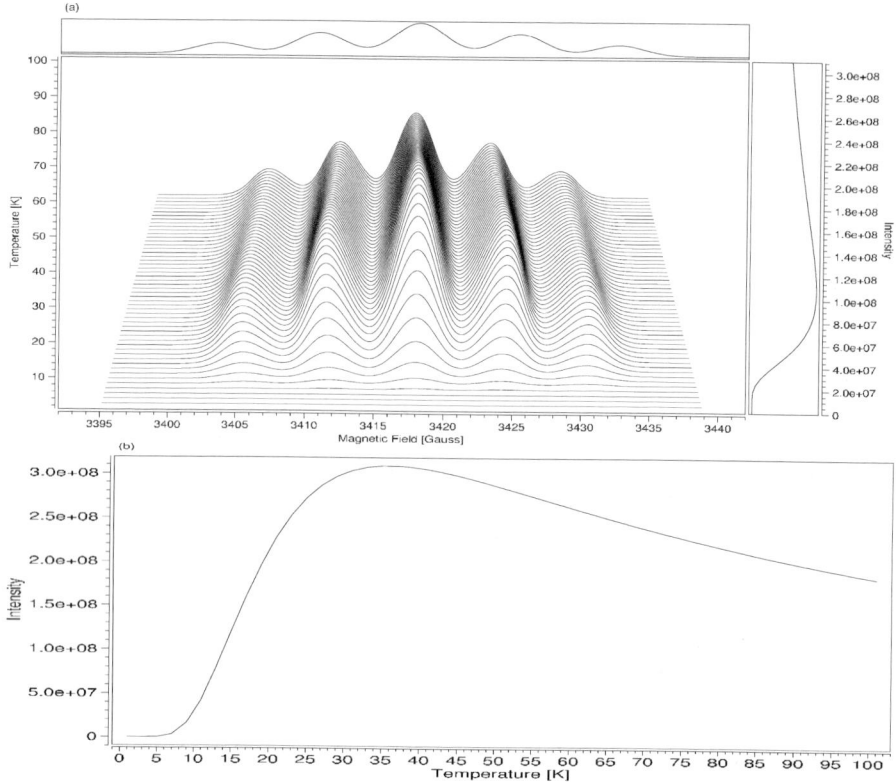

Figure 14. Variable temperature computer simulation of the nitroxide biradical. EPR parameters are given in the text. υ_c = 9.5962 GHz. The lower spectrum is a slice along the temperature axis at B= 3418 Gauss.

4.4 State mixing and the existence of energy level crossings, anticrossings and looping transitions

State Mixing occurs when one or more of the spin Hamiltonian interactions are of similar magnitude. This leads to energy level crossings and anticrossings. When the Nuclear Hyperfine and Quadrupole Interactions are of a similar magnitude state mixing can lead to the observation of formally forbidden $\Delta M_I = \pm$ 2 and 3 transitions. Computer simulation of these forbidden resonances allows the accurate determination of the quadrupolar coupling constants. At very low frequencies the Nuclear Hyperfine and Electron Zeeman Interactions become comparable leading to state mixing and complex spectra resulting in the apparent unequal spacings (A/gβ) between the hyperfine resonances. Conversely, state mixing can also occur for high spin systems when the zero field splitting or exchange interactions are of a similar magnitude to the electron Zeeman interaction. This not only leads to energy level crossings and anticrossings but also looping transitions (Figure 15 a,b). Increasing the microwave frequency (magnetic field) to Q- and W-band frequencies (Figure 15 c,d) leads to simpler spectra where the Electron Zeeman interaction dominates the Fine Structure interaction.

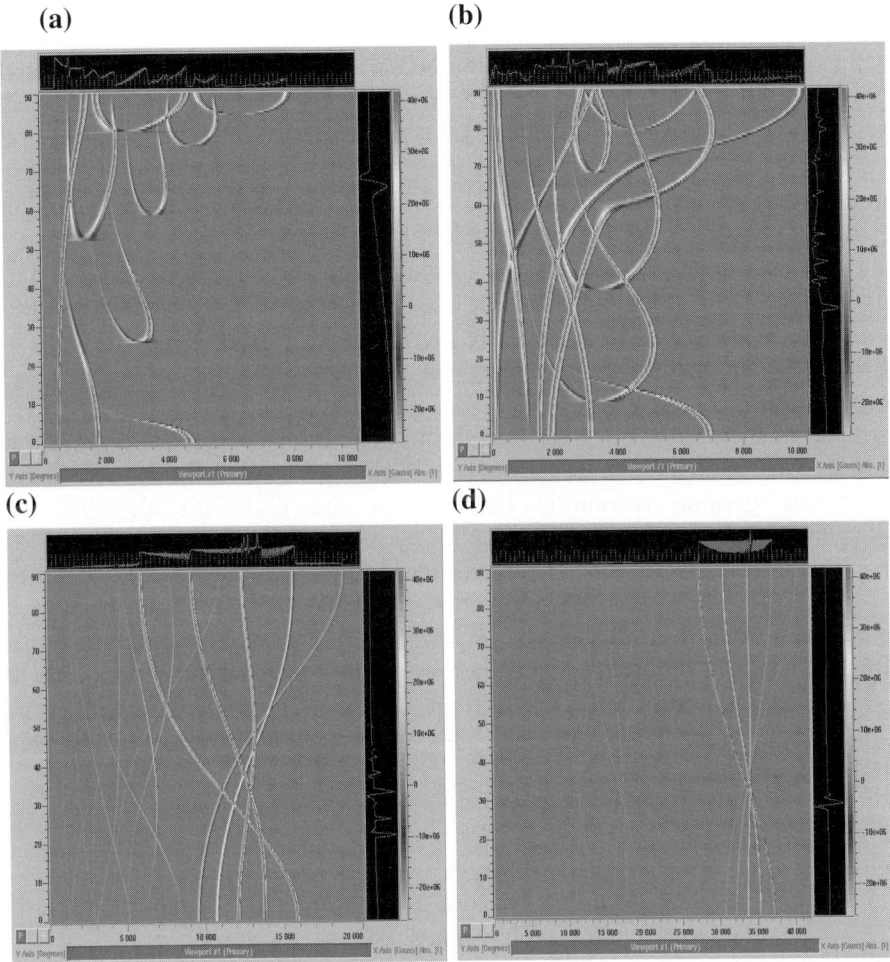

Figure 15. Single Crystal EPR spectra of a high spin Fe(III) spin system with D= - 4.5 GHz, E/D =0, and g = 2.0. (a) S-band spectrum, ν_c= 4.0 GHz, (b) X-band spectrum, ν_c= 9.75 GHz, (c) Q-band spectrum, ν_c= 35.0 GHz, (d) W-band spectrum, ν_c= 95.0 GHz.

State mixing also allows the observation of ΔM_S= ± 2 transitions in dipole-dipole coupled systems, for example binuclear copper(II) complexes. Analysis of CW EPR spectra containing these resonances allows the distance and orientation of the two copper(II) centres to be determined. The transition probability for these these resonances is inversely proportional to frequency and consequently their intenisty is maximised at lower microwave frequencies, such as S-band.

4.5 Distributions of parameters

Often metal centres in metalloproteins, transition metal complexes and glasses exist as a distribution of structures which can arise from interactions with solvents, poorly formed glasses in frozen solution samples and completely different coordination geometries/symmetries. These distributions of structures lead to distributions of various spin Hamiltonian parameters causing frequency dependent linewidth broadening. A correlated *g-A* strain model, originally developed by Froncisz and Hyde[41] and expressed in the frequency-domain (Eqn. [15]) has been incorporated into XSophe (Section 3.6) to account for the linewidth variations encountered in $S=1/2$ spin systems particularly in copper, low spin cobalt and molybdenum(V) complexes.[1,3,41,51] As the frequency is increased g value resolution increases (Eq. [20]) but the linewidth also increases (Figure 12, Eq. [20]). Conversely, lowering the frquency from X-band to S-band decreases g-value resolution but reduces the linewidth allowing the observation of superhyperfine coupling to ligating nuclei (Figure 12c). S-band EPR spectroscopy has been exploited in identifying ligating nuclei to copper(II) ions in Type II copper proteins,[3] oxomolybdenum(V) complexes[51] and oxomolybdenum proteins[52]. Importantly, the angular variation of this linewidth model (Eqn. [15]) does not accurately reproduce the linewidth of angular anomalies or off axis extrema. We are currently extending the approach developed by Wenzel and Kim[42] (Section 3.6) to all spin Hamiltonian parameters.

5. FUTURE DIRECTIONS FOR XSOPHE

The following areas are being actively pursued within the XSophe group to enhance the functionality of the XSophe-Sophe-XeprView software suite:
- Homotopy and the incorportaion of the Homotopy Segmentation algorithm,
- Improved linewidth models for incorporating distributions of parameters,
- Improved optimisation algorithms for extracting spin Hamiltonian parameters from a combination of CW and pulsed EPR spectra,
- Optimisation of multicomponent spectra,
- Pulsed EPR algorithms and

- An improved X windows interface.

6. CONCLUSIONS

The XSophe-Sophe-XeprView computer simulation software suite provides scientists with an easy-to-use research tool for determining the spin Hamiltonian parameters from isotropic, randomly orientated and single crystal CW EPR spectra. These spin Hamiltonian parameters can then be used to determine the electronic and geometric structure of the paramagnetic centre.

The output of one- and two-dimensional CW EPR spectra from the Sophe programme can be visualised in conjunction with the experimental spectrum in XeprView® or Xepr®. Computer simulation of single crystal spectra measured in a plane perpendicular to a rotation axis can be performed by defining the rotation axis and the beginning and end angles in the plane perpendicular to this axis. Energy level diagrams, transition roadmaps and transition surfaces aid interpretation of complicated randomly orientated EPR spectra and can be viewed with a netscape browser and an OpenInventor scene graph viewer for transition surfaces. The XSophe-Sophe-XeprView computer simulation software suite is available through Bruker biospin (http://www.bruker-biospin.de).

Acknowledgments *We would like to thank the Australian Research Council and the EPR Division of Bruker Biospin for financial support and past members of the Sophe group, including Andrae Muys and Dr. Deming Wang.*

7. REFERENCES

1. J. R. Pilbrow, "*Transition Ion Electron Paramagnetic Resonance*", Clarendon Press, Oxford, 1990.
2. F. E. Mabbs, D. C. Collison, "*Electron Paramagnetic Resonance of Transition Metal Compounds*", Elsevier, Amsterdam, 1992.
3. R. Basosi, W.E. Antholine, J.S. Hyde, "*Multifrequency ESR of Copper Biophysical Applications*" in "*Biological Magnetic Resonance*", L.J. Berliner, J. Reuben Eds., Vol. **13**, New York, Plenum Press, 1993.
4. G.R. Hanson, A.A. Brunette, A.C. McDonell, K.S. Murray, A.G. Wedd, *J. Amer. Chem. Soc.*, **103** (1981) 1953.
5. Y.S. Lebedev, *Appl. Magn. Reson.*, **7** (1994) 339.

6   L.C. Brunel, *Appl. Magn. Reson.*, **11** (1996) 417.
7   E.J. Reijerse, P.J. vanDam, A.A.K. Klaassen, W.R. Hagen, P.J.M. vanBentum, G.M. Smith, *Appl. Magn. Reson.*, **14** (1998) 153.
8   A. Abragam, B. Bleaney, B., *"Electron Paramagnetic Resonance of Transition Ions."* Clarendon Press, Oxford, 1970.
9   A. Bencini, D. Gatteschi, *"EPR of Exchange Coupled Systems"*, Springer-Verlag, Berlin, 1990.
10 T. D. Smith, J. R. Pilbrow, *Coord. Chem. Rev.*, **13** (1974) 173.
11 P. C. Taylor, J. F. Baugher, H. M. Kriz, *Chem. Rev.*, **75** (1975) 203.
12 (a) J. D. Swalen, H. M. Gladney, *IBM J. Res. Dev.*, **8** (1964) 515.
    (b) J. D. Swalen, T. R. L. Lusebrink, D. Ziessow, *Magn. Reson. Rev.*, **2** (1973) 165.
13 H. L. Vancamp, A. H. Heiss, *Magn. Reson. Rev.*, **7** (1981) 1.
14 B.J. Gaffney, H.J. Silverstone, *"Simulation of the EMR Spectra of the High-Spin iron in proteins"* in *"Biological Magnetic Resonance"*, (L.J. Berliner, J. Reuben Eds.) Vol. **13**, New York, Plenum Press, 1993.
15 S. Brumby, *J. Magn. Reson.*, **39** (1980) 1, *ibid* **40** (1980) 157.
16 D. Wang, G.R. Hanson, *J. Magn. Reson.* A, **117** (1995) 1.
17 D. Wang, G.R. Hanson, *Appl. Magn. Reson.*, **11** (1996) 401.
18 K.E. Gates, M. Griffin, G.R. Hanson, K. Burrage, *J. Magn. Reson.*, **135** (1998) 104.
19 (a) R. L. Belford, M. J. Nilges, *EPR Symposium 21st Rocky Mountain Conference*, Denver, Colorado, 1979.
    (b) A. M. Maurice. PhD thesis, University of Illinois, Urbana, Illinois, 1980.
    (c) M. J. Nilges, PhD thesis, University of Illinois, Urbana, Illinois, 1979.
20 M. C. M. Gribnau, J. L. C. van Tits, E. J. Reijerse, *J. Magn. Reson.*, **90** (1990) 474.
21 A.Kretter, J. Huttermann, *J. Magn. Reson.* **93** (1991) 12.
22 M. Griffin, A. Muys, C. Noble, D. Wang, C. Eldershaw, K.E. Gates, K. Burrage, G.R. Hanson, *Mol. Phys. Rep.*, **26** (1999) 60.
23 M. Heichel, P. Höfer, A. Kamlowski, M. Griffin, A. Muys, C. Noble, D. Wang, G.R. Hanson, C. Eldershaw, K.E. Gates, K. Burrage, *Bruker Report*, **148** (2000), 6.
24 XSophe Release Notes, 1.1.1(2002).
25 E. Anderson, Z. Bai, C. Bischof, J. Demmel, J. Dongarra, J. Du Croz, A. Greenbaum, S. Hammarling, A. McKenney, S. Ostrouchov, D. Sorensen, *"LAPACK Users' Guide"*, SIAM, Philadelphia, 1992.
26 J. R. Pilbrow, *J. Magn. Reson.*, **58** (1984) 186.
27 (a) G. R. Sinclair, PhD thesis, Monash University, 1988.
    (b) J.R. Pilbrow, G.R. Sinclair, D.R. Hutton, G.J. Troup, *J. Mag. Reson.*, **52** (1983) 386.
28 K.E. Gates, M. Griffin, G.R. Hanson, K. Burrage, *J. Magn. Reson.*, **135** (1998) 104.
29 M. Griffin, PhD thesis, The University of Queensland, 2002.
30 D. W. Alderman, M. S. Solum and D. M. Grant, *J. Chem. Phys.*, **84** (1986) 3717.
31 M. J. Mombourquette, J. A. Weil, *J. Magn. Reson.*, **99** (1992) 37.
32 G. van Veen, *J. Magn. Reson.*, **38** (1978) 91.
33 D. Nettar, N.I. Villafranca, *J. Magn. Reson.*, **64** (1985) 61.
34 M.I. Scullane, L.K. White, N.D. Chasteen, *J. Magn. Reson.*, **47** (1982) 383.
35 D.G. McGavin, M.J. Mombourquette, J.A. Weil, *"EPR ENDOR User's manual"*, University of Saskatchewan, Saskatchewan, Canada, 1993.
36 G.G. Belford, R.L. Belford, J.F. Burkhalter, *J. Magn. Reson.*, **11** (1973) 251.
37 B.-Q. Su, D.-Y. Liu, *"Computational geometry - Curves and Surface Modelling"*, Academic Press, Singapore, 1989.
38 C.J. Noble, G.R. Hanson, K.E. Gates, Unpublished Results.

39 D. Shaltiel, W. Low, *Phys. Rev.* **124** (1961) 1062.
40 (a) D. Kivelson, *J. Chem. Phys.* **33** (1960) 1094.
    (b) R. Wilson, D. Kivelson, *J. Chem. Phys.* **44** (1966) 154, *ibid* **44** (1966) 4440.
    (c) P.W. Atkins, D. Kivelson, *J. Chem. Phys.* **44** (1966) 169.
41 (a) W. Froncisz, J.S. Hyde, *J. Chem. Phys.* **73** (1980) 3123.
    (b) J.S. Hyde, W. Froncisz, *Ann. Rev. Biophys. Bioeng.* **11** (1982) 391.
42 R.F. Wenzel, Y.W. Kim, *Phys. Rev.* **140** (1965) 1592.
43 (a) S.K. Misra, *J. Mag. Reson.*, **23** (1976) 403.
    (b) S.K., Misra, *Mag. Reson. Rev.*, **10** (1986) 285.
    (c) S.K., Misra, *Physica*, **121B** (1983) 193.
    (d) S.K., Misra, S. Subramanian, *J. Physics* C, **15** (1982) 7199.
44 Spin Hamiltonian parameters are constrained to a portion of P-space as this will prevent the generation of a NULL spectrum.
45 R. Hooke, T.A. Jeeves, *J. Assoc. Computing Machinery*, **8** (1961) 212.
46 W. Spendley, G.R. Hext, F.R. Himsworth, *Technometrics*, **4** (1962) 441.
47 (a) D.M. Nicholson, A. Chowdhary, L. Schwartz, *Physical Review* B, **29** (1984) 1633.
    (b) I.O. Bohachevsky, M.E. Johnson, L.S. Myron, *Technometrics*, **28** (1986) 209.
    (c) A. Corana, M. Marchesi, C. Martini, S. Ridella, *ACM Trans. Math. Software*, **13** (1987) 262.
    (d) H. Heynderickx, H. De Raedt, D. Schoemaker, *J. Magn. Reson.*, **70** (1986) 134.
48 R. Basosi, G. Della Lunga, R. Pogni, *Appl. Magn. Reson.*, **11** (1996) 437.
49 G. Schenk, C.L. Boutchard, L.E. Carrington, C.J. Noble, B. Moubaraki, B., J. de Jersey, G.R. Hanson, S. Hamilton, *J. Biol. Chem.*, **276** (2001) 19084.
50 A. McCaleif, S. Bottle, G.R. Hanson, Unpublished Results.
51 G.R. Hanson, G.R. Wilson, T.D. Bailey, J.R. Pilbrow, A.G. Wedd, *J. Amer. Chem. Soc.*, **109** (1987) 2609.
52 (a) G.L. Wilson, R.J. Greenwood, J.R. pilbrow, J.T. Spence, A.G. Wedd, *J. Amer. chem. Soc.*, **113** (1991) 6803.
    (b) R.J. Greenwood, G.L. Wilson, J.R. pilbrow, A.G. Wedd, *J. Amer. chem. Soc.*, **115** (1993) 5385.

Chapter 6

THE CALCULATION OF THE HYPERFINE COUPLING TENSORS OF BIOLOGICAL RADICALS

Fuqiang Ban,[†] James W. Gauld,[‡] Stacey D. Wetmore[§] and Russell J. Boyd[†]
[†]Department of Chemistry, Dalhousie University, Halifax, Nova Scotia, Canada B3H 4J3
[‡]Department of Chemistry & Biochemistry, University of Windsor, Windsor, Ontario, Canada N9B 3P4
[§]Department of Chemistry, Mount Allison University, Sackville, New Brunswick, Canada E4L 1G8

Key words: Hyperfine coupling tensors, Density functional theory, Amino acid radicals.

Abstract: This chapter reviews the performance of density functional theory for the calculation of hyperfine coupling tensors and the development of computational methodology for biological radicals. Remarkable advances in density functional theory (DFT) have made it possible to study biological radicals as straightforwardly as closed-shell molecules. The conformational and charge dependence of the observed hyperfine coupling tensors can therefore be explored by employing reliable computational procedures. Although, hyperfine coupling tensors are highly dependent on the structure of the radical, temperature and environmental effects are typically smaller than 10%. As a consequence, hyperfine coupling tensor calculations on isolated biological radicals (at zero kelvin) closely reproduce the hyperfine coupling tensors of most radicals observed in condensed phases (at higher temperatures), and are appropriate for the purpose of theoretical assignment and interpretation. Often, precise identification of unknown radicals can be made by systematic comparison of observed experimental values with hyperfine coupling tensors calculated for all possible conformations of a radical and its protonated or deprotonated derivatives. The aim of this chapter is to review the ability of DFT methods to account for the hyperfine coupling tensors of a variety of biological radicals.

1. INTRODUCTION

DNA and protein oxidation is associated with various pathological conditions such as cancer, Alzheimer's, and many other diseases; thus, the study of the relevant biological radicals has become one of the most exciting topics in molecular biology.[1-4] It is also noteworthy that biological radicals play important roles in enzymatic catalysis[1] and in the mechanisms of many toxins[5] and hypoxia-selective antitumor drugs.[6-8] ESR spectroscopy and related techniques are powerful tools for investigating such radical species.[9] The key quantities measured by electron spin resonance spectroscopy are the g-tensors and the hyperfine coupling tensors, from which valuable information about the identity, the chemical environment and the spin density distribution of a radical can be deduced.

A general feature of radical species is their high reactivity and, therefore, short life-times. Indeed, many radicals live only long enough to be characterized by ESR spectroscopy at very low temperatures. Fundamental ESR experiments have concentrated on the identification of the products generated upon irradiation of model biological systems such as amino acids[10] or the components of DNA.[11] Due to the fact that experiments can often obtain the hyperfine coupling tensors for only some of the magnetic nuclei (such as protons) in biological radicals, ambiguous assignments of the observed hyperfine coupling tensors may occur. In addition, the conformational complexity of protein radicals and DNA radicals creates further difficulties for interpreting their ESR spectra. However, the characterization of new radicals may be assisted by comparisons of observed and computed hyperfine coupling tensors. Indeed, the increasing popularity of ESR techniques in probing the functionality of biological radicals *in vivo* demands reliable theoretical predictions on the relationship between hyperfine coupling tensors and molecular structure.

ESR experiments provide only indirect geometrical and electronic information. A more complete description of the molecular and electronic structures, and therefore the full spin density distribution and the associated hyperfine coupling tensors, is often provided by theory. In recent years, remarkable progress has been made in the calculations of hyperfine coupling tensors.[12-20ab] While accurate hyperfine coupling constants can be predicted by the highest levels of electron correlation theory (such as QCI, MRCI and CC methods) for small radical species,[21] comparable results can often be obtained from the methods based on density functional theory, DFT (such as B3LYP and PWP86).[14,16,18,22-24] Since the computational cost and memory requirements of DFT methods are considerably less than those of

6. Calculation of hyperfine coupling tensors

conventional correlated *ab initio* procedures, the number of basis functions, and hence atoms, is not nearly as limiting a factor at the DFT level. Furthermore, because of their ability to treat larger systems, DFT methods may be used to obtain more realistic descriptions of the interactions between radical systems and their surroundings by explicit consideration of the latter.[25-27ab] In addition to earlier assessments of DFT methods in combination with various basis sets, the performance of the SVWN, BLYP, B3LYP, BP86 and B3P86 methods and the Pople basis sets have recently been evaluated on the basis of gas-phase calculations performed on small radicals by Gauld *et al.*[18] Furthermore, the strategies for including environmental effects using current density functional theory methods have been reviewed by Barone and coworkers.[20a] In summary, DFT methods are now a serious alternative to conventional *ab initio* techniques for the computation of hyperfine coupling tensors.

Calculated hyperfine coupling tensors have been utilized as a probe for understanding the molecular structures and other properties of biological radicals. The hyperfine coupling tensors of peroxyl radicals,[22] the phenoxyl radical,[26] quinone radicals,[27ab] the radical cation and anion of chlorophyll,[27c] the bacteriopheophytin *a* radical anion,[27d] the β-carotene radical cation,[28] amino acid radicals,[29] radicals derived from various DNA components[30a-e] and radicals involved in B_{12}-dependent reactions[30fg] have been extensively studied at various levels of density functional theory. The results have shown that DFT methods work well, particularly for large organic radicals.

In this chapter, the relative performance of various levels of theory for the calculation of hyperfine coupling tensors is briefly reviewed. Methods based on density functional theory are highlighted, and the appropriateness of these techniques for investigating the hyperfine coupling tensors of biological radicals is discussed. Practical computational procedures of hyperfine coupling tensor calculations for biological radicals are illustrated using examples from our recent work on amino acid-derived radicals.

2. THEORETICAL BACKGROUND

The hyperfine splittings displayed in an electron spin resonance (ESR) spectrum are the fingerprints for radical species and can be reproduced by accurate theoretical calculations. Practical applications of theory, however, are limited by computer resources. The accuracy of hyperfine coupling calculations at a certain level of theory depends on the cancellation of errors arising from the deviation of geometrical structures from experiment, the truncation of electron correlation effects, and the neglect of environmental, temperature, and relativistic effects. Thus, it is critical to employ reliable computational procedures to obtain an unambiguous theoretical interpretation of experimental data. In this section, the theoretical description of hyperfine coupling tensors is briefly introduced and available theoretical methods are reviewed. The observed performance of various levels of theory on the basis of gas-phase calculations and the current methodologies for incorporation of environmental effects are also overviewed. Finally, a computational scheme of combined levels of theory is highlighted as being suitable for the hyperfine coupling calculations of large radical species.

2.1 Theoretical description of hyperfine coupling tensors

All radicals have one or more unpaired electrons with intrinsic spin. As the unpaired electron is delocalized in a radical, there is a non-zero probability that the unpaired electron exists at each nucleus. Hence, the magnetic moment of the electron can interact with the magnetic moment of each nucleus, and give rise to the hyperfine interaction, an intrinsic and unique property of radical species. Since the magnetic moment is a vector, the hyperfine interaction of two magnetic moments is described by a tensor, the so-called hyperfine coupling tensor. Experimentally, the hyperfine coupling tensors are determined by measuring the hyperfine structures detected by electron spin or paramagnetic resonance (ESR or EPR) spectroscopy.

For a given nucleus, the hyperfine interaction can be described by the hyperfine coupling tensor (A) in the Hamiltonian of the hyperfine interaction:

$$\hat{H}_{hf} = \hat{I} \cdot A \cdot \hat{S} \quad (1)$$

where $\hat{I}$ and $\hat{S}$ are the nuclear and electronic spin operators, and

$$A = \begin{bmatrix} A_{XX} & 0 & 0 \\ 0 & A_{YY} & 0 \\ 0 & 0 & A_{ZZ} \end{bmatrix} \quad (2)$$

The A_{ii} (i=X, Y and Z) are called the principal components of the hyperfine coupling tensor. The A tensor can be separated into two parts:

6. Calculation of hyperfine coupling tensors

$$A = A_{iso} \bullet \begin{bmatrix} 1 & 0 & 0 \\ 0 & 1 & 0 \\ 0 & 0 & 1 \end{bmatrix} + \begin{bmatrix} T_{XX} & 0 & 0 \\ 0 & T_{YY} & 0 \\ 0 & 0 & T_{ZZ} \end{bmatrix} \quad (3)$$

where A_{iso} is the isotropic hyperfine coupling constant (HFCC), which describes the magnitude of the hyperfine interaction, and T_{XX}, T_{YY} and T_{ZZ} ($T_{XX} + T_{YY} + T_{ZZ} = 0$) are the anisotropic hyperfine coupling constants, which reflect the asymmetry of the spin density about the nucleus.

For radicals with N magnetic nuclei, the hyperfine interaction Hamiltonian is given by:

$$\hat{H}_{hf} = \sum_{i=1}^{N} \hat{I}_i \bullet A_i \bullet \hat{S} \quad (4)$$

where the sum extends over all magnetic nuclei.

The isotropic hyperfine coupling constant of a given nucleus in a radical arises from a quantum mechanical contact of the unpaired electron and the nucleus, and can be calculated[9c] in hertz (Hz) by:

$$A_{iso} = \frac{2\mu_0}{3h} g\beta_e g_N \beta_N \rho^{\alpha-\beta}(0) \quad (5)$$

where $\rho^{\alpha-\beta}(0)$ is the spin density at the nucleus, μ_0 is the permeability constant in vacuum, h is Planck's constant, g is the electronic g factor, g_N is the nuclear g factor, β_e is the Bohr magneton, and β_N is the nuclear magneton. In many radicals, $g \approx g_e = 2.0023$ where g_e is the Zeeman splitting constant for the free electron. It should be noted that the "spin density" at a nucleus is often confused with the "Mulliken spin population" of the nucleus, as pointed out independently by Chipman[15] and Barone.[16]

The anisotropic hyperfine coupling constants result from the electron-nuclear magnetic-dipole interactions. The ij[th] component of the anisotropic coupling (in Hz) can be computed from:

$$T_{ij} = \frac{\mu_0}{4\pi h} g\beta_e g_N \beta_N \sum_{\mu\nu} \rho_{\mu\nu}^{\alpha-\beta} <\phi_\mu | r_{kN}^{-5}(r_{kN}^2 \delta_{ij} - 3r_{kN,i}r_{kN,j}) | \phi_\nu> \quad (6)$$

where $\rho_{\mu\nu}^{\alpha-\beta}$ is an element of the spin density matrix.

The hyperfine coupling constant is a field-independent property, while the isotropic coupling constant measures the difference of two resonance frequencies of the unpaired electron in magnetic atoms and molecules. Thus, the SI unit for hyperfine coupling constants is the hertz. However, hyperfine couplings have to be observed in an applied magnetic field B in ESR spectroscopy. In order to induce the electronic spin resonance of free electrons in EPR experiments, the resonance frequency and the applied magnetic field must be tuned to meet the requirement

$$h\nu = g_e \beta_e B \qquad (7)$$

where ν is the resonance frequency. EPR spectroscopists often give hyperfine coupling constants in gauss (G) or tesla (T) [1G=10^{-4}T], of the magnetic field. According to equation (7), the conversion factor from G to MHz is 2.8025.

2.2 A survey of available theoretical methods

Before undertaking a hyperfine coupling tensor calculation, it is important to carefully consider the levels of theory available, and their relative performance. In the present section, we survey the levels of theory currently available to calculate electronic properties. Subsequently, Section 2.3 provides a discussion of the relative success of these techniques for the calculation of HFCCs.

The Hartree-Fock (HF) method is the simplest *ab initio* model that can be employed for studying chemical problems. Inherent in the accuracy of Hartree-Fock calculations is the choice of the basis set used to describe the molecular orbitals. There are a number of different types of basis sets developed as a combination of Slater-type functions (orbitals) or Gaussian-type functions (orbitals). Slater functions ($e^{-\zeta r}$) closely resemble atomic orbtials; however, Gaussian functions ($e^{-\alpha r^2}$) are more computationally efficient. Therefore, typical basis sets use linear combinations of Gaussian functions to obtain the accuracy of Slater functions, but yet maintain the computational advantage of Gaussian functions. The smallest possible basis sets (minimal basis sets) use only one function to describe each occupied atomic orbital (*i.e.*, STO-3G). Improvements upon minimal basis sets are achieved by using a double-zeta (triple-zeta) split-valence basis set (denoted as DZ (TZ)), which distinguishes between the core and valence regions and uses twice (triple) the number of functions to describe the chemically important valence region (*i.e.*, 3-21G and 6-31G (DZ basis sets) or 6-311G

(TZ basis set)). Further improvements to the basis set include the addition of polarization functions (often denoted as *), which are functions with higher angular momentum that allow orbitals to distort within the molecular environment. These functions can be added to only the heavy atoms (*i.e.*, 6-31G* or 6-31G(d)) or also to hydrogen (*i.e.*, 6-31G** or 6-31G(d,p)). Diffuse functions (denoted as +), which are functions with small exponents (*i.e.*, very large orbitals), can also be added to heavy atoms (*i.e.*, 6-31+G(d,p)) or to all atoms including hydrogen (*i.e.*, 6-31++G(d,p)) in order to improve the description of systems with loosely bound electrons (*i.e.*, anions or hydrogen bonded systems). Pople and coworkers developed the basis sets mentioned thus far, however similar systematically designed basis sets developed by Dunning and coworkers (*i.e.*, cc-pVDZ, aug-cc-pVTZ) are also popular. Many other basis sets have also been designed to specifically calculate different properties to a high degree of accuracy.

There are two main methods based on HF theory used for open-shell molecules such as radicals. The first method, restricted open-shell HF (ROHF), uses a combination of doubly and singly occupied orbitals to describe the system. The second method, unrestricted HF (UHF), accounts for different interactions between an unpaired electron and the two paired electrons by using two different molecular orbital expansions which allow the paired electrons to occupy different regions in space. UHF generally provides a more accurate description of open-shell molecules than ROHF. However, a major drawback of UHF is that the wavefunction is not a pure spin state, a deficiency referred to as spin contamination.

The Hartree-Fock limit represents the best that can be done with a single electron configuration (*i.e.*, a HF calculation with an infinitely large basis set). However, quantitative predictions often require going beyond the Hartree-Fock level. The major deficiency of Hartree-Fock theory is its incomplete description of electron correlation. That is, Hartree-Fock theory predicts that the probability of finding two electrons in the same region of space is equal to the product of the individual probabilities. Clearly, this approximation does not hold for electrons of parallel spin since it is energetically favourable for these electrons to be far from each other (*i.e.*, the motion of electrons is correlated). Conventional electron correlation methods and methods based on density functional theory[31] are the two main approaches for incorporation of the effects of electron correlation.

A hierarchy of the conventional electron correlation methods, as well as a hierarchy of basis sets, is best illustrated by a Pople diagram,[32] such as shown in Figure 1.

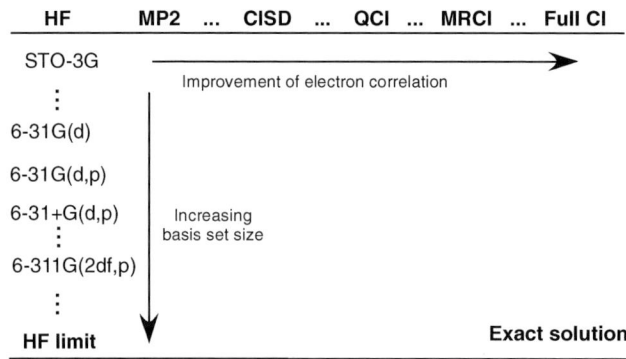

Figure 1. Hierarchy of conventional *ab initio* methods (adapted from ref. 32).

Following the hierarchy, the exact solution of the non-relativistic Schrödinger equation, and thus the exact spin density distribution for evaluation of hyperfine coupling tensors, may be approached in a systematic manner. Methods that approach the exact solution (*i.e.*, MRCI or Full CI with very large basis sets) may be highly accurate. However, they are often too computationally expensive, and therefore they are limited to small radicals (*i.e.*, less than five heavy atoms).

In comparison to the conventional correlated methods, density functional theory attempts to obtain all information, including the wave function, from the electron density of the system. In current DFT methods, the electron correlation is incorporated into the Hamiltonian by means of the exchange-correlation energy functional (E_{XC}). Unfortunately, the exact form of E_{XC} is unknown. Thus, various models, including the local spin density approximation (LSDA),[31c] and generalized gradient approximations (GGA),[33a-d] have to be employed for practical calculations. In the LSDA approximation, the explicit form of E_{XC} depends only on the electron density and is obtained on the assumption that E_{XC} can be separated into an exchange functional (E_X) and a correlation functional (E_C). One such functional combination is SVWN, which is constructed as a linear combination of the Slater exchange term[34] and the correlation functional fitted by Vosko, Wilk and Nusair.[35] In the GGA approximation, the spin gradients are introduced to correct for the non-uniformity of the electron density neglected by LSDA. The GGA-corrected exchange functionals PW86[33a] (Perdew and Wang) and B88[36a] (Becke), and the GGA-corrected correlation functionals P86[33bc] (Perdew), LYP[37] (Lee, Yang, and Parr), and PW91[33d] (Perdew and Wang) are in common use. To further correct for the deficiencies of GGA functionals, Becke developed hybrid functionals that combine DFT and Hartree-Fock methods. These functionals are written as a

6. Calculation of hyperfine coupling tensors

linear combination of Hartree-Fock, LSDA and B88 exchange contributions, together with LSDA and non-local correlation contributions (usually P86, PW91 or LYP). Three coefficients are used to fit the functional form to experimental data (atomization energies, ionization potentials and proton affinities). Indeed, Becke's three-parameter hybrid exchange functional combined with the LYP correlation functional (B3LYP)[36bc,37] has become the most popular E_{XC} functional. Basis sets typically used in Hartree-Fock based calculations (*i.e.* Pople or Dunning basis sets) are also employed in DFT calculations.

Since each DFT functional is constructed in a unique manner, the accuracy of density functional methods depends upon both the exchange-correlation functional and the basis set implemented. Furthermore, it is not possible to construct a hierarchical diagram for DFT methods. Therefore, there is no systematic way to study the convergence of DFT-based methods as in the conventional methods,[21] other than improving the basis set used. However, GGA functionals generally behave better than LSDA functionals for most properties. Extensive tests have shown that functionals like B3LYP can achieve chemical accuracy with some exceptions. Another important feature of well-behaved DFT methods for large molecules, such as B3LYP, is that the results are often less dependent on the size of the basis set compared with conventional techniques. Specific to open shell systems, spin contamination is less of a concern with DFT methods compared with conventional correlation techniques.

All DFT methods account for some of the effects of electron correlation and therefore DFT procedures are superior to HF methods. The principal advantages of DFT methods over more conventional *ab initio* electron correlation techniques are computational efficiency and small disk requirements. For example, the computational time required for QCISD(T) or CCSD(T) calculations scales roughly as N^7, where N is the number of orbitals, but typical DFT calculations scale approximately as N^3, which can be further reduced for very large molecules to N with modern linear scaling algorithms. Hence, DFT methods are the preferred choice for molecular systems that are too large to be treated by conventional electron correlation methods, such as biologically relevant systems.

2.3 An overview of the relative performance of various theoretical methods for HFCC calculations

Isotropic HFCCs are calculated by evaluating the spin density at the nucleus in question, a property that has been proven to be very difficult to calculate accurately. Since the anisotropic HFCCs are integrated over all space, rather than being evaluated at a single point, they are less sensitive to the quality of the method employed. Therefore, accurate anisotropic hyperfine coupling constants are also obtained when a level of theory suitable for the calculation of isotropic hyperfine coupling constants is implemented.

The performance of various levels of conventional theory for the calculation of isotropic hyperfine coupling constants has been summarized in reviews by Feller and Davidson,[13] Chipman,[15] Barone,[16] Engels et al.[17] and Eriksson.[19] Importantly, Chipman's spin polarization model[38] has made it possible to understand the factors involved in computing hyperfine coupling constants. As introduced in Section 2.2, the theoretical method and the basis set are the two variables that determine the quality of the calculated hyperfine coupling tensors. In summary, high recovery of electron correlation and the use of a well-balanced basis set, which accurately describes the spin density at the nucleus of interest (see, for example, refs 13, 15, 38), are both generally important for the quantitative prediction of isotropic HFCCs.

Basis set requirements for accurate HFCCs are much more involved than those for the accurate calculation of other electronic properties. Both the core and the valence space must be very well described for hyperfine coupling calculations (*i.e.*, a well-balanced basis set). In general, basis sets no smaller than triple zeta quality should be used to calculate HFCCs. In some instances, diffuse functions are very important; however, these drastically increase the computational cost.

Since the weight of electron correlation effects on A_{iso} varies from atom to atom, as well as with the electronic structures of different families of radicals, good results may be obtained for some radicals at the ROHF, UHF, and CIS levels of theory in conjunction with tailored basis sets due to fortuitous cancellation of errors. Therefore, these methods should not be relied upon if accurate data is desired. Among conventional *ab initio* approaches, it is essentially only multireference configuration interaction (MRCI), quadratic configuration interaction (QCI), and coupled-cluster (CC) techniques, in conjunction with large basis sets, that consistently generate HFCCs of high accuracy.[15,21,39-44] Basis sets of double-zeta plus polarization (DZP) quality or lower are found to be inadequate,[13-15,38] except for special cases where fortuitous cancellation of errors occurs. One problem with such approaches is that they are computationally quite expensive, even for moderately sized systems, and hence studies using these methods are restricted to relatively small systems.

As discussed in Section 2.2, an alternative approach for calculating hyperfine coupling tensors is represented by density functional theory (DFT). We must once again stress that the major deficiency of DFT is that a lower energy from one method does not guarantee that the functional leads to more accurate properties, and therefore all functional combinations must be carefully tested to determine the best DFT method for a particular property for a chosen class of molecules. In general, LSDA yields unacceptable isotropic HFCCs since the density is not localized, but GGA functionals lead to an improved description since an improved description of the core density is obtained. In particular, extensive assessments have shown that combinations of certain gradient-corrected functionals with appropriate basis sets such as (PBE0[45]/EPR-III[16]),[20a] (B3LYP/EPR-III and B3LYP/EPR-II),[16] (PWP86[33a-c]/IGLO-III[46]),[16,47] (PWP86/6-311G(2d,p))[24] and (B3LYP/6-311+G(2df,p)),[18] can provide reasonably accurate results. Noticeably, PWP86/6-311G(2d,p), B3LYP/EPR-III and B3LYP/EPR-II have been applied extensively to study the hyperfine coupling tensors of biological radicals.[27,29,30]

Since hyperfine coupling tensors are measured at a temperature above absolute zero (*i.e.*, the temperature routinely used for gas-phase calculations), the excitation of vibrational motions at higher temperatures may have effects on the molecular structure of the radical, and therefore on hyperfine coupling tensors. The features and applicability of theoretical methods for incorporating temperature or vibrational effects by means of vibronic coupling, vibrational averaging, and molecular dynamics have been summarized by Eriksson.[19] Barone and coworkers[20a] have more recently reviewed the procedures for including the vibrational averaging correction to hyperfine coupling tensors in detail. It should be noted that temperature effects on the hyperfine coupling constants are generally less than 10%, and therefore they are not included in our applied computational procedure.

2.4 Methodologies for incorporating environmental effects

One common feature of biomolecules in condensed phases is the existence of hydrogen bonding interactions with surrounding molecules. Inclusion of such interactions in calculations is important for a better understanding of environmental effects on the hyperfine coupling tensors. Static models for approximating surrounding molecules are (**a**) a finite supermolecular model, (**b**) a continuum with a dielectric constant characteristic of the solvent, and (**c**) a hybrid model in which the environment is represented by the combination of a few important surrounding molecules with a continuum for the remaining molecules. Currently it is often an overwhelming challenge to perform calculations on large supermolecular models (**a**) at reliable levels of theory (*i.e.*, including both the radical of interest and surrounding molecules in the geometry optimization and single-point calculations). However, calculations using models (**b**) and (**c**) can be routinely carried out using the self-consistent reaction field (SCRF) methods,[48] in which the target radicals of model (**b**), as well as some of the nearest neighbor molecules in model (**c**), are embedded in a cavity of the continuum of a dielectric. The advantage of model (**c**) over model (**b**) is the inclusion of specific solvent effects on the radical structure and therefore, on the hyperfine coupling tensors.

In the last ten years, tremendous progress has been made in the SCRF approach, for which the Onsager model[49,50] and a family of polarized continuum models (PCM) are currently available. The Onsager model and PCM models differ in how they define the cavity and the electric field,[48] where the PCM models use a more elaborate representation of the cavity. It has been shown that the Onsager model is very efficient for geometry optimizations,[50d] while PCM based methods, such as the self-consistent isodensity PCM (SCIPCM) often have convergence problems for geometry optimizations. Recently, Barone's group has implemented the conducter-like polarized continuum method (CPCM)[51] to describe the solvent effects on the hyperfine coupling tensors.[29i,52] Their results have shown that the CPCM method is very promising.

2.5 A practical computational scheme for the HFCCs in biological radicals

In principle, accurate hyperfine coupling tensors of all magnetic nuclei in a radical can be obtained at reliable levels of theory using a sufficiently large supermolecular model. However, as discussed in Section 2.4, such a supermolecular model is generally too expensive for geometry optimizations. On the basis of the fact that environmental effects on gas-phase optimized structures for radicals are typically minor, the use of isolated radicals for hyperfine coupling tensor calculations is widely employed in the literature. Indeed, since the major contribution to the hyperfine coupling constants arises due to spin delocalization and spin

6. Calculation of hyperfine coupling tensors

polarization within (isolated) radicals, calculations on isolated radicals often provide satisfactory results.

Predicting hyperfine coupling tensors for biological radicals begins with a conformational search that is generally extremely time-consuming and, therefore, computationally expensive. Since the molecular geometry depends only upon the first derivatives of the energy, reasonably accurate geometries can often be obtained at quite modest levels of theory (*i.e.*, MP2 or DFT with a small basis set). However, properties such as hyperfine coupling constants converge only at a very high level of theory. Thus, optimizing the geometry at a lower level of theory followed by a higher-level single-point calculation on the optimized structures represents a cheaper, yet reliable, computational scheme for hyperfine coupling tensor calculations.

It should be noted that there are cases where environmental effects must be considered to obtain the correct structures of certain radicals, for example, the zwitterionic forms of amino acid radicals in solution or in single crystals. In these instances, one must choose a suitable approach from those discussed in Section 2.4 on a case-by-case basis.

We have performed hyperfine coupling tensor calculations[29a-c] on a variety of isolated amino acid radicals derived from glycine, alanine and hydroxyproline (Scheme 1) at the PWP86/6-311G(2d,p)//B3LYP/6-31+G(d,p) level of theory. For zwitterionic forms of the amino acid radicals, the geometry optimizations were performed using the Onsager model at the B3LYP/6-31+G(d,p) level of theory (denoted by Onsager-B3LYP/6-31+G(d,p)). Thus, the hyperfine coupling tensor calculations were obtained from the PWP86/6-311G(2d,p)//Onsager-B3LYP/6-31+G(d,p) level. We note that the 6-31+G(d,p) basis set was chosen for the geometry optimizations in order to ensure that an accurate description is obtained for anions (*i.e.*, the diffuse functions) and hydrogen bonds (*i.e.*, the diffuse and polarization functions).

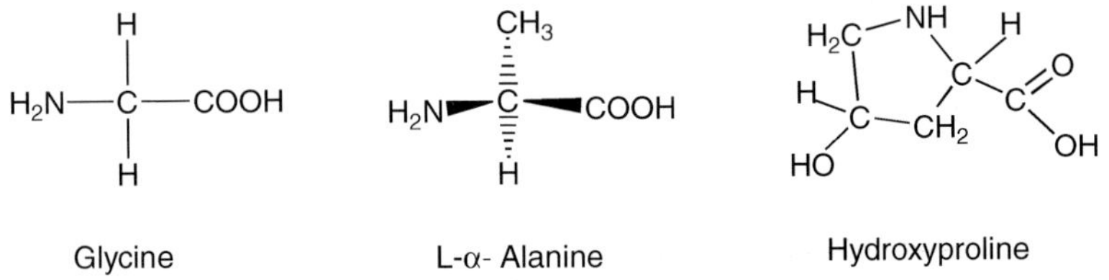

Scheme 1. Schematic illustration of the structures of glycine, L-α-alanine and hydroxyproline.

In the following section, we illustrate the procedures that have been employed in our studies of the relationship between conformations and hyperfine coupling tensors of amino acid radicals. For a complementary discussion of the radicals formed by radiation damage to the constituents of DNA, the reader is referred to a recent review.[20b]

3. THEORETICAL STUDIES OF AMINO ACID RADICALS

3.1 Calculations on isolated amino acid radicals

The structures of some amino acid radicals are scarcely affected by the environment of single crystals and solvents, and therefore can be readily obtained by gas-phase geometry optimizations. Comparison of the calculated hyperfine coupling tensors for each conformer of the isolated amino acid radicals with available experimental data may identify the specific conformations of the radical(s) that give rise to the observed hyperfine coupling tensors. For the first example, we illustrate how two observed (average) proton tensors (Table 1) of the –CH$_3$ moiety in the L-α-alanine radical R1: NH$_2$-C$^•$(CH$_3$)-COOH in the single crystals were assigned and interpreted by our calculations.

A B3LYP/6-31+G(d,p) conformational search found four possible conformers of R1 (Figure 2).

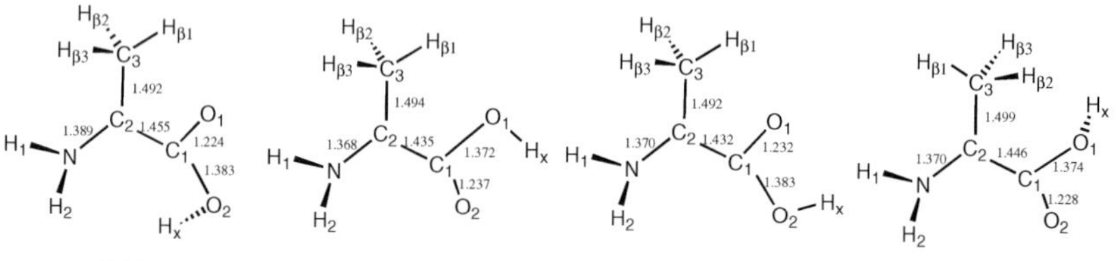

Figure 2. The optimized structures for the four conformations of the R1 radical of L-α-alanine.

The average of the calculated methyl proton isotropic HFCCs in the four conformers of R1 separates the conformers into two groups (Table 1). One group contains only R1-I with an averaged methyl proton isotropic HFCC (H$_{\beta(ave)}$) equal to 41.3 MHz, which is in excellent agreement with one of the experimental values (39.5 MHz). The second group consists of R1-II, R1-III, and R1-IV with an averaged methyl proton isotropic HFCC ($H_{\beta(ave)}$) of approximately 35 MHz, which is in excellent agreement with the second experimental value (33.1 MHz). In addition, all of the anisotropic components for both groups are in nearly perfect agreement with the corresponding experimental values. This leads to the conclusion that at least two conformers of R1 exist in the irradiated L-α-alanine crystals at 295 K. R1-I could be one conformer and the second conformer could be R1-II, R1-III or R1-IV.

Table 1. Comparison of calculated and experimental[10b] hyperfine couplings (MHz) of the R1 radical of L-α-alanine.

tensor	HFCC	R1-I	R1-II	R1-III	R1-IV	exptl	
$H_{\beta 1}$	A_{iso}	1.5	2.0	46.9	19.0		
	T_{xx}	-3.9	-3.4	-3.7	-3.5		
	T_{yy}	-2.7	-3.2	-2.7	-3.2		
	T_{zz}	6.6	6.6	6.4	6.6		
$H_{\beta 2}$	A_{iso}	61.1	61.2	56.0	71.7		
	T_{xx}	-3.5	-3.3	-3.5	-3.5		
	T_{yy}	-3.2	-3.0	-2.8	-2.9		
	T_{zz}	6.7	6.4	6.3	6.4		
$H_{\beta 3}$	A_{iso}	60.5	43.7	1.4	15.7		
	T_{xx}	-3.6	-3.5	-3.5	-3.3		
	T_{yy}	-2.9	-2.8	-2.7	-2.8		
	T_{zz}	6.6	6.3	6.2	6.1		
$H_{\beta(ave)}$	A_{iso}	41.3	35.6	34.8	35.4	39.5	33.1
	T_{xx}	-3.7	-3.4	-3.6	-3.4	-2.7	-2.3
	T_{yy}	-3.2	-3.0	-2.7	-3.0	-2.2	-2.3
	T_{zz}	6.6	6.3	6.3	6.4	5.0	4.6

The above theoretical assignment is supported by the correlation between the isotropic HFCCs and the structures. When the coplanarity of H_1, H_2, N and C_2 of R1 is examined by considering the magnitude of the sum of the three bond angles of the amino group (Table 2), the four conformers can be separated into the same two groups as isolated by examining $H_{\beta(ave)}$. The sum of the three bond angles in R1-I is at least 6.8° less than that of the second group. Thus, the N centre in R1-I is more pyramidal than those of the other three conformations, which is likely due to the specific repulsion between the amino group and the hydroxyl group in R1-I. The structural difference of the two groups of conformers results in distinct atomic spin populations (Table 3) and explains the difference in their proton isotropic HFCCs. More specifically, the more planar the amino group, the more easily the β spin electron of the lone pair of the N atom can be polarized by the unpaired α electron spin at the C_2 atom. As a consequence, the greater the spin polarization across the C_2—N bond, the lower the α (positive) spin population on the C_2 atom, and therefore the larger the α spin population on the N atom (Table 3), or the smaller the methyl proton isotropic HFCCs. Hence, the structural features of the two groups of conformers support the conclusion that R1-I possesses a larger spin population on the C_2 atom and therefore a larger methyl proton isotropic HFCC than R1-II, R1-III, and R1-IV.

Table 2. The three bond angles (degrees) and their sum of the amino group in the R1 radical of L-α-alanine.

structure	$\angle H_1NC_2$	$\angle H_2NC_2$	$\angle H_1NH_2$	sum
R1-I	116.1	117.4	112.4	345.9
R1-II	120.1	116.8	117.8	354.7
R1-III	119.6	117.1	117.1	355.4
R1-IV	119.9	117.3	117.3	352.7

Table 3. The PWP86/6-311G(2d,p) calculated spin populations of the C_2 and N atoms in the R1 radical of L-α-alanine.

atom	R1-I	R1-II	R1-III	R1-IV
C_2	0.597	0.519	0.521	0.537
N	0.162	0.232	0.222	0.235

In summary, the above example shows the dependence of HFCCs on spin polarization. Specifically, the H_β isotropic hyperfine coupling constants in an L-α-alanine derived radical depends on the extent of spin polarization across the N—C bond. Next, we will illustrate how theory may offer interpretations for observed differences in hyperfine coupling tensors of an amino acid radical based on a conformational study. In particular, we find that a radical derived from hydroxyproline leads to different HFCCs due to the conformational difference caused by intramolecular hydrogen-bonding interactions.

The hyperfine couplings of two hydroxyproline-derived radicals have been observed experimentally. These have been assigned to different conformations of the radical R2 (shown in Figure 3) which is formed upon deamination of the hydroxyproline radical anion. One conformation was observed at 77 K,[10c] denoted as exptl I (Table 4), while the second conformer was observed at 125 K,[10c] denoted as exptl II (Table 4). For temperatures between 77 and 125 K, a mixture of conformations was observed.[10c] Our calculations[29b] concluded that the observed radical is in a neutral form. Eight possible conformers R2-I to R2-VIII were found and are shown schematically in Figure 3. The conformers may be divided into two groups depending on whether or not they contain an intramolecular hydrogen bond between the carboxylic group and the –NH$_2$ moiety. The conformers R2-I,

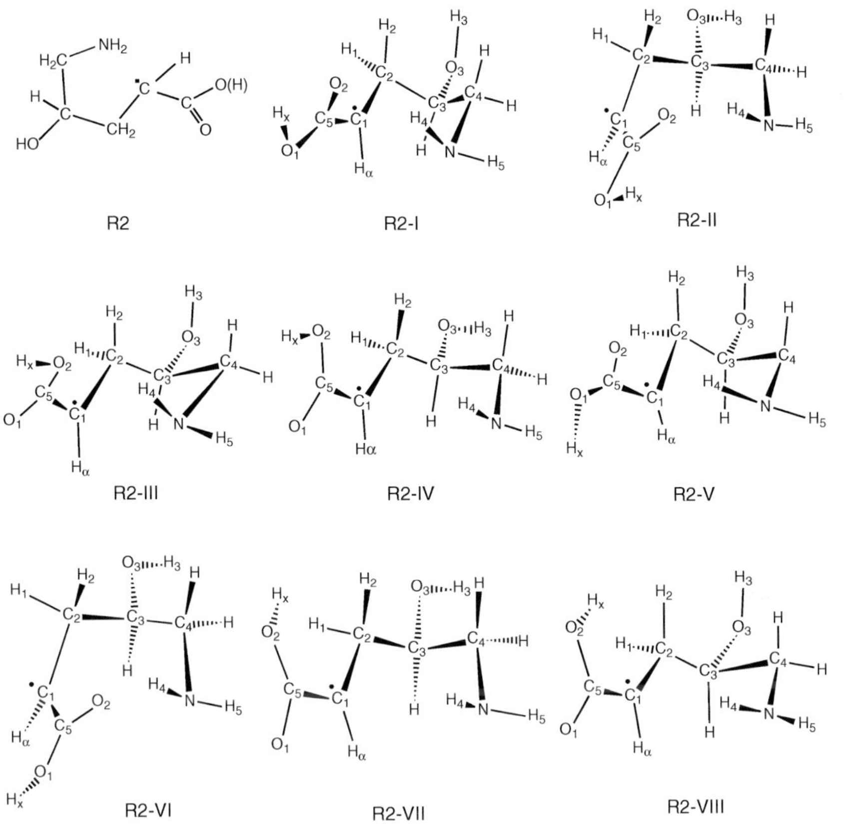

6. *Calculation of hyperfine coupling tensors*   255

Figure 3. Optimized structures for the eight conformers of the R2 radical of hydroxyproline.

Table 4. PWP86/6-311G(2d,p) calculated and experimental[10c] HFCCs (MHz) for eight conformers of the R2 hydroxyproline-derived radical.

system	H_α				H_1				H_2			
	A_{iso}	T_{xx}	T_{yy}	T_{zz}	A_{iso}	T_{xx}	T_{yy}	T_{zz}	A_{iso}	T_{xx}	T_{yy}	T_{zz}
R2-I	-47.5	-29.8	-1.9	31.6	72.4	-4.4	-3.2	7.6	3.7	-3.7	-2.8	6.5
R2-II	-49.0	-30.6	-1.9	32.6	79.1	-4.6	-3.2	7.8	3.3	-4.0	-2.8	6.8
R2-III	-47.0	-28.9	-2.3	31.2	78.3	-4.4	-3.2	7.6	2.5	-3.4	-3.3	6.7
R2-IV	-50.9	-30.5	-2.2	32.7	139.5	-5.4	-2.5	7.9	22.2	-3.9	-3.2	7.1
R2-V	-46.3	-29.8	-1.5	31.3	71.6	-4.4	-3.1	7.5	3.5	-3.7	-2.7	6.4
R2-VI	-47.5	-30.6	-1.6	32.2	80.2	-4.6	-3.1	7.7	2.9	-3.8	-2.8	6.7
R2-VII	-52.8	-30.8	-1.9	32.7	136.8	-5.2	-2.5	7.7	34.1	-3.9	-3.3	7.3
R2-VIII	-50.5	-30.4	-2.5	32.9	130.6	-5.1	-2.6	7.7	21.4	-3.7	-3.3	7.1
exptl I	-57.2	-32.4	1.9	30.6	78.0	-4.8	-3.4	8.3				
exptl II	-56.5	-32.4	1.3	31.1	123.1	-5.1	-2.2	7.4	21.4	-4.3	-3.2	7.6

R2-II, R2-III, R2-V and R2-VI, hereafter referred to as group I, contain a slightly shortened hydrogen bond between O_2 of the carboxylic group and H_4 of the $-NH_2$ moiety (2.316 Å< $r_{O...H}$ <2.431 Å). The conformers R2-IV, R2-

VII and R2-VIII, hereafter referred to as group II, contain no such hydrogen bond ($r_{O...H} > 3.661$ Å). Consequently, the structures of the group II conformers are more open than those of group I.

As can be seen in Table 4, these differences in the structures are also reflected in the calculated hyperfine coupling tensors. For the conformers in group I, the calculated H_α, H_1 and H_2 isotropic HFCCs range approximately from –46 to –49, from 70 to 80 and from 2 to 4 MHz, respectively. These coupling constants are in close agreement with those measured experimentally at 77 K (*i.e.*, exptl I). The small calculated coupling constants of H_2 support the experimental suggestion that the second β-hydrogen coupling may have been too small to be observed. Since the calculated hyperfine coupling tensors of the group I conformers are very similar to each other, it is not possible to determine if exptl I is due to one conformer or a mixture of the group I conformers. For conformers in group II, the H_α, H_1 and H_2 isotropic HFCCs range approximately from –50 to –53, from 130 to 140 and from 21 to 34 MHz, respectively. These coupling constants are in good agreement with the conformer observed at 125 K, i.e., exptl II, with those of R2-VIII in closest agreement. These results clearly show that differences in HFCCs at the two temperatures arise due to a change in conformation, where intramolecular hydrogen-bonding interactions are present at lower temperatures.

Through the above examples, it should be noted that the hyperfine coupling tensors for a particular radical significantly depend on its conformation. Therefore, a simple HFCC study based solely on the global minimum structure will not necessarily help the assignment of the experimentally observed hyperfine coupling tensors. Hyperfine coupling tensor calculations on the complete conformational space of the proposed radicals and its derivatives are required for the unambiguous assignment of the experimental data.

3.2   Calculations on the zwitterionic form of isolated amino acid radicals

Amino acids can exist as zwitterionic species in the crystalline state and in solution. Therefore, when irradiated, radicals in a zwitterionic form, such as R3: $^+NH_3^\bullet CHCOO^-$ (from glycine), can be formed. The zwitterionic structure of amino acids and their derived radicals has been a challenge for theoretical chemistry. *Ab initio* calculations on glycine[53] and its radical[29ij] have shown that their zwitterionic structures do not correspond to energy

6. Calculation of hyperfine coupling tensors

minima in the gas phase. Thus, environmental effects must be accounted for in order to explore the zwitterionic structures.

We have shown that the observed hyperfine coupling tensors of zwitterionic isomers of the amino acid radicals of glycine,[29b] alanine[29a] and hydroxyproline[29c] can be reproduced by PWP86/6-311(2d,p) single point calculations on the zwitterionic structures obtained by Onsager-B3LYP/6-31+G(d,p) optimizations. For example, the structure of R3 (Figure 4) was optimized using the Onsager model with an estimated cavity radius of 3.24 Å and the dielectric constant of water (ε=78.39). The resulting geometry possesses C_s symmetry with a planar radical centre. Thus, R3 is a typical π-radical. The computed full hyperfine tensors of R3, as well as the experimental values and other previously calculated values, are listed in Table 5.

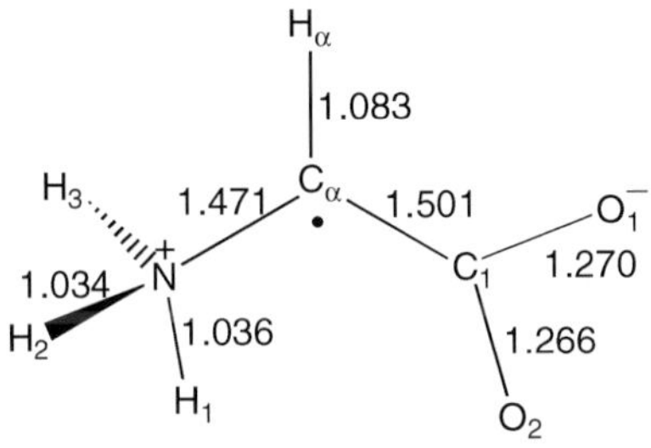

Figure 4. The optimized structure for the zwitterionic radical R3 of glycine.

Table 5. PWP86/6-311G(2d,p) calculated and experimental HFCCs (MHz) of the R3 of glycine.

tensor	A_{iso}	T_{xx}	T_{yy}	T_{zz}	A_{iso}^{exp}	T_{xx}^{exp}	T_{yy}^{exp}	T_{zz}^{exp}	A_{iso}^{e}
H_α	-58.96	-36.06	-1.46	37.51	-63.72[a]	-33.80[a]	1.85[a]	31.94[a]	-60.3
H_1	2.33	-5.39	-4.66	10.05	3.3[b]	-7.3[b]	-1.8[b]	9.2[b]	
H_2	77.06	-5.47	-4.52	9.99	62.91[a]	-6.60[a]	-4.07[a]	10.66[a]	
H_3	77.06	-5.47	-4.52	9.99	83.05[a]	-5.86[a]	-4.80[a]	10.65[a]	
H_{ave}	52.15	-5.44	-4.57	10.01	49.07[a]	-2.93[a]	-2.05[a]	4.97[a]	52.1
C_α	98.73	-76.29	-74.57	150.87	126.7[c]	-90.0[c]	-36.7[c]	126.8[c]	95.3
N	-6.96	-0.43	0.20	0.22	-8.72[d]	-0.98[d]	-0.76[d]	1.71[d]	-9.0

[a]Ref 10a. [b]Ref 54. [c]Ref 55. [d]Ref 56. [e]Ref 29i (B3LYP/EPR-II calculated values).

The H_α isotropic and anisotropic HFCCs are in good agreement with the experimental values. The anisotropic components of the hyperfine tensors of the three amino hydrogens (H_1, H_2 and H_3) are also in good agreement with the experimental values. The isotropic HFCCs of H_1, H_2 and H_3 are of similar magnitude as the experimental values with the largest differences of 14.15 and 5.99 MHz being observed for H_2 and H_3, respectively. The isotropic HFCC and anisotropic components of the nitrogen are in good agreement with the experimental values. It can be seen that the calculated C_α HFCC at the PWP86/6-311(2d,p) level is in fair agreement with the experimental value and slightly better than a previous B3LYP/EPR-II calculated value.[29i] The B3LYP/EPR-II study[29i] has shown that the deviation of C_α HFCC based on the optimized zwitterionic structure can be significantly corrected by including the effect of vibrational averaging. The HFCCs of H_α and N, however, are not sensitive to vibrational averaging.

The average (52.15 MHz) of the isotropic HFCCs of the three amino protons is in good agreement with the experimentally observed value of 49.07 MHz at 280 K, suggesting that the amino group rotates freely at this temperature. Furthermore, the results also suggest that at 100 K the orientation of the amino group in glycine crystals is constrained, giving rise to the three distinct hyperfine coupling tensors of the amino protons. Therefore, a further investigation was undertaken on the effects on the isotropic HFCCs of H_1, H_2, H_3, H_α, C_α and N of rotating the amino group about the N—C_α bond in R3. The rotation of the amino group was carried out by incrementally increasing the dihedral angle $\angle H_1NC_\alpha C_1$ by 30°, starting from $\angle H_1NC_\alpha C_1 = 0°$. The variation of the isotropic HFCCs as a function of the rotational angle is shown in Figure 5. From Figure 5 it can be seen that the isotropic HFCCs of H_1, H_2 and H_3 change dramatically, while the isotropic couplings of H_α, C_α and N are almost constant. It is noted that the difference in amplitude of the variation of the HFCC of H_1 and those of H_2 and H_3 is due to the fact that the geometrical parameters of the amino group have been constrained during rotation.

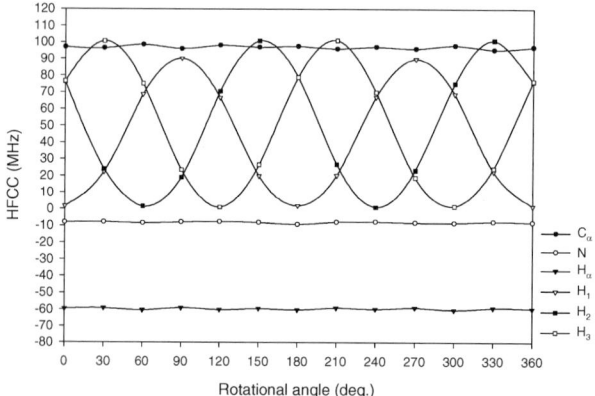

Figure 5. Variation of H_α, H_1, H_2, H_3, C_α, and N isotropic HFCCs with rotational angle of the amino group of the R3 radical of glycine.

6. Calculation of hyperfine coupling tensors 259

When the amino group is rotated by approximately 5°, the isotropic HFCCs of H_2 and H_3 are in good agreement with the experimental values of 62.91 and 83.05 MHz at 100 K, while the isotropic HFCC of H_1 is still less than 5 MHz. These results support the fact that only the hyperfine splittings of two of the three constrained amino protons are easily observed; while the third remains too small to be distinguished from the many lines observed in the ENDOR spectrum.[10a] Thus, from the hyperfine coupling tensor calculations on the isolated R3 radical it can be concluded that the crystalline environment has little direct effect on the ESR spectra of R3, similar to the conclusion of a previous study.[29i]

In summary, from the above example, it is clear that the PWP86/6-311G(2d,p)//Onsager-B3LYP/6-31+G(d,p) method provides a useful tool to study zwitterionic amino acid radicals. Furthermore, this example once again emphasizes the benefits of comparing theoretical and experimental HFCCs to obtain information about molecular structure.

3.3 Theoretical assignments versus experimental observations

We have shown in Sections 3.1 and 3.2 that the PWP86/6-311G(2df,p)//B3LYP/6-31+G(d,p) calculations on isolated amino acid radicals of both neutral and zwitterionic forms can reproduce the observed hyperfine coupling tensors. Thus, this validates our chosen computational scheme, and our procedure can now be used to study, and hopefully solve, more complex problems, in particular cases where experimental assignment is difficult.

To illustrate the usefulness of theory to understand experimental spectra and to aid spectral assignments, we will consider a hydroxyproline-derived radical. Upon irradiation of hydroxyproline, two H_β hyperfine coupling tensors were observed (the isotropic HFCCs (A_{iso}) are 23.9 and 61.0 MHz) which were assigned to two different conformations of the zwitterionic radical R^Z (Scheme 2), since no carboxylate hydrogen interactions were detected.[10c] However, our calculations suggest that this assignment is inaccurate.[29c]

Since proton-transfer may be involved in the formation of a radical in an irradiated hydroxyproline single crystal, we systematically examined the zwitterionic radical anion R^Z, its non-zwitterionic radical anion R^A and the protonated neutral radical R^P (Scheme 2). Two possible conformers of R^Z

were found at the Onsager-B3LYP/6-31+G(d,p) level, and four conformers of R^A and four conformers of R^P were found during a conformational search at the B3LYP/6-31+G(d,p) level.

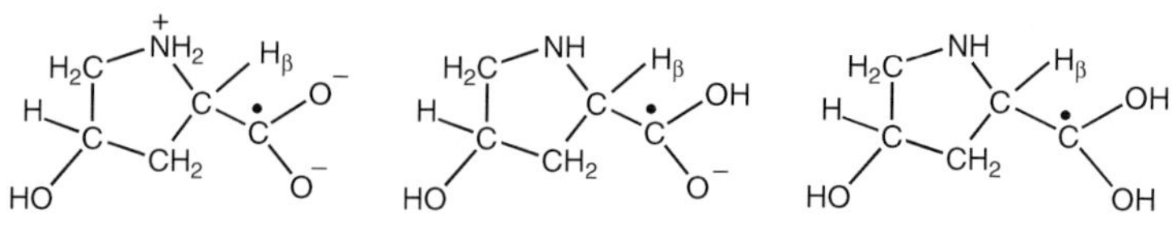

Scheme 2. Schematic illustration of the structures of the R^Z, R^A and R^P radicals of hydroxyproline.

The PWP86/6-311G(2d,p) calculated isotropic H_β HFCCs of R^Z, R^A and R^P are significantly different in magnitude. The calculated H_β isotropic HFCCs of R^Z are smaller than 5 MHz, which are very different from either of the two observed A_{iso}. The calculated H_β isotropic HFCCs of two of the four conformers of R^A are approximately 20 MHz, while the calculated H_β isotropic HFCCs for all four conformers of R^P are approximately 60 MHz.

Thus, the DFT calculations clearly illustrate that the two observed hyperfine coupling tensors are given by R^A and R^P and *not*, as previously proposed, by two conformations of R^Z. This example shows how theory can be used to aid experimental assignment in difficult cases.

4. CONCLUDING REMARKS

The present chapter has briefly reviewed the performance of theoretical methods available to predict the hyperfine coupling tensors for (bio)organic radicals. The computation of accurate isotropic hyperfine coupling constants requires levels of theory that recover a large portion of the total electron correlation and well-balanced basis sets. Suitable methods include conventional correlation theories, such as QCI, MRCI, CC, and density functional methods, such as B3LYP and PWP86. However, although QCI, MRCI, CC (in conjunction with appropriate basis sets) can provide accurate results for small radicals, they are prohibitively expensive for systematic studies of large biological radicals. In contrast, B3LYP and PWP86 methods in combination with Pople basis sets show promising accuracy with a dramatic reduction in computational time, and have been applied successfully to a range of large biological radicals.

6. Calculation of hyperfine coupling tensors

It is noteworthy that the spin density distribution, and therefore the hyperfine coupling constants, correlates with the conformations of the radical, and the correlation has been widely employed to rationalize the identity of the predominant biological radicals. Since many biological radicals are studied experimentally in solution or in the solid state, the environment can modify the gas-phase structure of the radicals. However, since modelling the conformation of biological systems with inclusion of the environmental effects arising from hydrogen bonding, van der Waals and steric interactions remains an overwhelming task (even for contemporary DFT levels of theory), isolated radicals are commonly used in hyperfine coupling tensor calculations under the assumption that the environmental and electronic effects of surrounding molecules on molecular structures and the hyperfine coupling tensor are minor.

By means of a few examples from our recent research on amino acid radicals, gas-phase PWP86/6-311(2d,p)//B3LYP6-31+G(d,p) hyperfine coupling tensor calculations have proven to be sufficiently accurate to complement experiment for the purpose of unambiguous characterization of most radical species. There are cases, however, for which the environmental effects can be large enough to transform the (gas-phase) structures of amino acids, and therefore their derived radicals, into different isomers (such as into the zwitterionic form) or to modify the coplanarity of the $-NH_2$ moiety.[29i] For zwitterionic amino acid radicals, we have shown that the Onsager model can account for the most dominant environmental effects, and reliable zwitterionic structures can be obtained using the Onsager-B3LYP/6-31+G(d,p) level of theory. Subsequently, PWP86/6-311(2d,p) gas-phase single-point calculations on the isolated zwitterionic structures of amino acid radicals appear to provide accurate proton hyperfine coupling constants. The results imply that the electronic effects of the environments on radical properties, such as HFCCs, are small.

Of course, the most accurate and reliable calculations of biological radicals should include the structural and electronic effects due to both the environment and the temperature. Although many practical calculations of hyperfine coupling constants of large biological radicals are heavily dependent on a cancellation of errors, considerable effort has been made by research groups, such as Barone and coworkers,[52] to examine the limit of current density functional theory in the prediction of hyperfine coupling constants. It can be expected that with further development of efficient algorithms and more powerful computers, more elaborate computational schemes will be able to provide more reliable descriptions of the relationship between structure and hyperfine coupling tensors. We are convinced that theoretical calculations will play an even more important role in identifying important biological radicals in the future.

Acknowledgements *We gratefully acknowledge the Natural Sciences and Engineering Research Council of Canada (NSERC) and the Killam Trusts for financial support. In addition, we thank Dr. Leif A. Eriksson for fruitful discussions and collaborations.*

5. REFERENCES

1. J. Stubbe, W. A. van der Donk, *Chem. Rev.* **98** (1998) 705.
2. W. M. Garrison, *Chem. Rev.* **87** (1987) 361.
3. W. K. Pogozelski, T. D. Tullius, *Chem. Rev.* **98** (1998) 1089.
4. (a) J. Stubbe, *Annu. Rev. Biochem.* **58** (1989) 257. (b) J. Stubbe, *J. Biochemistry* **27** (1988) 3893.
5. B. Halliwell, J. M. C. Gutteridge, *Free Radicals in Biology and Medicine*, Oxford University Press Inc.: New York, 1999.
6. W. A. Denny, W. R. Wilson, *Exp. Opin. Invest. Drugs* **9** (2000) 2889.
7. J. M. Brown, *Cancer Res.* **59** (1999) 5863.
8. J. M. Brown, *Br. J. Cancer,* **67** (1993) 1163.
9. (a) Jr. W. Weltner *Magnetic Atoms and Molecules*, Van Nostrand: New York, 1983. (b) N. M. Atherton, *Principles of Electron Spin Resonance,* Ellis Horwood Limited: England, 1993.
10. (a) Glycine: A. Sanderud, E. Sagstuen, *J. Phys. Chem.* B **102** (1998) 9353 and references therein. (b) Alanine: E. Sagstuen, E. O. Hole, S. R. Haugedal, W. H. Nelson, *J. Phys. Chem.* A **101** (1997) 9763 and references therein. (c) Hydroxyproline: W. H. Nelson, *J. Phys. Chem.* **92** (1988) 554. W. H. Nelson, C. R. Nave, *J. Chem. Phys.* **74** (1981) 2710 and references therein. (d) Tyrosine (in frozen solution): R. J. Hulsebosch, J. S. van der Brink, S. A. M. Nieuwenhuis, P. Gast, J. Raap, J. Lugtenburg, A. J. Hoff, *J. Am. Chem. Soc.* **119** (1997) 8685.
11. (a) Uracil: E. Sagstuen, E. O. Hole, W. H. Nelson, D. M. Close, *Rad. Res.* **149** (1998) 120. J. N. Herak, C. A. McDowell, *J. Chem. Phys.* **61** (1974) 1129. (b) Thymine: E. Sagstuen, E. O. Hole, W. H. Nelson, D. M. Close, *J. Phys. Chem.* **96** (1992) 1121. E. O. Hole, E. Sagstuen, W. H. Nelson, D. M. Close, *J. Phys. Chem.* **95** (1991) 1494. (c) Adenine: W. H. Nelson, E. Sagstuen, E. O. Hole, D. M. Close, *Radiat. Res.* **149** (1998) 75 and references therein. (d) Cytosine: E. Sagstuen, E. O. Hole, W. H. Nelson, D. M. Close, *J. Phys. Chem.* **96** (1992) 112 and references therein. (e) Guanine: E. O. Hole, W. H. Nelson, E. Sagstuen, D. M Close, *Rad. Res.* **129** (1992) 119 and references therein. E. O. Hole, W. H. Nelson, E. Sagstuen, D. M. Close, *Rad. Res.* **129** (1992) 119. (f) Sugar: D. M. Close, W. H. Nelson, E. Sagstuen, E. O. Hole, *Rad. Res.* **137** (1994) 300 and references therein. E. O. Hole, E. Sagstuen, *Rad. Res.* **109** (1987) 190. E. Sagstuen, *Rad. Res.* **84** (1980) 164. E. Sagstuen, *J. Magn. Reson.* **44** (1981) 518.
12. A. Lund, M. Lindgren, S. Lunell, J. Maruani, In *Molecules in Physics, Chemistry and Biology*; J. Maruani, Ed.; Kluwer Academic Publishers: Dordrecht, 1989; Vol. 3.

13. D. Feller, E. R. Davidson, In *Theoretical Models of Chemical Bonding*; Z. B. Maksić, Ed.; Springer-Verlag: Berlin, 1991; Part 3.
14. V. G. Malkin, O. L. Malkina, D. R. Salahub, L. A. Eriksson, In *Theoretical and Computational Chemistry*; P. Politzer, J. M. Seminario, Eds.; Elsevier: Amsterdam, 1995; Vol. 2.
15. D. M. Chipman *Magnetic Hyperfine Coupling Constants in Free Radicals* in *Quantum Mechanical Electronic Structure Calculations with Chemical Accuracy*, S. R. Langhoff Ed.; Kluwer Academic Publishers: The Netherlands, 1995.
16. V. Barone, In *Recent Advances in Density Functional Methods* Part I, D. P. Chong, Ed.; World Scientific Publishing Co. Pte. Ltd. 1996.
17. B. Engels, L. A. Eriksson, S. Lunell, *Adv. Quantum Chem.* **27** (1996) 297.
18. J. W. Gauld, L. A. Eriksson, L. Radom, *J. Phys. Chem.* A **101** (1997) 1352.
19. L. A. Eriksson, in *Encyclopedia of Computational Chemistry* 1998, John Wiley & Sons, Ltd.
20. (a) C. Adamo, M. Cossi, N. Rega, V. Barone, p467, and (b) S. D. Wetmore, L. A. Eriksson, R. J. Boyd, p409, In *Theoretical Biochemistry-Processes and Properties of Biological Systems*. L. A. Eriksson, Ed.; Elsevier: The Netherlands, 2001.
21. (a) J. Kong, R. J. Boyd, *J. Chem. Phys.* **107** (1997) 6270. (b) J. Kong, R. J. Boyd, L. A. Eriksson, *J. Chem. Phys.* **102** (1995) 3674. (c) K. Funken, B. Engels, S. D. Peyerimhoff, F. Grein, *Chem. Phys. Lett.* **172** (1990) 180. (d) B. Engels, M. Peric, W. Reuter, S. D. Peyerimhoff, F. Grein, *J. Chem. Phys.* **96** (1992) 4526.
22. S.D. Wetmore, R. J. Boyd, L. A. Eriksson, *J. Chem. Phys.* **106** (1997) 7738.
23. V. Barone, A. Bencini, A. di Matteo, *J. Am. Chem. Soc.* **119** (1997) 10831.
24. L. A. Eriksson, *Mol. Phys.* **91** (1997) 827.
25. E. Pauwels, V. van Speybroeck, P. Lahorte, M. Waroquier, *J. Phys. Chem.* A **105** (2001) 8794.
26. (a) D. M. Chipman, *J. Phys. Chem.* A **103** (1999) 11181. (b) D. M. Chipman, *J. Phys. Chem.* A **104** (2000) 11816.
27. (a) P. J. O'Malley, *J. Phys. Chem.* A **101** (1997) 6334. (b) P. J. O'Malley, *J. Phys. Chem.* A **102** (1998) 248. (c) P. J. O'Malley, *J. Am. Chem. Soc.* **121** (1999) 3185. (d) P. J. O'Malley, *J. Am. Chem. Soc.* **122** (2000) 7798.
28. F. Himo, *J. Phys. Chem.* A **105** (2001) 7933.
29. (a) F. Ban, J. W. Gauld, R. J. Boyd, *J. Phys. Chem.* A **104** (2000) 5080. (b) F. Ban, J. W. Gauld, R. J. Boyd, *J. Phys. Chem.* A **104** (2000) 8583. (c) F. Ban, S. D. Wetmore, R. J. Boyd, *J. Phys. Chem.* A **103** (1999) 4303. (d) F. Himo, A. Gräslund, L. A. Eriksson, *Biophys. J.* **72** (1997) 1556. (e) F. Himo, L. A. Eriksson, *J. Phys. Chem.* B **101** (1997) 9811. (f) F. Himo, G. T. Babcock, L. A. Eriksson, *Chem. Phys. Lett.* **313** (1999) 374. (g) F. Himo, L. A. Eriksson, M. R. A. Blomberg, P. E. M. Siegbahn, *Int. J. Quantum. Chem.* **76** (2000) 714. (h) F. Himo, *Chem. Phys. Lett.* **328** (2000) 270. (i) N. Rega, M. Cossi, V. Barone, *J. Am. Chem. Soc.* **120** (1998) 5723. (j) V. Barone, C. Adamo, A. Grand, R. Subra, *Chem. Phys. Lett.* **242** (1995) 351.
30. (a) S. D. Wetmore, R. J. Boyd, L. A. Eriksson, *J. Phys. Chem.* B **102** (1998) 5369. (b) S. D. Wetmore, F. Himo, R. J. Boyd, L. A. Eriksson, *J. Phys. Chem.* B **102** (1998) 7484. (c) S. D. Wetmore, R. J. Boyd, L. A. Eriksson, *J. Phys. Chem.* B **102** (1998) 9332. (d) S. D. Wetmore, R. J. Boyd, L. A. Eriksson, *J. Phys. Chem.* B **102** (1998) 10602. (e) S. D. Wetmore, R. J. Boyd, L. A. Eriksson, *J. Phys. Chem.* B **102** (1998) 7674. (f) S. D. Wetmore, D. M. Smith, L. Radiom, *J. Am. Chem. Soc.* **123** (2001) 8678. (g) S. D. Wetmore, D. M. Smith, B. T. Goldling, L. Radiom, *J. Am. Chem. Soc.* **123** (2001) 7963.
31. (a) P. Hohenberg, W. Kohn, *Phys. Rev.* B **136** (1964) 864. (b) W. Kohn, L. J. Sham, *Phys. Rev.* A **140** (1965) 1133. (c) R. G. Parr, W. Yang, *Density-Functional Theory of*

Atoms and Molecules, Oxford University Press, 1989. (d) W. Koch, M. C. Holthausen, *A Chemist's Guide to Density Functional Theory (2nd Edition),* Wiley-VCH, 1999.
32. W. J. Hehre, L. Radom, P. v. R. Schleyer, J. A. Pople, *Ab initio Molecular Orbital Theory,* John Wiley and Sons, 1986.
33. (a) J. P. Perdew, Y. Wang, *Phys. Rev.* B **33** (1986) 8800. (b) J. P. Perdew, *Phys. Rev.* B **33** (1986) 8822. (c) J. P. Perdew, *Phys. Rev.* B **34** (1986) 7406. (d) J. P. Perdew, Y. Wang, *Phys. Rev.* B **45** (1992) 13244.
34. J. C. Slater, *Quantum Theory of Molecules and Solids,* McGraw Hill: New York, 1974.
35. S. H. Vosko, L. Wilk, M. Nusair, *Can. J. Phys.* **58** (1980) 1200.
36. (a) A. D. Becke, *Phys. Rev.* A **38** (1988) 3098. (b) A. D. Becke, *J. Chem. Phys.* **98** (1993) 5648. (c) P. J. Stephens, F. J. Devlin, M. J. Frisch, C. F. Chabalowski, *J. Phys. Chem.* **98** (1994) 11623.
37. C. Lee, W. Yang, R. G. Parr, *Phys. Rev.* B **37** (1988) 785.
38. D. M. Chipman, *Theor. Chim. Acta.* **82** (1992) 93.
39. (a) B. Engels, *Chem. Phys. Lett.* **179** (1991) 398. (b) B. Engels, S. D. Peyerimhoff, *Mol. Phys.* **67** (1989) 583. (c) S. P. Karna, F. Grein, B. Engels, S. D. Peyerimhoff, *Int. J. Quantum Chem.* **36** (1989) 255. (d) S. P. Karna, F. Grein, B. Engels, S. D. Peyerimhoff, *Mol. Phys.* **69** (1990) 549. (e) K. Funken, B. Engels, S. D. Peyerimhoff, F. Grein, *Chem. Phys. Lett.* **172** (1990) 180. (f) D. Feller, E. R. Davidson, *J. Chem. Phys.* **88** (1988) 7580. (g) B. Engels, S. D. Peyerimhoff, E. R. Davidson, *Mol. Phys.* **62** (1987) 109. (h) B. Engels, S. D. Peyerimhoff, *J. Phys.* B **21** (1988) 3459.
40. (a) D. Feller, *J. Chem. Phys.* **93** (1990) 579. (b) D. Feller, E. Glendening, E. A. Jr. McCullough, R. J. Miller, *J. Chem. Phys.* **99** (1993) 2829.
41. V. Barone, C. Adamo, A. Grand, R. Subra, *Chem. Phys. Lett.* **242** (1995) 351.
42. (a) I. Carmichael, *J. Phys. Chem.* **95** (1991) 6198. (b) I. Carmichael, *J. Phys. Chem.* **95** (1991) 108. (c) I. Carmichael, *J. Chem. Phys.* **93** (1990) 863. (d) D. M. Chipman, I. Carmichael, D. Feller, *J. Phys. Chem.* **95** (1991) 4702. (e) I. Carmichael, J. Phys. Chem. 98 (1994) 5044. (f) I. Carmichael, *J. Phys. Chem.* **99** (1995) 6832.
43. H. Sekino, R. J. Bartlett, *J. Chem. Phys.* **82** (1985) 4225.
44. S. A. Perera, J. D. Watts, R. J. Bartlett, *J. Chem. Phys.* **100** (1994) 1425.
45. (a) J. P. Perdew, K. Burke, M. Ernzerhof, *Phys. Rev. Lett.* **77** (1996) 3865. (b) J. P. Perdew, K. Burke, M. Ernzerhof, *Phys. Rev. Lett.* **78** (1997) 1396. (c) C. Adamo, V. Barone, *Chem. Phys. Lett.* **298** (1998) 113.
46. (a) W. Kutzelnigg, U. Fleischer, M. Schindler, In *NMR-Basic Principles and Progress*; Springer-Verlag: Heidelberg, 1990; Vol. **23**. The IGLO-III basis set consists of an (11s7p2d/6s2p) primitive set contracted to [7s6p2d/4s2p].
47. (a) L. A. Eriksson, O. L. Malkina, V. G. Malkin, D. R. Salahub, *J. Chem. Phys.* **100** (1994) 5066. (b) M. A. Austen, L. A. Eriksson, R. J. Boyd, *Can. J. Chem.* **72** (1994) 695. (c) L. A. Eriksson, J. Wang, R. J. Boyd, S. Lunell, *J. Phys. Chem.* **98** (1994) 792. (d) J. M. Martell, L. A. Eriksson, R. J. Boyd, *J. Phys. Chem.* **99** (1995) 623. (e) L. A. Eriksson, J. Wang, R. J. Boyd, *Chem. Phys. Lett.* **235** (1995) 422. (f) J. Kong, L. A. Eriksson, R. J. Boyd, *Chem. Phys. Lett.* **217** (1994) 24.
48. J. B. Foresman, Æ. Frisch, *Exporing Chemistry with Electronic Structure Methods* Gaussian, Inc.: Pittsburgh, PA, 1996.
49. L. Onsager, *J. Am. Chem. Soc.* **58** (1936) 1486.
50. (a) M. W. Wong, M. J. Frisch, K. B. Wiberg, *J. Am. Chem. Soc.* **113** (1991) 4776. (b) M. W. Wong, K. B. Wiberg, M. J. Frisch, *J. Am. Chem. Soc.* **114** (1992) 523. (c) M. W. Wong, K. B. Wiberg, M. J. Frisch, *J. Am. Chem. Soc.* **114** (1992) 1645. (d) M. W. Wong, K. B. Wiberg, M. J. Frisch, *J. Chem. Phys.* **95** (1991) 8991.
51. V. Barone, M. Cossi, *J. Phys. Chem.* A **102** (1998) 1995.

52. C. Adamo, M. Heitzmann, F. Meilleur, N. Rega, G. Scalmani, A. Grand, J. Cadet, V. Barone, *J. Am. Chem. Soc.* **123** (2001) 7113.
53. Y. Ding, K. Krogh-Jespersen, *Chem. Phys. Lett.* **199** (1992) 261.
54. M. A. Collins, D. H. Whiffen, *Mol. Phys.* **10** (1966) 317.
55. J. R. Morton, *J. Am. Chem. Soc.* **86** (1964) 2325.
56. A. Hedberg, A. Ehrenberg, *J. Chem. Phys.* **48** (1968) 4822.

Chapter 7

AB INITIO AND DENSITY FUNCTIONAL CALCULATIONS OF ELECTRONIC G-TENSORS FOR ORGANIC RADICALS

Martin Kaupp
Institut für Anorganische Chemie, Universität Würzburg, Am Hubland, D-97074 Würzburg, Germany

Keywords: ab initio quantum chemistry, density functional theory, g-tensors, nitroxide radicals, phenoxyl radicals, semiquinone radical anions, spin-orbit coupling.

Abstract: Recent development and validation of quantum chemical methods for the calculation of electronic g-tensors is reviewed. The emphasis is on ab initio and density functional methods, whereas semi-empirical methods are covered only briefly. Methodological differences and the relative performance of various approaches are discussed critically, in particular regarding the treatment of spin-orbit coupling and of electron correlation. First applications to biologically relevant radicals are reviewed. Examples range from phenoxyl radicals via semiquinone radical anions to nitroxide spin labels.

1. INTRODUCTION

While hyperfine coupling tensors provide information about the detailed interactions between electronic spin density and certain nuclei within a given radical, the electronic g-tensor is a property of the entire molecule. It reflects the general spin density distribution and often is characteristically influenced by certain bonding features, and by interactions with the environment, *e.g.* in a protein binding site or in the cavities of a zeolite. It thus provides important spectroscopic information that may help in characterizing paramagnetic species in solids.[1,2]

In routine X-band (9 GHz) EPR of organic radicals, it is difficult to resolve accurately the individual components of the g-tensor, unless single-crystal data are available. The appearance of the spectra is usually dominated by partly resolved hyperfine interactions. Increased spin-orbit interactions lead to a larger spread of the tensor and thus allow closer analysis at low magnetic fields only when significant spin density is located onto heavier atoms, or when very low-lying excited states exist. During the past 10 years, however, the development of high-field high-frequency EPR spectroscopy (HF-EPR) has provided an increasing amount of accurate g-tensors of organic radicals, in particular regarding species of biological relevance.[3,4,5] At higher magnetic fields, the g-tensor anisotropy frequently dominates the solid-state spectra and may thus be resolved.

The g-tensors measured may hold key information about structure and bonding of radicals, and on their specific environment. Due to the relatively complicated nature of the g-tensor, its analysis by quantum chemical means is indispensable if we want to relate it to structure and bonding. Detailed models to rationalize the measured g-tensors have been developed early on in EPR history, for transition metal complexes,[6] and for organic π-radicals[7,8] (cf. below). Given the improved spectral information available from HF-EPR experiments, it has become increasingly desirable to be able to calculate g-tensors more quantitatively. Significantly enhanced interest in g-tensors of organic radicals by EPR spectroscopists, combined with improvements in reliable quantum chemical programs and computer hardware, have contributed to a tremendous development in quantitative calculations. This chapter focusses mainly on this progress during the past seven years on the side of ab initio and density functional methods. Previous semi-empirical approaches will be touched upon relatively briefly.

2. THE HAMILTONIAN

The link between the quantum chemical treatment and the actual EPR spectrum is provided by the effective spin Hamiltonian approach. In particular, the g-tensor parametrizes the Zeeman interaction between external magnetic field **B** and an effective spin **S** of the molecule:[2]

$$\hat{H}_{spin} = \mu_B \mathbf{B} \mathbf{g} \mathbf{S}, \qquad (1)$$

(where $\mu_B = e\hbar/2m_e$ is the Bohr magneton). Actually, the EPR experiment does not measure the unsymmetric tensor g, but the diagonal elements of the

7. Calculation of electronic g-tensors

symmetric tensor $G = g \cdot g^T$. We will in the following concentrate on spatially non-degenerate Kramers doublet states, although the perturbation theoretical treatment is also applicable to higher spin multiplicities (see below). Other spin-Hamiltonian terms like zero-field splittings or $\Delta M_S = \pm 2$ transitions will not be covered here.[1,6]

It is necessary to identify in our quantum chemical treatment those terms that are represented by the spin Hamiltonian for the Zeeman interaction. The g-tensor is dominated by spin-orbit (SO) coupling and is thus intrinsically a relativistic property. Therefore, the relativistic four-component Dirac equation in the presence of an external magnetic field provides a suitable starting point for the derivation of the relevant terms in the Hamiltonian.[2,7,9] However, to date all practical calculations of g-tensors have used either a) nonrelativistic (or scalar relativistic) wave functions, with SO coupling and other relativistic terms, as well as the influence of the external magnetic field, added by perturbation theory ("one-component approach"), or b) "two-component methods", in which SO coupling is treated variationally, but the positronic degrees of freedom (which are also described by the Dirac equation) have been eliminated.

In this review, we focus on organic radicals that contain only relatively light atoms. Then a perturbation theoretical treatment of SO coupling (one-component approach) is expected to be sufficient, and the following brief exposition of the theoretical background of quantum chemical g-tensor calculations will concentrate on the perturbation approach, based on the Breit-Pauli (BP) Hamiltonian. The detailed derivation of the terms that occur in the BP Hamiltonian of an open-shell system in the presence of an external magnetic field, starting from the Dirac equation, may be found, *e.g.*, in the book of Harriman[2] and will not be described here in detail. Briefly, the derivation requires transformation from the four-component Dirac equation to two- or one-component form, *e.g.* by the free-particle Foldy-Wouthuysen transformation.[2] Due to singularities in this transformation for systems in the presence of a Coulomb potential (electron-electron repulsion and electron-nucleus attraction), application of the BP Hamiltonian is only appropriate to first order in SO coupling (*i.e.* in a one-component approach with perturbational treatment of SO coupling), and for compounds containing only relatively light atoms. However, regularized transformations like the "zero-order regular approximation" (ZORA) method[10] or the Douglas-Kroll-Hess (DKH) transformation[11] are available that allow a variational, two-component treatment. A DFT-ZORA implementation of g-tensor calculations is already in wide use (see below; cf. ref 12 for more details). Interestingly, the earlier derivation by Harriman[2] is closely related to the

ZORA method.[12] During the preparation of this article, a two-component DKH-DFT implementation of g-tensor calculations has been reported.[13]

We will in the following refer to g-shift tensors, Δg, defined by

$$g = g_e \mathbf{1} + \Delta g, \tag{2}$$

i.e. deviations from the free-electron value, $g_e = 2.002319$. The g-shifts will often be given in ppm, *i.e.* in units of 10^{-6}. Restricting the Breit-Pauli Hamiltonian to those terms that are linear or bilinear in **B** and **S**, and employing Rayleigh-Schrödinger perturbation theory up to second order, we end up with the relevant terms contributing to Δg:[2]

$$\Delta g = \Delta g_{SO/OZ} + \Delta g_{GC} + \Delta g_{RMC}. \tag{3}$$

The "paramagnetic" second-order spin-orbit/orbital Zeeman cross term, $\Delta g_{SO/OZ}$ (eq. 4), dominates the g-shift tensor, except for the smallest Δg-values. It has already been introduced in Pryce's classical perturbation treatment[14] and arises from the joint action of the external magnetic field (orbital Zeeman operator, eq. 5) and the SO operator (BP SO Hamiltonian in eq. 6) in second-order perturbation theory:[2,15]

$$\Delta g_{SO/OZ,uv} = \frac{g_e \alpha^2}{2S} \left[\sum_n \frac{\langle \Psi_0^{(0)} | H_{SO,v} | \Psi_n^{(0)} \rangle \langle \Psi_n^{(0)} | H_{OZ} | \Psi_0^{(0)} \rangle}{E_0^{(0)} - E_n^{(0)}} + c.c \right], \tag{4}$$

$$H_{OZ,u} = -\sum_i l_{iO,u}, \tag{5}$$

$$H_{SO,v} = \sum_i \left(\sum_N \frac{Z_N \mathbf{L}_{iNv}}{r_{iN}^3} - \sum_{j \neq i} \frac{\mathbf{L}_{jv}^i}{r_{ij}^3} - 2\sum_{j \neq i} \frac{\mathbf{L}_{iv}^j}{r_{ij}^3} \right) s_i. \tag{6}$$

In eqs. 4-6 (which are in atomic units), α denotes the fine structure constant, u and v are Cartesian coordinates, $\Psi_o^{(0)}$ the unperturbed ground-state wavefunction, and $\Psi_n^{(0)}$ the unperturbed wavefunction of the n'th excited state (with eigenenergies $E_o^{(0)}$ and $E_n^{(0)}$, respectively). l_{io} represents orbital angular momentum, and S is the magnitude of total spin of the molecule (it has been introduced to make equations 4, 7, and 8 suitable also for high-spin systems[16,17]). Within the SO operator H_{SO} (eq. 6), Z_N is the charge of nucleus N, $\mathbf{L}_{iN}$ the angular momentum of electron i relative to the position of nucleus N, $\mathbf{L}^i_j$ the angular momentum of electron i relative to the position of electron j, $\mathbf{r}_{iN}$ and $\mathbf{r}_{ij}$ are relative electron-nucleus and electron-electron position vectors, and s_i is a spin operator for electron i. The first term in eq. 6 is the one-electron SO operator, in which the interaction

7. Calculation of electronic g-tensors

between spin and angular momentum of an electron is mediated by the potential due to the charge of the nuclei. The second term is the two-electron SO operator (largely due to the screening of nuclear charge around nucleus N by the core electrons), and the third term is the so-called spin-other-orbit (SOO) term (see 3.3).

The "paramagnetic" spin-Zeeman gauge correction terms, Δg_{GC} (eq. 7), reflect the magnetic-field dependence of the SO operators. They have been considered first by Stone to obtain a properly gauge-invariant theory of the g-tensor,[7] but they are usually much smaller than the $\Delta g_{SO/OZ}$ contributions. In most implementations, the two-electron contributions to Δg_{GC} are either neglected or taken into account only approximately (cf. Table 1).

$$\Delta g_{GC,uv} = -\frac{\alpha^2 g_e}{4S}\left\langle \Psi_o^{(0)} \left| (\sum_N Z_N \sum_i \frac{\mathbf{r}_{iN} \cdot \mathbf{r}_i \delta_{uv} - r_{iNu} r_{iv}}{r_{iN}^3}) s_i + (\sum_{ij} \frac{\mathbf{r}_{ij} \cdot (2\mathbf{r}_j - \mathbf{r}_i)\delta_{uv} - (2\mathbf{r}_j - \mathbf{r}_i)_v (\mathbf{r}_{ij})_u}{r_{ij}^3}) s_i \right| \Psi_o^{(0)} \right\rangle$$

(7)

Finally, the relativistic mass correction to the spin-Zeeman term, Δg_{RMC} (eq. 8), arises from field-dependent kinematic relativistic effects and provides another relatively small, isotropic contribution to the Δg-value:[2,18]

$$\Delta g_{RMC-SZ,uv} = -\frac{\alpha^2 g_e}{2S}\delta_{uv}\left\langle \Psi_o^{(0)} \left| \sum_i (-\frac{1}{2}\nabla_i^2) s_i \right| \Psi_o^{(0)} \right\rangle$$

(8)

Without this contribution, it would be impossible to explain the frequently negative Δg_{zz} contributions in organic π-radicals.[18] Δg_{GC} and Δg_{RMC} may be calculated in a relatively straightforward manner as expectation values of the unperturbed ground state wave function. Note that, in contrast to a two-component approach, in the perturbation theoretical one-component treatment S is still considered to be a good quantum number.

3. QUANTUM CHEMICAL APPROACHES

While the calculation of hyperfine tensors does already have an appreciable history of first-principles theoretical treatments,[19] the massive development of methods to calculate electronic g-tensors by ab initio quantum chemistry, and by modern approaches of density-functional theory (DFT), started only in the mid 1990's (earlier, approximate DFT implementations have already been reported in the 1980's,[20] and very early ab initio calculations date back even further[21]). In spite of this relatively short period of time, the variety of alternative approaches available is already appreciable and may appear confusing to the non-expert interested in applications of such quantum chemical calculations. We will in the following attempt to mark the main distinguishing features of various methods. Before turning to ab initio methods, we briefly summarize the main points of the earlier semiempirical approaches.

3.1 Semi-empirical calculations

Semiempirical MO treatments use the orbitals and orbital energies obtained in approximate semiempirical calculations (originally at the simple Hückel level,[7,8] later at extended-Hückel,[22] CNDO,[23] INDO[18,24-30] or NDDO[31-33] levels of approximation) to evaluate the dominant $\Delta g_{SO/OZ}$ terms in eq. 4. At the simplest level, the energy denominator was simply replaced by the energy differences between occupied and virtual MOs. This would correspond to an uncoupled Hartree-Fock approach. However, Hartree-Fock orbital energies (also in semi-empirical variants) are not well suited to describe excited states, as the virtual MOs are too diffuse and experience shielding due to all occupied MOs. As a next step, Coulomb-shift terms have been added (this corresponds to the use of configuration-state energies rather than orbital energies).[24] Due to the presence of Hartree-Fock exchange terms at CNDO, INDO, or NDDO levels, a full variational perturbation theory treatment would require the iterative solution of the coupled-perturbed Hartree-Fock equations (see below).

Further approximations were often applied in the semi-empirical approaches. In particular, the matrix elements in eq. 4 were usually restricted to their one-center contributions. While this appears to be a reasonable approximation for the matrix elements of the very short-ranged SO operator (Stone's first approximation[7]), the OZ operator is less localized, and the one-center approximation is doubtful in this case (Stone's second approximation[7]). Later calculations thus dropped the second approximation. It should be noted, however, that the minimal basis sets employed in semi-empirical MO approaches, and the often ill-suited nature of the virtual MOs, do already severely limit the accuracy achievable in such calculations. A further potential source of uncertainty stems from the semiempirical, effective one-electron SO operators employed (but see below[21]).

7. Calculation of electronic g-tensors

3.2 Ab initio approaches

Early Hartree-Fock calculations of g-tensors by Moores and McWeeny[21] still suffered from the small basis sets and limited wave functions available at the time (see also ref 24). The first modern ab initio calculations are due to Lushington *et al.*, who developed restricted-open-shell Hartree-Fock (ROHF)[34] and multi-reference configuration-interaction (MRCI) approaches.[15,35] They used explicitly a sum-over-states (SOS) approach (over a limited number of states) to calculate the $\Delta g_{SO/OZ}$ terms of eq. 4, and they included the other two terms in eq. 3 as well. The lack of electron correlation and spin polarization in the ROHF calculations is corrected for at the MRCI level, where the ground-state wavefunction consists already of a superposition of multiple configuration-state functions. Lushington *et al.* calculate explicitly all matrix elements of the full Breit-Pauli SO Hamiltonian (eq. 6). This, and the effort involved in the configuration-interaction approach, limits the calculations to small molecules. However, there the method provides benchmark results of high accuracy (cf. section 4). Recently, Lushington evaluated the use of a more compact CI expansion.[36] As only those excited states contribute to $\Delta g_{SO/OZ}$, which couple magnetically with the ground state, a smaller configuration space can be devised, which only includes these states. This allows the inclusion of all excitations within this space ("closed-form" CI expansions, CFCI approach). The first test calculations of the CFCI method indicate reduced computational effort for symmetrical, small molecules, with an accuracy comparable to that of the more expensive MRCI treatment.[36]

In their alternative approach, Vahtras *et al.* use linear response theory at the ROHF and multi-configuration self-consistent field (MCSCF) levels of ab initio theory.[37] Here the explicit SOS expansion of Lushington *et al.* is replaced by analytical linear response functions. This has the advantage of an implicit summation over all excited configurations. The MCSCF treatment, with a relatively small number of configuration-state functions, allows the inclusion of the major non-dynamical electron-correlation contributions. Again, while the ROHF approach is also applicable to somewhat larger systems, the correlated MCSCF treatment is limited to smaller molecules. The active orbital space within the MCSCF treatment is also restricted by the available computational resources, and it is usually not possible to cover much dynamical electron correlation. The initial implementation calculated explicitly the full set of one- and two-electron integrals over the Breit-Pauli SO Hamiltonian. Meanwhile, the more economic atomic-meanfield approximation (AMFI, cf. below) has been implemented into the code.[38]

The advantage of the ab initio treatments is that their accuracy may in principle be improved systematically towards the exact result (for a given Hamiltonian and level of perturbation theory), by extending the one-particle basis set and the number of configurations included in the CI or MCSCF treatment. Generally, however, the ab initio approaches mentioned up to now are most suitable for small molecules, and they serve best as methods to benchmark more economic, approximate treatments. Still on the ab initio side, we should mention the two-component UHF approach of Jayatilaka.[39] Notably, in the two-component treatment only first-order perturbation theory is required (see also below for the ZORA treatment). One-component unrestricted coupled-perturbed Hartree-Fock treatments have recently been reported as a side result of two different hybrid-DFT approaches[40,41] (see below).

3.3 Approaches based on density functional theory

A big step towards the reliable calculation of g-tensors for larger systems has been achieved by using DFT methods. DFT implicitly includes electron-correlation effects at much lower cost than the abovementioned post-Hartree-Fock methods.[42] A principal disadvantage of DFT is the lack of a way for systematic improvement towards an exact limit, as the exact exchange-correlation functional is not known. DFT calculations of properties like g-tensors thus require careful validation for a given functional. A further general, as yet unsolved question in DFT calculations of magnetic response properties is the dependence of the exchange-correlation potential on the paramagnetic currents induced by the magnetic field. The problems related to the current dependence have been discussed in detail in the field of NMR chemical shift calculations,[43] but the problem applies equally to calculations of g-tensors. As appropriate and reliable current-density functionals are lacking, all implementations presently neglect the current dependence. Some open questions pertain also to the proper treatment of SO coupling within a DFT framework (see below and ref 41). In spite of a significant number of open basic theoretical questions, DFT methods have become increasingly important in g-tensor calculations. This is mainly due to the overall high quality of the results combined with computational efficiency.

One of the two first modern DFT approaches (see below; cf. ref 20 for more approximate, older implementations at the X_α level) has been reported by van Lenthe et al.[12] and is implemented in the widely used ADF code.[44] It differs fundamentally from most of the other methods discussed below, as it is a two-component method. The zero-order regular approximation

7. Calculation of electronic g-tensors

(ZORA[10]) is employed, a regularized version of the elimination-of-the-small-component-method for the transformation of the Dirac equation. The main advantage of the approach is the variational treatment of SO coupling which is important mostly for compounds of heavier elements, or for systems with very close-lying excited states. A limitation in the current implementation is the use of a spin-restricted treatment, due to the difficulty of describing spin polarization within a two-component DFT approach. The approximate SO operators used are a further potential source of error (cf. discussion below). The very recent two-component DKH-DFT approach of Neyman et al.[13] is conceptually similar to the ZORA approach. It also currently uses a spin-restricted formalism and approximate SO operators.

The first modern one-component DFT approach reported by Schreckenbach and Ziegler (SZ)[45] has been followed by two alternative but related implementations, by us[46] and later by Neese.[40] These three approaches do all employ second-order perturbation theory based on unrestricted Kohn-Sham calculations. Apart from some more technical aspects summarized further below, the three implementations differ mainly in the treatment of the SO operators in eq. 4. During the preparation of this article, another DFT method has been reported by Pickard and Mauri.[47] It differs from the other approaches by employing periodic boundary conditions and augmented plane wave basis sets. It thus is particularly well suited for studies on extended solids.

A proper relativistic DFT treatment of SO coupling requires relativistic density functionals that include, among other things, the Breit interaction (transverse interaction), which derives from relativistic corrections to electron-electron repulsion. However, none of the existing implementations uses relativistic functionals, which introduce additional complications. Schreckenbach and Ziegler[45] (and van Lenthe et al. in their ZORA approach[12]) approximate the SO contributions in the following way: In addition to the potential due to the charge of the nuclei (one-electron SO term), the electrons experience shielding due to the other electrons, mainly due to the core electrons (two-electron SO term). Within Kohn-Sham theory, this shielding is represented by the relevant contributions from an approximate effective Kohn-Sham potential (Coulomb and X_α-exchange terms). The Breit interaction is not included in this treatment, and the so-called "spin-other-orbit" (SOO) terms are thus lacking. As the SOO terms are of opposite sign than the dominant one-electron SO terms, their lack leads overall to an overestimate of the SO/OZ contributions (see Section 4).

In our own implementation,[46] we step outside the conventional Kohn-

Sham treatment and use three different types of approaches to the SO matrix elements: i) The full treatment of all one- and two-electron integrals over the full BP SO operator (using explicitly the Kohn-Sham determinant as approximation to the wavefunction[41]), as done in the abovementioned ab initio methods. This approach includes the SOO terms (as well as some exact-exchange-type SO contributions[41]) but becomes prohibitively expensive for larger molecules. ii) The atomic meanfield approximation[48] (as implemented in the AMFI code[49]) replaces the explicit treatment of the integrals by integrals over a sum of effective atomic SO operators (replacing the full molecular density matrix by a superposition of one-center density matrices). This reduces the computational effort significantly relative to the full BP treatment, with very little loss of accuracy.[46] The AMFI approximation provides thus a very powerful means of treating the SO matrix elements in eq. 4 (including the SOO terms) accurately for larger molecules. It has also been used successfully in many other types of applications, ranging all the way from SO-effects on NMR chemical shifts[50] to photophysical transition probabilities.[51] iii) SO pseudopotentials (SO-ECPs) provide a further simplification by treating only the valence electrons explicitly. SO-ECPs are used in conjunction with scalar relativistic pseudopotentials inserted at the Kohn-Sham stage of the calculation.[46,52] This approximation is well-suited for g-tensor calculations and becomes particularly useful for compounds of heavy elements.[46] We will not discuss it further here. As another possibility, Neese uses semi-empirical, effective one-electron SO operators in his recent DFT implementation[40] (by introducing an effective nuclear charge, as done in all semi-empirical and early ab initio or DFT treatments). Pickard and Mauri apparently include an approximate treatment of the SOO terms.[47] However, the few data reported as yet are very close to those of SZ, suggesting that the SOO contributions are underestimated.

All initial DFT implementations used the local density approximation (LDA) or the generalized gradient approximation (GGA).[42] For these "pure" DFT approaches, the local exchange operators and lack of a current dependence of the functional (see above) make any coupling terms in the second-order perturbation theory treatment vanish. We thus end up with an uncoupled set of equations (uncoupled DFT approach, UDFT[53]), in which the energy denominator in eq. 4 may be replaced by the occupied and virtual Kohn-Sham orbital energies (and the sum over excited states is replaced by a double sum over all occupied and virtual orbitals). Fortunately, the Kohn-Sham orbitals include electron correlation effects implicitly. This is also reflected in the orbital energies. The UDFT approach is therefore a much better approximation than one might expect from what was said further

7. Calculation of electronic g-tensors

above about the poor quality of uncoupled Hartree-Fock methods. In particular, the virtual Kohn-Sham MOs experience the field of only n-1 electrons and are thus a much better basis for perturbation theory (and for the description of electronically excited states) than are virtual Hartree-Fock MOs.[42,53,54] The one-component DFT perturbation approach with LDA or GGA functionals is thus particularly easy to implement, computationally very efficient, and surprisingly accurate (see section 4). This is a major reason for its successful application to larger molecules.

The UDFT approach with currently available LDA or GGA functionals is less successful in the calculation of g-tensors for transition metal complexes.[45,46] Therefore, both we[41] and Neese[40] have recently extended our respective codes to allow the use of so-called hybrid functionals that include a part of exact Hartree-Fock exchange. This "exact-exchange mixing" requires the iterative solution of coupled-perturbed Kohn-Sham (CPKS) equations. While hybrid functionals have indeed been shown to provide a possibility to improve the performance for transition metal systems, they do not appear to lead to improved g-tensors for organic radicals.[40,41] In view of the larger computational effort involved in the CPKS treatment, hybrid functionals will thus not be discussed further here. Recently, we have also validated two so-called meta-GGA functionals for EPR parameter calculations.[55] No significant improvement over GGA results has been found for transition metal complexes. The g-shifts of main group radicals are slightly smaller than obtained with GGA functionals.[55]

Table 1. Characteristics of four different DFT approaches for the calculation of g-tensors

	van Lenthe[a]	SZ[b]	Malkina et al.[c]	Neese[d]
program	ADF	ADF	deMon-EPR	ORCA
general method	2-component ZORA	1-component BP	1-component BP	1-component BP
basis sets	STO	STO	GTO	GTO
functionals	LDA, GGA	LDA, GGA	LDA, GGA, hybrid (HF)	LDA, GGA, hybrid (HF)
SO integrals	effective Kohn-Sham	effective Kohn-Sham	explicit BP, AMFI, SO-pseudopotentials	semi-empirical eff. 1-el.
gauge treatment	GIAO, common	GIAO, common	IGLO, common[e]	common
first-order terms	RMC	RMC	RMC	RMC
	GC eff. 1-el.	GC eff. 1-el.	GC 1-el.	GC eff. 1-el.

[a]Cf. ref 12. [b]Cf. ref 45. [c]Cf. ref 46. Extensions to hybrid functionals, see ref 41. [d]Cf. ref 40.
[e]GIAOs have recently been implemented.

Due to the presence of an external magnetic field in the perturbation expressions (cf. eq. 1), g-tensor calculations are in principle subject to the so-called "gauge problem", that has hampered NMR chemical shift calculations for a long time. In calculations with a finite basis set, the result depends on the choice of gauge origin of the magnetic vector potential. In the field of NMR chemical shifts, distributed-gauge methods like gauge-including-atomic-orbitals (GIAO[56]) or individual-gauges-for-localized-orbitals (IGLO[57]) have been developed, which allow the gauge- and basis-set-dependence to be minimized. These approaches have also been extended to g-tensor calculations.[12,45-47] Fortunately, however, g-tensors are much less gauge-dependent than chemical shifts. Thus, calculations with a common gauge origin, e.g. at the center of charge or center of nuclear charges of the molecule, typically do already provide sufficiently accurate results.

Given the appreciable technical differences in the various DFT implementations, Table 1 summarizes the major characteristic features of four different available methods.

4. PERFORMANCE OF AB INITIO AND DFT METHODS, VALIDATION STUDIES

Initial validation studies of ab initio methods focussed on small main group radicals, for which either gas-phase microwave data or EPR spectra in inert matrices were available. This is expected to minimize environmental effects and to allow the evaluation of the intrinsic accuracy of the quantum chemical methods. The first studies were carried out at ROHF and MRCI levels by Lushington et al.[15,34-36] Subsequently developed methods have often been gauged initially for similar sets of small radicals. Table 2 shows a collection of results for some of these systems with Lushington's MRCI results and a range of different DFT methods (all using the BP86 GGA functional for better comparison). The comparison has been taken from the recent work of Neese,[40] except for a few corrections, and for missing entries with our own approach (AMFI), which we have added. We focus only on methods that incorporate electron correlation. Results at the overall less reliable Hartree-Fock level of theory may be found in the comparative study of Neese.[40]

Table 2. Comparison of different DFT approaches for the calculation of g-shift components (in ppm) for small radicals (BP86-UDFT results)[a]

molecule		expt.	AMFI	SZ	Neese	ZORA	MRCI
CO^+	$\Delta g_\perp$	-2400[b]	-2458	-3129	-2622	-3464	-2674
CN	$\Delta g_\perp$	-2000[b]	-1939	-2514	-2033	-2701	
BO	$\Delta g_\perp$	-1700[b]	-1796	-2298	-1924	-2455	-1899
BS	$\Delta g_\perp$	-8100,-8900	-9043	-9974	-9928	-11743	-8449
MgF	$\Delta g_\perp$	-1300	-1869	-2178	-2433	-1967	-1809
AlO	$\Delta g_\perp$	-1800 to -1900[b]	-1991	-222	-2920	855	-2284

7. Calculation of electronic g-tensors

H_2O^{+c}	Δg_x	200	-142	103	-185	-780	-292
	Δg_y	18800	10205	13824	11475	46527	16019
	Δg_z	4800	3702	5126	4412	7765	4217
HCO^c	Δg_x	1500	2275	2749	2183	3095	
	Δg_y	0	-224	-270	-307	-175	
	Δg_z	-7500	-7476	-9468	-7891	-12196	
NF_2^c	Δg_x	-100	-617	-738	-636	-318	
	Δg_y	6200	6288	7619	6970	10576	
	Δg_z	2800	3928	4678	4264	6254	
NO_2^c	Δg_x	3900	3400	4158	3281	5000	3806
	Δg_y	-11300	-11229	-13717	-11270	-16000	-10322
	Δg_z	-300	-688	-760	-706	-600	-235
O_2H^c	Δg_x	-800	-221		-305	-2150	
	Δg_y	39720	22629		24300	85158	
	Δg_z	5580	4879		5120	5127	
CO_2^{-c}	Δg_x	700	1086	1522	490	2236	
	Δg_y	-4800	-5420	-7210	-5151	-8056	
	Δg_z	-500	-769	-803	-969	-624	
O_3^{-c}	Δg_x	200, 1300	-429	-554	-476	-439	
	Δg_y	16400, 14700	15312	19380	15323	22475	
	Δg_z	10000, 9700	8552	10542	8736	12710	
H_2CO^{+c}	Δg_x	200	62	76	18	-286	1296
	Δg_y	-800	-1067	-1220	-678	-1131	-30
	Δg_z	4600	4837	6231	5507	8046	5510
$C_3H_5^c$	Δg_x	0	-65	-115	-79	-152	
	Δg_y	400	497	769	634	854	
	Δg_z	800	603	660	731	1023	

[a]Cf. ref 40 for more data and further discussions. [b]Best gas phase estimate.[58] [c]For C_{2v} symmetrical triatomic radicals, g_x is perpendicular to the molecular plane, g_z is parallel to the bisector of the B-A-B angle, and g_y is in-plane tangential. The orientation for the C_s symmetrical triatomic radicals is analogous. In H_2CO^+, g_x is oriented in-plane, perpendicular to the C-O bond, g_y is parallel to the C-O bond, and g_z is out-of-plane. In C_3H_5, g_x is in-plane, perpendicular to the central C-H bond, g_x is out-of-plane, and g_z is parallel to the C-H bond.

It should be noted that only part of the experimental data are from gas-phase microwave measurements, where $\Delta g_\perp$ is estimated from spin-rotation constants, via Curl's equation.[59] The other data are matrix-isolation EPR results. In the latter case, some of the smaller values (in particular $\Delta g_\parallel$ for some linear radicals, and some values for charged species) may be influenced significantly by environmental effects and should thus be viewed with caution. In general, very small values also tend to have significant relative contributions from first-order terms. We disregard in the following the $\Delta g_\parallel$ values for linear radicals. But even the more reliable data may easily exhibit error ranges of ±500 ppm.

Keeping these restrictions in the experimental data base in mind, it is clear that the MRCI approach provides rather accurate results, except for Δg_x

in H_2CO^+, where it differs both from experiment and from all DFT methods. As ROHF and UHF calculations give ca. 3000 ppm for this value,[40] it appears possible that the limited CI treatment in this case was not able to correct for the severe Hartree-Fock errors. In most cases, we may take the MRCI data as best benchmark data to gauge the DFT results. When doing so, it becomes clear that in the majority of the relevant components, DFT results with atomic meanfield SO operators (AMFI) tend to be in reasonable agreement with MRCI or experiment, or their magnitude is slightly too large. Exceptions are Δg_y in H_2O^+, and in O_2H. In these two, somewhat unusual cases, DFT results with gradient-corrected functionals (BP86) tend to significantly underestimate the corresponding g-shift. Interestingly, local density functionals provide larger values for these systems.[40,41] In the case of H_2O^+, this has been traced back to an unusually large dependence of the HOMO and SOMO energies on the functional, due to extremely pronounced spin polarization.[41]

To obtain a meaningful statistical comparison that allows us to judge the influence of the different SO operators and treatment of spin polarization, we select in the following the larger, presumably more reliable components of a number of first- and second-row radicals. We exclude H_2O^+ and O_2H, due to their exceptional behavior (cf. above). AlO is also excluded, as here the data of SZ are very different from all other DFT results, for unknown reasons. Table 3 shows the resulting statistical analysis of the correlation between the different DFT methods and experiment. Figure 1 shows the corresponding plot for the AMFI results. It should be kept in mind, that a different selection of examples could lead to somewhat larger or smaller slope or scatter.

We argue that the AMFI SO operator is an excellent approximation to the full Breit-Pauli SO Hamiltonian, and thus the remaining errors reflect the intrinsic deficiencies of the given exchange-correlation functional.[41,46] If we thus take the AMFI data as reference DFT values, we may obtain estimates of the influence of technical differences of the other three DFT methods. Neese's GTO implementation differs from our method mainly in the use of semi-empirical SO operators.[40] The g-shifts are close to the AMFI results in most cases, with a tendency to be somewhat higher (at most a few percent). The STO implementation of SZ[45] uses the abovementioned effective Kohn-Sham-potential approach to the two-electron SO contributions. It gives consistently ca. 25% larger g-shifts than the AMFI results, and thus somewhat inferior agreement with experiment or MRCI results for these light main group radicals (except for systems like H_2O^+ or O_3^-). About half of the discrepancy between AMFI and SZ results has been attributed to the lack of SOO contributions for the latter.[46] Another part is accounted for by

7. Calculation of electronic g-tensors

Table 3. Results of linear regression analyses for comparisons between theory and experiment for selected g-shift tensor components of first-row radicals[a]

	AMFI	SZ	Neese	ZORA
intercept B[a]	-1	4	73	359
slope A[a]	1.039	1.264	1.094	1.553
regression coefficient	0.996	0.995	0.992	0.995
standard deviation (in ppm)	489	603	680	808

[a] $\Delta g_{ii}(\text{calc.}) = A\, \Delta g_{ii}(\text{exp.}) + B$ (A and B in ppm). The data include the 16 presumably most reliable values: $\Delta g_\perp$ of CO^+, CN, BO, BS, MgF, as well as Δg_x and Δg_z of HCO, Δg_y and Δg_z of NF_2, H_2CO^+, and C_2H_5, Δg_x and Δg_y of NO_2, and Δg_y of CO_2^-. All calculations used the BP86 GGA functional and extended basis sets.

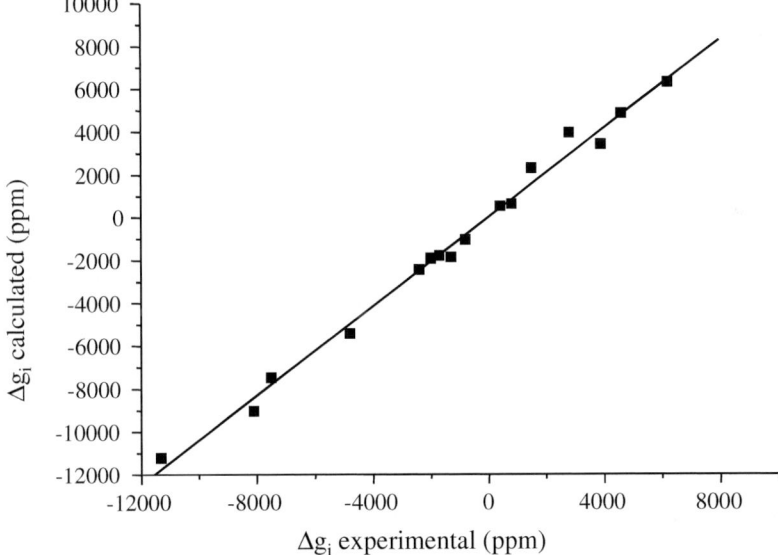

Figure 1. Comparison of DFT results with AMFI SO operators for selected g-tensor components of main group radicals (UDFT-BP86 data, cf. Tables 2,3)

exact-exchange-type SO integrals that appear in the AMFI (or full Breit-Pauli) implementation[41] but not in SZ's method. The good correlation coefficient suggests that the larger SO values are systematic. However, the relative magnitude of the different SO contributions changes for systems with atoms from other parts of the periodic table.[46]

The ZORA approach is a two-component method that includes SO coupling variationally and thus differs from the other three methods.[12] However, the two-component treatment is not expected to lead to major differences for the present light-atom radicals, as SO effects are small. The ZORA method and the SZ method are both implemented in the ADF code.[44]

They use the same types of STO basis sets and very similar SO operators. The major difference is the spin-restricted nature of the current ZORA implementation. This appears to lead to significantly larger g-shifts in most cases, *i.e.* the ZORA values typically overestimate experiment or MRCI results by ca. 55%. One notable exception is $\Delta g_\perp$ in the MgF radical, where the ZORA result is in the same range as the AMFI or SZ data. The SOMO in MgF is a magnesium 3s-orbital with slight $3p_z$ admixture, which polarizes the orbital away from the fluorine ligand. Spin polarization effects in this radical are known to be very small, and thus the spin-restricted approach gives close agreement with the spin-unrestricted methods. Consequently, the ZORA calculations overestimate Δg_y by more than 100% for systems like H_2O^+ and O_2H, where spin polarization is very large (*cf.* above). In a number of calculations, van Lenthe *et al.* used the one-component approach of SZ in a spin-restricted way for comparison to the ZORA calculations.[12] Then the agreement between ZORA and perturbational results was good, whereas the g-shift were generally larger than in an unrestricted treatment. This suggests that spin polarization tends to reduce g-shifts in general.

The above discussion used data obtained only with the BP86 GGA functional. However, available calculations with other GGA functionals gave very similar results. LDA functionals tend to provide somewhat larger g-shifts, which leads to overall inferior agreement with experiment for light main-group radicals (but to slightly improved agreement for many transition-metal complexes). The few available calculations with hybrid functionals indicate similar results as obtained at the GGA level.[40,41] Our first test calculations with meta-GGA functionals (which include dependencies on the Laplacian of the density or on local kinetic energy density) provided somewhat smaller g-shifts than GGA functionals.[55]

Calculations on larger organic π-radicals (see also section 5) confirm the small-radical results discussed here. UDFT calculations with GGA functionals and AMFI SO operators tend to provide good correlation with experimental data but apparently overestimate the largest g-shift components systematically by ca. 5-10%.[46,60] Correlated ab initio data are scarce in this case and consist essentially of some MCSCF calculations on the simplest aromatic π-radicals.[61-63] As these calculations had to be limited to relatively small basis sets and active orbital spaces, they are probably not suitable as high-level benchmark data. DFT results thus may only be judged against experiment. This requires usually an adequate treatment of environmental effects (hydrogen bonding, dielectric effects, etc.). Moreover, low-energy vibrational or rotational motion may also have to be considered for the larger systems. Examples will be mentioned in section 5.

Fortunately, it has been found that the basis-set requirements of DFT g-tensor calculations are moderate.[45,46] Typically, polarized valence double-zeta basis sets do already provide quite reasonable results, and further basis-set extension gives only small effects. However, it should be noted that in aromatic π-radicals polarization p-functions on the ring hydrogen atoms have been found necessary.[46] Many DFT programs improve the scaling of computational cost with system size by using auxiliary basis sets to fit the charge density (and sometimes the exchange-correlation potential). This introduces a further approximation that needs to be controlled carefully.

We note in passing, that recent DFT calculations on high-spin main group radicals provided an accuracy comparable to previous results on doublet states.[16] As expected, the accuracy deteriorated for systems containing transition metals. MCSCF calculations on diatomic triplet radicals had already demonstrated the feasibility of calculations on nondegenerate high-spin states.[17]

5. APPLICATIONS TO BIOLOGICALLY RELEVANT RADICALS

The more quantitative methods for g-tensor calculations discussed above have only become available recently. Therefore, the number of detailed applications to biologically relevant radicals is as yet limited. Among the persistent radicals of importance in biological electron transfer processes,[64] tyrosyl radicals, and in particular semiquinone radical anions have been the primary targets of the first g-tensor calculations by ab initio or DFT methods. Other studies have dealt with nitroxide spin labels, with sulfur-based radicals of potential biological interest, or with tryptophan radicals.

5.1 Phenoxyl Radicals

The tyrosyl radical, a substituted phenoxyl radical derived from the amino acid tyrosine, is of central importance in many single-electron transfer processes in living organisms, from photosynthesis to cell replication. It has been investigated by HF-EPR in a considerable variety of proteins, and in vitro.[31,65,66] Its g-tensor and that of related phenoxyls have thus also been studied quantum chemically, initially by semi-empirical calculations,[31,66] more recently also by ab initio[61-63] and DFT[46] methods.

Stone's perturbation model[7,8] predicts positive Δg_x and Δg_y components for the phenoxyl radical, with the largest (Δg_x) oriented parallel to the carbonyl C-O bond (Figure 2). The second largest component, Δg_y, is expected to be oriented within in the phenoxyl plane, perpendicular to the C-O bond. Essentially no $\Delta g_{SO/OZ}$ contributions are predicted for Δg_z, perpendicular to the plane (Figure 2). These expectations have been confirmed experimentally,[31,65,66] by semi-empirical MO calculations,[31,66] and by recent, more quantitative calculations.[46,61-63] The Δg_x component is dominated by a $^2B_1 \rightarrow {}^2B_2$ SO/OZ contribution (excitation from the n(b_2)-HOMO to the π(b_1)-SOMO), whereas Δg_y arises mainly from interaction between the π $^2B_1$ ground state and the σ $^2A_1$ excited state.[61]

It has been shown[61] that quantitative calculations of phenoxyl g-tensors require a particularly adequate treatment of nondynamical electron correlation effects. Ab initio ROHF results overestimate both Δg_x and Δg_y dramatically. MCSCF calculations provided much better results, but due to their large computational requirements these could only be applied to small models like unsubstituted phenoxyl.[61] A significant dependence of the g-shift components on the C-O bond length has also been noted. Good agreement with experiment has been obtained in UDFT calculations with AMFI SO operators and GGA functionals.[46] Typically, Δg_x is overestimated by ca. 5-10% at this computational level. This has been shown by comparison with experimental data for the tyrosyl radical (in the absence of hydrogen bonding), and for the 2,4,6-tris-*t*-Butyl-phenoxyl radical (no experimental data are available for phenoxyl itself).

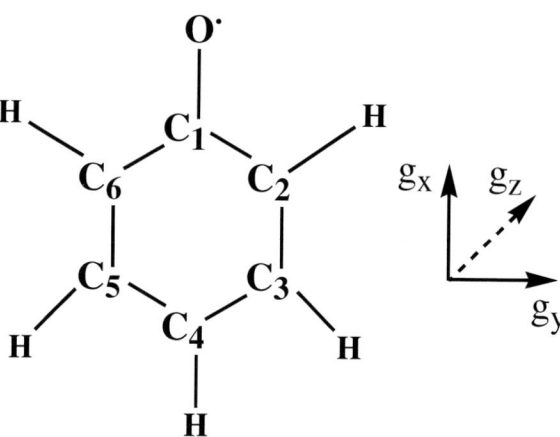

Figure 2. Conventional g-tensor orientation for phenoxyl radicals, and atom labelling

7. Calculation of electronic g-tensors

It had always been assumed that the g-tensor of phenoxyl radicals is dominated by the large spin density and SO coupling constant of the carbonyl oxygen atom. This was later confirmed by the MCSCF and DFT studies. However, our UDFT study, which used the AMFI approximation and thus allowed a break-down of the g-shift components into atomic contributions, showed that the ortho carbon atoms (C_2, C_6) provide small but nonnegligible negative contributions (ca. −360 ppm) to Δg_x and positive contributions (ca. +280 ppm) to Δg_y.[46] The para carbon atom (C_4) contributes another ca. +150 ppm to both Δg_x and Δg_y.

Substituents in 2- or 6-position of the phenoxyl radical have a rather significant influence on the g-tensor, in particular on Δg_x. For example, Δg_x in the 2,4,6-tris-*t*Butyl-phenoxyl is reduced by ca. 35% (Δg_y by ca. 15%) compared to phenoxyl.[46] This is due to hyperconjugation of the alkyl substituents with the delocalized π-system of phenoxyl, which reduces the spin density on oxygen. Similar substituent effects have been computed for semiquinones (see below). In contrast, differences between tyrosyl and unsubstituted phenoxyl are below 5%. This indicates that the "alanyl" substituent in 4-position of the tyrosyl radical has only a small influence on the g-tensor, and thus phenoxyl (or even better 4-methyl-phenoxyl) is a reasonable model for g-tensor calculations on the tyrosyl radical.[46]

Effects of thioether substituents in tyrosyl radicals have been of interest in the context of the function of galactose oxidase (GO). In this enzyme complex, which catalyzes the two-electron oxidation of primary alcohols to aldehydes,[64,67] a tyrosine residue is covalently cross-linked in *ortho*-position to the thioether function of a cysteine residue.[68] Whether this link has a crucial effect on spin-density distributions and g-tensors of the tyrosyl radical created, has been discussed controversially, from the experimental[69,70] and theoretical[63,70] side. EPR measurements indicate a reduced Δg_x (by ca. 25%) and enhanced Δg_y (by ca. 100%) compared to unsubstituted tyrosyl, leading to an almost axially symmetrical tensor.[69,70] Gerfen et al.[70] used Kohn-Sham orbitals obtained in LDA calculations to compute the $\Delta g_{SO/OZ}$ contributions to g-shifts for cresyl and 2-methylthiocresyl model radicals and found very large differences between the two systems. However, these calculations involved severe approximations, in particular the one-center approximation for both SO and OZ matrix elements in eq. 4 (Stone's first and second approximation,[7] cf. 3.1) and did therefore not provide quantitatively correct results. Recent MCSCF calculations of the g-tensor for the *ortho*-SH substituted phenoxyl radical[63] instead provided very small substituent effects, a ca. 13% decrease in Δg_x but only a 7% increase in Δg_y (the principal axis of g_x was rotated by ca. $9°$ from the C-O bond, compared to $23°$ in the DFT calculations[70]).

Our own recent DFT calculations[71] on systems like 2-methylthiophenoxyl provide a more accurate picture, which is intermediate between these extreme viewpoints: Compared to unsubstituted phenoxyl, Δg_x is reduced by ca. 30%, and Δg_y is increased by ca. 110%, in reasonable agreement with the experimental observation of a nearly axial tensor (Table 4). Probably, the active space in the MCSCF calculations was insufficient to provide quantitative results. The break-down of the DFT g-tensor results into atomic contributions (Table 5) indicates that a) substitution reduces the contribution from the phenoxyl fragment to both Δg_x and Δg_y, due to withdrawal of spin density from the phenoxyl oxygen atom, and b) the sulfur SO contribution is significant. It increases Δg_x somewhat and Δg_y dramatically (Table 5). A relatively small amount of spin density on the

Table 4. Comparison of contributions to Δg principal components of phenoxyl and 2-methylthiophenoxyl (in ppm)[a]

phenoxyl	Δg_x	Δg_y	Δg_z
Δg_{GC}(1el)	188	273	194
Δg_{RMC}	-206	-206	-206
$\Delta g_{SO/OZ}$	8808	2220	-8
total	8790	2286	-20
2-methylthiophenoxyl	Δg_x	Δg_y	Δg_z
Δg_{GC}(1el)	240	260	202
Δg_{RMC}	-207	-207	-207
$\Delta g_{SO/OZ}$	6109	4792	-56
total	6142	4845	-61

[a]UDFT/BP86 results with DZVP basis set, AMFI SO approximation, and common gauge origin at the center of mass.

Table 5. Fragment analysis of $\Delta g_{SO/OZ}$ contributions (in ppm) in *ortho*-SCH$_3$- phenoxyl[a]

fragment	Δg_{xx}	Δg_{xy}	Δg_{xz}	Δg_{yx}	Δg_{yy}	Δg_{yz}	Δg_{zx}	Δg_{zy}	Δg_{zz}
phenoxyl	4721	127	-1	-1	1658	-1	-1	0	-10
S	1312	-519	-1	-212	3201	-1	0	-1	-45
CH$_3$	6	7	0	6	3	0	0	0	-2
Σ	6037	-385	-2	-205	4862	-2	-1	-1	-57
$\Delta g_{SO/OZ}$	6039	-386	-2	-206	4861	-3	-1	-1	-56

[a]Contributions in general axis system, as shown in Figure 2. Fragment contributions are obtained by separate perturbation calculations, in which only the AMFI-SO operators on a given fragment are employed.

heavy sulfur atom (ca. 0.13 a.u.) is sufficient to produce significant SO contributions to the g-tensor of the radical. The comparably small substituent effect on Δg_x is due to a partial compensation between changes in the phenoxyl contribution and the direct sulfur contribution. Our systematic

7. Calculation of electronic g-tensors

DFT studies indicate furthermore an influence of hydrogen bonding on the g-tensor of the modified tyrosyl radical in apo-GO, and even much larger effects of heavier substituent atoms like selenium or tellurium.[71]

Both g_x and g_y of the tyrosyl radical are reduced by hydrogen bonding to the phenoxyl oxygen atom. Indeed, a reduction of g_x from ca. 2.0087-2.0089 in systems without significant hydrogen bonding to ca. 2.0066-2.0076 by hydrogen bonding is usually taken as characteristic probe of the nature of the protein environment.[31,65,66] Very similar behavior holds for semiquinones and for nitroxide radicals (see below). The reduced Δg_x is mainly due both to an increased energy denominator (stabilization of the "in-plane" b_1 HOMO by hydrogen bonding) and reduced SO and OZ matrix elements for the decisive $^2B_2 \to {}^2B_1$ excitation. As the electrostatic interactions are optimal for a direct alignment of the hydrogen bond with the HOMO, the reduction of Δg_x is expected to be most pronounced for an in-plane coordination. Δg_y is somewhat less affected by hydrogen bonding.

The qualitative features of these effects of hydrogen bonding have early on been accounted for within Stone's model.[8] Sun *et al.* used semi-empirical (PM3-MNDO) calculations to model the electrostatic influence of hydrogen bonding on the g-tensors.[31,66] Both point charge models and an explicit treatment of hydrogen bonding by water or acetic-acid complexes were employed. A Taylor expansion of the electrostatic interactions was used to evaluate the dependence of the effect on g_x on the distance to the hydrogen bond donor or to a positively charged metal center. From the comparison between models for the tyrosyl radical in ribonucleotide reductase (which does not exhibit hydrogen bonding), the one in photosystem II (Tyr-D^0), and tyrosyl in solid tyrosine hydrochloride, a structural model for hydrogen bonding to the Tyr-D^0 radical was suggested.[31] While the earliest calculations used relatively crude estimates for bond lengths, later studies employed fully DFT optimized structures for the model complexes.[66]

Engström *et al.* recently employed MCSCF wavefunctions and a phenoxyl-(H_2O) model.[63] They confirmed and quantified the findings of the earlier, semi-empirical studies. See below for detailed and systematic DFT studies of hydrogen-bonding effects on g-tensors of semiquinones and nitroxides.

5.2 Semiquinone Radical Anions

The electronic structures of semiquinone radical anions are closely related to those of phenoxyl radicals, and the qualitative features of the g-tensor have been accounted for by the Stone model.[8] As for phenoxyls, Δg_x is dominated by the coupling between the in-plane n-type HOMO and the out-of-plane π-type SOMO (similarly, Δg_y arises again mainly from a σ→π* transition). However, the spin density is distributed more symmetrically, and both carbonyl oxygen atoms contribute to the SO/OZ terms. Due to the larger energy gap between HOMO and SOMO, Δg_x is less positive than for phenoxyl. In contrast, Δg_y is more positive than for phenoxyl. Due to the resulting smaller difference between Δg_x and Δg_y, the resolution of g_x and g_y in EPR experiments is more demanding for semiquinones than for the tyrosyl radical.

The role of semiquinones in biological electron transfer processes is probably even more remarkable than that of tyrosyl radicals. Typically, quinones are reduced in one-electron steps, leading first to the semiquinones (either as radical anion or protonated as neutral radical), and often ultimately in a second step (accompanied by protonation) to the hydroquinones.[72] The intermediately produced semiquinone radical anions are essential spin probes in EPR experiments. This holds, *e.g.*, for photosynthetic reaction centers, and for several processes in respiration.[73] Some of the most important quinones involved in such processes are ubiquinone-10 (in reaction centers of purple bacteria, and in many other processes), plastoquinone-9 (in photosystem II, PS-II), and phylloquinone (Vitamin K_1; in PS-I). Similar to the discussion above for tyrosyl radicals, the g-tensors of semiquinones are influenced both by substituent effects and by hydrogen bonding. They are therefore often characteristic for the type of quinone involved in a given electron transfer process, as well as for the protein environment around the semiquinone. Semiquinones in proteins have thus been central targets of HF-EPR studies.[74] In addition, a considerable number of semiquinone radical anions have also been studied in isotropic solution. In particular, W-band (95 GHz) experiments on a large variety of semiquinones in frozen isopropanol, and some studies in other solvents, have provided a valuable database against which theoretical approaches can be validated.[75,76]

Semi-empirical calculations by Knüpling *et al.*[32] used a UHF-PM3 *ansatz* and modelled the solute-solvent interactions either by point charges or by explicit inclusion of two or four methanol molecules hydrogen-bonded to the semiquinone oxygen atoms. Given the rather severe approximations inherent in these calculations, as well as significant uncertainties in the structures employed, the final results with four methanol molecules are in surprisingly good agreement with experiment. However, these calculations would predict hydrogen bonding to reduce Δg_x only by ca. 200-400 ppm, *i.e.* only ca. 4-9%

relative to the results for the free radical anions. This is much less than the ca. 20-30% obtained in DFT calculations[60] (see below). The good agreement of the PM3 results with experiment thus appears to be partly due to cancellation of rather large errors. Due to their very approximate nature, these types of calculations are probably better suited for qualitative modelling than for quantitative predictions.

We have recently studied[60] the g-tensors of a very similar series of semiquinones by our DFT approach,[46] using gradient-corrected functionals (BP86), DZVP basis sets, and accurate AMFI SO operators. Hydrogen-bonding interactions with the solvent were modelled by explicit coordination with a varying number of water or isopropanol molecules. The structures were fully optimized at DFT level. The calculations suggested complexes with two hydrogen bonds to each of the carbonyl oxygen atoms to provide the most likely model for the first solvation shell, augmented by one additional hydrogen bond to each of the two methoxy groups in models for ubisemiquinone-10 (Figure 3).[60] The calculations indicated a nonnegligible

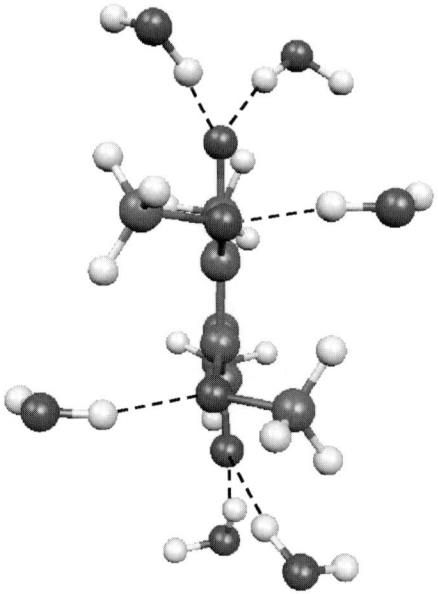

Figure 3. UQ-M $\cdot^-$ (H$_2$O)$_6$ model for solvation of ubisemiquinone in isotropic protic solution.[60] Note the extra two hydrogen bonds to the methoxy substituents. The structure of the UQ-M$\cdot^-$ (iPrOH)$_6$ model complex is similar.[60]

dependence of Δg_x (also partly of Δg_y) on the orientation of the hydrogen bonds, as well as on the conformation of substituents in the substituted semiquinones (again in particular of methoxy substituents in UQ-M). Moreover, the two types of motion appear to proceed along very shallow potential energy surfaces, and they are coupled to each other. This constitutes one of the major limitations in the accuracy achievable in the g-tensor calculations. In the absence of accurate molecular dynamics information, a reasonable selection of low-lying configurations is required.[60]

When taking this into account and using appropriate hydrogen-bonded model complexes Q$^-$(iPrOH)$_4$ (and UQ-M$^-$(iPrOH)$_6$), a remarkable correlation with experimental W-band data in frozen 2-propanol was obtained (Figure 4).[60] After scaling of the computed Δg_x values by a factor of 0.92, all results were within ca. ±130 ppm of experiment. Given the estimated experimental uncertainties of at least ca. ±100 ppm, this means quantitative agreement between theory and experiment. Quantitative agreement, even without the need of scaling, was found for Δg_y (Figure 5). Given the accuracy of the SO operators employed, it is thought that the systematic overestimate of Δg_x reflects inherent deficiencies in the currently used exchange-correlation functionals. Improved functionals may well eliminate the need for scaling. However, even at the present level of accuracy, the results allow quantitative predictions, making DFT calculations of g-tensors of semiquinone radical anions an extremely valuable tool to be used in conjunction with HF-EPR studies. We note in passing that unsolvated semiquinone models provide much too large g-shifts

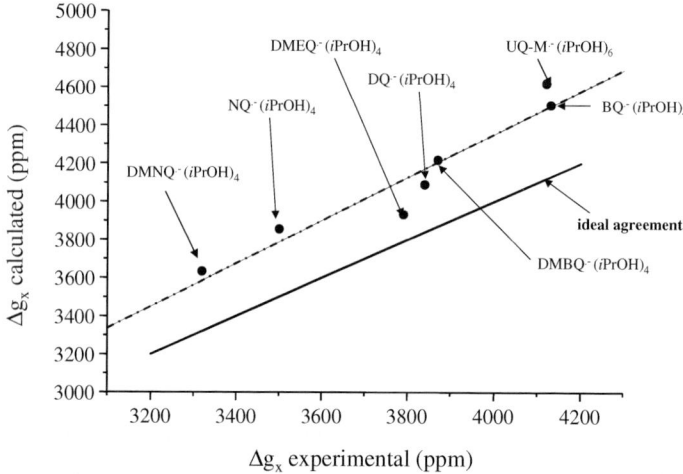

Figure 4. Comparison of DFT calculations (UDFT-BP86 results with AMFI SO operators) of explicitly solvated semiquinone model complexes with experimental Δg_x values in frozen 2-propanol.[60] Abbreviations: BQ (1,4-benzoquinone), DMBQ (2,3-dimethyl-1,4-benzoquinone), DMEQ (2,3-dimethyl-5-ethyl-1,4-benzoquinone), DQ (duroquinone: 2,3,5,6-tetramethyl-1,4-benzoquinone), TMQ (2,3,5-trimethyl-1,4-benzoquinone), UQ-M (2,3-dimethoxy-5,6-dimethyl-1,4-benzoquinone), NQ (1,4-naphthoquinone), DMNQ (2,3-dimethyl-1,4-naphthoquinone).

7. Calculation of electronic g-tensors

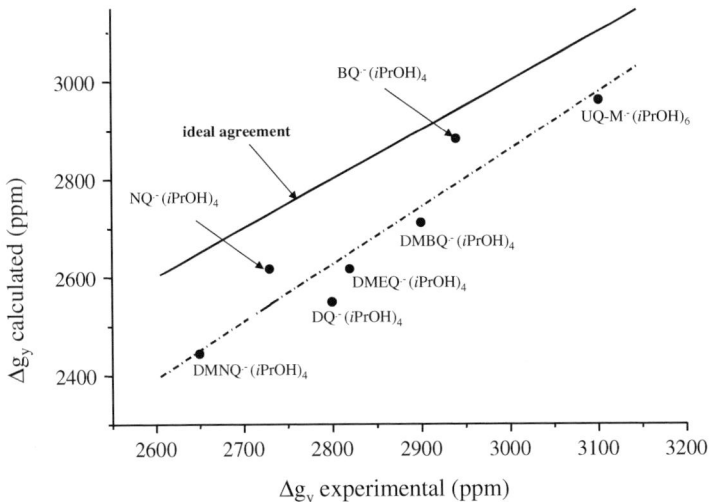

Figure 5. Comparison of DFT calculations (UDFT-BP86 results with AMFI SO operators) of explicitly solvated semiquinone model complexes with experimental Δg_y values in frozen 2-propanol.[60] See foonotes to Figure 4 for abbreviations.

and poor correlation with the experimental data in protic solution. However, these values are good estimates for the results expected in the absence of hydrogen bonding or other solvent interactions.[60]

This suggests that DFT methods should also be useful for studies on semiquinones in protein environments. Our first application[77] has focussed on the A_1 semiquinone binding site of PS-I. The unusual g-tensors of so-called "reconstituted" preparations of PS-I[78] caught our attention, where smaller quinones like DQ or NQ replace the native phylloquinone. HF-EPR data have indicated a different orientation of the semiquinone than in native PS-I. Moreover, both g_x *and* g_y were significantly larger than measured in protic solvents, or when reconstituted into reaction centers of purple bacteria (bRC).[78] Our calculations had indicated that even in the complete absence of hydrogen bonds (*i.e.* in the free gas-phase radical anions), Δg_y in $DQ^{\cdot-}$ or $NQ^{\cdot-}$ should never be significantly larger than ca. 3000 ppm (Table 6). In contrast, the reconstituted PS-I preparations exhibited values of ca. 4000 ppm with $DQ^{\cdot-}$, and of ca. 3700 ppm with $NQ^{\cdot-}$.[78] Thus, it seemed likely that previously neglected interactions could be responsible for the unusual g-tensors in these systems.

Table 6. Calculated g-shift tensors (in ppm) for various model systems of duro- and naphthosemiquinone compared to experimental data in different environments[a]

	Δg_{iso}	Δg_{xx}	Δg_{yy}	Δg_{zz}
DQ$^{\bullet-}$				
Free	2785	5347	3021	-14
+(H$_2$O)$_4$	2270	4200	2807	-195
+(iPrOH)$_4$	2071	3932	2549	-267
exp. (in iPrOH)[75,76]	2160	3790	2800	-100
exp. (in MTHF)[75,76]	2380	4380	2910	-140
exp. (in Zn-bRC)[b]	2200	3800	2800	0
exp. (in PS-I)[78]	3000	5000	4000	0
NQ$^{\bullet-}$				
Free	2795	5464	2896	-25
+(H$_2$O)$_4$	2233	4064	2632	3
+(iPrOH)$_4$	2167	3856	2617	29
exp. (in iPrOH)[75,76]	2060	3500	2730	-40
exp. (in Zn-bRC)[b]	2100	3500	2800	0
exp. (in PS-I)[78]	2700	4400	3700	0

[a]Cf. ref 60. [b]In zinc-substituted reaction centers of purple bacteria.[78]

We therefore started to investigate interactions of quinones and semiquinones with tryptophan (Trp) residues.[77] A variety of evidence had pointed to π-π interactions between phylloquinone and a Trp residue in the A$_1$ binding site. In the course of our computational study, the X-ray crystallographic analysis of the PS-I complex of *S. elongatus* at 2.5 Å resolution confirmed a π-stacked interaction between Q$_K$-A and Trp A697 (as well as a very similar one between Q$_K$-B and Trp B677 in the B-branch).[79]

Interestingly, our extensive structure optimizations of semiquinone-indole model complexes at the MP2 level of theory (this ab initio level was necessary to account for potential dispersion contributions to the π-stacking interactions) indicated that π-stacked arrangements are actually rather *unfavorable* for the anionic complex, and the optimizations converged to a T-stacked arrangement (Figure 6) with significant hydrogen bonding from the indole N-H hydrogen to the π-system of the semiquinone. In contrast, the π-stacked arrangement (Figure 7) was confirmed to provide the most favorable interaction for the neutral quinone-indole complexes.[77]

Subsequent DFT calculations of the g-tensors for the free semiquinones, their indole complexes in various structural arrangements, and models incorporating an additional hydrogen bond (to account for hydrogen bonding to the Leu A722 residue indicated by X-ray analysis), provided interesting insights: π-stacked semiquinone-indole interactions do not alter g_x or g_y

7. Calculation of electronic g-tensors

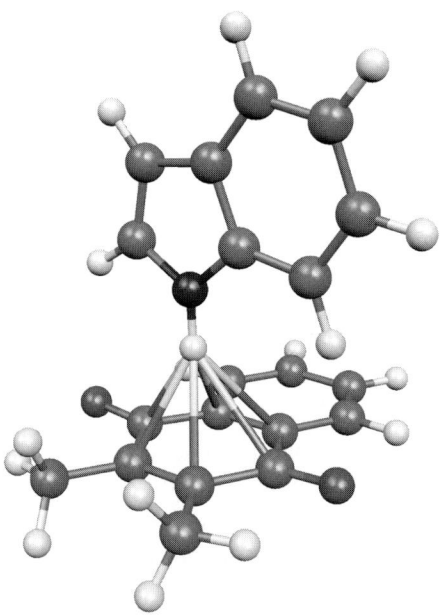

Figure 6. Fully MP2-optimized structure of the complex between 2,3-dimethylnaphthosemiquinone and indole. Note the T-stacked arrangement.[77]

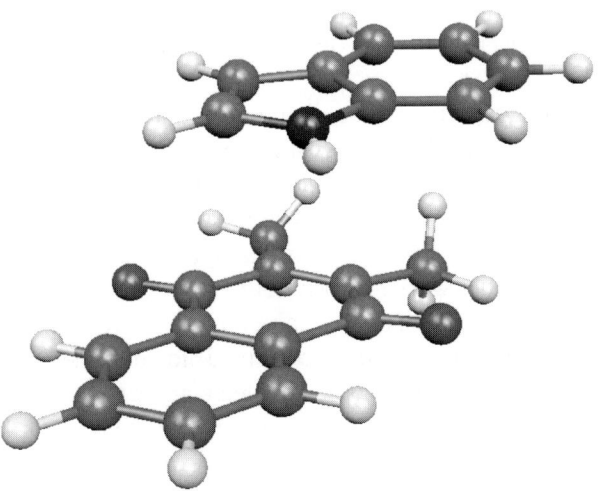

Figure 7. Fully MP2-optimized structure of the complex between neutral 2,3-dimethylnaphthoquinone and indole. Note the π-stacked arrangement.[77]

noticeably and certainly cannot account for the large g_y values of $DQ^{-\cdot}$ or $NQ^{-\cdot}$ reconstituted into PS-I. Interestingly, the only interactions we found to enhance *both* g_x *and* g_y relative to the free semiquinones were T-stacked hydrogen bonds to the semiquinone π-system.[77]

This suggested the following situation: Small quinones without extended side chain like DQ or NQ may reorient within the reconstituted A_1 site upon reduction to the semiquinone. This leads to a significant stabilization of the semiquinone relative to the neutral quinone, and thus to an increased redox potential. In these reconstituted preparations, the semiquinone should not be able to transfer its excess electron to the next acceptor (the iron-sulfur cluster F_x) within the electron transfer chain of PS-I, and thus photoreduction of $NADP^+$ will be interrupted. This agrees with experimental observation. In contrast, the native phylloquinone (Q_K) acceptor has an extended side chain in 3-position of the semiquinone, which prohibits reorientation of the semiquinone. In the enforced π-stacked orientation relative to the Trp residue, the semiquinone is destabilized relative to the quinone, and the redox potential is tuned more negative, as needed for forward electron transfer.

The increase of g_y in various T-stacked semiquinone model complexes was smaller than the observed enhancement in the reconstituted reaction centers, suggesting more than one T-stacked hydrogen bond to the semiquinone. It appears likely, that this type of T-stacked hydrogen bonding to semiquinone radical anions may generally be an important ingredient in biological electron transfer processes involving quinones.

In a recent combined experimental and theoretical EPR study of PS-I, DFT optimizations of a model A_1 binding site were combined with ZORA calculations of the g-tensor of the semiquinone.[80] For unknown reasons, the optimization of the semiquinone state did not reveal any reorientation relative to the π-stacked Trp residue, even though the model structure employed appears to provide sufficient flexibility. As expected from the comparison in section 4, the ZORA calculations gave significantly overestimated g-shifts, and it suggested extremely large changes in g-tensor due to semiquinone-protein interactions.[80] This may in part be due to the neglect of spin-polarization effects (cf. section 4). During the revision of this review, Neyman and coworkers also applied their two-component DKH approach to some semiquinone-water models.[81] The tendency is the same as found with ZORA (*i.e.* a significant overestimate of the g-shifts and unrealistically large solvent effects) and likely for the same reasons.

5.3 Nitroxide Spin Labels

Artificial spin labels are frequently introduced into proteins to be able to study structure and dynamics by EPR spectroscopy.[82,83] Spin labels may be, e.g., paramagnetic transition metal ions, or stable organic radicals. Typically, organic nitroxides are attached to cystein residues of the protein via a linker. Their advantage is the comparably high stability and the large sensitivity of their EPR parameters (A- and g-tensors) to the local environment. Examples of such a spin label and of a model system studied computationally are shown in Figure 8. Analogous to the abovementioned phenoxyls and semiquinones, nitroxides are π-radicals, and they are influenced in a similar way by hydrogen bonding and dielectric effects of the solvent or the protein environment.[84]

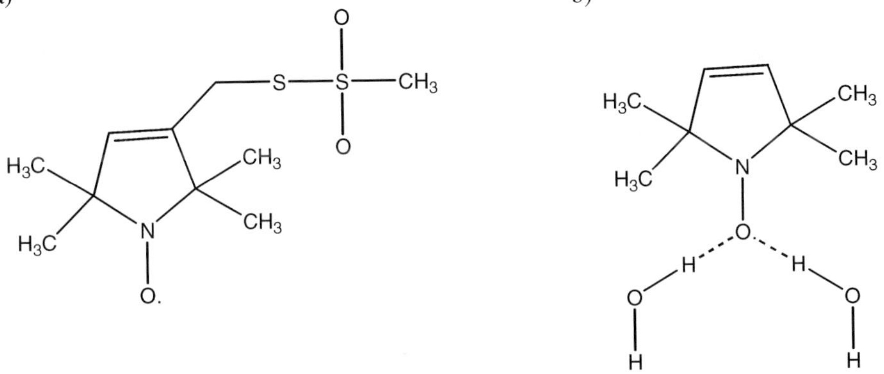

Figure 8. Structure of the MTSSL nitroxide spin label, and of the smaller MSL model studied computationally.

Early calculations on nitroxide spin labels by Mustafaev and Schastnev,[85] using a semiempirical (INDO) method, indicated a significant effect of the NO bond length but small effects of remote substituents on the g-tensor. In the context of recent multi-frequency EPR studies,[86] the EPR parameters of nitroxide spin labels like MTSSL have been the focus of ab initio and DFT calculations by Engström et al..[86-89] The g-tensor of a smaller model (MSL; cf. Figure 8) has initially been studied by ab initio ROHF linear-response calculations,[87,88] employing atomic meanfield (AMFI) SO operators (see section 3). The dependence of the g-tensor components on the NO bond length and on C-N-O-C dihedral angles within the spin labels, as well as the influence of hydrogen bonding, were studied. Hydrogen bonding leads to a significant reduction of Δg_x, and to a slight reduction of Δg_y. Models with one or two water or methanol molecules hydrogen-bonded to the nitroxide oxygen atom were investigated. The results resemble those obtained for phenoxyl and semiquinone radicals.[60,61] Methanol as solvent model leads to slightly shorter H-bonds and thus to a slightly larger reduction in Δg_x. Compared to experimental data, it appeared that the ROHF method exaggerated the effect of hydrogen bonding on the g-tensor.

As extensive MCSCF calculations were considered unreasonable for a system of this size, subsequent studies[87,86,89] used DFT methods,[46] with the AMFI approximation to treat the BP SO operators. Based on DFT-optimized structures, both A- and g-tensors were calculated, and solvent effects on the two quantities were studied both experimentally and theoretically. The AMFI approximation to the SO operators allowed an atomic break-down[46] of the g-shift components. While the spin density was found to be almost evenly distributed over the nitrogen and oxygen atoms of the NO group, the larger SO coupling of oxygen accounts for about 80% of the contributions to Δg_x. Hydrogen bonding to the nitroxide oxygen atom a) shifts spin density from oxygen to nitrogen and b) increases the energy gap for the relevant n-π^* excitation. Both effects reduce g_x. The computed changes in the g-tensor components at the DFT level were judged to be more realistic than the previous ROHF results, due to the implicit inclusion of electron correlation.

The relatively inexpensive DFT approach also allowed modelling of the protein environment. Up to six nearest amino acid residues were included in model studies[87,89] of the A- and g-tensor of a nitroxide spin label in the tissue-factor/factor-VIIa protein complex, which was also studied by X- and W-band EPR.[86] The protein environment was found to influence the g-tensor of the nitroxide spin label in particular via a change in the N-O bond length, but also by direct modifications of the $\Delta g_{SO/OZ}$ term due to hydrogen bonding. Effects of more remote residues were considered small. The effects of the different hydrogen-bonding interactions were nonadditive, due to a redistribution of spin density, as found also for semiquinones[60] (cf. 5.2).

5.4 Miscellaneous Systems

Various sulfur-containing radicals have been of interest in biological systems, *e.g.* radicals derived from amino acids like cysteine (see also 5.1 for thiolate-substituted tyrosyl). As sulfur is an element of the third period, its SO coupling is larger than, *e.g.*, that of oxygen or nitrogen. In cases with significant spin density on sulfur, appreciable deviations of the g-tensor components from g_e are thus expected. A recent MCSCF linear response study of Engström *et al.*[90] dealt with the g-tensors (and hyperfine couplings) of a variety of small sulfur-based radicals. Comparison of ROHF and MCSCF results for CH$_3$S· indicated very large correlation effects. The largest component was significantly reduced by hydrogen bonding. Disulfide anions and cations were also studied. Possible assignments of experimentally detected sulfur-containing radicals in biological systems were discussed in light of the calculations.[90]

Recent DFT calculations on tryptophan radicals in ribonucleotide reductase by Bleifuss et al.[91] augmented W-band EPR studies on these systems. The calculations used van Lenthe's ZORA approach[12] at the restricted Kohn-Sham level and thus overestimated the g-shifts. However, the authors were mostly interested in the relative orientation of the tensors of tryptophan and tyrosyl radicals. Based on the calculations, some structural suggestions were made.

An intermediate situation between typical organic π-radicals and transition metal complexes is provided by a number of binuclear complexes, in which the spin density is mainly, but not exclusively, localized on the bridging 2,2'-azobispyridine (abpy) or 2,2'-azobis(5-chloropyrimidine) ligands.[92] Recent HF-EPR studies of two rhenium complexes have shown that the g-anisotropy is larger than for organic radicals but smaller than for typical transition metal complexes with metal-centered spin density. DFT calculations with AMFI-SO operators and SO pseudopotentials have been used to analyze the g-tensors. The larger g-anisotropy for the abcp system was confirmed and traced back to larger metal SO contributions, due to more extensive metal-ligand covalency. The influence of changing the metal from Re to Tc, and of varying the metal-bound halide ligands, was studied, and the free ligands were investigated as well.[92]

6. CONCLUSIONS AND OUTLOOK

Compared to the mid 1990's, the situation in the field of quantum chemical g-tensor calculations has changed fundamentally. Before that period, calculations on organic radicals were essentially restricted to semiempirical methods. Today, a variety of ab initio and DFT methods are available. Hartree-Fock methods may be applied to relatively large systems but are not always reliable, due to the neglect of electron correlation. Sophisticated MRCI or MCSCF methods are applicable only to relatively small systems but serve as important benchmark methods to gauge more approximate approaches. At present, DFT methods undoubtedly provide the best compromise between accuracy and computational effort. While the deviations from experiment are still appreciable for transition metal complexes, the accuracy achievable for organic radicals is remarkable. As has been discussed in detail in section 4, differences between various DFT approaches arise mainly from the treatment of the spin-orbit matrix elements, as well as from the treatment of spin polarization. The latter is still a problem within two-component approaches like the ZORA method. The atomic meanfield approximation (AMFI) has been shown to provide a very accurate approximation to the full one- and two-electron Breit-Pauli spin-orbit matrix elements. DFT calculations with AMFI SO treatment and gradient-corrected functionals[46] are therefore recommended as a particularly accurate method in the calculation of g-tensors for organic radicals. Typically, the largest components of the g-shift tensor may be overestimated by at most ca. 10%. Appropriately chosen semiempirical SO operators may also provide adequate accuracy.[40]

The accuracy of the DFT approaches combined with their computational efficiency makes these methods very useful in applications in a variety of fields. We have concentrated here on studies of biologically relevant radicals, either in isotropic solution or in their protein environment. Examples have ranged from phenoxyl and tyrosyl radicals via nitroxide spin labels to semiquinone radical anions. In the latter case, the DFT calculations have even led to a novel proposal regarding the function of photosystem I. We expect that further computational studies will aid to achieve rapid progress in our understanding of the interrelation between structure, environment, and g-tensors of biologically relevant radicals.

Another field, which should benefit greatly from quantitative g-tensor calculations, but which we have not been able to cover here in detail, are paramagnetic defects in solids. These may be paramagnetic atoms or small ions encapsulated in cages, or defects related to vacancies or doping. Some of the smaller species have been studied as isolated molecules by MRCI calculations (see, *e.g.*, refs 15,35,36,58). The large expense of the calculations did not allow a detailed inclusion of the effect of the host crystal. Certainly, the more economical DFT approaches should allow more extensive scrutiny of the influence of the environment. Localized defects may be treated by an appropriate cluster model. Our group is currently interested, *e.g.*, in trapped hydrogen atoms in various environments.[93] The treatment of more extensively delocalized defects should be particularly well suited for the augmented plane-wave DFT approach of Pickard and Mauri. Their first applications were, however, restricted to the rather localized E'_1 and P_2 defects in α quartz.[47]

An aspect that needs closer investigation, is the influence of motional effects on the g-tensors. For example, the g-tensors of semiquinone radical anions have been found to depend significantly on a) the orientation of hydrogen bonds and b) the conformation of the substituents. Both motions proceed along shallow potential energy surfaces, and they are interdependent.[60] An accurate description of this situation calls for a combination of the new methods for g-tensor calculations with molecular dynamics approaches. In view of the open-shell character of the species involved, classical MD simulations do not appear very promising. Some type of ab initio molecular dynamics procedure (or a combined quantum mechanics/molecular mechanics treatment), followed by g-tensor calculations at selected snapshots from molecular dynamics trajectories, are needed. Work along these lines is presently underway in our laboratory.

Acknowledgments *I am very grateful for the important contributions by J. Vaara, O. L. Malkina, V. G. Malkin, and B. Schimmmelpfennig to our own entry into the field of g-tensor calculations, with further contributions due to A. Arbouznikov, R. Reviakine, I. Ciofini, M. Munzarová, C. Remenyi, C. Urban, T. Gress, J. Asher, and S. Schlund. Helpful comments on the manuscript are due to I. Ciofini, J. Vaara, A. Arbuznikov, and V. G. Malkin. Funding from Deutsche Forschungsgemeinschaft (Priority Program "High-Field EPR in Biology, Chemistry, and Physics) and Fonds der Chemischen Industrie is also gratefully acknowledged.*

7. REFERENCES

1. See, e.g.: J. A. Weil, J. R. Bolton, J. E. Wertz, *Electron Paramagnetic Resonance: Elementary Theory and Practical Applications*, Wiley & Sons, New York, 1994; W. Weltner Jr., *Magnetic Atoms and Molecules*, Van Nostrand, New York, 1983.
2. J. E. Harriman, *Theoretical Foundations of Electron Spin Resonance*, Academic Press, New York, 1978.
3. K. Möbius, in *Biological Magnetic Resonance*, Vol. 13 (Eds. L. J. Berliner, J. Reuben) Plenum Press, New York, 1993, pp. 253-274.
4. T.F. Prisner, in *Advances in Magnetic and Optical Resonance*, Vol. 20 (Ed. W. Warren) Academic Press, New York, 1997, pp. 245-299.
5. S. Un, P. Dorlet, A. W. Rutherford, *Appl. Magn. Reson.*, **21** (2001) 341.
6. Cf., e.g.: A. Abragam, B. Bleaney, *Electron Paramagnetic Resonance of Transition Ions*, Clarendon Press, Oxford, 1970. M. C. R. Symons, *Chemical and Biochemical Aspects of Electron-Spin Resonance Spectroscopy*, Van Nostrand, New York, 1978. N. M. Atherton, *Principles of Electron Spin Resonance*, Prentice Hall, New York, 1993.
7. A. J. Stone, *Proc. R. Soc. A*, **271** (1963) 424.
8. A. J. Stone, *Mol. Phys.*, **6** (1963) 509; A. J. Stone, *Mol. Phys.*, **7** (1964) 311.

9. R. McWeeny, *Methods of Molecular Quantum Mechanics*, Academic, London, 1992.
10. E. van Lenthe, E. J. Baerends, J. G. Snijders, *J. Chem. Phys.*, **99** (1993) 4597.
11. R. Samzow, B. A. Hess, G. Jansen, *J. Chem. Phys.*, **96** (1992) 1227.
12. E. van Lenthe, P. E. S. Wormer, A. van der Avoird, *J. Chem. Phys.*, **107** (1997) 2488.
13. K. M. Neyman, D. I. Ganyushin, A. V. Matveev, V. A. Nasluzov, *J. Phys. Chem. A*, **106** (2002) 5022.
14. M. H. Pryce, *Proc. Phys. Soc.*, **63** (1950) 25.
15. G. H. Lushington, PhD thesis, The University of New Brunswick, Canada, 1996.
16. S. Patchkovskii, T. Ziegler, *J. Phys. Chem. A*, **105** (2001) 5490.
17. M. Engström, B. Minaev, O. Vahtras, H. Ågren, *Chem. Phys.*, **237** (1998) 149.
18. R. Angstl, *Chem. Phys.*, **132** (1989) 435.
19. See, e.g.: B. Engels, L. A. Eriksson, S. Lunell, *Adv. Quantum Chem.*, **27** (1996) 297.
20. See, e.g.: P. J. Geurts, P. C. P. Bouten, A. J. van der Avoird, *Chem. Phys*, **73** (1980) 1306; P. Belanzoni, E. J. Baerends, S. van Asselt, P. B. Langewen, *J. Phys. Chem.*, **99** (1995) 13094; J. Swann, T. D. Westmoreland, *Inorg. Chem.*, **36** (1997) 5348.
21. W. H. Moores, R. McWeeny, *Proc. Roy. Soc. A*, **332** (1973) 365. See also: M. Ishii, K. Morihashi, O. Kikuchi, *THEOCHEM*, **235** (1991) 39.
22. See, e.g.: C. P. Keijzers, H. H. M. de Vries, A. van der Avoird, *Inorg. Chem.*, **11** (1972) 1338; E. Dalgård, J. Linderberg, *Int. J. Quantum Chem.*, **9** (1975) 269; K. C. Mishra, S. K. Mishra, J. N. Roy, S. Ahmad, T. P. Das, *J. Am. Chem. Soc.*, **107** (1985) 7898; J. N. Roy, K. C. Mishra, S. K. Mishra, T. P. Das, *J. Phys. Chem.*, **93** (1989) 194.
23. N. D. Chuvylkin, G. M. Zhidomirov, *Mol. Phys.*, **25** (1973) 1233.
24. R. Angstl, *Chem. Phys.*, **145** (1990) 413.
25. T. Morikawa, O. Kikuchi, *Theor. Chim. Acta*, **22** (1971) 224. D. C. McCain, D. W. Hayden, *J. Magn. Reson.*, **12** (1973) 312.
26. M. Plato, K. Möbius, W. Lubitz, in *Chlorophylls* (Ed. H. Scheer) CRC Press, Boca Raton, FL, 1991, p1015. B. J. Hales, *J. Am. Chem. Soc.*, **97** (1975) 5993.
27. Y.-W. Hsiao, M. C. Zerner, *Int. J. Quantum Chem.*, **75** (1999) 577.
28. B. N. Plakhutin, G. M. Zhidomirov, K. I. Zamaraev, *J. Struct. Chem.*, **24** (1983) 3.
29. For an INDO-CI approach, see: F. Neese, *Int. J. Quantum Chem.*, **83** (2001) 104; F. Neese, E. I. Solomon, *Inorg. Chem.*, **37** (1998) 37.
30. S. A. Mustafaev, V. G. Malkin, P. V. Schastnev, *J. Struct.Chem.*, **28** (1987) 667.
31. S. Un, M. Atta, M. Fontcave, A. W. Rutherford, *J. Am. Chem. Soc.*, **117** (1995) 10713.
32. M. Knüpling, J. T. Törring, S. Un, *Chem. Phys.*, **219** (1996) 291.
33. J. T. Törring, S. Un, M. Knüpling, M. Plato, K. Möbius, *J. Chem. Phys.*, **107** (1997) 3905.
34. G. H. Lushington, F. Grein, *Theor. Chim. Acta*, **93** (1996) 259.
35. P. Bruna, G. H. Lushington, and F. Grein, *Chem. Phys.*, **225** (1997) 1.
36. G. H. Lushington, *J. Phys. Chem. A*, **104** (2000) 2969.
37. O. Vahtras, B. Minaev, H. Ågren, *Chem. Phys. Lett.*, **281** (1997) 186.
38. O. Vahtras, M. Engström, B. Schimmelpfennig, *Chem. Phys. Lett.*, **351** (2002) 424.
39. D. Jayatilaka, *J. Chem. Phys.*, **108** (1998) 7587.
40. F. Neese, *J. Chem. Phys.*, **115** (2001)11080.
41. M. Kaupp, R. Reviakine, O. L. Malkina, A. Arbuznikov, B. Schimmelpfennig, V. G. Malkin, *J. Comput. Chem.*, **23** (2002) 794.
42. See, e.g.: W. Koch, M. C. Holthausen, *A Chemist's Guide to Density Functional Theory*, Wiley-VCH, Weinheim, 2000.
43. See, e.g.: Chr. v. Wüllen, *J. Chem. Phys.*, **102** (1995) 2806; A. M. Lee, N. C. Handy, S. M. Colwell, *J. Chem. Phys.*, **103** (1995) 10095, and references therein.

44. See, e.g.: G. Te Velde, F. M. Bickelhaupt, E. J. Baerends, C. Fonseca Guerra, S. J. A. van Gisbergen, J. G. Snijders, T. Ziegler, *J. Comput. Chem.*, **22** (2001) 931, and references therein.
45. G. Schreckenbach and T. Ziegler, *J. Phys. Chem.* A, **101** (1997) 3388.
46. O. L. Malkina, J. Vaara, B. Schimmelpfennig, M. L. Munzarová, V. G. Malkin, M. Kaupp, *J. Am. Chem. Soc.*, **122** (2000) 9206.
47. C. J. Pickard, F. Mauri, *Phys. Rev. Lett.*, **88** (2002), article 086403.
48. B. A. Hess, C. Marian, U. Wahlgren, and O. Gropen, *Chem. Phys. Lett.*, **251** (1996) 365.
49. B. Schimmelpfennig, *Atomic Spin-Orbit Meanfield Integral Program*, Stockholms Universitet, Sweden 1996.
50. O. L. Malkina, B. Schimmelpfennig, M. Kaupp, B. A. Hess, P. Chandra, U. Wahlgren, and V. G. Malkin, *Chem. Phys. Lett.*, **296** (1998) 93.
51. K. Ruud, B. Schimmelpfennig, H. Ågren, *Chem. Phys. Lett.*, **310** (1999) 215.
52. J. Vaara, O. L. Malkina, H. Stoll, V. G. Malkin, M. Kaupp, *J. Chem. Phys.*, **114** (2001) 61.
53. See, e.g.: V. G. Malkin, O. L. Malkina, M. E. Casida, D. R. Salahub, *J. Am. Chem. Soc.*, **116** (1994) 5898.
54. V. G. Malkin, O. L. Malkina, L. A. Eriksson, D. R. Salahub, in *Modern Density Functional Theory: A Tool for Chemistry; Theoretical and Computational Chemistry* (Eds. J. M. Seminario, P. Politzer), Vol. 2; Elsevier, Amsterdam 1995, pp. 273-347.
55. A. Arbuznikov, M. Kaupp, R. Reviakine, O. L. Malkina, V. G. Malkin, *Phys. Chem. Chem. Phys.* **4** (2002) 5467.
56. See, e.g.: R. Ditchfield, *Mol. Phys.*, **27** (1974) 789; K. Wolinski, J. F. Hinton, P. Pulay, *J. Am. Chem. Soc.*, **112** (1990) 8251.
57. W. Kutzelnigg, U. Fleischer, M. Schindler, in *NMR-Basic Principles and Progress*, Vol. 23, Springer, Heidelberg, 1990, pp.165-262.
58. P. J. Bruna, F. Grein, *Int. J. Quant. Chem.*, **77** (2000) 324; P. J. Bruna, F. Grein, *J. Phys. Chem.* A, **105** (2001) 3328.
59. R. F. Curl, *Mol. Phys.*, **9** (1965) 585; L. B. Knight, W. J. Weltner, *J. Chem. Phys.*, **53** (1970) 4111.
60. M. Kaupp, C. Remenyi, J. Vaara, O. L. Malkina, V. G. Malkin, *J. Am. Chem. Soc*, **124** (2002) 2709.
61. M. Engström, O. Vahtras, H. Ågren, *Chem. Phys.*, **243** (1999) 263.
62. M. Engström, F. Himo, A. Gräslund, B. Minaev, O. Vahtras, H. Ågren, *J. Phys. Chem.* A, **104** (2000) 5149.
63. M. Engström, F. Himo, H. Ågren, *Chem. Phys. Lett.*, **319** (2000) 191.
64. See, e.g.: J. Stubbe, W. A. van der Donk, *Chem. Rev.*, **98** (1998) 705.
65. See, e.g.: P. Allard, A. L. Barra, K. K. Andersson, P. P. Schmidt, M. Atta, A. Gräslund, *J. Am. Chem. Soc.*, **118** (1996) 895; P. J. van Dam, J.-P. Willems, P. P. Schmidt, S. Pötsch, A.-L. Barra, W. R. Hagen, B. M. Hoffman, K. K. Andersson, A. Gräslund, *J. Am. Chem. Soc.*, **120** (1998) 5080; S. Un, C. Gerez, E. Elleingard, M. Fontecave, *J. Am. Chem. Soc.*, **123** (2001) 3048; A. Ivancich, P. Dorlet, D. B. Goodin, S. Un, *J. Am. Chem. Soc.*, **123** (2001) 5050; A. Ivancich, T. A. Mattioli, S. Un, *J. Am. Chem. Soc.*, **121** (1999) 5743, and references cited in these works.
66. A. Ivancich, T. A. Mattioli, S. Un, *J. Am. Chem. Soc.*, **121** (1999) 5743.
67. J. P. Klinman, *Chem. Rev.*, **96** (1996) 2541.
68. N. Ito, S. E. V. Phillips, C. Stevens, Z. B. Ogel, M. J. McPherson, J. N. Keen, K. D. S. Yadaf, P. F. Knowles, *Nature*, **350** (1991) 87.
69. G. T. Babcock, M. K. El-Deeb, P. O. Sandusky, M. M. Whittaker, J. W. Whittaker, *J. Am. Chem. Soc.*, **114** (1992) 3727.

70. G. J. Gerfen, B. F. Bellew, R. G. Griffin, D. J. Singel, Ch. A. Ekberg, J. W. Whittaker, *J. Phys. Chem.*, **100** (1996) 16739.
71. M. Kaupp, T. Gress, R. Reviakine, O. L. Malkina, V. G. Malkin, *J. Phys. Chem. B.*, in press.
72. See, e.g.: S. Patai, *Chemistry of Quinoid Compounds*, Interscience, New York, 1974. *Function of Quinones in Energy Conserving Systems* (Ed. B. L. Trumpower) Academic Press, New York, 1982. R. A. Morton, *Biochemistry of Quinones*,Academic Press, New York, 1965.
73. See, e.g.: J. A. Pedersen, *EPR Spectra from Natural and Synthetic Quinones and Quinoids*, CRC Press, Boca Raton, FL, 1985.
74. See, e.g.: W. Lubitz, G. Feher, *Appl. Magn. Reson.*, **17** (1999) 1; H. Levanon, K. Möbius, *Annu. Rev. Biophys. Biomol. Struct.*, **26** (1997) 495.
75. O. Burghaus, M. Plato, M. Rohrer, K. Möbius, F. MacMillan, W. Lubitz, *J. Phys. Chem.*, **97** (1993) 7639. M. Rohrer, M. Plato, F. MacMillan, Y. Grishin, W. Lubitz, K. Möbius, *J. Magn. Reson.*, **116** (1995) 59.
76. O. Nimz, F. Lendzian, C. Boullais, W. Lubitz, *Appl. Magn. Reson.*, **14** (1998) 255.
77. M. Kaupp, *Biochemistry*, **41** (2002) 2895.
78. I. Sieckmann, A. van der Est, H. Bottin, P. Sétif, D. Stehlik, *FEBS Lett.*, **284** (1991) 98; A. van der Est, I. Sieckmann, W. Lubitz, D. Stehlik, *Chem. Phys.*, **194** (1995) 349.
79. P. Jordan, P. Fromme, H. T. Witt, O. Klukas, W. Saenger, N. Krauß, *Nature*, **411** (2001) 909.
80. C. Teutloff, W. Hofbauer, S. G. Zech, M. Stein, R. Bittl, W. Lubitz, *Appl. Magn. Reson.*, **21** (2001) 363.
81. K. M. Neyman, D. I. Ganyushin, Ž. Rinkevičius, N. Rösch, *Int. J. Quantum Chem.*, **90** (2002) 1404.
82. L. J. Berliner, J. Reuben, *Biological Magnetic Resonance, Vol. 8, Spin Labeling – Theory and Applications*, Plenum Press, New York, 1989.
83. W. L. Hubbell, A. Gross, R. Langen, M. A. Lietzow, *Curr. Opin. Struct. Biol.*, **8** (1998) 649.
84. See, e.g.: T. Kawamura, S. Matsunami, Y. Yonezawa, K. Fukui, *Bull. Chem. Soc. Jpn.*, **38** (1965) 1935; V. I. Krichnyi, O. Ya. Grinberg, V. R. Bogatyrenko, G. I. Likhtenstein, Ya. S. Lebedev, *Biophysics*, **30** (1985) 233; V. I. Krichnyi, O. Ya. Grinberg, Ye. I. Yudanova, Ye. V. Lyubashevskaya, L. I. Antsiferova, G. I. Likhtenstein, Ya. S. Lebedev, *Biophysics*, **32** (1987) 229; M. A. Ondar, O. Ya. Grinberg, A. A. Dubinskii, Ya. S. Lebedev, *Sov. J. Chem. Phys.*, **3** (1985) 781.
85. S. A. Mustafaev, P. V. Schastnev, *J. Struct. Chem.*, **30** (1989) 582.
86. R. Owenius, M. Engström, M. Lindgren, M. Huber, *J. Phys. Chem. A*, **105** (2001) 10967.
87. M. Engström, PhD Thesis, Linköpings Universitet, Sweden, 2001.
88. M. Engström, R. Owenius, O. Vahtras, *Chem. Phys. Lett.*, **338** (2001) 407.
89. M. Engström, J. Vaara, B. Schimmelpfennig, H. Ågren, *J. Phys. Chem. B*, **106** (2002) 12354.
90. M. Engström, O. Vahtras, H. Ågren, *Chem. Phys. Lett*, **328** (2000) 483.
91. G. Bleifuss, M. Kolberg, S. Pötsch, W. Hofbauer, R. Bittl, W. Lubitz, A. Gräslund, G. Lassmann, F. Lendzian, *Biochemistry*, **40** (2001) 15362.
92. S. Frantz, H. Hartmann, N. Doslik, M. Wanner, W. Kaim, H.-J. Kümmerer, G. Denninger, A.-L. Barra, C. Duboc-Toia, J. Fiedler, I. Ciofini, C. Urban, M. Kaupp, *J. Am. Chem. Soc.*, **124** (2002) 10563.
93. M. Kaupp, J. Asher, A. Arbuznikov, A. Patrakov, *Phys. Chem. Chem. Phys.*, **4** (2002) 5466.

Chapter 8

RADIOLABELLED RADICALS DERIVED FROM VOLATILE ORGANIC COMPOUNDS (VOCS) SORBED ON REACTIVE SURFACES

Implications for atmospheric chemistry and pollution control

Christopher J. Rhodes
School of Pharmacy and Chemistry, Liverpool John Moores University, Byrom St., Liverpool L3 3AF, U.K.

Key words: Volatile Organic Compounds, VOC, Radiolabelling, Atmosphere, Radicals, Troposphere, Oxidation, Heterogeneous, Surface, Pollution, Remediation, Environment.

Abstract: Free radicals have been formed by the addition of muonium (a radioactive hydrogen atom with a positive muon as its nucleus) to Volatile Organic Compounds (VOCs), sorbed on the surfaces of solid materials, representative of those found in the atmosphere, specifically: clays, porous carbon, silica-gel, zeolites and ice. Using techniques of transverse-field muon spin rotation (TF-MuSR) and longitudinal-field muon spin-relaxation (LF-MuSRx), motional correlation times for reorientation of molecular radicals on these surfaces were determined, and activation energies were estimated from their temperature dependences. The results are interpreted in terms of sorption within the pores and the interaction with specific surface sites in these materials.

1. INTRODUCTION

Until only recently, the majority of processes occurring in the atmosphere were viewed as gas-phase reactions, but it is now thought that many of them may in fact occur on the surfaces of suspended particles. The energy required for the activation of molecules in the atmosphere is provided by ultra-violet light. The essential chemistry in the troposphere is oxidation, a scene in which hydroxyl radicals are principal players, since they can intercept the molecules of volatile organic compounds (VOCs), converting them to free radicals.[1-3] The oxidation of organic compounds in the atmosphere leads both to an increased burden of greenhouse gas (CO_2) and to the formation of carbonaceous aerosol, thus contributing a solid surface which on which further oxidation can occur. We have suggested previously[4] that the surfaces of solid particles might provide the "third-body", often invoked in atmospheric chemistry[1] as a sink for the excess energy released when chemical bonds are formed, enabling molecular transformations to occur. There are few methods available to study radicals sorbed on surfaces; Electron Spin Resonance (ESR) mostly lacks the sensitivity that is required for reactive radicals under reactive conditions - *especially* on surfaces - and so we have imported from the suite of techniques known as "MuSR"[5,6], in which radicals are labelled with *muonium* - a radioactive hydrogen atom with a positive muon as its nucleus. These methods are tremendously sensitive, in part because they use *single-particle counting*; specifically, we have employed transverse-field muon spin rotation (TF-MuSR) and longitudinal-field muon spin-relaxation (LF-MuSRx) in these investigations, which are described below. Our particular motivation is to determine the reorientational diffusion of radicals derived from VOCs, sorbed on silica, clay, zeolite, carbon and ice surfaces, representative of those present in the atmosphere.[1-3]

2. METHODS

The positive muon, the radioactive nucleus of muonium, is relatively short-lived, having a *mean lifetime* of 2.2×10^{-6} s. This corresponds to a *half life* of 1.52×10^{-6} s, and a *radioactive decay constant* of 0.455×10^{6} s^{-1}, thus setting a "microsecond" timescale over which kinetic processes may be studied using muons. The decay of the muon produces a positron e^+ (equation 1) plus two neutrinos; it is, of course, the positrons which are counted.

$$\mu^+ \rightarrow e^+ + \nu_e + \nu_\mu \quad (1)$$
$$\pi^+ \rightarrow \mu^+ + \nu_\mu \quad (2)$$
$$\mu^+ + e^- \rightarrow Mu\bullet \quad (3)$$

Although they occur naturally as a consequence of cosmic radiation striking the nuclei of light atoms (*e.g.* O_2, N_2) in the atmosphere, muons are needed in high fluxes for research purposes. Thus they are produced using a particle accelerator, by which means a beam of energised protons is caused to impinge on a beryllium or a carbon target. Among the products of the ensuing nuclear reactions are pions (binding components of nuclei), which decay on a nanosecond timescale to form muons (equation 2).

8. Radiolabelled organic radicals on reactive surfaces

Muonium atoms (μ^+e^-; Mu) may be formed *in situ* in a range of liquid, solid and gas phase samples, according to (equation 3), where e^- is a radiolytic electron. Muonium is equivalent to a normal protium atom (p^+e^-) and indeed shows the chemical properties of a *light* hydrogen atom (Mu has a mass 1/9 that of a protium atom); so if the sample contains unsaturated organic molecules, Mu can undergo an addition reaction. The method is highly specific for the study of free radicals, of which many types may be so formed. Some examples are indicated in [equations (4)-(6)]:

$$R_2C=CR_2 \quad +Mu\bullet \quad \rightarrow \quad R_2C(Mu)CR_2\bullet \quad (4)$$
$$R_2C=CR-CR=CR_2 \quad + \quad Mu\bullet \quad \rightarrow \quad R_2C(Mu)-C(R)\bullet-CR=CR_2 \quad (5)$$
$$R_2C=O \quad +Mu\bullet \quad \rightarrow \quad R_2C\bullet-OMu \quad (6)$$

2.1 Transverse-field muon spin rotation spectroscopy (TF-MuSR)[5,6]

Though the process outlined above may be thought of in analogy with other isotopic labelling techniques, information equivalent to that obtainable from magnetic resonance experiments[7] is also available, because spectroscopic (hyperfine) magnetic interactions (couplings) are revealed by their influence on the positron count rate in designated detectors. Using high magnetic fields, applied transverse (at 90°) to the muon beam direction, the Transverse-Field Muon Spin Rotation (TF-MuSR) technique characterises each radical by a single pair of lines, which represent the -1/2, +1/2 m_S electron spin combination with the muon (m_μ) states. These occur at the precession frequencies from muons which experience the sum of the applied and (-1/2,+1/2 m_S) hyperfine magnetic fields. A classical picture is of the muon spin *rotating* around the axis of the applied magnetic field; rather as the spin of a similar magnetic nucleus, following a 90° *pulse* of radiofrequency radiation in an "FT-NMR" experiment.[8] The muon-electron hyperfine coupling constant is obtained from the difference between the high (v_2) and low (v_1) frequencies for each radical: $A_\mu = v_2 - v_1$. As the coupling increases, for a given magnetic field, the frequency v_2 increases, while concomitantly that v_1 first decreases, reaches zero and then increases once more, due to a sign change in the transition; the coupling is now obtained from the sum of the frequencies: $A_\mu = v_1 - (-v_2)$. A representative example [4] of an *actual* such *TF-MuSR* spectrum is that of *1,1-dichloroethyl* radicals (MuCH$_2$CCl$_2\bullet$), derived from 1,1-dichloroethene sorbed in kaolin (Figure 1). Given that *single-particle counting* methods are employed, and the muon spins are nearly 100% polarised (as compared with the Boltzmann factor, on which EPR and NMR depend), techniques involving muons are extremely sensitive (one *single* molecular radical is detected at a time by TF-MuSR).

2.2 Longitudinal-field muon spin relaxation (LF-MuSRx)

This is a relatively new technique, but shows promise in the study of the reorientational rates of radicals sorbed in porous materials, which so far include: zeolites[9-11], activated carbon[12-14], silica-gel[4], clays[4,15] and a highly dispersed ice-surface[16]. Though it does not (certainly in its present simple form) approach the level of detail regarding determining molecular reorientation that is possible with (Avoided Level Crossing) ALC-MuSR,[17,18] the method does provide a relatively direct estimate of motional correlation times and so of the activation energy associated with a particular kind of motional process. The underlying theory appears fairly well understood,[19,20] and other applications have been found in determining the intramolecular dynamics of radicals formed by muonium addition to Ph_4X (X=C, Si, Ge, Pb),[21,22] and to metallocenes and benzene-metal π-complexes,[21,22] in which the muon acts as a spin-probe of torsional motion of the phenyl groups or of the overall motion of the cyclopentadienyl or benzene ring about the metal atom. Similar torsional dynamics have been measured in samples of solid dipeptides[23], oleoyl esters[24] and polymers[25]. The restricted overall molecular reorientation of the muonium adduct of the C_{60} fullerene was also measured using LF-MuSRx.[26]

The physical basis of the method is one of resonance. In general, when the frequency of a particular molecular motion approaches that of the dominant spectral transition (ω) in the muon-electron coupled system,[19,20,26] there is an increase in the relaxation rate (λ) of the muon spins, as measured in a longitudinal magnetic field (LF). This reaches a maximum when the frequencies are equal. Different motional regimes may be identified, and in some cases two maxima are measured (*e.g.* Figure 2), which may reflect two distinct sorbed fractions, each with its own motional characteristics. Motional correlation times (τ) may be estimated from the LF-relaxation rates (λ), *via* (equation 7):

$$\lambda = (2\pi\delta A)^2 \bullet \tau/(1 + \omega^2 \tau^2) \qquad (7)$$

In (equation 7), δA reflects the modulation in the hyperfine frequency during the motional event, which relaxes the muon spin; ω is often assumed as the frequency of the <1| ←→ |2> transition, since it is strongly induced by the mechanism of motional partial averaging of the anisotropic (dipolar) muon hyperfine coupling, as the radical tumbles.[19,20,26] Since $\lambda = 1/T_1$, the maximum in the relaxation rate corresponds to a "T_1 minimum" familiar[7,8] in NMR and ESR spectroscopy. Simple estimates of activation parameters, *viz* inverse-frequency factors (τ_∞) and activation energies (E_a), for internal motion in molecules and for their overall motion in sorbed states, may be made by measuring the muon spin relaxation rate (λ) as a function of temperature. The activation terms (τ_∞) and (E_a) are obtained by fitting the τ values, determined over a particular temperature range, to (equation 8).

$$\bar{\tau} = \tau_\infty \bullet \exp(E_a/RT) \tag{8}$$

2.2.1 Additional relaxation effects

A number of influences might cause relaxation of the muon spin. At the outset, we note that there will be contributions from transitions other than that between the $<1| \leftrightarrow |2>$ levels; but since this particular transition is strongly induced through the dipolar modulation mechanism which is a feature of molecular reorientation, it is convenient to assume that its frequency is ω especially for reorientation on a surface. The choice of frequency will affect the values obtained for the motional correlation times (τ), and hence the inverse-frequency factor, (τ_∞) defined in (equation 8), but not the activation energy (E_a). Therefore, a plot of $-\ln(\tau)$ vs $1/T$ should provide a meaningful activation energy when a single motional mode is dominant.

Other than the effect of molecular reorientation, the degrees of freedom available to a rigid molecule like cyclohexadienyl are unlikely to affect significantly the muon spin relaxation rate (λ). This is not necessarily true in more flexible structures, which are fairly common in radicals derived by muonium addition to the molecules of VOCs: *e.g.* ring-inversion, out-of-plane vibrations of a bound muon, and rotation (or torsion) of $MuCH_2$-groups all modulate the coupling of the muon and may cause its relaxation. Paramagnetic centres in the host material are also potential contributors to (λ).

In the series of cation-exchanged zeolite X samples containing benzene, we used the inverse-frequency factors (τ_∞) to calculate activation entropies ($\Delta S^\ddagger$), from transition-state theory; we believe that a self-consistent set of data is thus provided for the rigid cyclohexadienyl radical in this medium, reflective of the varying nature of the cation. For the reasons mentioned above, we have less confidence in this procedure for some of the less-rigid radicals, when we have confined our discussion mainly to the activation energies.

3. ZEOLITES AND CLAY SURFACES

Until only recently atmospheric processes were viewed as being purely gas-phase reactions,[1] but it is now thought that atmospheric aerosols, clouds and dust particles can provide surfaces on which reactions can occur with great efficiency.[27] Indeed, the presence of a background atmospheric dust load, evenly spread through the atmosphere, has been established.[28] The quantity of atmospheric dust blown skywards by the desert winds is thought to amount to perhaps *5-10 billion tonnes* annually.[2,3,29] Clay particles, particularly, are a major component of airborne dust, and both clay-minerals and zeolites feature centrally in pollution control and remediation strategies;[33] additionally, the catalytic properties of zeolites[30,31] and clays[32] are well established. We believe that such properties of sorption and catalysis by zeolites and clays are likely to influence the environmental impact of VOCs.[34] Indeed, a number of catalytic applications have been found for montmorillonite,[32] which are in many respects similar to those for zeolites, and it seems highly probable that clays as airborne materials might contribute in the outcome of atmospheric pollution chemistry.

A growing body of evidence implicates organic free radicals as intermediates in many reactions catalysed by these materials;[30-32,35] however, free radicals are mostly highly reactive and have normally only a fleeting existence. ESR is a principal means for the detection and identification of radicals, but since they are present only in low concentrations, and often have fairly complex spectra, surface-sorbed radicals are generally thus undetectable. Some success *has* been found in using ESR to detect radicals on surfaces, but such measurements are limited to low temperatures,[31] since under catalytically relevant conditions, the spectral signature of the radicals is rapidly lost through their diffusion and consequent termination reactions; the extreme sensitivity of MuSR has a profound advantage in such studies. The first example of a TF-MuSR study of a free radical (1,1,2-trimethylallyl) sorbed on a surface (fumed silica) was reported about 15 years ago.[36] Later, TF-MuSR was used to provide the first detection *by any method* of a reactive radical (cyclohexadienyl) in a zeolite (NaX) *at ambient temperature*,[37] and was shortly extended[38] to other hydrocarbon radicals sorbed both in NaX, and in another zeolite, Na-mordenite.

All these studies show that sorption on a solid surface causes pronounced line-broadening: this is because the dipolar coupling between the muon and the unpaired electron is not fully averaged-out in a radical whose tumbling-rate is slowed by its interaction with a surface.[36-38] The linewidth is proportional to the time the molecule takes to reorient, nominally termed the "rotational correlation time" (τ). It is possible to extract this quantity from the linewidths of a series of TF-MuSR spectra measured as a function of temperature, and to estimate a frequency factor and an activation energy for the motional process, using an expression similar to the (equation 8). However, LF-MuSRx measurements may be made down to temperatures at which the corresponding TF-MuSR spectral lines become too broad for reliable analysis.

3.1 1,1-dichloroethyl radicals sorbed in clay and in silica-gel

To assess the possibility of detecting radiolabelled radicals sorbed on a clay-mineral surface, a study was made of 1,1-dichloroethyl radicals (MuCH$_2$CCl$_2$•) sorbed in silica-gel and in kaolin;[4] this was chosen to exemplify the kind of radical that might be derived from a halogenated VOC. An activation energy of 10.6±1.4 kJ/mol was determined for the reorientational diffusion of 1,1-dichloroethyl radicals sorbed on silica gel using TF-MuSR; this was supported by LF-MuSRx measurements, which gave an identical value of 10.9±0.7 kJ/mol. TF-MuSR measurements provided an activation energy of 11.0±1.3 kJ/mol for the same radical sorbed in kaolin powder, which also agrees, within experimental error, with the value obtained from an LF-MuSRx study (12.3±0.8 kJ/mol). An important difference is that the LF-MuSRx method revealed the presence of an additional, more mobile, fraction of 1,1-dichloroethyl radicals in both silica and kaolin with respective activation energies of 2.6±0.4 and 2.4±0.2 kJ/mol: no such second fraction was detected by TF-MuSR. As shown below, an explanation may be provided by considering the particular details of the two methods.

3.1.1 TF-MuSR

The major advantage of the TF-MuSR method[5,6] is that it allows an identification of radicals formed in a given system (so long as they are formed within the timescale set by the muonium hyperfine frequency, ca 10^{-9} -10^{-10} s) on the basis of the muon-hyperfine coupling. In Figure 1 is shown the spectrum recorded at 255 K from a sample of 1,1-dichloroethene sorbed in kaolin, revealing a clear pair of signals from a single radical species; the isotropic muon-electron hyperfine coupling constant is 238 MHz. Since there are two possible sites for addition, leading to either radicals of type $MuCH_2CCl_2\bullet$ or $MuCCl_2CH_2\bullet$, we then appeal to ESR data for the corresponding proton-equivalent radicals $HCH_2CCl_2\bullet$ and $HCCl_2CH_2\bullet$, in order to make comparisons. The latter radical is apparently unknown, but a coupling of 19.7 G was reported for the $CH_3CCl_2\bullet$ radical:[39] by scaling this value with the muon/proton magnetic moment ratio (3.1833) and converting to MHz, a coupling of 176 MHz is obtained; however, β-muon couplings are always larger than those of the protons in the corresponding radicals, generally by a factor of *ca* 1.4,[40] and so the $MuCH_2CCl_2\bullet$ radical is highly implicated.

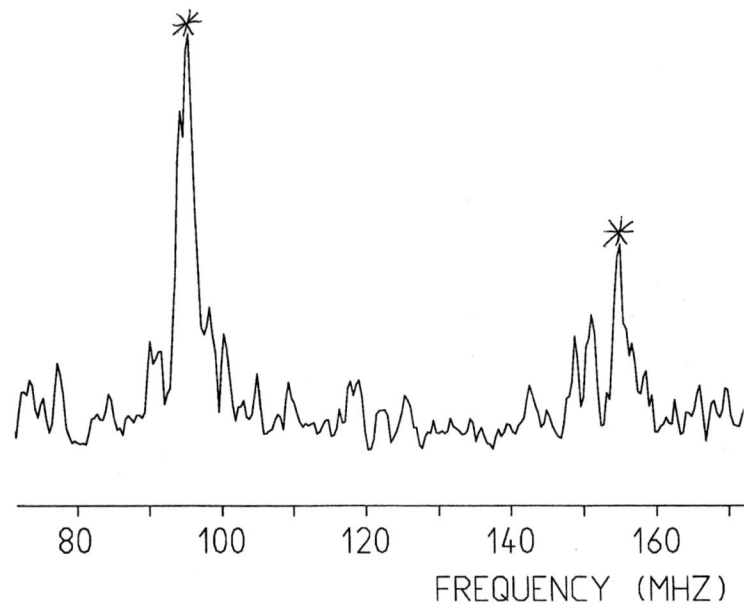

Figure 1. TF-MuSR spectrum recorded at 255 K from $MuCH_2Cl_2\bullet$ radicals sorbed in kaolin (see text).

As in a previous TF-MuSR study of radicals sorbed in zeolites,[38] the motional correlation times (τ) were determined from the temperature dependent linewidths (λ) in the $MuCH_2CCl_2\bullet$ radical; the results are shown in Table 1.

Arrhenius type plots of $-\ln\tau$ vs $1/T$ were made, according to (equation 8), so arriving at the values shown in the Table 3 which permit comparison

8. Radiolabelled organic radicals on reactive surfaces

between the activation parameters obtained from the TF-MuSR and LF-MuSRx measurements.

Table 1. Muon coupling constants (A) and motional correlation times (τ) for MuCH$_2$CCl$_2\cdot\bullet$ radicals sorbed in silica and in kaolin, as determined from TF-MuSR measurements.[4]

	T/K	A/MHz	τ/s
Silica Gel	255	243.9	5.64x10^{-9}
	265	242.4	5.21x10^{-9}
	275	240.8	4.46x10^{-9}
	285	239.3	3.33x10^{-9}
	295	237.9	2.98x10^{-9}
Kaolin	255	244.0	5.57x10^{-9}
	265	242.6	5.14x10^{-9}
	275	241.0	4.38x10^{-9}
	285	240.4	3.38x10^{-9}
	295	238.3	2.85x10^{-9}

3.1.2 LF-MuSRx

For both samples of 1,1-dichloroethene (in silica gel and in kaolin) motional correlation times (τ) were also determined (Table 2) from the measured longitudinal relaxation rates (λ). In each case, there are two maxima which correspond to two distinct sorbed fractions, each with its own mobility (Figure 2). From plots of -lnτ vs 1/T, the activation parameters shown in Table 3 were obtained, in clear agreement with the values determined from the TF-MuSR measurements.

Table 2. Motional correlation times (τ) determined from LF-MuSRx measurements on MuCH$_2$CCl$_2\cdot\bullet$ radicals sorbed in silica gel and in kaolin.[4]

	T/K	τ/s		T/K	τ/s
Silica	80	4.97x10^{-9}	Kaolin	75	5.67x10^{-9}
	100	3.66x10^{-9}		90	3.28x10^{-9}
	120	2.12x10^{-9}		115	2.12x10^{-9}
	135	9.35x10^{-10}		130	1.08x10^{-9}
	150	1.15x10^{-9}		145	1.12x10^{-9}
	165	6.18x10^{-10}		160	8.25x10^{-10}
	180	1.36x10^{-8}		190	6.06x10^{-9}
	195	8.90x10^{-9}		205	3.96x10^{-9}
	210	5.67x10^{-9}		220	2.12x10^{-9}
	225	3.82x10^{-9}		235	1.60x10^{-9}

240	2.67x10⁻⁹	250	8.27x10⁻¹⁰
255	2.12x10⁻⁹	265	6.39x10⁻¹⁰
270	1.60x10⁻⁹	280	5.97x10⁻¹⁰

Table 3. Motional activation parameters measured for MuCH$_2$CCl$_2$• radicals sorbed in silica gel and in kaolin, using LF-MuSRx and TF-MuSR methods.[4]

	E/(kJ/mol)	τ$_\infty$/s
Silica Gel	10.9±0.7(LF)	(1.1±0.3)x10⁻¹¹
	10.6±1.4(TF)	(4.0±1.8)x10⁻¹¹
	2.6±0.4(LF)	(1.2±0.4)x10⁻¹⁰
Kaolin	12.3±0.8(LF)	(1.6±0.5)x10⁻¹²
	11.0±1.3(TF)	(3.3±1.4)x10⁻¹¹
	2.4±0.2(LF)	(1.5±0.3)x10⁻¹⁰

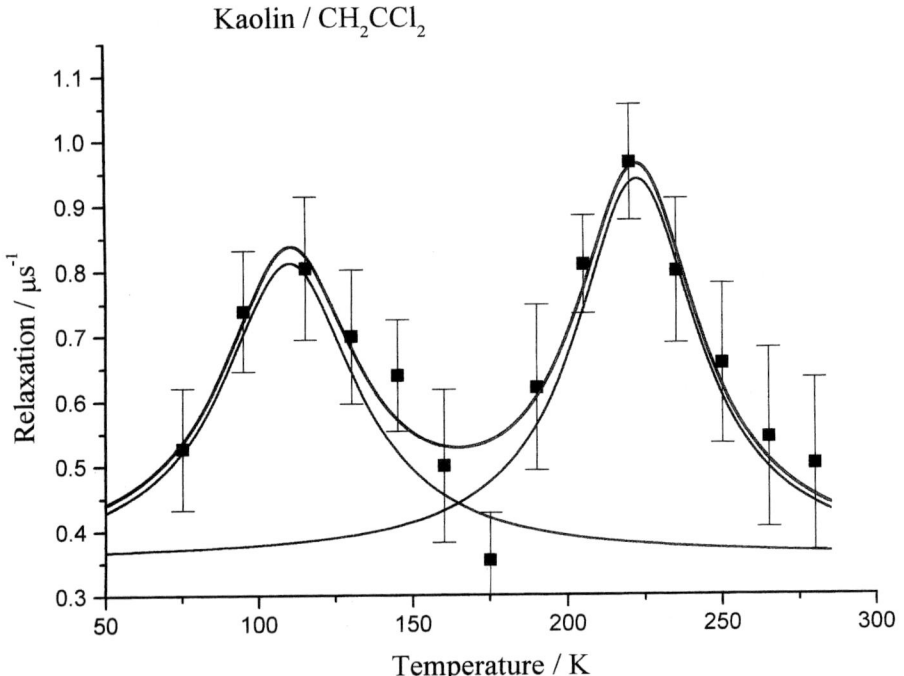

Figure 2. LF-MuSR plot of MuCH$_2$CCl$_2$• radicals sorbed in kaolin (see text).

3.1.3 Comparison between TF-MuSR and LF-MuSRx measurements

The correlation times listed in Table 1 and Table 2 are in reassuring agreement as determined by the two methods, and are very similar for both the silica and the kaolin samples. As already commented on, the major difference is that a fraction with a completely separate mobility is detected by the LF-MuSRx method (Figure 2), but is not evident in the TF-MuSR spectra (Figure 1). This fraction was measured in the LF spectra in the temperature range 75-165 K, which is much lower than that used for the TF measurements (255-295 K); however, extrapolation of the correlation times to this higher temperature range indicates a mobility some 4-5x greater than that of the fraction which is actually being measured. Since this would correspond to a relatively sharp signal superimposed on a broader one, it should be quite evident, particularly at the lowest temperature used for the TF measurements (255 K), where the differential line-broadening between the two fractions would be greatest; however there is nothing.

The relaxation times T_1 and T_2 correspond strictly to the two experimental arrangements,[5,6] since the muon beam is 100 % spin polarised (*i.e.* all the spins are aligned along the beam direction) with respect to the applied magnetic field, and so LF-MuSR and TF-MuSR measure separately these *longitudinal* and *transverse* components. The LF relaxation rate corresponds to muon spin-flips directly, along the beam direction, and measures the spin-lattice interaction, hence the appearance of Figure 2, which corresponds to "T_1 minima" in conventional magnetic resonance,[7,8] sensing the motional frequency - the minimum occurring when motional and transition frequencies are matched; the TF measurement is also sensitive to relaxation caused by motional processes, but also to additional processes such as chemical reactions.[5,6] In fact, the good agreement between the correlation times and activation parameters obtained from both types of measurement *indicates that factors extraneous to those of molecular reorientation are of little importance*; therefore, the lack of observation of the more mobile fraction by TF-MuSR cannot be explained in terms of differential T_1/T_2 relaxation effects.

We conclude that the explanation is actually kinetic, and refers to the rate of formation of radicals from 1,1-dichloroethene sorbed in different concentrations, in separate regions of each sorbant. In order to observe the TF-MuSR spectrum of a radical, it must be formed on a timescale set by the muonium hyperfine frequency (4.5 GHz), *i.e.* 10^{-9}-10^{-10} s,[5,6] which is close to the diffusion controlled limit for reactions of atomic muonium.[41] Therefore, for dilute fractions, the muonium may not be able to add fast enough for the resulting radical to be observed. In LF experiments, there is no such restriction other than that the radical must be formed within the lifetime of the muon itself, (2.2×10^{-6} s), and so radicals can, in principle, be detected even in concentrations 100-1000x lower than is possible in a TF experiment. Thus, the major fraction detected in LF-MuSRx corresponds to the (only) fraction detected by TF-MuSR, the more mobile fraction in the LF spectra being from a minor, more weakly sorbed, component of 1,1-dichloroethene. *Direct evidence that this is the case is provided by our observation that it could be readily pumped from the cell* - leaving the major, less mobile component - in accord with the low activation energy for its molecular reorientation (and diffusion), which indicates weak adsorption.

One possibility is that the minor fraction is located in a "fluid" type state; however, this should require a loading of the organic sorbate in excess of that equivalent to monolayer coverage. Using standard atomic radii,[42] we deduce that the surface area of one 1,1-dichloroethene molecule is 36 Å^2, and so 14 mmoles would cover an area of 3.04×10^{23} Å^2, which is far greater than the surface area of the kaolin but is only 83 % that of the silica sorbant. Therefore, on the basis of the BET surface areas determined by dinitrogen adsorption, the existence of a putative fluid fraction is entirely reasonable in kaolin, but not in silica; nonetheless, the similarities in activation parameters measured in both sorptive media would suggest its common nature, and we stress that the adsorption behaviour of dinitrogen may differ from that of 1,1-dichloroethene.

We believe that the actual surface adsorption is weak because there are fewer OH groups present in the external surface regions. Therefore, the radicals can easily break-free from the surface via a "sticking" collision with one or more 1,1-dichloroethene molecules. When a stronger surface (----HO) bond has to be broken simultaneously, the overall loss of entropy is correspondingly less since the initial, adsorbed, state of the radical is more restricted (ordered); this also accords with the large $\Delta G^{\ddagger}$ values, which are dominated by the enthalpies. Given the similarities between all parameters for each fraction in both silica gel and kaolin, we conclude that the sorptive properties of these materials are actually very similar.

It has been proposed[43,44] that, in the absence of viscous friction, the inverse-frequency factor for the reorientation time is expected to approach that of a classical free rotator: $\tau_{FR} = 2\pi(I/81kT)^{1/2}$. Assuming standard bond lengths and atomic masses for the MuCH$_2$CCl$_2\bullet$ radical, I=3.93×10^{-45} kg m^2, which gives 7.45×10^{-13} s at 250 K and 1.08×10^{-12} s at 120 K. These values are in agreement by a factor of 3 or 8 for the more strongly bound fraction in kaolin or silica, respectively, but are in excess of 70x faster for the weakly sorbed component, indicating that it is behaving as a viscous liquid. This may by taken with the large, negative activation entropies to indicate greater molecular association in the reorientation step.

3.2 Zeolites

3.2.1 LF-MuSRx measurements of cyclohexadienyl radicals sorbed in cation-exchanged zeolite X

A study[9,10] was reported of benzene sorbed in cation-exchanged (Li^+, Na^+, K^+, Mg^{2+}, Ca^{2+}, Sr^{2+}, Ba^{2+}) zeolite X and (Na^+) zeolite A, as is now discussed in some detail. Zeolite X has the essential structure of naturally occurring faujasite, which is a three-dimensional internal channel system with supercages: the supercages are some 13 Å in diameter, with access windows of 7.4 Å in cross-section, and there are 8 cages per unit cell.[45] The access windows readily admit molecules as small as benzene (kinematic diameter 5.5 Å) into the zeolite micropores. In contrast, the pore system of zeolite A consists of a cubic array of relatively large (*ca* 11.2 Å free diameter) cages, interconnected through 8-membered oxygen windows of only 4.3 Å diameter,[45] and so this structure admits only molecules with very small cross-sections, and definitely excludes benzene. Changing the nature of the counter-cation changes the diffusivity of molecules in zeolite A by relatively blocking or unblocking the already restrictive window sites; this is less important in zeolite X (and in faujasites generally),[45,46] but changes in the mobility of adsorbates and in their enthalpies of adsorption have been found, for instance in Na^+, K^+, Cs^+ exchanged zeolite X.[47] The aim of the study[9,10] was to determine the influence of the exchange cation on the reorientational mobility of cyclohexadienyl radicals in zeolite X.

While the analysis of the relaxation data is fairly straightforward for LiX, NaX, KX, MgX and 4A, at 18 wt% benzene loading, since there is in each case a single maximum, the situation for the other zeolites is more complex. At a 40 wt% loading in KX, the data clearly describe two well displaced maxima, and this also pertains in CaX, even at 18 wt% of benzene; for SrX and BaX the data are more ambiguous, since it is possible to fit each case either as a single or as two components. The abrupt and clear difference in behaviour of CaX from the group 1 series does, however, point to a potentially differing influence of the dipositive ions on the sorption of benzene into the micropores of zeolite X, and this is supported by the fitting-curves obtained, which each indicate the presence of two benzene components. On this basis, two sets of activation parameters are determined, as shown in Table 4. The data show that each member of the group 2 series, other then MgX, contains an additional, more restricted fraction, with similar activation parameters to those found in the heavily loaded KX and in the intrinsically pore-restricted 4A zeolites. The TF-MuSR method was used previously to study the effect of substrate (benzene) loading on the mobility of cyclohexadienyl radicals therefrom formed in NaX:[48] the results show that only for loadings above 50 wt% is a clear "fluid-like" phase observed; even at 50 % loading, the lines remained broad, indicating considerable restriction in molecular reorientation. According to nitrogen adsorption data,[30] the zeolite is at twice its fully saturated level; however, the nitrogen BET measures bulk diffusion phenomena. At the molecular level, which is probed by MuSR, reorientational motion is apparently impeded even under such conditions of bulk saturation.

Table 4. Activation parameters and effective Muon hyperfine anisotropies for reorientation of benzene in cation exchanged zeolite X and in other hosts.[10]

Medium.	E_a/(kJ/mol)	τ_∞/s	δA/MHz
LiX(18 wt% C_6H_6)	9.1±1.1	(1.4±0.4)x10^{-12}	5.7
NaX(18 wt% C_6H_6)	6.9±0.7	(5.3±1.4)x10^{-12}	4.7
KX(18 wt% C_6H_6)	4.0±0.5	(6.1±2.1)x10^{-11}	5.9
KX(40 wt% C_6H_6)	12.9±1.5	(1.0±0.3)x10^{-12}	5.4
4A(18 wt% C_6H_6)	12.7±1.3	(2.7±1.1)x10^{-13}	5.0
MgX(18 wt% C_6H_6)	7.7±0.8	(1.6±0.6)x10^{-12}	5.5
CaX(18 wt% C_6H_6)	6.9±0.6	(3.7±0.7)x10^{-12}	4.8
	12.4±1.3	(3.5±1.5)x10^{-13}	4.2
SrX(18 wt% C_6H_6)	6.4±0.7	(8.1±3.7)x10^{-12}	5.9
	12.8±1.4	(3.6±1.1)x10^{-13}	4.6
BaX(18 wt% C_6H_6)	12.0±1.1	(5.4±2.7)x10^{-13}	5.1
	12.3±1.4	(1.1±0.5)x10^{-12}	4.4
Carbon Pores d >20 Å	5.9±0.5	(6.2±2.2)x10^{-12}	5.9
Bulk Benzene	6.6±0.5	(2.4±0.7)x10^{-11}	4.7

We note that the values of δA are in the range 4.2-5.9 MHz, which is consistent with the major reorientational motion occurring about the ring plane axis, since the nature of the hyperfine anisotropy in the muonated cyclohexadienyl radical requires a variation of 5.6 MHz during a step of this kind, assuming a 90° "jump".[13,49]

3.2.2 Energetics for reorientation of cyclohexadienyl radicals in cation-exchanged zeolite X

Transition-state theory was used to determine free energies ($\Delta G^\ddagger$), enthalpies ($\Delta H^\ddagger$) and entropies ($\Delta S^\ddagger$) of activation for cyclohexadienyl radicals (and by inference, benzene molecules) sorbed in the range of group 1 and group 2 cation exchanged zeolite X. The results are shown in Table 5 along with those determined for benzene in zeolite 4 Å, in the large pores (d > 20 Å) of porous carbon and in bulk benzene, and from which some apparent trends emerge. For the group 1 series (LiX, NaX, KX), all at 18 wt% loading of benzene, $\Delta G^\ddagger$ has an almost common value of *ca* 10 kJ/mol, but the relative contribution made to it by $\Delta H^\ddagger$ and $T\Delta S^\ddagger$ varies, with the entropy term becoming increasingly dominant as the cation radius increases ($r_{K+} > r_{Na+} > r_{Li+}$). This reflects the decreasing importance of cyclohexadienyl/benzene-cation interactions, since the $\Delta H^\ddagger$ value falls from 7.9 kJ/mol in LiX to only 2.8 kJ/mol in KX, and is in order with previous measurements of the isosteric enthalpies of adsorption of benzene in Na^+, K^+, Cs^+ exchanged zeolite X.[47] In both NaX and KX, the motional process is dominated by the entropy change, and in fact, the values for $\Delta H^\ddagger$ and $\Delta S^\ddagger$ are quite similar to those determined for bulk benzene and for benzene sorbed in large-pores in activated carbon.[13] Thus, we conclude that the activation process, which causes the molecular motion we are measuring, is dominated by interactions/collisions between almost free benzene molecules, certainly in NaX and in KX. In all cases, $\Delta S^\ddagger$ is negative, and it has been argued that this represents a "sticking" collision between the cyclohexadienyl radical and one or more benzene molecules in the activation step,[4,9,10] since reorientation of an effectively *isolated* molecule is expected to show a near zero $\Delta S^\ddagger$;[50] such a collision is expected to lose degrees of freedom from the initial state of the molecule.

In the case of benzene sorbed in LiX, the large $\Delta H^\ddagger$ indicates far stronger associations with cations, and so it is reasonable that the initial state is more "ordered"; therefore, the overall loss of degrees of freedom in making the "sticking" collision with other benzene molecules is accordingly reduced, hence the less negative $\Delta S^\ddagger$ value. In support of this explanation, the parameters $\Delta H^\ddagger$ and $\Delta S^\ddagger$ are in clear interdependence in the series LiX → KX, so reflecting the relative "ordering" in the initial state compared with the activated state.

The $\Delta H^\ddagger$ and $\Delta S^\ddagger$ parameters for the fraction of lower E_a in the series MgX → BaX indicate a similar situation: MgX appears much like LiX, and we would expect strong interactions between cyclohexadienyl radicals/benzene molecules and both Li^+ and Mg^{2+} cations, as we found for cation-PhCH•-OMu radical complexes in these zeolites;[51] an interdependency of $\Delta H^{\ddagger}$ and $\Delta S^{\ddagger}$ is apparent for benzene molecules sorbed in CaX and in SrX, implying their relatively weakening association with Ca^{2+} and Sr^{2+} cations, and that achieving the activated motional state in CaX is similar to doing so in bulk benzene. As in the series LiX → KX, it is expected that the cyclohexadienyl radicals/benzene molecules will associate (bind) less strongly with the larger cations, simply because the positive charge *density* is reduced; however BaX contains no equivalent mobile component. This observation does not fit the above model.

Some insight, both to this apparent anomaly and the motional behaviour of benzene molecules in the pores of zeolite X, generally, may be provided by a comparison with the activation parameters for bulk benzene;[13] the exchange cations can be considered mainly to act in perturbing the essential ordering of a small cluster of benzene molecules, located in one supercage. We note that the series of group 2 cation exchanged zeolites contain only half the number of cations that the group 1 series does, and having a greater positive charge, they are expected to associate more strongly with the zeolite framework, so leaving a greater "volume" *centred* in the cage, into which benzene molecules can be accommodated. Although, the supercages are filled with benzene molecules, the overall cage geometry will limit the possibilities for their mutual orientation, compared with the bulk material.

In CaX and SrX, the $\Delta H^{\ddagger}$ values are slightly reduced, being similar to that found in bulk benzene,[13] but the $\Delta S^{\ddagger}$ parameters are nearly identical. This supports our notion of a motional behaviour which is that of a small number (*ca* 6-8) of benzene molecules clustered in the zeolite X supercage,[45] interacting very weakly with the Ca^{2+} or Sr^{2+} cations, and is equivalent to the molecular reorientation occurring in the bulk material. The relaxations which are measured, we believe to represent an *average* molecular motion, or at least a *distribution* of mobilities, since at any instant, some molecules may be coordinated with cations while others remain free, and presumably characteristic activation parameters apply for each situation.

At high loadings (40 wt%) of benzene in KX, a second fraction is identified, to which is provided a more restricted molecular environment. Since this loading represents more than twice the saturation capacity of the supercages, the excess benzene must be located elsewhere, and it appears reasonable that it is resident on the external surface and in the extragranular voids: this idea is favoured by the near identity of the E_a and τ_{∞} parameters, and consequently those of $\Delta H^{\ddagger}$, $\Delta S^{\ddagger}$ and $\Delta G^{\ddagger}$, found with an 18 wt% loading

8. Radiolabelled organic radicals on reactive surfaces

of benzene in 4A zeolite. The small access pores[45,46] in 4A (4.1 Å) definitely exclude benzene molecules, given their kinematic diameter of 5.5 Å, which have no alternative than to occupy (external) non-cage sites.

The motional processes pertinent to benzene molecules in these non-cage regions are dominated by their enthalpies of activation; the relatively less negative $\Delta S^{\ddagger}$ values implying a relatively *ordered* initial state. This seems strange, as one might envisage the benzene molecules either lying flat on the granular surface, or forming clusters in the extragranular voids, and these arrangements surely would behave much as the benzene molecules sorbed on the walls of large carbon pores or in bulk benzene, respectively; yet these latter have significantly reduced motional activation enthalpies, and more negative $\Delta S^{\ddagger}$ values for their molecular reorientation. We make, therefore, the following suggestion, which is that the benzene molecules become "trapped" in pockets (niches) proximate to the external surface, which normally provide initial access for molecules into the zeolite grain; in KX, saturation of the inner structure (pores) of the grain prevents these "external" molecules making further inroads to the supercages/channel system, while in 4A, the intrinsic internal dimensions of the zeolite provide a physical barrier to them. Such "trapping" of the benzene molecules in the near-surface niches would tend to *order* them relative to the bulk material, so resulting in a less negative entropy change in their reorientation processes.

Table 5. Free energies (kJ/mol), enthalpies (kJ/mol), entropies (J/mol/K) of activation and entropy of activation terms at mid-temperature range (kJ/mol) for each series.[10]

	$\Delta G^{\ddagger}$	$\Delta H^{\ddagger}$	$\Delta S^{\ddagger}$	$T\Delta S^{\ddagger}$
LiX(18 wt% C_6H_6)	10.7(1.5)	7.9(1.1)	-20.1(3.0)	-2.8(0.4)
NaX(18 wt% C_6H_6)	10.5(1.1)	5.7(0.7)	-31.7(2.6)	-4.8(0.4)
KX(18 wt% C_6H_6)	10.3(0.9)	2.8(0.5)	-51.7(3.0)	-7.5(0.4)
KX(40 wt% C_6H_6)	15.5(2.0)	11.2(1.5)	-20.6(2.4)	-4.3(0.5)
MgX(18 wt% C_6H_6)	9.4(1.3)	6.6(0.8)	-20.8(3.9)	-2.8(0.5)
CaX(18 wt% C_6H_6)	9.7(1.0)	5.7(0.6)	-28.2(3.2)	-4.0(0.4)
	13.2(1.7)	10.9(1.3)	-10.6(2.1)	-2.3(0.4)
SrX(18 wt% C_6H_6)	10.3(1.2)	5.2(0.7)	-35.0(3.1)	-5.1(0.5)
	12.9(1.8)	10.9(1.4)	-11.0(2.3)	-2.0(0.4)
BaX(18 wt% C_6H_6)	13.1(1.8)	10.5(0.9)	-14.0(3.8)	-2.5(0.7)
	15.0(2.0)	11.2(2.0)	-21.2(3.1)	-4.4(0.6)
4A (18 wt% C_6H_6)	12.6(2.1)	11.2(1.3)	-8.2(4.3)	-1.4(0.8)
Carbon Pores d >20 Å	8.9(0.9)	4.8(0.5)	-31.8(3.0)	-4.1(0.4)
Bulk Benzene	14.3(1.1)	5.0(0.5)	-46.6(3.0)	-9.3(0.6)

3.2.3 Cyclohexadienyl and 2,5-dimethylcyclohexadienyl radicals in ZSM5, silicalite and mordenite

We have also used LF-MuSRx to investigate the sorption and mobility of cyclohexadienyl radicals and 2,5-dimethylcyclohexadienyl radicals (derived from benzene and from *p*-xylene) in Na-ZSM5, in silicalite and in Na-mordenite.[11] Unlike the faujasites, zeolites X and Y, which contain cages, ZSM5 and mordenite both contain only channels.[45] For cyclohexadienyl radicals in ZSM5, silicalite and mordenite, a fraction was detected with a common reorientational activation energy of *ca* 5 kJ mol^{-1}; however, in both ZSM5 and silicalite there appeared a secondary fraction with an activation energy of *ca* 14 kJ mol^{-1}. This was absent in mordenite. Mordenite contains a single kind of linear channel, which has a cross-section of *ca* 7 Å; therefore, there is only one broad location for sorbed molecules, namely within these channels, and the single distribution represents this. In ZSM5 and silicalite, the straight channels have a cross-section of *ca* 5.5 Å; there is also a *zig-zag* channel system which provides a lateral interconnection of the straight channels.[45] It is well known that, at low loadings hydrocarbon molecules tend to locate at the intersections of these channels, but as the loading increases, molecules are forced to occupy more restrictive sites within the channels themselves: since loadings close to the saturation capacity of the zeolites were employed, we believe the two distinct motional distributions represent the channel intersection and channel locations, the latter having the higher activation energy. The results for 2,5-dimethylcyclohexadienyl radicals are rather similar, but indicate slightly reduced activation energies. This is probably because more limited molecular excursions are permitted for these larger molecules within channels that are narrower than the long-axis of *p*-xylene; for benzene, free rotation is allowed. These conclusions are in accord with those drawn from ^{2}H-quadrupole-NMR measurements.[52-54]

3.2.4 PhCH•-OMu radicals sorbed in Zeolite X

It was shown that electron withdrawing substituents act to reduce the isotropic muon coupling in PhCH•-OMu radicals.[51,55] The question then arose, of whether association between the aromatic ring and a cation exchanged into the zeolite might provide a similar electron withdrawing influence. To explore this possibility, benzaldehyde was sorbed into a series of cation-exchanged zeolite X samples. Changes in the isotropic muon couplings were indeed found, which accord with the formation of such o-complexes: in each series, Li$^+$, Na$^+$, K$^+$ and Mg^{2+}, Ca^{2+}, Sr^{2+}, Ba^{2+}, the strength of this complexation was found to decrease with the increasing radius of the cation, the coupling being smallest for Li$^+$ and Mg^{2+} cations.[51] While representing an advance in the study of molecular sorption by zeolites, the effect noted is indirect, being one of perturbation on the π-electrons of the radical, and does not determine the influence of the cation on the reorientational motion of the sorbed species. Nonetheless, logic would suggest that such motion might be impeded, and increasingly so, by complexation of increasing strength. This was investigated, and the reorientational activation energies for PhCH•-OMu radicals sorbed in cation-exchanged zeolite X were measured using LF-MuSRx.[56]

In all cases, other than MgX, two clear maxima were identified. These correspond to PhCH•-OMu radicals confined in two distinct motional regimes. The nearly common activation energy of *ca* 3 kJ/mol (Table 6) measured for the *low-temperature range* among all the samples is striking, and since it is the same as that attributed to benzaldehyde sorbed in the larger pores of porous carbon[12] - *i.e.* in an effectively *bulk* phase - we believe that some proportion of the benzaldehyde is also sorbed as bulk in zeolite X. This fraction could not be detected positively in MgX. In contrast, the activation energies measured in the *higher temperature range* vary appreciably according to which exchange-cation is present. In both series, Li$^+$ → K$^+$ and Mg^{2+} → Sr^{2+}, the activation energy (E_a) falls as the cation radius increases, in accord with the reduced polarising-power and hence weaker π-complexation to the cation, as was inferred from the change in the isotropic muon couplings measured in these samples. The relatively large E_a value measured in BaX is anomalous, and can be ascribed to a fraction excluded from the supercages by large Ba^{2+} cations which obstruct their access windows. A similar result for benzene sorbed in BaX has been alluded to, *vide supra*.

Table 6. Activation energies (E_a) as determined for the muonium adduct radicals formed from benzaldehyde, PhCH•-OMu, sorbed in cation-exchanged zeolite X, according to equation 8, showing two distinct motional regimes.[56]

Zeolite	(E_a)/kJ/mol	(E_a)/kJ/mol
LiX	3.4±0.4	11.4±1.7
NaX	4.2±0.5	11.0±0.7
KX	2.1±0.4	8.0±0.7
MgX	---------	12.2±1.4
CaX	3.1±0.2	8.3±0.5
SrX	3.1±0.0	6.2±0.9
BaX	3.1±0.3	13.6±1.2
C$_{(s)}$	2.9±0.3	11.5±0.4

4. CARBON PARTICLES

In the lower atmosphere the higher concentration of oxygen makes very significant contributions to its chemistry, and so the troposphere is dominated by oxidation reactions, mediated largely by •OH radicals. The plant kingdom emits enormous quantities of hydrocarbons into the atmosphere, especially terpenes from forests. Indeed the "Blue-Ridge-Mountains-of-Virginia", of Laurel and Hardy fame, appear blue behind the haze caused by the tropospheric oxidation of pinenes and other terpenes.[2,3] There are also anthropogenic sources. The oxidation of these and many other organic compounds (VOCs) increases the atmospheric burden of greenhouse gases (CO_2) but also leads to the formation of carbonaceous aerosol in the atmosphere.[2,3] Thus they may be self-promoting in providing a solid surface on which further oxidations of VOCs could occur, involving organic radicals as intermediates. We have tried to explore the utility of MuSR methods in studying the reorientational diffusion of radicals derived from VOCs sorbed in porous carbon, as a simple model of a carbon-rich surface.[12,13] Benzene was chosen since it is forms a relatively rigid cyclohexadienyl radical, and should give reliable activation entropies as well as energies and enthalpies, it is also a significant VOC in the atmosphere, as is toluene, partly from human activities; benzaldehyde is a principal oxidation product of toluene in the troposphere.[3]

4.1 Benzene sorbed in porous carbon

When benzene is sorbed in porous carbon,[13] the resulting LF-MuSRx plot describes two maxima, peaking at *ca* 125 K and *ca* 200 K. Using (equation 7) motional correlation times (τ) were extracted from the λ values, and were plotted *vs* 1/T according to (equation 8), with its typical Arrhenius form. From the linear plots taken over the two temperature ranges, the thermodynamic activation parameters listed in Table 7 were obtained. We have suggested[13] that the anomalous $\Delta S^{\ddagger}$ value of +52.3 J/mol/K reflects benzene molecules sorbed in the smaller (d < 10 Å) micropores of porous carbon, and which undergo rapid collisions with the pore walls; its positive sign was proposed to represent the acquisition of degrees of freedom as such initially confined molecules escape into more spacious pores. We imagine the effect on a benzene molecule as it squeezes-out through a narrow access window (slit) in the small pore, as being akin to an elastic disk, held in the hand, whose edges are suddenly compressed: the disk is thus propelled rapidly, gaining both translational and reorientational freedom. It literally "pops-out" from the small pores into more spacious regions.

That no such *positive* $\Delta S^{\ddagger}$ value is found for toluene or benzaldehyde (*vide infra*) strongly implies that these molecules are unable to penetrate the very small carbon pores (*i.e.* those toward the low-end of the 4 - 20 Å range, which designates the "micropore" region). This accords with our view[13] that the access holes to the micropores are very narrow, and only just admit the molecules of unsubstituted benzene; we suggest that increasing the molecular dimension by even a CH_3 or CHO substituent is sufficient to discourage admission of the substrate. The second fraction, measured for sorbed benzene, presents activation parameters which are all very similar to those measured in bulk-benzene, which is reasonable if it is resident in the meso- and macro-pores of the carbon structure (*i.e.* those of dimension, 20 > d < 10,000 Å). Benzene sorbed in porous silica[44] was shown to behave rather like bulk benzene, although a range of melting temperatures was inferred, according to the differing sizes of molecular clusters that had formed within the distribution of pore-sizes present in porous silica. We note that the distributions of pore sizes in activated carbons[45,57] show clear maxima, both in the micropore and macropore regions, with a relatively smaller mesopore fraction, and that the majority of the (high) surface area in the material used in this study (*ca* 1600 m^2/g) lies within its micropores. The loading used corresponds to a nominal 30-40 % coverage of the carbon surface for the three substrates.

In our previous studies of sorbed radicals using LF-MuSRx,[4,9,10,12-14] we ascribed *negative* $\Delta S^{\ddagger}$ values to reorientational events that were not those of *single* molecules, but rather involved "sticking-collisions" between molecules, so increasing the molecular "ordering" as the activated state for the process is achieved. We believe this is the situation for benzene, toluene and benzaldehyde sorbed in porous carbon. The latter two cases are now considered specifically.

4.2 Toluene sorbed in porous carbon

Toluene reveals two maxima in the LF-MuSRx plot, which we would normally attribute to two types of sorbed fraction. We have already discounted that toluene molecules manage to penetrate the very small micropores, since both $\Delta S^{\ddagger}$ values are negative. From Table 7 it is clear that the fraction characterised by the very small activation energy is dominated by its activation entropy, $\Delta H^{\ddagger}$ being close to zero (1.5±0.4 kJ/mol). The alternative toluene distribution has a very similar $\Delta H^{\ddagger}$ to that of "bulk" benzene,[13] but its $\Delta G^{\ddagger}$ value is enhanced by the more negative $\Delta S^{\ddagger}$ term, which again dominates over $\Delta H^{\ddagger}$. That toluene should require a more specific molecular ordering between its molecules for reorientation to occur than benzene does is not surprising, given the presence of the methyl group. Some mutual orientations of toluene molecules will be sterically less favourable than those which minimise eclipsing of methyl groups between molecules; hence a greater "ordering" is achieved in the activated state. This is reflected in the more negative $\Delta S^{\ddagger}$ values than apply for benzene.

Table 7. Free energies (kJ/mol), enthalpies (kJ/mol), entropies (J/mol/K) of activation and entropy of activation terms at mid-temperature range (kJ/mol) for each series of benzene, toluene and benzaldehyde sorbed in porous carbon.[12]

Substrate.	$\Delta G^{\ddagger}$	$\Delta H^{\ddagger}$	$\Delta S^{\ddagger}$	$T\Delta S^{\ddagger}$
Low-temperature ranges.				
Benzene (100-160 K)	8.9±0.9	4.8±0.5	-31.8±3.0	-4.1±0.4
Toluene (85-140 K)	8.5±0.7	1.5±0.4	-55.7±2.5	-7.0±0.3
Benzaldehyde (85-140 K)	11.4±0.6	1.8±0.3	-76.4±2.0	-9.6±0.3
Higher-temperature ranges.				
Benzene (175-235 K)	13.4±2.8	24.1±2.3	+52.3±2.4	+10.7±0.5
Toluene (160-220 K)	13.2±1.2	4.3±0.7	-47.0±2.5	-8.9±0.5
Benzaldehyde (160-220 K)	17.4±0.7	9.9±0.4	-39.5±1.7	-7.5±0.3

The lower symmetry of toluene molecules may also change the particular nature of the molecular reorientation. From the change in the muon hyperfine anisotropy (δA) which occurs during the reorientation of "bulk" benzene molecules in porous carbon, we deduced[13] that the most probable mechanism was a process involving "90° jumps" about the molecular axis. In toluene, three isomeric radicals are expected, according to *transverse-field* (TF-MuSR) measurements on liquid toluene,[58] formed by Mu addition at the 2-, 3- and 4- positions (the *ipso* isomer was not detected in the TF study). We are reminded also of a study of anisole (methoxybenzene) sorbed on silica surfaces,[59] made using the *Avoided Level Crossing* (ALC)-MuSR method. The results from this study show that the 2- and 3- Mu adducts of anisole have similar activation energies, close to ca 10 kJ/mol and similar to the value for the Mu adduct of benzene sorbed on 7 nm silica grains, while the 4-Mu adduct of anisole shows a near-zero activation energy. This result leaves us to speculate that rather than having two distinct fractions of toluene sorbed in porous carbon, there is a mainly bulk phase present in the meso- and macro-pores, but that the different radical isomers themselves exhibit differential motional behaviour. In view of the prior result for anisole, we make the tentative suggestion that it is the 4-Mu adduct of toluene which has the lowest E_a (2.5±0.4 kJ/mol), while the 2- and 3-Mu adducts display a mean value of ca 6 kJ/mol, similar to that for benzene.[13]

8. Radiolabelled organic radicals on reactive surfaces 325

Despite the difference in E_a, the relaxation maxima yield nearly identical δA values, in the region of 7-8 MHz, which is close to that expected (*ca* 6 MHz) for cyclohexadienyl radicals undergoing a 90° jump motion about the ring-plane axis:[13,49] the apparent increase in δA, even though there is some spin-delocalisation onto the methyl group (as shown by the fall[58] in the isotropic muon hyperfine coupling from that in the benzene-Mu adduct, cyclohexadienyl radical) suggests that the reorientational motion of methyl-substituted cyclohexadienyl radicals is more complex; accordingly, the entropy requirements appear greater too (*ca* -50 J/mol/K *c.f.* -30 J/mol/K for unsubstituted cyclohexadienyl radicals) too, suggesting that a more ordered activated state occurs.

4.3 Benzaldehyde sorbed in porous carbon

TF-MuSR studies have shown that Mu addition to benzaldehyde forms dominantly the radical PhCH•-OMu, with only relatively minor amounts of the ring-adducts:[51,55] we assume, therefore, that the measured relaxation stems mainly from the molecular reorientation of this radical species. Previously, we discovered that the isotropic muon coupling in PhCH•-OMu. is sensitive both to effects of its medium and to temperature, and have used the value measured in solution in cyclohexane in our calculation of the transition frequency (ω), since this was the most weakly perturbing medium found.[51,55] Proceeding as described earlier, motional correlation times (τ) were estimated and Arrhenius-type plots were made for both temperature ranges. Clearly, the τ values are smaller by a factor of *ca* 10 than those determined for toluene in the fraction exhibiting a nearly identical activation energy (*ca* 3 kJ/mol), and we envisage that "hydrogen-bonding" effects involving the muon in PhCH•-OMu are important (the viscosity[60] of PhCHO is only *ca* twice that of toluene, but that of PhCH$_2$OH is greater by a factor of *ca* 10, encouraging the view that "H-bonding" effects will be influential in such a system). Indeed, the very large and negative $\Delta S^{\ddagger}$ value of -76 J/mol/K estimated for benzaldehyde (Table 7) indicates that an appreciable molecular ordering is achieved in the activated state, consistent with "H(Mu)-bond" formation; we are cautious about the absolute value, however (see section 2.2.1). Since the process is dominated by the entropy, it appears that critical molecular orientations are required to form the "Mu-bond"; *$\Delta H^{\ddagger}$ being close to zero*.

The δA values for both benzaldehyde fractions are smaller than those found for benzene or toluene sorbed in porous carbon. This is surprising since the muon hyperfine anisotropy is expected to be larger in PhCH•-

OMu. than in cyclohexadienyl type radicals. In a single crystal study of related Ph$_2$C•-OMu radicals,[61] formed by Mu addition to benzophenone, dipolar couplings of approximately -15, -6, +21 MHz were determined; in comparison with which, values of *ca* 2 MHz for δA indicate that the type of reorientational motion for sorbed PhCH•-OMu radicals can only involve fairly small-angle jumps, whose amplitude is limited by such Mu-bonding as proposed earlier. In contrast, Me$_2$C•-OMu radicals apparently undergo extensive averaging of the muon hyperfine anisotropy, as sorbed in the zeolite NaX at 298 K.[62]

The larger E_a measured for benzaldehyde over the higher temperature range (160-220 K) most likely represents the motion of effectively "isolated" PhCH•-OMu molecules on the carbon surface, in contrast to the "clusters" that are envisaged to account for the lower temperature fraction. This explanation follows those advanced previously by us to account for the similar behaviour shown by benzene sorbed in cation-exchanged zeolite X samples[10] and by 1,1-dichloroethene sorbed in silica and in kaolin.[4] We note that the $\Delta S^\ddagger$ values are very similar for all such "isolated" molecules, and are more positive (*i.e.* less negative) than is found in molecular clusters. The apparent small-angle jumps indicate that there is Mu-bonding to the carbon surface, which is normally fairly rich in oxygen-containing functional groups.[57]

4.4 Viscosity effects for toluene and benzaldehyde sorbed in carbon

Using standard bond lengths, bond angles and atomic masses, we have estimated the moment of inertia for benzaldehyde (I = 2.1 x 10^{-45} kgm^2) and for toluene (I = 1.8 x 10^{-45} kgm^2). These values were used to estimate the reorientation time for a free rotor,[43,44] $\tau_{FR} = 2\pi (I/81kT)^{1/2}$, for both benzaldehyde and toluene. Since the inverse-frequency factor (τ_∞) is expected to approach τ_{FR}, in the absence of *viscous friction*,[43,44] the ratio τ_∞/τ_{FR} provides an indication of the effective viscosity for a given sorbed fraction.[10] The greatest departure from the predicted τ_{FR} value is found for the fraction of sorbed benzaldehyde which we believe to be in a sorbed cluster-phase; this accords with a substantial viscosity as measured by PhCH•-OMu radicals which are Mu-bonded within this sorbed medium. Lesser viscosities are apparent for the "isolated" PhCH•-OMu. radicals, despite their higher reorientational activation energy, and for all benzene and toluene fractions. [The highly accelerated motion of benzene molecules sorbed in small carbon micropores was discussed previously.[13]]

Another check on the viscosity of sorbed molecules is provided by comparing the motional correlation time (τ) measured at (or extrapolated to) 300 K (τ_{300}) with that expected for molecules present in the corresponding organic liquids.[10] For such non-viscous liquids as benzene and toluene, typical τ values are close to 5×10^{-12} s;[50] a value of ca 10^{-11} s is probably more appropriate for benzaldehyde since its bulk viscosity is higher.[60] Once again, the actual τ_{300} values are far greater than those expected, demonstrating that molecules when sorbed in porous carbon experience an effective viscous impedance to their reorientation compared with a liquid sample; this is consistent with the general finding that pore-confinement often provides a resistance to molecular motion.[45]

4.5 Terpenes sorbed in activated carbon

Terpenes, especially the pinenes, are a principal component of naturally emitted (biogenic) hydrocarbons in the atmosphere.[2,3] Using LF-MuSRx, we have attempted to investigate the formation of radicals from a variety of terpenes and their interaction with a porous carbon material.[14] The results for α-pinene and β-pinene are representative of this study. Addition of Mu to either pinene isomer will yield very similar radicals, differing only in the relative position of the muon and a proton. It might be expected, therefore, that their reorientational dynamics would also be very similar. Both samples reveal two distinct motional regimes, corresponding to radicals located at different sorption sites within the carbon micropores. The activation parameters are shown in Table 8 and it is clear that they are identical within error for both fractions for both isomers. From our LF-MuSRx results for benzene sorbed in activated carbon[13] we propose that the fraction of lower activation energy corresponds to molecules that occupy the meso- and macro- pores (*i.e.* those of dimensions > 20 Å), while the fraction at higher activation energy is sorbed within the micropores (4-20 Å). However, the inverse-frequency factors indicate there is little increase in freedom (entropy) for the pinene derived radicals, as they escape from the small pores, in contrast with that which is apparent for benzene (*vide supra*). This may be because the larger pinene molecules are denied access to the very small (< 10 Å) pores, but occupy those which are somewhat larger in their size distribution.

We stress that the radical derived by muonium addition to α-pinene has a quite rigid structure, which lends confidence that the experiment is measuring true reorientation events, and the identical results for β-pinene provide no indication that the rotation of the $MuCH_2$- group in its radical is contributing appreciably to the muon spin relaxation process. Furthermore, we have estimated, using ESR spectroscopy, the spin concentration in the unloaded porous carbon to be *ca* one unpaired electron per 10,000 carbon atoms, which we do not believe will influence the muon spin relaxation unduly.

Table 8. Activation parameters measured for α-pinene and β-pinene sorbed in activated carbon powder.

	τ_∞^{-1}/s^{-1}	E_a/kJ mol^{-1}	τ_∞^{-1}/s^{-1}	E_a/kJ mol^{-1}
α-pinene	$(8.1\pm2.4) \times 10^{11}$	6.0 ± 0.3	$(4.8\pm1.3) \times 10^{12}$	13.1 ± 1.7
β-pinene	$(4.5\pm1.1) \times 10^{11}$	6.6 ± 0.9	$(8.1\pm2.5) \times 10^{12}$	16.0 ± 1.2

5. ONGOING FURTHER STUDIES

5.1 Clays

One example - of 1,1-dichloroethyl radicals sorbed in kaolin - has already been mentioned.[4] We have recently undertaken preliminary experiments using LF-MuSRx to distinguish between the effects of sorption of radicals in kaolin and in montmorillonite. These are fundamentally different sorbent materials, and are characterised by specific surface areas in the region of 40 m^2/g and up to 800 m^2/g, respectively, in consequence of a relatively rigid arrangement of aluminosilicate sheets for kaolin, but a more flexible structure in montmorillonite, whose layers are able to separate at the behest of molecular sorption to expose a far more extensive surface.[32,46] Indeed, a number of catalytic applications have been found for montmorillonite,[32] which are in many respects similar to those for zeolites, and it seems reasonable that clays as airborne materials might contribute in the outcome of atmospheric pollution chemistry.

The probe radicals were MuCH(Cl)CCl$_2\bullet$ and Me$_2$C$\bullet$-OMu, chosen to represent reactive intermediates in atmospheric halocarbon photolysis and in tropospheric hydrocarbon oxidation.[1-3] Two sorbed fractions could be identified for both radicals in both types of clay: MuCH(Cl)CCl$_2\bullet$ revealed activation energies of 4.6 and 8.5 kJ/mol in kaolin and of 1.9 kJ/mol and 10.5 kJ/mol in montmorillonite; the values measured for Me$_2$C$\bullet$-OMu were not greatly different, at 1.9 and 7.4 kJ/mol in kaolin and 2.7 and 9.3 kJ/mol in montmorillonite. In analogy with some of the more detailed examples already discussed, the low-energy fraction can probably be ascribed to a sorbed essentially *liquid* fraction in all cases, in analogy with measurements of MuCH$_2$Cl$_2$• sorbed in both kaolin and silica [4] and for PhCH•-OMu sorbed in porous carbon[12] and in zeolite X[56] (all *ca* 2-3 kJ/mol); the other fractions are attributed to isolated surface-sorbed radicals. The activation energy for the surface sorbed fraction is rather greater in montmorillonite than in kaolin for both MuCH(Cl)CCl$_2$• and Me$_2$C•-OMu, which might be explained in terms of a lateral penetration between the layers of montmorillonite by both CHCl=CCl$_2$ and Me$_2$C=O substrates, and hence an association between the resulting probe radicals and cations present in the clay.

5.2 ICE

Ice surfaces, prepared at low temperatures, have been used to model tropospheric clouds; however, on annealing above *ca* 200 K, an abrupt reduction in the surface-area occurs, although there is little change in density or particle size.[63] We have attempted some exploratory measurements on an ice-surface. In order to maintain a high surface area throughout sample measurements made over a wide temperature range, an ice surface was prepared by deposition of water from the vapour-phase into silica gel cooled to 195 K: the resulting bimolecular "ice" layer had a surface-area of *ca* 300 m^2/g. A subsequent monolayer coverage each of benzene and α-pinene was sorbed in separate samples, also at 195 K. LF-MuSRx measurements were made in the temperature range 90-300 K; both α-pinene and benzene showed a fraction with an activation energy to surface reorientation of 10 kJ/mol, but the sorbed benzene sample contained an additional fraction with an activation energy of 25 kJ/mol. Since the latter is manifest in the region of the melting-point of normal ice, it is tempting to ascribe the underlying dynamic process as involving a melting process of the ice component; however, the effect was not observed from sorbed α-pinene, and more likely represents benzene molecules which are trapped in localised "pockets" of the ice surface. In support of this, we recall that a nearly identical activation energy was measured from benzene in the small pores of porous carbon.[13]

6. CONCLUSIONS

In all cases discussed, a good straight-line plot of -ln(τ) *vs* 1/T was obtained, according to (equation 8), which encourages our confidence that a single dominant activation process is involved. When we have been able to compare results obtained using both TF-MuSR and LF-MuSRx for a given system, (*e.g.* 1,1-dichloroethyl radicals sorbed in silica-gel and in kaolin, and benzene sorbed in porous carbon), the agreement is excellent, and strongly suggests that the process being measured is indeed a molecular reorientation.

Of further significance are the results obtained for α-pinene and β-pinene sorbed in porous carbon. The radical formed from β-pinene has a rotating MuCH$_2$- group, whereas the muonium atom is bound in an essentially "fixed" position by addition to α-pinene. Nonetheless, both E$_a$ and τ_∞^{-1} values are identical, within experimental error (Table 8), for both motional regimes (sorbed fractions). Since the unloaded porous carbon samples reveal only a low concentration of paramagnetic sites (*ca* one unpaired electron per 10,000 carbon atoms) we do not believe they influence the muon relaxation rate to any great extent. Therefore, the model of molecular reorientation is also supported in the pinene/carbon samples.

A contribution to (λ) in the Me$_2$C•-OMu radical, formed in the solid, liquid and gas-phases of acetone, was attributed to the "out-of-plane" oscillation of the O-Mu bond, and this may also apply to the PhCH•-OMu radical. However, the only *difference* between the samples of benzaldehyde sorbed in cation-exchanged zeolite X is the type of cation which is present. The activation energies (Table 6) are found to decrease as the cation radius increases in each series Li$^+$ -> K$^+$ and Mg^{2+} -> Sr^{2+}. This accords with the formation of increasingly weak o-complexes between the PhCH•-OMu radicals and the exchange-cations, as deduced previously from changes in the isotropic muon hyperfine couplings measured in these samples. Once again, the results are more readily explained by a dominant reorientational process rather than one of intramolecular dynamics.

The results show that a diversity of VOCs can be converted into radicals bearing the muonium radiolabel, and so may be investigated on a wide range of solid surfaces, representative of the various particles present in the atmosphere.

6.1 "implications for atmospheric chemistry and pollution control"

"MuSR" would not normally be described as an "analytical" method, since it generally requires the prior introduction of substrate molecules to a given system to "catch" muonium atoms and form free radicals. One is not normally looking at chemically unperturbed materials, nor naturally occasioned samples, *e.g.* as taken directly *from* the atmosphere. If this is a criticism, it is one that could be levelled at most "surface science" techniques, which rarely examine the complex conditions of operating catalytic processes, but rather provide tractable models for them. In this perspective, with the specific intention of providing a fundamental appreciation of radicals in their interaction with surfaces, the incisive aspect of MuSR is apparent.

During the previous decade, heterogeneous chemistry has taken-on a greater role in our musings over how the atmosphere "works"; *e.g.* the role of clouds in ozone depletion is a topical feature of enquiry. We also know that there are staggering quantities of suspended particles present, of the kind discussed in this chapter, which must provide an extensive and reactive surface to promote atmospheric transformations, many of which are mediated by radicals. So MuSR might contribute insight from studies of model systems, such as we have just considered. Greater insight might be possible through studies of actual atmospheric samples, but the generally small quantities available will test the limits of the sensitivity of these methods. As future details of heterogeneous processes are disclosed perhaps by MuSR and through other strategies, their *true* "implications for Atmospheric Chemistry and Pollution Control" will become clear.

Acknowledgements *I thank my colleagues, Ivan Reid, Tim Dintinger, Harry Morris, Chantal Hinds, Estelle Butcher, Chris Scott, Ulrich Zimmerman, Steve Cox, Brian Webster and Jas Jayasooriya either for their actual participation in the experiments described, or for their helpful thoughts or both! I further acknowledge financial support from the Leverhulme Trust, the Paul Scherrer Institute, the European Union, the Engineering and Physical Sciences Research Council (EPSRC) of the United Kingdom, Unilever Research and The Royal Society of Chemistry (for a J.W.T.Jones Travelling Fellowship).*

7. REFERENCES

1. R.P. Wayne, *Chemistry of Atmospheres*, Clarendon Press, Oxford, 1985.
2. R.P. Turco, *Earth Under Siege*, Oxford University Press, Oxford, 1997.
3. B.J. Finlayson-Pitts and J.N. Pitts, Jr., *Chemistry of the Upper and Lower Atmosphere*, Academic Press, London, 2000).
4. C.J. Rhodes, T.C. Dintinger, I.D. Reid and C.A. Scott, *Magn. Reson. Chem.* **38** (2000) 281.
5. D.C. Walker, *Muon and Muonium Chemistry*. Cambridge University Press, Cambridge, 1983.

6. E. Roduner, *The Positive Muon as a Probe in Free Radical Chemistry. Lecture Notes in Chemistry*. Springer, Heidelberg, 1988.
7. A. Carrington and A.D. McLachlan, *Introduction to Magnetic Resonance*. Chapman and Hall, London, 1979.
8. R.K. Harris, *Nuclear Magnetic Resonance Spectroscopy*. Pitman, Melbourne, 1983.
9. C.J. Rhodes, T.C. Dintinger and C.A. Scott, *Magn. Reson. Chem.* **38** (2000) 62.
10. C.J. Rhodes, T.C. Dintinger and C.A. Scott, *Magn. Reson. Chem.* **38** (2000) 729.
11. T.C. Dintinger, *Ph.D Thesis*. Liverpool John Moores University, 2000.
12. C.J. Rhodes and I.D. Reid, *Spectrochimica Acta A*, in press.
13. C.J. Rhodes, T.C. Dintinger, I.D. Reid and C.A. Scott, *Magn. Reson. Chem.* **38** (2000) S58.
14. C.J. Rhodes, H. Morris, I.D. Reid and U. Zimmermann, *PSI Annual Report*, 2000.
15. C.J. Rhodes, I.D. Reid and U. Zimmermann, to be published.
16. C.J. Rhodes, I.D. Reid and U. Zimmermann, to be published.
17. E. Roduner, *Chem. Soc. Rev.* **22** (1993) 337.
18. E. Roduner, M. Schwager and M. Shelley, *Radicals on Surfaces*, p.259. A. Lund and C.J. Rhodes, Eds., Kluwer, Dordrecht, 1995.
19. S.F.J. Cox, Solid State *Nucl. Magn. Reson.* **11** (1998) 103.
20. S.F.J. Cox and D.S. Sivia, *Appl. Magn. Reson.* **12** (1997) 213.
21. U.A. Jayasooriya, J.A. Stride, G.M. Aston, G.A. Hopkins, S.F.J. Cox, S.P. Cottrell and C.A. Scott, *Hyperfine Interact.* **27** (1997) 106.
22. U.A. Jayasooriya, G.M. Aston and J.A. Stride, *Appl. Magn. Reson.* **13** (1997) 165.
23. C.J. Rhodes, T.C. Dintinger, H. Morris and C.A. Scott, *Magn. Reson. Chem.*, in press.
24. Clayden NJ, Jayasooriya UA, Cottrell SP, *Physical Chemistry Chemical Physics*. **1** (1999) 4379.
25. Clayden NJ, Jayasooriya UA, Stride JA, King P, *Polymer*. **41** (2000) 3455.
26. C. Christides, S.F.J. Cox, W.I.F. David, R.M. Macrae and K. Prasides, *J. Chim. Phys.* **90** (1993) 663.
27. R. van Dingenen, N.R. Jensen, J. Hjorth and F. Raes, *J. Atmospheric Chemistry*. **18** (1994) 211.
28. D.J. Wuebbles, *J. Geophys. Res.* **88** (1983) 1433.
29. I. Tegen and I. Fung, *J. Geophys. Res.* **99** (1994) 22897.
30. *Properties and Applications of Zeolites*. R.P. Townsend, Ed., Chemical Society, London, 1980.
31. *Radicals on Surfaces*, A. Lund and C.J. Rhodes, Eds., Kluwer, Dordrecht, 1995.
32. P. Laszlo, *Science*. **233** (1987) 1473.
33. P.H. Jacobs and U. Forstner, *Water Res.* **33** (1999) 2083.
34. *Volatile Organic Compounds in the Atmosphere*, ed. R.H. Hester and R.M. Harrison, Royal Society of Chemistry, Cambridge, 1995.
35. J.K.A. Clarke, R. Darcy, B.F. Hegarty, E. O'Donoghue, V. Amir-Ebrahimi and J.J. Rooney, *J. Chem. Soc., Chem. Commun.* (1986) 425.
36. M. Heming and E. Roduner, *Surf. Sci.* **189** (1987) 535.
37. C.J. Rhodes, E. Roduner and I.D. Reid, *J. Chem. Soc., Chem. Commun.* (1993) 512.
38. C.J. Rhodes, E.C. Butcher, H. Morris and I.D. Reid, *Magn. Reson. Chem.* **33** (1995) S134.
39. J.K. Kochi, Advances in *Free Radical Chemistry*. **5** (1975) 189.
40. C.J. Rhodes and M.C.R. Symons, *J. Chem. Soc., Faraday Trans.* 1, **84** (1988) 1187.
41. D.C. Walker, *J. Chem. Soc., Faraday Trans.* **94** (1998) 1.
42. J.E. Huheey, *Inorganic Chemistry, Principles of Structure and Reactivity*; Harper and Row, New York, 1978.

43. H.J.V. Tyrrell and K.R. Harris, *Diffusion in Liquids*, Butterworths, London, 1984.
44. E. Roduner, M. Shwager, P. Tregenna-Piggott, H. Dilger, M. Shelley and I.D. Reid, *Ber. Bunsenges. Phys. Chem.* **99** (1995) 1338.
45. J. Karger and D.M. Ruthven, *Diffusion in Zeolites - and other microporous solids*. Wiley, New York, 1992.
46. R.M. Barrer, *Zeolites and Clay Minerals*, Academic Press, London, 1978.
47. M.A. Hepp, V. Ramamurthy, D.R. Corbin and C. Dybowski, *J. Phys. Chem.* **96** (1992) 2629.
48. C.J. Rhodes, C.S. Hinds and I.D. Reid, unpublished results.
49. E. Roduner, *Hyperfine Ineractions.* **65** (1990) 857.
50. S.R. Harrison, P.S. Pilkington and L.H. Sutcliffe, *J. Chem. Soc., Faraday Trans.1,* **80** (1984) 669.
51. C.J. Rhodes, C.S. Hinds and I.D. Reid, *J. Chem. Soc., Faraday Trans.* **92** (1996) 4265.
52. J.B. Nagy, E.G. Derouane, H.A. Resing and G.Ray Miller, *J. Phys. Chem.* **87** (1983) 833.
53. R.R. Eckman and A.J. Vega, *J. Phys. Chem.* **90** (1986) 4679.
54. I. Kustanovich, D. Fraenkel, Z. Luz and S. Vega, *J. Phys. Chem.* **92** (1988) 4134.
55. C.J. Rhodes, I.D. Reid and R.A. Jackson, *Hyperfine Interactions.* **106** (1997) 193.
56. C.J. Rhodes, I.D. Reid and U. Zimmermann, to be published.
57. R.C. Bansal, J.-B. Donnet and F. Stoeckli, *Active Carbon*. Marcel Dekker, Inc., New York, 1988.
58. E. Roduner, G.A. Brinkman and P.W.F. Louwrier, *Chem. Phys.* **73** (1982) 117.
59. I.D. Reid, T. Azuma and E. Roduner, *Hyperfine Interactions.* **65** (1990) 879.
60. *CRC Handbook of Chemistry and Physics 66th Edition*, ed. R.C. West, CRC Press, Florida, 1985.
61. G.M. Aston, J.A. Stride, U.A. Jayasooriya and I.D. Reid, *Hyperfine Interactions.* **106** (1997) 157.
62. C.J. Rhodes and B.C. Webster, *J. Chem. Soc., Faraday Trans.* (1993) 1283.
63. L.F. Keyser and M.T. Leu, *J. Colloid Interface Sci.* **155** (1993) 137.

PART II: TRENDS IN APPLICATIONS

Chapter 9

EPR STUDIES OF ATOMIC IMPURITIES IN RARE GAS MATRICES

Henrik Kunttu and Jussi Eloranta
Department of Chemistry, University of Jyväskylä, P.O. Box 35, FIN-40351 Jyväskylä, Finland

Key words: EPR, matrix, isolation.

Abstract: In this article we give an overview of the matrix isolation technique combined with electron paramagnetic resonance (EPR) detection for embedded atomic impurities in solid rare gases. A special emphasis is put on impurity - matrix coupling effects combining both experimental and theoretical approaches.

1. INTRODUCTION

Chemical dynamics in condensed phases forms a rapidly evolving field of research dealing with a wide range of phenomena from simple point defect diffusion to the broad field of chemical reaction dynamics. In the strict sense such dynamics in dense media is always dictated by interactions with strong many-body nature, thus detailed understanding of even the most elementary events at a molecular level remains a challenge. A convergence of quantitative modelling and interpretation of various processes is attained only by synergy between experiment and theory. Experimental methods at one's disposal range from optical absorption and emission based techniques to magnetic spectroscopies among which electron paramagnetic resonance (EPR) has a long tradition as a versatile tool for interrogation of open shell species.[1] Due to the rapidly increasing computational resources and development of new efficient algorithms, microscopic description of the observed spectroscopy, optical and magnetic, can now be pursued at the first-principles level in highly multi-dimensional models approaching the size and complexity of a realistic chemical system.

Rare gases (Rg) solidify in cryogenic conditions and form more or less ordered structures, namely, glassy crystalline or strongly light scattering powder like substances which are commonly called matrices. The inertness of the Group VIII elements against chemical reactivity is reflected in their other name, the noble gases. As such, they provide a versatile medium for studies of elementary physico-chemical processes. In particular, doping these solids with atomic or small molecular impurities, *i.e.* matrix isolation, is ideally suited for spectroscopic observations of highly reactive transient species in inert frozen environment. This was, in fact, the original goal of the matrix isolation technique.[2] Besides providing spectral reference data for investigations in related fields such as atmospheric and combustion chemistry,[3] understanding of trapping, transport, and reactivity in these conceptually simple solids has also more practical applications. Chemical energy storage in cryogenic solids by trapping reactive atoms, radicals or chemical intermediates, and controlling their subsequent recombination by annealing is an example of such developments.[4] Moreover, matrix isolation in solid hydrogen is actively used is research aiming at development of advanced propellants called "high energy density materials".[5,6] Finally, besides their role in isolation, Rg matrices can be used as excellent media to investigate chemical species containing rare gas atoms.[7,8]

The information gathered from experiments in solids is subject to perpetual averaging processes of dynamical and structural origin. Therefore, as the spectroscopic methods become blunted in this respect, the need for theoretical aid in interpretation becomes evident. On the other hand, cryogenic solids provide an ideal testing ground for theoretical treatments. Either way around, the theoretical approaches encounter challenging obstacles due to the immense number of interacting species and degrees of freedom that need to be considered. This is why purely quantum mechanical approaches, that is, solving the time-dependent Schrödinger equation by some affordable way are rarely used in large-scale simulations and, instead, various approximations of semi-classical nature are usually employed.[9]

EPR spectroscopy has proven to be a powerful experimental technique for elucidating details of the electronic structure of open shell molecules and ions in crystals and solutions. Because EPR transitions (i) are dominated by the properties of the electronic ground state, (ii) exhibit very narrow line widths in dilute samples, and (iii) are very sensitive to small changes in the environment of the paramagnetic species under investigation, this method seems conceptually ideally suited for studies of atomic trapping and dynamics in low temperature matrices. However, detection of atoms with orbitally degenerate ground states has turned out to be difficult, and most of the available EPR data on atoms or ions in Rg matrices have concerned cases with isotropic g values and hyperfine interaction.[1] Even for these spherical atoms theoretical treatment of the isotropic hyperfine coupling (IHC) is a relatively problematic task for modern *ab initio* methods. The challenge here is the local nature of the spin density, and the isotropic component of the hyperfine interaction, which may also be composed of indirect effects such as spin polarization and electron correlation.[10,11,12] A proper theoretical approach would therefore necessitate a high quality basis set combined with a substantial effort in treatment of electron correlation. Since a typical matrix shift for the IHC of an atom is some tens of MHz, the accuracy needed for such computation is exceedingly high. Consequently, a combined quantum-classical approach with an assumption of pair additive hyperfine coupling is by far the only choice for computation of magnetic properties of atoms trapped in matrix environments.[13,14,15]

In this review we briefly summarize some of the previous EPR spectroscopic observations of atomic impurities isolated in rare gas matrices and introduce some of the theoretical tools applicable for analysis of the obtained spectra.

2. EXPERIMENTAL TECHNIQUES

Here we consider some of the special experimental aspects related to matrix EPR measurements. For a detailed overview of the experimental techniques we refer to 1 and 16. For description of the basic EPR instrumentation we refer to 17.

Most of the matrix EPR experiments that have been carried out below 10 K have employed a liquid He bath cryostat. For example, the first study of Bowers *et al.* on atomic impurities in rare gas matrices employed this type of arrangement.[18] For studies above 10 K, a standard closed-cycle He cryostat is a convenient choice. An example of such setup is shown in Fig. 1, where both EPR and optical measurements can be performed from the same sample. In both approaches the cryostat is used to cool down a copper or sapphire substrate (cold target) on which the thin matrix film is grown at low temperature. During the measurement the substrate resides in the microwave cavity of a EPR spectrometer, and thus care must be taken in designing the geometry and proper positioning of the substrate. For example, a copper substrate interacts strongly with the electrical component of the microwave radiation field, and even small mechanical vibrations will greatly increase noise in the EPR signal. In the case of a standard TE102 microwave resonator the signal-to-noise ratio can be improved by making the copper substrate flat (for example, 4 mm x 30 mm x 0.4 mm) and placing it carefully in the nodal plane of the electric field. Mechanical vibrations are not usually a problem when He bath cryostats are used but most of the closed cycle cryostats suffer from this problem.

Once the cold substrate is at a suitable temperature, a gaseous mixture can be slowly sprayed on it. In magnetic measurements the optical quality of the matrix is not important, and thus low temperature (4 – 10 K) deposition, which usually results in very "snowy" looking and strongly scattering solids, can be applied. In some cases low temperature deposition has a great advantage over high temperature deposition (20 – 50 K) as it prevents diffusion of the impurity atoms or molecules during the freezing stage of the matrix. The impurities can be premixed with the rare gas before deposition, or they can be deposited from separate inlets as shown in Fig. 1.

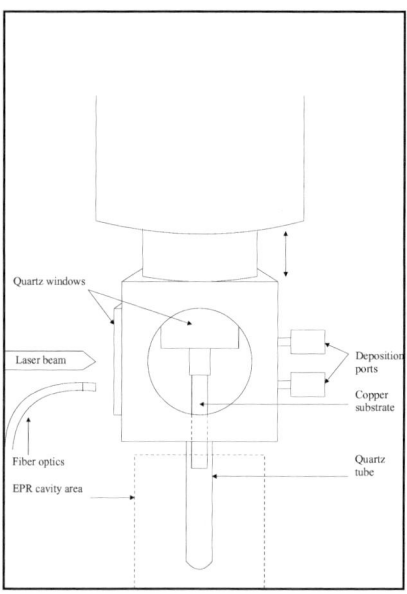

Figure 1. Sketch of the matrix EPR/luminescence apparatus after Ref. 19 is shown. The flat copper substrate can be moved from the area with optical access to the quartz tube fitting the microwave cavity of the EPR spectrometer.

Reactive species, like most atoms, can not be premixed in the rare gas due to practical limitations. Hence, various alternative methods have been developed for producing such species in the rare gas matrices. The most common techniques include *in situ* photolysis of a suitable parent molecule

9. EPR studies of atomic impurities in rare gas matrices

in the matrix, laser ablation, various thermal vaporization sources, chemical reactions in the matrix, and discharge based methods. For example, photodissociation of hydrogen halides will yield hydrogen atoms and halogen atoms in the matrix.[20] Comparison of the two preparation techniques, laser ablation and a Knudsen oven, in generation of atomic species is given in Ref. 21. A sketch of a laser ablation apparatus is shown in

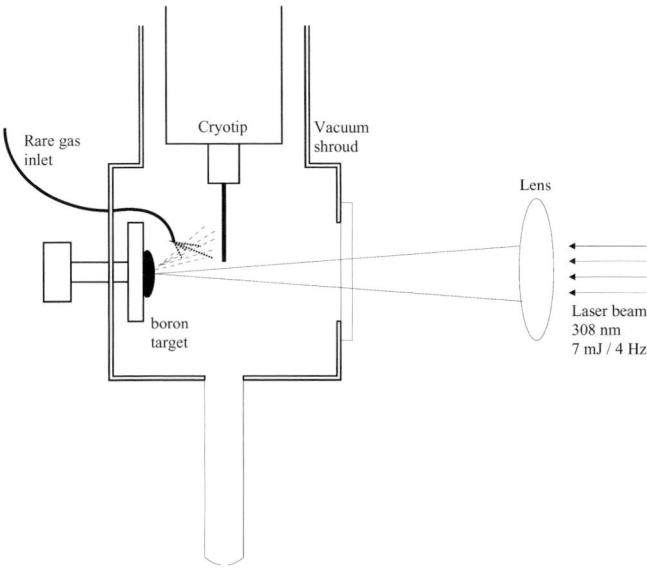

Figure 2. Laser ablation / EPR apparatus of Refs. 15 and 21 is shown. To preserve laser ablation efficiency the copper target was rotated periodically.

3. ATOMIC IMPURITIES IN RARE GAS MATRICES

3.1 ^2S state atoms

For ^2S state atoms trapped in rare gas matrices the EPR spectrum is highly isotropic. Depending of the impurity, the spectrum may contain resonances due to multiple trapping sites of different size and symmetry, and somewhat different g and/or A values. For example, in the case of hydrogen atom doped argon matrix, two distinct trapping sites occur as shown in Fig. 3.[18,22]

These sites have been identified as singly substitutional (12 nearest neighbors) and interstitial octahedral sites (6 nearest neighbors) in the face-centered-cubic (FCC) lattice.[13,23] This assignment is based on a theoretical treatment which will be discussed in section 4. An empirical and relatively well justified rule of thumb is that when the interaction pair potential between the impurity atom and the rare gas atom is much more repulsive than the rare gas - rare gas potential, then a distribution of trapping sites with varying number of lattice vacancies is usually observed. An example of such situation in rare gas matrices is provided by the alkali metal atoms, where the atoms may reside in sites involving multiple lattice vacancies.[14,21] In this case, thermally activated dynamics, characterized by activation barrier of few meV, within the trapping cavity has been reported.[24] Due to their different structures, different trapping sites have usually characteristic thermal stabilities, which allows one to use sample annealing at suitably high temperatures to simplify the observed spectrum significantly. In most favourable cases EPR resonances due to a single trapping configuration is observed after annealing.

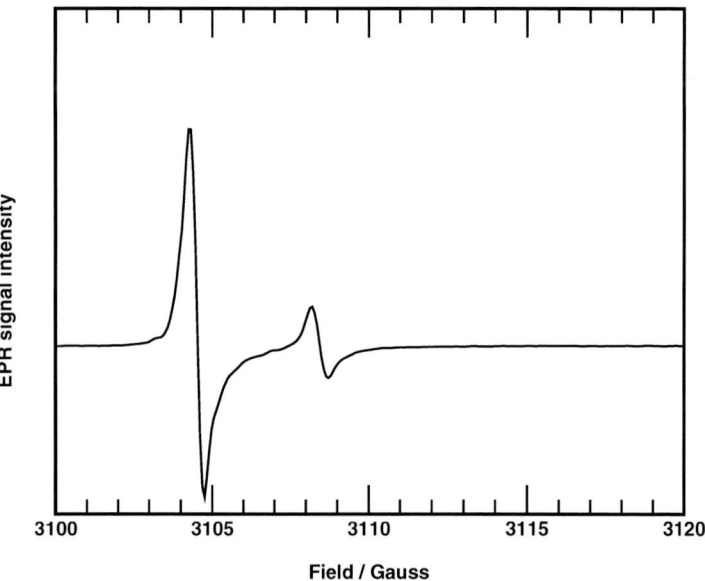

Figure 3. Low field resonance lines of H atom trapped in Ar matrix as produced by UV photodissociation of HCl precursor. Two distinct trapping site resonances (interstitial octahedral and substitutional) are separated by ~ 4 G. In this sample the interstitial octahedral

9. EPR studies of atomic impurities in rare gas matrices

site is more populated than the substitutional site. Center of the spectrum is located at *ca.* 3350 Gauss.

The resulting EPR spectra can be simulated by standard methods using isotropic g and hyperfine coupling matrix A. In should, however, be carried out with caution since atomic species may have large A and therefore higher than first order corrections become very important. An interesting example of such case is provided by Cu atoms trapped in rare gas matrices as this species shows rather peculiar EPR spectrum in which only two lines are seen for a particular Cu isotope.[25] The origin for this rather odd spectrum can be understood by inspection of the energy level diagram of Fig. 4. In order to carry out a rigorous theoretical analysis of these systems, one must obtain both g and A matrices with proper accuracy. For this reason, it is suggested that spectrum simulation with iterative parameter fitting with highly accurate evaluation of the exact spin Hamiltonian is used (see section 4.4). Furthermore, the transition moments should be evaluated based on the proper eigenvectors in order to get the intensities reliably. In the previous Cu atom example both dominantly electron spin and dominantly nuclear spin transitions are observed. Summary of experimental magnetic parameters for ^{2}S state atoms trapped in solid rare gases is shown in Table 1.

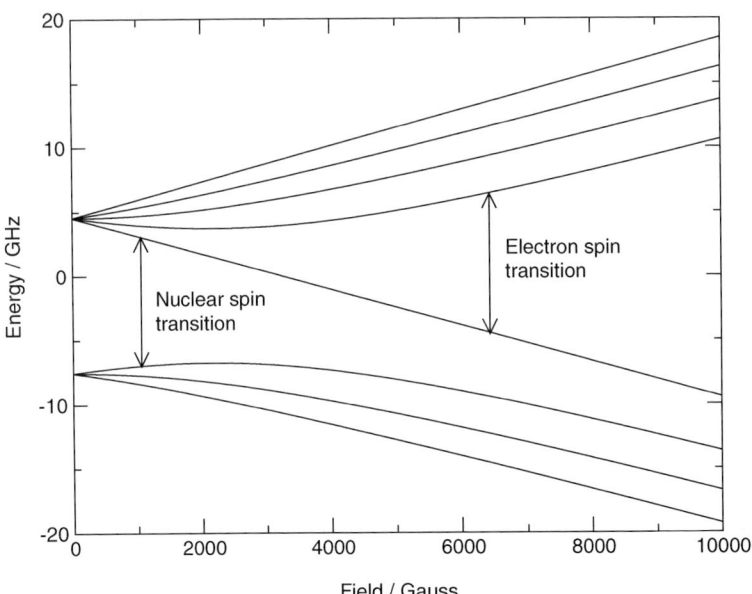

Figure 4. Energy level diagram as a function of magnetic field strength for a single Cu atom. At X-band only two transitions are seen below 10000 G which can be approximately classified as being electron spin and nuclear spin types by numerically evaluating the transition moments.

Table 1. Summary of relevant magnetic parameters of ^2S state atoms in solid rare gases is shown. Atom preparation method is denoted as: (1) photodissociation, (2) thermal source (Knudsen oven), or (3) laser ablation. Hyperfine coupling shifts with respect to the gas phase values (labelled as free) are expressed in MHz. Data for most abundant isotopes are shown.

Atom	I	Matrix	g_{iso}	A_{iso}	Site	Ref.
H		Ar(1)	2.0012	+17	Int.	13,18,22,23
H		Ar(1)	2.0017	-6	Subst.	13,18,22,23
H		Kr(1)	1.9992	+8	Int.	13,18,22,23
H		Kr(1)	2.0013	-11	Subst.	13,18,22,23
^{1}H	1/2	Free	2.00226	1420.4		26
Li		Ar(2)	2.0018	-7	12 vac.	14,27
Li		Ar(2)	2.0010	+12	8 vac.	14,27
Li		Ar(3)	1.9984	+56	Subst.	14,21
Li		Ar(3)	1.9998	+14	8 vac.	14,21
Li		Kr(2)	1.9987	-7	12 vac.	14,27
Li		Kr(2)	1.9966	+9	6 vac.	14,27
Li		Kr(3)	1.9860	+44	Subst.	14,21
Li		Xe(2)	1.9914	-5	12 vac.	14,27
^{7}Li	3/2	Free	2.00231	401.7		26
Na		Ar(2)	Broad	distribution	of sites	21,27
Na		Ar(3)	Broad	distribution	of sites	21,27
Na		Kr(2)	1.9917	-14	12 vac.	14,21,27
Na		Kr(2)	1.9872	+11	6 vac.	14,21,27
Na		Kr(3)	1.9832	+115	Subst.	14,21
Na		Xe(2)	1.9925	-12	12 vac.	14,21,27
Na		Xe(3)	1.9925	+97	Subst.	14,21
^{23}Na	3/2	Free	2.00231	885.81		26
K		Ar(2)	Broad	distribution	of sites	27
K		Kr(2)	1.9964	-3	--	27
K		Kr(2)	1.9849	+15	--	27
K		Xe(2)	1.9857	+4	--	27
^{39}K	3/2	Free	2.00231	230.86		26
Rb		Ar(2)	Broad	distribution	of sites	27
Rb		Kr(2)	Broad	distribution	of sites	27
Rb		Xe(2)	1.9821	-16	--	27
^{85}Rb	5/2	Free	2.00241	1012		26
Cs		Ar(2)	2.0051	+11	--	27
Cs		Kr(2)	2.0015	-21	--	27
^{133}Cs	7/2	Free	2.00258	2298		26
Cu		Ar(2)	1.9994	+282	--	25a
Cu		Kr(2)	1.9955	+176	--	25a
Cu		Kr(3)	1.996	+181	--	25b
Cu		Xe(2)	1.9942	+29	--	25a
^{63}Cu	3/2	Free	2.0023	5867		26
Ag		Ar(2)	1.9998	+98	--	25a

Atom	I	Matrix	g_{iso}	A_{iso}	Site	Ref.
Ag		Kr(2)	1.9942	+65	--	25a
Ag		Xe(2)	1.9922	+17	--	25a
^{107}Ag	1/2	Free	2.0023	-1713		26
Au		Ar(2)	2.0012	+85	--	25a
Au		Kr(2)	1.9962	+43	--	25a
Au		Xe(2)	1.9970	-27	--	25a
Au	3/2	Free	2.0023	3053		26

Trapped atoms may be thermally activated provided that the barrier for diffusion can be overcome with thermal energies. By measuring the time decay profiles of the EPR signals at specified temperatures one can obtain kinetic information of the impurity atom diffusion. Thermal behaviour usually depends on the site structure around the impurity as was already noted in the context of sample annealing. For example, this method has been used to determine the diffusion rates of atomic hydrogen in rare gas matrices.[22] In this case the substitutional sites are thermally stable up to the matrix evaporation temperatures, whereas the octahedrally trapped atoms are mobilized at low temperatures (in Ar at 16 K and in Kr at 24 K). No new resonances are observed as the final products are not paramagnetic (H_2 and thermally formed rare gas compounds[7,8]). In the case of metal atoms (Li[28], Na[29], K[29,30], Cu[31], Ag[32]) formation of paramagnetic clusters has been observed. However, the alkali metal clusters were formed during sample deposition within a semiliquid interface between the solid and vacuum, and Cu and Ag were mobilized thermally in solid neon only. Therefore, these atoms do not exhibit real diffusion behaviour in solid Ar, Kr, or Xe. It appears that only hydrogen exhibits diffusion mediated propagation in these solids, which suggests the importance of tunneling in the mechanism. This is consistent with the observed different thermal behaviour of H and D atoms.[22] When the initial concentration of the H/D atoms is high, interesting spin-paired species can be observed without thermal activation.[33] Based on the magnetic parameters it was estimated that the distances between the impurity centres was greater than 7 Å in the matrix.

3.2 High spin S state atoms

Atoms having half-filled np shells (*e.g.* N (^{4}S), P (^{4}S), As (^{4}S)) all have isotropic electron distribution and hence the rare gas - atom interaction potentials are isotropic. The matrix - impurity interaction ground state potentials dictate the trapping site symmetry and for purely substitutional trapping the site symmetry will be spherical. Thus, the EPR spectra are expected to be isotropic. If multiple substitutional trapping occurs then deviation from the spherical symmetry results and the electron spin-spin interaction may be observed. For rare gas matrices this can not be seen but, for example, N atoms in N_2 matrix shows signs of this interaction.[34] Both g and hyperfine coupling experience deviation from their gas phase values and similar analysis as presented for the ^2S atoms may be applied. In general, it has been observed that the relative matrix induced shifts for these atoms are larger than for the ^2S atoms indicating stronger impurity - matrix coupling. Data for some ^4S atoms are listed in Table 2. The simple theory based on shifts in the isotropic hyperfine coupling constants predict that these atoms occupy sites that resemble closely substitutional sites in the FCC lattice.

Table 2. Summary of relevant magnetic parameters for ^4S state atoms in solid rare gases is shown. Atom preparation method is denoted as: (1) photodissociation or (2) γ-irridation of suitable molecular species. Hyperfine coupling shifts with respect to gas phase values (labelled as free) are expressed in MHz. Data for most abundant isotopes are shown.

Atom	I	Matrix	g_{iso}	A_{iso}	Site	Ref.
N		Ar(1,2)	2.0020	+1.6	Subst.	19,35
N		Kr(1,2)	2.0019	+1.9	Subst.	19,35
N		Xe(1,2)	2.0019	+1.9	Subst.	19,35
^{14}N	1	Free	2.0022	10.45		26
P		Ar(2)	2.0012	+26	Subst.	35
P		Kr(2)	2.0001	+28	Subst.	35
P		Xe(2)	--	~ +31	Subst.	35
^{31}P	½	Free	2.0019	55.06		26
As		Ar(2)	1.9960	-31	Subst.	35
As		Kr(2)	1.9951	-35	Subst.	35
As		Xe(2)	1.9943	--	--	35
^{75}As	3/2	Free	1.9965	-66.20		26

Thermal mobility of ^4S state atoms has not been reported. The experimental data of Ref. 19 show that at least nitrogen atoms, produced in a photolysis, were stable up to the evaporation temperatures of the matrix. The N – Ar pair potential (minimum ~ at 3.6 Å with depth of 8 meV)[36] is comparable to the H - Ar (minimum at 3.6 Å with 4 meV well depth)[13] pair potential and yet no thermal mobility is observed. However, only one trapping site was observed for nitrogen atoms and therefore it is quite probable that N atoms occupy pure substitutional sites in the lattice. As mentioned, hydrogen atoms trapped in substitutional cavities are also stable and this thermal stability can also be related to the site structure. No reports for diffusive cluster formation in this group of atoms has been reported. In highly concentrated matrices N atoms show spin-pairing spectra similar to the H atoms.[37] Other observed high spin atoms in matrices include Cr (^7S) and Mn (^6S), which yield very close to isotropic EPR spectra.[38]

3.3 Superhyperfine interaction

Both ^{83}Kr (11.6 % of I = 9/2) and ^{129,131}Xe (26.4 % of I = 1/2; 21.2 % of 3/2) atoms have magnetic isotopes, which may affect the observed EPR spectra. If the interaction between impurity and the matrix atom is sufficiently strong (*i.e.* there is sufficient mixing of matrix and impurity orbitals) then the EPR spectrum may exhibit additional structure due to electron spin (impurity) - nuclear spin (matrix) coupling. This effect is usually called as superhyperfine coupling. An interesting demonstration of this effect was given by Morton *et al.* who isolated hydrogen atoms in solid isotopically enriched ^{129}Xe (37 %).[39] The authors were able to simulate the resulting anisotropic EPR spectrum by assuming octahedral trapping site and proper statistical distribution of magnetic matrix atoms. A similar approach was used in explaining the observed superhyperfine structure for sodium atoms trapped in solid Xe.[21] Experimental and simulated spectra for one of the sodium lines are shown in Fig. 5. Accurate EPR spectrum simulations can be effectively used to predict the number of magnetic matrix atoms around the impurity and therefore conclusions on the trapping site structure can be made.

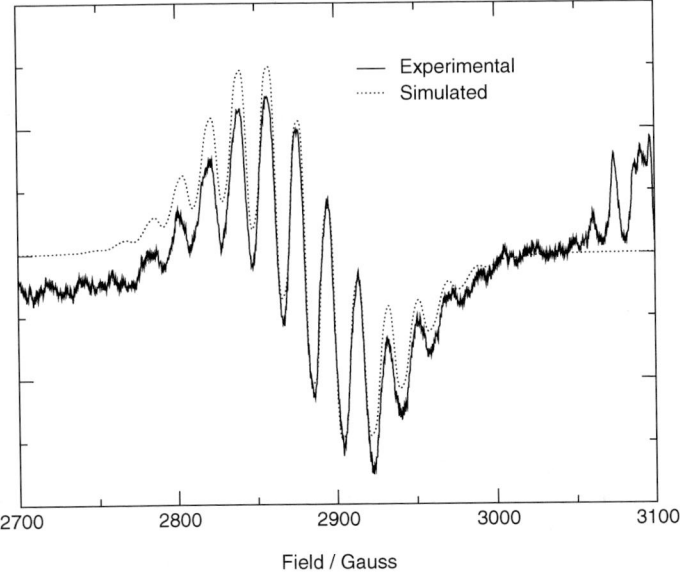

Figure 5. Superhyperfine structure of Na atoms trapped in solid Xe (only single Na resonance line is shown). Simulation assumes statistical distribution of 12 Xe atoms with their natural

isotope distribution around the alkali metal center. Only isotropic hyperfine interaction (A_{iso} ~ 115 MHz) has been included in calculation. For details see Ref. 21.

3.4 ^{2}P state atoms

Detection of ^{2}P atoms in rare gas matrices has turned out to be a difficult task.[1] Atoms with a single electron in their outer p-shell (B, Al, and Ga) show powder spectra with axial symmetry and, consequently, EPR detection sensitivity is greatly reduced.[15,40-42] As will be justified in section 4, the axial symmetry is caused by locking of the unpaired electron orbital to a preferential direction with respect to an origin placed on the impurity. On the other hand, the sample will contain atoms with random distribution of p-orbital orientations with respect to the direction of the external magnetic field. The other extreme, a hole in the p-orbital frame (*e.g.* halogens), is even more difficult to detect by EPR.[1] The only clean observation of this group of atoms concerns F atom in a Kr matrix, which appear to form relatively tightly bound tri-atomic Kr$_2$F molecule, again in axial symmetry.[43] Other halogen atoms have only been observed in strongly quenched states.[44] However, in these few observed cases, the EPR spectrum revealed detailed information about the lattice structure around the impurity. Simulation of the EPR spectra of ^{2}P state atoms involves powder integration along axial symmetry, and depending on the atom, higher order corrections may be necessary to include. For loosely bound atoms (*i.e.* B, Al, and Ga), by explicitly including spin-orbit coupling, orbital Zeeman, and crystal field interactions, one can obtain the associated parameters from the spectrum. This has been carried out in detail for B atom in Ar, Kr, and Xe matrices by Kiljunen *et al.*[15] In this case the Hamiltonian was written as:

$$H = \beta_e H \cdot (L + g_e S - g_N I) - \xi L \cdot S - S \cdot A \cdot I + H_{tetr}(L)$$
where $H_{tetr}(L) = \alpha_t [3L_d^2 - L(L+1)\mathbf{1}], d = x, y, z$

and was solved numerically in uncoupled representation. Note that $\alpha_t = -\Delta/2$ where Δ is the crystal field splitting parameter. By performing parameter fitting between simulated and experimental powder EPR spectra, values for spin-orbit and crystal field parameter α_t can be obtained. As explained in more detail in the theoretical part of the text, the external heavy atom effect can be included in this model by artificially reducing the value of the spin-orbit coupling constant. By way of an example experimental and simulated EPR spectra of boron atoms trapped in solid Ar and Kr are shown in Fig. 6.

9. EPR studies of atomic impurities in rare gas matrices

The external heavy atom effect is seen clearly as exchange in the magnitude of $g_{xx,yy}$ and g_{zz} as the EPR spectrum changes its overall phase completely. A summary of magnetic parameters for selected atoms in solid rare gases is shown in Table 3.

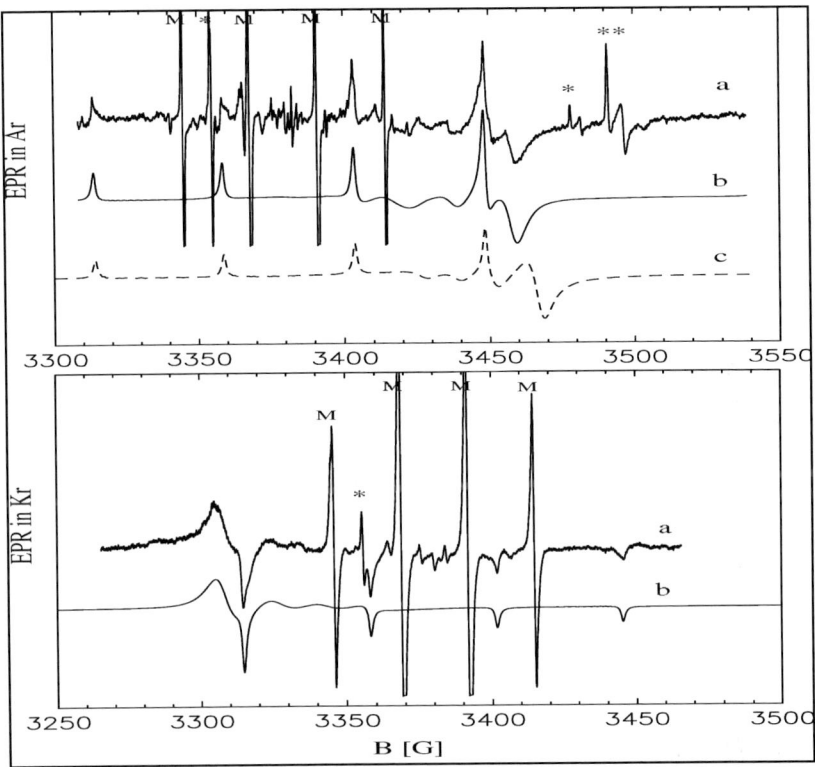

Figure 6. Experimental and simulated EPR spectra of boron atoms in solid Ar and Kr: (a) experimental, (b) axially symmetric simulation (Ref. 15), (c) axially symmetric simulation (Ref. 41). The methyl radical impurities are marked by M, ^{10}BO by *, and ^{11}BO by **.

Table 3. Summary of relevant magnetic parameters for ^{2}P state atoms in solid rare gases is shown. Atom preparation method is denoted as: (1) thermal source or (2) laser ablation of the solid. Hyperfine coupling constant shifts with respect to gas phase values (labelled as *free*) are expressed in MHz. Data for most abundant isotopes are shown. A_{ii} ($ii = xx, yy, zz$) refers to the anisotropic components of the hyperfine coupling, ξ is the effective spin-orbit coupling constant (cm^{-1}), and Δ is the crystal field splitting parameter as defined in eqn. (1) (cm^{-1}). When clearly distinguishable sites for Al and Ga exist then the data shown is the average over these sites.

Atom	I	Matrix	g_{zz}	$g_{xx,yy}$	A_{zz}	$A_{xx,yy}$	ξ	Δ	Ref.
^{11}B		Ar	2.0012	1.9645	126	-33	10.7	380	41
^{11}B		Ar	2.0014	1.970	126.1	-46.9	7.8	361	15

Atom	I	Matrix	g_{zz}	$g_{xx,yy}$	A_{zz}	$A_{xx,yy}$	ξ	Δ	Ref.
^{11}B		Kr	2.0018	2.032	122.3	-44.9	-4.7	224	15
^{11}B		Xe	2.002	2.10	100	-36	-11.4	168	15
^{11}B	3/2	free	--	--	11.6	11.6	10.7	--	26
^{27}Al		Ar	2.000	1.954	143.0	-101.5	74.9	2100	42
^{27}Al		Kr	2.001	1.989	135.8	-89.9	74.9	7540	42
^{27}Al		Xe	1.997	1.962	173.9	-75.8	74.9	--	42
^{27}Al		free	--	--	-4.6	-4.6	74.9	--	26
^{69}Ga		Ar	1.943	1.591	433	-598	550.8	2068	42
^{69}Ga		Kr	1.956	1.688	394	-536	550.8	2643	42
^{69}Ga		Xe	1.968	--	342	--	550.8	3460	42
^{69}Ga		free	--	--	-146	-146	550.8	--	26

The theoretical analysis of the boron atom has accounted for external heavy atom effect, whereas the Al and Ga were treated essentially by assuming the gas-phase values for the spin-orbit coupling constants. As already demonstrated in the B atom case, this assumption fails for heavier rare gases. It is thus expected that the crystal field splitting in Table 3 for Al and Ga is too large for Kr and Xe. This was noted in Ref. 42, where the calculations indicated that the crystal field splitting is infinite in Xe matrix. Since the experimental EPR spectrum can only yield information on the ratio between the crystal field splitting and the spin-orbit coupling constant, additional theoretical calculations are required to determine either of the free parameters.

As was noted in Refs. 15 and 41, diffusion of B atoms can be thermally activated, and in some cases formation of B_2 as well as impurity related products H_2BO (presumably originating from H_2O + B reaction) and BO (from O_2 + B) are observed. The existence of these species clearly demonstrate that the matrix environment is well suited for studying reactions between molecular species and B atoms. No information of thermal stability of Al and Ga atoms is available.

4. THEORETICAL TREATMENTS FOR ATOMIC IMPURITIES

The first theoretical analysis of matrix induced shifts on the isotropic g and A values was presented by Adrian.[23] In the case of isotropic hyperfine coupling shifts, the impurity atom - matrix interaction was assumed to be pair-wise additive and to consist of two types of interactions: van der Waals attraction and Pauli repulsion. The van der Waals interaction causes expansion of the impurity atom electron cloud and, in simple cases like H atom, yields reduced spin density at the nucleus. This is the origin of the negative shift in the isotropic hyperfine coupling constant. The Pauli repulsion has an opposite effect for the shift since it causes compression of the unpaired electron orbital. The experiments on atoms in rare gas matrices measure the delicate balance between these two effects. For this reason different trapping sites exhibit distinct isotropic hyperfine coupling constants. The origin for g-value shift is more difficult to visualize but essentially it is caused by mixing of the outer atomic orbitals of the rare gas into the wavefunction of the impurity by exchange interaction. This effect increases the effective spin-orbit coupling of the atom, and hence we expect negative shift in isotropic g-value. This is, indeed, observed for ^2S impurities as shown in Table 1. In the following paragraphs we will consider theoretical calculation of these effects using a pair-wise additive model.

4.1 Calculation of isotropic hyperfine coupling constants

A number of theoretical studies have demonstrated that *ab initio* calculation of the isotropic hyperfine coupling constant is very sensitive to proper treatment of electron correlation and quality of the basis set. The same conclusion applies for calculation of the pair potentials, which similarly consist of van der Waals interaction and Pauli repulsion. To a good approximation the Pauli repulsion can be described by single determinant calculations (Hartree-Fock), whereas the van der Waals part requires inclusion of electron correlation. For the case of atomic hydrogen, the maximum negative IHC shifts have been calculated to occur at slightly larger internuclear distances than the van der Waals potential energy minimum. The simple interpretation of van der Waals expansion / Pauli compression fits into these results very well.

As a representative case we will consider the pair interaction between atomic hydrogen and Ne, Ar, Kr, and Xe atoms. For Ne and Ar Dunning's augmented basis set aug-cc-pVQZ[45] was applied, whereas for the heavier rare gases the effective core potential basis set of the Stuttgart group[46] with additional augmentation had to be used. In the latter case the effective core potential basis set reduces the number of explicit electrons and makes the calculation computationally more affordable. In both cases the diffuse basis functions are of great importance for obtaining accurate results. The isotropic hyperfine coupling constant a can be obtained from the Fermi contact analysis:[47]

$$a = \frac{2}{3h}\mu_0 g_e g_N \mu_B \mu_N |\psi(0)|^2$$

where μ_0 is the vacuum permeability, g_e and g_N are the electron and nuclear g values, and μ_B and μ_N are the Bohr and nuclear magnetons, respectively. The spin density $|\psi(R)|^2$ can be obtained from the single particle density matrices by standard methods.[47] Efficient methods for including electron correlation are provided by the coupled cluster (CC) and Møller-Plesset (MP) theories.[48] For evaluation of the pair potentials, methods such as CCSD(T) (single, double, and perturbative triple excitations) and MP4 have proven to be very accurate. In the hydrogen atom - rare gas case the calculated van der Waals minima account for about 80 % of the binding energy when compared to experimental results. Unfortunately, most of the available quantum chemistry programs allow evaluation of the required single particle density matrix only at the less accurate CCD and MP4(SDQ) levels.

Since finite basis expansion calculations are prone to artificial basis set superposition errors, one must use the counterpoise procedure of Boys and Bernardi[49] for both pair potentials and isotropic hyperfine coupling shifts. In both cases it is not necessary to obtain fully converged results for the asymptotic region (*e.g.* infinite nuclear separation) since we are only interested in relative energies and shifts (Δa). In the latter case the following equation is applied:[13]

$$\Delta a(R) = a_{RGH}(R) - a_{RG''H}(R)$$

where $a_{RGH}(R)$ is the calculated isotropic hyperfine coupling constant shift for H-RG system and $a_{RG''H}(R)$ is the same quantity when the rare gas atom is included as "ghost atom". For ghost atoms the calculation involves all basis functions for the center but no electrons. By way of an example, the calculated isotropic hyperfine coupling shifts for {Kr, Xe} - H as function of internuclear distance are shown in Fig. 7. All the obtained curves show sudden collapse at short distances. The onset for the Xe - H case can be seen in Fig. 7, whereas for Kr - H the turn over occurs at shorter bond lengths. At short distances it is expected that contribution of the diatomic Rydberg states increases dramatically and this, consequently, causes shielding of the unpaired electron from the impurity nucleus. Most of the available basis sets are not sufficient in describing these Rydberg states properly, and the collapse in Δa may in fact occur even at larger distances than the calculations have indicated so far. This may have, for example, considerable effect in line broadening in the Xe - H system. In fact, experiments show that sensitivity for detecting hydrogen atoms in Xe matrix via EPR is strongly reduced. Although the lack of sensitivity is partially caused by broadening caused by the superhyperfine interaction, it does not explain the situation fully. Finally, we note that the anisotropic hyperfine coupling will exhibit small shifts as well, but comparison with the experimental observations is not easy because of the poor resolution in the powder spectra.

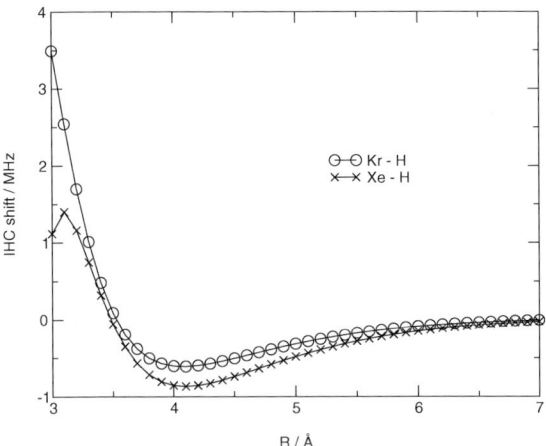

Figure 7. The distance dependence of IHC shifts for {Kr, Xe}-H systems (MP4(SDQ) calculations of Ref. 13) is shown.

4.2 Calculation of the g shift and anisotropy

The behaviour of the g shift may be estimated by calculating the induced spin-orbit coupling change by the approaching rare gas atom. In the case of free hydrogen atom, no spin-orbit interaction exists, and thus the rare gas atom is solely responsible for spin-orbit interaction in this case. The spin-orbit matrix elements in a diatomic system can be conveniently evaluated by a method presented in Ref. 50 by using multi-reference configuration interaction wave functions. Both regular and effective core potential basis sets can be applied in these calculations. This type of calculation has been carried out in detail for boron atom in rare gas matrices where anisotropic interactions also occur.[15] The g-shift can be approximated by the sum over states bilinear formula:[47,51]

$$g_{ab} = g_e \delta_{ab} + g_e \sum_{n \neq 0} \frac{<0|L_a S_a|n><n|L_b|0>}{E_n - E_0}$$

where a, b = x, y, z, L_aS_a is the spin-orbit operator, L_b is the orbital Zeeman operator (gauge dependent) and summation is taken over the excited states. Provided that the matrix element of L_b would not change much for different impurity - rare gas combinations, then it can be seen that it is the spin-orbit operator that is responsible for mixing with the excited states to a varying extent. These excited states have considerable contributions from the rare gas atoms and, in the case of Xe, the contribution of ionic charge transfer state (B^- Xe^+) is also significant. In this case it was possible to account for the isotropic g shift by reducing the effective value of the spin-orbit coupling constant as shown in Table 3.

9. EPR studies of atomic impurities in rare gas matrices

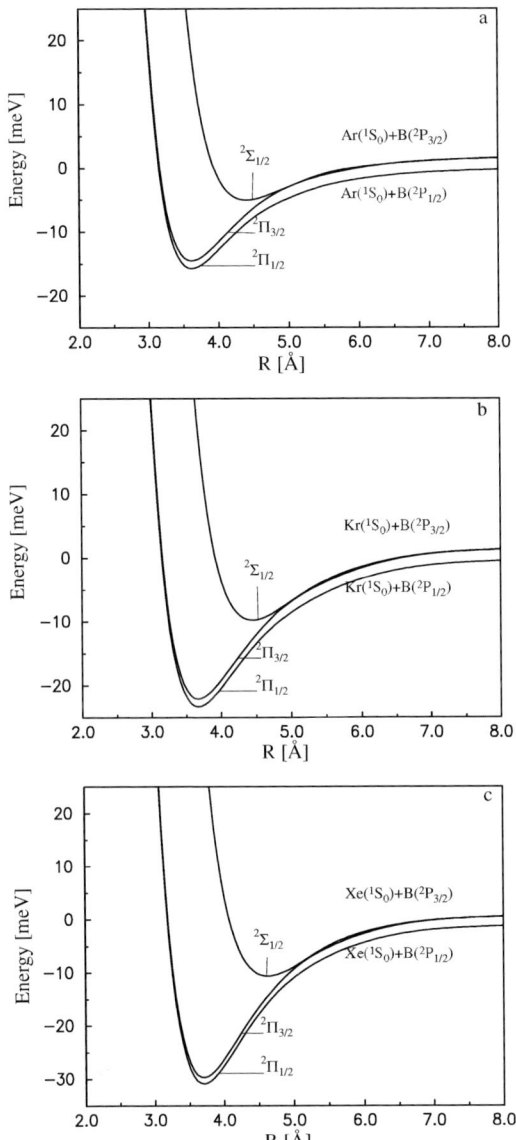

Figure 8. The potential energy curves for B - Rg pairs including spin-orbit coupling (Ref. 15) are shown.

In addition to the isotropic g shifts, the ^{2}P state atoms experience strong splitting of the p-orbital frame by the surrounding crystal field. For example, boron atom trapped in rare gas matrices produce axially symmetric powder spectrum. This effect can be understood in terms of the Σ and Π pair potentials for B – Rg, which are sufficiently different to produce axially symmetric trapping site (see Fig. 6). In terms of the crystal field theory, this means that the surrounding matrix removes the orbital degeneracy and therefore quenches the orbital angular momentum. It can be shown that simple relation for crystal field splitting energy (ΔE) and $\Delta g_{xx,yy}$ can be obtained:

$$\Delta g_{xx,yy} = g_e - 2\frac{\xi}{\Delta E}$$

In most theoretical analysis the spin-orbit (ξ) coupling is taken as a constant corresponding to the free atom limit. However, as discussed earlier, the external heavy atom effect of the matrix atoms may modify the spin-orbit coupling, and therefore the above expression will yield increasingly incorrect results when heavier rare gas matrices are treated. This behaviour was in fact observed by Ammeter *et al.* in their study of crystal field splitting in Al and Ga atoms embedded in rare gas matrices (see Table 3), where too large crystal field splitting was observed.[42] Furthermore, this expression does not allow a change of sign in the g shift and this effect is already clearly seen in the case of B atoms in Kr matrix. Thus a careful analysis of the effective spin-orbit coupling constant must always be carried out when using this equation.

4.3 Molecular dynamics simulations

Once the pair-wise additive potentials and isotropic hyperfine coupling shifts have been calculated, trapping of the impurity in the rare gas lattice can be simulated. Although light atoms such as H and B have considerable zero-point spread associated with them, to first approximation we can treat them classically. Classical molecular dynamics simulations for these systems have been carried out in Refs. 13 and 15. Electronic and nuclear degrees of freedom were separated in these calculations by assuming that the electronic part adiabatically follows the nuclear motion. For hydrogen atoms these simulations are able to yield correct trapping sites (*i.e.* they produce the experimentally observed isotropic hyperfine coupling shifts as shown in Table 1). This method was the basis in identifying the trapping sites for atomic hydrogen. Simulation of ^2P state atoms requires more complex way of evaluating the potential energy surface used in the molecular dynamics simulations because the B - Rg Σ and Π potentials are different from each other. For this purpose a minimal diatomics-in-molecules (DIM) description is well suited. The DIM Hamiltonian is given as:[15]

9. EPR studies of atomic impurities in rare gas matrices

$$H^{DIM} = \sum_{i<j} E(i,j) \otimes \mathbf{1} + \sum_{i} \begin{pmatrix} E_i(^2\Pi) + x_i^2\Delta_i & x_i y_i \Delta_i & x_i z_i \Delta_i \\ x_i y_i \Delta_i & E_i(^2\Pi) + y_i^2\Delta_i & y_i z_i \Delta_i \\ x_i z_i \Delta_i & y_i z_i \Delta_i & E_i(^2\Pi) + z_i^2\Delta_i \end{pmatrix}$$

where $\Delta_i = E_i(^2\Sigma) - E_i(^2\Pi)$, x_i, y_i, z_i denote the direction cosines between the ^2P atom and matrix atom i, $E(i,j)$ is the Rg - Rg pair potential, **1** is a 3x3 unit matrix. The first summation is over ^1S matrix atom pairs and the second over the surrounding matrix atoms. Numerical diagonalization of this real symmetric matrix can be efficiently performed, for example, by DSYEV LAPACK routine.[52] This diagonalization yields three eigenstates and the effective orientation of the unpaired electron orbital is given by the eigenvectors. It should be noted that atoms that have a hole in their p-orbital frame must be treated differently as the ionic states have major contributions to the ground state. By including vacancies around the boron atom it was observed that in the lowest state the p-orbital direction became locked in space. Based on the experimental EPR spectrum this is the correct behaviour. The calculated crystal field splitting (*i.e.* the difference between the eigenvalues) correspond very nicely with the experimental observations. Spatial locking of the p-orbital can be conveniently represented by the orbital autocorrelation function of the ith state:

$$C_i(t) = <c_i(t) \cdot c_i(0)>$$

where c_i is the eigenvector of the ith state. By way of an example, the autocorrelation function of the unpaired electron orbital of boron in various Rg hosts is shown in Fig. 9.

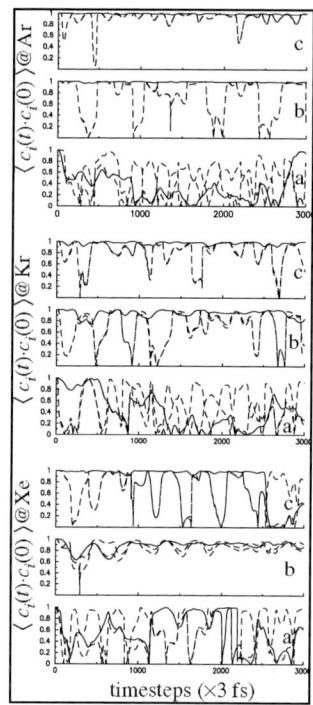

Figure 9. The p-orbital autocorrelation function for B atom in Ar, Kr, and Xe matrices after Ref. 15 are shown. Case (a) no vacancies, (b) one vacancy, and (c) two vacancies. The solid lines correspond to the lowest eigenstate (z) and the dashes lines to nearly degenerate excited eigenstates (xy).

The calculated crystal field splitting by this method are listed in Table 3. From the simple g shift expression in 4.2 it can be seen that experiments can only yield the ratio between the effective spin-orbit and the crystal field splitting. In practice this means that the theoretical calculations are mandatory in order to obtain proper results. The DIM model can be easily extended to include interactions that are important for EPR spectroscopy (DIM-SH). When spin-orbit interaction is included the Hamiltonian becomes complex 6x6 matrix since electron spin has to be added to the basis set. It then depends on the problem weather Cartesian, uncoupled, or coupled basis representation is the most efficient one. One should observe that a consistent phase convention is used in transformations between different schemes.[53]

The DIM-SH method combined with molecular dynamics calculations is expected to be very useful, for example, in modelling of the spin dynamics in solids. Ensemble averages for spins can be performed by averaging over multiple trajectories with different initial conditions. Even inhomogenities can be included by introducing different trapping sites in the averaging

9. EPR studies of atomic impurities in rare gas matrices

process. However, long simulation times and extensive ensemble averaging may lead to excessive computation times.

4.4 Simulation of EPR spectra

Simulation of EPR spectra for ^{2}S atoms in rare gas matrices can be carried out by standard methods using an effective isotropic g value. The spectrum is completely isotropic which further simplifies the spectrum simulation. However, the ^{2}P state case requires some more complex treatment and the spin Hamiltonian of section 3.4 must be solved numerically. Furthermore, this simulation involves powder integration with axial symmetry. For this purpose we have implemented a generic spin engine, which can deal with arbitrary number of spins and operations on them in uncoupled representation. Given the spin Hamiltonian of 3.4 this routine builds the required spin Hamiltonian matrix efficiently, diagonalizes it, and calculates the transition moments between the states in question. In order to mimic a real EPR experiment, which involves sweeping the magnetic field, this procedure is repeated at various values of external magnetic field strength. For generating EPR spectrum linear interpolation is used in obtaining intermediate transition energies and transition moments. Since the obtained energy levels may cross, it is important to correlate correct states with each other. For this purpose, for state i we correlate state k that maximizes the overlap:

$$\max_k |<c_i(B_j)|c_k(B_{j+1})>|^2$$

where B_j are the values of magnetic field where the spin Hamiltonian has been diagonalized and c_i are the eigenvectors.

When the powder spectrum integration was carried out a cubic spline interpolation of the transition energies and moments was applied. This greatly reduces the number of orientations at which the spin Hamiltonian has to be diagonalized. After powder integration procedure we finally convolute in the EPR lineshape by using the fast Fourier transform (FFT) method where the convolution is carried out in the Fourier space. For the boron case we have used different linewidths for parallel and perpendicular transitions, which complicates the calculation by forcing the FFT convolution separately for each single crystal spectrum. In this particular case it was observed that the perpendicular lines were 7 - 9 times broader than the parallel ones. Coupling of the p_z orbital with $p_{x,y}$ by the external magnetic field is the most likely source for this effect. The x-y frame is more strongly coupled to the lattice and hence its presence causes severe line broadening. This type of effect is in fact quite general and it has been observed in more complex systems as well.[54]

The simulated EPR spectrum can be fitted to the experimental spectrum by using least squares minimization process. In the existing simulation code any of the parameters of 3.4 Hamiltonian can be included. Since the analytic gradients are very difficult to obtain, a method that does not require gradients is preferred. Three such methods were used in the simulations: Monte Carlo, Simplex, and Marquardt.[55] Restrictions on parameters can be included by the penalty terms, which essentially increase the object function value if some parameters are out of their preferred range.

All the described spectral simulation methods have been implemented in Xemr program, which is available for free download through Internet.[56]

5. REFERENCES

1. W. Weltner, Jr., *Magnetic Atoms and Molecules*, Dover Publications, New York, 1983.
2. E. Whittle, D. A. Dows, and G. C. Pimentel, *J. Chem. Phys.* **22** (1943) 1954.
3. M. E. Jacox, *Vibrational and Electronic Energy Levels of Polyatomic Transient Molecules*, http://webbook.nist.gov/chemistry/polyatom/.
4. G. C. Pimentel, in *Frormation and Trapping of Free Radicals*, edited by A. M. Bass and H. P. Broida, Academic Press, New York, 1960.
5. P. Palaszewski, L. S. Ianovski, P. Garrick, *J. Propul. Power*, **14** (1998) 641.
6. S. Tam, M. Macler, M. E. DeRose, and M. E. Fajardo, *J. Chem. Phys.* **113** (2000) 9067.
7. M. Pettersson, J. Lundell, and M. Räsänen, *Eur. J. Inorg. Chem.* (1999) 729.
8. L. Khriachtchev, M. Pettersson, N. Runeberg, J. Lundell, and M. Räsänen, *Nature* **406** (2000) 874.
9. V. A. Apkarian and N. Schwentner, *Chem. Rev.* **99** (1999) 1481.
10. D. M. Chipman, *Theor. Chim. Acta*, **82** (1992) 93.
11. D. Feller, *J. Chem. Phys.* **93** (1990) 579.
12. D. Feller, E. D. Glendening, E. A. Mc. Cullough, Jr., and R. J. Miller, *J. Chem. Phys.* **99** (1993) 2829.
13. T. Kiljunen, J. Eloranta, and H. Kunttu, *J. Chem. Phys.* **110** (1999) 11814.
14. J. Ahokas, T. Kiljunen, J. Eloranta, and H. Kunttu, *J. Chem. Phys.* **112** (2000) 7475.
15. T. Kiljunen, J. Eloranta, J. Ahokas, and H. Kunttu, *J. Chem. Phys.* **114** (2001) 7144.
16. I. R. Dunkin, *Matrix-Isolation Techniques: A practical approach*, Oxford University Press, New York, 1998.
17. J. A. Weil, J. R. Bolton, and J. E. Wertz, *Electron Paramagnetic Resonance: Elementary theory and practical applications*, John Wiley & Sons, New York, 1994.
18. S. Foner, E. L. Cochran, V. A. Bowers, and C. K. Jen, *J. Chem. Phys.* **32** (1960) 963.
19. J. Eloranta, K. Vaskonen, H. Häkkänen, T. Kiljunen, and H. Kunttu, *J. Chem. Phys.* **109** (1998) 7784.

20. J. Eloranta, K. Vaskonen, and H. Kunttu, *J. Chem. Phys.* **110** (1999) 7917.
21. K. Vaskonen, J. Eloranta, and H. Kunttu, *Chem. Phys. Lett.* **310** (1999) 245.
22. K. Vaskonen, J. Eloranta, T. Kiljunen, and H. Kunttu, *J. Chem. Phys.* **110** (1999) 2122.
23. F. J. Adrian, *J. Chem. Phys.* **32** (1960) 972.
24. A. Schrimpf, R. Rosendahl, T. Bornemann, H.-J. Stöckmann, F. Faller, and L. Manceron, *J. Chem. Phys.* **96** (1992) 7992.
25. P. Kasai and D. McLeod, Jr., *J. Chem. Phys.* **55** (1971) 1566; J. Eloranta, Unpublished results.
26. Gas phase atomic data.
27. C. K. Jen, V. A. Bowers, E. L. Cochran, and S. N. Foner, *Phys. Rev.* **126** (1962) 1749.
28. D. A. Garland and D. M. Lindsay, *J. Chem. Phys.* **80** (1984) 4761.
29. G. A. Thompson, F. Tischler, and D. M. Lindsay, *J. Chem. Phys.* **78** (1983) 5946.
30. G. A. Thompson and D. M. Lindsay, *J. Chem. Phys.* (1981) 959.
31. R. J. Van Zee and W. Weltner, Jr., *J. Chem. Phys.* **92** (1990) 6976.
32. S. B. H. Bach, D. A. Garland, R. J. Van Zee, and W. Weltner, Jr., *J. Chem. Phys.* **87** (1987) 869.
33. L. B. Knight, Jr., W. E. Rice, L. Moore, E. R. Davidson, and R. S. Dailey, *J. Chem. Phys.* **109** (1998) 1409.
34. T. Cole and H. McConnell, *J. Chem. Phys.* **29** (1958) 451.
35. G. Jackel, W. Nelson, W. Gordy, *Phys. Rev.* **176** (1968) 453.
36. J. Eloranta, unpublished results: *BSSE corrected CCSD(T)/AVQZ calculation*.
37. L. B. Knight, Jr., B. A. Bell, D. P. Cobranchi, and E. R. Davidson, *J. Chem. Phys.* **111** (1999) 3145.
38. P. Kasai, *Phys. Rev. Lett.* **21** (1968) 67.
39. J. R. Morton, K. F. Preston, S. J. Strach, F. J. Adrian, and A. N. Jette, *J. Chem. Phys.* **70** (1979) 2889.
40. L. B. Knight, Jr. and W. Weltner, Jr., *J. Chem. Phys.* **55** (1971) 5066.
41. W. R. Graham and W. Weltner, Jr., *J. Chem. Phys.* **65** (1976) 1516.
42. J. H. Ammeter and D. C. Schlosnagle, *J. Chem. Phys.* **59** (1973) 4784.
43. A. R. Boate, J. R. Morton, and K. F. Preston, *Chem. Phys. Lett.* **54** (1978) 579.
44. S. V. Bhat and W. Weltner, Jr., *J. Chem. Phys.* **73** (1980) 1498; M. Iwasaki, K. Toriyama, and H. Muto, *J. Chem. Phys.* **71** (1979) 2853.
45. D. E. Woon and T. H. Dunning, Jr., *J. Chem. Phys.* **100** (1994) 2975.
46. A. Nicklass, M. Dolg, H. Stoll, and H. Preuss, *J. Chem. Phys.* **102** (1995) 8942.
47. R. Mcweeny, *Methods of Molecular Quantum Mechanics*, Acadenic, London, 1992.
48. F. Jensen, *Introduction to Computational Chemistry*, Wiley, New York, 1999.
49. S. F. Boys and F. Bernardi, *Mol. Phys.* **19** (1970) 553.
50. A. Berning, M. Schweizer, H-J. Werner, P. Knowles, *Mol. Phys.* **98** (2000) 1823.
51. J. E. Harriman, *Theoretical Foundations of Electron Spin Resonance*, Academic, New York, 1978.
52. E. Anderson, Z. Bai, C. Bischof et al., *LAPACK Users' Guide* (3rd ed.), Society for industrial and applied mathematics, Philadelphia, PA, 1999.
53. E. Condon and G. Shortley, *The Theory of Atomic Spectra*, Cambridge University Press, Cambridge, 1970.
54. J-L. Du, G. R. Eaton, and S. S. Eaton, *J. Magn. Reson.* A **117** (1995) 67.
55. B. Kirste, *Anal. Chim. Acta* **265** (1992) 191.
56. J. Eloranta, http//epr.chem.jyu.fi/xemr

Chapter 10

ORGANIC RADICAL CATIONS AND NEUTRAL RADICALS PRODUCED BY RADIATION IN LOW-TEMPERATURE MATRICES

Vladimir Feldman
Karpov Institute of Physical Chemistry, 10 Vorontsovo Pole Str., Moscow 105064,Russia; Institute of Synthetic Polymeric Materials of RAS, 70 Profsoyuznaya Str., Moscow 117393, Russia;
Department of Chemistry, Moscow State University, Moscow 119992, Russia

Key words: EPR; radiation chemistry; radical cations; radicals; matrix isolation; low-temperature chemistry; macromolecules.

Abstract: Paramagnetic species produced by ionising radiation in organic materials have been studied extensively by EPR for fifty years. Using low temperature matrices made it possible to characterise a wide class of highly reactive radicals. More recently, the focus was shifted to the investigations of ionised molecules (radical cations). Several approaches based on frozen solution technique, trapping in porous media and rigorous matrix isolation method has been developed up to 1990s. This chapter presents a review of recent progress in the field with special attention to the EPR studies of radical cations and radicals generated by high-energy irradiation of moderate-size organic molecules in solid rare gas matrices. The following aspects are discussed: (i) trap-to-trap positive hole transfer between organic solute molecules in low-temperature matrices; (ii) matrix effects on trapping and reactivity of organic radical cations in rigid inert media; (iii) the role of excess energy in the reactions of ionised organic molecules in solids; (iv) the nature of selectivity of the primary bond rupture in organic molecules and macromolecules. The prospects and problems of EPR and combined spectroscopic studies of the radiation-induced species in organic systems are outlined.

1. INTRODUCTION

Organic radicals and radical ions are the key reactive intermediates in a wide variety of chemical and biological processes induced by oxidising agents, heat, light, ionising radiation, etc. Unstable radicals produced in thermal chemical reactions normally occur in very small concentrations, and their lifetimes are quite short (especially, in liquid phase), which limits direct detection of these species by EPR. Meanwhile, in the case of irradiation of organic solids and polymers, various-type radicals and radical ions are easily produced and trapped in high concentrations, which makes it possible to investigate their structure and properties in detail. Probably, the first EPR study of organic radicals generated by high-energy radiation (X-rays) was reported in 1951,[1] so now we can celebrate fifty years of using EPR for characterisation of the radiation-induced damage in organic materials. It is worthwhile noting that both EPR spectroscopy and radiation chemistry benefited greatly from these studies. Indeed, ionising radiation has proved to be the most common tool to obtain various organic radicals, including highly reactive «high-energy» species, which can be hardly produced by other techniques. Many basic results concerning the structure and dynamics of free radicals were derived from the studies of irradiated systems. As an illustrative example, one can mention classical works on determination of proton hyperfine coupling tensors for aliphatic radicals trapped in irradiated single crystals.[2] On the other hand, high sensitivity and unique structural informativity of EPR spectroscopy gave invaluable help in elucidating the radiation-chemical mechanisms, and since the 1960s EPR was accepted as one of the basic experimental tools of radiation chemistry.[3]

Extensive studies carried out during the first thirty years (1951 - 1980) resulted in dramatic progress in the field. In particular, the radicals trapped in different organic systems and polymers irradiated at 77 K were identified unambiguously, and the radiation-chemical yields were determined for a number of systems. A comprehensive review of the early work can be found in the book by Pshezhetskii et al.[4] An important problem addressed (but not solved) in these studies is concerned with the possibility to differentiate primary radiation-induced chemical events from secondary reactions. At first, it was supposed that the secondary processes should be completely blocked in rigid media at 77 K. However, further experiments revealed that some reactions activated by local molecular motion in organic solids and polymers could occur well below this point.[5-10] Furthermore, although the EPR data on the radiolysis of organic systems at cryogenic temperatures (below 77 K) are still rather limited, it is clear that some primary radical species cannot be trapped even at lowest attainable temperatures. In general, it is suggested that the radical reactions occurring at very low temperatures are due to tunnelling phenomena. Meanwhile, in the case of radiation-induced radicals, one should also bear in mind possible involvement of «hot» or «unrelaxed» species.

At this point, it should be stressed out that even detection of «primary» neutral radicals resulting from dissociation of organic molecules is not sufficient to establish the basic mechanisms of radiation-induced damage. Actually, the *primary* event induced by radiation is ionisation or electronic excitation rather than chemical bond rupture. The ionisation process specific for high-energy radiation is generally favoured in condensed phases. An important role of ionic processes is clearly demonstrated by the observations of trapped electrons and radical anions in a number of irradiated organic systems[4]. Thus, from fundamental viewpoint, it is crucial to characterise the structure and reactivity of the primary ionised molecules, or radical cations. The radical cations are paramagnetic species, so they should be detectable by EPR. Nevertheless, most aliphatic radical cations were not observed in early studies. An obvious reason is extremely high reactivity of ionised aliphatic molecules, which may undergo recombination with electrons, fast reactions with neighbouring neutral molecules, and «hot» fragmentation. In fact, none of these processes requires molecular diffusion, so the reactions of the primary radical cations cannot be stopped at low temperatures. To avoid the ion—molecule reactions, it is practicable to use matrix isolation, whereas the ion—electron recombination may be ceased in the presence of electron scavengers. A simple and efficient solution for the EPR studies of reactive radical cations, the so-called «Freon matrix technique», was suggested by Shida and Kato in 1979.[11] The method is based on irradiation of frozen solutions of organic substances in fluorinated halocarbons (mainly freons) at reasonably low temperatures (typically, 77 K). Detailed scheme, advantages and problems of the method will be discussed below. Extensive studies of a wide class of organic cations were carried out using this approach,[12-14] and the up-to-date knowledge on electronic structure, geometry, and chemical properties of these species relies essentially on the data obtained in halocarbon matrices.

As a whole, up to early 1990ths, the basic information on the structure of the radiation-induced paramagnetic species (radicals and radical ions) was available for many organic compounds and a number of important polymers (including macromolecules of biological interest). Nevertheless, general understanding of the factors controlling primary radiation-induced events in organic molecular materials was still lacking. In particular, the following issues could be defined as major unresolved problems:

(a) *The problem of localisation.* Typically, molecular solids and polymers consist of molecules or units, which are chemically similar, but physically not exactly equivalent, because of conformation difference, variations in molecular interaction, packing, etc. The role of this dispersion in localisation of primary radiation-induced events is not clear.

(b) *The problem of matrix effects.* Even in the case of chemically inert environment, the primary radiation-induced effects in solid media can be sensitive to the matrix nature, especially if we consider the relaxation and reactions of strongly interacting primary ionised molecules. However, the data on matrix effects on the early processes are rather ambiguous.

(c) *The problem of excess energy* (involvement of «vibrationally hot» species). In general, it is assumed that the high-energy reaction paths involving «hot» species (vibrationally excited radical cations or radicals) are less significant in solids than in the gas phase. However, the formation of fragment radicals in the low-temperature radiolysis may imply effect of excess energy of the primary species. The conclusions on this problem are speculative and are not based on any direct experiment.

(d) *The problem of selectivity.* Despite the intuitive reasoning on non-selectivity of the primary bond rupture in «high-energy processes», the EPR data suggest non-random mode of radical formation. The interpretation is not straightforward since it is not easy to discriminate between primary and secondary events. In fact the «selection rules» in the radiation chemistry of molecules in solids are still not established.

It is to be noted that EPR spectroscopy is the most suitable experimental tool, which could shed a bit of light on these issues. Indeed, EPR spectra are quite sensitive to chemical structure and conformation of organic radicals and radical ions; matrix effects can be often visualised in magnetic parameters of the radical species. Although the EPR studies of the radiation-induced species during the past decade were not so extensive as in early period, the results provide some important clues for basic problems outlined above. This chapter will present a review of recent development with specific impact on experimental studies carried out in our laboratories.

2. EXPERIMENTAL APPROACHES AND OVERVIEW OF RESULTS

2.1 Low-temperature techniques

Most commonly, the low-temperature EPR studies of radiation-induced radicals in organic solids are carried out at 77 K using standard equipment for irradiation and measurements, which is commercially available for tens of years. Meanwhile, as mentioned above, some chemical reactions of active species and local physical processes in matrices cannot be stopped at this temperature. Using "helium-range" temperatures (below 77 K) may be significant both for trapping of highly reactive intermediates and for visualisation of spatial distribution of the primary chemical events. Early EPR studies of the radiation damage in organic materials at very low temperatures (down to 1.5 K) were focused mainly on spatial distribution of the primary radicals, in particular, on the structure and dynamics of radical pairs produced by irradiation. Linear alkane crystals and polyethylene were the most popular objects in these works.[5-10,15-17] In a few cases, it was possible to identify the radicals, which are unstable at 77 K (*e.g.*, methoxy radicals in methanol[18]). Also, extensive work was made on DNA crystals irradiated at 4 K.

Considering early studies from experimental viewpoint, one should note that the irradiation of organic crystals in ampules with X- or γ-rays was typically carried out in a liquid-helium bath (Dewar) at 4.2 K (or at even lower temperature, if the pump-out procedure was used[5]). The EPR measurements were made in the same Dewar at 4.2 K, or at higher temperatures (using liquid nitrogen Dewar or nitrogen-flow cryostat for the measurements at 77 K and above). Such an approach is somewhat limited, especially for detailed studies at intermediate temperatures (between 4 and 77 K); furthermore, it is not suitable for matrix isolation studies. A more versatile approach is based on using continuous-flow or closed-cycle helium cryostats operating in a wide temperature range. At present, the general-purpose helium-flow cryostats for the EPR measurements are commercially available and widely used in many laboratories. However, the cryostats for high-energy irradiation and EPR measurements under the conditions strictly avoiding intermediate heating of the sample are still custom-made devices used by a few groups. To my knowledge, most of these devices make use of X-ray radiation.[19-21] A different approach using fast electron irradiation produced by a Van-de-Graaf accelerator (typically, 1-1.2 MeV) has been developed at Karpov Institute.[22-24] One advantage of using fast electrons results from easy tunability of the beam parameters, i. e., electron energy and beam current. Thus, it is possible to tune the penetration depth and to vary the dose rate over a wide range. Generally, the irradiation time in electron-beam experiment is shorter than in the case of X-ray irradiation.

The EPR cryostat for matrix isolation studies (Fig. 1) contains a cylindrical (H_{011} mode) vacuum resonator cavity. A sample is obtained by slow controlled deposition of gaseous mixture onto the tip of sapphire rod inserted into the cavity. The rod is a 4-mm cylinder with truncated conical tip, which corresponds naturally to the symmetry of the resonator cavity. If the deposition is slow enough and the nozzle comes close to the tip, the shape of the growing sample follows the tip shape. In certain experiments, relatively large deposited samples were visualised as small "bulbs" (semi-spheres). In this case, the paramagnetic species produced in the sample show essentially random orientation in macroscopic scale. For this reason, the effect of preferential orientation often observed in the spot-like samples obtained on a flat rod[25] is of minor importance for the samples prepared on the truncated conical tip. After the deposition is complete, the sample is irradiated with fast electrons through a special aluminium foil window, and then the cavity is connected to the microwave bridge of spectrometer.

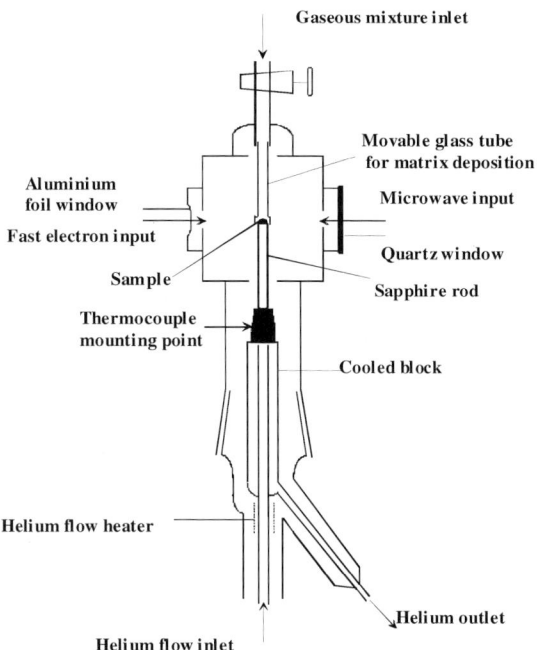

Figure 1. Cryostat with vacuum resonator cavity for matrix isolation EPR studies of species generated by fast electron irradiation

Another-type EPR cryostat was designed for the studies of paramagnetic species produced in polymers. In this case, the sample is kept in cold helium flow coming out of the capillary. A small polymeric disk with the hole is mounted on the capillary close to its outlet. The capillary is placed into a quartz Dewar to be inserted into a standard resonator cavity of the EPR spectrometer. Using the disks obtained from oriented polymeric films makes it possible to study the angular dependence of the EPR spectra of macroradicals by rotating the stretch axes in the external magnetic field.

Both cryostats operate at the temperatures down to 8 K. Other details of experimental techniques used at Karpov Institute are given elsewhere.[22-24]

2.2 Matrix isolation for radiation chemistry

Matrix isolation is a widely used experimental approach for the studies of highly reactive intermediates. In general, it implies trapping of a reactive species in a rigid, chemically inert environment at low temperature. Classic procedure makes use of deposited matrices obtained mainly from rare gases;

however, "compromise" techniques using dilute frozen solutions are also often referred to as matrix isolation methods. Meanwhile, the matrices used for the EPR studies of radiation-induced radicals and radical cations should meet a number of specific requirements. First, the matrix substance should have relatively high ionisation and/or excitation potential. Indeed, the high-energy radiation is absorbed primarily by matrix, and the species from guest molecules are produced only by positive hole or excitation transfer. Second, as mentioned above, trapping of the primary radical cations is possible only in the presence of electron scavengers, which blocks ion—electron recombination. Finally, the EPR spectra of paramagnetic species produced by matrix radiolysis should not overlap with the spectra of radicals under study (ideally, the matrix should yield no paramagnetic species under irradiation). These requirements result in significant restriction in the choice of suitable matrices for the radiation-chemical studies. Nevertheless, a number of approaches have been developed to overcome the above-mentioned limitations. Recent development in this field, unresolved problems and future prospects are briefly analysed below.

2.2.1 Low-temperature organic glasses

Organic glasses stable at 77 K (*e.g.*, branched alkanes, MTHF, methylcyclohexane, alcohols and alkyl halides) have been used as matrices for optical spectroscopic studies of structure and photochemistry of organic radicals produced by different methods.[26] Using glassy matrices was found to be especially valuably for investigations of trapped electrons and radical anions.[14] In general, these media are less suitable for the EPR studies of radical cations and neutral radicals produced by radiolysis, because the matrix radicals yield quite intense and complicated EPR signals. Among a few examples, it is possible to mention identification of the EPR signals of radical cations of tetramethylethylene and some long-chain alkanes with relatively low ionisation potentials in 3-methylpentane and 3-methylhexane glasses.[27, 28] The problem of background signal can be partially solved by using perdeuterated matrices. Indeed, the hyperfine coupling constants for deuterons are 6.51 times smaller than the corresponding values for protons, so the EPR spectra of perdeuterated radicals show much smaller total spread. For this reason, the outer lines of the species produced from protiated solute molecules can be easily detected in perdeuterated glassy solvents. Such an approach was used for the studies of hydrogen atom abstraction from guest alkane molecules in glassy perdeuterated alcohol matrices.[29]

2.2.2 Freon matrices

Freons are chemically inert fluorinated halocarbons widely used in general-purpose technologies. Actually, the "Freon matrix technique" is a modification of the halocarbon method for optical spectroscopic studies of radical cations suggested by Shida and Hamill.[30] As mentioned in Introduction, application of freons to the EPR studies of radiation-induced radical cations has been uniquely successful. General scheme of production of organic radical cations in the most popular matrix, Freon-11 (fluorotrichloromethane), may be represented as follows:

$$CFCl_3 \;\text{-}/\!\!\bigwedge\!\!\text{-}\!\!> CFCl_3^{+\cdot} + e^- \qquad (1)$$

$$CFCl_3^{+\cdot} + RH \rightarrow CFCl_3 + RH^{+\cdot} \qquad (2)$$

$$CFCl_3 + e^- \rightarrow (CFCl_3^{\cdot\cdot-}) \rightarrow CFCl_2^{\cdot} + Cl^- \qquad (3)$$

Here symbol "-/\/\->" denotes the action of ionising radiation, and RH is a solute organic molecule. The problem of background EPR signal from matrix radicals is not crucial, because fluorinated radicals yield very broad and weak signals in macroscopically disordered media due to large anisotropy of ^{19}F hyperfine coupling. Other popular matrix substances are the isomers of trifluorotrichloroethane (Freon-113), mainly $CFCl_2CF_2Cl$ and CF_3CCl_3. The former matrix is especially useful for the studies of ion - molecule reactions, whereas the latter one may be helpful for investigation of unimolecular transformations of the radical cations. The mixture of CF_2BrCF_2Br (Freon-114B2) and Freon-11 suggested for optical studies[31] was also used in combination with EPR. In addition to freons, some other fluorinated compounds (in particular, perfluoroalkanes and SF_6) were applied for the EPR studies of radical cations.

The "golden age" in Freon matrix studies is associated with the 1980ths, and comprehensive reviews in the field are available.[12-14] Here I have to mention in short some recent studies (not covered in early reviews).

Identification of a large number of hydrocarbon radical cations was made in early works; meanwhile, new data became available during the past decade. Despite the extensive studies of linear alkane radical cations in halocarbon matrices in the previous period, the discussion on the structure of these species is not closed, and some additional experimental and theoretical arguments have been presented.[32,33] The peculiarities of ion - molecule reactions of the linear alkane radical cations occurring in alkane aggregates in frozen halocarbon solutions were studied by Ceulemans and co-workers.[34] EPR spectra of radical cations of some branched alkanes have been characterised in various freon matrices.[33,35] Several studies were concerned with the structure and reactions of the radical cations of long-chain linear alkenes (pentenes, hexenes and octenes).[36-39] The radical cation of vinyl cyclopropane and other C_5H_8-related species has been studied extensively in various Freon matrices.[40-42] The primary species resulting from ionisation at vinyl group may exist as *gauche* or *anti* conformer, which easily convert to two kinds of distonic ring-open radical cations. The reactions of primary and distonic radical cations are quite sensitive to environment, which leads to a very diverse chemistry, depending on the experimental conditions. A number of papers reported EPR and ENDOR studies of structure and rearrangements of bicyclic and polycyclic hydrocarbon radical cations in halocarbon matrices.[43-47] The application of the Freon matrix technique for general organic chemistry may be illustrated well by recent studies of trimethylene methane radical cation produced from methylene cyclopropane.[48]

In the case of aromatic hydrocarbons, the main problem is precise determination and assignment of small hyperfine couplings. Recent progress is associated with the application of ENDOR. The latter technique made it possible to characterise in detail the structure of both monomeric and dimeric cations.[49-52] Also to be mentioned, recent EPR studies revealed dramatic effects of halocarbon matrix nature on the conformation of some alkyl benzene radical cations.[53, 54]

A number of recent investigations using halocarbon matrices were related to heteroatomic radical cations. An interesting example is given by characterisation of structure and reactions of methyl *tert*-butyl ether (MTBE) radical cation.[55] This cation shows a major hyperfine coupling of $a(3H) = 3.3$ mT due to methoxy group protons, which is substantially smaller than the corresponding value for dimethyl ether ($a(6H) = 4.2 - 4.3$ mT[56, 57]). Unlike the radical cations of linear ethers, the MTBE radical cation easily undergoes methane loss at the temperatures above 100 K. Several papers reported the studies of the radical cations of vinyl monomers and related compounds, including vinyl ethers,[58] dihydrofuranes,[59] and acrylates.[60] The latter species appear to be of special interest in view of practical significance of acrylates and confusing data of early studies. It was shown that ionisation of acrylates occurred from the non-bonding orbital of carbonyl oxygen, which implies small hyperfine couplings in the primary cation yielding a broad unresolved singlet in the EPR spectra. The primary species easily undergo intramolecular H transfer (in the case of ethyl acrylate, the transformation starts at 40 K). Further reactions depend crucially upon the matrix used. Also to be mentioned, several new studies were made on the radical cations of organometallic compounds.[61, 62]

Specific deuterium labelling was used to analyse the low-temperature dynamics of the radical cations[63-65] in halocarbon matrices as discussed in detail in a separate chapter of this book.

Recent development in the field of "Freon matrix technique" also included extensive application of this method to the photochemistry of the radical cations. Early investigations of the phototransformations of organic radical cations in solid halocarbons used mainly optical absorption spectroscopy,[66,67] whereas the application of EPR was restricted to the qualitative "photobleaching" experiments and photochemical studies of some hydrocarbon cations.[12-14] Systematic quantitative determination of the kinetic parameters of the photochemical reactions of a series of aliphatic radical cations based on combination of EPR and electronic absorption spectroscopy was made in the past decade. The optical characteristics, quantum yields of decay and reaction products were determined for the radical cations of ethers and acetals,[68-70] amides,[71] alkanes,[72] acetone[73] and acetaldehyde[74] (a mini-review was published recently[75]). In general, these studies reveal three kinds of photoreactions of the radical cations: (i) "trivial pathway", i. e., photostimulated charge transfer to matrix followed by charge recombination, (ii) deprotonation, and (iii) specific reactions. The first channel is typically characterised by high quantum yields. It occurs in different freons, but shows maximum efficiency in a polycrystalline Freon-11 matrix. The proton loss occurring with much lower quantum efficiency (except for tetrahydrofuran and dimeric radical cations) can be revealed clearly in a sulphur hexafluoride matrix with high ionisation potential, because the photoinduced charge transfer in this matrix is energetically unfavourable due to large "IP gap" (> 5.5 eV). Such a process also may occur in some other media (e. g., in Freon-113); the nature of proton acceptor is not fully clear. Specific reactions are unique for each system studied: fragmentation for alkanes,[72] intramolecular H transfer for amides and acetaldehyde,[71,74] ring cleavage for 1,3-dioxolane.[70] An unusual photochemical reaction, methyl group migration to oxygen, was found recently for the methyl *tert*-butyl ether radical cation.[55] An important feature of photochemical transformations of radical cations in halocarbons and related matrices is concerned with strong matrix effects, which can be described as matrix-controlled selection of reaction channels.[75]

In summary, the "Freon matrix technique" has played a very important role in the studies of radiation-induced radical cations, and it is still of certain potential value (some examples will be shown below). Meanwhile, this approach suffers from several significant limitations:

(a) In fact, this method is *not a true matrix isolation technique*. Indeed, the typical procedure makes use of frozen solutions. The microstructure of these samples is unknown, and the aggregation of solute molecules cannot be excluded. Furthermore, the concentrations of guest molecules used in halocarbon matrix studies are often quite high (typically, 1 mol %, or even higher), so, in general, the assumption of isolation is invalid.

(b) The assumptions of "matrix inertness" and "low disturbance" for freons are questionable. Although irreversible chemical reactions of radical cations with these matrices were not observed, the formation of strong matrix—cation complexes was detected for in a number of cases.[12]

(c) Frozen halocarbon solutions, which seem to be excellent media for the EPR studies, are not so attractive for UV/VIS absorption studies because of strong scattering (except for glassy Freon mixture matrix); they are even less suitable for IR spectroscopy due to intense absorption in the low-frequency region.

In order to overcome some of these limitations, we suggested to use matrix deposition technique for EPR and optical spectroscopic studies in halocarbon matrices.[23] However, in any case, one should look for some alternative (or complementary) approach to answer a number of basic issues.

2.2.3   Zeolites and other porous media

The idea of using inorganic sorbents as matrices for the EPR studies of paramagnetic species produced from organic molecules by ionising radiation was first tested for benzene adsorbed on silica gel some thirty-five years ago,[76] which led to identification of monomeric and dimeric benzene radical cations. However, the application of silica gel was probably limited to the studies of simple aromatic radical cations.[77,78] An important step was turning to specific cavity-type hosts, namely, synthetic zeolites[79,80] In principle, the scheme of formation of radical cations in zeolites is similar to that given above for the halocarbon matrices; the nature of electron traps in this case is not fully clear. The basic difference between the radical cations trapped in halocarbon matrices and zeolites results from the fact that the cages (trapping sites) in zeolites have regular, well-defined geometry. Both trapping and reactions of radical cations in zeolites occur in a completely rigid environment. Extensive EPR studies of structure and reactions of

10. Organic radical cations and radicals in matrices

hydrocarbon radical cations produced by γ-irradiation in zeolites[§] in the past decade were made by the Argonne group,[82] recent reviews are available.[83, 84] The most popular matrices used for these studies were zeolites of ZSM family (mainly, ZSM-5). Other hosts tested were X, Y and Beta zeolites, mordenite, silicalite, and MCM-41 molecular sieve. Also to be mentioned, the studies of dynamics of amine radical cations produced by radiolysis in zeolites have been reported recently.[85]

In general, zeolites are less suitable for specific studies of the structure of organic radical cations than halocarbons. For example, the radical cations of n-hexane and n-octane were stabilized in a ZSM-5 zeolite only at 4 K,[79] whereas smaller alkane cations were not detected at all. The same problem occurs with the radical cations of small alkenes.[84] Furthermore, an attempt to observe the EPR spectra of radical cations produced by irradiation of many organic molecules (e. g., ethers or esters) in zeolites failed, probably, due to secondary reactions.[84] The stability of radical cations and neutral radicals produced in zeolites depends strongly on the size of guest species. Small paramagnetic species often decay rapidly even at 77 K, whereas larger radicals and radical cations may be observed at 200 K or above.[82] On the other hand, large cations cannot be accommodated in the pores of ZSM-5, and their trapping requires using the zeolite hosts with larger cage size (e. g., mordenite).[83,84] These "size effects" resulting from size-dependent molecular diffusion and geometrical constraints reveal both limitations and advantages of zeolites. As to the latter, one should note that the EPR spectra of radicals in zeolites typically show better resolution than in halocarbon hosts. This result is understandable in view of larger pore size in zeolites, which allows more rotational freedom for small guest species. Geometrical constraints also play an important role in selection of reaction channels in zeolites. Thus, zeolites can be described as "microreactors" with tunable pore size, polar interactions and acidity.[83,84]

Formation of radical cations is not the only process occurring upon radiolysis of hydrocarbon molecules adsorbed in zeolites. Recent studies of the Argonne group has also demonstrated the importance of formation of H adducts for olefins, dienes and aromatic molecules.[86]

Generally speaking, zeolites are fascinating matrices for the EPR studies of structure and dynamics of some radical cations and neutral radicals. Furthermore, radiolysis of adsorbed molecules in zeolites provides

---

[§] It should be noted that radical cations and radicals in zeolites can be also produced from some organic molecules due to chemical or photochemical one-electron oxidation[81]; these processes will be not considered in the present review.

a powerful tool for designing "spin probes" of adsorption and valuable models for heterogeneous catalysis.[84] However, a wide-scale application of zeolites for basic studies of the radiation-induced processes in solids is questionable because of a lot of complications resulting from inhomogeneous adsorbate distribution, strong chemical interactions, *etc*.

2.2.4 Solid rare gas matrices

Solid rare gases are classical media for matrix isolation studies, which have been used extensively for spectroscopic characterisation of highly reactive intermediates for several decades. In particular, organic radical cations (mainly, of aromatic and conjugated systems) were widely studied in rare gas matrices by optical absorption spectroscopy.[66,67,87] A number of neutral organic radicals produced by different techniques (i. e., photolysis, pyrolysis, glow discharge, or chemical reactions) were characterised in solid argon and neon by EPR since the 1960s.[88] Meanwhile, up to recently, the application of rare gas matrices to the EPR investigations of paramagnetic species produced by the solid-state radiolysis was limited. First EPR studies of methane radiolysis in solid argon, krypton and xenon at 4.2 K reported by Bouldin and Gordy[89] used frozen solution technique ("ampule method") rather than the classical matrix isolation procedure. Several other groups applied the same technique to radiolysis of larger hydrocarbons in solid rare gases.[90] Obvious limitations of this method are concerned with aggregation of solute molecules, which is especially important for rare gas matrices since the solute—solute interactions are much stronger than the solute—matrix interactions. More recently, Qin and Trifunac reported the EPR spectrum of 1,1,2,2-tetramethylcyclopropane radical cation in a frozen xenon solution containing Freon-113 as an electron scavenger at 77 K.[91] To my knowledge, it was the only EPR observation of a large organic radical cation in solid rare gas matrices before 1996.

Rigorous matrix isolation technique was applied by Knight and co-workers for the studies of inorganic and small organic radical cations in neon matrices (see Ref. 92 for review; recent results may be found in Refs. 20, 25, 93-95). Early studies of this group used a number of different techniques for generation of radical ions *during* the matrix deposition (in-situ photoionisation, electron bombardment, pulsed laser and discharge treatment)[92]. Meanwhile, it was found later that the X-ray radiolysis of solid deposited matrices was quite effective for producing of organic radical cations in high concentrations.[20,25,94] Using neon as a matrix material made it possible to characterise the radical cations generated from small molecules with high IP (e. g., methane and methanol), which cannot be produced in halocarbon matrices. In some other cases (e. g., for acetaldehyde), the neon matrix provides a benefit of superior resolution of the EPR spectra (in comparison with freons). However the matrix isolation studies of organic radical cations in neon were restricted to a few small species (maximum, two carbon atoms). Other applications of neon and argon matrices for EPR studies of the species produced by solid-state radiolysis included investigations of small organometallic compounds.[95,96]

Somewhat surprisingly, up to recently, the radiation-induced radical cations were not characterised by EPR in solid argon, which is the most common medium for matrix isolation studies. Attempts to produce very small cations in argon were reported to be unsuccessful,[20,25] whereas this matrix was not applied for the studies of larger species.

Several years ago we started an experimental program aimed at characterisation of intermediates resulting from fast-electron irradiation of various organic molecules in solid rare gas matrices. In contrast with the work of other groups, our main interest was focused on chemical aspects rather than spectroscopic or molecular dynamics problems. The experimental technique and apparatus used for these studies are described above.

The first experiments with heptane in a xenon matrix[23,24,97] revealed that irradiation of solid deposited mixtures (mole ratio of 1:400 to 1:1000) yields nearly balanced amounts of trapped hydrogen atoms and alkyl radicals. Addition of an electron scavenger results in dramatic drop in the yield of hydrogen atoms, whereas the spectrum of neutral alkyl radicals is replaced by the spectrum of radical cation (known from previous studies in halocarbon matrices[12-14]). These observations may be rationalised in the frame of simple scheme:[24,97]

$$RG \xrightarrow{\wedge\wedge} RG^{+\cdot} + e^{-} \qquad (4)$$

$$RG^{+\cdot} + RH \rightarrow RG + RH^{+\cdot} \qquad (5)$$

$$RH^{+\cdot} + e^{-} \rightarrow RH^{*} \rightarrow R^{\cdot} + H^{\cdot} \qquad (6)$$

Here RG denotes a rare gas atom, and RH^{*} is an excited organic molecule. In the presence of an electron scavenger S, the ion—electron recombination (6) is ceased, and the radical cation is trapped in the matrix:

$$S + e^- \rightarrow S^{-\cdot} \tag{7}$$

$$RH^{+\cdot} \rightarrow RH^{+\cdot}(tr) \tag{8}$$

Formal meaning of reaction (8) is *trapping (stabilisation)* of the radical cation in matrix. It implies that the primary radical cation resulting from the positive hole transfer is in *"unrelaxed"* state (the sense of this difference will become clear from the later discussion). A very strong effect of electron scavenger clearly suggests that the main primary process is positive hole transfer rather than excitation transfer. Further studies[98-102] have shown that the scheme given above should be basically valid for different organic molecules in solid rare gases. Indeed, in the absence of an electron scavenger, the yields of trapped radicals and hydrogen atoms were nearly balanced (in the case of xenon) or, at least, comparable (for argon and krypton matrices)**. In all the cases studied, addition of electron scavengers resulted in drastic decrease in the yield of trapped hydrogen atoms (by one or two orders of magnitude). This effect may be used as criterion of electron scavenging (with freons, high efficiency was achieved at quite low scavenger concentration, typically 0.1-0.2 mol %). Using freons as electron scavengers is beneficial in view of spectroscopic reasons mentioned above. It should be noted that addition of scavenger was found to be vitally important for observation of the radical cations (unlike to the case of neon matrix studies of Knight group). Meanwhile, even in the presence of scavengers, the relative yields of trapped radical cations for some systems were low (or even zero). As the "hydrogen atom criterion" was met, the lack of radical cations should not be attributed to low efficiency of electron scavenging. Furthermore, in most cases, the composition of radicals resulting from organic molecules is changed in the presence of electron scavengers. Thus, the most reasonable explanation implies reactions of "unrelaxed" radical cations before trapping, or in competition with trapping:

$$RH^{+\cdot} \rightarrow products \tag{9}$$

In summary, matrix isolation in solid rare gases in combination with EPR detection offers wide opportunities for basic studies of the primary radiation-induced processes in solids. It is worthwhile noting that physical characteristics or the rare gas matrices (ionisation potential, polarisability, rigidity) vary over a wide range when turning from neon to argon, krypton

---

** In principle, dissociation of excited molecules may also yield the products of skeleton bond rupture. However, the probability of escape from matrix cage for heavier fragments is much lower.

10. Organic radical cations and radicals in matrices 379

and xenon, which allows one to follow the most general matrix effects. These gains may justify complex experimental procedure and relatively high cost of the equipment for matrix isolation studies.

2.3 Combination with other spectroscopic methods

Obviously, as we are concerned about paramagnetic species, EPR is a very powerful structural method and a valuable kinetic tool. However, important information on the whole radiation-induced process is missed, if EPR is used as the only spectroscopic probe. Indeed, EPR is "silent" about the diamagnetic species (both neutral and charged). In particular, it gives no information on the state of parent molecule in the matrix prior to irradiation, which may be quite important for solid-state processes. Furthermore, one cannot estimate the overall efficiency of the radiation-induced process from EPR data, so there is a chance that the observed formation of paramagnetic species represents only a minor channel.

The most widely used complementary method for the low-temperature studies of the radiation-induced species is electronic absorption (UV/VIS) spectroscopy. In the case of good absorbers, the sensitivity of this method is comparable to EPR. A combination of EPR and UV/VIS spectroscopy has been applied extensively to identification of radical cations and investigation of the photochemical reactions of the radiation-induced radicals (both neutral and charged).[26,67,75] However, it should be noted that the electronic absorption spectra of aliphatic radical cations in solid matrices typically exhibit broad featureless bands, which give no detailed structural information. Furthermore, the absorptions of neutral aliphatic radicals are often not characteristic, and the parent aliphatic molecules typically absorb only in the far UV region. Thus, being a valuable kinetic method (especially, for photochemical studies of the radiation-induced intermediates), UV/VIS spectroscopy is not very helpful for the problems outlined above.

The IR (vibrational) absorption spectroscopy offers an opportunity of getting rich structural information on different-type species, regardless of their magnetic properties. Using this method may allow one to characterise the state of parent molecule in the solid matrix in great detail, including conformation, molecule—matrix interactions and association. Often it is supposed that the main drawback of IR spectroscopy as applied to the studies of intermediate species is relatively low sensitivity of this method. However, this problem is not so crucial when using modern FTIR spectrometers, which provide high signal-to-noise ratio and fast scan speed.

In particular, we have shown that it is possible to obtain the EPR and IR spectroscopic characteristics of radiolytic intermediates using the same dose range, which makes valid direct comparison of the results.

In general, a combination of EPR and FTIR spectroscopy is an important part of our experimental strategy in the studies of the radiation-induced processes in low-temperature matrices.[23] First, we applied such an approach to the studies in halocarbon matrices.[23,38,103] This made it possible to obtain direct evidence for strong interaction between freon and solute molecules and gave indications of the IR features of the radiation-induced cations. The main problem is concerned with the meaning of "combination". In early studies[103] we just used a comparison between IR spectra obtained for deposited matrices and EPR data obtained by a conventional "ampule" (frozen solution) technique. Certainly, the validity of this approach is not evident, so later we turned to rigorous matrix isolation procedure (deposition technique) for both EPR and IR studies. In particular, this method was applied for the studies in rare gas matrices.[23] At present, the "combination" used in our laboratory implies obtaining the matrix samples in two cryostats from the same gaseous mixture with the same deposition system followed by irradiation with fast electrons from the same source. Our cryostat for FTIR studies of electron irradiated matrix samples is described in detail elsewhere.[22,23] Using FTIR spectroscopy allowed us to obtain the first direct estimate of overall efficiency of the radiation-induced transformation of guest molecules in rare gas matrices by measuring the intensity of IR absorptions of the parent molecules before and after irradiation. It was found[98,104] that the total radiation-chemical yields of consumption of organic molecules in solid argon and xenon were between 1.7 and 2.9 molecules per 100 eV of energy absorbed by the matrix. These values are quite high, especially if one notes that they probably represent the lower limits.[104] Thus, the overall "energy transfer" (i. e., positive hole and excitation transfer) from rare gas matrices to the guest molecules is very effective, even at high dilution. Also, the FTIR studies revealed significance of non-radical paths in the radiation-induced degradation of organic molecules in solid rare gases resulting in formation of stable diamagnetic products.[24,104,105] In certain cases, we tried to assign the vibrational features of aliphatic radical cations on the basis of comparison between the EPR and FTIR data,[23] however, unambiguous identification is still lacking, and the work is in progress now.

3. POSITIVE HOLE MIGRATION AND TRAPPING

The localisation of the primary radiation-induced events in solids is controlled by migration and trapping of positive hole and excitation. In fact, the significance of distant positive hole transfer in low-temperature matrices (e. g., frozen halocarbons and solid rare gases) is well illustrated by efficient formation of guest organic radical cations under matrix isolation conditions. A case of specific interest is concerned with the situation, when different traps with relatively close ionisation energies are distributed in a rigid matrix. This seems be a typical model of organic solid or polymer taking into account occurrence of different-type structural defects, chemical impurities, etc.; it can be also applied to heterogeneous systems. The questions addressed to this model are as follows: (i) what is the minimum "driving force" for a distant trap-to-trap hole transfer ? and (ii) what is the role of specific solid-state effects (i. e., the difference in conformation, molecular interactions, etc.) ? Actually the trap-to-trap hole transfer in a number of irradiated glassy systems was studied previously by optical absorption spectroscopy;[106] however, this work aimed at kinetic aspects. Meanwhile, application of EPR in halocarbon matrices makes it possible to investigate a much wider class of trap pairs and to characterise the trapped hole in detail, which is the main focus of the present consideration.

3.1 Two-trap model

An indication of the occurrence of trap-to-trap hole transfer between organic solute molecules upon radiolysis in solid halocarbons was reported by Toriyama and Okazaki.[33a] A simple quantitative model for analysis of this phenomenon was suggested recently by Werst et al.[107] It implies irradiation of the frozen halocarbon solutions containing simultaneously two kinds of dissolved organic molecules at 77 K. If the EPR spectra of the radical cations resulting from the two solutes are substantially different, it is possible to determine the relative contributions of these species from computer simulation or additive least-square analysis. The bias in final population of the two "hole traps" R_{mn} can be expressed as follows:[107]

$$R_{mn} = (S_m/S_n) \cdot (N_n/N_m)$$

Here indices *m* and *n* are related to the two solutes, $S_{m,n}$ denote the corresponding integrated intensities of the EPR signals (proportional to concentrations of the radical cations), and $N_{m,n}$ are the concentrations of parent neutral molecules. In principle, the bias in favour of the low-energy trap ($R_{mn} > 1$) may result either from different efficiency of the primary hole trapping or from trap-to-trap hole transfer. The former factor can be estimated by comparison of the radiation-chemical yields of the radical cations obtained in a usual ("single-trap") experiment. In general, the yields of radical cations in halocarbons were reported to vary by a factor of 3, as a maximum,[108] however, these values are rather close for solutes of similar chemical nature. The latter is most probably true for olefins and dienes used for the two-trap studies.[107] Anyway, the bias found for some pairs in a $CFCl_3$ matrix was so high (up to 10^3), that the dominating role of the trap-to-trap transfer was quite evident. The study of the concentration dependence made it possible to estimate the characteristic distance of hole transfer as 2 to 4 nm, which is supposed to be due to a single-step tunnelling.[107]

In addition to halocarbon matrices, the trap-to-trap hole transfer was also revealed in double loaded zeolites.[107] Meanwhile, in this case, the interpretation is complicated because of the effects of inhomogeneous adsorbate distribution and possible difference in the site energy.

3.2 "Fine tuning" effects

Following the approach outlined above, recently we have applied the two-trap model to the analysis of the trap-to-trap hole transfer between benzene derivatives in different Freon matrices.[54,109] Small difference in the gas-phase IP values (< 0.5 eV) and similarity of the chemical structure of the traps warranted similar efficiency of the primary hole transfer. The total concentration of solute molecules was kept constant (1 mol %), whereas relative concentrations of the two traps varied by a factor of 5-10.

In the case of pair benzene/toluene, the results obtained clearly reveal distant trap-to-trap transfer from benzene to toluene. Qualitatively similar result (hole transfer to a low-energy trap) was obtained for the toluene/*para*-xylene pair. Meanwhile, in the case of pair *metha*-xylene/*para*-xylene, bias in the trap population was not found ($R_{mn} \sim 1$). These data make it possible to estimate roughly the threshold value of the IP difference (ca. 0.2-0.3 eV), which results in effective trap-to-trap transfer.

The most interesting result was obtained for the pair toluene/ethyl benzene.[109] In this case, the gas-phase IP difference is quite small (ca. 0.06 eV), so it cannot provide sufficient "driving force" for the hole transfer. On the other hand, our studies revealed that ethyl benzene radical cation could be trapped in different conformations, depending on the freon matrix used[††].

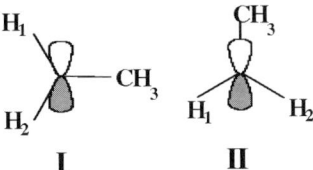

Figure 2. Conformations of ethyl benzene radical cation.

Conformer I with the hyperfine coupling constants of $a(2H) = 2.85$ mT and $a(1H) = 1.28$ mT is observed in a $CFCl_3$ matrix, in agreement with earlier data.[110] In view of well-known "$\cos^2\theta$ rule",[111] in this case, methyl group lies in the plane of phenyl ring ($\theta_1 = \theta_2 = 30°$; θ_i is the dihedral angle between the C_β--H_i bond and unpaired electron orbital axis). Conformer II observed mainly in a CF_3CCl_3 matrix (with small contribution from I). This species exhibits the hyperfine coupling constants of $a(2H) = 0.95$ mT and $a(1H) = 1.25$ mT, which implies $\theta_1 = \theta_2 = 60°$. Both conformers were observed in a $CFCl_2CF_2Cl$ matrix.[54] Note that conformer II corresponds to the most stable configuration of the parent neutral molecule because of minimum steric repulsion between methyl and phenyl groups (this intuitive argument is supported by the quantum-chemical calculations[112]). Thus, conformer II of the radical cation may result from vertical ionisation of ethyl benzene molecule, and it can be trapped, if the matrix does not allow relaxation to occur. Early INDO data predicted that the same conformer should be more stable for the radical cation,[113] so trapping of conformer I in a $CFCl_3$ matrix was attributed to a "matrix effect". However, recent DFT calculation suggest higher stability of conformer I for the radical cation.[112] If it is the case, conformer I can be described as "relaxed" conformer. Apparently, the ionisation energy for the two conformers should be different. In fact, we observed inversion in the direction of positive hole transfer for the pair toluene/ethyl benzene in different matrices. This result may imply that the ionisation energy of toluene just falls in the "gap" between the ionisation energies of the two conformers of ethyl benzene in matrix. At this stage, we refrain from quantitative conclusions since the actual ionisation energy of the conformers depends on the matrix interactions, which may be somewhat different. This observation of the conformation-controlled hole transfer appears to be the first evidence of the "fine tuning" effects in distant charge migration in a solid matrix.

[††] Strong matrix effect on the conformation of 1,4-diethylbenzene radical cation in various freons has been reported previously[53]

Table 1. Positive hole transfer between alkyl benzene molecules in frozen halocarbons[54,109]

Solute molecule pair	Matrix	Direction of transfer	ΔIP_{gas}, eV[a]	Controlling factor
Benzene/toluene (B/T)	$CFCl_2CF_2Cl$	B → T	0.43	ΔIP_{gas}
Toluene/ethyl benzene (T/EB)	$CFCl_3$	T → EB	0.06	Conformation
-"-	CF_3CCl_3	EB → T	- 0.06	Conformation
Toluene/*para*-xylene (T/p-X)	$CFCl_2CF_2Cl$	T → p-X	0.38	ΔIP_{gas}
Para-xylene/*metha*-xylene (p-X/m-X)	$CFCl_2CF_2Cl$	no	0.12	-

[a] The data were taken from Ref. 114.

The results of our studies of the trap-to-trap hole transfer are summarised in Table 1. In conclusion, our findings suggest that at relatively large IP differences (> 0.2-0.3 eV) a distant positive hole transfer should lead to charge trapping and localisation of the primary radiation-induced chemical event at the low-energy trap. In particular, this is valid for the systems doped with "hole traps" (molecules with low ionisation potential). Meanwhile, in the region of small IP gaps (typical for undoped systems with "natural" trap dispersion, *e.g.*, polymers), specific effects (conformation, formation of weak dimers, matrix interactions) may be significant. Further work should be done to determine the actual role of the "fine tuning" effects in positive hole transfer and trapping in different-type media.

4. MATRIX EFFECTS ON TRAPPING AND REACTIONS OF RADICAL CATIONS

Generally speaking, matrix environment has strong and rather complex influence on different stages of the radiation-induced chemical transformations of molecules in solids. Some of these effects are common for different-type species (*e.g.*, effects of matrix rigidity or cage geometry). Meanwhile, the primary radical cations resulting from ionisation of molecules are uniquely sensitive to the medium effects, because of strong electrostatic interactions of these species with surrounding molecules. "Matrix effects" for radical cations may imply a wide range of observations from spectroscopic effects to matrix-controlled and matrix-assisted chemistry. In the case of molecular matrices, the nature of the effects may be very complicated, so, in most cases, it is difficult to rationalise them in clear physical terms. This section will analyse mainly the basic effects of chemically rare gas matrices on trapping and properties of the radiation-induced radical cations as revealed by recent EPR studies in our laboratory.

4.1 Spectroscopic effects: rare gas matrices vs. freons

As mentioned above, up to recently, only a few organic radical cations were characterised by EPR in solid neon. However, at present, we have succeeded in obtaining the EPR spectra of a number of larger cations in argon and xenon matrices, so a wider comparison is possible.

The available data on isotropic hyperfine couplings in the radical cations obtained in solid rare gas and halocarbon matrices are given in Table 2.

Table 2. Comparison of the hyperfine coupling constants for organic radical cations trapped in solid rare gas and halocarbon matrices pair.

Radical cation	Matrix	Isotropic hfc, mT	Reference	Isotropic hfc in halocrbon matrices, mT
$CH_2O^{+\cdot}$	Neon	13.29 (2H)	115	13.97 (2H) ($CFCl_3$, Ref. 116)
$CH_3CHO^{+\cdot}$	Neon	12.88 (1H)	93	13.7 (1H) ($CFCl_3$, Ref. 93)
$CH_3OCH_3^{+\cdot}$	Argon	4.3 (6H)	98	4.3 (6H) ($CFCl_3$, Ref. 56)
$CH_3OCH_2OCH_3^{+\cdot}$	Argon	14.0 (2H); 3.43 (2H)	98, 101	13.6 (2H); 3.13 (2H) ($CFCl_3$, Ref. 117)
$THF^{+\cdot}$	Argon	8.9 (2H); 4.0 (2H)	118	8.9 (2H); 4.0 (2H) ($CFCl_3$, Ref. 57)
$CH_3COCH_3^{+\cdot}$	Argon	< 0.25 (unresolved)	118	$\leq$ 0.15 (CCl_4, Ref. 119)[a]
$C_6H_6^{+\cdot}$	Argon	0.64 (4H)[b]	99	0.82 (2H); 0.24 (4H)[c] ($CFCl_3$, Ref. 120)
$C_6H_5CH_3^{+\cdot}$	Argon	1.9 (3H); 1.3 (1H)	100	2.0 (3H); 1.25 (1H) (CF_3CCl_3, Ref. 121)
$n-C_7H_{16}^{+\cdot}$	Xenon	3.1 (2H)	23, 24	3.0 (2H) ($CFCl_2CF_2Cl$, Ref. 122)

[a] ENDOR data; [b] $^2B_{1g}$ state; [c] $^2B_{2g}$ state.

In most cases, the coupling constants in rare gas and halocarbon matrices are rather close. An interesting exception is benzene radical cation. This cation is a typical Jahn-Teller species. The two distorted states denoted as $^2B_{1g}$ (acute minimum) and $^2B_{2g}$ (obtuse minimum) show essentially different spin density distribution, so they can be easily distinguished by EPR:

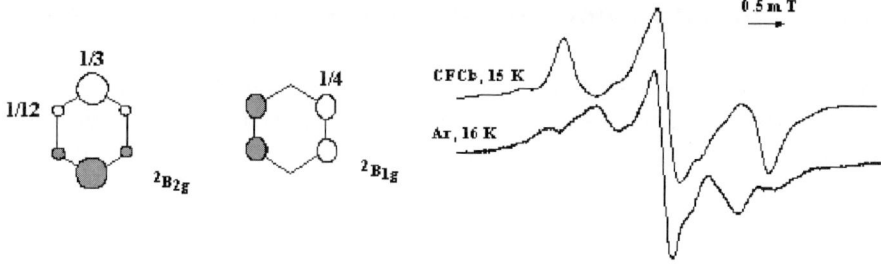

Figure 3. Spin density distribution in the Jahn-Teller states of benzene radical cation and EPR spectra observed in Freon and argon matrices (see Refs. 99,100 for detailed analysis).

Basic energy separation between these states is quite small (the $^2B_{1g}$ state lies 8 cm^{-1} lower as revealed by recent ZEKE studies[123]). Therefore, stabilisation of a specific structure in the solid state should be determined mainly by the matrix effects. Previous studies gave clear evidence for preferential stabilisation of the $^2B_{2g}$ state in a CFCl$_3$ matrix.[50,120] Meanwhile, the results of our matrix isolation studies suggest trapping of the $^2B_{1g}$ state in an argon matrix.[99] This difference can be explained by various-type matrix interactions:[100] localised asymmetrical interaction with a Freon molecule leads to stabilisation of the $^2B_{2g}$ state, whereas more symmetrical delocalised interaction in an argon matrix may favour stabilisation of the $^2B_{1g}$ state.

It should be noted that superior resolution of the EPR spectra observed for small cations in neon matrices was not found in our argon matrix studies of larger organic species. Main reason of relatively poor resolution may be anisotropic line broadening for randomly oriented species. In conclusion, from spectroscopic point of view, using solid rare gas matrices may be justified in two cases: (i) for small radical cations, which cannot be trapped in halocarbon media or (ii) for the radical cations with nearly degenerate states, which can be extremely sensitive to the matrix effects.

4.2 Matrix-assisted deprotonation of primary radical cations in xenon

As stated in section 2.2.3, in certain cases, the yields of the primary organic radical cations produced by irradiation in solid rare gas matrices in the presence of freons were found to be quite low. In particular, we failed to detect the trapped radical cations of simple ethers, acetals and acetaldehyde in a xenon matrix.[98,101,102] Instead of this, we observed large yields of the radicals resulting from specific C-H bond rupture, which corresponds formally to proton loss in the primary radical cations.

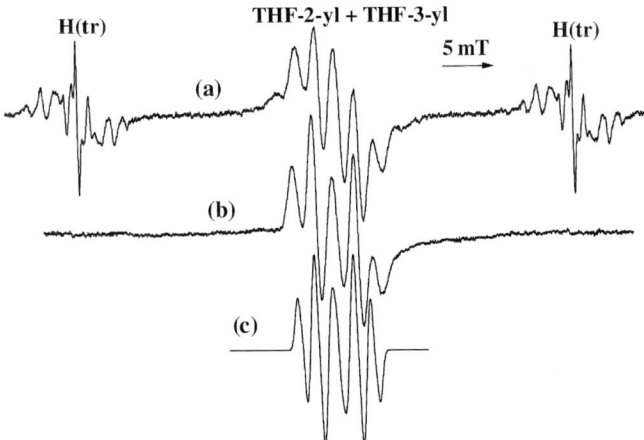

Figure 4. EPR spectra of the radicals resulting from irradiation of THF in a xenon matrix at 16 K (a) in absence of electron scavenger and (b) in presence of Freon-11; (c) simulated EPR spectrum of THF-2-yl radical (details are given elsewhere[114]).

Fig. 4 illustrates this conclusion for the case of tetrahydrofuran (THF) radical cation.[102] In the absence of Freon, THF yields a mixture of THF-2-yl and THF-3-yl radicals in balance with comparable amount of trapped hydrogen atoms. Most reasonably, these species result from dissociation of excited tetrahydrofuran molecules formed upon ion - electron recombination (reaction (6)). Meanwhile, only THF-2-yl radicals are observed in the presence of electron scavenger. Note that the formation of hydrogen atoms is strongly suppressed, which indicates effective electron scavenging (see section 2.2.3); thus, the observed radicals should result from reaction of the THF radical cation occurring before trapping (formally written as (9)).

Similar results were obtained for a number of other oxygen-containing compounds listed in Table 3. It is worthwhile noting that, in all the cases studied, we observed formation of only one specific radical for each parent cation. This means that deprotonation of the primary radical cations is a *regioselective* process (in contrast with the C-H bond cleavage in excited molecules observed in the absence of electron scavengers). The selectivity of this process correlates with spin density distribution in the radical cation, i. e., proton loss occurs at the maximum spin density position. The implication of this correlation will be discussed in more detail in section 5.1.

Table 3. Radicals observed upon irradiation of organic molecules in a xenon matrix in the presence of Freon-11 at 16 K.

Parent radical cation	Observed radical
$CH_3OCH_3^{+\cdot}$	$CH_3OCH_2^{\cdot}$
$CH_3OCH_2OCH_3^{+\cdot}$	$CH_3O\dot{C}HOCH_3$
$CH_3CHO^{+\cdot}$	$CH_3\dot{C}O$
$THF^{+\cdot}$	THF-2-yl
1,3-dioxolane$^{+\cdot}$	1,3-dioxolane-2-yl

Concerning the mechanism of proton transfer, the main problem is assignment of the proton acceptor site in a xenon matrix. At first glance, deprotonation might occur in dimers or larger molecular aggregates. However, such a possibility can be ruled out in view of the following arguments: (i) only deprotonation products were found even at high dilutions (above 1:1000), when major part of guest molecules should be in monomeric form; (ii) under similar conditions, isolated monomeric radical cations were observed in an argon matrix, and there is no reason to suggest that aggregation of organic molecules in xenon is much stronger than in argon (in fact, the opposite is probably true for some systems). Thus, the observed reaction involves essentially isolated molecules, and the matrix plays an active role. Actually, the proton affinity of a xenon atom is rather high (5.4 eV), that is, comparable to proton affinities of simple organic molecules and neutral radicals[114] This means that, even in a gas-phase approximation, direct proton transfer from some highly acidic radical cations to xenon should be only slightly endothermic process. Meanwhile, in the case of solid xenon matrix, additional stabilisation of proton results from specific collective solvation (formation of protonated xenon clusters of the Xe_nH^+ type) and long-range medium polarisation. In particular, the formation of linear centrosymmetrical cation $XeHXe^+$ in solid xenon is well documented.[124] Thus, direct deprotonation of radical cations to matrix may be justified reasonably (at least, from qualitative viewpoint). The reaction scheme for ether-type radical cations can be written as follows:

$$R_1CH_2OR_2^{+\cdot} \text{ (Xe)} \rightarrow R_1\dot{C}HOR_2 + XeHXe^+ \qquad (10)$$

Verification of this scheme could be made by direct observation of the protonated species due to its characteristic vibrational spectrum[124] (progression in a low-frequency region with the strongest band at 731 cm^{-1}). In fact, we observed such an absorption by FTIR spectroscopy in some experiments; however, up to now, we were unable to establish the correlation between formation of $XeHXe^+$ and radical products of deprotonation. For this reason, it is of value to consider other possible mechanisms.[75,101,102]

An alternative explanation can be based on recent theoretical finding,[125] which shows that relatively polarisable rare gas atoms (in particular, xenon) may facilitate intramolecular rearrangement (H transfer to oxygen atom) in methanol radical cation due to formation of transition-state complex. In this case, the role of xenon is lowering potential barrier for the reaction, so the effect may be described as *matrix catalysis*. If such a model is applicable to the ether-type radical cations (and other oxygen-containing species), the observed transformation may be represented by the scheme:

$$R_1CH_2OR_2^{+\cdot} \; (Xe) \rightarrow R_1\dot{C}H(OH^+)R_2 \qquad (11)$$

Note that the EPR spectra of the distonic radical cations resulting from reaction (11) may be indistinguishable from the spectra of the corresponding neutral radicals (deprotonation products) since the OH proton coupling should be small. Formally, reaction (11) is a hydrogen atom shift rather than proton transfer; however, in fact, the process is accompanied by substantial redistribution of positive charge. Theoretical analysis[125] shows that the catalytic effect directly correlates with the proton affinity of matrix atom, so the analogy with deprotonation is reasonable. Verification of this mechanism implies observation of the OH group in the radical product, which could come from IR spectroscopic studies or ENDOR measurements.

Finally, one may consider a combination of the two possibilities discussed above, namely matrix-assisted intermolecular proton transfer to oxygen atom of a distant molecule ("matrix pseudocatalysis"). In this case, the role of xenon matrix is providing a system of shallow traps ("conducting chain") for proton transport to a deeper trap (organic molecule).

4.3 "Hot fragmentation" in argon: effect of excess energy

Deprotonation of some primary radical cations was also observed in a krypton matrix,[98,101] however, this was not the case for argon. On the other hand, the radical cations of linear alkanes,[23] methyl *tert*-butyl ether and 1,3-dioxolane[118] were not trapped in an argon matrix, and methylal radical cations were found only in trace amounts.[101] In this case, the observed radicals result mainly from the skeleton bond fragmentation (*i. e.*, cleavage of C—C or C—O bond). Formation of these products was attributed to the effect of excess energy resulting from high exothermicity of the positive hole transfer in the case of argon matrix.[23,75,98,101] In first approximation, one can estimate this excess energy from the difference in the IP values between matrix atom and guest organic molecule (IP gap). Typical IP values for simple organic molecules are ca. 9-10 eV, so, in the case of argon, the IP gap is around 6 eV. This value definitely exceeds the energy of chemical bonds in the resulting radical cations. Taking into account inefficient dissipation of excess energy to the argon lattice, one can conclude that the "hot" fragmentation should be highly probable. In fact, an indication of such process was obtained in early studies of the radiolysis of alkanes in argon and krypton.[90c] Our recent studies make it possible to analyse the effect of excess energy on various organic radical cations, as shown in Table 4.

Table 4. Radicals resulting from fragmentation of the primary radical cations upon irradiation of organic molecules in an argon matrix in the presence of Freon-11 at 16 K.

Parent radical cation	Fragment radicals	Relative yield of fragmentation	Reference
$CH_3OCH_3^{+\cdot}$	$CH_3^{\cdot}$	Low	75, 98
$CH_3OCH_2OCH_3^{+\cdot}$	$CH_3^{\cdot}$, $CH_3O^{\cdot}$	High	101
$n-C_7H_{16}^{+\cdot}$, $n-C_5H_{12}^{+\cdot}$	$CH_3^{\cdot}$	Moderate high	23
$CH_3CHO^{+\cdot}$	$CH_3^{\cdot}$	Moderate high	118
$CH_3COCH_3^{+\cdot}$	$CH_3^{\cdot}$	Low	118
$(CH_3)COCH_3^{+\cdot}$	$CH_3^{\cdot}$	Very high	118
1,3-dioxolane$^{+\cdot}$	$ROCH_2^{\cdot}$ (?)	High	118

The efficiency of "hot" fragmentation of the radical cations in solid argon varies strongly, depending on the molecular structure of the guest species, even for similar values of the IP gap. For example, dimethyl ether, heptane and methylal all have the IP values very close to 10 eV; meanwhile, the former radical cation is mainly trapped in argon, whereas the cations produced form the two latter molecules show large yields of fragmentation products. In general, our observations may be summarised as follows:
(a) Radical cations of saturated molecules containing alkyl groups other than methyl exhibit C(1)—C(2) bond cleavage to yield methyl radicals.
(b) Radical cations of linear and cyclic acetals containing -O—CH_2—O- moiety undergo efficient fragmentation with the C—O bond cleavage, which is not the case for simple monoethers.
(c) Radical cations produced from aromatic molecules (*e.g.*, alkyl benzenes) show no "hot" fragmentation, despite the excess energy may be quite high (ca. 7 eV in the case of toluene).

These findings clearly suggest an important role of intramolecular relaxation of excess energy in the radical cations produced by highly exothermic positive hole transfer in argon. As revealed by the FTIR spectroscopic studies, the fragmentation in solid argon also leads to formation of molecular products, *e.g.*, methane and vinyl-type olefins from alkane radical cations[23,104] and CO in the case of carbonyl compounds.[105] It should be noted that the reactivity of "hot" radical cations generated in argon is quite different from the behaviour of electronically excited cations resulting from photoexcitation in halocarbon matrices.[75] It seems to be likely that indirect ionisation in argon leads to population of high vibrational levels, which cannot be reached in photoprocesses. If it is the case, using argon matrices may provide unique information on the behaviour of vibrationally excited radical cations; however, further work is necessary to verify this assumption.

In addition to fragmentation, the "high-energy" pathways realised in an argon matrix may result in rearrangement of the primary radical cations. The processes of this kind were observed in optical spectroscopic studies of hydrocarbon radical cations.[66] Recently, we obtained an indication of intramolecular rearrangement for ethyl acrylate cation in argon.[60]

4.4   "Matrix switching" between reaction channels

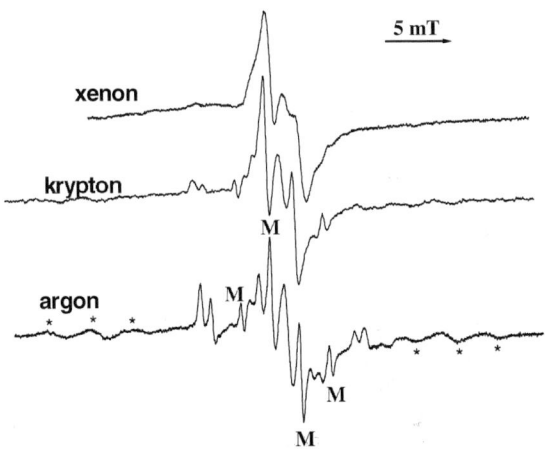

Figure 5. Matrix switching between reaction channels for methylal radical cation: radical products resulting from irradiation of methylal in rare gas matrices in the presence of $CFCl_3$ at 16 K. Symbol M shows the lines from methyl radicals; asterisks indicate the components of the outer triplets from methylal radical cations (see Ref. 101 for details).

One of the most interesting findings is concerned with observation of both reaction channels for the same species, when the yields of the primary radical cations are low for both xenon and argon matrices, but the observed products are quite different. An illustrative example of such effect of nearly complete *"matrix switching between reaction channels"* (deprotonation to fragmentation) is given by methylal radical cation.[101]

In the case of xenon, the EPR spectra show dominating contribution from the $CH_3O^\bullet CHOCH_3$ radicals (anisotropic doublet), which corresponds to selective deprotonation of the primary cation. Mainly fragmentation products ($CH_3^\bullet$ and $CH_3O^\bullet$, probably in the form of complex [$CH_3O^\bullet$ ...$^+CH_2OCH_3$]) were found in argon.[101] Both processes occur in krypton.

General consideration of the nature of matrix effects in chemically simplest and apparently inert environment should be addressed to the basic physical characteristics of the matrices used. As shown in Table 5, IP and polarisability of rare gas atoms vary over a rather wide range. While going from neon to xenon, the IP value decreases by more than 9 eV, whereas the polarisability increases roughly by an order of magnitude. In fact, the electronic characteristics of xenon are rather close to molecular matrices.

Table 5. Electronic characteristic of the rare gas atoms used as matrix materials

Rare gas atom	IP, eV (Ref. 114)	Polarisability $\cdot 10^3$, nm^3 (Ref. 126)	Proton affinity, eV (Ref. 114)
Ne	21.56	0.40	2.08
Ar	15.75	1.64	3.8
Kr	14.0	2.48	4.5
Xe	12.13	4.16	5.4

High polarisability implies relatively strong interaction with the radical cations and appreciable proton affinity (basicity) of the xenon matrix.

As discussed above, the fragmentation of the primary radical cations in argon results from high IP value of the matrix used. From this point of view, one could be rather pessimistic about obtaining high yields of radiolytically produced complex organic radical cations in neon, because the excess energy (ΔIP) in this case is too large. The fragmentation of some radical cations was also found in a krypton matrix,[98,101] but not in xenon. Thus, formally it is possible to estimate the threshold excess energy for "hot" fragmentation as 2.5 to 4 eV. Note, however, that matrix polarisability may also affect the probability of fragmentation, because energy dissipation to the matrix lattice becomes more efficient in krypton and especially in xenon.

Deprotonation of the studied radical cations becomes less important in krypton and does not occur in an "electronically rigid" (low-polarisable) argon matrix. Meanwhile, for the most "acidic" small radical cations (*e.g.*, methane or methanol), even argon probably may act as a proton acceptor. In this case, neon should be the best choice because of its extremely low polarisability and proton affinity. In summary, trapping of organic radical cations in solid rare gas matrices is probably a matter of compromise between different electronic characteristics of the matrix used.

5. SELECTIVITY OF THE PRIMARY RADIATION-INDUCED CHEMICAL EVENTS

The problem of selectivity of the primary chemical effects induced by ionising radiation in molecular materials is of primary significance from both basic and practical points of view. In general, it implies determination of the factors, which control the reaction pathways of primary ionised and excited molecules in solid media. EPR studies in low-temperature matrices are especially valuable for model purposes, because they may reveal the mode of initial chemical bond rupture in various molecular systems.

5.1 Site-selective reactivity of organic radical cations

The phenomenon of site-selective reactivity has been first clearly demonstrated by Toriyama and co-workers for deprotonation of linear alkane radical cations in halocarbon matrices, SF_6 and zeolites.[79,127,128] It was shown that ion-molecule reactions of linear alkane cations in the extended (planar zigzag) conformation resulted in selective proton loss from the chain-end position to yield terminal alkyl radicals:

$$CH_3(CH_2)_nCH_3^{+\cdot} + RH \rightarrow CH_3(CH_2)_nCH_2^{\cdot} + RH_2^+ \qquad (12)$$

Here RH denotes neutral alkane molecule. The reactions of this type occur either in pre-existing dimers[79] and larger aggregates[34] or in the complexes formed due to molecular diffusion at intermediate temperatures.[127, 128] This process was also invoked to explain preferential formation of terminal alkyl radicals in the irradiated C_{10} — C_{25} linear alkane crystals.[17] Meanwhile, in the case of *gauche*-C_2 conformers of linear alkane cations, deprotonation occurs at the C_2 position to yield penultimate radicals.[34, 127, 128] Thus, in both cases, the most "acidic" proton is an in-plane proton, which bears maximum spin density in σ-delocalised alkane radical cations. In other words, it implies a correlation between the isotropic proton hyperfine coupling constant and probability of deprotonation.

The correlation of this kind was also found for linear alkene radical cations.[39] Meanwhile, extensive studies of alkyl radical formation in halocarbon matrices containing the parent radical cation of various alkyl-substituted cyclohexanes did not reveal any direct relationship between the location of high hydrogen spin density in the cation and the site of the resulting π–type alkyl radicals.[128a] The results obtained for ether and acetal radical cations in halocarbon matrices were also not so definitive.[129,130] However, the studies in xenon matrices considered in section 4.2, clearly reveal the same trend for all the radical cations examined. Certain difference between the data obtained in halocarbon and xenon matrices can be easily understood, if one takes into account that the ion-molecule reactions observed in halocarbons may imply not only proton transfer from the radical cation to neutral molecule, but also hydrogen atom transfer in the reverse direction. In the latter case, the radical cation acts similar to any neutral radical abstracting an H atom from neutral molecule. The selectivity of such process should be determined by the dissociation energies of specific C—H bonds in a neutral molecule; in the case of small difference, the relative yields of radicals may be essentially controlled by statistical and steric factors. In particular, this may be the reason for formation of relatively large amounts of the $^{\bullet}CH_2OCH_2OCH_3$ radicals upon ion-molecule reactions of the methylal radical cation occurring at high methylal concentrations or at elevated temperatures in halocarbons.[130] As shown in the previous section, deprotonation in a xenon matrix yields only $CH_3O^{\bullet}CHOCH_3$ radicals. Thus, one may conclude that deprotonation of the radical cations controlled by *electronic factors* is, in general, much more selective than thermal hydrogen atom abstraction.

Although the correlation between spin density distribution and selectivity of deprotonation of the radical cations appears to be rather well established, its theoretical meaning was not clear. An attempt of qualitative interpretation of this correlation in terms of specific "bond weakening" was made in recent MNDO-UHF studies of a wide class of organic radical cations.[131] Actually, the values of relative bond weakening upon ionisation (in percents) calculated from the double-centered integrals[131] may be considered as specific reactivity indices. These values for the C-H bonds indeed show reasonable correlation with the corresponding isotropic proton hyperfine coupling constants. Conclusive verification of this correlation and obtaining the quantitative data requires more rigorous theoretical treatment.

One could note that "bond weakening" may also imply another possibility, namely hydrogen atom loss (dissociation of $RH^{+\bullet}$ to R^+ and $H^{\bullet}$). This process may be favourable in gas phase; however, in the case of condensed phase, deprotonation should be favoured because of large gain in polarisation energy. In fact, H atoms were never found as significant products of reactions of organic cations in halocarbons or solid rare gases.

Deprotonation is probably the most important and the most widely studied site-selective process among the reactions of organic radical cations. Nevertheless, the concept of selective bond weakening is also applicable to the skeleton bonds in the radical cations. In particular, large elongation and weakening of specific C-C bonds upon ionisation has been proved for branched alkanes[33] Experimental and theoretical evidences for the C-O bond weakening were reported for the radical cations of linear and cyclic acetals (1,1-diethers).[30-132] In conclusion, the effect of strong differentiation in the chemical bond energy upon ionisation of organic molecules should be of key significance for understanding of specific selectivity of the primary events in the radiation chemistry of molecular systems and macromolecules.

5.2 Selectivity of other primary processes

Formation of primary ionised molecules (radical cations) is a specific feature of the processes induced by high-energy radiation, so the selective reactivity of these species is of particular interest for radiation chemistry. Meanwhile, the radicals observed in molecular solids also result from dissociation of neutral excited molecules. Controversial results concerning the selectivity of the primary C-H bond rupture were reported for linear alkane crystals.[16,17,133] In general, a mixture of different-type radicals (*i.e.*, terminal, penultimate and interior alkyl radicals) is observed after irradiation of crystalline alkanes at 4.2 K.[16,17] As deprotonation of the primary cations yields selectively terminal radicals (see above), other processes of radical formation are probably not so selective. In recent studies, we tried to estimate the mode of C-H bond rupture in isolated excited heptane molecules produced in a xenon matrix.[24,134] The experiment revealed formation of a mixture of penultimate and interior radicals, the former being predominated (detailed quantitative analysis was not made). Terminal alkyl radicals are not formed, as was demonstrated in the studies of selectively deuterated heptane $CD_3(CH_2)_5CD_3$. Thus, the formation of radicals from neutral excited alkane molecules also appears to be non-random; however, the mode of C-H bond rupture is not so specific as in the case of deprotonation of the primary cations.

Little is known about the reactions of neutral excited molecules produced in the radiolysis of other simple aliphatic molecules in solid phase. As shown in section 4.2 (Fig. 5 and related discussion), dissociation of excited THF molecules in solid xenon yields a mixture of THF-2-yl and THF-3-yl radicals in roughly comparable concentrations, which is characteristic of a non-selective process.

In addition to reactions of the primary radical cations and neutral excited molecules, the composition of radicals resulting from the low-temperature solid-state radiolysis is also determined by the reactions of hydrogen atoms. Strictly speaking, it is not a primary process; however, H atoms produced upon radiolysis are not trapped in solid alkanes (except for methane) and other organic systems even at 4 K, because they abstract hydrogen atoms from molecules via tunnelling mechanism.[7] Thus, the radicals produced by the reactions of hydrogen atoms are often treated as the "primary" radiolysis products. According to the work of Ichikawa and Yoshida,[29] the selectivity of reactions of H atoms with alkane molecules in the solid phase depends strongly on the matrix physical state. In the case of crystalline linear alkanes, the abstraction yields both penultimate and interior radicals, whereas only penultimate radicals are produced selectively in glassy matrices of perdeuterated alcohols.[29a] This effect was attributed to impeded C-C-C bending motion in low-temperature glasses; similar data were reported for glassy branched alkanes.[29b]

5.3 Application to macromolecules

From practical point of view, studying the mode of radiation-induced damage in macromolecules and complex polymeric systems is one of the most important tasks of the solid-state radiation chemistry. It is also a challenging basic problem since it addresses to a number of issues, i. e., validity of local molecular models for description of the long-range effects, role of conformational defects, significance of molecular packing, etc. Numerous data on the radicals trapped in various irradiated polymers are available,[4] however, the primary distribution of the radiation-induced events is not known, even for most widely studied macromolecules. A specific problem is associated with structural and chemical inhomogeneity of real polymers, which complicates the interpretation to a great extent.

Here main focus will be made on recent results concerning the role of structural defects in the radiolysis of polyethylene (PE), chemically the simplest and practically the most important polymer. For linear polyethylene (if one disregards branching and chemical impurities), the most important type of defects is conformational defects, *i. e. gauche-trans* (GT) conformers. Although theoretical predictions for ideal macromolecules indicated possible role of the conformational defects in localisation of primary events,[135] up to recently, there was no experimental evidence for the effects of this type. It is well known that irradiation of PE at low temperatures leads to formation of interior-type alkyl radicals ~CH$_2$˙CHCH$_2$~,[4] and only the radicals resulting from *trans-trans* (TT) conformers were detected in early EPR studies at 77 K.[136] However, the observed distribution of radicals may be affected by secondary processes, namely, local radical site migration, which was found to occur below 77 K.[5-10] Also, to make definite conclusions, it is important to deal with a chemically pure, well-organised and well-characterised polymer. Taking into account these points, we have reinvestigated the initial mode of radical formation in high-density linear PE irradiated at 15 K.[137,138] In order to get an unequivocal test of the role of structural defects occurring in small concentrations, we used PE with extended chain crystals (ECC PE) obtained by high-temperature annealing of linear high-density PE under high pressure. The ECC PE samples are characterised by extremely high crystallinity degree (95-98%) and very low concentration of conformational defects. In addition, using oriented samples made it possible to obtain high-quality EPR spectra and to get more information from the angular dependence.

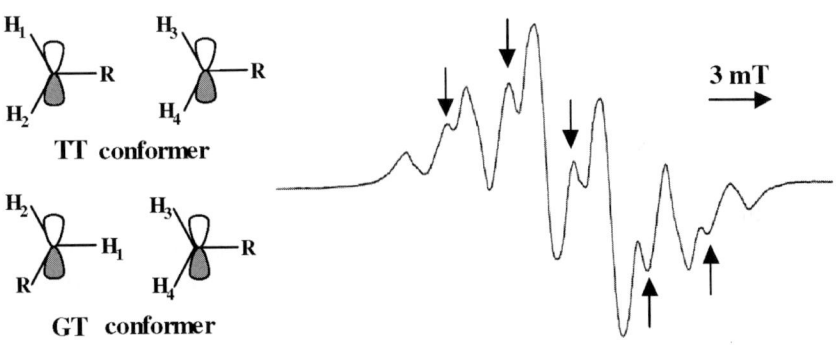

Figure 6. EPR spectrum of ECC PE irradiated at 15 K (see Refs. 137, 138 for details).

The EPR spectrum shown in Fig. 6 clearly reveals the presence of substantial amounts of radicals resulting from GT conformers (the lines marked with arrows). Indeed, when external magnetic field is applied parallel to the draw axis, the TT conformer (dihedral angles $\theta = 30^0$ for all the four β-protons) yields a nearly perfect sextet spectrum: $a(H_\alpha) \approx a(4H_\beta) \approx 3.3$ mT. An admixture of the GT conformer ($\theta_1 = 90^0$, $\theta_2 = \theta_3 = \theta_4 = 30^0$) is easily detected since in the latter case one β-proton takes a position in the nodal plane in respect to unpaired electron orbital axes, and the corresponding hyperfine coupling is lost. Detailed analysis using computer simulation[138] shows that the fraction of the radicals resulting from GT conformers is ca. 30%, despite the concentration of such conformers is very small. Thus, we have obtained the first direct experimental evidence for highly preferential localisation of the primary radiation-induced events at conformational defects in macromolecules. It is worthwhile noting that annealing of the irradiated samples at 80-90 K results in irreversible conversion of the defect-localised radicals to "normal" TT conformers, probably due to local "radical hopping".[138] This result accounts for failure to observe the GT conformers in earlier studies carried out at 77 K.

To explain the nature of conformational selectivity in PE, it is logical to turn back to selective effects observed for the prototype molecules (linear alkanes). As stated above, deprotonation of the primary alkane radical cations occurs selectively, either at terminal methyl group (for extended all-*trans* conformers) or at *gauche* position (for *gauche*-C_2 conformers). This rule remains valid for rather long molecules (at least, up to C_{25}[17]). Thus, a similar consideration may be applied to the PE macromolecules: deprotonation of the primary positive hole may occur selectively at the defect position (again, the in-plane proton is lost preferentially).

In other words, this model implies that the *conformational defects act as "effective chain-ends"* for hole delocalisation. Interestingly, the fraction of radicals localised at conformational defects in PE is very close to the fraction of terminal alkyl radicals resulting from selective deprotonation of the long-chain linear alkane cations.

Detailed information on the selectivity of early events in other polymer systems is lacking. It seems probable that the processes of long-range positive hole migration are basically important for the radiation chemistry of a wide class of polymeric systems.[134,137] In particular, in the case of polystyrene, the trap-to-trap hole migration may result in favourable localisation of the radiation-induced events at specific conformers or dimeric associates of aromatic rings.[137] In the case of microheterogeneous systems, the interphase migration of positive hole and electron may cause specific non-additive effects in the formation of radicals, especially, if the electronic properties of the components are different.[139] Some consequences of selective and long-range effects for the radiation chemistry of polymers are discussed in more detail in recent reviews.[134,137]

6. CONCLUSIONS AND OUTLOOK

In summary, EPR studies of the radiation-induced organic species in low-temperature matrices in the past decade has led to significant progress in understanding of basic mechanisms of the radiation effects in molecular solids and polymers and revealed some new trends.

The structure and reactivity of the primary ionised molecules (radical cations) remains in a focus of basic radiation chemistry and some other areas of physical chemistry. Using halocarbon matrices for the EPR characterisation of organic radical cations has made a sort of "revolution" in the field some twenty years ago. Probably, the basic resource of this method for wide-scale structural studies has been used. Further progress may be associated with the studies of some specific radical cations (e. g., of bifunctional molecules of "bridged" structure and other complex systems with nearly degenerate orbitals) and with investigation of the very nature of the positive hole transfer and trapping. On the other hand, application of matrix isolation in solid rare gases to the EPR studies of relatively large organic radical cations is just being developed. Somewhat unexpectedly, this method has proved to be especially valuable for elucidation of chemical aspects rather than for spectroscopic studies. Indeed, up to now, little new information on the EPR parameters of aliphatic cations was derived from the experiments in solid argon and xenon; meanwhile, the effects of matrix electronic characteristics on trapping and reactions of the radical cations were found to be quite strong and specific. I believe that the matrix isolation studies will bring the key to basic understanding of the matrix-assisted and matrix-controlled chemistry of ionised organic molecules in solids.

Regarding the radicals produced in complex organic systems and polymers, the most important finding is specific selectivity of the radiation-induced processes. In fact, it was clearly demonstrated that the primary chemical events induced by high-energy radiation were far more selective than it might be expected from formal energetic reasons; furthermore, the radiation damage was found to be sensitive to very "subtle" effects (e. g., molecule conformation or weak association). To a large extent, this selectivity is probably determined by long-range hole migration and specific bond weakening in the ionised molecules. There are still quite a few studies for complex systems, and more work should be done. The implications of "selective radiation chemistry" may be significant for a number of fields, including radiation modification of polymers and radiation biology.

Important new information on the structure and reactivity of the radiation-induced paramagnetic species can be obtained using a combination of EPR with other spectroscopic techniques. In particular, IR spectroscopy is useful since it yields detailed structural data concerning parent molecules, primary paramagnetic species and diamagnetic products of their reactions. The most challenging issue is probably concerned with obtaining vibrational spectra of simple aliphatic radical cations. These characteristics may provide essential information on chemical bonding in the radical cations, which should be crucial for testing the concept of selective bond weakening.

Recent development in theoretical methods made it possible to describe the geometry and magnetic resonance parameters of relatively large organic radicals with reasonably high accuracy. Meanwhile, the theoretical treatment of reactivity of radicals and, especially, radical cations is not so extensive. In view of observed correlation between spin density distribution and specific reactivity of organic radical cations, in principle, it seems possible to predict the main reaction channel from the magnetic resonance parameters of the parent cation. In a wider context, it means developing of an approach, which could allow one to predict the distribution of the reaction channels for complex ionised molecules and macromolecules on the basis of EPR data and quantum-chemical calculations made for model systems. Probably, the most difficult problem is concerned with correct theoretical description of the effect of matrix environment on the reaction profile. Different ideas, including a novel concept of "matrix catalysis", should be considered.

Generally speaking, the results presented in this chapter have to illustrate the potential of using EPR spectroscopy for elucidation of the most challenging basic problems of radical chemistry in solids (in particular, radiation chemistry). Some other aspects of the EPR studies of radiation-induced species related to radical dynamics, radiation damage in biomolecules and dosimetry are considered in other chapters of this book.

Acknowledgments *I am indebted to all my collaborators and students, who took part in our research of the radiation-induced radicals in low-temperature matrices and polymers during the past decade. I would especially acknowledge invaluable contribution of Dr. Sukhov into designing of original versions of matrix isolation technique and apparatus used for these studies. The work in our laboratories was continuously supported by the Russian Foundation for Basic Research and INTAS (projects no. 96-03-32949, 00-03-32041, IR-95-0008, IR-97-1262, and INTAS 2000-0093). Grant for talented young researchers from the Science Support Foundation* *is gratefully acknowledged. I appreciate a valuable comment by Prof. Shiotani regarding the selectivity of deprotonation.*

7. REFERENCES

1. E. E. Schneider and M. J. Day, G. Stein, *Nature* **168** (1951) 645.
2. J. R. Morton, *Chem. Rev.* **64** (1964) 453.
3. E. Henly and E. Johnson, *The Chemistry and Physics of High Energy Reactions.* University Press, Cambridge, 1969.
4. S. Ya. Pshezhetskii, A. G. Kotov, V. K. Milinchuk, V. A. Roginskii and V. I. Tupikov, *EPR of Free Radicals in Radiation Chemistry.* John Wiley & Sons, NY, 1974.
5. K. Nunome, H. Muto, K. Toriyama and M. Iwasaki. *Chem. Phys. Lett.* **39** (1976) 542.
6. M. Iwasaki, K. Toriyama, H. Muto and K. Nunome. *J. Chem. Phys.* **65** (1976) 596.
7. K. Toriyama, H. Muto, K. Nunome, M. Fukaya and M. Iwasaki. *Radiat. Phys. Chem.* **18** (1981) 1041.
8. V. I. Feldman, S. M. Borzov, F. F. Sukhov and N. A. Slovokhotova. *Khim. Fiz.* **6** (1987) 477; *Sov. J. Chem. Phys.* **6** (1990) 875 (English translation).
9. V. I. Feldman, S. M. Borzov, F. F. Sukhov and N. A. Slovokhotova. *Khim. Fiz.* **7** (1988) 781 ; *Sov. J. Chem. Phys.* **7** (1990) 1303 (English translation).
10. V. I. Feldman, S. M. Borzov, F. F. Sukhov and N. A. Slovokhotova. *Khim. Fiz.* **8** (1989) 949.
11. T. Shida and T. Kato. *Chem. Phys. Lett.* **68** (1979) 106.
12. M. C. R. Symons. *Chem. Soc. Rev.* **13** (1984) 393.
13. M. Shiotani. *Magn. Res. Rev.* **12** (1987) 333.
14. *Radical Ionic Systems. Properties in Condensed Phases.* (Ed. A. Lund and M. Shiotani). Kluwer, Dordrecht, 1991.
15. T. Gillbro and A. Lund. *Chem. Phys. Lett.* **34** (1975) 375.
16. T. Gillbro and A. Lund. *Int. J. Radiat. Phys. Chem.* **8** (1976) 625.
17. M. Iwasaki, K. Toriyama, M. Fukaya, H. Muto and K. Nunome. *J. Phys. Chem.* **89** (1985) 5278.
18. K. Toriyama and M. Iwasaki. *J. Am. Chem. Soc.* **101** (1979) 2516.
19. A. J. McKinley and J. Michl. *J. Phys. Chem.* **95** (1991) 2674.
20. L. B. Knight, Jr., K. Kerr, M. Villanueva, A. J. McKinley and D. Feller. *J. Chem. Phys.* **97** (1992) 5363.
21. S. Yamada, K. Komaguchi, M. Shiotani, N. P. Benetis and A. R. Sornes. *J. Phys. Chem. A.* **103** (1999) 4823.
22. F. F. Sukhov. Dissertation Doc. Sci. (Chem.). Karpov Institute of Physical Chemistry, Moscow, 1988.
23. V. I. Feldman. *Acta Chem. Scand.* **51** (1997) 181.
24. V. I. Feldman, F. F. Sukhov, N. S. Nekhoroshev, V. K. Ivanchenko and N. A. Shmakova. *Khim. Vys. Energ.* **32** (1998) 15; *High Energy Chem.* **32** (1998) 15 (English translation).
25. L. B. Knight, Jr., G. M. King, J. T. Petty, M. Matsushita, T. Momose and T. Shida. *J. Chem. Phys.* **103** (1995) 3377.
26. M. Ya. Mel'nikov and V. A. Smirnov. *Handbook of Photochemistry of Organic Radicals.* Begell House Inc. Publishers. NY, 1996.
27. T. Ichikawa and P. K. Ludwig. *J. Am. Chem. Soc.* **91** (1969) 1023.
28. T. Ichikawa and N. Ohta. *J. Phys. Chem.* **91** (1987) 3244.

29. T. Ichikawa and H. Yoshida. *J. Phys. Chem.* **96** (1992) (a) 7656; (b) 7661.
30. T. Shida and W. H. Hamill. *J. Chem. Phys.* **44** (1966) 2369.
31. A. Grimpson and G. A. Simpson. *J. Phys. Chem.* **72** (1968) 1776.
32. Y.-J. Liu and M. B. Huang. *Chem. Phys. Lett.* **321** (2000) 89.
33. (a) K. Toriyama and M. Okazaki. *J. Phys. Chem.* **96** (1992) 6986; (b) *Acta. Chem. Scand.* **51** (1997) 167.
34. (a) G. Luyckx and J. Ceulemans. *J. Chem. Soc., Faraday Trans.* **87** (1991) 3499; (b) D. Stienlet and J. Ceulemans. *J. Chem. Soc., Perkin Trans.* 2 (1992) 1449; (c) G. Luyckx and J. Ceulemans. *Radiat. Phys. Chem.* **41** (1993) 567; (d) D. Stienlet and J. Ceulemans. *J. Phys. Chem.* **97** (1993) 8595; (e) A. Demeyer and J. Ceulemans. *J. Phys. Chem.* **104** (2000) 4004.
35. N. Ohta, M. Shiotani and T. Ichikawa. *J. Chem. Soc. Faraday Trans.* **87** (1991) 3869.
36. L. Sjökvist, M. Shiotani and A. Lund. *Chem. Phys.* **141** (1990) 417.
37. L. A. Erikson, L. Sjökvist, S. Lunnel, M. Shiotani, M. Usui and A. Lund. *J. Am. Chem. Soc.* **115** (1993) 3244.
38. E. A. Ulyukina, V. I. Feldman, S. M. Borzov, F. F. Sukhov and N. A. Slovokhotova. *Khim. Fiz.* **8** (1990) 1053; *Sov. J. Chem. Phys.* **8** (1991) 1779 (English translation).
39. V. I. Feldman, E. A. Ulyukina, F. F. Sukhov and N. A. Slovokhotova. *Khim. Fiz.* **12** (1993) 1613.
40. I. Yu. Shchcapin, V. I. Feldman and V. N. Belevskii. *Dokl. Akad. Nauk.* **334** (1994) 338; *Doklady Chemistry* **334** (1994) 28 (English translation).
41. I. Yu. Shchapin, V. I. Feldman, V. N. Belevskii, N. A. Donskaya and N. D. Chuvylkin. (a) *Izv. Akad. Nauk. Ser. Khim.* (1994) 11; *Russ. Chem. Bull.* **43** (1994) 1 (English Translation); (b) *ibid* (1995) 212; *Russ. Chem. Bull.* **44** (1995) 203 (English Translation).
42. V. N. Belevskii and I. Yu. Shchapin. *Acta Chem Scand.* **51** (1997) 1085.
43. G-F. Chen and F. Williams. *J. Chem. Soc., Chem. Communs* (1992) 670.
44. A. Faucitano, A. Butttafava, F. Martinotti, R. Sustman and H. Korth. *J. Chem. Soc., Perkin Trans.* 2 (1992) 865.
45. F. Gerson. *Acc. Chem. Res.* **27** (1994) 63.
46. (a) T. Bally, L. Truttman, J. T. Wang and F. Williams. *J. Am. Chem. Soc.* **117** (1995) 7923; (b) *Chem. Phys. Lett.* **212** (1993) 141; (c) T. Bally, L. Truttman, S. Dai and F. Williams. *J. Am. Chem. Soc.* **117** (1995) 7916.
47. (a) I. Yu. Shchapin, V. I. Feldman, V. N. Belevskii, A. V. Khoroshutin and A. A. Bobylyova. *Radiat. Phys. Chem.* **55** (1999) 559; (b) I. Yu. Shchapin, S. I. Belopushkin, D. A. Tyrin, B. I. No, G. M. Butov and V. M. Mokhov. *Dokl. Akad. Nauk.* **372** (2000) 60.
48. K. Komaguchi, M. Shiotani and A. Lund. *Chem. Phys. Lett.* **265** (1997) 217.
49. R. Erickson, N. P. Benetis, A. Lund and M. Lindgren. *J. Phys. Chem. A.* **101** (1997) 2390.
50. R. M. Kadam, R. Erickson, K. Komaguchi, M. Shiotani and A. Lund. *Chem. Phys. Lett.* **290** (1998) 371.
51. Y. Itagaki, N. P. Benetis, R. M. Kadam and A. Lund. *Phys. Chem. Chem. Phys.* **2** (2000) 2683.
52. Y. Itagaki, A. Lund, M. Shiotani and A. Hasegawa. *Trends in Chemical Physics.* **7** (1999) 278 and references cited therein.
53. Y. Kubozono, Okada, T. Miyamoto, M. Ata, Y. Gomodo, M. Shiotani and S. Yasutake. *Spectrochimica Acta. Part A.* **48** (1992) 213.
54. A. A. Zezin and V. I. Feldman. *Dokl. Akad. Nauk.* **370** (2000) 481; *Doklady Chemistry* **370** (2000) 21 (English translation).

55. D. A. Tyurin and V. N. Belevskii. *Khim. Vys. Energ.* **35** (2001) 442; *High Energy Chem.* **35** (2001) 404. (English translation).
56. J. T. Wang and F. Williams. *J. Am. Chem. Soc.* **103** (1981) 6994.
57. H. Kubodera, T. Shida and K. Shimokoshi. *J. Phys. Chem.* **85** (1981) 2583.
58. W. Knolle, I. Yanovsky, S. Naumov and R. Mehnert. *Radiat. Phys. Chem.* **55** (1999) 625.
59. W. Knolle, I. Yanovsky, S. Naumov and R. Mehnert. *J. Chem. Soc., Perkin Trans.* 2. (1999) 2447.
60. W. Knolle, V. I. Feldman, I. Yanovsky, S. Naumov, R. Mehnert, H. Langguth, F. F. Sukhov and A. Yu. Orlov. *J. Chem. Soc., Perkin Trans.* 2. (2002) 687.
61. V. N. Belevskii, Zh. Kh. Urtaeva, S. I. Belopushkin and Yu. A. Okhlobystin. *Dokl. Akad. Nauk.* **325** (1992) 973.
62. V. N. Belevskii, S. I. Belopushkin, M. Ya. Mel'nikov and V. I. Feldman. *Acta Chem. Scand.* **52** (1998) 903.
63. K. Komaguchi and M. Shiotani. *J. Phys. Chem. A.* **101** (1997) 6983.
64. P. Wang, M. Shiotani and S. Lunell. *Chem. Phys. Lett.* **292** (1998) 110.
65. M. Shiotani, N. Isamoto, M. Hayashi, T. Fängström and S. Lunell. *J. Am. Chem. Soc.* **122** (2000) 12281.
66. T. Bally. In *Radical Ionic Systems. Properties in Condensed Phases.* (Ed. A. Lund and M. Shiotani). Kluwer, Dordrecht. (1991) 3.
67. T. Shida. *Electronic Absorption Spectra of Radical Ions.* Elsevier, Amsterdam, 1988.
68. M. Ya. Mel'nikov, D. V. Baskakov, I. A. Baranova, V. N. Belevskii and O. L. Mel'nikova. *Mendeleev Commun.* (1998) 2.
69. M. Ya. Mel'nikov, O. L. Mel'nikova, V. N. Belevskii and S. I. Belopushkin. *Khim. Vys. Energ.* **32** (1998) 39; *High Energy Chem.* **32** (1998) 34 (English translation).
70. D. V. Baskakov, I. A. Baranova, V. I. Feldman and M. Ya. Melnikov. *Dokl. Akad. Nauk.* **375** (2000) 56.
71. M. Ya. Mel'nikov, E. N. Seropegina, V. N. Belevskii, S. I. Belopushkin and O. L. Mel'nikova. *Khim. Vys. Energ.* **31** (1997) 281; *High Energy Chem.* **31** (1997) 250 (English translation).
72. D. A. Tyurin, M. Ya. Mel'nikov and V. N. Belevskii. *Khim. Vys. Energ.* **35** (2001) 266; *High Energy Chem.* **35** (2001) 236 (English translation).
73. M. Ya. Mel'nikov, E. N. Seropegina, V. N. Belevskii, S. I. Belopushkin and D. V. Baskakov. *Mendeleev Commun.* (1996) 183.
74. M. Ya. Mel'nikov, I. A. Baranova and O. L. Mel'nikova. *Dokl. Akad. Nauk.* **381** (2001) 214.
75. V. I. Feldman and M. Ya. Mel'nikov. *Khim. Vys. Energ.* **34** (2000) 279; *High Energy Chem.* **34** (2000) 236 (English translation).
76. O. Edlund, P.-O. Kinell, A. Lund and A. Shimizu. *J. Chem. Phys.* **46** (1967) 3679.
77. T. Komatsu and A. Lund. *J. Phys. Chem.* **76** (1972) 1727.
78. R. Erickson, A. Lund, M. Lindgren, A. Lund and L. Sjökvist. *Colloids Surf. A.* **72** (1993) 207.
79. K. Toriyama, K. Nunome and M. Iwasaki. *J. Am. Chem. Soc.* **109** (1987) 4496.
80. X.-Z. Qin and A. D. Trifunac. *J. Phys. Chem.* **94** (1990) 4751.
81. For review: A. M. Volodin, V. A. Bolshov and T. A. Konovalova. In *Radicals on Surfaces* (Eds. A. Lund and C. Rhodes). Kluwer, Dordrecht (1995) 201.
82. (a) M. V. Barnabas and A. D. Trifunac. *Chem. Phys. Lett.* **187** (1991) 565; (b) *ibid* **193** (1992) 1008; (c) *J. Chem. Soc., Chem. Commun.* (1993) 813; (d) M. V. Barnabas, D. W. Werst and A. D. Trifunac. *Chem. Phys. Lett.* **204** (1993) 435; *ibid* (e) **206** (1993) 21; (f) D. W. Werst, E. A. Piocos, E. E. Tartakovsky and A. D. Trifunac. *Chem. Phys. Lett.*

229 (1994) 421; (g) K. R. Cromack, D. W. Werst, M. V. Barnabas and A. D. Trifunac. *Chem. Phys. Lett.* **218** (1994) 485; (h) D. W. Werst, E. E. Tartakovsky, E. A. Piocos and A. D. Trifunac. *J. Phys. Chem.* **98** (1994) 10249; (i) E. A. Piocos, D. W. Werst, A. D. Trifunac and L. A. Eriksson **100** (1996) 8408.
83. D. W. Werst and A. D. Trifunac. *Acc. Chem. Res.* **31** (1998) 651.
84. D. W. Werst and A. D. Trifunac. *Magn. Res. Rev.* **17** (1998) 163.
85. W. Liu, P. Wang, K. Komaguchi, M.Shiotani, J. Michalik and A. Lund. *Phys. Chem. Chem. Phys.* **2** (2000) 2515.
86. D. W. Werst, P. Han, S. C. Choure, E. I. Vinokur, L. Xu and A. D. Trifunac. *J. Phys. Chem. B.* **103** (1999) 9219.
87. L. Andrews. In *Radical Ionic Systems. Properties in Condensed Phases.* (Ed. A. Lund and M. Shiotani). Kluwer, Dordrecht (1991) 55.
88. *Chemistry and Physics of Matrix Isolated Species.* (Ed. L. Andrews and M. Moskovits). Elsveier, Amsterdam, 1989.
89. W. V. Bouldin and W. Gordy. *Phys. Rev.* **135** (1964) A806.
90. (a) D. Bhattachrya and J. E. Willard. *J. Phys. Chem.* **85** (1981) 154; (b) H. Muto, K. Toriyama, K. Nunome and M. Iwasaki. *Radiat. Phys. Chem.* **19** (1982) 201; (c) K. Gotoh, T. Miyazaki, K. Fueki and K.-P. Lee. *Radiat. Phys. Chem.* **30** (1987) 89.
91. X.-Z. Qin and A. D. Trifunac. *J. Phys. Chem.* **94** (1990) 3188.
92. (a) L. B. Knight, *Jr. Acc Chem. Res.* **19** (1986) 313; (b) L. B. Knight, Jr. In *Radical Ionic Systems. Properties in Condensed Phases.* (Ed. A. Lund and M. Shiotani). Kluwer, Dordrecht (1991) 73.
93. L. B. Knight, Jr., B. W. Gregory, S. T. Cobranchi, F. Williams and X.-Z. Qin. *J. Am. Chem. Soc.* **110** (1988) 327.
94. L. B. Knight, Jr., B. W. Gregory, D. W. Hill, C. A. Arrington, T. Momose and T. Shida. *J. Chem. Phys.* **94** (1991) 67.
95. (a) L. B. Knight, Jr., A. J. McKinley, R. M. Babb, d. W. Hill and M. Morse. *J. Chem. Phys.* **99** (1993) 7376; (b) L. B. Knight, Jr., G. C. Jones, G. M. King, R. M. Babb and A. J. McKinley. *J. Chem. Phys.* **103** (1995) 497.
96. E. Karakyriakos, J. R. Davis, C. J. Wilson, S. A. Yates, A. J. McKinley, L. B. Knight, Jr., R. Babb and D. J. Tyler. *J. Chem. Phys.* **110** (1999) 3398.
97. V. I. Feldman, F. F. Sukhov and A. Yu. Orlov. *Chem. Phys. Lett.* **280** (1997) 507.
98. V. I. Feldman. *Radiat. Phys. Chem.* **55** (1999) 565.
99. V. I. Feldman, F. F. Sukhov and A. Yu. Orlov. *Chem. Phys. Lett.* **300** (1999) 703.
100. V. I. Feldman, F. F. Sukhov, A. Yu. Orlov, R. Kadam, Y. Itagaki and A. Lund. *Phys. Chem. Chem. Phys.* **2** (2000) 29.
101. V. I. Feldman, F. F. Sukhov, A. Yu. Orlov and N. A. Shmakova. *J. Phys. Chem. A.* **104** (2000) 3792.
102. V. I. Feldman, F. F. Sukhov, A. Yu. Orlov and N. A. Shmakova. *Khim. Vys. Energ.* **35** (2001) 352; *High Energy Chem.* **35** (2001) 319 (English translation).
103. V. I. Feldman, S. M. Borzov, F. F. Sukhov and N. A. Slovokhtova. *Khim. Fiz.* **5** (1986) 510; *Sov. J. Chem. Phys.* **5** (1990) 788 (English translation).
104. V. I. Feldman, F. F. Sukhov, N. A. Slovokhotova and V. P. Bazov. *Radiat. Phys. Chem.* **48** (1996) 261.
105. V. I. Feldman, F. F. Sukhov and A. Yu. Orlov, unpublished result.
106. A. Kira, Y. Nosaka and M. Imamura. *J. Phys. Chem.* **84** (1980) 1882.
107. D. W. Werst, P. Han and A. D. Trifunac. *Chem. Phys. Lett.* **269** (1997) 333.
108. V. N. Belevskii. Disseratation Doc. Sci. (Chem.). Moscow State University, 1990.
109. A. V. Egorov, A. A. Zezin and V. I. Feldman. *Khim. Vys. Energ.* **35** (2001) 437; *High Energy Chem.* **35** (2001) 399 (English translation).

110. D. N. Ramakrishna Rao, H. Chandra and M. C. R. Symons. *J. Chem. Soc., Perkin Trans. 2.* (1984) 1201.
111. C. Heller and H. M. McConnell. *J. Chem. Phys.* **32** (1960) 1535.
112. M. Ya. Mel'nikov, K. I. Marushkevich, I. A. Baranova, O. L. Mel'nikova and D. A. Tyurin. *Khim. Vys. Energ.*, submitted.
113. H. Chandra, M. C. R. Symons and A. Hasegawa. *J. Chem. Soc., Faraday Trans.1.* **83** (1987) 759.
114. *Bond Energies. Ionisation Potentials and Electron Affinities.* (Ed. V. N. Kondrat'ev). Moscow, Nauka, 1974.
115. L. B. Knight, Jr. and J. Steadman. *J. Chem. Phys.* **80** (1984) 1018.
116. C. J. Rhodes and M. C. R. Symons. *J. Chem. Soc., Faraday Trans. 1.* **84** (1988) 4501.
117. L. D. Snow, J. T. Wang and F. Williams. *J. Am. Chem. Soc.* **104** (1982) 2062.
118. V. I. Feldman, F. F. Sukhov, A. Yu. Orlov and I. V. Tyulpina. *Phys. Chem. Chem. Phys.* To be submitted.
119. P. J. Boon, L. Harris, M. T. Olm, L. Wyatt and M. C. R. Symons. *Chem. Phys. Lett.* **106** (1984) 408.
120. M. Iwasaki, K. Toriyama and H. Nunome. *J. Chem. Soc., Chem. Commun.* (1983) 320.
121. M. Tabata and A. Lund. *Z. Naturforsch., A.* **38** (1983) 428.
122. K. Toriyama, K. Nunome and M. Iwasaki. *J. Phys. Chem.* **85** (1981) 2149.
123. R. Linder, K. Muller-Dethlefs, E. Wedum, K. Haber and E. R. Grant. *Science.* **271** (1996) 1698.
124. (a) A. A. Karatun, F. F. Sukhov and N. A. Slovokhotova. *Khim. Vys. Energ.* **15** (1981) 471; (b) H. Kunttu, J. Seetula, M. Räsänen and A. Apkarian. *J. Chem. Phys.* **96** (1992) 5630.
125. T. D. Fridgen and J. M. Parnis. *Int. J. Mass Spectrom. Ion Processes.* **190-191** (1999) 181.
126. L. Pauling. *General Chemistry.* W. H. Freeman & Co., San-Francisko, 1970.
127. K. Toriyama, K. Nunome and M. Iwasaki. *J. Chem. Phys.* **77** (1982) 5891.
128. K. Toriyama, K. Nunome and M. Iwasaki. *J. Phys. Chem.* **90** (1986) 6836.
128a.M. Shiotani, M. Lindgren, F. Takahashi and T. Ichikawa, *Chem. Phys. Lett.* **170** (1990) 201.
129. F. Williams and X.-Z. Qin. *Radiat. Phys. Chem.* **32** (1988) 299.
130. I. A. Baranova, V. N. Belevskii, S. I. Belopushkin and V. I. Feldman. *Khim. Vys. Energ.* **25** (1991) 536; *Hugh Energy Chem.* **25** (1991) 450 (English translation).
131. (a) V. N. Belevskii, S. I. Belopushkin and N. D. Chuvylkin. *Khim. Vys. Energ.* **32** (1998) 202; *High Energy Chem.* **32** (1998) 171 (English translation); (b) V. N. Belevskii, D. A. Tyurin and N. D. Chuvylkin. *ibid.* **32** (1998) 424; *High Energy Chem.* **32** (1998) 381 (English translation).
132. I. A. Baranova, V. I. Feldman and V. N. Belevskii. *J. Radioanal. Nucl. Chem., Lett.* **126** (1988) 39.
133. W. J. Chappas and J. Silverman. *Radiat. Phys. Chem.* **16** (437) 1980.
134. V. I. Feldman. *Vestn. Mosk. Univ. Ser. 2.: Khimiya.* **42** (2001) 194.
135. R. H. Partridge. In *The Radiation Chemistry of Macromolecules*, vol. 1 (Ed. M. Dole). Academic Press, NY, 1972.
136. S. Shimada, H. Kashiwabara, J. Sohma and S. Nara. *Jap. J. Appl. Phys.* **8** (1969) 145.
137. V. I. Feldman, F. F. Sukhov and N. A. Slovokhotova. *Vysokomolek. soedin.* B. **36** (1994) 519; *Polym. Sci.* B. **36** (1994) 420 (English translation).
138. V. I. Feldman. *Appl. Radiat. Isot.* **47** (1996) 1497.
139. A. A. Zezin and V. I. Feldman. *Radiat. Phys. Chem.* **63** (2002) 75.

Chapter 11

MOLECULE-BASED EXCHANGE-COUPLED HIGH-SPIN CLUSTERS

Conventional, high-field/high-frequency and pulse-based electron spin resonance of molecule-based magnetically coupled systems

Takeji Takui, Hideto Matsuoka, Kou Furukawa, Shigeaki Nakazawa, Kazunobu Sato, Daisuke Shiomi
Departments of Chemistry and Materials Science, Graduate School of Science, Osaka City University, Osaka 558-8585, Japan

Key words: molecule-based exchange-coupled systems, molecular high-spin clusters, low-dimensional molecule-based magnetic materials, high-field/high-frequency ESR, 2D electron spin transient nutation spectroscopy, pulsed ESR, fine-structure ESR, off-principal-axis extra lines, hybrid eigenfield approach, perturbation approach, spin quantum tunneling, long-time-tail transverse relaxation, biological high-spin clusters.

Abstract: Syntheses and magnetic functionalities of exchange-coupled magnetic systems in a controlled fashion of molecular basis have been the focus of the current topics in chemistry and materials science; particularly extremely large spins in molecular frames and molecular high-spin clusters have attracted much attention among the diverse topics of molecule-based magnetics and high spin chemistry. Magnetic characterizations of molecule-based exchange-coupled high-spin clusters are described in terms of conventional as well as high-field/high-frequency ESR spectroscopy. Off-principal-axis extra lines as a salient feature of fine structure ESR spectroscopy in non-oriented media are emphasized in the spectral analyses. Pulse-ESR-based two-dimensional electron spin transient nutation spectroscopy applied to molecular high-spin clusters is also dealt with, briefly. Solution-phase fine-structure ESR spectroscopy is reviewed in terms of molecular magnetics. In addition to finite molecular high-spin clusters, salient features of molecule-based low-dimensional magnetic materials are dealt with. Throughout the chapter, electron spin resonance for high-spin systems is treated in a general manner in terms of theory. Hybrid eigenfield method is formulated in terms of direct products, and is described as a powerful and facile approach to the exact numerical calculation of resonance fields and transition probabilities for molecular high spin systems. Exact analytical expressions for resonance fields of high spin systems in their principal orientations are for the first time given.

1. INTRODUCTION

Magnetic-ion-based exchange-coupled systems and their fine-structure ESR spectral analyses have been long-standing issues in electron magnetic resonance spectroscopy. Many standard or advanced ESR textbooks have been published, which have dealt with the issues, more or less.[1.1] Among these, the most comprehensive one devoted to the issues appeared in 1987.[1.1f] Recently, syntheses of exchange-coupled magnetic systems in a controlled fashion of molecular basis have been the focus of the current topics in chemistry and materials sciences; particularly extremely high spins in molecular frames and high-spin molecular clusters, or large nonvanishing angular momentum systems have attracted much attention among the diverse topics of molecule-based magnetics and high spin chemistry.[1.2] This is partly due to the potential applications of the quantum nature of spins and orbital angular momenta controlled in well-designed molecular frames, emphasizing molecular designs for such as spin-mediated memory devices, spin magnetization oscillations, single-spin (extremely large S) detection and its dynamics, and the utilizations of dynamic (transverse) phase transitions of spin magnetizations; these issues are termed molecular spinics in future science and technology. Organic molecular systems give exceptional diversity as subjects of novel quantum magnetic phenomena or functionalities. Organic super high-spin magnetics utilizing through-bond approaches is closely related to conceptual advances in magnetics which underlie novel molecular functionality devices such as genuine liquid-phase magnets and magnetic spin quantum well effects. In this context, spin dynamics for superparamagnets or extremely large molecular spins is expected to develop.

Also, from the theoretical side, molecule-based magnetics underlain by high spin chemistry is an important testing ground for a variety of theoretical models, whether they are established or not. In favorable conditions, electron magnetic resonance of molecular magnetic materials plays a crucial role to understand novel aspects of their magnetic property. For example, organic ferrimagnetics demonstrate a breakdown of classical and conventional pictures for ferrimagnetics.[1.3] Experimentally direct detection of molecular systems with large fine-structure constants due to spin-orbit couplings has been challenged in the recent progress with high-field/high-frequency ESR

11. Exchange-coupled high-spin clusters

spectroscopy. Findings of the quantum tunneling of spin magnetization from exchange-coupled transition metal ion clusters have affected research trends in interdisciplinary fields between chemistry and physics, emphasizing general interests in single-molecule magnets (SMM's) exhibiting stepwise magnetic hystereses of the magnetization at low temperatures. Also, magnetic-ion clustering systems of biological importance have been elucidated, attracting general attention.

In a magnetically ordered molecular substance, the macroscopic magnetization as cooperative property is described in terms of dynamics of microscopic details (molecular spins). There are a variety of magnetic excitations, where the excitation does not remain localized at a given spin site but propagates in the form of coherent waves. The collectivized coherency (collective excitation) typically originates in quantum-mechanical exchange interactions, forming the simplest type of magnetic excitation as spin waves or magnons. The exchange interaction is of short range. There is another type of spin waves, which is driven by an electron-spin dipolar interaction, i.e., a Walker mode. In terms of spin carriers, all documented spin waves are not molecule-based, but atom-based. These conventional spin waves are standing waves along the direction perpendicular to the thin film. Recently, novel types of quantum modes for standing spin waves have appeared which arise from artificial superlattices of ferromagnetic thin films on a micron size. One of the novel quantum modes is a lateral Walker's mode, and another is a dipolar spin wave from gigantic magnetic moments of one- or two-dimensionally arrayed ferromagnetic islands. Experimental identifications of the quantum modes have been carried out by electron spin (ferromagnetic) resonance spectroscopy.[1,4] Those artificial superlattices of ferromagnetic microstructures are an intermediate substance between bulk magnetic materials and molecular magnetic clusters with extremely high spins. In this context, magnetic resonance phenomena on a semi-microscopic scale are new applications of electron spin resonance spectroscopy. The spin dynamics of both magnetic substance on a semi-microscopic scale and molecular superparamagnets with internal spin (polar) structures are one of the current issues in molecule-based magnetics and spin science.

This chapter surveys the recent progress in fine-structure ESR spectroscopy of exchange-coupled molecular systems with finite spin multiplicities, both experimentally and theoretically, focusing on the documentation appeared recently after the book by Bencini and Gatteschi.[1,1f] The molecular systems include mainly inorganic ones, and genuinely organic high-spin clusters of chemical and spectroscopic importance are dealt with as well. Throughout the present chapter, molecular high-spin clusters mean intermolecularly exchange-coupled spin systems in chemistry terms. Thus, high-spin oligopolynitrenes or high-spin hydro-carbons such as oligopolycarbenes[1.5] or cyclopentanediyl-based hydro-carbons[1.5] with large high-spin multiplicities, whose high spins arise from the topological symmetry of their π-conjugation network of chemical bonding, are not dealt with in the present chapter except otherwise in theoretical terms. They are grouped into intramolecularly exchange-coupled systems and are dealt with in the chapter by Baumgarten. In terms of effective spin-Hamiltonian approach, both inter- and intramolecularly exchange-coupled systems can be treated on the same theoretical background, but in microscopic terms both systems show remarkably different magnetic properties and they require intrinsic molecular designs. In harmony with the book title, the issues treated in this chapter are associated with solid states. Nevertheless, in terms of spin dynamics in high-spin ESR spectroscopy and future technology, solution ESR spectroscopy for stable high-spin molecular systems with hyperfine interactions of comparable order to fine structure ones is of particular importance. The chapter spares pages for this issue.

Referred to electron magnetic resonance in ordered regime such as low dimensional magnetic systems, superparamagnetic ones and spin frustration ones, readers are recommended to consult recent reviews and monographs.[1.1f, 1.2, 1.6] Magnetic properties of magnetic materials such as low dimensional magnetic systems appearing on a macro- or semi-macroscopic scale are characterized by invoking both magnetic resonance and magnetic susceptibility measurements. Approaches in terms of both micro- and macroscopic magnetic measurements are complementary for molecular exchange-coupled systems. Methodological establishments for molecular systems with nonvanishing and sizable orbital angular momenta and their assemblages are one of the current topics in the field of molecule-based magnetics and related fields.[1.2a, 1.2f, 1.7-8] Electron spin resonance spectroscopy applied to such molecular magnetic systems and their exchange-coupled assemblages in the crystal is immature, where analyses include magnetic interactions between orbital angular momenta. Treatments are beyond conventional Heisenberg-Dirac types exchange couplings.

Electron-nuclear multiple resonance spectroscopy devoted to molecule-based high-spin systems is not included here, although it gives crucially important microscopic details such as spin density distributions, direct determination of spin sublevels involved in the ESR transitions, and thus signs of fine structure constants.[1.9] Electron-nuclear multiple resonance spectroscopy in solid-state oriented media can afford direct evidence of inverted large negative molecular spins anticipated for antiferromagnetically exchange-coupled hetero-spin systems.[1.9] Applications of the multiple resonance technique to molecular high-spin systems in solid states are still premature.[1,10]

2. THEORETICAL BACKGROUND

The most striking feature of electron magnetic resonance phenomena due to molecule-based magnetically coupled clusters is that a variety of high spin states arising from the strength of exchange couplings give rise to a diversity of anisotropic ESR spectra. In this context, single-crystal fine-structure ESR spectroscopy is apparently the most powerful method for giving us anisotropic information on various magnetic tensors, which can be related to crystallographic structural data and electronic spin structures of molecular high-spin clusters. Nevertheless, single-crystal work is not always feasible simply because well-defined and magnetically diluted molecular systems are not available for most cases. In order to characterize magnetic properties of molecules or molecular clusters themselves, magnetically diluted molecular systems are required to suppress intermolecular exchange interactions, yielding anisotropic information on microscopic details with high accuracy. Thus, fine-structure ESR spectroscopy in non-oriented media is particularly important from the experimental side. Magnetically concentrated high-spin molecular clusters (molecule-based multinuclear high-spin clusters) are intriguing targets for fine-structure ESR spectroscopy in view of molecularly controlled exchange couplings, intramolecular or intermolecular. In most magnetically concentrated high-spin nuclear clusters, hyperfine structures are smeared out in the ESR spectra. ESR spectroscopy gives unique microscopic information on their electronic and molecular structures with the help of their bulk magnetic properties based on magnetic susceptibility.

Generally, the fine structure spectroscopy in non-oriented media requires spectral simulations to acquire spin Hamiltonian parameters with high accuracy from observed fine-structure spectra for most cases otherwise except special cases with $S = 1$. Then, facile and easy-to-access interpretive approaches for high-spin identification in the ESR spectroscopy are useful. This chapter aims to give those approaches to readers who do not specialize in high-spin ESR spectroscopy and which carry them to extract spin Hamiltonian parameters from observed fine-structure spectra. Under favorable conditions, the analytical exact expressions for resonance fields described in this chapter help to extract the parameters without spectral simulations.

It should be noted that spectral simulations with the help of theoretical considerations for the electronic molecular structures of systems under study can afford us much information on magnetic properties of both high-spin molecules and molecular high-spin clusters. The theory-based or theoretically oriented spectral simulations are applicable to intramolecularly exchange-coupled high-spin systems under some restrictions, e.g., high-spin oligonitrenes whose high spin alignments are governed by the topological symmetry of π-spin polarizations.[1.5] In high-spin nitrene chemistry, there has been serious controversial issues between the documented fine-structure constants and quantum-chemistry-based spin structures. Those apparently puzzling issues have been disclosed by theory-based simulations.[1.5] Such a theoretical approach has been underlain by tensor-based analyses for the fine-structure spin-spin interaction, exemplifying a novel organic-radical-based molecule in Section 2.3. The approach is applicable to metal-based dinuclear high-spin clusters when spin sites are apart in the range of 0.3 to 0.8 nm and a classical magnetic dipole-dipole interaction is operative.

In terms of anisotropic ESR spectroscopy, another striking feature is the appearance of off-principal-axis lines, which correspond to the stationary points with the static magnetic field B_0 along the off-principal-axis of the fine structure tensor. Off-principal-axis absorption peaks in ESR spectroscopy are called extra lines (or off-axis extra lines). The occurrence of extra lines is inherent in fine-structure ESR spectra due to high spin molecular systems with $S > 1$. Referred to triplet-states, an extra line appears only in the region for forbidden transitions[1.10b, 1.11]; the extra resonance line in fine-structure ESR spectra from random orientation has been called B_{min}. The occurrence of extra lines complicates fine-structure spectra from molecular high-spin systems, but correct identifications give a rationale for experimentally determined sets of spin Hamiltonian parameters with high accuracy.[1.10b, 1.11]

Molecule-based infinite systems of exchange couplings feature low dimensionality of the systems. A long-time transverse relaxation occurring in the magnetic assemblages with extremely high purity depends on the dimensionality of the systems under study. This issue will be treated in a later section of this chapter.

2.1 Effective spin Hamiltonian approach to exchange-coupled systems; Tensorial analyses underlying theoretical spectral simulations

Spectral and theoretical analyses in fine-structure ESR spectroscopy for molecular exchange-coupled high-spin clusters require the following two-site spin Hamiltonian to start,

11. Exchange-coupled high-spin clusters

$$\hat{H}^{spin} = \hat{H}_A^{spin} + \hat{H}_B^{spin} + (-2J_{AB})\mathbf{S}_A \cdot \mathbf{S}_B + \mathbf{S}_A \cdot \mathbf{D}_{AB} \cdot \mathbf{S}_B + \mathbf{d}_{AB} \cdot \mathbf{S}_A \times \mathbf{S}_B \quad (1)$$

$$\hat{H}_\alpha^{spin} = \beta \mathbf{B}_0 \cdot \mathbf{g}_\alpha \cdot \mathbf{S}_\alpha + \mathbf{S}_\alpha \cdot \mathbf{D}_\alpha \cdot \mathbf{S}_\alpha + \sum_k \mathbf{I}^k \cdot \mathbf{A}_\alpha^k \cdot \mathbf{S}_\alpha \quad (\alpha = A, B) \quad (2)$$

where we simply assume an isotropic exchange interaction between spins, $\mathbf{S}_A$ and $\mathbf{S}_B$, and also we neglect nuclear Zeeman terms, nuclear quadrupolar interactions and group-theoretically allowed quartic or higher-order fine-structure terms of even numbers such as $B_0 S_m^3$, $S_m^2 S_n^2$, and $S_m^3 S_n^3$ (n, m = x, y, z). The term of BS_m^3 is required for some cases with high symmetry such as distorted tetrahedral or octahedral symmetry and high spin multiplicity. In the strong exchange-coupling limit, the isotropic exchange interaction term, $(-2J_{AB})\mathbf{S}_A \cdot \mathbf{S}_B$ dominates and the other terms are considered to be perturbation ones. The z-component of $\mathbf{S} = \mathbf{S}_A + \mathbf{S}_B$ and $\mathbf{S}^2$ commute with $(-2J_{AB})\mathbf{S}_A \cdot \mathbf{S}_B$, giving the common eigenstate belonging to $(-2J_{AB})\mathbf{S}_A \cdot \mathbf{S}_B$ and to $\mathbf{S}^2$ and S_z. The resultant spin quantum number S is given as $|S_A - S_B| \le S \le S_A + S_B$, and the corresponding eigenenergy, $E(S)$ is as follows;

$$E(S) = -J_{AB}[S(S+1) - S_A(S_A+1) - S_B(S_B+1)] \quad (3)$$

with $|S_A - S_B| \le S \le S_A + S_B$.

In the strong exchange-coupling limit, no ESR transition occurs between different spin multiplicity states, giving superimposed fine-structure spectra with weights of Boltzmann distributions at a given temperature. It is to be noted that particular relationships of resonance fields and transition probabilities between the different spin states hold when superimposed fine-structure spectra appear from the same molecular magnetic origins. For example, the transition probabilities inherent in the quintet and triplet states arising from an $S_A = S_B = 1$ coupled system with an equivalent $\mathbf{D}_\alpha$, assuming $J \gg |D_S|$, strongly depend on the ratio of $|D_S|/h\nu$, where D_S stands for the fine structure constant for the quintet or triplet state and $h\nu$ is the microwave transition energy. For the experimental condition of the ratio much smaller than unity, the transition probability for the $|S=1; M_S=\pm1\rangle \Leftrightarrow |S=1; M_S=0\rangle$ allowed transitions becomes, to first order, one half of that for the $|S=2; M_S=\pm2\rangle \Leftrightarrow |S=2; M_S=\pm1\rangle$ transitions. This is one of the reasons why a thermally accessible triplet state originating in intramolecularly interacting triplet-triplet systems has been difficult to identify experimentally in non-oriented media such as organic glasses. In general, high-field/high-frequency resonance conditions are most favorable for identifying experimentally the different spin states from the same magnetic origin. The fact that the resulting resonance fields and transition probabilities from the different spin states in exchange-coupled systems are interrelated has been overlooked in the documentation so far except for a few cases.[1,12] The interrelation is crucially important for comprehensive analyses of the electronic structures of molecular exchange-coupled systems based on their fine-structure spectra.

Fine structure spectra due to any spin state with a well-defined spin quantum number S whose energy is given by Eq.(3) are described by the resultant spin Hamiltonian, $\hat{H}_S^{spin}$ as follows;

$$\hat{H}_S^{spin} = \beta \mathbf{B}_0 \cdot \mathbf{g}_S \cdot \mathbf{S} + \mathbf{S} \cdot \mathbf{D}_S \cdot \mathbf{S} + \sum_k \mathbf{I}^k \cdot \mathbf{A}_S^k \cdot \mathbf{S} \quad (4)$$

where the similar neglect of tensor terms as in Eq.(2) is made. Intrinsic relationships between a set of spin Hamiltonian parameters, $\mathbf{g}_S$, $\mathbf{D}_S$, $\mathbf{A}_S^k$ and those of the A and B components, $\mathbf{g}_\alpha$, $\mathbf{D}_\alpha$, $\mathbf{A}_\alpha^k$ (α = A, B) are acquired by invoking Wigner-Eckart theorem with irreducible tensor operators and using Clebsch-Gordon coefficients as follows;[1.1f, 1.2f, 1.13]

$$\hat{H}_S^{spin} = \beta [\, \mathbf{B}_0 \cdot (1/2)(\mathbf{g}_+ + c\mathbf{g}_-) \cdot \mathbf{S}\,] + \mathbf{S} \cdot (1/2)[\,(1 - c_+)\,\mathbf{D}_{AB} + c_+\mathbf{D}_+ + c\mathbf{D}_-\,] \cdot \mathbf{S}$$
$$+ \sum_k \mathbf{I}^k \cdot (1/2)[\,(1 + c)\,\mathbf{A}_A^k + (1 - c)\,\mathbf{A}_B^k\,] \cdot \mathbf{S}$$
$$- J_{AB}\,[\,S(S + 1) - S_A(S_A + 1) - S_B(S_B + 1)\,]\,, \quad (5)$$

where the term of $\mathbf{d}_{AB} \cdot \mathbf{S}_A \times \mathbf{S}_B$ is neglected. A set of the coefficients, c, c_+, and c_-, are expressed as functions of S, S_A, and S_B only. Formulas for the functions are given in Table 1.[1.1f, 1.2f] The relationships between the resultant tensors in the resultant spin Hamiltonian and those in the component spin Hamiltonians are given by

$$\begin{array}{ll} \mathbf{g}_S = c_1 \mathbf{g}_A + c_2 \mathbf{g}_B & (\mathbf{g}_\pm = \mathbf{g}_A \pm \mathbf{g}_B) \quad (6a) \\ \mathbf{D}_S = d_1 \mathbf{D}_A + d_2 \mathbf{D}_B + d_{12}\mathbf{D}_{AB} & (\mathbf{D}_\pm = \mathbf{D}_A \pm \mathbf{D}_B) \quad (6b) \\ \mathbf{A}_S^k = c_1 \mathbf{A}_A^k + c_2 \mathbf{A}_B^k & \quad (6c) \\ c_1 = (1 + c)/2, \quad c_2 = (1 - c)/2 & \\ d_1 = (c_+ + c_-)/2, \quad d_2 = (c_+ - c_-)/2, \quad d_{12} = (1 - c_+)/2 & \end{array}$$

All the relationships above are given in terms of tensor-based expressions in their own coordinate systems. There have been seen many documentations which commit errors in the expressions and the understanding of the relationships. Thus, it is worthwhile noting that unitary transformation procedures are required for theoretical calculations of the resulting tensors from the component tensors.[1.12] For example, a given component tensor $\mathbf{D}_i$ in a convenient reference frame (e. g., a common coordinate-axis system such as molecular principal-axis system) is given by

$$\mathbf{D}_i = {}^t\mathbf{U}_i \cdot \mathbf{D}_i^d \cdot \mathbf{U}_i \quad (i = A, B, AB), \quad (7)$$

11. Exchange-coupled high-spin clusters

Table 1. Formulas for the coefficients, c, c_+, and c_- for a given set of S_A and S_B.[a]

$$c = \{S_A(S_A + 1) - S_B(S_B + 1)\}/S(S + 1)$$
$$c_+ = [3\{S_A(S_A + 1) - S_B(S_B + 1)\}^2 + S(S + 1)\{3S(S + 1) - 3 - 2S_A(S_A + 1)\}]/(2S + 3)(2S - 1)S(S + 1)$$
$$c_- = [4S(S + 1)\{S_A(S_A + 1) - S_B(S_B + 1)\} - 3\{S_A(S_A + 1) - S_B(S_B + 1)\}]/(2S + 3)(2S - 1)S(S + 1)$$

[a] The coefficients are taken to be zero for spin-singlet states ($S = 0$), and also the coefficients should be zero for the vanishing denominator.

where D_i^d denotes the given component tensor in the diagonal (principal axis and local molecular) frame and U_i is a rotation (unitary) matrix in which three rows are constructed by three set of the corresponding direction cosines for three principal values of D_i^d, and ^tU_i stands for the transposed matrix of U_i. During this procedure, the local molecular structure relevant to D_i^d, theoretically or empirically, should be known. The resultant D_S (e.g., Eq. (6b)) in terms of the convenient reference frame is diagonalized to give D_S^d, i.e., the principal values and corresponding direction cosines of D_S in the reference frame. In order to produce theory-based spectral simulations, all the resultant theoretical tensors are transformed to the field-based coordinate axes defined by both B_0 and B_1 (microwave oscillating field).

In the weak exchange-coupling limit and for intermediate exchange-coupling cases, general and group-theoretical treatments have been made also by Bencini and Gatteschi,[1.1f] and analytical expressions for spin Hamiltonian parameters are given in their book together with examples of ESR spectra. Extended applications of the above approach to three- or four-spin exchange-coupled systems in the strong coupling limit are feasible. In the case of intermediate exchange coupling, remarkable change in fine structure ESR spectra take place due to spin quantum mixing between interrelated spin states. The changes depend on the ratio of $|D_S/2J|$ and the group-theoretical symmetry of the systems under study. Figure 1 exemplifies a quantum spin mixing occurring in an exchange-coupled triplet pair,

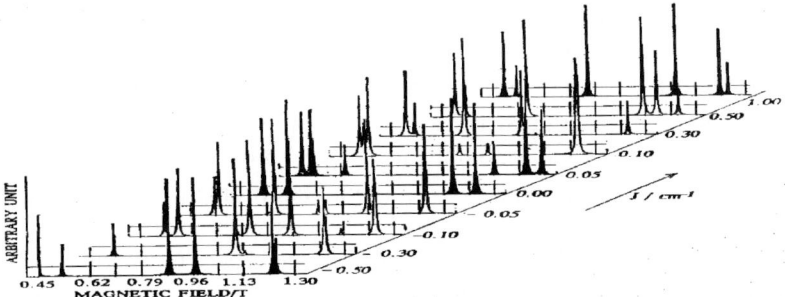

Figure.1. Fine-structure ESR transitions from an exchange-coupled triplet pair undergoing spin quantum mixing. A K-band microwave frequency is assumed with the fine structure constants comparable to $|2J|$.

showing typical fine-structure ESR spectra at K-band (~25 GHz) calculated for the static magnetic field B_0 along a given direction. For symmetry arguments, $D_A = D_B$ is assumed in the calculation.[1.14] As the spin quantum mixing with $|2J|$ comparable to $|D_S|$ grows, new ESR transitions arise with intensity borrowing and a considerable amount of the shifting of the resonance fields occurs. In the case of complete mixing, the characteristic spectral features arising from the quintet and triplet states disappear, giving the appearance of only two triplet-state fine-structure spectra. They are not independent, but interrelated. In this example, a singlet-quintet complete mixing takes place due to the group-theoretical symmetry requirement for a pair of the equivalent $S_i = 1$ spins (i = A, B), while the resultant triplet state is isolated because of symmetry requirements. The permutation symmetry for Bose ($S_i = 1$) particles is symmetric, preventing the mixing between the triplet and the other spin multiplicities. Thus, remarkable spectral changes are anticipated when complete spin quantum mixing occurs with a nearly vanishing exchange coupling. In conventional X-band ESR spectroscopy, the transition probabilities which gain intensity owing to the high-field approximation valid in high-frequency ESR spectroscopy are reduced a great deal. As a result, the characteristics of intensity distributions in the high-field approximation are weakened. In this sense, the practical advantages of X-band ESR spectroscopy are limited for the cases of spin quantum mixing. High-field/high-frequency ESR spectroscopy is desirable for the detection of spin quantum mixing in terms of the transition probability of intermediate spin states. For these reasons, molecular designs for the fine-tuning of on- and-off quantum spin mixing with potential device are the current topics of molecular spin science. Effects of the quantum mixing in hetero-spin systems are shown more remarkably for molecular exchange-coupled systems.[1.15] Under favorable conditions, relative signs of the fine structure constants for the resultant spin states can be determined.

Referred to the higher-order terms of fine structure tensors in spin Hamiltonians, Eqs. (2) and (4), the group-theoretical arguments apparently allow the inclusion of them for molecular high spins of relatively low molecular symmetry ($S = 2, 3$). Indeed, for even high-spin hydrocarbons in the electronic ground state the inclusion may yield apparent better agreement between observed fine-structure spectra and simulated ones, where different spin-state species gave complicated superimposed fine-structure spectra.[1.16] Nevertheless, additionally derived spin Hamiltonian parameters cannot give a rationale for such high-spin hydrocarbons with small spin-orbit couplings of carbon atoms. The obtained result has been misleading in terms of physical meaning of the derived terms. A refined set of spin Hamiltonian parameters have produced much better agreement between the observed spectra and simulated ones by invoking similar spectral simulation methods without incorporating the higher-order, quartic terms.[1.17] Due care is necessary to incorporate the higher-order fine-structure terms in spin Hamiltonians for genuinely organic high-spin systems to avoid artifacts and over-parameterization in spectral simulations.

2.2 Appearance of off-principal-axis lines in fine-structure spectra in molecular high spin systems

One of the salient features appearing in the fine structure spectra due to high spin systems is the appearance of off-principal-axis absorption peaks, called extra lines (or off-axis extra lines), in fine-structure ESR spectroscopy from random orientation. Extra lines inevitably appear when high spin systems have sizable fine structure constants compared with the magnitude of microwave excitation energy. In general, lineshapes of forbidden transition peaks are anomalous, compared with those of the allowed ones. This is because the angular dependence (anisotropy) of forbidden transitions is constrained and thus undergoes angular anomaly, which corresponds to stationary points with $\mathbf{B}_0$ pointing off the principal axis of the fine-structure tensor. As described above, a typical example is B_{min} appearing in the fine-structure spectra from triplet states in non-oriented media. The B_{min} anomaly underlies the stationary behavior of the extra lines from high spins ($S > 1$).[1.10b, 1.11] Appearance of off-axis extra lines is general for anisotropic ESR spectroscopy, and physical origins of "hyperfine-structure" extra lines have been elucidated by several authors.[1.1g, 1.18] The hyperfine-structure extra lines originate in large anisotropic g- and hyperfine tensors.

The first extensive study of fine-structure extra lines has been made in terms of higher-order (second- and third-order) perturbation treatments, giving general conditions and formulas for the appearance of extra lines for arbitrary high spins as well as recipes for the analyses of fine-structure extra lines.[1.11] The formulas in literature[1.11] also enable us to identify off-principal-axis orientations within the frame work of the perturbation theory. In order to complete the analyses of all the fine-structure stationary peaks including extra lines, it is desirable to invoke exact numerical diagonalization approaches which execute the calculation of resonance fields, transition probabilities and their angular dependence with high accuracy. Such a good example is illustrated in Figure 2, where a quartet ground state from m-phenylenebis(phenyl-methylene) monoanion in an organic glass is shown. High spin states with odd spin quantum numbers show strong characteristic peaks of extra lines from the $M_S = -1/2 \Leftrightarrow M_S = +1/2$ transition, as illustrated in Figure 2. The inevitable occurrence of the angular anomaly corresponding to this transition arises from the disappearance of the first-order fine-structure term in the resonance field in terms of perturbation theory. High-field/high-frequency ESR spectroscopy affords to simplify fine structure spectra, weakening the importance of extra line analyses for spin multiplicities of odd numbers. Nevertheless, the simplification is hampered by anisotropic nature intrinsic to molecular systems with both even spin multiplicities and sizable fine-structure constants.

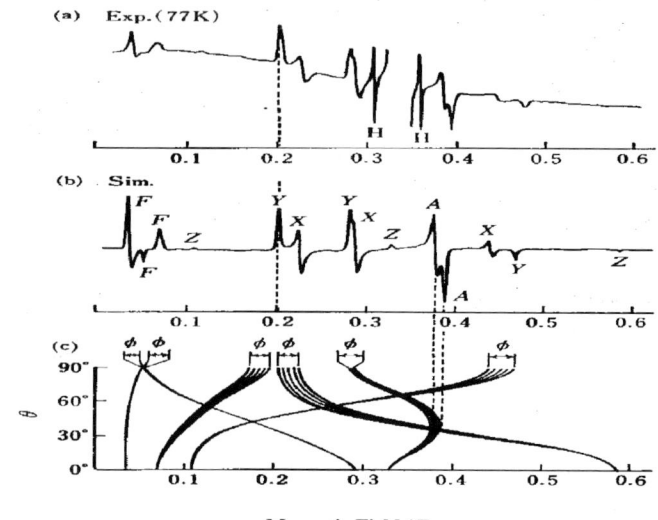

Figure 2. Appearance of extra lines in fine-structure ESR spectroscopy from random orientation; Fine-structure spectra from *m*-phenylenebis(phenylmethylene) monoanion in the quartet ground state.[2,2,1] (a) Observed spectrum; H's denote the absorption lines due to hydrogen atom. (b) Simulated one; X, Y and Z denote the absorption lines corresponding to the canonical orientations. F's denote the lines from $\Delta M_S = \pm 2$ and ± 3 forbidden transitions. A's denote the extra lines.

For smaller fine-structure constants, first-order perturbation treatments have been invoked for the sake of simplicity in order to extract fine-structure constants and *g*-values. It has turned out that most of the documented first-order-perturbation analyses for molecular high spins in the ground state, where π-π spin-spin interactions dominate in the contribution to the fine-structure parameters, do not fulfill the appearance conditions for extra lines. During the simulation procedure, readers are recommended to check the appearance conditions by substituting any trial fine-structure parameters and the microwave frequency employed into the formulas given in literature.[1,11] Whether or not extra lines appear also depends on the line width of the single ESR transition used for the simulation. If the difference in resonance fields between extra lines and principal-axis lines is comparable to or smaller than the line width, the extra line will not be distinctly observed. Instead, asymmetry of the line shape or intensity anomaly will be appreciable.

2.3 Theoretical spectral simulation of molecular high-spin clusters; direct comparison between theoretical simulations and observed spectra of a triplet-state cluster

This section gives a theoretical spectral simulation approach to molecular high-spin clusters, exemplifying a direct comparison of simulated fine-structure spectra with the observed one from a novel exchange-coupled triplet-state molecule, calix[4]arene-based biradical with two nitroxide spin sites.[2.3.1] The system is composed of two NO-based radical centers within the molecule. Thus, spectral simulations require the incorporation of hyperfine tensors due to two nitrogen atoms. The crystal and molecular structures of the exchange-coupled triplet molecule are not known, but the X-ray crystal structure analysis of the corresponding precursor of a new nitroxide-based biradical has been made. The local molecular structures of the two nitroxides have attracted attention. With the help of theoretical spectral simulations, the molecular geometry of the biradical of the two-site spin exchange-coupling can be evaluated, as given below.

For the exchange-coupled system under study, we adopt spin Hamiltonians as follows;

$$\hat{H} = \hat{H}_1 + \hat{H}_2 + \hat{H}_{12} \tag{8}$$
$$\hat{H}_1 = \beta \mathbf{B}_0 \cdot \mathbf{g}_1 \cdot \mathbf{S}_1 + \mathbf{I}_1 \cdot \mathbf{A}_1 \cdot \mathbf{S}_1 \tag{9}$$
$$\hat{H}_2 = \beta \mathbf{B}_0 \cdot \mathbf{g}_2 \cdot \mathbf{S}_2 + \mathbf{I}_2 \cdot \mathbf{A}_2 \cdot \mathbf{S}_2 \tag{10}$$
$$\hat{H}_{12} = \mathbf{S}_1 \cdot \mathbf{J}_{12} \cdot \mathbf{S}_2 , \tag{11}$$

where $\mathbf{g}_i$, $\mathbf{A}_i$ ($i =1, 2$) and $\mathbf{J}_{12}$ stand for the g tensor, hyperfine coupling tensor for the i-th nitroxide site and the exchange coupling tensor, respectively. Here an anisotropic exchange coupling interaction is assumed for theoretical arguments. In the case of biradicals where the interacting spins are sufficiently far from each other, magnetic dipole-dipole interactions dominates the exchange interaction. The present two-nitroxide-based biradical is the case. The classical magnetic dipole-dipole interaction between two spins is given by

$$\mathbf{J}_{12} = \frac{\mu_0}{4\pi} \left[\frac{\boldsymbol{\mu}_{e1} \cdot \boldsymbol{\mu}_{e2}}{r^3} - \frac{3(\boldsymbol{\mu}_{e1} \cdot \mathbf{r})(\mathbf{r} \cdot \boldsymbol{\mu}_{e2})}{r^5} \right]$$
$$= \frac{\mu_0}{4\pi} \beta^2 \left[\frac{\mathbf{g}_1 \cdot \mathbf{g}_2}{r^3} - \frac{3(\mathbf{g}_1 \cdot \mathbf{r})(\mathbf{r} \cdot \mathbf{g}_2)}{r^5} \right] \tag{12}$$

with $\boldsymbol{\mu}_{ei} = -\beta \mathbf{g}_i \cdot \mathbf{S}_i$, noting $\beta = \mu_B$ throughout this section.

This equation indicates that the interacting tensor based on the classical magnetic dipolar interaction depends on both the anisotropy of the g tensor of the two spins and the distance between the spins. Relationships between the anisotropic g tensors reflect the molecular geometry of the biradical. J_{12} is a dyadic, and it can be decomposed into the sum of a scalar product, a symmetric traceless tensor and an antisymmetric tensor. Then, $\hat{H}_{12}$ can be rewritten as

$$\hat{H}_{12} = \mathbf{S}_1 \cdot \mathbf{J}_{12} \cdot \mathbf{S}_2 = J_{12}\,\mathbf{S}_1 \cdot \mathbf{S}_2 + \mathbf{S}_1 \cdot \mathbf{D}_{12} \cdot \mathbf{S}_2 + \mathbf{d}_{12} \cdot \mathbf{S}_1 \times \mathbf{S}_2, \qquad (13)$$

where J_{12} is a scalar, $\mathbf{D}_{12}$ is the symmetric traceless tensor, and $\mathbf{d}_{12}$ is a polar vector which is constructed by the off-diagonal elements of the antisymmetric tensor part of $\mathbf{J}_{12}$. Each term is termed an isotropic, anisotropic, and antisymmetric spin-spin interaction, respectively. Relationships between $\mathbf{J}_{12}$ and the decomposed terms are given by the following equations,

$$J_{12} = (1/3)\,\mathrm{Tr}(\mathbf{J}_{12}) \qquad (14)$$

$$\mathbf{D}_{12} = (1/2)(\mathbf{J}_{12} + {}^t\mathbf{J}_{12}) - J_{12} \qquad (15)$$

$$d_{12,x} = (1/2)(\mathbf{J}_{12} - {}^t\mathbf{J}_{12})_{yz}, \qquad d_{12,y} = (1/2)(\mathbf{J}_{12} - {}^t\mathbf{J}_{12})_{zx},$$
$$d_{12,z} = (1/2)(\mathbf{J}_{12} - {}^t\mathbf{J}_{12})_{xy}, \qquad (16)$$

where ${}^t\mathbf{J}_{12}$ stands for the transposed matrix of $\mathbf{J}_{12}$.

Considering the spin Hamiltonian parameters of the coupled spin system in terms of the coupled eigenbase of S_{1z} and S_{2z}, the exchange interaction term reflects a relative orientation of the two radical sites. The isotropic term makes the energies of all eigenstates shift, not affecting the spectrum at all in ESR spectroscopy. The antisymmetric interaction arises from a second-order term of the spin-orbit coupling in terms of perturbation treatments. The antisymmetric vector given by Eq.(16) is relevant to the symmetry of the coupled spin system. When the spin system has an inversion center, the vector vanishes. For most of the organic biradicals, the antisymmetric term is safely assumed to be zero because of their small spin-orbit interaction in the ground state. Returning to the effective spin Hamiltonian approach, we recall Eqs. (4) and (5), as given below

$$\hat{H}_S^{spin} = \beta \mathbf{B}_0 \cdot \mathbf{g}^{eff} \cdot \mathbf{S} + \mathbf{S} \cdot \mathbf{D}^{eff} \cdot \mathbf{S} + \mathbf{I}_1 \cdot \mathbf{A}_1^{eff} \cdot \mathbf{S} + \mathbf{I}_2 \cdot \mathbf{A}_2^{eff} \cdot \mathbf{S} \qquad (17)$$

with $\mathbf{S} = \mathbf{S}_1 + \mathbf{S}_2$. $\mathbf{g}^{eff}$ denotes the effective g tensor of the exchange-coupled system with $S = 1$, and $\mathbf{D}^{eff}$ the effective fine-structure tensor for the triplet state. $\mathbf{A}_1^{eff}$ and $\mathbf{A}_2^{eff}$ designate the hyperfine coupling tensors of nitrogen nuclei 1 and 2 in the effective spin framework. Applying Eqs.(6a-c), we obtain the following relationships;

$$g^{eff} = (1/2)\,g_1 + (1/2)\,g_2, \tag{18a}$$
$$D^{eff} = (1/2)\,D_{AB} \tag{18b}$$
$$A_i^{eff} = (1/2)\,A_i \quad (i = 1, 2) \tag{18c}$$

It is worthwhile again noting that the relationships above in Eqs. (18a-c) are tensor-based and should be expressed in terms of a common reference molecular frame, as generally described in Section 2.1.

Figure 3 shows a fine-structure ESR spectrum of the calix[4]arene-based biradical with two nitroxide radical sites observed in an organic glass together with calculated spectra. The observed ESR spectrum consists of

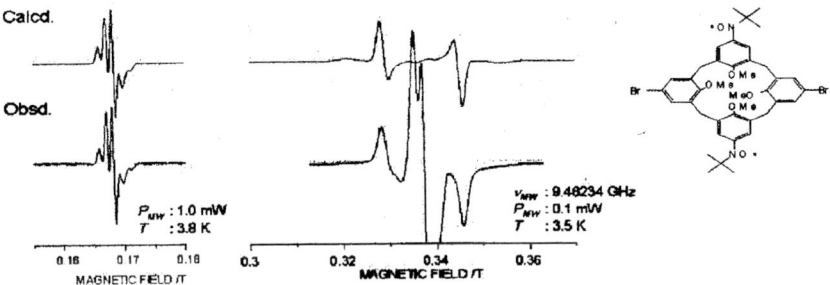

Figure 3. Observed and calculated triplet-state ESR spectra from a calix[4]arene-based biradical with two nitroxide radical centers.

ESR transitions in both the target biradical and the monoradical in which hydroxyamino groups were partially oxidized. In the field range of forbidden transitions with $|\Delta M_S=2|$, an asymmetric hyperfine structure pattern due to interactions between electron spin and two nitrogen nuclei was observed. Absorption peaks assignable to the ESR transitions due to the Z canonical orientations for a triplet state are too broad. These salient features should be reproduced by theoretical spectral simulations, giving electronic molecular structures of the novel calixarene-based biradical. The theoretical spectral simulations based on the effective spin Hamiltonian approach have successfully reproduced the observation. The obtained spin Hamiltonian parameters are summarized in Table 2. The observed spectral features are well interpreted by considering the anisotropic hyperfine coupling tensors of both the nitrogen nuclei. The broadened Z components are due to large hyperfine splittings of the nitrogen nuclei in the principal Z axis of the fine structure tensor, as shown in Figure 4. It shows that the two π-orbitals localized on the nitrogen nuclei point head to head each other in a σ-type bonding. The derived molecular structure agrees with the fact that the triplet state is thermally accessible. The distance between the two spin centers can also be evaluated.

Figure 4 shows the molecular structural dependence of the theoretically simulated spectra for the calixarene-based biradical in non-oriented media.

Table 2. Spin Hamiltonian parameters of a calix[4]arene-based biradical ($S = 1$). Only the principal values are given. A principal Z axis is parallel to the direction connecting the two nitroxide radical sites.

	g^{eff}	D^{eff}/cm^{-1}	A_1^{eff}/cm^{-1}	A_2^{eff}/cm^{-1}
XX	2.008	0.0053	0.00025	0.00025
YY	2.005	0.0053	0.00025	0.00025
ZZ	2.003	-0.0106	0.00135	0.00135

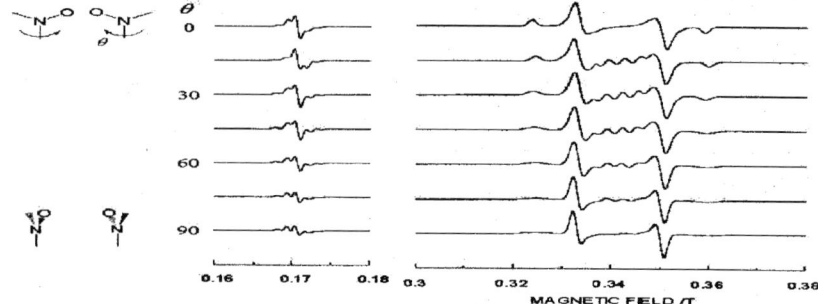

Figure 4. Dependence of the fine-structure spectra on the relative orientation between the tensors in the calixarene-based biradical with the two nitroxide radical sites.

When both the z axes of the nitrogen hyperfine coupling tensors are parallel to the principal Z axis of the fine structure tensor D^{eff}, i.e., $\theta = \pi/2$ (see Fig. 4 for the definition of θ), the Z canonical peaks are broadened because of the large hyperfine splittings due to the nitrogen nuclei in the direction. In the intermediate orientations of the z axes, absorption peaks with appreciable amounts of intensity appear in the magnetic field range of $g \sim 2$. The appearance of these peaks is due to the disagreement of the principal axes of the fine-structure tensor, hyperfine coupling ones and the g tensors. A salient lineshape of the half-field transitions is also affected by the relative orientations of the tensors. Although the ESR transitions due to the partially oxidized nitroxide monoradical masked the central part of the fine-structure ESR spectrum of the biradical, the molecular structural features for the novel biradical are probable. A pure triplet-state spectrum from the biradical gives much more precise information on the molecular conformation.[2.3.2] In addition, the molecular structure with one hydrogen bond due to partial oxidation gives a clue to molecular magnetic functionality controlled by intramolecular hydrogen bonding.[2.3.1]

Tensor-based theoretical spectral simulation approaches are also useful for extracting spin Hamiltonian parameters for molecular high-spin systems whose spin sites are composed of transition metal ions and organic high spins. A subtle issue during the simulation procedure is to evaluate the interaction term D_{AB} for hybrid high-spin systems with extremely delocalized spins over the whole system or moiety. For such cases, if thermally accessible spin states with different S's are observed, D_{AB} can be estimated under favorable conditions. Otherwise, quantum chemical computations are invoked for the estimation of D_{AB}. In exchange-coupled systems with sizable molecular frames, the contribution of the interacting terms in the ground or nearby excited states decreases rapidly with increasing molecular sizes. The molecular size effect, however, does not hold for electronic excited high-spin states with extreme spin delocalization, and high spin chemistry in electronic excited states of inter- and intra-molecularly exchange-coupled systems are the focus of current issues in chemistry and spin science. It is worth noting that semi-empirical tensor-based spectral simulation gives a rationale for molecular structural determination of relatively small molecular clusters such as triplet-state dimeric $(NO)_2-(Na^+)_x$ adsorbed in zeolites with the help of theoretical estimation for the triplet-state fine structure tensor, in which accurate quantum chemical calculation should be invoked because of a large amount of spin densities on the nitrogen and oxygen sites.[2.3.3] An absolute sign of the principal fine-structure D-value determined experimentally (by high-field/high-frequency ESR or ENDOR experiments) gives crucial information on the molecular and electronic spin structures of the systems under study.

2.4 Microscopic spin dynamics underlying magnetization dynamics in infinite systems of exchanged couplings

One of the crucial issues in electron magnetic resonance spectroscopy applied to "ordered" magnetic assemblages with exchange couplings is to understand their magnetization dynamics in terms of microscopic spin dynamics.[2.4.1] The magnetization dynamics has been treated by both Kubo-Tomita[2.4.2] and Mori (and Tokuyama-Mori)[2.4.3] theories in terms of spin assemblages in microscopic detail. The former theory describes the transverse relaxation by invoking a relaxation function $\phi(t)$ of a macroscopic magnetization as

$$\phi(t) = <M_x(t) M_x(0)>/ k_B T, \qquad (19)$$

where $<M_x(t) M_x(0)>$ stands for a time correlation function of M_x under thermal fluctuation and therefore the symbol $\phi(t)$ describes the magnetization, classically. Isotropic exchange interactions expressed as $-2J_{ij}\mathbf{S}_i \cdot \mathbf{S}_j$ are commutable with $\mathbf{S} = \mathbf{S}_i$, thus giving no contribution to the transverse relaxation. Noncommutable local magnetic fluctuations are expressed in terms of a time correlation function $\varphi(t)$ for the local magnetic field. The relaxation function $\phi(t)$ for the macroscopic magnetization can be expressed by using $\varphi(t)$. Noting that the phase memories of individual spins S_i's are vanishing in nearly $h/4J$ because of mutual exchange interactions with surrounding spins, the motion of **M**, i.e., that of **S** is conserved for a long time because **M** commutes with isotropic exchange interactions. Dynamics of S_i can be expressed in terms of a spin-time-correlation function $\Phi(t)$ as

$$\Phi(t) = 3<S_{xi}(t)\, S_{xi}(0)>/\, S(S+1) . \tag{20}$$

In the rotating frame, the first three terms of $\Phi(t)$ are approximated to be

$$\Phi(t) = \exp[-(t/\tau_e)^2/2] \quad (t < \tau_e) \tag{21}$$

with $\tau_e = (h/4J)\{3/[2qS(S+1)]\}^{1/2}$, where J stands for an exchange integral of a pair of nearest-neighboring spins and q for the number of the nearest-neighboring spins. For a long time, $t > \tau_e$,

$$\Phi(t) \propto [S(S+1)/3]\, t^{-d/2} \tag{22}$$

holds, where d stands for the dimension of the spin assemblages. According to Eq.(22), $\Phi(t)$ decays slowly for low-dimensional spin assemblages ($d = 1, 2$), leading to the occurrence of a long transverse relaxation called a long time tail (LTT). An LTT arises from increases in the contributing weights of the zero-mode Fourier component of the wave vector in $\Phi(t)$ for smaller d. Compared with three-dimensional spin assemblages, whose spin correlations decay in times much shorter than periods of microwave frequency $\omega_0/2\pi \sim 10 - 10^2$ GHz, free induction decays with transverse relaxations characteristic of LTT or satellite peaks due to spin dipolar interactions appear for low-dimensional organic spin assemblages in pulsed ESR experiments. In this context, high-field/high-frequency ESR spectroscopy such as W-band ESR spectroscopy is potentially capable of extending vital information on magnetization dynamics in microscopic detail. Particularly, it is worth noting that microscopic spin dynamics characteristic of low dimensionality in organic magnetics reflects the contribution of nearly-zero-mode Fourier components as T approaches T_c (or T_N). The difference in the contributions originates from the difference in temperature evolution of short-range interactions between ferromagnetics and antiferromagnetics. Pulsed-FT-based time domain ESR spectroscopy is direct and sensitive to the difference.

3. SPECTRAL SIMULATION BASED ON A HYBRID EIGENFIELD METHOD AND PERTUBATION TREATMENTS

Electron spin resonance spectroscopy from random orientation of high-spin molecular systems gives us a variety of fine structure spectra. It is generally tough to extract the underlying physical parameters of paramagnetic species from spectra in non-oriented media in a straightforward manner except for simple cases. Difficulties arise from the spectroscopic methodology that cw-ESR spectroscopy measures resonance with an electromagnetic irradiation field by sweeping a static magnetic field $B_0(=B)$. A spectroscopic coordinate in ESR spectra is not easy to be intuitively transformed to an energy difference between electronic spin states involved in the resonance transitions, contrary to most other spectroscopies. Because of the many types of spin Hamiltonian terms, ESR spectral simulations[3.1-17] are an important procedure for quantitatively interpreting observed ESR spectra and extracting physically meaningful parameters with high accuracy. A failure of successful spectral simulations weakens the usefulness of random orientation ESR spectroscopy. This section is intended to introduce to readers a general and powerful procedure for simulating ESR spectra from random orientation, exemplifying fine structure spectra. Methods given in this section are free from notorious problems of non-convergence and save computation time. The program softwares based on the present methods are available, which are written in terms of Mathematica-based and other languages.[3.12]

In order to reproduce overall fine-structure spectra with low-field absorption peaks from forbidden transitions, the second- or higher-order perturbation treatments in terms of both Rayleigh-Schroedinger and Brillouin-Wigner types have been frequently used and all the mathematical expressions required for spectral simulations are available in analytical forms with respect to arbitrary coordinate axes as well as arbitrary spin quantum numbers.[3.18-20] The perturbation approach has been developed to the third-order stage in terms of fine structure terms in spin Hamiltonians.[3.11] Fourth order treatments for the analyses of fine-structure spectra in an arbitrary coordinate-axis system are available.[3.21] Program packages of second-order treatments for random orientation ESR spectroscopy, including hyperfine terms, were intended to be distributed by various authors at early time.[3.18] Recently, a program package for ESR spectral simulations based on a second-order perturbation theory has been commercially distributed, its efficiency has been hampered by a failure to simulate forbidden transition peaks and the forced assumption of collinearity between *g* and fine structure tensors.[3.22] The perturbation approaches are efficient in terms of computation time and they are free from a non-convergence problem due to energy crossings, but a weakness is that they fail to reproduce low-field peaks and give out-of–phase in conventional X-band fine-structure spectra. The appearance of such somewhat peculiar-looking lines strongly suggests a possible breakdown of the perturbation treatments. Nevertheless, it is apparent that the perturbation approach programs are useful for the cases of $|D|/h\nu \ll 1$ under the experimental conditions of X-band or high-field/high-frequency spectroscopy (K-, Q-, W-, and higher bands).

The breakdown of the perturbation approaches based on general analytical expressions can be avoided by invoking exact numerical diagonilizations of spin Hamiltonian matrices, either the n × n eigenenergy matrix or the $n^2 \times n^2$ eigenfield matrix, where n = 2S + 1. The eigenenergy approach suffers from notorious non-convergence problems while matching $h\nu$ to the difference between the energies involved in the transitions, if the calculation involves avoided energy crossings. Particularly, this difficulty takes place in the simulation procedure for the transitions appearing in the low-field region of X-band fine-structure spectra.

On the other hand, the eigenfield approach is free from the non-convergence problem but needs much more computation time, which increases in proportion to the third power of the matrix dimension. Thus, the eigenfield approach becomes impractical when the dimension of an original eigenenergy matrix is large. An enormous amount of computation time inherent in the original eigenfield approach[3.9-10] should be eliminated for the practical use of the methodology. A hybrid eigenfield approach here gives a practical solution for this problem. In addition, some important technical problems with eigenfield matrices should be solved in order to acquire physically meaningful parameters. It should be recollected that the exact analytical solutions of the eigenfield for triplet states have been derived in an arbitrary orientation of B_0 (= B).[3.23] Not only a weakness of the eigenfield method is the sizable dimension of the $n^2 \times n^2$ eigenfield matrix, but also the method requires solving generalized eigenvalue problems, which give rise to imaginary eigenfield values. Technically, elaborate mathematical techniques are necessary during the numerical convergence procedure. There have been a commercially available program package[3.24] and a home-made one[3.12, 3.16] which fulfil the above requirements. The latter unites the above requirements with the shortening the computation time by invoking a methodologically hybrid method between the original eigenfield theory and conventional eigenenergy method.[3.12, 3.16]

3.1 Fine-structure ESR spectroscopy from random orientations in non-oriented media

The effective spin Hamiltonian[3.2-3] for analyses of ESR spectra is generally expressed by

$$\hat{H} = \hat{H}_{eZ} + \hat{H}_D + \hat{H}_{hf} + \hat{H}_{nZ} + \hat{H}_Q$$
$$= \beta \mathbf{S} \cdot \mathbf{g} \cdot \mathbf{B} + \mathbf{S} \cdot \mathbf{D} \cdot \mathbf{S} + [\mathbf{S} \cdot \mathbf{A}_k \cdot \mathbf{I}_k - \beta_n \mathbf{B} \cdot \mathbf{g}_n^k \cdot \mathbf{I}_k + \mathbf{I}_k \cdot \mathbf{Q}_k \cdot \mathbf{I}_k], \quad (23)$$

where $\hat{H}_{eZ}$, $\hat{H}_D$, $\hat{H}_{hf}$, $\hat{H}_{nZ}$, and $\hat{H}_Q$ stand for the electron Zeeman, fine structure, hyperfine coupling, nuclear Zeeman, and quadrupole terms, respectively. The electron and nuclear Zeeman terms are magnetic-field dependent, and the others are independent. A single-crystal ESR spectrum, i.e., the one at a single given orientation of the magnetic field $\mathbf{B}$ (= $\mathbf{B}_0$) is constructed from resonance fields B^{res} and transition probabilities P calculated from the spin Hamiltonian for $\mathbf{B}$. Hereafter we omit the suffix 0 for the static magnetic field $\mathbf{B}_0$. When assuming an appropriate function $f(B-B^{res}, \Delta B_{1/2})$ for the ESR signal, the spectrum pattern $s(\theta,\varphi,\psi,B)$ is described by

$$s(\theta, \varphi, \psi, B) = P_i(\theta, \varphi, \psi) \times f[B - B_i^{res}(\theta, \varphi), \Delta B_{1/2}] \quad (24)$$

where θ, φ, and ψ stand for Euler angles and $\Delta B_{1/2}$ for the line width at half height. Euler angles are required for the calculation of resonance fields and transition probabilities in the laboratory frame referred to both $\mathbf{B}$ and $\mathbf{B}_1$ (called Zeeman coordinate systems). When a given component tensor such as $\mathbf{D}_i$ in a convenient reference frame is transformed to the laboratory frame, Euler angles are defined referred to the convenient reference frame. The summation runs over all the transitions i's. The first derivative of a normalized Gaussian or Lorentzian function is commonly applied as the function for field-swept and field-modulation scheme detection.

An overall ESR spectrum from powdered states or non-oriented media such as organic glasses is composed of the single-crystal like ESR transitions due to all the randomly oriented molecules in the laboratory frame. Thus, the overall ESR spectrum $S(B)$ from random orientation is constructed by integrating single-crystal spectrum patterns, i.e.,

$$S(B) = \iiint_{\theta,\varphi,\psi} s(\theta,\varphi,\psi,B) \sin\theta\, d\theta\, d\varphi\, d\psi \quad (25)$$

One of the most time-consuming processes in spectrum simulation is a numerical integration given in Eq.(25). Several methods for improving the numerical integration, such as angular grid methods, application of interpolation techniques, statistical calculations based on random sampling, and so on, have been developed so far in order to reduce computation time in the simulation procedure.[3,4-6] A new method based on a 'sophe' partition and interpolation scheme has been developed by Hanson and coworkers.[3,7-8] Their simulation software achieves high efficiency and accuracy for any parameters in the spin Hamiltonian by solving the eigenvalue problem with the help of homotopy. Application of homotopy is a solution that can avoid mathematical difficulties in the vicinity of anti-level crossings and looping transitions.

3.2 Eigenfield method and hybrid eigenfield approach as an improved accessible method

One of the "tricky" manipulations in the ESR spectroscopic analysis is to convert transition frequencies obtainable from energy eigenvalue problems to external (applied) static magnetic fields satisfying the resonance condition, because current ESR spectroscopy is magnetic field-swept in contrast to ordinary spectroscopy which is frequency-swept. In the direct calculation, exact resonance fields are acquired by an iteration procedure using the numerical diagonalization of the spin Hamiltonian matrix with B included. However, we sometimes meet a problem that the iteration does not converge. In addition, even though convergence takes place within a certain accuracy, a method employed under a certain algorithm does not always assure whether calculated resonance fields are complete and exact. The eigenfield method[3,9-13] formulated and developed by Belford et al. and by Hatfield is free from such numerical "breakdown or pitfall". It is worth while noting that Brillouin-Wigner types of higher-order perturbation treatments require iterations for numerical convergence in contrast to Rayleigh-Schroedinger ones, but the former ones do converge quickly and give more accurate resonance fields than the latter ones.[3,13b]

In the following, we derive the eigenfield method so that the readers can easily understand to encode programs. Since the spin Hamiltonian consists of a magnetic field dependent term FB and an independent term G, we rewrite the spin Hamiltonian as given by

$$H = FB + G. \tag{26}$$

The time-independent Schrödinger equation and its complex conjugate are written as follows,

$$H|\psi_i> = w_i|\psi_i> \tag{27a}$$

11. Exchange-coupled high-spin clusters

$$H^*|\psi_j^*\rangle = w_j|\psi_j^*\rangle, \quad (27b)$$

where i denotes a particular energy eigenvalue w_i and the corresponding eigenvector $|\psi_i\rangle$. $|\psi_i\rangle$ and $|\psi_j^*\rangle$ are explicitly defined in terms of a basis set $\{|\phi_n\rangle\}$ as

$$|\psi_i\rangle = \sum_n a_{in}|\phi_n\rangle \quad (28a)$$

$$|\psi_j^*\rangle = \sum_n a_{jn}^*|\phi_n^*\rangle, \quad (28b)$$

where we assume another solution denoted by j. Making the direct product space which is constructed by both $|\psi_i\rangle$ and $|\psi_j^*\rangle$, we can rewrite Eqs.(27a) and (27b) as

$$H \otimes \mathbf{E}|\psi_i,\psi_j^*\rangle = \omega_i \mathbf{E} \otimes \mathbf{E}|\psi_i,\psi_j^*\rangle \quad (29a)$$

$$\mathbf{E} \otimes H^*|\psi_i,\psi_j^*\rangle = \omega_j \mathbf{E} \otimes \mathbf{E}|\psi_i,\psi_j^*\rangle, \quad (29b)$$

where $|\psi_i,\psi_j^*\rangle \equiv |\psi_i\rangle \otimes |\psi_j^*\rangle$ and $\mathbf{E}$ is the $(2S+1)\times(2S+1)$ identity matrix. By subtracting Eq.(29b) from Eq.(29a), we obtain

$$\left(H \otimes \mathbf{E} - \mathbf{E} \otimes H^*\right)|\psi_i,\psi_j^*\rangle = (\omega_i - \omega_j)\mathbf{E} \otimes \mathbf{E}|\psi_i,\psi_j^*\rangle. \quad (30)$$

The eigenvalue $(\omega_i - \omega_j)$ of Eq.(30) is identical to the energy difference between the $|\psi_i\rangle$ and $|\psi_j^*\rangle$ states. Since resonance occurs when the difference is equal to a microwave enegy $h\nu$, we obtain

$$\left(H \otimes \mathbf{E} - \mathbf{E} \otimes H^*\right)|\psi_i,\psi_j^*\rangle = h\nu \mathbf{E} \otimes \mathbf{E}|\psi_i,\psi_j^*\rangle. \quad (31)$$

By substituting Eq.(26) into Eq.(31), we have the following eigenfield equation which affords resonance magnetic fields (eigenfields) as generalized eigenvalues.

$$(h\nu \mathbf{E} \otimes \mathbf{E} - \mathbf{G} \otimes \mathbf{E} + \mathbf{E} \otimes'\mathbf{G})\mathbf{Z} = B(\mathbf{F} \otimes \mathbf{E} - \mathbf{E} \otimes'\mathbf{F})\mathbf{Z}, \quad (32)$$

where $\mathbf{Z} = |\psi_i,\psi_j^*\rangle = \sum_{m,n} a_{im} a_{jn}^*|\phi_m,\phi_n^*\rangle$.

By solving this eigenfield equation, we have the resonance field B as the generalized eigenvalue and corresponding eigenvector $\mathbf{Z}$. The $\mathbf{Z}$ eigenvector in the eigenfield equation is composed of a direct product of two energy eigenstates associated with a particular transition on resonance. The $\mathbf{Z}$ vector includes all the information on the corresponding transition.

The transition moment μ_{ij} between $|\psi_i\rangle$ and $|\psi_j\rangle$ and the corresponding transition probability is calculated according to

$$\mu_{ij} = \langle \psi_j | \beta \mathbf{S} \cdot \mathbf{g} \cdot \mathbf{B}_1 | \psi_i \rangle = \langle \psi_j | H_1 | \psi_i \rangle$$
$$= \sum_{m,n} a_{im} a_{jn}^* \langle \phi_n | H_1 | \phi_m \rangle \qquad (33)$$
$$= \sum_{m,n} (H_1)_{nm} \times a_{im} a_{jn}^* = \boldsymbol{\mu} \cdot \mathbf{Z},$$

where H_1 is the transition moment operator and $\boldsymbol{\mu}$ is a vector composed of all the rows of H_1, as defined by

$$\boldsymbol{\mu} = \sum_{m,n} (H_1)_{nm} \langle \phi_m, \phi_n^* |. \qquad (34)$$

The transition probability $|\mu_{ij}|^2$, therefore, is given by

$$|\mu_{ij}|^2 = |\boldsymbol{\mu} \cdot \mathbf{Z}|^2. \qquad (35)$$

The expectation values for S_Z of the states involved in the transition are also derived in the eigenfield formalism as follows:[3,12]:

$$\langle S_Z^i \rangle = \langle \psi_i | S_Z | \psi_i \rangle = \langle \psi_i, \psi_j^* | S_Z \otimes \mathbf{E} | \psi_i, \psi_j^* \rangle$$
$$= \widetilde{\mathbf{Z}}^* S_Z^i \mathbf{Z} \qquad (36a)$$

$$\langle S_Z^j \rangle = \langle S_Z^j \rangle^* = \langle \psi_j^* | S_Z | \psi_j^* \rangle = \langle \psi_i, \psi_j^* | \mathbf{E} \otimes S_Z | \psi_i, \psi_j^* \rangle$$
$$= \widetilde{\mathbf{Z}}^* S_Z^j \mathbf{Z}, \qquad (36b)$$

where $S_Z^i = S_Z \otimes \mathbf{E}$, $S_Z^j = \mathbf{E} \otimes S_Z$, and $\widetilde{\mathbf{Z}}^* = \langle \psi_i, \psi_j^* |$.

Numerical computations and encoding programs by invoking the use of the expressions derived in this section are rather straightforward.[3,12] All the resonance fields and transition probabilities in the eigenfield method are directly calculated without any iteration procedure for searching the resonance field. Thus, the method eliminates the difficulties that the iteration does not converge in the vicinity of anti-level crossing and looping transitions. It is, however, necessary to completely solve the generalized eigenvalue problem of a sizable dimension compared with the energy eigenvalue problem. Although the generalized eigenvalues are calculated with enough accuracy, the calculation procedure sometimes loses a great deal in obtaining the eigenvectors in terms of accuracy even if applying the latest sophisticated program package for generalized eigenvalue problems. In order to avoid such numerical inconvenience, we have extended a hybrid approach which unites the original eigenfield method with conventional eigenenergy calculations. The approach is termed a hybrid eigenfield method. In the hybrid eigenfield method, resonance fields are acquired by the eigenfield method and the corresponding eigenvectors required for transition probabilities are calculated by numerically diagonalizing the ordinary eigenenergy spin-Hamiltonian matrices with the eigenfields substituted. The hybrid method has been successfully applied to organic and inorganic high-spin molecular systems so far.[3.11-15] Besides the numerical convergence problem, the hybrid method saves computation time a great deal.

3.3 Exact analytical treatment of the spin Hamiltonian; Exact analytical solutions for fine-structure resonance fields and transition probabilities

3.3.1 General arguments

Analytical expressions for resonance fields and transition probabilities are useful for interpreting ESR spectra in any levels, and perturbation treatments are employed for this purpose and predicting unknown spectral features. Perturbation-based analytical expressions for a given effective spin in an arbitrary direction of the static magnetic field have been documented.[3.18-21] Electron-nuclear multiple magnetic resonance frequencies in terms of second-order perturbation theory are also given in a general and comprehensive form.[3.19] This section gives "exact analytical formulae" derived from the eigenfield method. The formulae can be neither general nor for an arbitrary spin, but are particular for spin quantum numbers and spin Hamiltonian terms. Exact analytical approaches for $S \geq 3/2$ are feasible for fine-structure terms of axial symmetry.[3.25-26] The treatment here does not assume symmetry and is enough to be general in this context.[3.12]

We start with the spin Hamiltonian for an arbitrary effective spin S as given by Eq.(23). We treat only the electron Zeeman and fine-structure terms, ignoring the others for simplicity. Once exact eigenvectors for the electron spin part are acquired, details such as hyperfine interactions in terms of electron magnetic resonance can be obtained with good accuracy. For an arbitrary direction of the magnetic field, we have to solve an eigenvalue problem of the $(2S+1) \times (2S+1)$ spin-Hamiltonian matrix based on Eq.(23) to obtain the energy eigenvalues. Then, the corresponding secular equation for the energy eigenvalues is given as a $2S+1$ degree polynomial equation of the energy. As analytical general solutions for the polynomial equations less than a quintic are available, the eigenenergy expressions for spin states less than a quintet ($S = 2$) can be derived. It was proven by two mathematicians, Ruffini in 1799 and Abel in 1826, that it is not possible to derive an explicit analytical solution for the general quintic equation. To obtain resonance fields for an arbitrary orientation of the magnetic field one needs to solve the eigenfield equation of the $(2S+1)^2$ dimension. As the eigenfield equation formally gives zero and negative fields, the secular equation for the resonance fields is reduced to an $S(2S+1)$ degree polynomial equation of the function of the squared resonance field B^2, as shown in Eq.(37). The general analytical expressions for the resonance fields, therefore, do not exist for high-spin states more than $S = 3/2$.

$$\begin{aligned} f &= \det[h\nu \mathbf{E} \otimes \mathbf{E} - \mathbf{G} \otimes \mathbf{E} + \mathbf{E} \otimes \mathbf{G}^* - B(\mathbf{F} \otimes \mathbf{E} - \mathbf{E} \otimes \mathbf{F}^*)] \\ &= \det[h\nu \mathbf{E} \otimes \mathbf{E} - (\mathbf{G} + B\mathbf{F}) \otimes \mathbf{E} + \mathbf{E} \otimes (\mathbf{G}^* + B\mathbf{F}^*)] \\ &= \det(h\nu \mathbf{E} \otimes \mathbf{E} - \mathbf{H} \otimes \mathbf{E} + \mathbf{E} \otimes \mathbf{H}^*) \\ &= B^{2S+1} f_0(B^2) \end{aligned} \qquad (37)$$

In addition, it is not easy to describe general solutions for ESR transitions derived from the spin Hamiltonian in an arbitrary orientation of the static magnetic field. Useful expressions for the analysis of powder-pattern fine structure ESR spectra, however, are available when assuming that the principal axes of the magnetic tensors in the spin Hamiltonian coincide. The expressions allow us to reproduce the canonical peaks, i.e., the observed resonance fields corresponding to **B** along the principal axes, without time-consuming sophisticated computation, once salient features of the observed spectral patterns are characterized.

In the following argument, we assume that the principal axes of the **g** and **D** tensors coincide for convenience. This assumption is not a requirement for the present approach and can be eliminated if necessary. In the Zeeman coordinate system where the direction of the magnetic field is taken parallel to the z axis, choosing the eigenfunctions $\{|S, M_S\rangle\}$ of S_z as a basis set, we can rewrite Eq.(3.2-8) in the following;

$$H = \beta g_{zz} S_z + \frac{1}{2} g_1 S_- + \frac{1}{2} g_1^* S_+ + D_0 \left[S_z^2 + \frac{1}{4}(S_+ S_- + S_- S_+) \right]$$
$$+ \frac{1}{2} D_1 (S_z S_- + S_- S_z) + \frac{1}{2} D_1^* (S_z S_+ + S_+ S_z) + \frac{1}{4} D_2 S_-^2 + \frac{1}{4} D_2^* S_+^2 , \qquad (38)$$

with $g_1 = g_{zx} + ig_{zy}$,

$D_0 = -D_{zz}$, $D_1 = D_{zx} + iD_{zy}$ and $D_2 = D_{xx} - D_{yy} + 2iD_{xy}$.

In the case where one of the axes of the Zeeman coordinate system coincides with one of the principal axes of the **D** tensor, both g_1 and D_1 vanish because the **g** and **D** tensor become diagonal. Therefore, the spin Hamiltonian matrix

11. Exchange-coupled high-spin clusters

is reducible to two sub blocks in some cases, as shown below; where n_1 and n_2 are $2S$ and $2S+1$ for half-integral spins, and $2S+1$ and $2S$ for integral spins, respectively. We can reduce the matrix size of the spin Hamiltonian. Then, we do not necessarily solve the eigenvalue problem with the full-size Hamiltonian matrix, solving the eigenvalue problem with each

$$H = \begin{pmatrix} H_{1,1} & 0 & H_{1,3} & & & & & & 0 \\ 0 & H_{2,2} & 0 & H_{2,4} & & & & & \\ H_{3,1} & 0 & H_{3,3} & & \ddots & & & & \\ & H_{4,2} & & \ddots & & & H_{2S-2,2S} & & \\ & & & & & H_{2S-1,2S-1} & 0 & H_{2S-1,2S+1} & 0 \\ & & & & H_{2S,2S-2} & 0 & H_{2S,2S} & 0 \\ & & & & & H_{2S+1,2S-1} & 0 & H_{2S+1,2S+1} \end{pmatrix}$$

$$= \begin{pmatrix} \begin{pmatrix} H_{1,1} & H_{1,3} & & \\ H_{3,1} & \ddots & & \\ & & H_{n_1,n_1} \end{pmatrix} & & 0 \\ & \begin{pmatrix} H_{2,2} & & \\ & \ddots & H_{n_2-2,n_2} \\ & H_{n_2,n_2-2} & H_{n_2,n_2} \end{pmatrix} \\ 0 & \end{pmatrix} = H_1 \oplus H_2$$

(39)

block matrix. For instance, for a quintet state ($S = 2$) it is enough to solve secular equations from 2×2 and 3×3 spin Hamiltonian matrices. This reduction means that for the canonical orientations with **B** along the principal axes of the **D** tensor it is possible under certain conditions to obtain exact analytical expressions for energy eigenvalues even for high-spin states. The corresponding eigenfield equation is also reducible and solved by a block diagonalization. The eigenfield matrix giving the secular equation is expressed by

$$\mathbf{C} = \begin{pmatrix} \mathbf{C}_{11} & & & 0 \\ & \mathbf{C}_{12} & & \\ & & \mathbf{C}_{21} & \\ 0 & & & \mathbf{C}_{22} \end{pmatrix} = \mathbf{C}_{11} \oplus \mathbf{C}_{12} \oplus \mathbf{C}_{21} \oplus \mathbf{C}_{22}$$

(40)

with $\mathbf{C}_{ij} = h\nu \mathbf{E}_i \otimes \mathbf{E}_j - \mathbf{H}_i \otimes \mathbf{E}_j + \mathbf{E}_i \otimes \mathbf{H}_j^*$ $(i, j = 1, 2)$.

The secular equation of the determinant of **C** is given by a polynomial in B^2. In the case of an integral spin, the total determinant is factorized into the product of four polynomials in B^2, corresponding to the determinant of each block matrix. In the case of a half-integral spin, the determinant of each block matrix is a polynomial in B, even though the total determinant is a polynomial in B^2.

For the quintet state the 4×4 block matrix due to $\mathbf{C}_{22}$ provides a linear equation in B^2, and all the remaining ones provide cubic equations in B^2. Thus, we can obtain the exact analytical solutions for principal-axis resonance fields. We can also calculate the transition probabilities using Eq.(35) when obtaining the eigenvectors of the spin Hamiltonian. However, since the simulation of the powder-pattern ESR spectrum requires integration over all the transitions arising from all the orientations distributed in three-dimensional space, the transition probability for only a particular orientation is of no crucial importance. On the other hand, it is more useful to know salient features for the canonical for most cases. The above arguments give the possible expression of the exact analytical eigenenergy and eigenfield, as summarized in Table 3.

3.3.2 Exact analytical formulae for resonance fields by eigenfield method

Exact analytical formulae for only three particular high-spin states derived by the present treatment of the eigenfield equation are given in this section, for space limitation. The other exact analytical formula for a given S summarized in Table 3 are available.[3.12, 3.16] These expressions are correct for an anisotropic g value under the assumption of the colinearity between the $\mathbf{g}$ and $\mathbf{D}$ tensors. The formulae given here are their energy eigenvalues and resonance fields with the magnetic field $\mathbf{B}$ parallel to the Z principal axis of the $\mathbf{D}$ tensor. The formulae with $\mathbf{B}$ parallel to the X and Y axes are straightforwardly obtained by the cyclic permutation with respect to X, Y, and Z, as follows;

X → Y, Y → Z, and Z → X for the X principal-axis orientation, or
X → Z, Y → X, and Z → Y for the Y principal-axis orientation.

Throughout the present description, the fine structure parameters D and E are defined as

$$D = (3/2)D_{ZZ},$$
$$\text{and } E = (D_{XX} - D_{YY})/2. \tag{41}$$

The expressions can be converted into the D- and E-based ones with the traceless relation $D_{XX} + D_{YY} + D_{ZZ} = 0$.

Since the analytically derived complete formulae for the eigenfield solution of a cubic or quartic equation are complicated, only the original cubic or quartic equations are explicitly presented instead, for space limitation. These equations can exactly and analytically be solved using, e.g., Cardano's formula for the cubic equation or Ferrari's formula for the quartic equation without mathematical difficulty. Resonance fields for double

11. Exchange-coupled high-spin clusters

quantum transition are also obtained by changing the resonance condition involving $2h\nu$ absorption in the expressions. In the calculation, only non-negative eigenfields are valid. The solutions given in literature include all the forbidden transitions with **B** // Z and the cyclic permutation procedure (thus, with the other canonical orientations), predicting that in favorable cases the $\Delta M_S = \pm 2$ forbidden transitions show a marked fine-structure anisotropy in fine-structure ESR spectra ($S \geq 3/2$) from random orientation.

The exact analytical expressions for $S = 1$ and $3/2$ are compact, but not given here because of space. All the other exact analytical formula for a given S larger than $3/2$ are quite long, but manageable to calculate resonance fields and they are available also in electronic media.[3.12, 3.16]

Table 3. Possibility for analytically obtaining the eigenvalue of the spin Hamiltonian (see the text) and the ESR resonance field. The circle, triangle, and cross denote exactly soluble, partially soluble, and insoluble, respectively.

	Arbitrary Direction		Directions of principal axes	
	Eigenvalue	Resonance field	Eigenvalue	Resonance field
1/2	○	○	○	○
1	○	○	○	○
3/2	○	×	○	○
2	×	×	○	○
5/2	×	×	○	×
3	×	×	○	—
7/2	×	×	○	×
4	×	×	△	×
≥9/2	×	×	×	×

3.3.3 Parallel microwave polarization excitation spectroscopy combined with the hybrid eigenfield method

In high spin chemistry, parallel microwave polarization (parametric) excitation spectroscopy under the experimental condition of $B//B_1$ in X band has attracted considerable attention, recently. In high-field/high-frequency ESR experiments, parametric excitation techniques are not so important in practice as in X-band, simply because group-theoretically allowed transitions dominate in high-field/high-frequency fine-structure spectra. In conventional X-band spectroscopy, forbidden transitions from molecular high-spin systems with intermediate or sizable D values sometimes dominate in the fine structure spectra and afford us key peaks in determining the spin Hamiltonian parameters. Exact eigenvectors obtained by the present hybrid eigenfield method reproduce enhanced "forbidden" (= allowed) transition probabilities by the use of the expressions Eqs. (33-35) in Section 3.2.[3.14,3.17] High spin systems arising from transition metal ions require the inclusion of higher-order fine-structure terms such as $S_i^2 S_j^2$ or $S_i^3 B$ in interpreting powder-pattern spectra.[3.14,3.17] The hybrid eigenfield method has been extended to lanthanoid ions. Conventional X-band parametric (parallel) excitation spectroscopy is strengthened with the help of the hybrid eigenfield method. Also, the transition moment calculated by the hybrid eigenfield method plays an essential role to identify the unequivocal transition assignment combined with theoretical spectral simulation of pulsed-based two-dimensional electron spin transient nutation spectroscopy, which is termed transition moment spectroscopy.

4. SOLUTION ESR SPECTROSCOPY FOR MOLECULAR HIGH-SPIN SYSTEMS WITH EXCHANGE INTERACTION COMPARABLE TO HYPERFINE INTERACTIONS

4.1 Introductory remarks

Exchange interactions between paramagnetic centers are important for some biological molecular systems and magnetic materials. The electron-electron exchange has been studied by magnetic susceptibility measurements when the magnitude of the exchange interaction is comparable to, or larger than, the thermal energy k_BT of the temperature examined in the susceptibility experiments. The exchange interactions much weaker than k_BT are not detectable in conventional susceptibility measurements. This limitation can be overcome by the use of ESR spectroscopy. This section concerns the determinations of exchange interactions by ESR spectroscopy. In the aspect of the exchange interactions, ESR spectroscopy has been used to characterize metal-containing proteins, spin-labeled biomolecules,[4.1.1-.2] and building blocks for magnetic materials. Our focus in this section is on organic polyradicals in which two or more unpaired electrons are exchange-coupled in a single molecule. As for systems in which two unpaired electrons are nonequivalent, i.e., one from transition metal ions and the other from organic radicals such as nitroxides, lots of studies have been reported and textbooks and reviews are available.[4.1.3-5]

Hyperfine splitting patterns for a given set of nuclear spins can be analyzed in terms of the chemical equivalence of the nuclear spins, or chemical environment in a molecule, and hyperfine coupling constants. One can determine the electron spin density distribution from the obtained hyperfine coupling constants and elucidate the electronic spin structure of the molecule, as found in many text books.[4.1.6-8] In addition, one can detect the exchange interaction from the splitting patterns when it is comparable to the hyperfine coupling. The exchange interaction is usually described by the scalar product of two vector operators of electron spins $S_1 = S_2 = 1/2$, as $H = -2J\mathbf{S}_1 \cdot \mathbf{S}_2$. The scalar parameter J represents the strength of the exchange interaction between $\mathbf{S}_1$ and $\mathbf{S}_2$. In Section 4.2, a quantum mechanical and quantum chemical description of the Heisenberg exchange interaction above is described only briefly. Section 4.3 exemplifies a spectral simulation for organic biradicals with two unpaired electrons which are coupled by an intramolecular exchange interaction. A large number of superficially different molecules are classified as biradicals.[4.1.9-10] We restrict ourselves here to nitroxide-based biradicals consisting of two radical fragments since they are sufficiently stable and widely used both as spin labels and building blocks for molecule-based magnetic materials.

The spin Hamiltonian suitable for such nitroxide biradicals is written as

$$H = -2J\mathbf{S}_1 \cdot \mathbf{S}_2 + g\mu_B B(S_1^Z + S_2^Z) + A(S_1^Z I_1^Z + S_2^Z I_2^Z). \tag{42}$$

The first term represents the Heisenberg exchange interaction with the exchange parameter J. The second and the third terms denote the electronic Zeeman and the hyperfine interaction for nitrogen nuclei, respectively. The hyperfine term in the spin Hamiltonian (42) does not commute with the exchange term, as demonstrated in Section 4.3. Therefore, the energy eigenvalues and the spin eigenfunctions of the spin Hamiltonian (42), and hence the resonance field and intensity of ^{14}N hyperfine ESR transitions, depend on the relative magnitudes of J and A. As shown below, extreme limits of $|J| \gg |A|$ or $|J| \ll |A|$ give simple hyperfine splitting patterns in ESR spectra reflecting the relative magnitude, while intermediate cases of $|J| \sim |A|$ give rise to complicated hyperfine splitting patterns as compared with monoradicals of $S = 1/2$ with one unpaired electron. The magnitude of hyperfine interactions $|A|$ falls within the order of 10^1 Oe (1 mK or 10^{-3} cm^{-1} for $g = 2$) for nitrogen nuclei in stable nitroxide radicals. We are able to determine the magnitude of exchange interaction in this range of energy. The hyperfine splitting patterns of biradicals are a spectroscopic "probe" for intramolecular exchange interactions. The energy eigenvalues and the resonance fields analytically obtained from the spin Hamiltonian (42) will be shown in Section 4.3 in order to allow nonexperts to simulate ESR spectra.

Hyperfine splitting patterns reflecting the magnitude of exchange interaction can be distorted by dynamical effects. Time-dependent perturbations affecting the ESR spectra of biradicals are only briefly discussed in Section 4.4. Finally, an application of hyperfine ESR spectroscopy to building blocks for molecule-based magnets is presented in

Section 4.5. The building block molecules have characteristics of two types of intramolecular exchange interactions; one is much larger than hyperfine interactions and the other is comparable to them.

4.2 The Heisenberg exchange coupling

Coulombic energies of electrons relevant to magnetic resonance of molecules or magnetism of molecular assemblages in solid states are described by exchange interactions between electron spins. As outlined below, the spin-dependent part of the coulombic energy between electrons, which are assumed to spatially localized on atoms or molecules, are termed "exchange interaction". The spin Hamiltonian of the exchange-coupled spins are given by $H_0 = -2J_{ij}\mathbf{S}_i \cdot \mathbf{S}_j$, where $\mathbf{S}_i$ and $\mathbf{S}_j$ denote vector operators of two electron spins ($S_i = S_j = 1/2$). The scalar parameter J_{ij} represents the strength of the exchange interaction between $\mathbf{S}_i$ and $\mathbf{S}_j$. The basic description of the exchange interaction has been given by Heisenberg,[4.2.1] Dirac,[4.2.2-3] and Van Vleck.[4.2.4-5] The spin Hamiltonian for many-body systems are written as

$$H = \sum_{i,j} -2J_{ij}\mathbf{S}_i \bullet \mathbf{S}_j . \qquad (43)$$

The formalism based on the inner product of spin operators as given above or Eq. (43) has been utilized to analyze the magnetic coupling within and between open-shell molecules. We have to recall, however, that the Heisenberg Hamiltonian above or Eq.(43) is phenomenological. On the theoretical side, accurate quantum chemical computations are necessary for J_{ij}. The electronic energy of the two spin system is $E_A = E_0 + K_{12}^0 - J_{12}^0$ for the triplet ($S = 1$) state and $E_A = E_0 + K_{12}^0 + J_{12}^0$ for the singlet ($S = 0$) state in usual notation. Therefore, the electronic energt is written as

$$E = E_0 + K_{12}^0 - \frac{1}{2}J_{12}^0 - 2J_{12}^0\mathbf{S}_1 \cdot \mathbf{S}_2 . \qquad (44)$$

Omitting the constants independent of the spin variables, the Hamiltonian of the two-spin system is written as $H_{12} = -2J_{12}^0\mathbf{S}_1 \cdot \mathbf{S}_2$. H_{12} or H_0 above is customarily called the Heisenberg Hamiltonian. Expanding the description to systems of $S > 1/2$ spins and to those containing more than two spin variables,[4.2.3-5] we reach the Heisenberg-Dirac-Van Vleck (HDV) Hamiltonian (43). The applicability of the Hamiltonian for nonorthogonal orbitals has also been established.[4.2.6-7] It should be noted that a single parameter J_{ij} contains many nontrivial integrals other than J_{ij}^0 in general cases of many-spin systems. Furthermoe, the Hamiltonian H_0 does not mean any "actual magnetic coupling" between the magnetic moments $g\mu_B\mathbf{S}_i$ and $g\mu_B\mathbf{S}_j$; it describes only in a phenomenological way the electronic or coulombic interaction between two open-shell atoms, molecules, or molecular fragments.

4.3 Biradicals composed of two radical fragments with a time-independent exchange coupling

Let us consider a simple system of two unpaired electrons in a single molecule. The system is informative to extend solution hyperfine ESR spectroscopy so as to be applicable to molecule-based magnetics. The molecule is either in a triplet or a singlet state. The energy separation between the states, the singlet-triplet energy gap, is designated by $2J$, as discussed in Section 4.2. We assume that the two radical fragments are identical, each having one nucleus of nitrogen with the nuclear spin quantum number $I = 1$. This type of the two-spin system is frequently found in nitroxide radicals which are used as spin labeling reagents and building blocks for molecule-based magnetic materials.

The spin Hamiltonian of the system is written by Eq.(42), as shown above. In the two terms S_i^Z and I_i^Z stand for the spin operators for the z components of the electron and nuclear spins in the ith fragment. The averaged g-factor g is adopted here in solution ESR spectroscopy. In Eq.(42) the hyperfine interactions are assumed to be small as compared with the electronic Zeeman interaction and nonsecular terms of the hyperfine coupling are neglected. The parameter A denotes the isotropic hyperfine coupling constant at the nitrogen nuclei. Since the Hamiltonian (42) commutes with the z-component of the total electron spin, $S_T^Z = S_1^Z + S_2^Z$ ($[H, S_T^z]=0$), the Hamiltonian is set up in the ket space spanned by the set of direct product defined as

$$S_i^z \left| m_1^S, m_2^S, m_1, m_2 \right\rangle = \pm \frac{1}{2} \left| m_1^S, m_2^S, m_1, m_2 \right\rangle, \quad (i = 1, 2), \tag{45}$$

where m_i^S and m_i are the quantum numbers for the z-components of the electron and nuclear spins in the ith fragment, respectively. The matrix representation of the spin Hamiltonian is given as

$$H = \begin{bmatrix} -\frac{J}{2} + \frac{A}{2}(m_1 + m_2) + g\mu_B B & 0 & 0 & 0 \\ 0 & \frac{J}{2} + \frac{A}{2}(m_1 - m_2) & -J & 0 \\ 0 & -J & \frac{J}{2} - \frac{A}{2}(m_1 + m_2) & 0 \\ 0 & 0 & 0 & -\frac{J}{2} - \frac{A}{2}(m_1 + m_2) - g\mu_B B \end{bmatrix} \begin{matrix} \left| +\frac{1}{2}, +\frac{1}{2} \right\rangle \\ \left| +\frac{1}{2}, -\frac{1}{2} \right\rangle \\ \left| -\frac{1}{2}, +\frac{1}{2} \right\rangle \\ \left| -\frac{1}{2}, -\frac{1}{2} \right\rangle \end{matrix}$$

$$\tag{46}$$

where the electron spin part of the basis ket $|m_1^S, m_2^S\rangle$ is shown in the right side of the matrix. The energy eigenvalues of the spin Hamiltonian are obtained by diagonalizing the matrix (46). The results are straightforward. The transition field $B(i \leftrightarrow j)$ and the relative transition intensity $P(i \leftrightarrow j)$ associated with the ith and the jth states for the allowed ESR transition with $\Delta M_S = \pm 1$ are calculated as

$$h\nu = E(i) - E(j) \equiv \Delta E(i \leftrightarrow j) \quad (\nu: \text{the microwave frequency}) \quad (47)$$

$$\delta\Delta E(i \leftrightarrow j) \equiv \Delta E(i \leftrightarrow j) - g\mu_B B \quad (48)$$

$$B(i \leftrightarrow j) = B_0 - \delta\Delta E(i \leftrightarrow j)/g\mu_B \quad (49)$$

$$P(i \leftrightarrow j) = \left|\langle i|S_T^X|j\rangle\right|^2 . \quad (50)$$

In Eq. (50) S_T^X denotes the spin operator of the x-component of the total electron spin, assuming the configuration of the oscillating microwave field perpendicular to the static field in conventional experiments. The central field B_0 in Eq. (49), $B_0 = h\nu/g\mu_B$ is the single resonance field appearing when both J and A are vanishing. We have four allowed transitions at most for each nuclear spin configuration $|m_1, m_2\rangle$ in the two-spin system, which are listed in Table 4 for readers' convenience. The resonance fields $\delta\Delta E(i \leftrightarrow j)/g\mu_B$ are measured from the central field B_0. Spectral simulations are made by applying an appropreate function such as a Lorentzian and a line width for all the allowed transitions of $4 \times (2I+1)^2 = 36$ and accumulating them. The overall distribution of the intensity is simulated by weighting the transitions with the relative intensity $P(i \leftrightarrow j)$ and the Boltzmann factor, $|\exp(-E(i)/k_B T) - \exp(-E(j)/k_B T)|$.

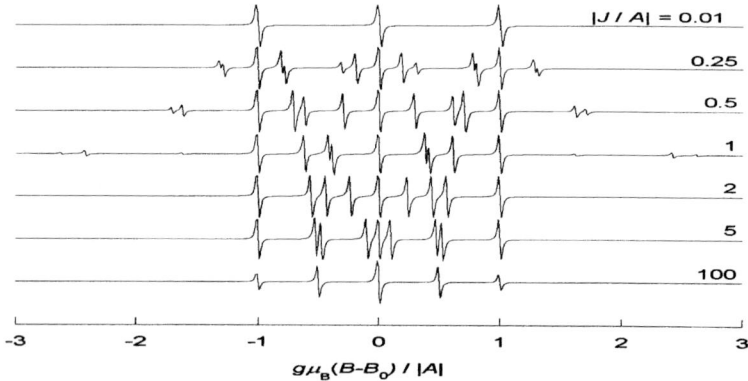

Figure 5. Simulated spectra of biradicals consisting of two identical radical fragments for various ratios of |J/A| as a function of the reduced magnetic field $g\mu_B(B-B_0)/|A|$. The Lorentzian line shapes are adopted with the line width of 0.01|A|.

11. Exchange-coupled high-spin clusters

In Figure 5 are shown the simulated spectra for various ratios of $|J/A|$ as a function of the reduced magnetic field $g\mu_B(B-B_0)/|A|$. When the exchange interaction between the two unpaired electron spins is weak as compared with the hyperfine coupling ($|J/A| = 0.01$), the resultant hyperfine splitting pattern consists of three equally weighted lines with a spacing of $|A|$. This pattern is the same as that expected for an $S = 1/2$ radical with one nitrogen nucleus. The weak exchange interaction makes each electron stay at one particular $I = 1$ nucleus such that a fluid solution of the biradical with such a weak J value is equivalent to that of an $S = 1/2$ radical with a doubled concentration. In contrast, if $|J| >> |A|$ ($|J/A| = 100$ in Fig.5), an electron spin interacts with the two nitrogen nuclei in the two fragments owing to the strong exchange interaction, resulting in five lines with the intensity ratio of 1:2:3:2:1 and the spacing of $|A|/2$. When the exchange interaction $|J|$ falls within the same order of magnitude as the hyperfine coupling $|A|$, the hyperfine splitting pattern is complicated as depicted in Figure 5. In some favorable cases, however, one can determine the strength of a specific exchange interaction directly from the splitting patterns of an ESR spectrum. An example is found in literature.[4.4.3] Another example by the present authors is presented in Section 4.5. It should be noted that the commutator of the Hamiltonian and the total electron spin

$$\mathbf{S}_T^2 \equiv (\mathbf{S}_1 + \mathbf{S}_2)^2 \tag{51}$$

$$[H, S_T^2] = \begin{bmatrix} 0 & 0 & 0 & 0 \\ 0 & 0 & A(m_1 - m_2) & 0 \\ 0 & -A(m_1 - m_2) & 0 & 0 \\ 0 & 0 & 0 & 0 \end{bmatrix} \tag{52}$$

has nonvanishing elements in the second and the third rows or columns, which correspond to the states with the z components of the total spin $M_S = 0$. This incommutability mixes the states of $M_S = 0$ with each other. The well-known spin wavefunctions for $M_S = 0$ ($S = 1, 0$) are no longer valid for $A \neq 0$.

Table 4. Resonance fields $B(i \leftrightarrow j) = B_0 - \delta\Delta E(i \leftrightarrow j)/g\mu_B$ and relative intensities of allowed ESR transitions for the Hamiltonian (42).[a]

i	j	$\delta\Delta E(i \leftrightarrow j)$ [b]	Relative intensity [b]
1	3	$-\frac{1}{2}\Delta + J + \frac{1}{2}A(m_1 + m_2)$	$\frac{1}{4} + \frac{J}{2\Delta}$
1	4	$\frac{1}{2}\Delta - J + \frac{1}{2}A(m_1 + m_2)$	$\frac{1}{4} + \frac{J}{2\Delta}$
2	3	$\frac{1}{2}\Delta + J + \frac{1}{2}A(m_1 + m_2)$	$\frac{1}{4} - \frac{J}{2\Delta}$
2	4	$-\frac{1}{2}\Delta - J + \frac{1}{2}A(m_1 + m_2)$	$\frac{1}{4} - \frac{J}{2\Delta}$

[a] $B_0 \equiv h\nu/g\mu_B$. [b] $\Delta = \sqrt{4J^2 + A^2(m_1 - m_2)^2}$.

442   Chapter 11

4.4 Effects of time-dependent interactions

A five-line hyperfine splitting pattern with an intensity ratio of 1:2:3:2:1 is expected for biradicals with two nitrogen nuclei in the strong exchange limit $|J| \gg |A|$, whereas a simple three-line pattern is found for the weak exchange limit $|J| \ll |A|$, as described in Section 4.3. One finds many examples of biradicals exhibiting such ESR spectra. Typical examples of nitroxide biradicals are given in literature.[4.4.1-3] It is worth noting that these hyperfine splitting patterns reflecting the exchange interactions $|J|$ as compared with the hyperfine coupling $|A|$ are valid only when the exchange interaction $|J|$ is independent of time. When $|J|$ is modulated by time-dependent perturbations, the intensity ratio of the five-line spectra is distorted even for $|J| \gg |A|$. In Figure 6 is given the solution ESR spectrum of a glutarate biradical,[4.4.4-5] which contains five lines implying $|J| \gg |A|$. An alternation of line widths is, however, found in the spectrum: The two lines between the central and the outermost lines are broadened with a seemingly weaker intensity. This alternation of line width is a historically important example of molecular dynamics affecting ESR spectra, which has been rationalized by Luckhurst[4.4.4-5] on the basis of. the relaxation matrix method developed first by Redfield.[4.4.6-8] The rationalization is outlined as follows: The matrix elements $R_{ii',jj'}$ of the the relaxation matrix **R** are given from the matrix

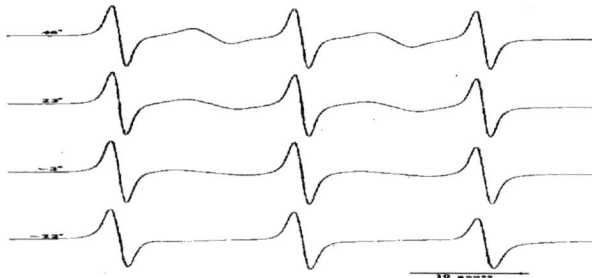

Figure 6. Hyperfine spectra of a glutarate biradical.[4.4.5] The spectra were observed at 45 °C, 23 °C, -3 °C, and -22 °C from top to bottom.

elements of $H'(t) = [J(t) - J_{av}] \mathbf{S}_1 \cdot \mathbf{S}_2$.[4.4.4-5] The time average of the exchange interaction J_{av} is expressed as $J_{av} = (\tau_a J_a + \tau_b J_b)/(\tau_a + \tau_b)$ in the simplest case, where an interconversion, or a jumping, is assumed to occur between two molecular conformations with the lifetimes τ_a and τ_b, and with the exchange interactions J_a and J_b. The eigenvalue of **R** multiplied by -1 for a hyperfine transition with the nuclear quantum numbers m_1 and m_2 is

$$T_2^{-1} = \frac{A^2(m_1 - m_2)^2}{4J_{av}^2} j(J_{av}), \qquad (53)$$

where $j(J_{av})$ is the spectral density for the fluctuation at an angular frequency $J_{av}/\hbar$. Equation (53) shows that the hyperfine transitions with $m_1-m_2 = 0$ are not affected by the fluctuation of J; the outermost lines ($m_1 = m_2 = \pm 1$) and one of the overlapping central lines ($m_1 = m_2 = 0$) remain sharp. The other lines are broadened, as observed in the experimental spectra. Only the above three lines ($m_1 = m_2 = \pm 1, 0$) with an equal weight appear with the spacing of $|A|$ in the limit of extreme broadening, which are just alike those for biradicals in the weak exchange limit $|J| \ll |A|$ or those for monoradicals with one $I = 1$ nucleus. As for an organic triradical, a line width broadening depending on nuclear quantum numbers m's has been reported.[4.4.9] The broadening has been explained in the same way as for the biradical.[4.4.9] Hyperfine ESR spectra reflecting dynamic phenomena of chemical importance besides the fluctuation of $|J|$ have been analyzed in the same way as described here. Examples are *cis-trans* isomerism, restricted rotations, ring inversion, proton exchange, and ion pair formation. An extensive review[4.4.10] has been published on dynamical effects on ESR line shapes, which includes dynamic effects other than the line width alternation.

An alternative way of analyzing ESR line shapes affected by some chemical dynamics is to use a general line shape equation, which has been derived in terms of the density matrix theory in the Liouville representation.[4.4.11-17] Kinetics for the conformational dynamics of a symmetrical triradical has also been analyzed assuming an interconversion among conformers with strong or weak exchange interactions in isosceles and equilateral triangle symmetries.[4.4.17]

4.5 An application to models for organic molecule-based magnets

An application of solution hyperfine ESR spectroscopy to building blocks for molecule-based magnets is described in this section. Molecule-based magnets and other molecular functionality magnetics have received great interests in recent years. For reviews of molecule-based magnetism, see References.[4.5.1-7] More than thirty crystalline ferromagnets have been discovered in genuinely organic molecule-based materials[4.5.1-7] since the discovery of the first organic ferromagnet.[4.5.8-10] Ferrimagnets have also attracted attention as one of the facile approaches to organic magnets since the first proposal by Buchachenko in 1979.[4.5.11] Ferrimagnetic spin ordering is conventionally regarded as an antiferro-magnetic ordering of different spin quantum numbers, e.g., $S = 1$ and $S = 1/2$, giving net and bulk magnetization in the solid state. This picture has been initiated by Néel's mean field theory.[4.5.12] The concept of organic ferrimagnetics and the seemingly practical feasibility are based on the tendency for organic open-shell molecules to have antiferromagnetic intermolecular interactions in their assemblages. The antiferromagnetic interactions would bring about antiparallel spin alignment between neighboring molecules with different magnetic moments to result in a possible ordered state in the mixed crystalline assemblages. Genuinely organic ferrimagnets composed of two discrete kinds of organic open-shell molecules, however, have not been discovered yet and is a challenging issue in spin chemistry and materials science. The possibility of the occurrence of the ferrimagnetic spin alignments has been examined by the authors both by experiments and theoretical calculations in an elaborate fashion.[4,5,13-21]

One of the practical difficulties in constructing molecule-based ferrimagnetics is co-crystallization of molecules with different spin quantum numbers. Generally, co-crystallization of distinct molecules in a crystal lattice gives rise to a decrease in entropy, which prevents them from packing in a structurally ordered fashion. Some purposive molecular designing is needed to overcome the separative crystallization driven by entropy. As one of such a purposive molecular designing for purely organic molecule-based ferrimagnets, the authors have proposed a strategy of "single-component ferrimagnetics",[4,5,22-23] which is schematically shown in Figure 7. When a π-biradical with an $S = 1$ ground state and a π-monoradical with $S = 1/2$ are

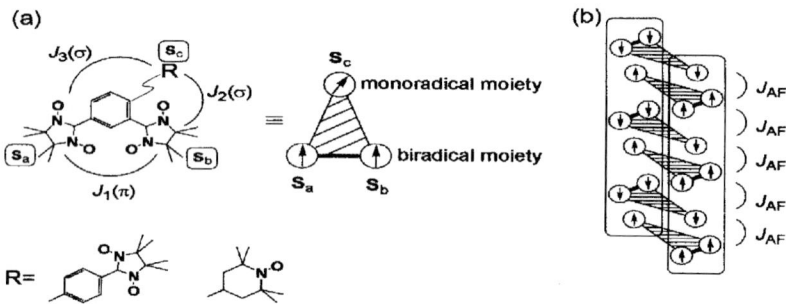

Figure 7. Schematic pictures of single-component ferrimagnetics. (a) A triradical consisting of weakly coupled biradical and monoradical moieties. $J_1(\pi)$ denotes the intramolecular exchange interaction within the biradical moiety with the spin S_a+S_b ($S_a = S_b = 1/2$), which is coupled with the monoradical moiety with $S_c = 1/2$ by $J_2(\sigma)$ and $J_3(\sigma)$ ($|J_2(\sigma)|\sim|J_3(\sigma)|<<|J_1(\pi)|$). (b) An example of the ferrimagnetic chain based on the intermolecular antiferromagnetic interaction J_{AF} in the crystalline solid. The rounded rectangles represent the alternating molecular chain of the biradical and the monoradical.

united by σ-bonds, the π-conjugation between the biradical and the monoradical moieties should be effectively truncated in the resultant triradical. The triradical has magnetic degrees of freedom for both $S = 1$ and

11. Exchange-coupled high-spin clusters 445

$S = 1/2$ in the single molecule, serving as a building block for ferrimagnets. Intermolecular π-orbital overlaps between an $S = 1$ moiety of one molecule and an $S = 1/2$ moiety of the adjacent molecule mimic molecular complexation of π-biradicals and π-monoradicals in a crystalline solid state. The chemical bonding between the biradical and the monoradical moieties within the single triradical molecule plays a role of binding only; the antiferromagnetic interactions underlying the ferrimagnetic spin alignment have a chance to undergo an *intermolecular* π-π orbital overlap of π-SOMO's of the constituent biradical and monoradical moieties. Thus, in this single-component ferrimagnetics it is important whether or not the additional intramolecular interactions through the σ-bonds $J_2(\sigma)$ and $J_3(\sigma)$ are negligible as compared with the intramolecular ferromagnetic interaction $J_1(\pi)$ in the biradical moiety. In this section, ESR spectroscopic determination of the intramolecular exchange interaction in organic triradicals is described.

The authors have designed and have synthesized triradicals **7** and **8** from benzoic acid monoradicals **9** and **10** and phenol-substituted biradical **11** (Fig.8) as building blocks of "single-component ferrimagnetics". The phenol biradical **11** is known to have an intramolecular ferromagnetic interaction of $J_1(\pi)/k_B = 13$ K.[4,5,19] The experimental spectra of the triradicals **7** and **8** are shown in Figures 9 and 10. The hyperfine splitting patterns of the triradicals are analyzed in the same way as those of biradicals in Section 4.3.

Fgiure 8. Stable nitroxide radicals as building blocks of "single-component ferrimagnetics".

The spin Hamiltonian for the triradicals is written as

$$H = g\mu_B B(S_a^z + S_b^z + S_c^z) - 2J_1(\pi)\mathbf{S}_a \cdot \mathbf{S}_b - 2J_2(\sigma)\mathbf{S}_b \cdot \mathbf{S}_c - 2J_3(\sigma)\mathbf{S}_c \cdot \mathbf{S}_a$$
$$+ A_{ab}(I_a^z S_a^z + I_b^z S_b^z) + A_c I_c^z S_c^z, \tag{54}$$

which consists of the electronic Zeeman, the Heisenberg exchange, and the hyperfine coupling terms. The three-centered Heisenberg exchange couplings are described with the inner products of the three vector operators. This is derived[4.5.24] in the same way as that of the two-centered system described in Section 4.2. A generalized formulation of the exchange couplings for molecular systems with more than two electrons is found in literature.[4.5.25-26] Since the nitronyl nitroxide radical group has two nitrogen nuclei which are chemically equivalent, the nuclear spin operators in Eq.(54) are given by

$$I_a^z \equiv I_{a1}^z + I_{a2}^z, \quad I_b^z \equiv I_{b1}^z + I_{b2}^z \tag{55}$$

$$I_c^z \equiv I_{c1}^z + I_{c2}^z \text{ (Triradical } \mathbf{7}), \quad I_c^z \equiv I_{c1}^z \text{ (Triradical } \mathbf{8}). \tag{56}$$

The Hamiltonian is set up in the ket space spanned by the set of direct product

$$S_I^z \left| m_a^S, m_b^S, m_c^S, m \right\rangle = \pm \frac{1}{2} \left| m_a^S, m_b^S, m_c^S, m \right\rangle, \quad (I=a,b,c) \tag{57}$$

where m is the collective index of a nuclear spin configuration of

$$m = \{m_{a1}, m_{a2}, m_{b1}, m_{b2}, m_{c1}(, m_{c2})\}. \tag{58}$$

The Hamiltonian is represented by a $2^3 \times 2^3$ matrix. The energy eigenvalues and eigenvectors can be obtainned analytically. They are, however, quite complicated, hampering their practical usability. Instead, from the exact, numerical diagonalization of the spin Hamiltonian, the resonance fields and the transition probability are calculated in the same way as Eq.(47) through Eq.(50). We have fifteen allowed transitions, i.e., fifteen pairs of the spin states (five within the multiplets and ten across the multiplets) with $\Delta M_S = \pm 1/2$ for one set Eq.(58) of nuclear configuration. We have simulated the hyperfine ESR spectra for the triradicals **7** and **8** by superposing at most $15 \times (2I+1)^5 = 3645$ or $15 \times (2I+1)^6 = 10935$ of the transitions.

The simulated spectra for triradical **7** are shown in Figure 9. In the weak exchange limit of $|J_2(\sigma)/A| \ll 1$ and $|J_3(\sigma)/A| \ll 1$, the simulated spectrum is a simple superposition of the spectrum of the biradical and that of the monoradical. The biradical spectrum consists of nine lines with equal spacings of $|A|/2$ due to five nitrogen nuclei, while the monoradical has the contribution of five lines attributable to two nitrogen nuclei. The biradical is in the strong exchange limit of $|J_{ab}/A| \gg 1$. Thirteen lines with equal spacing of $|A|/3$ show up in the simulated spectra due to the strong exchange limit of $|J_2(\sigma)/A| \gg 1$ and $|J_3(\sigma)/A| \gg 1$. The experimental spectra of the triradical **7** are reproduced by assuming $|J_2(\sigma)/A_a| \sim |J_3(\sigma)/A_a| \geq 10$, i.e., $|J_2(\sigma)|/k_B \sim$

11. Exchange-coupled high-spin clusters

$|J_3(\sigma)|/k_B \geq 10$ mK for $|A|/g\mu_B = 0.75$ mT. The simulated hyperfine splitting patterns are insensitive to the difference between $|J_2(\sigma)|$ and $|J_3(\sigma)|$: The deviation of molecular symmetry from the isosceles triangle is not detectable in the strong exchange limit, $|J_2(\sigma)/A_a| \sim |J_3(\sigma)/A_a| \geq 10$. The spectral simulation gives the lower limit of the hyperfine coupling constant in triradical **7**. The exchange interactions in **7** are much larger than the hyperfine couplings. The π-conjugation through the phenyl group of the monoradical moiety in **7** is not completely truncated. An estimate for the exchange interactions $J_2(\sigma)$ and $J_3(\sigma)$ in **7** has been obtained from paramagnetic susceptibility measurements in the crystalline solid state;[4,5,22] $|J_2(\sigma)|/k_B$ and $|J_3(\sigma)|/k_B$ are in the order of 0.1 K.

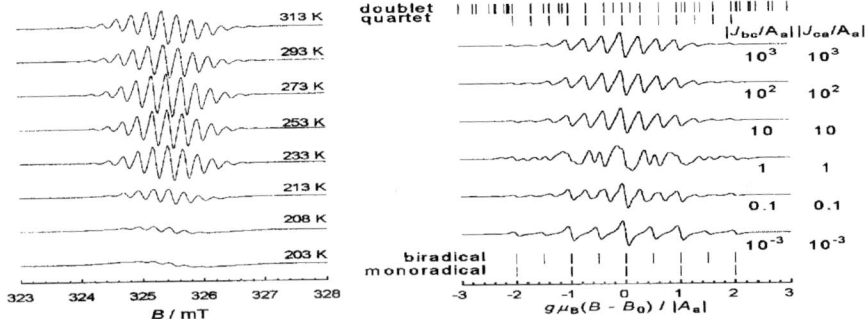

Figure 9. ESR spectra of **7**. (left) Observed in a toluene solution. (right) Simulated spectra. B_0 is the central field, $B_0 = h\nu/g\mu_B$. The sticks in the upper portion indicate the resonance field for the quartet and doublet states of the triradical in the strong exchange limit ($|J_{bc}/A_a| \sim |J_{ca}/A_a| >> 1$), while in the lower portion are shown those of the constituent biradical and monoradical in the weak exchange limit ($|J_{bc}/A_a| \sim |J_{ca}/A_a| << 1$). See text for the experimental conditions and the parameters employed for the simulation.

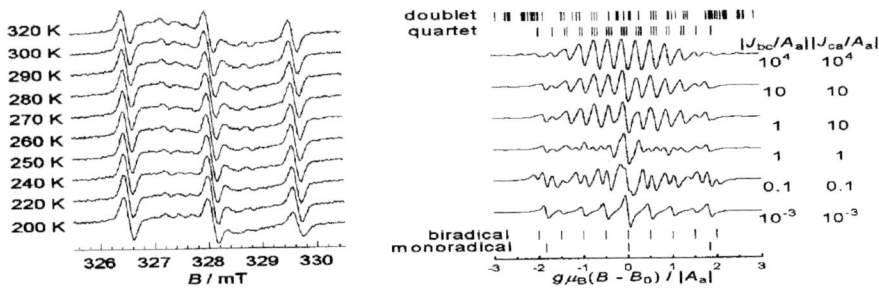

Figure 10. ESR spectra of **8**. (left) Observed in a toluene solution. (light) Simulated spectra. B_0 is the central field; $B_0 = h\nu/g\mu_B$. The sticks in the upper portion indicate the resonance field for the quartet and doublet states of the triradical for the strong exchange limit ($|J_{bc}/A_a| \sim |J_{ca}/A_a| >> 1$), while in the lower portion are shown those of the constituent biradical and monoradical for the weak exchange limit ($|J_{bc}/A_a| \sim |J_{ca}/A_a| << 1$). See text for the experimental conditions and the parameters employed for the simulation.

The solution ESR spectra of triradical **8** are shown in Figure 10. An intense triad of signals with spacings of the ^{14}N hyperfine coupling of the TEMPO monoradical feature in the spectra. The other signals between the triad decrease in intensity on lowering the temperature, as shown in the figure. The spectral simulations for **8** were carried out in the same way as **7** except that $|A_c|/g\mu_B$ =1.38 mT and I_{c2}=0. The simulated spectra with the spectrum for $|J_2(\sigma)/A_a| = |J_3(\sigma)/A_a| = 10^{-3}$ (at the bottom on the right) that is equivalent to the simple Lorentzian ($\Delta B = 0.04$ mT) are shown in Figure 10. The spectrum simulated for $|J_2(\sigma)/A_a| = |J_3(\sigma)/A_a| = 10^4$ demonstrates the strong limit of intramolecular exchange interaction. On the other hand, the simulated superposition of the spectra of the biradical and the monoradical, representing the weak limit of intramolecular exchange interaction. The experimental spectra at room temperature are interpreted in terms of neither the strong nor the weak exchange limit. Thus, the exchange interactions between the biradical and the monoradical moieties are estimated to be in the same order of magnitude as the hyperfine couplings; $|J_{bc}/A_a| \sim |J_{ca}/A_a| \sim 1$. Satisfactory agreement between the experimental and the simulated spectra, however, is not obtained. The disagreement suggests the fluctuation of J_{bc} and J_{ca} due to conformational interconversion of the molecule **8** in solution.

Effects of molecular dynamics on ESR spectra in solution have not been studied so much for triradicals as compared with those for biradicals. ESR spectra of a triradical with three nitroxide groups containing three nitrogen nuclei have been analyzed in terms of the relaxation matrix[4,4,9] and the composite Liouville space formalism.[4,4,17] We present here an alternative perturbation treatment approach,[4,5,27] assuming that the exchange interaction within the biradical moiety $|J_1(\pi)|$ is much larger than those between the biradical and the monoradical moieties $|J_2(\sigma)|$, $|J_3(\sigma)|$. This approach gives the assignment of specific signals in complicated hyperfine splitting patterns, explaining the appearance of the triad signals for **8**. Such an assignment is difficult to obtain by the exact diagonalization described above. The spin Hamiltonian (54) is split into two parts

$$H^{(0)} = g\mu_B B(S_a^z + S_b^z + S_c^z) - 2J_1(\pi)\mathbf{S}_a \cdot \mathbf{S}_b, \tag{59}$$

$$H^{(1)} = -2J_2(\sigma)\mathbf{S}_b \cdot \mathbf{S}_c - 2J_3(\sigma)\mathbf{S}_c \cdot \mathbf{S}_a + A_{ab}(I_a^z S_a^z + I_b^z S_b^z) + A_c I_c^z S_c^z. \tag{60}$$

The unperturbed Hamiltonian (59) is block-diagonalized according to the z-component of the total electron spin M_S

$$(\sum_I S_I^z)|m_a^S, m_b^S, m_c^S, m^I\rangle = M^S|m_a^S, m_b^S, m_c^S, m^I\rangle, \; (M_S = \pm 3/2, \pm 1/2) \tag{61}$$

11. Exchange-coupled high-spin clusters

$$\mathbf{H}^{(0)} = \begin{bmatrix} \frac{3}{2}g\mu_B B - \frac{1}{2}J_1(\pi) & \mathbf{O} & \mathbf{O} & 0 \\ \mathbf{O} & \mathbf{H}^{(0)}_{234} & \mathbf{O} & \mathbf{O} \\ \mathbf{O} & \mathbf{O} & \mathbf{H}^{(0)}_{567} & \mathbf{O} \\ 0 & \mathbf{O} & \mathbf{O} & -\frac{3}{2}g\mu_B B - \frac{1}{2}J_1(\pi) \end{bmatrix}. \tag{62}$$

The four block submatrices correspond to M_S = +3/2, +1/2, -1/2, and -3/2. The submatrices for $M_S = \pm 1/2$ are given as follows;

$$\mathbf{H}^{(0)}_{234} = \begin{bmatrix} \frac{1}{2}g\mu_B B + \frac{1}{2}J_1(\pi) & -J_1(\pi) & 0 \\ -J_1(\pi) & \frac{1}{2}g\mu_B B + \frac{1}{2}J_1(\pi) & 0 \\ 0 & 0 & \frac{1}{2}g\mu_B B - \frac{1}{2}J_1(\pi) \end{bmatrix}, \tag{63}$$

$$\mathbf{H}^{(0)}_{567} = \begin{bmatrix} -\frac{1}{2}g\mu_B B + \frac{1}{2}J_1(\pi) & -J_1(\pi) & 0 \\ -J_1(\pi) & -\frac{1}{2}g\mu_B B + \frac{1}{2}J_1(\pi) & 0 \\ 0 & 0 & -\frac{1}{2}g\mu_B B - \frac{1}{2}J_1(\pi) \end{bmatrix}. \tag{64}$$

By diagonalizing the matrix (62), we obtain the zeroth order energy eigenvalues and the eigenvectors. The energies are listed in literature.[4,5,28] The eight eigenvectors $\{\mathbf{x}_i^{(0)}\}$ are represented in columns of a matrix $\mathbf{U}$;

$$\mathbf{U} = [\mathbf{x}_1^{(0)} \mathbf{x}_2^{(0)} \mathbf{x}_3^{(0)} \mathbf{x}_4^{(0)} \mathbf{x}_5^{(0)} \mathbf{x}_6^{(0)} \mathbf{x}_7^{(0)} \mathbf{x}_8^{(0)}] = \begin{bmatrix} 1 & 0 & 0 & 0 & 0 & 0 & 0 & 0 \\ 0 & -\frac{1}{\sqrt{2}} & \frac{1}{\sqrt{3}} & -\frac{1}{\sqrt{6}} & 0 & 0 & 0 & 0 \\ 0 & \frac{1}{\sqrt{2}} & \frac{1}{\sqrt{3}} & -\frac{1}{\sqrt{6}} & 0 & 0 & 0 & 0 \\ 0 & 0 & \frac{1}{\sqrt{3}} & \frac{2}{\sqrt{6}} & 0 & 0 & 0 & 0 \\ 0 & 0 & 0 & 0 & -\frac{1}{\sqrt{2}} & \frac{1}{\sqrt{3}} & -\frac{1}{\sqrt{6}} & 0 \\ 0 & 0 & 0 & 0 & \frac{1}{\sqrt{2}} & \frac{1}{\sqrt{3}} & -\frac{1}{\sqrt{6}} & 0 \\ 0 & 0 & 0 & 0 & 0 & \frac{1}{\sqrt{3}} & \frac{2}{\sqrt{6}} & 0 \\ 0 & 0 & 0 & 0 & 0 & 0 & 0 & 1 \end{bmatrix}. \tag{65}$$

The perturbed vectors and the state energies to the first order are given as

$$\mathbf{x}_n = \mathbf{x}_n^{(0)} + \sum_{i \neq n} \frac{H_{in}^{(1)}}{E_n^{(0)} - E_i^{(0)}} \mathbf{x}_i^{(0)}, \quad E_n = E_n^{(0)} + H_{nn}^{(1)}, \tag{66}$$

where $H_{in}^{(1)}$ is the matrix element of the first order submatrices

$$\mathbf{U}^{-1}\mathbf{H}^{(1)}\mathbf{U} = \begin{bmatrix} -J(\sigma)+\frac{1}{2}[A_b(m_a+m_b)+A_m m_c] & 0 & 0 & 0 \\ 0 & (\mathbf{U}^{-1}\mathbf{H}^{(1)}\mathbf{U})_{234} & 0 & 0 \\ 0 & 0 & (\mathbf{U}^{-1}\mathbf{H}^{(1)}\mathbf{U})_{567} & 0 \\ 0 & 0 & 0 & -J(\sigma)-\frac{1}{2}[A_b(m_a+m_b)+A_m m_c] \end{bmatrix}$$

(67)

$$(\mathbf{U}^{-1}\mathbf{H}^{(1)}\mathbf{U})_{234} = \begin{bmatrix} \frac{1}{2}A_m m_c & \frac{1}{\sqrt{6}}A_b(m_a-m_b) & -\frac{1}{2\sqrt{3}}A_b(m_a-m_b) \\ \frac{1}{\sqrt{6}}A_b(m_a-m_b) & -J(\sigma)+\frac{1}{6}[A_b(m_a+m_b)+A_m m_c] & \frac{1}{3\sqrt{2}}[A_b(m_a+m_b)-2A_m m_c] \\ -\frac{1}{2\sqrt{3}}A_b(m_a-m_b) & \frac{1}{3\sqrt{2}}[A_b(m_a+m_b)-2A_m m_c] & 2J(\sigma)+\frac{1}{6}[2A_b(m_a+m_b)-A_m m_c] \end{bmatrix},$$

(68)

$$(\mathbf{U}^{-1}\mathbf{H}^{(1)}\mathbf{U})_{567} = \begin{bmatrix} -\frac{1}{2}A_m m_c & -\frac{1}{\sqrt{6}}A_b(m_a-m_b) & \frac{1}{2\sqrt{3}}A_b(m_a-m_b) \\ -\frac{1}{\sqrt{6}}A_b(m_a-m_b) & -J(\sigma)-\frac{1}{6}[A_b(m_a+m_b)+A_m m_c] & -\frac{1}{3\sqrt{2}}[A_b(m_a+m_b)-2A_m m_c] \\ \frac{1}{2\sqrt{3}}A_b(m_a-m_b) & -\frac{1}{3\sqrt{2}}[A_b(m_a+m_b)-2A_m m_c] & 2J(\sigma)-\frac{1}{6}[2A_b(m_a+m_b)-A_m m_c] \end{bmatrix}.$$

(69)

where the two exchange interactions $J_2(\sigma)$ and $J_3(\sigma)$ are assumed to be identical for simplicity; $J_2(\sigma) = J_3(\sigma) \equiv J(\sigma)$. Two pairs of the vectors $\{\mathbf{x}_3^{(0)}, \mathbf{x}_4^{(0)}\}$ and $\{\mathbf{x}_6^{(0)}, \mathbf{x}_7^{(0)}\}$ are degenerate in energy to the zeroth order. Diagonalizing the perturbation matrices (68) and (69) with respect to these degenerate states, the state energies $E(i)$ and vectors $\{\mathbf{x}_i\}$ to the first order are obtained. The energies are listed in Table 4.5.2. We have fifteen allowed transitions with $\Delta M_S = \pm 1/2$, as mentioned above. Since the intramolecular exchange interaction within the biradical moiety $J_1(\pi)/k_B = 13$ K is much larger than all other interactions, one of the two doublet states D1 lies far apart above the other doublet (D2) and the quartet (Q) states, as depicted in Figure 11. Therefore, we can exclude contribution from allowed ESR transitions across the multiplets, i.e., those between D1 and Q or between D1 and D2. One has only nine pairs of states, five within the multiplets and four across the multiplets of Q and D2. The resonance fields $B(i \leftrightarrow j)$ are calculated in the same way as Eqs.(47) – (50) for the $S = 1/2$ radicals. In literature[4.5.27-28] are listed the first-order resonance fields as measured from the central field B_0.

11. Exchange-coupled high-spin clusters

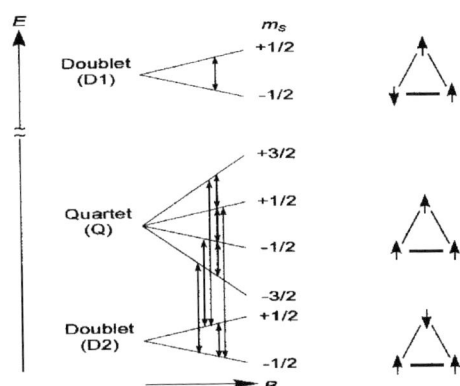

Figure 11. Energy diagram for the triradicals **7** and **8**. The arrows indicate the allowed transitions. The hyperfine sublevels are omitted for clarity. In the right side are shown the schematic representations of electron spin configurations for the quartet (Q) and doublet (D1, D2) states.

It should be noted that the resonance field $\delta \Delta E(2 \leftrightarrow 5) /g\mu_B$ is independent of the exchange interaction $J(\sigma)$ ($J_2(\sigma)$ and $J_3(\sigma)$). Thus, the transition $\Delta E(2 \leftrightarrow 5)$ is little affected by the fluctuation of the $J(\sigma)$ values. Furthermore, the hyperfine splitting of this transition is $|A_m|$, that is the hyperfine coupling constant of the monoradical. The intense triad of the signals with the spacing of $|A_m|$ in the observed ESR spectra is assigned to the transition of $\Delta E(2 \leftrightarrow 5)$ in D1. Resonance fields of other transitions have a contribution of $J(\sigma)$, which fluctuates in a solution. The appearance of the intense triad signals and the rest of the broadened signals indicate that the exchange interactions through the σ bondings of the ester group are quite small as compared with that within the biradical moiety:

$$J_1(\pi) >> |J_2(\sigma)|, |J_3(\sigma)| \approx |A_{ab}|, |A_c|. \tag{70}$$

This indicates that the triradical **8** has the potential to be building blocks of single-component organic ferrimagnets.

Some of the perturbation-based useful expressions derived in this section are given in literature.[4,5,27-28]

5. HIGH SPIN CHEMISTRY OF VARIOUS MOLECULAR CLUSTERS; UTILIZATION OF HIGH-FIELD/HIGH-FREQUENCY ESR AND PULSED ESR SPECTROSCOPY

This section deals with the recent documentation, focusing on important issues in terms of electron magnetic resonance spectroscopy with the emphasis of recent important progress in chemistry, materials science and physics of molecule-based high-spin clusters. Readers are recommended to refer to the monographs "EPR of Exchanged Coupled Systems" by Bencini and Gatteschi[5.1.1] and "Molecular Magnetism" by Kahn[1.2b] for comprehensive treatises on metal-ion-based molecular high-spin multi-clusters in terms of electron magnetic resonance and materials science, respectively.

5.1 Inorganic molecule-based metal high-spin clusters including dinuclear triplet-state clusters

In this section recent progress on pure inorganic molecule-based metal high-spin clusters is summarized. Readers are recommended to refer to other chapters for inorganic molecule-based non-transition metal clusters with high spin multiplicities, where important chemical species such as dimeric $(NO)_2(Na^+)x$ in the triplet state adsorbed on zeolites are included.

5.1.1 Exchange-coupled dinuclear clusters

In isolated high-spin systems composed of dinuclear clusters, the spin Hamiltonian given in Section 2.1 is generally applicable. Throughout this chapter, dinuclear clusters are termed metal-ion-based molecular bi-clusters. The total spin Hamiltonian can include all the terms and any resulting effective spin Hamiltonian can be expressed in terms of component spin Hamiltonian terms. If the component spin Hamiltonian parameters in tensor-based terms are known, experimentally or theoretically, the spin Hamiltonian parameters for the exchange-coupled dinuclear cluster can be derived, reproducing the corresponding ESR fine-structure spectrum. This procedure is phenomenological, but applicable to either homo-spin or hetero-spin molecular systems. During the spectral simulation procedure, molecular structural parameters are required for unitary transformation into desired molecular principal-axes systems. One can assume the molecular structures, theoretically, or derive them from X-ray structural analyses.

In dealing with the exchange-coupled spin systems in terms of ESR spectroscopy, a crucial point is to evaluate the interacting terms such as the exchange coupling J^{AB}, the magnetic dipolar coupling D^{AB} and the asymmetric terms d^{AB} between the component spins A and B. In the approximation given in Section 2, the interacting terms can be derived from dinuclear clusters of two effective spin centers. For multi-centers, this is not the case, generally. Only for a series of well controlled molecule-based high-spin clusters, semi-empirical estimations for the interacting terms can be acquired. Otherwise, for most molecular clustering systems, quantum chemical computations which enable us to consider nearby excited electronic states with high accuracy are desirable to estimate the interacting terms, the

11. Exchange-coupled high-spin clusters

computations for high spin systems with non-vanishing angular momenta are formidable. For most cases, the asymmetric terms are regarded to be negligible, although they are important in terms of group theoretical arguments for the systems under study. In order to understand molecular high-spin clusters in the crystalline state, it is worth noting that ESR spectroscopy should be supported with the help of magnetic susceptibility measurements of the bulk magnetic properties.

The oxalato-briged bi-clusters (dimers) of Cu(II) have been thoroughly studied in magnetochemistry because they provide model systems to give testing grounds for theories describing magnetic interactions on a molecular orbital basis.[5.1.2-4] Focusing on other metal-based oxalato-bridge homo-spin pairs, examples are known with Ti(III),[5.1.5] Fe(III),[5.1.6-7] and Cr(III).[5.1.8] Also, an oxalato-bridged hetero-pair Cr(III)-Fe(III) was reported.[5.1.9-10] The high-spin-based hetero-spin bi-cluster is particularly intriguing in terms of molecular exchange couplings, although the crystal structure was not reported. Triki et al. compared the hetero-spin pair (Cr(III)-Fe(III)) with the homo-spin ones (Cr(III)-Cr(III) and Fe(III)-Fe(III)). From the magnetic susceptibility measurements for the homo- and hetero-spin pairs, the exchange coupling parameters were obtained. In Table 5 are summarized their adjustable magnetic parameters in terms of bulk magnetic properties. The table shows that in the homo-spin pairs antiferromagnetic interactions between the spins were observed, while the ferromagnetic exchange interaction between the spins was observed in the hetero-spin pair. The ESR spectra for these dinuclear clusters qualitatively agree with the temperature dependence of the magnetic susceptibilities. Figure 12 shows the temperature-dependent ESR spectra for the Cr(III)-Cr(III), Fe(III)-Fe(III), and Cr(III)-Fe(III) pairs. In the Cr-Cr clusters, only one peak around $g = 2.00$ was observed and an asymmetric large signal around 0.2 T was observed, indicating that the feature is typical of ESR spectra for high-spin systems with large zero-field splittings. The accurate spin Hamiltonian parameters have neither been determined in terms of ESR spectroscopy and mechanisms of the exchange coupling nor interpreted yet. Possible applications of high-field/high-frequency ESR spectroscopy and parallel microwave excitation spectroscopy at liquid helium temperature are desirable to give direct information about the high-spin ground or intermediate state of the homo- and hetero-spin pairs. Particularly, pulse-based two-dimensional electron spin nutation spectroscopy with high time-resolution for dinuclear molecular bi-clusters at low temperature gives reliable spin identification in a straightforward manner. Recently, many authors have reported various types of molecule-based exchange-coupled dinuclear clusters.[5.1.11-25] For most dinuclear clusters, the exchange interaction is larger than the energy of the X-band microwave. Thus, high-field/high-frequency ESR technique is a powerful tool.[5.1.26-27]

Table 5. Adjustable magnetic parameters for Cr(III)-Cr(III), Fe(III)-Fe(III), and Cr(III)-Fe(III) pairs.

Compounds	J/cm^{-1}	G	θ/cm^{-1}	ρ/ %	$N\alpha$ emu mol^{-1}
Cr-Cr^a	-3.23	1.989			610 x 10^{-6}
Fe-Fe^b	-3.84	2.00		0.5	
Cr-Fe^c	+1.10	1.97	-0.97		547 x 10^{-6}

^{a} $\chi_m = \chi_D + N\alpha$.
^{b} $\chi_m = (1-\rho)\chi_D + \rho(35N\beta^2 g^2/(12kT))$.
^{c} $\chi_m = \chi_D T/(T-\theta) + N\alpha$, where the dimer susceptibility χ_D can be obtained for any isotropic dimer.

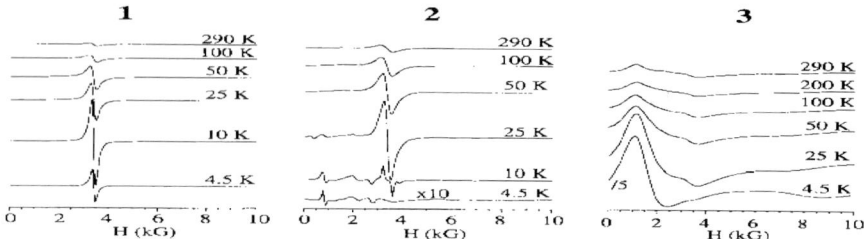

Figure 12. ESR spectra of dinuclear-pair compounds in solids at various temperatures.[5.1.10] 1:Cr-Cr, 2:Fe-Fe, 3:Cr-Fe.

5.1.2 Exchange-coupled trinuclear clusters; spin frustration and mixed valence states

There has been much interest in the spin frustration of equilateral triangular tri-clusters composed of paramagnetic metal-ions. A crucial point of the spin frustration systems is the occurrence of the antiferromagnetic interaction between the spin centers. Figure 13 shows the schematic model of the spin frustration systems for trinuclear clusters. J^{AB}, J^{BC}, and J^{CA} denote the exchange interaction between the A-B, B-C, and C-A spin sites, respectively. The frustration systems are important issues in spin chemistry,

Figure 13. Schematic models of the spin frustration systems for trinuclear clusters.

11. Exchange-coupled high-spin clusters

and related structural chemistry and magnetics in quantum terms because the whole systems are not described by simple spin schemes for the ground stabilized state. In Figure 14, a simplified model for trinuclear clusters is given on the left side, where the three spin sites are composed of Cu(II) ($S_i = S = 1/2$) with $J^{AB} = J^{CA} = J_1$ and $J^{BC} = J_2$, noting that exchange coupling terms in spin Hamiltonians should be expressed by coefficients $-2J_i$ or $-2J^{ij}$, i.e., $-2J^{ij}S_i \cdot S_j$. Spin structures representing typical ground states are described for the cases of $2J_1 \neq 2J_2$. When the spin states are represented by $\mathbf{S} = \mathbf{S}^A + \mathbf{S}^B + \mathbf{S}^C$, the resulting states are two doublets and one quartet. The ground state for the isosceles three spin system is described by a ket $\mid S, S' \rangle$ with the definition of the total spin $S = S_1 + S_2 + S_3$ and $S' = S_2 + S_3$. For $2J_1 > 2J_2$, the ground state is given by $\mid 1/2, 1 \rangle$ and for $2J_1 < 2J_2$, the ground state by $\mid 1/2, 0 \rangle$. For both cases, the highest spin state $\mid 3/2, 1 \rangle$ is above the two doublet states by $-2(J_1/2 + J_2/4)$. For $2J_1 = 2J_2$, the two doublet states are degenerate, giving rise to trinuclear spin frustration. The spin frustration associated with vibronic interaction is expected to cause rapid spin fluctuation. Many authors have studied spin-frustration systems until recently.[5.1.28-41]

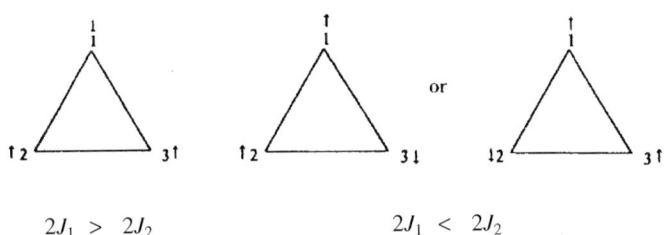

$2J_1 > 2J_2$ $2J_1 < 2J_2$
Figure 14. Spin frustration modes of the ground spin state for two cases.

Recently, So et al. have reported ESR studies of the Cu(II)-based equilateral triangular clusters **1**:Na$_9$[β-SiW$_9$O$_{37}${Cu(H$_2$O)}$_3$][5.1.42] and **2**:K$_{12}$[As$_2$W$_{18}$O$_{66}$Cu$_3$(H$_2$O)$_2$]·11H$_2$O,[5.1.43] revealing the possible occurrence of the trinuclear spin frustration. The crystal structure for **1** has not been reported. The presumed molecular structure for 1 and the crystal structure of **2**, however, have been reported by Robert et al.,[5.1.44-45] showing that three Cu(II) ions form a nearly equilateral triangle. From the temperature dependence of the magnetic susceptibility, the antiferromagnetic exchange interactions between the Cu(II) spins ($S = 1/2$) were confirmed: $2J_1 = 2J_2 = -7.8$ cm^{-1}. The sizable magnitude of the exchange coupling indicates that the through-bond exchange interactions occur. If the bridging oxygen and tungsten atoms play a role of superexchange interactions between the Cu(II) spin sites, the lowest doublet states are expected to be dynamically stabilized also by vibronic coupling or a vibrational spin-orbit one. Such spin frustration systems undergoing pseudo-rotations are expected to be ESR silent in terms of conventional ESR spectroscopy at low temperatures. They measured conventional ESR spectra of **2**, in which the crystal structure is known, in order to obtain information on the frustration effects. Figure 15 shows the observed and simulated ESR fine-structure spectra for **2** at room temperature (left) and 77 K (right). A second-order-perturbation-based simulation was carried out assuming axial symmetry: $S = 3/2$, $g_x = g_y = 2.226$, $g_z = 2.062$, $D = 0.0223$ cm^{-1}, and $E = 0$ cm^{-1} at 77 K: the smaller D value of 0.0189 cm^{-1} was determined at room temperature. No doublet state was explicitly detected, indicating the occurrence of rapid spin fluctuation. In terms of tensor-based exchange-coupling analyses for the observed D values the inter-distance between the Cu(II) sites was determined to be 0.420 nm at 77 K and 0.437 nm at room temperature, suggesting the possible vibronic contraction of the nuclear distance. Also, it was indicated that the doublet-quartet energy gap is enough smaller than the thermal energy at room temperature in **2**. For the complete analysis, solving the puzzles of this molecular spin frustration system, high-field/high-frequency ESR spectroscopy at liquid helium temperatures is required. The relaxation anomaly due to the pseudo rotation of the trinuclear sites should be hampered at temperatures below 2 K. Another model system **1** for the triangular spin frustration was ESR silent at ambient temperature, suggesting an anomalously rapid spin relaxation occurring in the excited high-spin quartet state. In addition, the temperature dependence of the magnetic susceptibility of **1** in the range of 5-300 K has given another complicated puzzle,[5.1.42] to which single-crystal high-field/high-frequency ESR spectroscopy finds a clue.

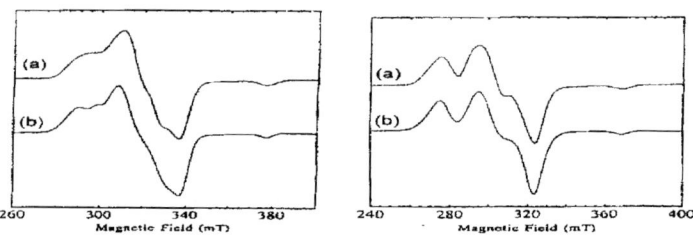

Figure 15. ESR spectra observed for **2**:$K_{12}[As_2W_{18}O_{66}Cu_3(H_2O)_2]\cdot 11H_2O$ (polycrystalline) at room temperature (left) and 77 K (right).[5.1.43] (a) and (b) denote the observed and simulated spectra ($S = 3/2$), respectively.

Sandwich-type triangular clusters with various transition metals were characterized by ESR spectroscopy.[5.1.46] They reveal the antiferromagnetic exchange interactions between the metals from the temperature dependence of the signal intensity of high-field/high-frequency ESR spectroscopy. Similar clusters were synthesized and studied by many authors.[5.1.46-48]

However, detail information on the electronic spin structures due to the frustration has not been clear yet. A crucial issue of the frustration systems in terms of X-band ESR spectroscopy is the enormous line broadening due to rapid spin relaxations intrinsic to the spin degeneracy or rapid relaxation assisted by vibronic quantum spin mixing between the ground and excited spin states. For some cases of spin frustration systems, high-field/high-frequency ESR techniques have proven to be useful.[5.1.36, 5.1.49-54]

Okubo et al. have examined a triangular-*Kagome* antiferromagnet $Cu_9X_2(cpa)_6$ by high-field/high-frequency EPR spectroscopy, where cpa and X denote carboxypentonic acid and halogen atoms (F, Cl, and Br).[5.1.49] They have obtained the temperature dependence of the line width of the ESR spectra for the *Kagome* lattice.

McCusker et al. have reported the μ_3-oxide trinuclear mixed-valence manganese clusters of $Mn(II)Mn(III)_2O$.[5.1.55-59] The reported structures of this type were $[Mn_3O(acetato)_6(pyr)_3]pyr$,[5.1.56] $[Mn_3O(acetato)_6(3Cl-pyr)_3]$,[5.1.57] $[Mn_3O(benzoato)_6(pyr)_2(H_2O)]1/2MeCN$,[5.1.58] and $[Mn_3O(X-benzoato)_6L_3]$ (X = 2-F, 2-Cl, 2-Br, 3-F, 3-Cl, 3-Br; L = pyridine or water).[5.1.59] A remarkable feature of these mixed-valence clusters are the variety of their ground-state. For their ESR analysis, Vincent et al. have applied the similar spin Hamiltonian of isotropic exchange-couplings used for the triangular spin frustration systems,[5.1.58] given by

$$H = -2J^{12}\mathbf{S}^1 \cdot \mathbf{S}^2 - 2J^{13}\mathbf{S}^1 \cdot \mathbf{S}^3 - 2J^{23}\mathbf{S}^2 \cdot \mathbf{S}^3$$
$$= -2J(\mathbf{S}^1 \cdot \mathbf{S}^2 + \mathbf{S}^1 \cdot \mathbf{S}^3) - 2J^*(\mathbf{S}^2 \cdot \mathbf{S}^3) \quad , \quad (71)$$
$$= -J(\mathbf{S}^{T2} - \mathbf{S}^{*2}) - 2J^*\mathbf{S}^{*2}$$

where $J = J^{12} = J^{13} = J_1$, $J^* = J^{23} = J_2$, $\mathbf{S}^T = \mathbf{S}^1 + \mathbf{S}^*$, and $\mathbf{S}^* = \mathbf{S}^2 + \mathbf{S}^3$. McCusker et al. have calculated the J/J^* dependence of the energy. The energy diagram reveals that there are various spin multiplicities for the ground state of the mixed-valence μ_3-oxide trinuclear Mn cluster. Ribas et al. have shown from magnetic susceptibility measurements and ESR spectroscopy that the variety of the ground states is governed by ligands.[5.1.59]

Referred to mixed-valence clusters, one of the important processes in nature occurs in the oxygen-evolving complex (OEC) of photosystem II (PSII), where the four-electron oxidation of water to molecular oxygen is believed to be catalyzed by a cluster composed of four manganese ions. Mixed-valence homo- and hetero-trinuclear linear clusters containing the [tris(dimethylglyoximato)-metalate(II)]$^{4-}$ anions as bridging ligands have been examined. From the X-ray diffraction, magnetic susceptibility, and

ESR spectra, the exchange interaction between the nucleuses have been discussed.[5.1.60-67] Manganese clusters of biological importance will be only briefly dealt with in Section 6 of this chapter. Sizable organometallic triangular clusters of the transition metal [(Cp*)(dppe)Fe(III)-]$^+$ units bridged by 1,3,5-triethynylbenzene spacers have been studied, emphasizing that triangular topology is important and the ferromagnetic coupling occurs at nanoscale distances between the metal spin carriers.[5.1.68]

5.2 Inorganic molecule-based high-spin large clusters revealing quantum spin tunneling: Single-molecule magnets

This section deals with molecule-based high-spin metallic multi-clusters which reveal apparent magnetic hystereses associated with molecular superparamagnetic entities at low temperature. These large clusters have emerged recently and they have been the focus of the current topics in the field of molecule-based magnetism and high spin chemistry. Their microscopic details have most successfully been characterized by invoking high-field/high-frequency ESR spectroscopy.

5.2.1 Single-molecule magnets

In last decades, impressive progress has been made in molecular design and synthesis in the field of molecule-based magnetism, leading to various types of high-spin polynuclear transition-metal clusters. Also, exotic magnetic properties have emerged. For example, a cyano-bridged extremely high-spin cluster, [Mn$^{II}_9$(μ-CN)$_{30}$Mo$^V_6$], with an $S = 51/2$ ground state has recently been synthesized and magnetically characterized by Larionova et al.[5.2.1] In addition, nano-scale molecular clusters whose magnetic behaviors are superparamagnetic have emerged. These high spin multi-clusters, currently termed "Single-Molecule Magnets(SMM's)",[5.2.2] have attracted increasing interest in this field.[5.2.3] This is partially due to the fact that SMM's could potentially be used for storage of a large density of information in their well-defined nano-scale dimensions. Since SMM's exhibit intriguing quantum size effects such as quantum tunneling of the magnetization,[5.2.4-6] these materials are currently considered as a candidate for quantum computing devices.[5.2.7] Moreover, SMM's could potentially provide us with a deep insight into the interpretation of the transition from molecular paramagnetism (quantum mechanical phenomena) to bulk ferromagnetism (macroscopic "classical" phenomena). Magnetic characterizations of SMM's have extensively been carried out in terms of ESR,[5.2.8-21] NMR,[5.2.16, 5.2.22-29] neutron diffraction[5.2.30] or inelastic neutron scattering,[5.2.31] torque[5.2.32-33] or

11. Exchange-coupled high-spin clusters

micro-hall magnetometry,[5.2.34] specific heat[5.2.35-37] or heat capacity measurements,[5.2.38-40] Mössbauer,[5.2.15, 5.2.41] and magnetization measurements. In particular, ESR spectroscopy provides direct information about the magnetic anisotropy and energy levels of their ground states. In the following, single-molecule magnetism and applications of ESR spectroscopy to SMM's are surveyed, emphasizing SMM's of chemical implication in terms of materials science.

5.2.2 Mn_{12} clusters

A variety of SMM's have successfully been found so far: Mn_4, Mn_7, Mn_{10}, Mn_{12}, Mn_{30}, Fe_4, Fe_8, V_4 (see Table 6-9). The cluster appearing as a SMM, $[Mn_{12}O_{12}(CH_3COO)_{16}(H_2O)_4] \cdot 2CH_3COOH \cdot 4H_2O$ (abbreviated as $Mn_{12}Ac$), was first synthesized by Lis in 1980.[5.2.50] Among SMM's, Mn_{12} families are the most intensively investigated systems in terms of ESR spectroscopy.[5.2.8-11, 5.2.14-16] The magnetic core of $Mn_{12}Ac$ is schematically shown in the inset of Figure 16.[5.2.8a] The core is composed of an external ring of eight manganese(III) ions ($S = 2$) with an internal tetrahedron of four manganese(IV) ions ($S = 3/2$). ESR and magnetization measurements of $Mn_{12}Ac$ by Caneschi et al.[5.2.51] provided experimental evidence of the $S = 10$ ground state for $Mn_{12}Ac$. The high-spin ground state is generated by superexchange interactions between Mn(III) and Mn(IV) ions through oxygen bridges. As shown in Figure 16, the magnetic cluster exhibits a

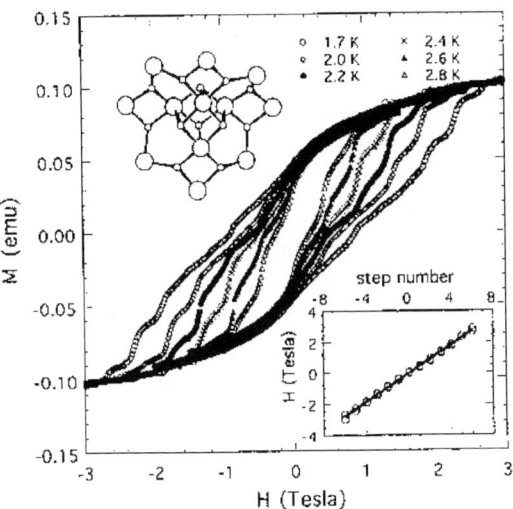

Figure 16. Plots of magnetization versus magnetic field for $Mn_{12}Ac$ (field sweep rate of 67 mT/min). The inset shows plots of the magnetic fields where steps occur versus step number. Based on a least square fit, the straight line with a slope of 0.46 T has been obtained.[5.2.4a]

hysteresis loop below a blocking temperature of about 3K, and also it shows slow exponential relaxation of the magnetization that obeys Eq.(72) down to 2.1 K. Any intermolecular interactions are considered to be negligible because the neighboring molecules are 7 Å apart at least. In addition, there was no evidence of three-dimensional magnetic order in magnetic susceptibility for frozen solutions and polymer-doped samples,[5.2.52] specific heat measurements,[5.2.53] and MCD spectra in solution.[5.2.54] Therefore, it is concluded that the hysteresis loop observed for $Mn_{12}Ac$ arises from the individual isolated molecules.

In general, SMM's can be regarded as a molecular cluster exhibiting very slow relaxation of the magnetization. The slow relaxation originates in a large S value of the ground spin state and large magnetic anisotropy associated with a large negative zero-field splitting constant. Figure 17 shows a potential energy diagram for a single-molecule magnet with an

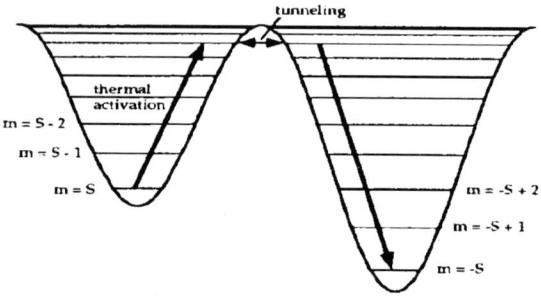

Figure 17. Plot of the potential energy versus magnetization direction for a SMM with an electron spin S and axial symmetry in an external magnetic field.[5.2.4b]

electron spin S in a certain field,[5.2.4b] where the lowest energy level with $m = S$ in the left well corresponds to the "spin-up" state and the level with $m = -S$ in the right well corresponds to the "spin-down" state. Assuming axial symmetry, the height of the energy barrier U in zero field is given by $S^2|D|$ and $(S^2-1/4)|D|$ for integer and half-integer spins, respectively. In order to invert the magnetization vector from "spin-up" to "spin-down", the spin system climbs over the potential energy barrier through a thermally activated process. Thus, the magnetization of SMM's relaxes slowly at sufficiently low temperature. In fact, the relaxation time of $Mn_{12}Ac$ is of the order of 2 months at 2K.[5.2.55] The slow magnetization relaxation results in the observation of out-of-phase AC (alternating current) magnetic susceptibility signals and magnetization hysteresis loops. The relaxation time follows the Arrhenius equation which governs elementary activation processes:

$$\tau = \tau_0 \exp\left(\Delta / k_B T\right), \tag{72}$$

where Δ and τ_0 stand for the height of the activation energy and the pre-exponential factor, respectively.[5.2.56] This behavior is analogous to that observed for superparamagnets.[5.2.55-57] In addition, it was found that at low temperature the relaxation time is given by

$$\tau = C \frac{S^6}{\Delta^3} \exp\left(\Delta / k_B T\right), \tag{73}$$

where C is a constant which depends on the phonon coupling and on Δ.[5.2.58] Consequently, it is indicated that a ground spin state with a large S value is a crucial factor to design and synthesize new SMM's.

At extremely low temperature (below about 2K), a striking deviation from the thermal activation process was also observed for $Mn_{12}Ac$. The hysteresis loop exhibits step-like features as shown in Figure 16. These behaviors at low temperature were interpretable as signs and magnitudes of fine-structure constants closely related to quantum tunneling of the magnetization (QTM).[5.2.59-60] The first explicit and quantitative evidence that the quantum tunneling through the barrier as well as thermal activation over the barrier occurs in $Mn_{12}Ac$ was demonstrated by Friedman et al.,[5.2.4] and several groups also confirmed such phenomena.[5.2.5] For simplicity, we discuss QTM here in terms of the following spin Hamiltonian:

$$\hat{H} = -DS_z^2 - g\beta_e \mathbf{S} \cdot \mathbf{B}, \tag{74}$$

where D stand for the axial zero-field splitting parameter corresponding to the anisotropy energy. The forth order terms are neglected because of their small contribution to resonance fields. When the static field $\mathbf{B}$ is applied along the easy axis, the eigenstates of the spin Hamiltonian are given as $|S, m\rangle$. S and m are the total spin and the magnetic spin quantum number, respectively. When the applied field is equal to

$$B_n = -\frac{Dn}{g\beta_e} \quad (n = 0, 1, 2 \cdots), \tag{75}$$

the eigenenergy of the $|S, m\rangle$ state in the left well coincides with that of the $|S, -m+n\rangle$ state in the right well. Under this conditions the pairs of the energy levels can be quantum mechanically admixed well, resulting in the sufficiently strong coupling between the states. Therefore, the observed steps

462   Chapter 11

in the hysteresis loop occur at intervals of field given by $B_n = 0.4n$ T. The resonance tunneling must be derived from off-diagonal terms in the spin Hamiltonian (a transverse magnetic anisotropy). This means that the spin Hamiltonian should contain a term that does not commute with S_z in order to observe QTM. $Mn_{12}Ac$, however, has tetragonal symmetry (axial symmetry), so that only the following fourth-order terms originating from the crystal field correspond to such anisotropy:

$$\sum_{k,q} B_k^q O_k^q = B_4^0 O_4^0 + B_4^4 O_4^4 \qquad (76)$$

$$O_4^0 = 35 S_z^4 + 30 S(S+1) S_z^2 + 25 S_z^2 - 6 S(S+1) + 3 S^2 (S+1)^2 \qquad (77)$$

$$O_4^4 = \left(S_+^4 + S_-^4 \right) / 2 , \qquad (78)$$

where B_k^q and O_k^q represent the crystal field constants and Steven's equivalent operators, respectively.[5.2.61] The spin Hamiltonian parameters including the higher order terms can precisely be determined by high-field/high-frequency ESR spectroscopy. Figure 18 shows fine-structure ESR

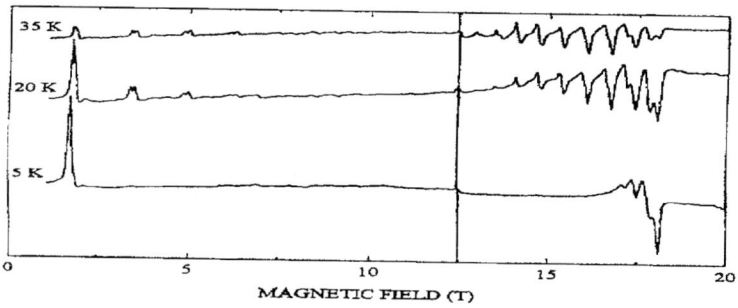

Figure 18. Observed ESR spectra for a pellet of $Mn_{12}Ac$ at low temperatures (ν = 349 GHz). The narrow signal arises from DPPH used for field calibration.[5.2.8b]

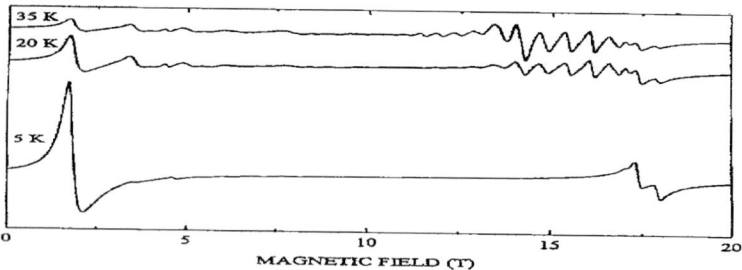

Figure 19. Theoretical ESR spectra for $Mn_{12}Ac$ with the following parameters: S = 10, $g_{//}$ =1.93, $g_\perp$ =1.96, D = - 0.472 cm^{-1}, B_4^0 = - 2.2×10^{-5} cm^{-1}, and B_4^0 =± 4.0×10^{-5} cm^{-1}.[5.2.8b]

spectra observed for a pellet of $Mn_{12}Ac$ at 349 GHz.[5.2.8] There are significant temperature dependences in the spectra. The theoretical (simulated) spectra using the following spin Hamiltonian parameters are shown in Figure 19, which are reasonably consistent with the observed spectra: $g_{//} = 1.93$, $g_\perp = 1.96$, $D = -0.472$ cm^{-1}, $B_4^0 = -2.2\times10^{-5}$ cm^{-1}, and $B_4^0 = \pm 4.0\times10^{-5}$ cm^{-1}. The fourth order terms allow only $\Delta m = \pm 4$ transitions, and prohibit all other ones. From the experimental data, however, it is indicated that the observed transitions are allowed ones. This suggests that the tunneling is induced by another small transverse magnetic field such as the hyperfine interaction. In fact, it has been reported that the assistance of the fluctuating fields driven by the magnetic nuclei contributes to the relaxation mechanism of Mn_{12} at very low temperature.[5.2.62] It was also indicated that half-integer-spin systems should not tunnel coherently in the absence of an applied field because of the spin parity effect related to Kramers degeneracy.[5.2.63-64] In fact, evidence of the spin parity effects has been shown by Gomes *et al.* for $Mn_{12}Bz$ clusters with $S = 19/2$.[5.2.65] On the other hand, one of Mn_{12} clusters with $S = 19/2$, $[PPh_4][Mn_{12}O_{12}(O_2CEt)_{16}(H_2O)_4]$, exhibits the step-like features on the hysteresis loop.[5.2.66] This is probably due to the nuclear spins in the molecule. A theoretical study of QTM occurring in half-integer-spin systems in terms of ESR spectroscopy is seen in literature.[5.2.67]

5.2.3 Fe$_8$ clusters

An octanuclear iron(III) cluster, $[Fe_8O_2(OH)_{12}(tacn)_6]Br_8\cdot 9H_2O$ (Fe$_8$), with an $S = 10$ ground state has also been extensively studied by many researchers.[5.2.18-20, 5.2.38, 5.2.68-74] An interesting point is that the Fe$_8$ cluster exhibits temperature-independent magnetic relaxation time below 0.36 K.[5.2.68] Under this condition, only the tunneling mechanism between the $m = \pm 10$ states alone contributes to the relaxation magnetization. Thus, $[Fe_8O_2(OH)_{12}(tacn)_6]^{+8}$ enables us to investigate the mechanism of QTM in detail. The molecular structure of $[Fe_8O_2(OH)_{12}(tacn)_6]^{+8}$ has been given in literature.[5.2.18-19, 5.2.68, 5.2.75] The ground state $S = 10$ originates from competing antiferromagnetic interactions between the eight $S = 5/2$ iron ions. The unpaired spin density has been obtained by polarized neutron experiments,[5.2.76] showing that the ground state has six up-spins and two down-spins. High-field/high-frequency ESR spectroscopy has also played a crucial role for the understanding of the magnetic properties of Fe$_8$ clusters.[5.2.18-20, 5.2.74] Single-crystal high-field/high-frequency ESR measurements have been carried out for Fe$_8$ by Barra *et al.*, providing information on the orientation of the principal axes in the molecular frame.[5.2.19] Figure 20 shows the single-crystal ESR spectra observed with the static field parallel to the easy axis. The principal axes of the magnetic anisotropy has been

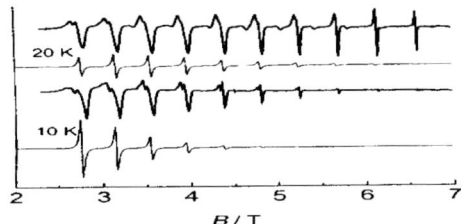

Figure 20. Single-crystal ESR spectra observed (bold) and calculated (plain) for $[Fe_8O_2(OH)_{12}(tacn)_6]^{8+}$ at 189.9541 GHz with the main magnetic field parallel to the easy axis.[5.2.19]

identified. The separations between the fine-structure transitions are not constant, indicating that higher-order fine-structure terms should be considered in order to reproduce the ESR spectra observed for the Fe_8 clusters. Only second- and fourth-order terms have been considered for the interpretation of the observed spectra:[5.2.19]

$$\hat{H} = \beta_e \mathbf{S} \cdot \mathbf{g} \cdot \mathbf{B} + D[S_z^2 - S(S+1)/3] + E(S_x^2 - S_y^2) + B_4^0 O_4^0 + B_4^2 O_4^2 + B_4^4 O_4^4 , \quad (79)$$

$$O_4^0 = 35 S_z^4 + 30 S(S+1) S_z^2 + 25 S_z^2 - 6 S(S+1) + 3 S^2 (S+1)^2 , \quad (80)$$

$$O_4^2 = [(7 S_z^2 - S(S+1) - 5)(S_+^2 + S_-^2) + (S_+^2 + S_-^2)(7 S_z^2 - S(S+1) - 5)]/4 , \quad (81)$$

$$O_4^4 = (S_+^4 + S_-^4)/2 . \quad (82)$$

The best fit parameters have been obtained as $g = 2.00$, $D = -0.205$ cm^{-1}, $E = 0.038$ cm^{-1}, $B_4^0 = 1.6 \times 10^{-6}$ cm^{-1}, $B_4^2 = -5 \times 10^{-6}$ cm^{-1}, and $B_4^4 = 8 \times 10^{-6}$ cm^{-1}. The calculated spectra have not completely reproduced the corresponding observed spectra. This might be due to the fact that the strong exchange limit is not fully achieved for the Fe_8 clusters, leading to an admixture of m states belonging to the different spin states. In fact, an energy gap of ca. 25 cm^{-1} to the excited states has been obtained from the magnetic susceptibility.[5.2.3e]

Magnetization measurements have been carried out for Fe_8 clusters at several temperatures (field sweeping rates: 0.14 T/s), showing magnetic hysteresis curves with the six equally spaced steps (see Fig. 21). The step-like feature originates from quantum tunneling of the magnetization (QTM), as described earlier. The temperature-independent hysteresis loops below 0.4 K indicate the purely quantum tunneling regime.[5.2.68] Therefore, the Fe_8 cluster seems to be the best suited system to investigate QTM in SMM's. Moreover, as shown below, Fe_8 clusters have given the first observation of the Berry phase in magnets.[5.2.69]

11. Exchange-coupled high-spin clusters

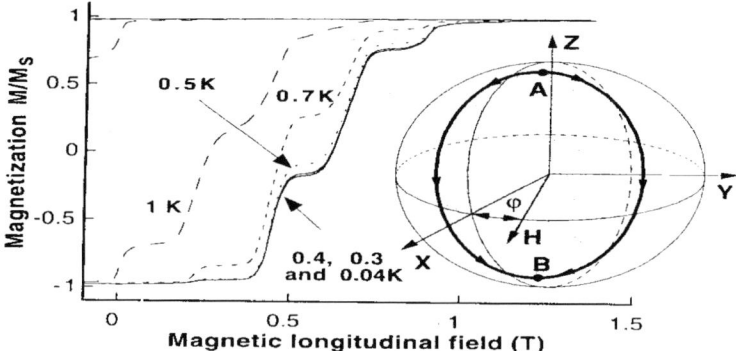

Figure 21. Magnetic hysteresis curves observed for [Fe$_8$O$_2$(OH)$_{12}$(tacn)$_6$] Br$_8$·9H$_2$O clusters at several temperatures (field sweeping rates of 0.14 T/s). The six equally separated steps originate from quantum tunneling of the magnetization. The temperature-independent hysteresis loops indicate the pure quantum tunneling regime. Degenerate minima (A and B) joined by two tunnel paths (heavy lines) are shown in unit sphere (Inset). The hard, medium, and easy axes correspond to x, y, and z directions, respectively.[5.2.69]

The energy diagram obtained for Fe$_8$ clusters is shown in Figure 22, where the static magnetic field is applied along the easy axis. The energy levels have been calculated by the numerical diagonalization of the spin Hamiltonian.[5.2.72] It should be noticed that for simplicity the higher order terms were neglected in this calculation. As shown in the inset of Figure 22, the avoided level crossing arising from transverse terms is seen around $H_z = 0$. The energy gap corresponds to the so-called tunneling splitting Δ. The inset of Figure 22 shows degenerate minima (A and B) joined by two tunneling paths (heavy lines)[5.2.68] In zerofield, the giant magnetic momentum

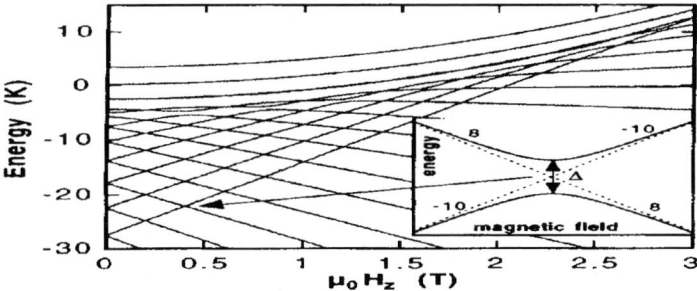

Figure 22. Energy versus the magnetic field of [Fe$_8$O$_2$(OH)$_{12}$(tacn)$_6$]Br$_8$·9H$_2$O clusters with an $S = 10$ ground state. The field is applied to the direction of the easy axis (see Figure XX-5.2.10). The energy levels are labeled by magnetic quantum numbers $m = \pm 10, \pm 9, \cdots, 0$ from the bottom to the top. The level crossing occurs at the field equal to $n \times 0.22$ T ($n = 1, 2, 3, \cdots$). The higher the gap Δ, the stronger the tunnel rate.[5.2.72]

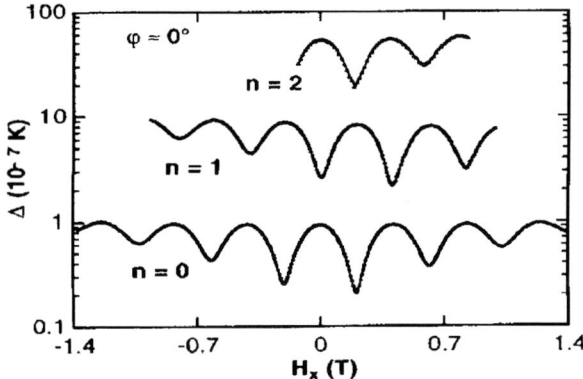

Figure 23. Experimentally determined tunneling splitting Δ versus transverse field H_x (parallel to the hard axis) for quantum tunneling between $m = -10$ and $(S - n)$. The parity effects are observed.[5.2.72]

reverses via the interference of two quantum spin paths of opposite windings in the easy anisotropy (yz) plane. Based on a semi-classical description, the observed oscillations are considered as constructive or destructive interference of quantum spin phases associated with the two tunneling paths. In fact, it has been found experimentally that a magnetic field applied along the hard axis can periodically change the splitting Δ as shown in Figure 23. This is the first direct observation of the topological part of the quantum spin phase (Berry phase) in magnets.[5.2.69] The tunneling splitting Δ can be directly obtained in terms of the Landau-Zener model.[5.2.69, 5.2.77] Following the model, the tunneling probability is given by the following expression when the longitudinal field H_z is swept over the avoided level crossing at a constant:

$$P = 1 - \exp\left[-\frac{\pi \Delta^2}{4\hbar g \mu_B S dH/dt}\right] \quad (83)$$

where dH/dt is the field sweeping rates. For the application of the Landau-Zener model, the following measurement has been carried out by Wernsdorfer et al. for the Fe_8 clusters.[5.2.69-72] First, the magnetic cluster was saturated in a field of $H_z = -1.4$ T. Then, the applied field was swept at a constant rate over one of the resonance transitions, and the fraction of molecules that reversed their spin was measured. As a result of this procedure, the tunneling rate P related with the tunneling splitting Δ was obtained. This technique enables us to directly measure very small tunneling splittings of the order of 10^{-8} K, which is not accessible in terms of resonance techniques at the moment.

11. Exchange-coupled high-spin clusters

Table 6. The list of SMM's documented so far and their spin Hamiltonian parameters determined by magnetization measurements

Complex	S	g	D / cm^{-1}	U/K	Ref.
[Mn$_{12}$O$_8$Cl$_4$(O$_2$CPh)$_8$(hmp$^-$)$_6$]	7	1.79	-0.6	42.3	42
(PPh$_4$)$_2$[Mn$_{12}$O$_{12}$(O$_2$CCH$_2$Cl)$_{16}$(H$_2$O)$_3$]	10	1.94	-0.26	37.4	43
(PPh$_4$)$_2$[Mn$_{12}$O$_{12}$(O$_2$CCH$_2$Cl)$_{16}$(H$_2$O)$_4$]	10	2.00	-0.27	38.8	43
(PPh$_4$)$_2$[Mn$_{12}$O$_{12}$(O$_2$CC$_6$F$_5$)$_{16}$(H$_2$O)$_4$]	10	2.01	-0.28	40.3	43
(PPh$_4$)$_2$[Mn$_{12}$O$_{12}$(O$_2$CC$_6$H$_3$-2,4-(NO$_2$)$_2$)$_{16}$(H$_2$O)$_4$]	10	1.94	-0.28	40.3	43
[Mn$_{12}$O$_{12}$(O$_2$CCHCl$_2$)$_8$(O$_2$CCH$_2$Bu)$_8$(H$_2$O)$_3$]	10	1.89	-0.45	64.7	25
[Mn$_{12}$O$_{12}$(O$_2$CHCl$_2$)$_8$(O$_2$CEt)$_8$(H$_2$O)$_3$]	10	1.83	-0.42	60.4	25
(PPh$_4^+$)[Mn$_{12}$O$_{12}$(O$_2$CC$_6$H$_4$F(-o))$_{16}$(H$_2$O)$_4$]	21/2	1.96	-0.39	61.7	44
[Fe(C$_5$Me$_5$)$_2$]$^+$[Mn$_{12}$O$_{12}$(O$_2$CC$_6$H$_4$F(-o))$_{16}$(H$_2$O)$_4$]	19/2	1.96	-0.375	48.6	44
[Co(C$_5$Me$_5$)$_2$]$^+$[Mn$_{12}$O$_{12}$(O$_2$CC$_6$H$_4$F(-o))$_{16}$(H$_2$O)$_4$]	19/2	1.95	-0.396	51.7	44
Mn$_{30}$O$_{24}$(OH)$_8$(O$_2$CCH$_2$C(CH$_3$)$_3$)$_{32}$(H$_2$O)$_2$(CH$_3$NO$_2$)$_4$	7	1.97	-0.79	55.7	45
[Mn$_4$O$_2$(MeO)$_3$(O$_2$CPh)$_2$(L)$_2$(MeOH)]$^{2+}$	7/2	1.8	-0.7	12.1	46
[Mn$_4$(O$_2$CMe)$_2$(Hpdm)$_6$][ClO$_4$]$_2$	8	1.85	-0.25	23	47
[Mn$_4$O$_3$Cl$_4$(O$_2$CMe)$_3$(py)$_3$]	9/2	1.95	-0.32	9.2	2
[Mn$_4$O$_3$(N$_3$)(O$_2$CMe)$_3$(dbm)$_3$]	9/2	1.95	-0.38	10.9	2
[Mn$_4$O$_3$(NCO)(O$_2$CMe)$_3$(dbm)$_3$]	9/2	1.84	-0.27	7.8	2
[V$_4$O$_2$(O$_2$CEt)$_7$(bpy)$_6$][ClO$_4$]	3	1.93	-1.52	19.7	23
[Net$_4$][V$_4$O$_2$(O$_2$CEt)$_7$(Pic)$_2$]	3	2.00	-1.50	19.4	23
[V$_4$O$_2$(O$_2$CEt)$_7$(bpy)$_6$][ClO$_4$]	3	1.93	-1.52	19.7	23
[Net$_4$][V$_4$O$_2$(O$_2$CEt)$_7$(Pic)$_2$]	3	2.00	-1.50	19.4	23
[V$_4$O$_2$(O$_2$CEt)$_7$(bpy)$_2$](ClO$_4$)	3	1.93	-1.52	19.7	23
[Fe$_4$(sae)$_4$(MeOH)$_4$]	8	2.316	-0.31	28.5	48

Table 7. The list of SMM's documented so far and their spin Hamiltonian parameters determined by non-oriented ESR spectroscopy.

Complex	S	g	D / cm^{-1}	E / cm^{-1}	B_4^0 / cm^{-1}	Ref.
[Fe(C$_5$Me$_5$)$_2$][Mn$_{12}$O$_{12}$(O$_2$CC$_6$F$_5$)$_{16}$(H$_2$O)$_4$] · 2H$_2$O	21/2	1.908	-0.351		-3.6×10^{-7}	17
[Mn$_{12}$O$_{12}$(NO$_3$)$_4$(O$_2$CCH$_2$Bu^t)(H$_2$O)$_4$]	10	1.99	-0.46		-2.0×10^{-5}	16
(PPh$_4$)[Mn$_{12}$O$_{12}$(O$_2$CEt)$_{16}$(H$_2$O)$_4$]	19/2	1.97	-0.61		-4.8×10^{-6}	15
[Mn$_4$(hmp)$_6$Br$_2$(H$_2$O)$_2$]Br$_2$. 4H$_2$O	9	1.999	-0.346	0.086	1.19×10^{-5}	11

Table 8. The list of SMM's documented so far and their spin Hamiltonian parameters determined by inelastic neutron scattering measurements.

Complex	S	D / cm^{-1}	E / cm^{-1}	B_4^0 / cm^{-1}	Ref.
[Mn$_4$O$_3$Br(OAc)$_3$(dbm)$_3$]	9/2	-0.502	0.017	-5.1×10^{-5}	31
[Mn$_4$O$_3$Cl(OAc)$_3$(dbm)$_3$]	9/2	-0.529	0.022	-6.5×10^{-5}	31
[Mn$_4$O$_3$(OAc)$_4$(dbm)$_3$]	9/2	-0.469	0.017	-7.9×10^{-5}	31
[Mn$_4$O$_3$F(OAc)$_3$(dbm)$_3$]	9/2	-0.379		-11.1×10^{-5}	31

Table 9. The list of SMM's documented so far and their spin Hamiltonian parameters determined by single-crystal ESR spectroscopy.

Complex	S	g_x	g_y	g_z	D/cm⁻¹	E/cm⁻¹	B_4^0/cm⁻¹	B_4^2/cm⁻¹	B_4^4/cm⁻¹	Ref.
[Fe$_8$O$_2$(OH)$_{12}$(tacn)$_6$]Br$_8$·9H$_2$O	10	2.00	2.00	2.00	-0.205	0.04	1.6×10⁻⁶	-5×10⁻⁶	-8×10⁻⁶	19
[Fe$_4$(OCH$_3$)$_6$(dpm)$_6$](Isomer 1)	5	1.995	1.997	2.009	-0.206	0.01	-1.1×10⁻⁵	-0.8×10⁻⁵	-0.4×10⁻⁵	21
[Fe$_4$(OCH$_3$)$_6$(dpm)$_6$](Isomer 2)	5			2.009	-0.19		-1.6×10⁻⁵			21
[Fe$_4$(OCH$_3$)$_6$(dpm)$_6$](Isomer 3)	5			2.009	-0.175		-1.6×10⁻⁵			21

5.3 Hydrogen-bonded molecule-based high-spin clusters

Recently, hydrogen-bonded molecule-based high-spin clusters have emerged, where two- or three-dimensional hydrogen network plays crucial roles for constructing cooperative magnetic behaviors.[5.3.1] For some cases, magnetic sites of transition metal ions are in high symmetry such as tetrahedron or octahedron, giving models for molecular magnetic assemblages with nonvanishing orbital angular momenta. In these exchange-coupled systems, isotropic Heisenberg types of exchange interactions are not enough to interpret their bulk magnetic properties. The models of well defined molecular structures are suitable for establishing molecular magnetism with sizable orbital momenta. Sophisticated theoretical treatments have also been documented recently.[1.2f] Fine-structure ESR spectroscopy for molecule-based exchange-coupled systems having non-vanishing orbital angular momenta has not fully been established in terms of both effective and theoretical spin Hamiltonians. In addition, fine-structure ESR spectroscopy for such molecular systems has neither been established yet. From a methodological point of view, analyses for magnetic susceptibility of molecular assemblages with sizable orbital angular momenta require reliable g-tensors of the systems under study as well as fine-structure tensors for the lowest and nearby excited multiplets.[5.3.2]

Referred to roles of hydrogen bonding in high spin clusters or infinite molecular assemblages, magnetic functionality properties are modulated by changes in the bonding scheme reflecting temperature variation and the modulation effects occur in a wide range of temperature from ambient to low temperature. The functionality changes arise mainly from vibronic and rotational modulations of magnetic tensors of paramagnetic metal ions. Single-crystal high-field/high-frequency ESR spectroscopy and conventional ESR spectroscopy below 2 K and at ultra-low temperature are powerful tools for identifying such microscopic changes.[5.3.3]

5.4 Genuinely organic molecule-based high-spin clusters: Spin identification by pulse-ESR-based electron spin transient mutation spectroscopy

Genuinely organic high-spin molecular multi-clusters can date back to Hirota's pioneering work on alkaline-metal-ion bridged aromatic ketone-based dianion in the triplet state.[5.4.1] *Meta*-connected oligoketones have pseudo-degenerate π-LUMO's near zero-energy in units of resonance integral β, and the extended π-conjugation network of the LUMO's undergoes robust dynamic spin polarization upon reduction by excess electrons. *Meta*-connected-oligoketone-based inter-molecular high-spin clusters and their magnetic characterization have been a long standing issue in high spin chemistry and materials science. Recently, such inter-molecular pluri-anionic high-spin clusters have been unequivocally identified by invoking pulsed ESR based 2D electron spin nutation spectroscopy applied to non-oriented media.[5.4.2] The nutation spectroscopy is a novel spectroscopy which is termed transition moment spectroscopy and also applicable to transition assignments of hyperfine allowed and forbidden spectra in a 2D representation.[5.4.3] The nutation spectroscopy applied to intramolecularly exchange-coupled high-spin systems is dealt with in the following chapter. The nutation technique is the most powerful for identifying molecular spins of the spin mixture in non-oriented media in a straightforward manner.

Figure 24 shows observed (top) and simulated (bottom) fine-structure ESR spectra due to various high-spin states derived from 1,3-dibenzoylbenzene upon chemical reduction in 2-MTHF solution.[5.4.3] Complete simulation has been made by evaluating their contributing weights and identifying their spin multiplicities, experimentally. Agreement between experiment and theoretical simulation is satisfactory. The experimentally derived spin Hamiltonian parameters are as follows; $g = 2.001$, $D = 0.00585$ cm^{-1}, $E = 0$ for $S = 2$ tetra-anionic dimer species, $g = 2.001$, $D = 0.0066$ cm^{-1}, $E = 0$ for $S = 3/2$ tri-anionic dimer species, and $g = 2.001$, $D = 0.0115$ cm^{-1}, $E = 0$ for $S = 1$ dianionic dimer species. The most crucial point to determine these spin Hamiltonian parameters with high accuracy has been to identify the contributing high spin states in the observed complex fine-structure ESR spectra. Figure 25 demonstrates the usefulness of nutation spectra in the 2D representation. The nutation frequency peaks ω_a, ω_b, ω_c, ω_d denoted by a, b, c, and d in the slice spectra correspond directly to the transitions $S = 1/2$; $m = +1/2 \leftrightarrow -1/2>$, $S = 3/2$; $m = \pm 3/2 \leftrightarrow \pm 1/2>$, $S = 2$; $m = \pm 2 \leftrightarrow \pm 1>$ and $S = 3/2$; $m = +1/2 \leftrightarrow -1/2>$, and $S = 2$; $m = \pm 1 \leftrightarrow 0>$ with the ratios of ω_a: ω_b: ω_c: ω_d = 1: 1.80: 2.06: 2.51. The ratios of the theoretical values are 1: $\sqrt{3}$: 2: $\sqrt{6}$ in terms of the first-order transition

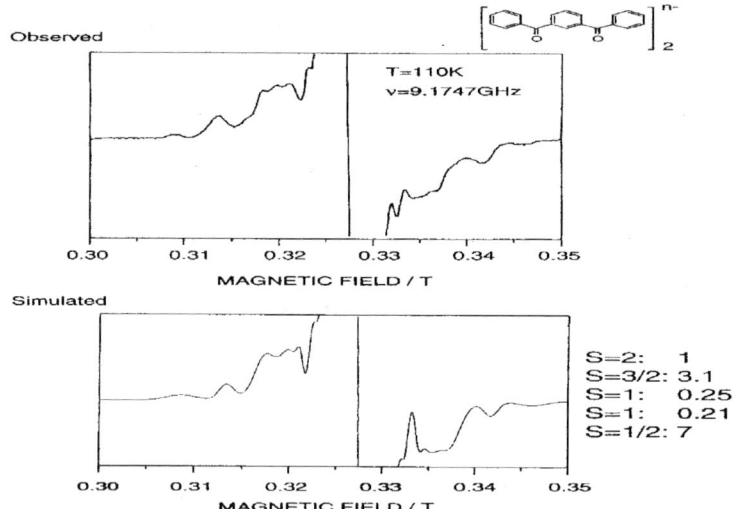

Figure 24. Observed (top) at 110 K and simulated (bottom) fine-structure X-band spectra composed of inter-molecularly exchange-coupled triplet, quartet, and quintet pluri-anionic dimers derived from chemical reduction of 1,3-dizenzoylbenezene in 2-MTHF solution. The contributing weights are given on the right hand side of the simulated spectrum.

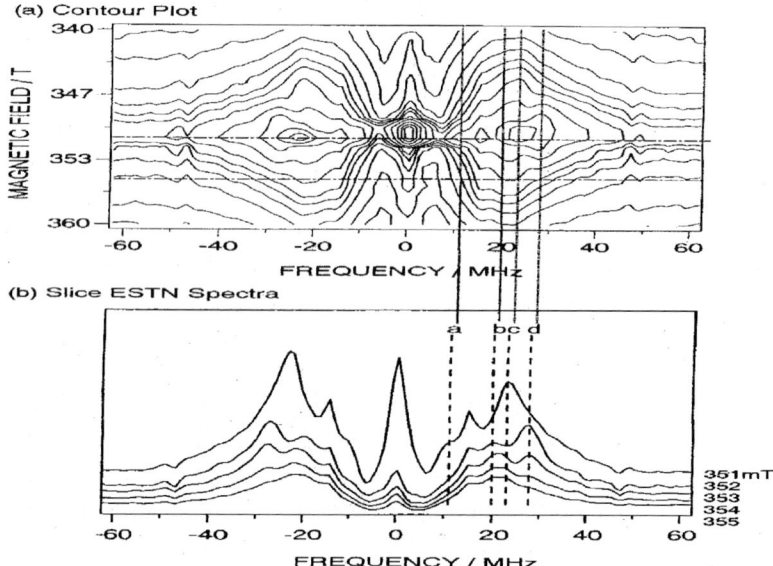

Figure 25. Contor plot of nutation spectra in the 2D representation and slice spectra observed for high spin species derived from chemical reduction of 1,3-dibenzoylbenzene in 2-MTHF. The nutation frequency peaks denoted by a, b, c, and d correspond to the doublet, quartet, quartet and quintet, and quintet state, respectively. The 1 mT-step slice spectra are from 351 to 355 mT.

moment, assuming that the g anisotropy of the organic high-spin clusters is small. Agreement between experiment and theory is satisfactory. Temperature dependence of the fine structure spectra have shown that all the inter-molecular high-spin clusters are in the ground state. Throughout the tensor-based analyses for their probable molecular structures of the oligoketone-based dimeric clusters, the intramolecularly exchange-coupled 1,3-dibenzoylbenzene dianion in the triplet state has been assumed and the corresponding fine-structure constants have been predicted.

Benzoylenebenzene, decacyclene and C_{60} fullerene also give intermolecularly exchange-coupled high-spin clusters upon chemical reduction by alkaline metal ions.[5.4.4] They are stable in polar solutions at ambient temperature. On the basis of experimental fine-structure constants, their probable molecular structures via alkaline metal ions can be determined with the help of tensor-based calculations for the fine structure tensors, as given in section 2. It should be noted that C_{60} fullerene upon chemical reduction gives a variety of intermolecularly exchange-coupled high-spin clusters depending on the reduction stage.[5.4.5] Both the intramolecular high-spin pluri-anionic C_{60} species and intermolecular high-spin pluri-anionic C_{60} clusters give useful models for formation of molecular high-spin clusters in crystalline solid of alkaline-doped C_{60} fullerene. It is worth noting that γ-ray irradiations of oligoketones at organic glasses generate solvent-containing ionic molecular clusters.[5.4.6] Neutral organic hydrocarbon-based high-spin clusters in crystalline solids are classified into two types in terms of the packing motif of intermolecular assemblages. One is a motif of herringbone types, and the other the columnar motif. Documented molecular systems of these categories have been rare. Recently, the latter one has appeared,[5.4.7] where gable *syn*-dimers of a stable neutral 1,3-diazaphenalenyl radical are stacked in a columnar motif, and the triplet state of the dimer is thermally accessible. The columnar motif has been established by elaborate molecular designs which are influential in molecular packing. In contrast, homoatomic phenalenyl radicals undergo π-dimerization in the crystal and the triplet-state dimmers are stacked with the herringbone motif.[5.4.8]

5.5 Low-dimensional molecule-based exchange-coupled assemblages

Interacting magnetic moments in the crystal experience the dipolar and exchange fields in addition to the Zeeman interaction in the presence of the applied magnetic field. A fundamental theoretical description of the electron spin resonance phenomena for exchange-coupled systems is given as

$$\text{Re}(\chi) = \text{Re}\left(\omega V/k_B T \int <M_x(t) M_x(0)> e^{-i\omega t} dt\right)$$
$$= \chi''(\omega) = \omega V/k_B T \int <M_x(t) M_x(0)> \cos\omega t\, dt \quad (84)$$

where the resonance is analyzed by using a linear response theory,[5.5.1] and the symbols denote the usual meanings. The intensity of the ESR-absorbed power $\int \chi''(\omega)\, d\omega$ is proportional to the static susceptibility χ. A relaxation function $\phi(t) = <M_x(t) M_x(0)>/k_B T$ of macroscopic magnetization is defined by a time correlation function $<M_x(t) M_x(0)>$, as given also in Section 2.4. Equation (84) is derived from the fluctuation-dissipation theorem. The resonance field is dominated by the molecular g-tensor, which is determined by the resonance condition of $h(\omega/2\pi) = g\beta B_0$, where B_0 stands for the strength of the static applied field $\mathbf{B_0} = B_0\mathbf{h}$ with $g^2 = \mathbf{h}\cdot g\cdot g\cdot \mathbf{h}$. The quantization axis of effective S points in the direction of a unit vector, $\mathbf{u} = g\cdot\mathbf{h}/g$. Experimental determination of the molecular g-tensor is the most crucial part in single-crystal ESR spectroscopy for the exchange-coupled systems, and the g-tensor gives the essential magnetic nature of the intermolecular interaction with due theoretical arguments and crystallographic molecular data from X-ray measurements. This approach applies successfully to uniform molecular spin chains.[5.5.2] The approach has been applied to more complicated molecular hetero-spin ($S = 1/2$ and $S = 1$ molecular chain assemblages coupled antiferromagnetically in the crystal) systems, where two kinds of the g-tensors with different magnetic behaviors were observed, showing the formation of magnetic supramolecules.[5.5.3]

Temperature dependence of the g-tensor gives crucial information on evolution of inter- as well as intra-molecular magnetic interactions such as internal fields arising from demagnetization. Temperature dependences of the g-shift due to the magnetic dipolar field were studied theoretically[5.5.4] and experimentally.[5.5.5] The pioneering theoretical work by Nagata and Tazuke showed that a significant modification of the T^{-1}-dependence of the g-shift is established for magnetic short-range interactions along the chains occurring in quasi 1D Heisenberg antiferromagnets.[5.5.4] Resonance give clues for kinds of the intermolecular magnetic interactions occurring in molecular assemblages; the resonance lines of homogeneous broadening are responsible for dipolar fields, while those of narrowing undergo exchange fields. The purely dipolar broadening gives Gaussian, while the exchange narrowing Lorentzian line shapes. Departures from Lorentzian line shapes give rationales for the possible occurrence of additional spin-related mechanisms such as spin diffusions at high temperatures.[5.5.6] Angular variations of the line shapes with line widths as a function of the static magnetic field $\mathbf{B_0}$ for single-crystal molecular magnetic materials, under favorable conditions, yield such magnetic relaxation phenomena as closely related to the dimensionality of magnetic interactions.[5.5.7] The magnetic

11. Exchange-coupled high-spin clusters

relaxations reflect spin dynamics and magnetic interactions in microscopic detail, which is based on the crystal and electronic molecular structures of component magnetic molecules. Anomalies appearing in magnetic relaxations are vital for organic magnetics because of the low dimensionality inherent in organic molecule-based magnetic materials.

The anomalies originate in dimension d and the low symmetry of spin interactions. Related ESR phenomena are described in terms of a relaxation function for spin magnetization, $\Phi(t) = <M_+(t)M_-(0)>/<M_+M_->$ (a spin self-time-correlation function for transverse magnetization) and its Fourier transform. Time dependence of M_+ or M_- is expressed by the slow modulation due to H', which stands for two-center spin interactions such as spin dipolar ones. It is to be noted that H_{ex} for intermolecular exchange interactions between molecular spins and H' undergo the restrictions of the dimensionality of magnetic systems under study. When H' stands for spin dipolar interactions ($H' = H_{dip}$), resonance absorption line shapes as a function of the angle θ depend on the dimensionality of the systems, where θ denotes the angle between the direction of a one-dimensional magnetic chain and B_0, or the normal axis of a two-dimensional magnetic plane and B_0 in two-dimensional square-planar systems. For reduced symmetric spin systems, the appearance of additional shifts in resonance fields is predicted; the shifts are due to the topology of J-connectivity (the connectivity of intermolecular exchange interactions) when the contribution to magnetic relaxation from a long transverse relaxation, i.e., a long-time tailing (LTT) occurs.

In Table 10 are briefly summarized the relations between ESR line shapes, line widths vs. dimensions of spin systems, and the spin relaxation functions, $\Phi(t)$. In terms of the magnitude of spin relaxations in low-dimensional magnetic systems, only cw-ESR spectroscopic studies have been documented. Recently, pulsed ESR techniques have been applied to directly identify the dimensionality of exchange-coupled molecular magnetic systems, whose relaxation times fall within the time resolution of spectrometers.[5,5,7]

Table 10. Relations between ESR line shapes, line widths vs. dimensions of spin systems, and the spin relaxation functions, $\Phi(t)$.

Dimension d	relaxation function $\Phi(t)$	line shape	line width $\Delta B_{1/2}(\theta) \propto$
1	$\exp(-\Gamma t^{3/2})$	FT of $\Phi(t)$	$(3\cos^2\theta - 1)^{4/3}$
2	$\exp(-\Gamma t \ln t)$	FT of $\Phi(t)$	$(3\cos^2\theta - 1)^2$
3	$\exp(-\Gamma t)$	FT of $\Phi(t)$	$(\cos^2\theta + 1)$

6. METAL HIGH-SPIN CLUSTERS OF BIOLOGICAL IMPORTANCE; MANGANESE CLUSTERS IN PHOTOSYSTEM II

This section briefly deals with transition-metal-ion high-spin clusters of biologically important photosystem II, emphasizing the use of high-field/high-frequency ESR and parallel microwave excitation techniques and contrasting with synthetic molecule-based metallic clusters treated in the other sections of this chapter. A comprehensive treatise on this subject in terms of chemistry and biophysics is given in a later chapter of this book.

6.1 Oxygen-evolving high-spin complexes

Most of the molecular oxygen in the atmosphere has been released as a by-product of water oxidation during photosynthesis in plants and algae. The photosynthetic oxidation of water to molecular oxygen is energetically driven by light-induced charge separations in the reaction center of photosystem II. Water is chemically the terminal electron donor for the electron transfer processes that constitute the light reactions of plant photosynthesis. Absorption of a photon results in a charge separation between a chlorophyll molecule (P680) and a pheophytin molecule. The pheophtytin anion transfers an electron to a quinone Q_A and P680$^+$ is reduced by a tyrosine residue, Tyr$_Z$. Interestingly, the oxidized Tyr$_z$ is in turn reduced by a nearby cluster of four Mn ions, often called the oxygen-evolving complex (OEC) because it catalyzes the oxidation of water. Photochemical charge separation is a one-electron process while water oxidation is a concerted process involving four electrons. Joliot and Kok showed that this can be described by a cyclic process involving five redox states,[6.1.1-2] termed S_{0-4} and that the presence of a number of metastable redox states can be demonstrated in chloroplasts following flash illumination. When the complex reaches the state S_4, molecular oxygen is released and the complex reverts to the S_0-state. Although many models for the oxygen-evolving process have been proposed,[6.1.3-6] no consensus has yet emerged.

S_i-states below the S_0-state are generated via reducing the Mn cluster of the OEC by hydrazine and hydroxylamine, which state are named S_{-1}, S_{-2}, and S_{-3}-state.[6.1.7-11] The detailed structure of the Mn$_4$ cluster of the OEC is still unknown. There is structural information obtained form X-ray spectroscopic experiments (EXAFS and XANES) and models for the structure have been developed. Extended X-ray absorption fine-structure (EXAFS) spectroscopy experiments provide evidence that in the S_1- and S_2-states the Mn cluster contains two 0.27-nm Mn-Mn distances and one 0.33-

11. Exchange-coupled high-spin clusters

nm Mn-Mn distance.[6.1.12-14] The possible structures of the manganese cluster in the OEC have been proposed by DeRose et al.[6.1.15] In this section, we review only the recent ESR results for the exchange-coupled Mn cluster in the S_{0-3}-states of the OEC in Photosystem II. An exhaustive review and in-depth discussions of relevant controversial issues are beyond the scope of this chapter, and readers are recommended to refer to the chapter by Kawamori.

6.2 The S_2-state of the OEC in Photosystem II

The light-induced S_2 oxidation state of the OEC is the most extensively characterized S state. The S_2-state is paramagnetic and gives rise to three characteristic ESR signals: (i) A multiline signal, (ii) $g = 4.1$ signal, (iii) $g = 10.6$ signals. Among them, the $g = 2$ multiline signal was first reported to provide direct evidence for the involvement of the Mn cluster.[6.2.1] The multiline ESR signal appearing near $g = 2$ is spread over about 180 mT and is made up of at least 18 lines, each separated by 8-9 mT originating from ^{55}Mn hyperfine structures.[6.2.2] The four effective ^{55}Mn hyperfine tensors (A_X, A_Y, A_Z) have been reported to be (-232, -232, -270), (200, 200, 250), (-311, -311, -270), (180, 180, 240) in units of MHz, which are derived from the simultaneously constrained simulations of the CW-ESR and ESE-ENDOR.[6.2.3] The multiline signal is ascribed to the ground state with $S = 1/2$.

Figure 26 shows the multiline ESR signal obtained by conventional X-band and Q-band ESR spectroscopy reported by Smith et al.[6.2.7] A $g = 4.1$

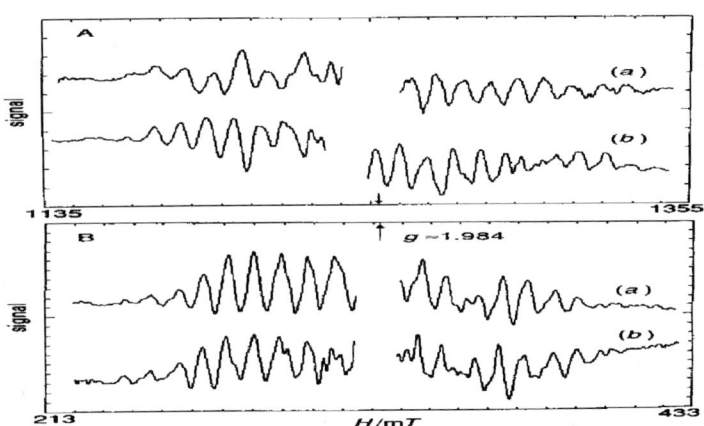

Figure 26. Multiline signals from Photosystem II at (A) Q-band and (B) X-band. (a) Minus alcohol, (b) plus alcohol. Illumination temperatures: (a) 200 K, (b) 220 K. Spectrometer conditions: X-band, 9.03 GHz, 2 mT modulation amplitude, 30 mW microwave power, 100 kHz modulation frequency, temperature 8 K. Q-band, 34.650 GHz, 0.5 mT modulation amplitude, 5 mW microwave power, 100 kHz modulation frequency, temperature 8 K.[6.2.7]

signal reveals a broad unstructured ESR spectrum with a peak-to-peak width of 35 mT at X-band. The $g = 4.1$ signal has been assigned to arise from the Mn cluster having an $S = 5/2$ state[6.2.4-5] or an $S = 3/2$ state.[6.2.6-7] The $g = 4.1$ signal has been observed by Q-band ESR spectroscopy, exhibiting two partially resolved components at $g = 4.34$ and $g = 4.14$. Boussac et al. have reported that the new signals at $g = 10$ and 6 can be observed by a conventional ESR spectroscopy. They are assigned to be $m_s = \pm 5/2$ and $\pm 3/2$ transitions in an $S = 5/2$ spin state.[6.2.8]

6.3   The S_0-state of the OEC in Photosystem II

Because the S_0-state is reduced by two further electrons form the S_2-state, it is expected to be an odd-electron or Kramers state observable with conventional ESR spectroscopy. The first observation of an ESR signal arising from the S_0-state was made by different methods. One is a chemical reduction with hydrazine of the S_1-state to the S_{-1}-state followed by photogeneration of the S_0-state at 273 K in the presence of DCMU (3-(3,4-dichlorophenyl)-1,1-dimethylurea), the other is a chemical reduction with hydroxylamine of the S_1-state to the S_0-state.[6.3.1] The S_0-state ESR signals have hyperfine structures similar to the multiline signal in the S_2-state except for a broader overall width.[6.3.1-2] The splittings of the hyperfine lines are more variable (7-11 mT) than those for the S_2-state. The total spectral breadth is in the range of 220-240 mT and the total number of peaks amounts to 24-26. The S_0 multiline gives rise to an asymmetry of this ESR signal, indicative of an average g value below $g = 2.0$. A temperature dependence of the S_0 multiline signal has shown that the $S = 1/2$ state responsible for the signal is an isolated ground state.[6.3.3-4]

6.4   The S_1-state of the OEC in Photosystem II

The S_1-state which is one-electron reduced state from the S_2-state is paramagnetic, but the S_1-state is of even electron number and a non-Kramers ESR signal is observed in parallel-polarized ESR spectroscopy. Recently, the multiline signal for the S_1-state has been detected from photosystem II particles isolated from the cyanobacterium Synechocytstis sp. PCC 6803[6.4.1] and spinach[6.4.2], showing the parallel polarization ESR spectrum composed of a well-resolved multiline pattern centered at $g \sim 12$ with at least 18 well-resolved hyperfine lines having an average splitting of 3.2 mT. This hyperfine pattern gives unambiguous evidence for the existence of a tri- or tetra nuclear exchange-coupled paramagnetic Mn cluster in the S_1-state of the OEC in photosystem II. Another parallel polarization ESR signal of the S_1-state of photosystem II particles isolated from spinach has been also observed, which appears at an effective g value of approximately 4.8 with a peak-to-peak width of roughly 60 mT.[6.4.3-4] It has been concluded that this parallel polarization ESR signal of the S_1-state arises from a thermally excited state with $S = 1$ which is located 2.5 K above the ground singlet state.

6.5 The S_3-state of the OEC in Photosystem II

Finally, we discuss the ESR studies for the S_3-state that is the third one-electron oxidation state from the S_0-state of the OEC in photosystem II. The S_3-state is expected to be an integer spin state because the S_2-state is characterized by the ESR signals from half-integer spin states. The parallel polarization ESR spectra of the S_3-state have been reported by Kawamori *et al.*, in which the ESR signals at $g = 12$ and 8 are observed in accompaniment with their peak-to-peak widths of about 30 and 20 mT, respectively.[6.5.1] The signals have been obtained after two turnovers from the S_1-state by flash illumination and its intensity has showed the periodicity of a four flash illumination oscillation, thus the observed signals have been rationalized to originate from the S_3-state. The spectral simulation for the ESR signals of the S_3-state has been performed by using the ordinary fine-structure spin Hamiltonian and numerical diagonalization of the spin Hamiltonian, giving $S = 1$, $D = \pm 0.435 \pm 0.005$ cm^{-1}, $E/D = -0.317 \pm 0.002$ cm^{-1}. The temperature dependence of the EPR signals at $g = 12$ and 8 has been studied, showing that the signals arise from a low-lying excited triplet state. The authors[6.5.1] have concluded that the manganese in the OEC might be oxidized during the transition $S_2 \rightarrow S_3$-state. The reason is that if an organic residue such as histidine is oxidized the interaction between the Mn cluster in the S_2-state with $S = 1/2$ and $g = 2$ and the organic residue with $S = 1/2$ and $g = 2$ would have the resonance line at $g_{ob} = 2$ (g_{ob}: the observed field position) but the experimental filed position is $g_{ob} = 12$ and 8 in their work. There have been researchers who insist that the oxidized site in the S_3-state is organic residues on the basis of the study of Mn Kβ X-ray emission spectroscopy.[6.5.2-3] The issue is the focus of the current topics in Photosystems II.

7. CONCLUSIONS

Recently, organic and inorganic hybrid high-spin systems are emerging in which both organic high-spin entities and inorganic metallic sites are composed of open-shell systems and both are connected via chemical bonding. They are classified into intramolecularly exchange-coupled hybrid high-spin systems. Molecular assemblages composed of the hybrid molecular entities have attracted attention in the field of molecule-based magnetism and molecular materials spin science. The reason is that magnetic orbitals in the entities are basically composed of transition-metal-ion based localized orbitals and organic spin-delocalized SOMO's. Spin alignments in the entities are governed by orbital and topological symmetry. Thus, magnetism-based functionalities are controllable in microscopic molecular terms in the hybrid systems. Molecular design and synthetic molecular engineering for exotic materials and chemical system materials are feasible. A typical example is high-spin carbene-based paramagnetic porphyrins.[7.1] The other examples are novel neutral radical-based metallo-oxophenalenyls with nonvanishing orbital angular momenta.[7.2] The above two categories are for organic moieties which are extremely spin-delocalized. For less spin-delocalized and weak spin-polarized organic parts, the concept of the hybrid high-spin systems is also applicable.[7.3-4] In the former two hybrid systems, highly delocalized spins from the organic entity give robust dynamic spin polarization network, establishing strong exchange-coupled molecular systems if transition-metal-ion sites are coordinated in such a way as dynamic spin polarization or spin delocalization mechanisms are operative. For the latter *N-tert*-butyl-*N*-aminoxyl-based manganese(II) and copper(II) complexes, asymmetric coordination of NO-based SOMO's and nitrogen sites of pyrimidines with the magnetic orbitals are elaborately performed, giving useful testing grounds for the spin alignment in the hybrid molecular systems.[7.3-4] Nevertheless, the symmetry reduction and the weak spin polarization occurring in the heterocyclic linkers such as pyrimidine give ferromagnetic-antiferromagnetic competing effects. In the dimerized hybrid exchange-coupled systems, through-space orbital interactions also participate in the competition.[7.3-4] Fine-structure and hyperfine-structure ESR spectroscopy plays a crucially important role to characterize novel magnetic properties in microscopic details. Ground and excited high-spin states originating in such novel hybrid molecular systems give intriguing exchange coupling schemes not documented so far. The research area is immature, but the relevant topics are important in terms of chemistry and molecular materials science. Nevertheless, the topics are not included in this chapter because of space. Genuinely organic molecule-based super high-spin intermolecular clusters appearing in the crystal edge are also an intriguing issue.[7.5]

Also, genuinely organic molecule-based exchange-coupled systems with high spin multiplicities are only briefly exemplified in Section 5.4. There have been potentially important examples for inter-molecularly exchange-coupled high-spin systems in oriented or non-oriented media. Among them, well identified chemical species are not many in terms of electronic-spin and molecular structures. In ESR terms, resonance peaks due to intermolecularly exchange-coupled organic high spins appear in the $g = 2$ region because of both small spin-orbit couplings in the ground state and apparently decreasing fine-structure constants: the decrease is due to projection factors even if the magnitude of the exchange coupling is kept constant. For these particular high-spin systems, 2D pulsed ESR-based electron spin nutation spectroscopy is the most useful in identifying chemical species and deriving spin Hamiltonian parameters. The latter procedure for quantitative evaluations of the parameters can be carried out by invoking spectral simulation of 2D nutation spectra as functions of resonance fields and transition moments in two dimensions.

The nutation and transition moment spectroscopic techniques are also useful for phosphorescence fine-structure ESR spectroscopy in the lowest excited triplet state as well as in the excited high-spin multiplet states. Even extremely small fine-structure parameters due to high symmetry of the molecular systems under study can be detctable.[7.6]

Acknowledgements *The authors acknowledge the Ministry of Education, Culture, Sports, Science and Technology, Japan and JSPS for the financial supports.*

8. REFERENCES

Section 1 and Section 2.1

1.1. Standard and advanced textbooks. (a) J. R. Bolton, J. E. Wertz, *Electron Paramagnetic Resonance. Elementary Theory and Practical Applications*, McGraw-Hill, Inc., New York (1972). (b) J. A. Weil, J. R. Bolton, J. E. Wertz, *Electron Paramagnetic Resonance. Elementary Theory and Practical Applications*, John Wiley & Sons, Inc., New York (1994). (c) N. M. Atherton, *Principles of Electron Spin Resonance*, Ellis Horwood Ltd., New York (1993); N. M. Atherton, *Electron Spin Resonance, Theory and Applications*, Ellis Horwood Ltd., Chichester (1973). (d) J. R. Pilbrow, T*ransition Ion Electron Paramagnetic Resonance*, Clarendon Press, Oxford (1990). (e) K. Kuwata and K. Itoh, *Introduction to Electron Spin Resonance*, Nankou-Dou, Inc., Tokyo (1978). (f) A. Bencini, D. Gatteschi, *Electron Paramagnetic Resonance of Exchange Coupled Systems*, Springer-Verlag (1990). (g) W. Weltner, Jr., *Magnetic Atoms and Molecules*, Scientific and Academic Editions, New York (1983). (h) M. Date, *Electron Spin Resonance*, Baifuh-Kan Tokyo (1978).

1.2. Monographs and books for molecule-based magnetism. (a) D. Gatteschi, O. Kahn, J. S. Miller, Editors, *Molecular Magnetic Materials*, Kluwer, Dortrecht (1991). (b) O. Kahn, *Molecular Magnetism*, Wiley-VCH, New York (1993). (c) T. Takui, *Organic Molecule-Based Magnetic Materials*, in *Handbook of Opto-Electronic Organic Functionality Materials*, Asakura Inc., Tokyo (1994). (d) P. M. Lahti, Editor, Magnetic Properties of Organic Materials, Marcel Dekker, Inc., New York (1999). (e) K. Itoh, M. Kinoshita,

Molecular Magnetism; New Magnetic Materials, Kodansha, Tokyo, Gordon & Breach, Amsterdam (2000). (f) R. Boca, *Theoretical Foundations of Molecular Magnetism, Elsevier*, Amsterdam (1999). Book (f) is the most comprehensive treatise on inorganic-molecule-based magnetism from the group-theoretical side after Ref. 1f. (g) T. Takui, K. Sato, D. Shiomi, K. Itoh, Chap. 11 in *Magnetic Properties of Organic Materials*, Paul M. Lahti, Editor, Marcel Dekker, Inc., New York (1999).

1.3. (a) D. Shiomi, M. Nishizawa, K. Sato, T. Takui, and K. Itoh, H. Sakurai, A. Izuoka, T. Sugawara, *J. Phys. Chem.*, **B 101**, 3342 (1997). (b) M. Nishizawa, D. Shiomi, K. Sato, T. Takui, K. Itoh, H. Sawa, R. Kato, H. Sakurai, A. Izuoka, T. Sugawara, *J. Phys. Chem.*, **B 104**, 503 (2000). (c) D. Shiomi, K. Sato, T. Takui, *J. Phys. Chem.*, **B 104**, 1961 (2000). (d) D. Shiomi, K. Sato, and T. Takui, *J. Phys. Chem.*, **B 105**, 2932 (2001). (e) D. Shiomi, K. Sato, T. Takui, *J. Phys. Chem.*, **A 106**, 2096 (2002). (f) D. Shiomi, T. Kanaya, K. Sato, M. Mori, K. Takeda, T. Takui, *J. Am. Chem. Soc.*, **123**, 11823 (2001).

1.4. K. Furukawa, D. Shiomi, K. Sato, H. Yamano, H. Takahashi, A. Maeda, T. Takui, to be published.

1.5. Intramolecularly exchange-coupled organic high-spin systems are exclusively dealt with in the following chapter.

1.6. B. Pilawa, *Ann. Phys. (Leipzig)*, **8**, 191 (1999).

1.7. (a) M. Drillon, R. Georges, *Phy. Rev.*, **B 24**, 1278 (1981). (b) M. Drillon, R. Georges, *Phy. Rev.*, **B 26**, 3882 (1982). (c) B. Leuenberger, H. U. Guedel, *Mol. Phys.*, **51**, 1 (1984).

1.8. (a) J. J. Borras-Almenar, J. M. Clemente-Juan, E. Coronado, A. V. Palii, B. S. Tsukerblat, *Chem. Phys.*, **274**, 131 (2001). (b) J. J. Borras-Almenar, J. M. Clemente-Juan, E. Coronado, A. V. Palii, B. S. Tsukerblat, *Chem. Phys.*, **274**, 145 (2001). (c) J. J. Borras-Almenar, J. M. Clemente-Juan, E. Coronado, A. V. Palii, B. S. Tsukerblat, *J. Chem. Phys.*, **114**, 1148 (2001). (d) Overparamerization in the analysis of magnetic susceotibility is a crucial problem. H. Matsuoka, H. Onishi, K. Kubono, K. Sato, D. Shiomi, K. Yokoi, T. Takui, to be published.

1.9. T. Takui, S. Kita, S. Ichikawa, Y. Teki, T. Kinoshita, K. Itoh, *Mol. Cryst. Liq. Cryst.*, **176**, 67 (1989).

1.10. (a) H. Kurreck, B. Kirste, W. Lubitz, *Electron Double Resonance Spectroscopy of Radicals in Solution; Application to Organic and Biological Chemistry*, VCH, Weinheim (1988). (b) T. Takui, Electron *Nuclear Multiple Resonance Spectroscopy (Chap. 8.5) in Molecular Spectroscopy III*, A Monograph Series of Experimental Chemistry, Fourth Ed. Vol. 8 (Chem. Soc. Jpn.), Maruzen, Tokyo (1993).

1.11. Y. Teki, T. Takui, K. Itoh, *J. Chem. Phys.*, **88**, 6134 (1988).

1.12. (a) K. Itoh, *Pure Appl. Chem.*, **50**, 1251 (1978). (b) T. Takui, *PhD. Thesis*, Osaka University, 1973.

1.13. B. R. Judd, *Operator Techniques in Atomic Spectroscopy*, Princeton University Press, Princeton (1998).

1.14. T. Takui, K. Itoh, unpublished.

1.15. H. Yagi, Y. Teki, T. Takui, K. Itoh, unpublished.

1.16. K. Tanaka, K. Sato, D. Shiomi, T. Takui, to be published.

1.17. K. Tanaka, K. Sato, D. Shiomi, M. Baumgarten, W. Adam, T. Takui, to be published.

1.18. (a) H. M. Gersmann, J. D. Swalen, *J. Chem. Phys.*, **36**, 3221 (1962). (b) L. D. Rollman, S. I. Chan, *J. Chem. Phys.*, **50**, 3416 (1969). (c) I. V. Ovchinnikov, V. N. Konstantinov, *J. Mag. Reson.*, **32**, 179 (1978). (d) K. Itoh, T. Takui, *ESR in Solids (Chap. 8.4) in Molecular Spectroscopy III*, A Monograph Series of Experimental Chemistry, Fourth Ed. Vol. 8 (Chem. Soc. Jpn.), Maruzen, Tokyo (1993).

11. Exchange-coupled high-spin clusters

Section 2.2

2.2.1. M. Matsushita, M. Momose, T. Shida, Y. Teki, T. Takui, K. Itoh, *J. Am. Chem. Soc.*, **112**, 4702(1990).

Section 2.3

2.3.1. Q. Wang, J.-S. Wang, Y. Li, G.-S. Wu, In *EPR in the 21st Century: Basics and Applications to Materials, Life and Earth Sciences (A. Kawamori, J. Yamauchi, H. Ohta, Eds.)*, Elsevier, 326 (2002).
2.3.2. R. P. Scaringe, D. J. Hodgson, W. E. Hatfield, *Mol. Phys.* **35**, 701 (1978).
2.3.3. S. Nakazawa, K. Sato, D. Shiomi, T. Takui, unbublished.

Section 2.4

2.4.1 (a) K. Nagata, Bussei, **13**, 149 (1972). (b) K. Nagata, *Physics for Random Systems*, Chapter 13, Bifuhkan, Tokyo, 1981.
2.4.2 R. Kubo, K. Tomita, *J. Phys. Soc. Japan*, **9**, 888 (1954).
2.4.3 (a) H. Mori, *Prog. Theor. Phys.*, **33**, 423 (1965). (b) *ibid.*, **34**, 399 (1966). (c) Tokuyama, H. Mori, *Prog. Theor. Phys.*, **55**, 411 (1976).

Section 3

3.1. A. Rockenbauer, J. Pilbrow (ed.), *Computer Simulations in EPR Spectroscopy*, *Mol. Phys. Reports* **26** (1999).
3.2. A. Abragam, B. Bleaney, *Electron Paramagnetic Resonance of Transition Ions*, Clarendon, Press, Oxford (1970).
3.3. J. R. Pilbrow, *Transition Ion Electron Paramagnetic Resonance*, Clarendon Press, Oxford (1990).
3.4. D. W. Alderman, M. S. Solum, D. M. Grant, *J. Chem. Phys.* **84**, 3717 (1986).
3.5. M. J. Mombourquette, J. A. Weil, *J. Magn. Reson.* **99**, 37 (1992).
3.6. S. Galindo, L. Gonzáles-Tovany, *J. Magn. Reson.* **44**, 250 (1981).
3.7. D. Wang, G. R. Hanson, *J. Magn. Reson. A* **117**, 1 (1995).
3.8. K. E. Gates, M. Griffin, G. R. Hanson, K. Burrage, *J. Magn. Reson.* **135**, 104 (1998).
3.9. G. G. Belford, R. L. Belford, and J. F. Brukhalter, *J. Magn. Reson.* **11**, 251 (1973).
3.10. K. T. McGregor, R. P. Scaringe, W. E. Hatfield, *Mol. Phys.* **30**, 1925 (1975).
3.11. Y. Teki, I. Fujita, T. Takui, T. Kinoshita, K. Itoh, *J. Am. Chem. Soc.* **116**, 11499 (1994).
3.12. K. Sato, Doctoral Thesis, Osaka City University, Japan (1994).
3.13. (a) T. A. Fukuzawa, K. Sato, A. S. Ichimura, T. Kinoshita, T. Takui, K. Itoh, P. M. Lahti, *Mol. Cryst. Liq. Cryst.* **278**, 253 (1996). (b) T. Takui, unpublished work.
3.14. H. Matsuoka, K. Sato, D. Shiomi, Y. Kojima, K. Hirotsu, N. Furuno, T. Takui, In *EPR in the 21st Century: Basics and Applications to Materials, Life and Earth Sciences (A. Kawamori, J. Yamauchi,H. Ohta, Eds.)*, Elsevier, 264 (2002).
3.15. N. Oda, T. Nakai, K. Sato, D. Shiomi, M. Kozaki, K. Okada, T. Takui, *Mol. Cryst. Liq. Cryst.* **376**, 501 (2002).
3.16. K. Sato, T. Takui, K. Itoh, Program Software Package (ESR Spectral Simulation and Exact Analytical Expressions Based on Hybrid Eigenfield Approach). Osaka City University.
3.17. H. Matsuoka, K. Sato, D. Shiomi, T. Takui, *Appl. Mag. Reson.*, in press.

3.18. (a) A. Rokenbauer, P. Simon, *J. Mag. Reson.*, **11**, 217 (1974). (b) M. Iwasaki, *J. Mag. Reson.*, **16**, 417 (1974). (c) K. Toriyama, M. Iwasaki, personal communication, 1974. (d) K. Shimokoshi, personal communication, 1975.
3.19. T. Takui, *Electron Nuclear Multiple Resonance Spectroscopy (Chap. 8.5) in Molecular Spectroscopy III*, A Monograph Series of Experimental Chemistry, Fourth Ed. Vol. 8 (Chem. Soc. Jpn.), Maruzen, Tokyo (1993).
3.20. (a) J. A. Weil, J. R. Bolton, J. E. Wertz, *Electron Paramagnetic Resonance. Elementary Theory and Practical Applications*, John Wiley & Sons, Inc., New York (1994). (b) N. M. Atherton, *Principles of Electron Spin Resonance*, Ellis Horwood Ltd., New York (1993).
3.21. T. Takui, unpublished work.
3.22. Program Package, Symfonia; Bruker Analytishe Messtehnik GMBH, D-7512 Rheinstetten, FRG.
3.23. E. Wasserman, L. C. Snyder, W. A. Yager, *J. Chem. Phys.*, **41**, 1763 (1964).
3.24. Program Package, X'Sophe; Bruker Analytishe Messtehnik GMBH, D-7512 Rheinstetten, FRG.
3.25. M. L. Matta, B. D. Sukheeja, M. L. Narchal, *J. Mag. Reson.*, **9**, 121 (1973).
3.26. K. Sato, D. Shiomi, T. Takui, unbublished work.

Section 4.1

4.1.1. L. J. Berliner (ed.), *Spin Labeling*, Academic Press, New York, 1976.
4.1.2. L. J. Berliner (ed.), *Spin Labeling II*, Academic Press, New York, 1979.
4.1.3. G. I. Likhtenstein, *Spin Labeling Methods in Molecular Biology*, Wiley Interscience, New York, 1976.
4.1.4. S. S. Eaton, G. R. Eaton, *Coord. Chem. Rev.*, **26**, 207 (1978).
4.1.5. G. R. Eaton, S. S. Eaton, *Acc. Chem. Res.*, **21**, 107 (1988).
4.1.6. J. A. Weil, J. R. Bolton, J. E. Wertz, *Electron Paramagnetic Resonance, Elementary Theory and Practical Applications*, Chapter 6, John Wiley and Sons, New York, 1994.
4.1.7. N. M. Atherton, *Principles of Electron Spin Resonance*, Ellis Horwood, Chichester, 1993.
4.1.8. G. E. Pake, T. L. Estle, *The Physical Principles of Electron Paramagnetic Resonance*, W.A. Benjamin, Massachusetts, 1973.
4.1.9. L. Salem, C. Rowland, *Angew. Chem. Int. Ed. Engl.*, **11**, 92 (1972).
4.1.10. W. T. Borden (ed.), *Diradicals*, John Wiley and Sons, New York, 1982.

Section 4.2

4.2.1 W. Heisenberg, *Z. Phys.*, **38**, 411 (1926).
4.2.2 P. A. M. Dirac, *Proc. Roy. Soc.*, **112A**, 661 (1926).
4.2.3 P. A. M. Dirac, *Proc. Roy. Soc.* **1929**, *123A*, 714.
4.2.4 J. H. Van Vleck, *Rev. Mod. Phys.*, **17**, 27 (1945).
4.2.5 J. H. Van Vleck, *The Theory of Electric and Magnetic Susceptibilities*, Chapter XII, Oxford University Press, Oxford, 1932.
4.2.6 T. Arai, *Phys. Rev.*, **126**, 471 (1962).
4.2.7 T. Arai, *Phys. Rev.*, **134A**, 824 (1964).

Section 4.4

4.4.1. R. M. Dupeyre, H. Lemaire, A. Rassat, *J. Am. Chem. Soc.*, **87**, 3771 (1965).

11. Exchange-coupled high-spin clusters

4.4.2. R. Briere, R. M. Dupeyre, H. Lemaire, C. Morat, A. Rassat, P. Rey, *Bull. Soc. Chim. France*, **1965**, 3290.
4.4.3. A. Nakajima, H. Ohya-Nishiguchi, Y. Deguchi, *Bull. Chem. Soc. Jpn.*, **45**, 713 (1972).
4.4.4. G. R. Luckhurst, *Mol. Phys.*, **10**, 543 (1966).
4.4.5. G. R. Luckhurst, G. F. Pedulli, *J. Am. Chem. Soc.*, **92**, 4738 (1970).
4.4.6. A. G. Redfield, *IBM J. Res. Develop.*, **1**, 19 (1957).
4.4.7. J. H. Freed, G. K. Frankel, *J. Chem. Phys.*, **39**, 326 (1963).
4.4.8. J. H. Freed, G. K. Frankel, *J. Chem. Phys.*, **41**, 699 (1964).
4.4.9. A. Hudson, G. R. Luckhurst, *Mol. Phys.*, **13**, 409 (1967).
4.4.10. A. Hudson, G. R. Luckhurst, *Chem. Rev.*, **69**, 191 (1969).
4.4.11. J. Heinzer, *Mol. Phys.*, **22**, 167 (1971).
4.4.12. J. Heinzer, QCPE No.209, 1972; Quantum Chemistry Program Exchange, Indiana University, Bloomington, Indiana.
4.4.13. G. Binsch, *Mol. Phys.*, **15**, 469 (1968).
4.4.14. G. Binsch, *J. Am. Chem. Soc.*, **91**, 1304 (1969).
4.4.15. D. A. Kleier, G. Binsch, *J. Magn. Reson.*, **3**, 146 (1970).
4.4.16. S. Sankarapandi, G. V. R. Chandramouli, C. Daul, P. T. Manoharan, *J. Magn. Reson. A*, **103**, 163 (1993).
4.4.17. C. Corvaja, M. DeMarchi, A. Toffoletti, *Appl. Magn. Reson.*, **12**, 1 (1997).

Section 4.5

4.5.1 D. Gatteschi, O. Kahn, J. S. Miller, F. Palacio (Eds.), *Molecular Magnetic Materials*; Kluwer Academic: Dordrecht, 1991.
4.5.2 H. Iwamura, J. S. Miller (Eds.), *Mol. Cryst. Liq. Cryst.*, **232, 233** (1993).
4.5.3 J. S. Miller, A. J. Epstein (Eds.), *Mol. Cryst. Liq. Cryst.*, **271-274** (1995).
4.5.4 K. Itoh, J. S. Miller, T. Takui, (Eds.), *Mol. Cryst. Liq. Cryst.*, **305, 306** (1997).
4.5.5 O. Kahn, (Ed.), *Mol. Cryst. Liq. Cryst.*, **334, 335** (1999).
4.5.6 P. M. Lahti (Ed.), *Magnetic Properties of Organic Materials*, Marcel Dekker: New York, 1999.
4.5.7 K. Itoh, M. Kinoshita (Eds.), *Molecular Magnetism*, Gordon and Breach: Amsterdam (Kodansha: Tokyo), 2000.
4.5.8 M. Kinoshita, P. Turek, M. Tamura, K. Nozawa, D. Shiomi, Y. Nakazawa, M. Ishikawa, M. Takahashi, K. Awaga, T. Inabe, Y. Maruyama, *Chem. Lett.*, **1991**, 1225.
4.5.9 M. Tamura, Y. Nakazawa, D. Shiomi, K. Nozawa, Y. Hosokoshi, M. Ishikawa, M. Takahashi, M. Kinoshita, *Chem. Phys. Lett.*, **186**, 401 (1991).
4.5.10 Y. Nakazawa, M. Tamura, N. Shirakawa, D. Shiomi, M. Takahashi, M. Kinoshita, M. Ishikawa, *Phys. Rev. B*, **46**, 8906 (1992).
4.5.11 A. L. Buchachenko, *Dokl. Phys. Chem.*, **244**, 107 (1979).
4.5.12 L. Néel, *Ann. Phys. Ser. 12*, **3**, 137 (1948).
4.5.13 D. Shiomi, M. Nishizawa, K. Sato, T. Takui, K. Itoh, H. Sakurai, A. Izuoka, T. Sugawara, *J. Phys. Chem. B*, **101**, 3342 (1997).
4.5.14 M. Nishizawa, D. Shiomi, K. Sato, T. Takui, K. Itoh, H. Sawa, R. Kato, H. Sakurai, A. Izuoka, T. Sugawara, *J. Phys. Chem. B*, **104**, 503 (2000).
4.5.15 D. Shiomi, K. Sato, T. Takui, *J. Phys. Chem. B*, **104**, 1961 (2000).
4.5.16 D. Shiomi, K. Sato, T. Takui, *J. Phys. Chem. B*, **105**, 2932 (2001).
4.5.17 D. Shiomi, K. Sato, T. Takui, *J. Phys. Chem. A*, **106**, 2096 (2002).
4.5.18 D. Shiomi, M. Nishizawa, K. Kamiyama, S. Hase, T. Kanaya, K. Sato, T. Takui, *Synth. Met.*, **121**, 1810 (2001).
4.5.19 S. Hase, D. Shiomi, K. Sato, T. Takui, *J. Mater. Chem.*, **11**, 756 (2001).

4.5.20 D. Shiomi, K. Ito, M. Nishizawa, K. Sato, T. Takui, K. Itoh, *Synth. Met.*, **103**, 2171 (1999).

4.5.21 D. Shiomi, M. Nishizawa, K. Kamiyama, S. Hase, T. Kanaya, K. Sato, T. Takui, *Synth. Met.*, **121**, 1810 (2001).

4.5.22 D. Shiomi, T. Kanaya, K. Sato, M. Mito, K. Takeda, T. Takui, *J. Am. Chem. Soc.*, **123**, 11823 (2001).

4.5.23 T. Kanaya, D. Shiomi, K. Sato, T. Takui, *Polyhedron*, **20**, 1397 (2001).

4.5.24 D. Shiomi, K. Sato, T. Takui, to be reported elsewhere.

4.5.25 R. McWeeny, *Spins in Chemistry*, Academic Press, New York and London, 1970.

4.5.26 R. McWeeny, B. T. Sutcliffe, *Methods of Molecular Quantum Mechanics*, Academic Press, New York and London, 1969.

4.5.27 C. Kaneda, D. Shiomi, K. Sato, T. Takui, *Polyhedron*, inpress.

4.5.28 D. Shiomi, C. Kaneda, K. Sato, T. T. Takui, *Appl. Mag. Reson.*, inpress.

Section 5.1

EN.REFLIST

Section 5.2

5.2.1 J. Larionova, M. Gross, M. Pilkington, H. Andres, H. Stoeckli-Evans, H. U. Güdel, S. Decurtins, *Angew. Chem. Int. Ed. Engl.*, **39**, 1605 (2000).

5.2.2 S. M. J. Aubin, M. W. Wemple, D. M. Adams, H.-L. Tsai, G. Christou, D. N. Hendrickson, *J. Am. Chem. Soc.*, **118**, 7746 (1996).

5.2.3 Recent developments in single-molecule magnetism are overviewed in the following literatures. (a) D. Gatteschi, L. Sorace, *J. Solid State Chem.*, **159**, 253 (2001). (b) D. Gatteschi, *J. Alloy Comp.*, **317-318**, 8 (2001). (c) D. N. Hendrickson, G. Christou, H. Ishimoto, J. Yoo, E. K. Brechin, A. Yamaguchi, E. M. Rumberger, S. M. J. Aubin, Z. Sun, G. Aromi, *Polyhedron*, **20**, 1479 (2001). (d) D. Luneau, *Current Opinion in Solid State and Materials Science*, **5**, 123 (2001). (e) D. Gatteschi, *J. Phys. Chem. B*, **104**, 9780 (2000). (f) G. Aromí, S. M. J. Aubin, M. A. Bolcar, G. Christou, H. J. Eppley, K. Folting, D. N. Hendrickson, J. C. Huffman, R. C. Squire, H.-L. Tsai, S. Wang, M. W. Wemple, *Polyhedron*, **17**, 3005 (1998).

5.2.4 (a) J. Friedman, M.P. Sarchik, J. Tejada, R. Ziolo, *Phys. Rev. Lett.*, **76**, 3838 (1996). (b) J. Friedman, M.P. Sarchik, J. Tejada, J. Maciejewski, R. Ziolo, *J. Appl. Phys.*, **79**, 6031 (1996).

5.2.5 (a) J. M. Hernandez, X.X. Zhang, F. Luis, J. Bartolome, J. Tejada, R. Ziolo, *Europhys. Lett.*, **35**, 301 (1996). (b) L. Thomas, F. Lionti, R. Ballou, D. Gatteschi, R. Sessoli, B. Barbara, *Nature*, **383**, 145 (1996). (c) J. M. Hernandez, X. X. Zhang, F. Luis, J. Tejada, J. Friedman, M. P. Sarachik, R. Ziolo, *Phys. Rev.*, **B55**, 5858 (1997).

5.2.6 Short review articles about (a): P. C. E. Stamp, *Nature*, **383**, 125 (1996). E. M. Chudnovsky, *Science*, **274**, 938 (1996).

5.2.7 M.N. Leuenberger, D. Loss, *Nature*, **40**, 789 (2001).

5.2.8 (a) A. Barra, D. Gatteschi, R. Sessoli, *Phys. Rev.* **B56**, 8192 (1997). (b) A. L. Barra, A. Caneschi, D. Gatteschi, and R. Sessoli, *J. Magn. Magn. Mater.*, **177-181**, 709 (1998).

5.2.9 S. Hill, J. A. A. J. Perenboom, N. S. Dalal, T. Hathaway, T. Stalcup, J. S. Brooks, *Phys. Rev. Lett.*, **80**, 2453 (1998).

5.2.10 K. Awaga, K. Takeda, and T. Inabe, *Mol. Cryst. and Liq. Cryst.*, **335**, 473 (1999).

5.2.11 S.M.J. Aubin, Z. Sun, L. Pardi, J. Krzystek, K. Folting, L.-C. Brunel, A.L. Rheingold, G. Christou, D. N. Hendrickson, *Inorg. Chem.*, **38**, 5329 (1999).

11. Exchange-coupled high-spin clusters

5.2.12 A.L. Barra, A. Caneschi, A. Cornia, F. F. De Biani, D. Gatteschi, C. Sangregorio, R. Sessoli, L. Sorace, *J. Am. Chem. Soc.*, **121**, 5302 (1999).

5.2.13 J. Yoo, E.K. Brechin, A. Yamaguchi, M. Nakano, J.C. Huffman, A.L. Maniero, L.-C. Brunel, K. Awaga, H. Ishimoto, G. Christou, D. N. Hendrickson, *Inorg. Chem.*, **39**, 3615 (2000).

5.2.14 R. Blinc, P. Cevc, D. Arcon, N. S. Dalal, R. M. Achey, *Phys. Rev.* **B63**, 212401/1 (2001).

5.2.15 T. Kuroda-Sowa, M. Lam, A. L. Rheingold, C. Frommen, W. M. Reiff, M. Nakano, J. Yoo, A. L. Maniero, L.-C. Brunel, G. Christou, D. N. Hendrickson, *Inorg. Chem.*, **40**, 6469 (2001).

5.2.16 P. Artus, C. Boskovic, J. Yoo, W. E. Streib, L.-C. Brunel, D. N. Hnedrickson, G. Christou, *Inorg. Chem.*, **40**, 4199 (2001).

5.2.17 J. Yoo, A. Yamaguchi, M. Nakano, J. Krzystek, W. E. Streib, L.-C. Brunel, H. Ishimoto, G. Christou, D.N. Hendrickson, *Inorg. Chem.*, **40**, 4604 (2001).

5.2.18 A. L. Barra, P. Debrunner, D. Gatteschi, C. E. Schulz, R. Sessoli, *Europhys. Lett.*, **35**, 133 (1996).

5.2.19 A.L. Barra, D. Gatteschi, and R. Sessoli, *Chem. Eur. J.*, **6**, 1608 (2000).

5.2.20 A.-L. Barra, F. Bencini, A. Caneschi, D. Gatteshi, C. Paulsen, C. Sangregorio, R. Sessoli, L. Sorace, *CHEMPHYSCHEM*, **2**, 523 (2001).

5.2.21 A. Bouwen, A. Caneschi, D. Gatteschi, E. Goovaerts, D. Schoemaker, L. Sorace, M. Stefan, *J. Phys. Chem. B*, **105**, 2658 (2001).

5.2.22 J. Dolinsek, D. Arcon, R. Blinc, P. Vonlanthen, J. L. Gavilano, H. R. Ott, R. M. Achey, N. S. Dalal, *Europhys. Lett.*, **42**, 691 (1998).

5.2.23 S. L. Castro, Z. M. Sun, C. M. Grant, J.C. Bollinger, D. N. Hendrickson, G. Christou, *J. Am. Chem. Soc.*, **120**, 2365 (1998).

5.2.24 T. Goto, T. Kubo, T. Koshiba, Y. Fujii, A. Oyamada, J. Arai, K. Takeda, K. Awaga, *Physica B*, **284-288**, 1227 (2000).

5.2.25 M. Soler, P. Artus, K. Folting, J. C. Huffman, D. N. Hendrickson, G. Christou, *Inorg. Chem.*, **40**, 4902 (2001).

5.2.26 N. Aliaga, K. Folting, D. N. Hendrickson, G. Christou, *Polyhedron*, **20**, 1273 (2001).

5.2.27 S.M. Aubin, Z. Sun, H. J. Eppley, E.M. Rumberger, I. A. Guzei, K. Folting, P. Gantzel, A.L. Rheingold, G. Christou, D.N. Hendrickson, *Inorg. Chem.*, **40**, 2127 (2001).

5.2.28 T. Kubo, T. Koshiba, T. Goto, A. Oyamada, Y. Fujii, K. Takeda, K. Awaga, *Physica B*, **284**, 310.

5.2.29 R. M. Achey, P. L. Kuhns, A. P. Reyes, W.G. Moulton, N. S. Dalal, *Solid State Commun.*, **121**, 107 (2001).

5.2.30 R. A. Robinson, P. J. Brown, D. N. Argyriou, D. N. Hendrickson, S. M. J. Aubin, *J. Phys. Condens. Matter.*, **12**, 2805 (2000).

5.2.31 H. Andres, R. Basler, H. U. Gudel, G. Aromi, G. Christou, H. Buttner, B. Ruffle, *J. Am. Chem. Soc.*, **122**, 12469 (2000).

5.2.32 A. Cornia, M. Affronte, A. G .M. Jansen, D. Gatteschi, A. Caneschi, R. Sessoli, *Chem. Phys. Lett.* **322**, 477 (2000).

5.2.33 A. Cornia, M. Affronte, D. Gatteschi, A. G. M. Jansen, A. Caneschi, R. Sessoli, *J. Magn. Magn. Mater.*, **226-230**, 2012 (2001).

5.2.34 L. Bokacheva, A.D. Kent, M. Walters, *Polyhedron*, **20**, 1717 (2001).

5.2.35 M. A. Novak, A. M. Gomes, R.E. Rapp, *J. Appl. Phys.*, **83**, 6943 (1998).

5.2.36 F. L. Mettes, G. Aromi, F. Luis, M. Evangelisti, G. Christou, D. N. Hendrickson, L. J. de Jongh, *Polyhedron*, **20**, 1459 (2001).

5.2.37 A. M. Gomes, M. A. Novak, W. C. Nunes, R. E. Rapp, *J. Magn. Magn. Mater.*, **226-**

5.2.38 F. Fominaya, P. Gandit, G. Gaudin, J. Chaussy, R. Sessoli, C. Sangregorio, *J. Magn. Magn. Mater.*, **195**, L253 (1999).

5.2.39 Y. Miyazaki, A. Bhattacharjee, M. Nakano, K. Saito, S. M. J. Aubin, H. J. Eppley, G. Christou, D. N. Hendrickson, M. Sorai, *Inorg. Chem.*, **40**, 6632 (2001).

5.2.40 A. Bhattacharjee, Y. Miyazaki, M. Nakano, J. Yoo, G. Christou, D. N. Hendrickson, M. Sorai, *Polyhedron*, **20**, 1607 (2001).

5.2.41 A. Caneschi, L. Cianchi, F. D. Giallo, D. Gatteschi, P. Moretti, F. Pieralli, G. Spina, *J. Phys. Condens. Matter.*, **11**, 3395 (1999).

5.2.42 C. Boskovic, E. K. Brechin, W. E. Streib, K. Folting, D. N. Hendrickson, G. Christou, *Chem. Commun.*, 467 (2001).

5.2.43 M. Soler, S. K. Chandra, D. Ruiz, J. C. Huffman, D. N. Hendrickson, G. Christou, *Polyhedron*, **20**, 1279 (2001).

5.2.44 T. Kuroda-Sowa, M. Nakano, G. Christou, D. N. Hendrickson, *Polyhedron*, **20**, 1529 (2001).

5.2.45 M. Soler, E. Rumberger, K. Folting, D. N. Hendrickson, G. Christou, *Polyhedron*, **20**, 1365 (2001).

5.2.46 E. C. Sanudo, V. Z. Grillo, J. Yoo, J. C. Huffman, J. C. Bollinger, D. N. Hendrickson, G. Christou, *Polyhedron*, **20**, 1269 (2001).

5.2.47 E. K. Brechin, J. Yoo, M. Nakano, J. C. Huffman, J. C. Bollinger, D. N. Hendrickson, and G. Christou, *Chem. Commun.*, 783 (1999).

5.2.48 (a) H. Oshio, N. Hoshino, T. Ito, *J. Am. Chem. Soc.*, **122**, 12602 (2000). (b) H. Oshio, N. Hoshino, and T. Ito, *Preprints of Symposium on Molecular Structure*, Chem. Soc. Japan, 522 (2000).

5.2.49 C. Benelli, J. Cano, Y. Journaux, R. Sessoli, G. A. Solan, R. E. P. Wingpenny, *Inorg. Chem.* **40**, 188 (2001).

5.2.50 T. Lis, *Acta Crystallogr.*, **B36**, 2042 (1980).

5.2.51 A. Caneschi, D. Gatteschi, R. Sessoli, A. L. Barra, L. C. Brunel, M. Guillot, *J. Am. Chem. Soc.*, **113**, 5873 (1991).

5.2.52 S. M. J. Aubin, N. R. Dilley, L. Pardi, J. Krzystek, M. W. Wemple, L. C. Brunel, M. B. Maple, G. Christou, D. N. Hendrickson, *J. Am. Chem. Soc.*, **120**, 4991 (1998).

5.2.53 M. A. Novak, R. Sessoli, A. Caneschi, D. Gatteschi, *J. Magn. Magn. Mater.*, **146**, 211 (1995).

5.2.54 M. R. Cheesman, V. S. Oganesyan, R. Sessoli, D. Gatteschi, A. J. Thomson, *Chem. Comun.*, 1677 (1997).

5.2.55 R. Sessoli, D. Gatteschi, A. Caneschi, M.A. Novak, *Nature*, **365**, 141 (1993).

5.2.56 R. Sessoli, H.L. Tsai, A. R. Schake, S. Wang, J. B. Vincent, K. Folting, D. Gatteschi, G. Christou, D. N. Hendrickson, *J. Am. Chem. Soc.*, **115**, 1804 (1993).

5.2.57 D. Gatteschi, A. Caneschi, L. Pardi, R. Sessoli, *Science*, **265**, 1054 (1994).

5.2.58 J. Villain, F. Hartman-Boutron, R. Sessoli, A. Rettori, *Europhys. Lett.*, **27**, 159 (1994).

5.2.59 C. Paulsen, J.-G. Park, B. Barbara, R. Sessoli, A. Caneschi,, *J. Magn. Magn. Mater.*, **140-144**, 379 (1995).

5.2.60 B. Barbara, W. Wernsorfer, L. C. Sampaio, J. G. Park, C. Paulsen, M. A. Novak, R. Ferre, D. Mailly, R. Sessoli, A. Caneschi, K. Hasselbach, A. Benoit, L. Thomas, *J. Magn. Magn. Mater.*, **140-144**, 1825 (1995).

5.2.61 A. Abragam, B. Bleaney, *Electron Paramagnetic Resonance of Transition Ions*, Dover Publications, New York, 1986.

5.2.62 W. Wernsdorfer, R. Sessoli, D. Gatteschi, *Europhys. Lett.*, **47**, 254, (1999).

5.2.63 D. Loss, D. P. Di Vincenzo, G. Grinstein, D. Awschalom, J. F. Smyth, *Physica B*, **189**, 189 (1993).

11. Exchange-coupled high-spin clusters

5.2.64   D. P. Di Vincenzo, *Physica B*, **197**, 109 (1994).
5.2.65   A. M. Gomes, M. A. Novak, W. Wernsdorfer, R. Sessoli, L. Sorace, D. Gatteschi, *J. Appl. Phys.*, **87**, 6004 (2000).
5.2.66   S.M. J. Aubin, S. Spagna, H.J. Eppley, R. Sager, G. Christou, D.N. Hendrickson, *Chem. Commun.*, 803 (1998).
5.2.67   S. Miyashita and N. Nagaosa, *Prog. Theoretical Phys.*, **106**, 533 (2001).
5.2.68   C. Sangregorio, T. Ohm, C. Paulsen, R. Sessoli, D. Gatteschi, *Phys. Rev. Lett.*, **78**, 4645 (1997).
5.2.69   W. Wernsdorfer and R. Sessoli, *Scienc*, **284**, 133 (1999).
5.2.70   W. Wernsdorfer, R. Sessoli, A. Caneschi, D. Gatteschi, A. Cornia, *Europhys. Lett.*, **50**, 552 (2000).
5.2.71   W. Wernsdorfer, I. Chiorescu, R. Sessoli, D. Gatteschi, D. Mailly, *Physca B*, **284-288**, 1231 (2000).
5.2.72   W. Wernsdorfer, R. Sessoli, A. Caneschi, D. Gatteschi, A. Cornia, D. Mailly, *J. Appl. Phys.*, **87**, 5481 (2000).
5.2.73   R. Sessoli, A. Caneschi, D. Gatteschi, L. Sorace, A. Cornia, W. Wernssdorfer, *J. Magn. Magn. Mater.*, **226-230**, 1954 (2001).
5.2.74   S. Maccagnano, R. Achey, E. Negusse, A. Lussier, M. M. Mola, S. Hill, N. S. Dalal, *Polyhedron*, **20**, 1441 (2001).
5.2.75   K. Wieghardt, K. Pohl, I. Jibril, G. Huttner, *Angew. Chem. Int. Ed. Engl.*, **23**, 77 (1984).
5.2.76   Y. Pontillon, A. Caneschi, D. Gatteschi, R. Sessoli, E. Ressouche, J. Schweizer, E. Lelievre-Berna, *J. Am. Chem. Soc.*, **121**, 5342 (1999).
5.2.77   S. Miyashita, *J. Phys. Soc. Jpn.*, **65**, 2734 (1996).

Section 5.3

5.3.1.   M. Sakai, J. Toyata, M. Mitsumi, K. Nakasuji, K. Furukawa, D. Shiomi, K. Sato, T. Takui, *Synthetic Metals*, **121**, 1776 (2001).
5.3.2.   H. Matsuoka, H. A. Onishi, T. Yoshida, K. Kubono, K. Sato, D. Shiomi, K. Furukawa, T. Kato, K. Yokoi, T. Takui, *Syn. Met.*, in press.
5.3.3.   M. Sakai, J, Toyata, K. Furukawa, M. Mitsumi, K. Sato, D. Shiomi, K. Nakasuji, T. Takui, to be published.

Section 5.4

5.4.1.   N. Hirota and S. I. Weissman, *J. Am. Chem. Soc.*, **86**, 2538 (1964).
5.4.2.   For pulse-ESR based electron spin transient nutation spectroscopy and applications to transition assignments of fine structure ESR spectra, refer to the following chapter and also see (a) J. Isoya, H. Kanda, J. R. Norris, J. Tang, M. K. Bowman, *Phys. Rev.*, **B41**, 3905 (1990). (b) A. V. Astashkin, A. Schweiger, *Chem. Phys. Lett.*, **174,** 595 (1990). (c) K. Sato, D. Shiomi, T. Takui, K. Itoh, T. Kaneko, E. Tsuchida, H. Nishide, *J. Spectrosc. Soc. Japan*, **43**, 280 (1994).
5.4.3.   For pulse-ESR based electron spin transient nutation spectroscopy applied to hyperfine fine-structure spectroscopy, see H. Matsuoka, K. Sato, D. Shiomi, T. Takui, *Appl. Mag. Reson.*, in press.
5.4.4.   (a) M. C. B. L. Shohoji, M. L. T. M. Franco, M. C. R. L. R. Lazana, K. Sato, S. Nakazawa, D. Shiomi, T. Takui, to be published. (b) S. Nakazawa, K. Sato, D. Shiomi, T. Takui, to be published.

5.4.5. S. Nakazawa, K. Sato, D. Shiomi, M. C. B. L. Shohoji, M. L. T. M. Franco, M. C. R. L. R. Lazana, T. Takui, to be published.
5.4.6. T. Nakamura, S. Nakazawa, K. Sato, D. Shiomi, T. Takui, T. Shida, K. Ihoh, unbublished work.
5.4.7. Y. Morita, T. Aoki, K. Fukui, S. Nakazawa, K. Tamaki, S. Suzuki, A. Fuyuhiro, K. Yamamoto, K. Sato, D. Shiomi, A. Naito, T. Takui, K. Nakasuji, *Angew. Chem., Int. Ed.*, **41**, 1793 (2002).
5.4.8. K. Goto, T. Kubo, K. Yamamoto, K. Nakasuji, K. Sato, D. Shiomi, T. Takui, M. Kubota, T. Kobayashi, K. Yakushi, J. Ouyang, *J. Am. Chem. Soc.*, **121**, 1619 (1999).

Section 5.5

5.5.1. R. Kubo, K. Tomita, *J. Phys. Soc. Japan*, **9**, 888 (1954).
5.5.2. (a) B. Pilawa, J. Ziegker, *Syn. Met.*, **82**, 53 !1996). (b) A. Wolter U. Frasol, . Jaeppelt, E. Dormann, *Phys. Rev.* **B54**, 12272 (1996).
5.5.3. (a) D. Shiomi, M. Nishizawa, K. Sato, T. Takui, K. Itoh, H. Sakurai, A. Izuoka, T. Sugawara, *J. Phys. Chem. B*, **101**, 3342 (1997). (b) M. Nishizawa, D. Shiomi, K. Sato, T. Takui, K. Itoh, H. Sawa, R. Kato, H. Sakurai, A. Izuoka, T. Sugawara, *J. Phys. Chem. B.*, **104**, 503 (2000).
5.5.4. K. Nagata, Y. Tazuke, *J. Phys. Soc., Japan*, **321**, 337 (1972).
5.5.5. Only typical examples are cited in the following: (a) K. Oshima, K. Okuda, M. Date, *J. Phys. Soc. Japan*, **41**, 475 (1976). (b) B. Pilawa, T. Pietrus, *J. Mag. Mag. Mat.*, **150**, 165 (1995). (c) J. –L. Stanger, J. J. Abdre, P. Turek, Y. Hosokoshi, M. Tamura, M. Kinoshita, P. Rey, J. Cirujeda, J. Veciana, *Phys. Rev.* **B55**, 8398 (1997).
5.5.6. H. Benner, J. P. Boucher, in Magnetic Properties of Layered Transition Metal Compounds (ed. by L. J. de Jongh), Kluwer Academic Pub., 1990.
5.5.7. P. Turek, *Mol. Cryst. Liq. Cryst.*, **233**, 191 (1993).

Section 6.1

6.1.1. P. Joliot, G. Barbieri , R. Chabaud, *Photochem. Photobiol.*, **10**, 309-329 (1969).
6.1.2. B. Kok, B. Forbush, M. McGloin, *Photochem. Photobiol.*, **11**, 457-475 (1970).
6.1.3. R. J. Debus, *Biochim. Biophys. Acta*, **1102**, 269 (1992).
6.1.4. H. T. Witt, *Ber. Bunsenges. Phys. Chem.*, **100**, 1923 (1996).
6.1.5. R. D. Britt, in *Oxygenic Photosynthesis: The Light Reactions*, D. R. Ort and C. F. Yocum (Eds.), Kluwer, Dordrecht, Netherlands (1996),137-164.
6.1.6. V. K. Yachandra, K. Sauer, M.P. Klein, *Chem. Rev.*, **96**, 2927 (1996).
6.1.7. Ö. Saygin, H. T. Witt, *Photobiochem. Photobiophys.*, **10**, 71-81 (1985).
6.1.8. W. F. Beck, G. W. Brudvig, *Biochemistry*, **26**, 8285-8295 (1987).
6.1.9. J. Messinger, G. Renger, *Biochemistry*, **32**, 9379-9386 (1993).
6.1.10. P. J. Riggs-Gelasco, R. Mei, C. F. Yocum. J. E. Penner-Hahn, *J. Am. Chem. Soc.*, **118**, 2387-2399 (1996).
6.1.11. J. Messinger, G. Seaton, T. Wydrzynski, U. Wacker, G. Renger, *Biochemistry*, **36**, 6862-6873 (1997).
6.1.12. V. K. Yachandra, K. Sauer, M. P. Klein, *Chem. Rev.*, **96**, 2927-2950 (1996).
6.1.13. H. Schiller, J. Dettmer, L. Iuzzolino, W. Dörner, W. Meyer-Klaucke, V. A. Solé, H. - F. Nolting, H. Dau, *Biochemistry*, **37**, 7340-7350 (1998).
6.1.14. J. E. Penner-Hahn, *Metal Sites in Proteins and Models: Redox Centres*; H. A. O. Hill, P. J. Sadler, A. J. Thomson (Eds.), Springer, Berlin (1998), pp1-36.

11. Exchange-coupled high-spin clusters

6.1.15. V. J. DeRose, I. Mukerji, M. J. Latimer, V. K. Yachandra, K. Sauer, M. P. J. Klein, *Am. Chem. Soc.*, **116**, 5239-5249 (1994).

Section 6.2

6.2.1. G. C. Dismukes, Y. Siderer, *Proc. Natl. Acad. Sci. USA*, **78**, 274 (1981).
6.2.2. Review Article; R. J. Debus, *Biochim. Biophys. Acta*, **1102**, 269-352 (1992).
6.2.3. J. M. Peloquin, K. A. Campbell, D. W. Randall, M. A. Evanchik, V. L. Pecoraro, W. H. Armstrong, R. D. Britt, *J. Am. Chem. Soc.*, **122**, 10926-10942 (2000).
6.2.4. A. Haddy, W. R. Dunham, R. H. Sands, R. Aasa, *Biochim. Biophys. Acta*, **1099**, 25-34 (1992).
6.2.5. A. V. Astashkin, Y. Kodera, A. Kawamori, *J. Magn. Reson.*, **105**, 113-119 (1994).
6.2.6. O. Hansson, R. Aasa, T. Vanngard, *Biophys. J.*, **51**, 825-832 (1987).
6.2.7. P. J. Smith, K. A. Åhling, R. J. Pace, *J. Chem. Soc., Faraday Trans.*, **I 89**, 2863-2868 (1993).
6.2.8. A. Boussac, S. Un, O. Horner, A. W. Rutherford, *Biochemistry*, **37**, 4001-4007 (1998).

Section 6.3

6.3.1. J. Messinger, J. H. A. Nugent, M. C. W. Evans, *Biochemistry*, **36**, 11055-11060 (1997).
6.3.2. K. A. Åhrling, S. Peterson, S. String, *Biochemistry*, **36**, 13148-13152 (1997).
6.3.3. J. Messinger, J. H. Robblee, W.O. K. Yu, Sauer, V. K. Yachandra, M. P. Klein, *J. Am. Chem. Soc.*, **119**, 11349-11350 (1997).
6.3.4. K. A. Åhrling, S. Peterson, S. Styring, *Biochemistry*, **37**, 8115-8120 (1998).

Section 6.4

6.4.1. K. A. Campbell, J. M. Peloquin, D. P. Pham, R. J. Debus, R. D. Britt, *J. Am. Chem. Soc.*, **120**, 447-448 (1998).
6.4.2. K. A. Campbell, W. Gregor, D. P. Pham, J. M. Peloquin, R. J. Debus, R. D. Britt, *Biochemistry*, **37**, 5039-5045 (1998).
6.4.3. S. L. Dexheimer, M. P. Klein, *J. Am. Chem. Soc.*, **114**, 2821-2826 (1992).
6.4.4. T. Yamauchi, H. Mino, T. Matsukawa, A. Kawamori, T. Ono, *Biochemistry*, **36**, 7520-7526 (1997).

Section 6.5

6.5.1. T. Matsukawa, H. Mino, D. Yoneda, A. Kawamori, *Biochemistry*, **38**, 4072-4077 (1999).
6.5.2. N. Ioannidis, V. Petrouleas, *Biochemistry*, **39**, 5246-5254 (2000)
6.5.3. J. Messinger, J. H. Robblee, U. Bergmann, C. Fernandez, P. Glatzel, H. Visser, R. M. Cinco, K. L. McFarlane, E. Bellacchio, S. A. Pizarro, S. P. Cramer, K. Sauer, M. P. Klein, V. K. Yachandra, *J. Am. Chem. Soc.*, **123**, 7804-7820 (2001).

Section 7

7.1 N. Koga, *et al.*, personal communications.

7.2 Y. Morita, S. Suzuki, K. Fukui, S. Nakazawa, K. Sato, D. Shiomi, T. Takui, and K. Nakasuji, *Polyhedron*, in press.
7.3 (a) Y. Ishimaru, K. Inoue, N. Koga, and H. Iwamura, *Chem. Lett.*, **1994**, 1693. (b) M. Kitano, Y. Ishimaru, K. Inoue, N. Koga, and H. Iwamura, *Inorg. Chem.*, **33**, 6012 (1994). (c) H. Iwamura and N. Koga, *Mol. Cryst. Liq. Cryst.*, **334**, 437 (1999).
7.4 L. M. Field, P. M. Lahti, and F. Palacio, *J. Chem. Soc. (Chem. Commn.)*, **2002**, 636.
7.5 K. Sato, D. Shiomi, and T. Takui, to be published.
7.6 D. Shiomi, K. Sato, T. Takui, and K. Itoh, unpublished work.

Chapter 12

HIGH SPIN MOLECULES DIRECTED TOWARDS MOLECULAR MAGNETS

Martin Baumgarten
Max Planck Institute for Polymer Research, Ackermannweg 10, D-55128 Mainz or PO Box 3148, D-55 021 Mainz, Germany

Key words: EPR, ESR, high spin molecules, stable radicals, biradicals, triradicals, tetraradicals, spin alignment, conjugation, topological and geometrical control, spin density distribution.

Abstract: EPR has often been used to define spin states of high spin molecules. This chapter outlines some meaningful considerations for the classification of interacting radicals and the determination of their ground spin multiplicity. Then a review on different classes of high spin molecules follows, sorted by their spin state, summarizing the application of EPR in that domain.

1. INTRODUCTION

Molecular magnetism is a part of material science chemistry, which strongly developed over the last 20 years. It may be viewed as a frontier research among inorganic, organic and their hybrid solids. Milestones include the detection of bulk ferromagnetism in the molecular complex of decamethylferrocene-tetracyanoethylene $Fe^{III}(C_5(CH_3)_5)_2$[TCNE] in 1987 by Miller and Epstein[1] and the first pure organic ferromagnet of p-nitrophenyl-nitronylnitroxide in 1991 by Kinoshita.[2] Besides use of inorganic metal complexes for molecular magnets large efforts went into the controlled synthesis of new stable organic high spin moieties and their in depth characterization. Therefore, by now a huge number of organic bi- and oligo-radicals of very different nature concerning their radical sites and the connectivity between them are available. Some of them are fairly old and are known since the beginning of the last century: ones like the Schlenk hydrocarbon[3] and Tschitschibabin's[4] or Yang's[5] and Coppinger's[6] biradicals

(Fig.1). Many more were just added over the last twenty years and most of them were synthesized and characterized or trapped as intermediates to clarify their electronic structure and ground state spin multiplicity. This has been performed especially in order to gain further insight into the factors controlling spin alignment and their possible use as building blocks for organic materials with magnetic ordering.

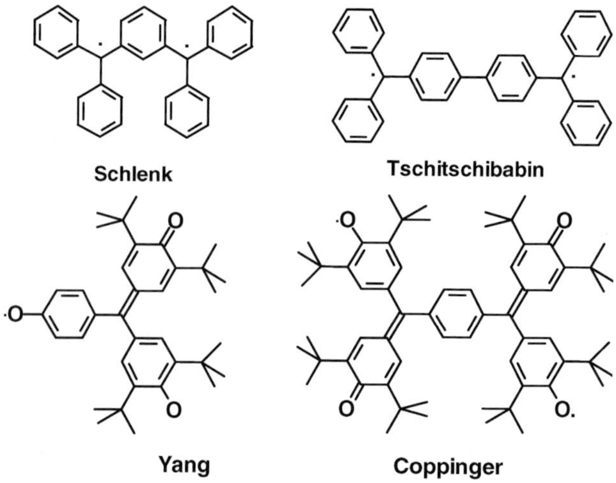

Figure 1. Some classical biradicals

Triplet state molecules, on the other hand, also play an important role as intermediates in bio-organics and bio-physics. In order to define the basic structure of this chapter we may first consider the classification of bi- and oligoradicals (see Fig. 2 for the energy levels of two spin systems in a magnetic field). While a monoradical is just represented by the spin up/spin down energy levels which lose their degeneracy in a magnetic field, a biradical can be described in many different ways: if there is a large distance (d ≥ 1 nm) and no conjugation between the radical sites the exchange coupling J is zero and the biradical may either be presented as two independent monoradicals or as an equilibrium mixture of a triplet and a singlet state (Fig. 2b,c).

If we deal with the excited states of neutral organic precursors, most often aromatic compounds, then the lowest triplet is below the lowest photoexcited singlet state and can be accessed by intersystem crossing from the latter. Such highly activated triplet states will not be considered in this review, although it might be an intriguing task to find even higher than triplet spin multiplicities after photoexcitation, *e.g.* by considering two phenyl bridged naphthalenes, where a quintet state may become accessible.

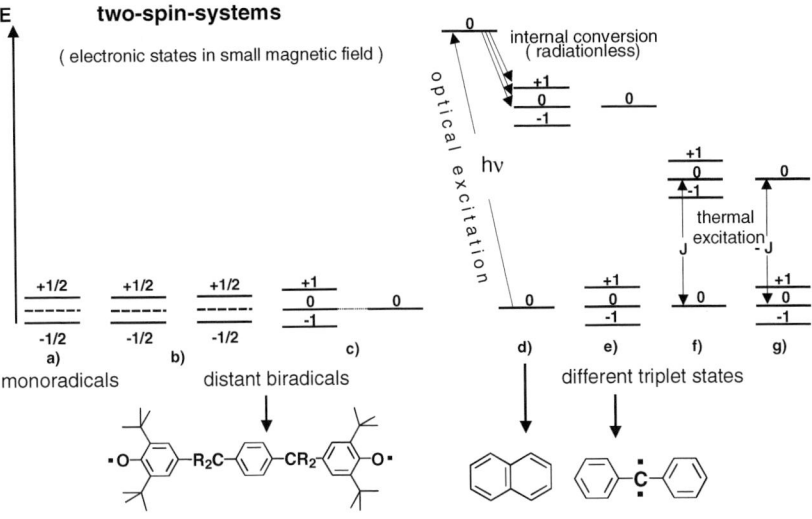

Figure 2. Different kinds of the biradicals and their energetic states compared to a) monoradical, b) or c) biradical with distant radical centers d) photoexcited triplet state, e) ground state triplet, and f) ground state singlet with thermally accessible triplet, and g) ground state triplet with thermally accessible singlet.

In this vein it was demonstrated very recently that quartet and quintet states are amenable upon photoexcitation of a 9,10-Diphenyl-anthracene **1** situated between stable nitronylnitroxide or oxo-verdazyl radicals.[7,8]

In the discussion we can delineate three cases in order of decreasing high-spin stability: a) a triplet ground state where no thermal activation to the lowest singlet is possible under ambient laboratory conditions (Fig.2e); b) a ground-state triplet with thermally accessible singlet state, and c) thermally activated triplet state above a singlet ground state. While biradicals at large distance connected through saturated spacers may easily be classified as b) or c) and photo excited triplet states d) are rarely long lived stable species, the last three groups (Fig. 2e, f, g) deserve more intimate characterization in order to define their ground state spin multiplicity unambiguously. This also holds for even higher spin states and the method of choice to do that in discrete molecules certainly is EPR spectroscopy.

1.1 Determination of the spin states and ground state multiplicities of high spin molecules

Over the last decade EPR has often been applied to identify many ambiguous spin states in high spin molecules and for characterizing new molecules. This was performed in order to fully understand the conditions and prerequisites for obtaining high spin molecules or parallel spin alignment in extended polymers with radical sites in the main or side chain with linear, branched, dendritic, or hyperbranched extensions. The basis for these approaches was the clean experimental elucidation of the spin states of the molecular building blocks by EPR spectroscopic methods and their theoretical understandings. Continuous wave X-band EPR and sometimes ENDOR studies have most often been employed, besides the use of high field high frequency and FT-pulsed EPR techniques. The latter especially transient nutation experiments developed fast during the last 10 years.[9-11]

The basic parameters for characterization of more than one unpaired electron are the electron electron spin-spin interactions H_{SS}, consisting of the isotropic exchange J and the dipolar exchange described by tensor **D**. The latter are also referred to as fine structure or zero field splitting (ZFS) parameters. The spin Hamiltonian H_{spin} for describing high spin molecules is then expressed by eq.(1), including further the electronic Zeeman $H_{ZE} = \beta\, \boldsymbol{B} \bullet \boldsymbol{g} \bullet \boldsymbol{S}$ and the hyperfine interaction $H_{SI} = \boldsymbol{S} \bullet \boldsymbol{A} \bullet \boldsymbol{I}$, while higher order terms for nuclear Zeeman $H_{nZ} = -g_n\beta_n \boldsymbol{I} \bullet \boldsymbol{B}$ and quadrupolar interactions $H_{II} = \boldsymbol{I} \bullet \boldsymbol{P} \bullet \boldsymbol{I}$ are often neglected.

$$H_{spin} = H_{ZE} + H_{SS} + H_{SI} [+ H_{nZ} + H_{II} + ...] =$$
$$\beta\, \boldsymbol{B} \bullet \boldsymbol{g} \bullet \boldsymbol{S} + \boldsymbol{S} \bullet \boldsymbol{D} \bullet \boldsymbol{S} + J\, \boldsymbol{S} \bullet \boldsymbol{S} + \boldsymbol{S} \bullet \boldsymbol{A} \bullet \boldsymbol{I}\, [-g_n\beta_n \boldsymbol{I} \bullet \boldsymbol{B} + \boldsymbol{I} \bullet \boldsymbol{P} \bullet \boldsymbol{I} +...] \quad (1)$$

In solid state EPR spectra the isotropic exchange J is not considered, since it just contributes a common constant to the energy of each high spin state. Since **D** is traceless, the fine structure is usually described only by two energetic parameters D and E, which for a triplet are $D = 3D_z/2$ and $E = (D_x - D_y)/2$, most often expressed in cm^{-1} as $D' = D/hc$ and $E' = E/hc$ or in magnetic field units $D' = D/g_e\beta$ and $E' = E/g_e\beta$. The EPR spectral analyses for isolated triplet state molecules are well established and described in standard textbooks.[12]

For high spin molecules usually it is assumed that the g anisotropy is small and that the zero field splitting |D| is not larger than the microwave transition energy (high field approximation), but exceeds the hyperfine interaction, which can then be considered as perturbation.

$$H_{spin} = \beta\, \boldsymbol{B} \bullet \boldsymbol{g} \bullet \boldsymbol{S} + D(S_z^2 - S(S+1)/3) + E(S_x^2 - S_y^2) \quad (2)$$

12. High spin molecules directed towards molecular magnets

This approximation has been encountered most widely in bridged bi- and oligoradicals leading to the "typical" high spin patterns with a maximum of 2Sx3 allowed fine structure transitions ($\Delta m_S = 1$). In addition $\Delta m_S = 2$ and sometimes $\Delta m_S = 3$ forbidden transitions can be observed. One should note that the relative intensities of the $\Delta m_S = 1$, $\Delta m_S = 2$, and $\Delta m_S = 3$ transition obey the ratio $1:(D'/B_o)^2:(D'/B_o)^4$, respectively. A typical spectral simulation for a quartet state (S = 3/2) at X-band is shown in Figure 3.

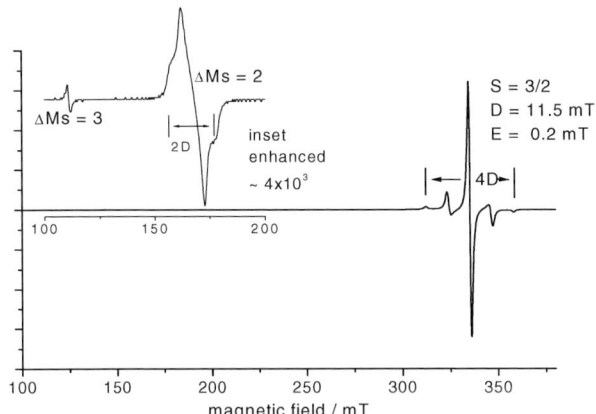

Figure 3. Simulated X-band EPR spectrum for S = 3/2 with D' = 0.01075 cm^{-1}, E' = 0.00019 cm^{-1}, and g = 2.003 at 9.4 GHz. (B_{res}= 335 mT).

Very complex patterns and additional absorption lines have been identified and analysed for high spin molecules with very large zero field splittings (e.g. carbenes and nitrenes) where $H_{SS} \geq H_{ZE}$. Such spectra were a true puzzle in the early 60's, when Wassermann et al. reported[13] about new powder spectra for *meta*-phenylene bridged dicarbenes and dinitrenes. They could not be explained until 1967,[14] when Itoh[15] submitted a single crystal study on *meta*-phenylene-bis(phenyl-methylene) **2**.

The breakdown of the perturbation approach in cases of very large zero field splitting parameters $|D| \geq h\nu$ was pointed out carefully by Takui et al.[16] leading to off axis extra absorption peaks. Those can be simulated by direct diagonalization of the spin Hamiltonian but not directly extracted from the observed spectra, which led to spurious documentation on spin-quintet dinitrenes.

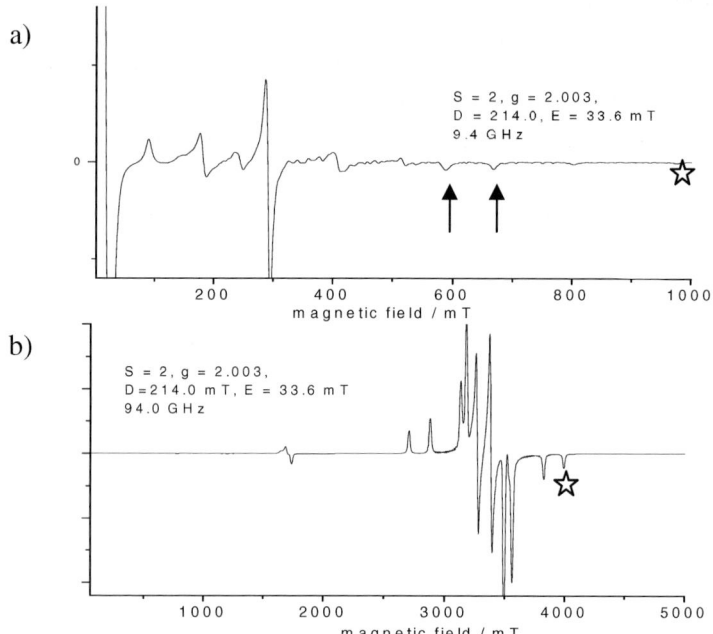

Figure 4. Simulated EPR spectra at X-band a) and at W-band b) for very large ZFS parameters D' = 0.200 cm^{-1}, E' = 0.0314 cm^{-1}. Broad star marks highest z-field component, the sharp arrows some extra absorption lines.

If these measurements, on the other hand, would be performed in W-band at 94 GHz, where the high field approximation still holds, they will give the "usual" pattern for a S = 2 spin state (Fig.4a,b). The standard cw EPR application is limited to discrete spin systems since with increasing number of spins the number of allowed transitions grows rapidly, producing high spectral density in the center with extremely weak intensity of the outermost signals. This was found already for a hexaradical in a septet state[17] and can be further demonstrated for an S = 5 state with D' = 0.0060 cm^{-1} and E' = 0.0002 cm^{-1}.

Another problem in determining the fine structures by EPR may occur when they are smaller or equal to the hyperfine splitting. In some cases with well defined hyperfine couplings as in phosphinyl radicals (^{31}P A$_\parallel$= 850 MHz, A$_{iso}$= 297 MHz while D' = 260 MHz), shown by Janssen *et al.*,[18] the well resolved transitions can still be fully analysed. Otherwise this situation may lead to unresolved EPR spectra in the Δm_S = 1 region but also to weak signals for the Δm_S = 2 transition due to small D'-value. In such cases high resolution ENDOR experiments can be very helpful to analyze hyperfine and sometimes exchange interaction.

12. High spin molecules directed towards molecular magnets

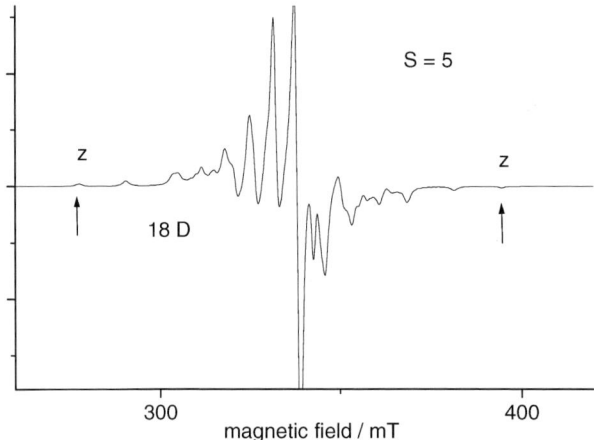

Figure 5. EPR spectrum for S=5, D=0.0060 and E=0.0002 cm-1 at 9.4 GHz and g=2. The outermost z components are separated by 18 D.

They have been applied for the identification of multispin states in solution and in rigid media,[19,20] but hardly for the identification of the ground state multiplicity itself. However, additional information on the hyperfine coupling scheme and thus spin density distribution can be obtained. If the absolute signs of the hyperfine coupling constants are determined, e.g. from isotopic labelling and triple resonance, the sign and the size of the isotropic exchange coupling J and of the dipolar ZFS component D' can be deduced as well.[20]

Other hurdles may be faced in large molecular or polymeric spin systems where the observed spectra are often composed of mixtures of different spin states and also conformers rendering the analysis very difficult.

New FT-EPR methods based on transient nutation therefore seem to be very promising for elucidating such spin states.[9,10] The transient nutation spectroscopy provides an extremely high resolution for spin multiplicities and magnitudes of ZFS parameters. It is based on the precession or nutation of the magnitization vector M around the effective magnetic field B_{eff} summed by the static field B_o and the microwave field B_1 described in the rotating frame. In the rotating frame under microwave irradiation this precession is no longer the Larmor frequency as in the laboratory frame but called "transient nutation", since the signal is observed for a finite time only after turning on the microwave radiation until a steady state is reached. When the excitation pulse is turned off, M undergoes the free induction decay (FID). The effective nutation frequency ω_{eff} is described in the rotating frame ω_{rot} as typical

Larmor precession $\omega_s = g\beta_e B_o/\hbar$ and the additional precession about the microwave field $\omega_1 = g\beta_e B_1/\hbar$.

$$\omega_{eff} = \sqrt{[(\omega_s - \omega_{rot})^2 + \omega_1^2]} \qquad (3)$$

In the resonant case $\omega_s = \omega_{rot}$ and $\omega_{eff} = \omega_1$. That is the resonant magnetization nutates with ω_1 around the rotating x-axis, whereas the off resonance magnetization nutates with frequency ω_{eff} around the effective field vector, which is deflected from the z-axis by the angle θ. The description of the spin nutation about the effective field follows the Hamiltonian including their spin states.

$$H = H_o + H_1 = (\omega_s - \omega_{rot})S_z + \omega_1 S_x \qquad (4)$$

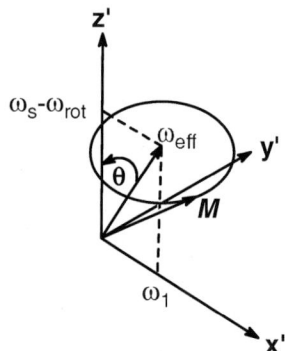

Figure 6. Nutation of the magnetization vector *M* in rotating frame.

In case of high spin molecules the nutation frequency is directly depending on the contributing spin states. The transition amplitudes depend on the spin quantum number S. If the microwave radiation excites only a single transition, the nutation frequency is given by ω_{nut}:

$$\omega_{nut}(m_S, m_S+1) = \omega_1 [(S(S+1) - m_S(m_S+1))]^{1/2}, \qquad (5)$$

where ω_1 can be calibrated for a standard spin state. If the microwave excitation is not transition sensitive, and in strong microwave field limit where all transitions are excited, a nutation of the magnetization along ω_1 is observed, which is independent of the spin states. For the intermediate situations the interpretation of the spectra may be more or less complicated. Nutation experiments can be conducted by either observing the FID or the electron spin echo, and the latter has most often been applied for characterizing high spin molecules, *e.g.* by Takui *et al.*[9,21]

12.High spin molecules directed towards molecular magnets 499

In order to define the ground state spin multiplicities and to separate them from thermally activated ones, the most usual characterization is to follow the temperature dependent variation of signal intensities of the given spin state. If the intensity changes linearly with the temperature ($I*T$ = const) the behavior is Curie like and the exchange interaction is either very large or very small. The activation energy for experimental determination is usually limited to the temperature range 2 ~ 300 K (~ 4 - 600 cal/mol) or even lower due to the thermal stability of the radicals or their matrix.

Especially for the standard cw-techniques, complications in the ground state determination for the bi- and oligoradicals may arise from overlapping signal contamination of other spin states, and very different saturation effects. Therefore these measurements should be checked for reversibility, temperature errors, and saturation effects, which become most important at lowest accessible temperature range. A deviation from linear dependence of signal intensity versus square root of microwave power showing the saturation effect may easily be implied upon lowering the temperature.

Besides experimental techniques for determining ground state spin multiplicities, where also magnetic susceptibility measurements have often been applied, many molecular systems have been considered theoretically in order to obtain a deeper understanding and further confirmation. Such quantum chemical studies on high spin states also enable the prediction of ground state multiplicities for molecules so far unknown as a guiding design.

Discrepancies between experimental results and theoretical predictions have often spurred more accurate characterisation of old or synthesis of novel high spin candidates. A number of high spin molecules have been designed from standard radicals attached to coupling units (CU) providing spin exchange by topological or geometrical control. Thus in the following paragraph we will pass through the common rules and exceptions, in order to get a more complete picture on the molecular level. Since this book focuses on EPR spectroscopy and its application, we will sort this chapter by the spin states and not by the type of radicals. A few exceptions may be apologized, for very closely related molecules. Research has been focused mainly on bi-, tri-, and tetra-radicals, while higher spin states have only been dealt with in a few cases.

Since the number of biradicals is huge we have to limit this outline to a choice of importance. For a more complete overview on all available organic high spin molecules and extended molecular based magnets further textbooks and review articles may be considered.[22-31]

2. BIRADICALS – THE TRIPLET STATE

The most often used approach for high spin molecules is the control of through bond exchange in conjugated molecules leading to so called non-Kekulé structures as in *m*-Xylene (1,3-Dimethylene-benzene), where no double bond between the unpaired electrons can be formed. In difference thereto, the *para* and *ortho* derivatives usually undergo spin pairing to the more stable quinoid structures in the Kekulé forms.

Non-Kekulé vs. Kekulé structures

The number of non bonding MO's (NBMO's) and thus the number of unpaired electrons S can easily be predicted by the rules of Longuet Higgins[32] [nNBMO = (N-2T); S = 0.5 (N-2T) with N number of π-centers and T number of double bonds] or according to the rule of Ovchinnikov[33] based on spin polarisation, by substracting the number of starred and unstarred π-centers S = 0.5 (n*-n°). The spin polarisation is thus used to explain high spin ground states for Schlenk's and Yang's biradicals, but low spin ground state for Tschitschibabins and Coppingers biradicals (Fig. 1).

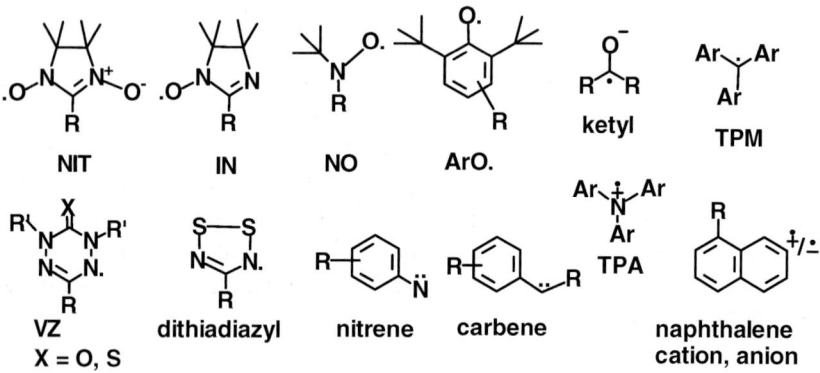

Figure 7. Common radical sites, nitronyl nitroxide **NIT**, iminonitroxide **IN**, tert.-but.-nitroxide **NO**, phenoxide **ArO**, ketyl radicals, triphenylmethyl **TPM**, verdazyl **VZ**, Dithiadiazyl, nitrenes, carbenes, Triphenylamin cation **TPA** or aromatic radical cation and anions like naphthalene.

The *m*-phenylene as coupling unit (**CU**) to connect radical sites has therefore been used in many high spin molecules. Some common radical sites are given in Fig. 7. Some of them are very stable and can be handled even in solution at room or higher temperatures (**NIT, IN..**). Others are moisture and air sensitive as the ketyl anion or aromatic radical anions and cations. The usual nitrenes and carbenes are very reactive and only stable below 100 K. Over the last years a major breakthrough was made with triplet carbenes being stable up to room temperatures. They could be synthesized with halides in the o-phenyl-positions close to the carbene as perchlorodiphenylcarbene, 2,2',4,4',6,6'-hexachlorodiphenylcarbene, or 2,2',6,6'-tetrabromo-4,4'-di*tert*-butyldiphenyl carbene.[34,35,11]

In the 90s, on the other hand, there were already known several exceptions of high spin states in non-Kekulé hydrocarbon structures, which were then addressed as "Violations of Hund's Rule in Non-Kekulé Hydrocarbons".[36] It was shown and predicted earlier by Borden[37,38] that another classification of non-Kekulé molecules is very important, *e.g.* the separation of molecules into those possessing non-disjoint and those possessing disjoint non-bonding MO's. Most easily it can be demonstrated by comparing Trimethylenemethane (**TMM**) and Tetramethyleneethane (**TME**). In **TMM** the NBMOs are distributed over the whole molecule (non-disjoint) while in **TME** and in the more rigid 2,3-Dimethylenecyclohexane-1,4-diyl (**DMC**) the NBMOs are localized on two subparts as separate moieties (disjoint). Thus spin polarization already predicts ground state singlets for **TME** and **DMC**.

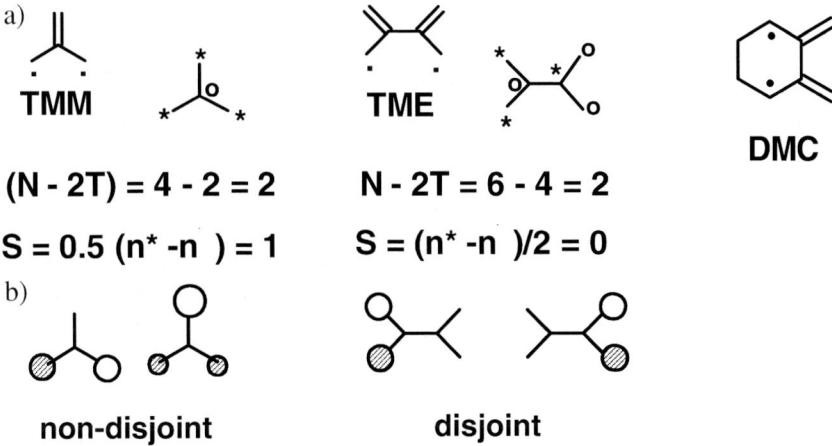

Figure 8. a) Structure and spin state of **TMM**, **TME**, and **DMC**, b) the NBMO's of **TMM** and **TME**.

Experimentally Dowd[39-41] could access all 3 molecules in frozen matrices showing similar biradical behavior concerning their EPR fine structure with ZFS's of D' = 0.024 cm^{-1} (**TMM**), and D' = 0.025 cm^{-1} (**TME**) and D'= 0.024 cm^{-1} (**DMC**). Although irradiations for nitrogen extrusion of the precursors were performed at 10 K, Curie like temperature dependence of EPR signal intensities was only plotted down to 20, 15, and 16 K, respectively, suggesting ground spin triplets for all three of them. While for **TMM** there were no doubts about, the ground spin multiplicities of **TME** and **DMC** were heavily discussed for quite some time, and many further theoretical calculations with different geometries and basis sets resulted in contradictory descriptions of **TME** and **DMC**.[42-46] In 1996 Iwamura's group[47] unambiguously evidenced that the ground state for **DMC** and **TME** is singlet, nearly degenerate with a thermally excited triplet above 10 K. This final experiment was even more important, since there exists a large number of similar biradicals and substituted derivatives which can be grouped into the "families" of **TMM** and **TME** possessing the same spin interaction core.[48-51]

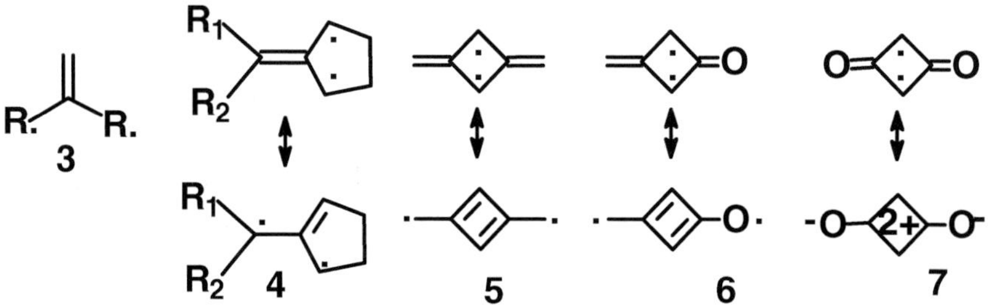

The "**TMM** family" has been widely tested for all kinds of radicals attached to 1,1'-ethenyl (**3**).[51,52] It comprises also cyclic structures like dihydropentafulvenes **4** and cyclobutenes **5-7**. Different symmetric and asymmetric substitutions R$_1$,R$_2$ have been used for **4** and for all of them the ground state was triplet.[49,53] Just the ZFS values were largest for electron donating groups like methoxy, while electron acceptors or extension of the π-system leads to a decrease of the zero field splitting. For the Cyclobutadion-diyl **7** only, no biradical could be found and a singlet state is predicted, also argued from the strong contribution of dipolar resonance structures, as shown below.

Within the "**TME** family" as for **TME** itself, there occurred many ambiguous papers, while theory predicts singlet ground states for them, even more in the cyclic planarized forms. For **8-10** first triplet states were assumed,[54,55] then Berson studied the tetramethylenebenzene **8** in more detail

12. High spin molecules directed towards molecular magnets

demonstrating ground state singlet.[56] Therefore it must be assumed that also Cyclopentadienes **9** and **10** should be further requested. Closely related is the cyclopentane-1,3-diyl **11** where a triplet ground state has been established.[57]

But the singlet triplet splitting should be small, since the two competing pathways through σ bond yield opposite signed spin polarization presented by α and β spin (or spin up and spin down) as given schematically.

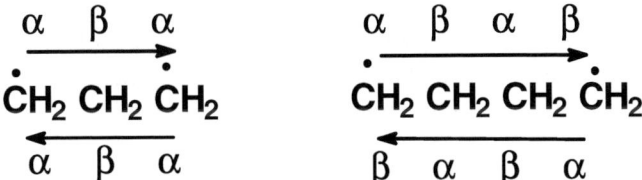

Figure 9. Spin polarisation through methylene and ethylene

The heteroatomic substitution in the five membered ring systems leading to Dimethylenethiophene **12**, Dimethylenefurane **13**, and Dimethylenepyrrol **14**, destabilizes the triplet state to such an extent that all of them are biradical singlets, even without thermal activation of the triplet state. Partially zwitterionic forms must be considered.[58] One exception seems the tosylated pyrrole **15** where a stable triplet was reported with thermally activated components of only 19 cal /mol.[59] This exception was explained with long lived spin isomerism and shown to be dependent on the excitation wavelength for nitrogen extraction leading to different species.

Also several exceptions for *meta*-phenylene bridging without high spin ground state multiplicities have been found in the 90's. While for a typical meta-phenyl bridged bisnitroxide **16** the triplet state was witnessed,[60] for the tris-methylated bisnitroxide **17** two isomers were found in their singlet state and the triplet excitation was reported to be in the range 2J = −(66 − 81) K.[61] This was also the first case where it could be shown that a corresponding *para*-phenylene bridged biradical is more stable triplet with 2J/k=-35K or ca 70 cal/mol than its meta form.[62] A very similar finding of thermally excited triplet bisnitroxides was reported by Iwamura[63] using methoxy groups neighboring the nitroxides **18** and yielded 2J/k = -7.0K for a diluted sample in solid PVC, where the intermolecular exchange is negligible, and -73.8 K for a crystalline material, where strong antiferromagnetic intermolecular interactions are enclosed. Thus not surprisingly also the three spins in a trisnitroxide with one more nitroxide radical between the methoxy groups are coupled antiferromagnetically to a doublet ground state.[64]

The reason for this unusual singlet ground states of **17** and **18** was further discussed by Borden.[65] On the basis of ab initio calculations considering twisted radical sites, singlet states for angles of torsion between 70- 90° were unequivocally predicted. A surprise was found for a bis-trifluormethyl derivative **19** reported by Rajca et al.[66] demonstrating that rapid cooling of the biradical leads to a singlet state, which slowly converts to its ground state triplet. Different temperature dependent conformer preferences were assumed to be responsible for the change.

Other exceptions of *m*-phenylene acting as ferromagnetic CU's when strong steric hindrances are involved were manifested by the diphenothiazine and diphenoxazine dication derivatives.[67,68] The zero field splittings of the biradical *para*-, *meta*-, and *ortho*- derivatives (**20-25**) formed in sulfuric acid nicely reflected their distance dependence with D' = 0.0043, 0.0060, and 0.0204 cm^{-1} (**20-22**) corresponding to r ~ 0.84, 0.76, and 0.50 nm, respectively. This dipolar approach with D'~ $3g_e^2\beta_e^2/4r^3$ estimation can be refined for distributed spin densities in delocalised biradicals, as well. The above finding is one of the very rare cases, where *o*-phenylene bridging gave EPR active species, what is based on the strong twisting and stabilization of the dication through the sulfonate counterion. In case of the phenoxazines the ZFS components are slightly larger, due to the higher spin density on the connecting nitrogens and lower charge on the oxygen compared to sulfur in phenothiazines. The temperature dependence indicated a singlet ground state for **21** with 2J = -28 cal/mol and here again more stable triplet states with Curie like behavior were identified for the corresponding *p*-phenylene bridged biradicals **20**.

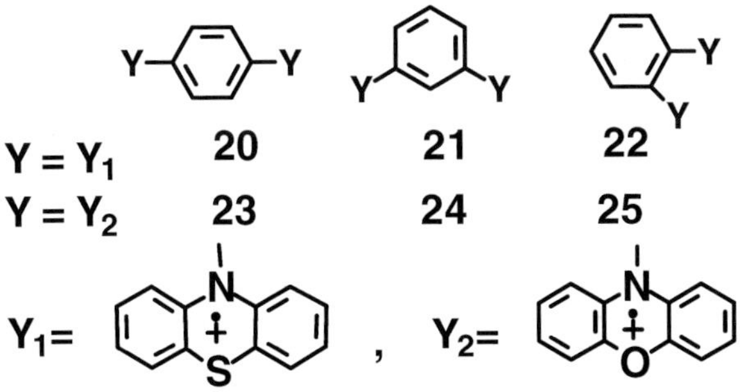

20, 21, 22 structures

Y = Y₁: 20, 21, 22
Y = Y₂: 23, 24, 25

$Y_1 =$ (phenothiazine-S), $Y_2 =$ (phenoxazine-O)

In principle strict orthogonal alignment of radicals may also open a way to high spin ground states (geometrical vs. topological approach).[23] In an early approach this was tested by Veciana et al.[69] with the perchlorobiphenyl- dication **26**, where the large chloro-substituents lead to a strong twisting of the biphenyl unit and also decrease the oxidation potential. But later it was shown that the biradical had singlet ground state. Since Bianthryl **27** is aligned perpendicularly and can be oxidized with $SbCl_5$ to the biradical dication (D' = 0.00225 cm^{-1}) or reduced with alkali metal to the biradical dianion (D' = 0.00177 cm^{-1}), also oligo(anthrylenes) **27-29** and the corresponding polymer were studied for this purpose.[70] Temperature dependent studies of the zero field splittings (zfs) in the frozen solution EPR

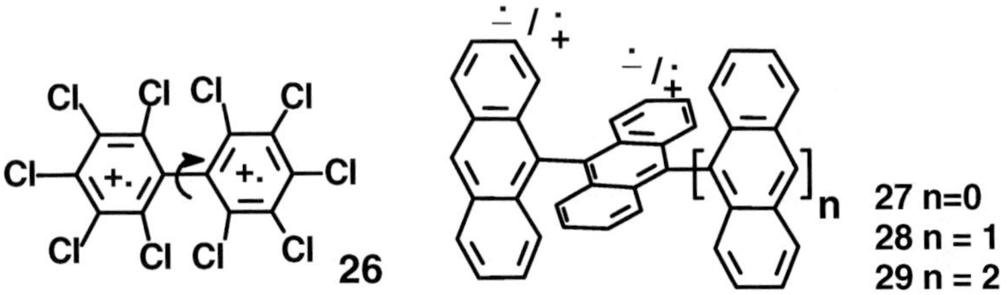

26 (perchlorobiphenyl dication)

27 n=0
28 n = 1
29 n = 2

spectra evidenced that the charged oligo(9,10-anthrylenes) **27-29** persist in low spin ground states. The low temperature measurements of the bianthryl dianion additionally revealed that the average orthogonal alignment is lost upon lowering of the temperature, with a clear orthorhombicity of the zero field splitting parameters D'= 0.00182 cm^{-1} and E' = 0.0010 cm^{-1} at 20 K which correlates to an angle θ of 81° derived from the two extremes $E_{min} = 0$

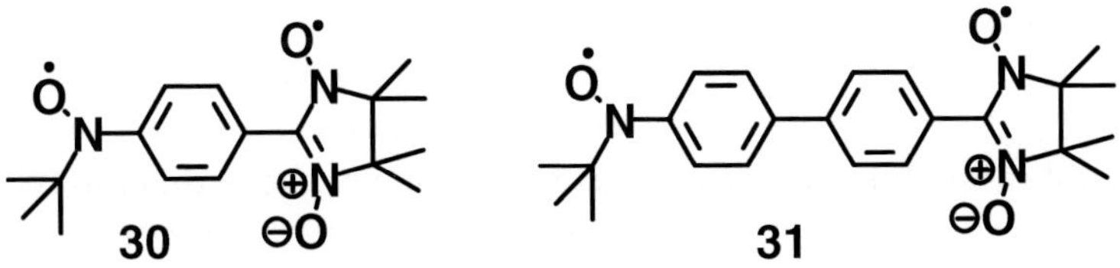

30, **31**

($\theta = 90°$) and $E_{max} = D/3$ ($\theta = 0°$). Also in other asymmetric biaryles with strong to orthogonal hindrance only thermally activated triplet states were found.[71]

Another way to overcome the standard topological rule was excellently demonstrated by Iwamura and Inoue[72,73] using asymmetric radical sites in **30, 31**. Then spin polarisation leads to stable triplet ground state biradicals with strong intramolecular ferromagnetic coupling.

In a similar way the topological rules for exchange coupling may be overcome upon use of delocalized radicals and radical ions.[74,75] A necessary property is large sign-alternating spin densities along the periphery, well known for the neutral phenalenyl- (32) and pyrenyl (33) cation or anion radicals. Bridging through so called antiferroagnetic CU's as single bond, 1,2-ethenyl, or *p*-phenylene will then lead to high spin formation if the radical ions are connected through positions of opposite signed spin densities. While the bridged 1,2'-biphenalene 34 was just considered as model theoretically[74] the 1,2'-bipyrene 35 and 2-(9-anthryl)-pyrene 36[75] should also form high spin ground state entities and were synthesized and charged to their biradical dianion.[76]

A third opportunity to obtain high spin molecules besides topology and geometry, is based on high symmetry. It has long been established for molecules having threefold or higher symmetry, that they possess doubly degenerate HOMO's or LUMO's which can be charged, leading to triplet biradicals. Classical examples are the pentadienyl cation **37**[77] and some cations of substituted derivatives[78,79] as well as the benzene dianion **38**,[80] so called antiaromatic annulenes. The perphenyl substituted derivatives **37, 38**, on the other hand, yielded singlet ground states or paramagnetic

12.High spin molecules directed towards molecular magnets 507

contributions only.[81,82] For hexaphenylbenzene **38** even a one step dehydrocyclization to hexabenzocoronene **46** is well established.[83]

Therefore also 1,3,5-triphenylbenzene **39** dianion and dication,[84] triphenylene dianion **40**,[85] and the dication of donor substituted derivatives **41**,[86] **42**[87] yield triplet ground states. Suitable derivatives with lowered reduction/oxidation potentials as **42**[88] and hexaazahydrocoronene **43**[89] were also deemed as building blocks for organic magnets by Breslow[88-92] in alternating donor acceptor stacks, in line with McConnell model II.[93] For **43** more detailed studies revealed the singlet ground state[94,95] which was also found for the dianion (D' = 0.054 cm^{-1}, ΔE_{ST} = 2.3 kcal/mol)[96] and dication (D' = 0.0591 cm^{-1}, ΔE_{ST} = 1.4 kcal/mol)[97] of coronene **44**. Due to Jahn-Teller distortions also other fused hydrocarbons like corannulene **45**[98] and hexabenzocoronene **46**,[99] loose their high symmetry upon charging, yielding singlet ground state dianions. For **46**, however the thermal activation of the triplet state is extremely small, nearly degenerate with the singlet (ΔE_{st} = 15 cal/mol). For higher charged molecules the use of an additional independent method besides EPR spectroscopy to characterize the charging process like optical absorption or potentially controlled generation has shown to be very powerful in many occasions. The latter holds also for the reduction of [60]fullerene which even possesses threefold degenerate LUMO's in the neutral form. Only for the dianion of C$_{60}$ new large thermally excited ZFS components were measured, being lost at trianion formation, additionally evidenced by absorption spectroscopy[100] and potentiometric control.[101] Without such control in higher concentrated samples even triradicals and higher spin states could be found,[102] while the single crystal study of the trianion supported its doublet state.[103]

3. TRIRADICALS - THE QUARTET STATE

Many triradicals have been evidenced by EPR spectroscopy, since the early work by Brickmann und Kothe[104] on a triradical in the quartet state **47** and the full analysis of its powder spectra.[105] At first glance some ambiguities from cw-EPR spectral analysis may easily occur, since the ZFS in the $\Delta m_S = 1$ region itself can be very similar to a mixture from mono- and biradicals. Additional information then come from the possible identification of a $\Delta m_S = 3$ forbidden transition and splittings of the size of D' in the $\Delta m_S = 2$ region (*e.g.* Figure 3).[106,70] Further hints can be derived in cases of stepwise formation upon charging and control of the relative signal intensities as well as by comparison with similarly structured triradicals.

47 R = phenyl
48 R = H
49 R = H, **50** R = OCH$_3$

In 1992 a triradical trication of Trisphenylaminobenzene (**49**) was reported,[107] but the ZFS components were very small (D = 0.0012 cm^{-1}), and the 5 line solution spectrum for three equivalent nitrogens unexplained. This forced Blackstock[108] to fully characterize the charging of the hexa-anisyl (**50** R=OCH$_3$) derivative. The electron donating methoxy groups certainly support the trication formation, where cyclic voltammetry at low temperature (-78 °C) yielded 3 reversible oxidation waves. The liquid solution EPR spectrum of the trication showed 10 lines for strong exchange coupled nitrogens. In frozen state a typical spectrum of a quartet was found (5 lines) with D' = 0.0046 cm^{-1}. This value is very close to the one given for the isoelectronic hexaphenyl-trimethylenebenzene **48** [109] (D' = 0.0049 cm^{-1}) and further supported the correct quartet state assignment. Thus it was

12. High spin molecules directed towards molecular magnets

assumed that in the earlier report some dimerization might have occurred, which can easily be found for triphenylamines and triphenylmethylenes.[110] The ground state quartet of **50** was further evidenced by 2D-ESTN spectroscopy,[111] where in a well defined trication no hint of contributing excited doublet states was found.

51 $R_1=R_2=C_6H_5$
52 $R_1=C_6H_5; R_2=CH_3$

53 $R_1=R_2=C_6H_5$
54 $R_1=C_6H_5; R_2=CH_3$

55 $R_1=R_2=C_6H_5$
56 $R_1=C_6H_5; R_2=CH_3$

The problems of dimerization can also be overcome in extended oligo(*m*-*p*-N-phenylaniline)s **51-56** reported by Janssen et al.[112-114] They indicated quartet ground state for all of them, but certainly with much smaller ZFS (D' = 0.0026, 0.0029 cm^{-1} **51, 52**) than for **50** due to the larger delocalisation. As seen from the D' values, the exchange of each phenyl group by a donating methyl substituent still does not strengthen much dipolar interactions.

57 $R_1=R_2=C_6H_5$
58 $R_1=R_2= pC_6H_4OCH_3$

Comparable triradicals with triazine skeleton obtained by Blackstock,[115] on the other hand, were found to deviate from Curie like behavior at low temperatures, and a low spin ground state was assumed. This is somewhat in contrast to other reports, claiming the disubstituted 1,3-triazines or trisubstituted 1,3,5-triazines to be a more favourable CU than di- or trisubstituted 1,3- or 1,3,5-benzene.[116,117] The s-triazines have also been shown to support high spin ground states for dinitrenes.[118]

Many more stable quartet triradicals have been reported, synthesized by attaching radical sites to topologically controlled positions of a π-network. Only a few will be discussed here focusing on EPR studies.

Triradicals with nitroxides as spin carrying units have been synthesized extensively by Iwamura et al.[119-122] Here the quartet states were proven for the phenylene bridging between the radical sites in **59-61**. They have also been established as valuable stable building blocks for molecular magnets upon further ligation with metal complexes through the nitroxide unit.[123] This is inherently combined with the need of very clean oligo-radicals (100%).

If a methyl center is used (**62**) a triplet ground state results, while for a central amine in **63** or borane in **64** only doublet ground states were found.[124,125] This was reasoned by electron excess in case of the amine leading to superexchange between two spin sites, and by electron deficiency in borane, stabilizing degenerate doublet states below the quartet.

Similarly, many nitronyl-nitroxides (**NIT**) and iminonitroxides (**IN**) have been attached to aromatic cores and used also in mixtures with nitroxide radicals and as building blocks for molecular magnets.[126] Typical

12. High spin molecules directed towards molecular magnets

examples are the 1,3,5-tris-NIT-benzene (**65**) by Shiomi et al.[127] and the tris-NIT **66** by Turek et al.[128]

While many publications just treated symmetrical radical sites with a 1,3,5-phenylene bridging, Matsuchita and Itoh et al.[129] tried to charge dicarbenes to the monoanion **67** and monocation **68**. First some deviation from Curie type behavior in the very low temperature range ~ 5K was reported, but since there was no obvious reason for spin frustration of the quartet state, the work was redone carefully also with ^{13}C labelling of the methylene.[130] The quartet ground state was evidenced thereby for both, **67** and **68**.[131]

Several other triradicals in the quartet state were also obtained by using mixtures of different radical sites - so called "heterospin molecules" - and modified topology as for instance nitrenes in combination with **NIT** or nitroxides.[132,133] Although a quinoid structure can be formed in **69** it remains a non-Kekulé form with 3 unpaired electrons and quartet ground state. The ZFS value (D' = 0.277 cm^{-1}, E' ~ 0) is smaller than for the *m*-isomer **70** (D' = 0.347 cm^{-1}, E' = 0.0045 cm^{-1}) which was found to be a general criterion for a range of different radicals, *e.g.* the nitroxide **71** (D' = 0.336 cm^{-1}, E' = 0.0045 cm^{-1}) yields nearly identical ZFS components as the NIT **70**. This similarity for the ZFS still makes sense although **70** actually should be low spin ground state judged from the topology (it was only reported at 77 K) while **71** should have high spin ground state.

The features of the nitrenes coupled with **NIT** or nitroxide radicals are structurally very close to those found earlier by Matsuda and Iwamura[134,135] for carbenes mixed with **NIT 72** (D = 0.113 cm^{-1}, E = 0.006 cm^{-1}) or nitroxide **73** (D' = 0.118-0.125 cm^{-1}, E' = 0.006 cm^{-1}). Just in the case of carbenes, conformational isomers occur again and the spin is partially delocalised in the outer phenyl ring.

As shown for the trisnitroxide **59** one diphenyl-nitroxide can also be used as bridging unit between radical sites. This was extended to bis-NIT **74**, bis-IN **75** and their mixed structure **76**.[136] Here the EPR spectra yielded no resolved ZFS components, only a strong central signal and some shoulders, together with non resolved $\Delta m_S = 2$ signal. The magnetic measurements further proved the quartet state and were fitted with nearly twice as large exchange coupling for the bis-NIT **74** (J = 231 cm^{-1}) than for bis-IN **75** (J = 127 cm^{-1}). For the asymmetric **76** an even more complex analysis yielded one very strong and a weak interaction. The elucidation of these effects on a cleanly separated sample seems worthwhile, since **NIT** can easily be contaminated with **IN** upon sample preparation.

12.High spin molecules directed towards molecular magnets 513

Sugawara demonstrated that the attachment of two **NIT** units to thianthrene in **77, 78** regardless of the substitution position in 2,7 or 2,8 lead to quartet ground states upon one electron oxidation of the thianthrene.[137]

The ZFS components also are nearly identical for both isomers D' = 0.012(5) , E' = 0.0009 cm^{-1}. The independence of the topology was reasoned by equally signed spin densities on the 7 and 8 position of the thianthrene cation. Several other redox switchable high spin molecules based on stable radicals and easily oxidized π-systems like tetrathiafulvalenes or triarylamines have been prepared by Sugawara et al.[138,139]

4. TETRARADICALS - THE QUINTET STATE

Many tetraradicals in the quintet state have been evidenced by EPR. Classical examples are the *m*-phenylene bridged dicarbenes and dinitrenes mentioned in the introduction and a 3,6-Dimethyleneanthracenediyl-1,7-dioxy **80** in its quintet state.[140] The carbenes and nitrenes have been applied as spin sources with many different CU's.

In 1991 Iwamura[141] evidenced that quintet spectra are formed for the diphenylacetylene and diphenyldiacetylene dinitrenes **81-84**. But stable ground state entities resulted from the *m,p'* isomers **81, 82**, only. The quintet *m,m'*-isomers **83, 84** gave maximum intensity at 50 and 28 K, according to antiferromagnetic exchange with J ~ -30 (**83**) and J~-14 cal/mol (**84**), respectively. The exchange couplings were derived from the fit of the intensity of the quintet state by assuming standard thermal excitation with ΔE_{Quin-S}= 6J and accounting for the intermediate triplet state ΔE_{S-T}=2J.

$$I \cdot T = C \cdot \{\exp(6J/RT)\}/\{1+3\exp(2J/RT) + 5\exp(6J/RT)\} \qquad (6)$$

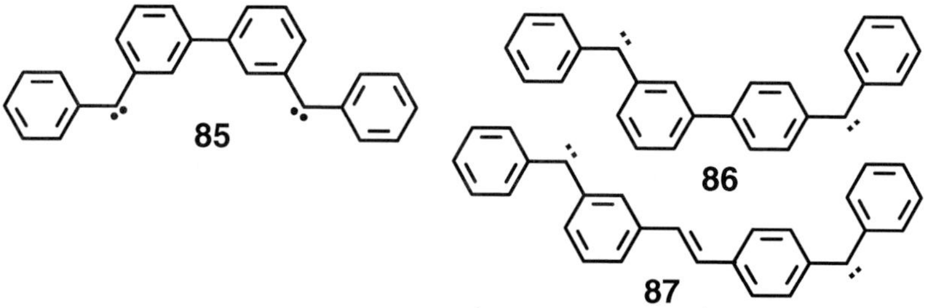

85

86

87

From the ZFS (D' = 0.169 cm^{-1}, E' = 0.040 cm^{-1}) it was concluded that the Z-isomers should be preferred. The overall result is similar to the reports of ground state singlets for the dicarbene **85**[142] and ground state quintet for **86**.[143] Several years before it had been shown that a *meta-para* linkage at a stilbene unit yields quintet ground states for **87**.[144]

Several other diphenyldinitrenes with 1,1-ethenyl[145,146] or carbonyl[146] bridges were also studied for obtaining deeper insight into the spin coupling mechanism. The *p,p'*-isomers **88, 91** were shown to be the only one to form ground state quintets, while for the *m,m'*- and *m,p'*-isomers **89, 90** thermally activated quintet states were defined.

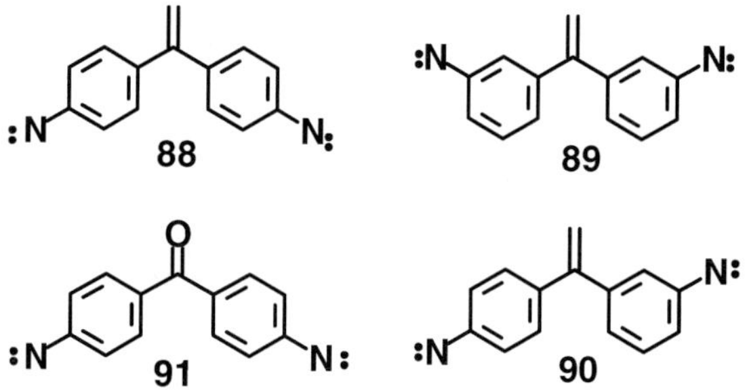

88

89

91

90

While spin polarization accounts for the low spin state of the *m,p'*-isomer it does not in case of the *m,m'*-isomer. The latter was concluded by Iwamura to be explained as a "doubly-disjoint" structure on basis of Borden's classification. A comparison with nitroxide spin centers instead of nitrenes at the same positions shown above, yielded the same signs for the exchange interaction, even though the dinitroxides lead to weaker spin polarization into the bridging units. Based on the nitrene results and Itoh's model[147] for weakly interacting triplets, the ZFS D_q of a quintet was estimated from the two triplets D_t from centers a and b (eq. 7). Lahti[148,149] proposed a vector model to explain the angular dependence of the ZFS in bridged dinitrenes. Thereby it clearly came out that a decent approximation is obtained but that it is still limited to planarized spin centers and π-

12. High spin molecules directed towards molecular magnets

network.[149] Overall the topology determines the ground state multiplicity as long as no large geometrical demands (strong torsions) occur.

$$D_q = (D_t^a + D_t^b)/6 + D_t^{ab}/3 \qquad (7)$$

In an elegant work using the **TMM**-approach with 2-alkylidene-1,3-cyclopentanediyl, Dougherty[150] tested the spin interaction through various spacers **92-95**. Even saturated CU's as cyclobutanediyl, cyclopentanediyl, and adamantyl were included. For each system Kekulé forms can be drawn towards the inner/outer units, indicating the bridging of triplets directly connected to the spacer. In **92** different R were also applied demonstrating the ability to control the interaction. For R = methyl still sufficient coupling to the quintet was found, but no longer for R = *tert.*-butyl where only triplets can be measured since the biradicals are twisted out of conjugation and become separated moieties.

93 and **94** were shown to possess quintet ground states, with D' = 0.021 and 0.018 cm^{-1}, respectively. For **94** thermally activated triplet components appeared around 40 K in the temperature dependent study, suggesting a small quintet triplet gap $\Delta E_{QT} = 200$ cal/mol ($\Delta E_{QT} = 4J$, J = 50 cal/mol). For the adamantane derivative **95** containing a 1,3-cyclohexanediyl fragment, no quintet was found upon photolysis of the diaza-precursor. This showed that the interaction between the triplet sites depends sensitively on the through bond coupling, the bond angle, and the distance.

Heterocycles have been considered further as substitutes for benzene, e.g. the aforementioned triazine[118] and pyridine.[151-154] Dougherty[151] described stable quintet ground state for all three pyridine isomers **96-98**. Only small changes in D' were found for the neutral tetraradicals and as assumed from model calculations **96** is closest to **92**, since the nitrogen is substituted in a so called "inactive" position (unstarred). Additional formation of the pyridinium cation by protonation led to singlet ground state for the 2,6-isomer **99** as predicted, but surprisingly not for the protonated 2,4-isomer **98**, where also an active position is involved.

Using phenyl carbenes as radical sites, other authors[152] suggested that only 3,5-pyridine bridging should yield high spin quintet states, while for the 2,6- and 2,4-pyridine bridging, the disturbance of active sites by heteroatomic substitution should lead to ground state singlets. This is somewhat contradictory to other reports where the 2,6-pyridine bridging is well acknowledged for high spin formation.[153,154] Also Lahti et al.[155] tested the interaction of di- and trinitrenes through 2,6-, and 2,4-, and 2,4,6- substituted pyridines. They further supported the finding of high spin ground states by ab initio and DFT calculation with 6-31G* basis sets. The pyridine seems thus to act as coupling unit very similar to benzene.

Rajca[156] synthesized and characterized many oligoarylmethanes which were used to get further control of spin interaction but also directed towards use in larger arrays, where a defect on one spin site would not hinder interaction between the others. One of these examples is the cyclic tetraradical **100**, which has been used as a core or a building block in many other polyradicals. The ZFS components of **100** (D' = 0.0033 cm^{-1}) are only slightly larger than those for the branched tetraradical **101**.[157]

A small central peak was attributed to half integral spin (S=1/2, 3/2) impurities estimated to contribute less than 10%. No thermally excited low spin components were envisaged up to 80 K, which are a problem for the even further extended high spin states.

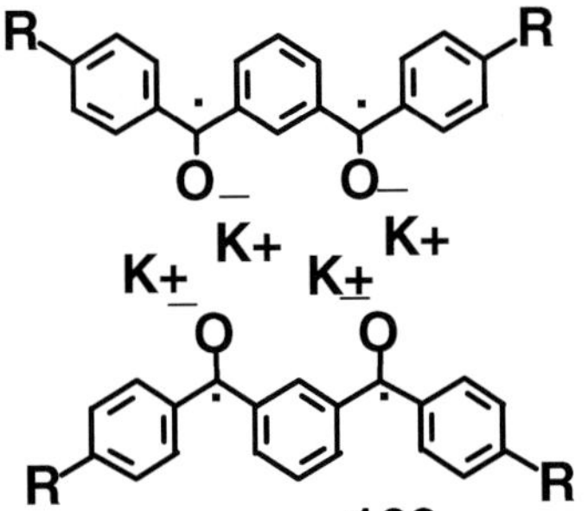

102

Finally it should be mentioned that in a very different approach well defined tetraradical formation could be proven for diketyl radical dimerizaion through metal bridging.[158,159] Based on very early examples of triplet formaion in benzophenone anions[160] dibenzoylketones **102** were shown to persist in their quintet ground state.

5. HIGHER SPIN STATES, S ≥ 5/2

Just a handful of papers are dealing with spin state detection by EPR of even higher spin multiplicities than the quintet, with the majority of reports for the heptet state. Two pentaradicals with S=5/2 state based on NO as in **103**[161] and arylmethyl **104**[157] radical sites of very similar structure have been identified. They were found in their sextet ground state from temperature dependent studies. As expected from further spin delocalisation into the outer aryl units in **104** the ZFS parameters are smaller (D' = 0.0027 cm, E' = 0.0009) than for the oligonitroxide **133** (D' = 0.0039, E' = 0.0013 cm^{-1}).

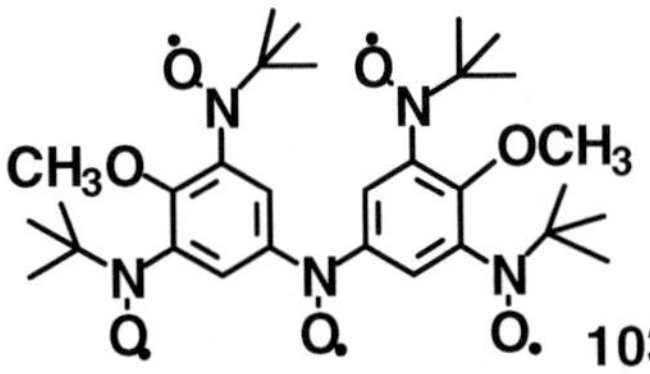

103 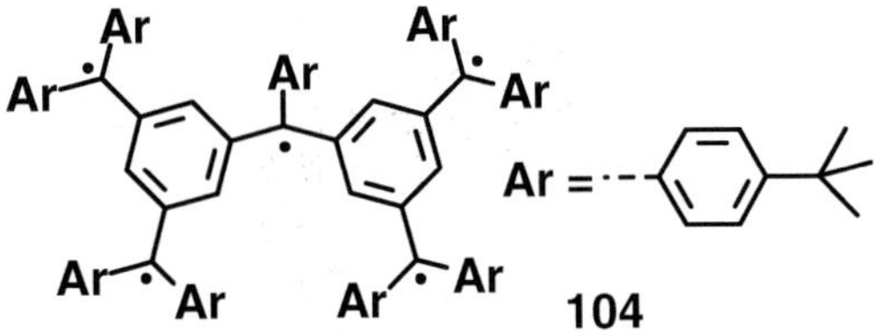   **104**

The heptet state was considered most often, since it is accessible through threefold 1,3,5-benzene substitution with S=1 radical units. The first prepared hexaradical of this type was a tricarbene reported in 1973[162] soon after the understanding of the dicarbene spectra. Further preparations on extended linear carbenes **105, 106** followed. They were studied by single crystal EPR spectroscopy and shown to persist in S = 4 and S = 5 state.[163,164]

The oligocarbenes are much more difficult to characterize in powders, since the number of possible conformers increases, all giving rise to somewhat different ZFS parameters.

105 n = 1, 106 n = 2

Adam et al[165] reported the hexaradical **107** based on the 1,3-cyclopentane-diyl unit as a spin carrier. It was shown that nitrogen extrusion takes place in a stepwise fashion, leading to a triplet, a quintet and finally a heptet ground state. The intermediately formed bi- and tetraradicals were shown to be identical to the ones found earlier **108, 109**.[166] The difficulty in analysing the hexaradical was its incomplete formation (yield ~ 35%). Although the identification with ZFS components of D' = 0.00907 cm^{-1} and a small E'= 0.000187 cm^{-1} was possible[166,167] such molecules then can not be considered as building blocks for organic magnets, since incomplete formation is a major problem in all extended organic high spin systems.

108

107

109

In the early 90s Rajca[167] synthesized and characterized by cw-EPR spectroscopy star shaped polyarylmethylradicals **110, 111** achieving S= 7/2 (D' = 0.00163, E ~ 0) and S = 5 (D' = 0.0012, E ~ 0), which could still be resolved due to symmetry (E ~ 0) and quite clean preparations of the spin states with only small contributions from lower spin moieties. Similar carbene structures with S = 6 (**112**) and S = 9 (**113**) were reported,[168,169] which at that time have been the highest spin states from organic high spin molecules. However EPR spectroscopy was applied only up to S = 6.

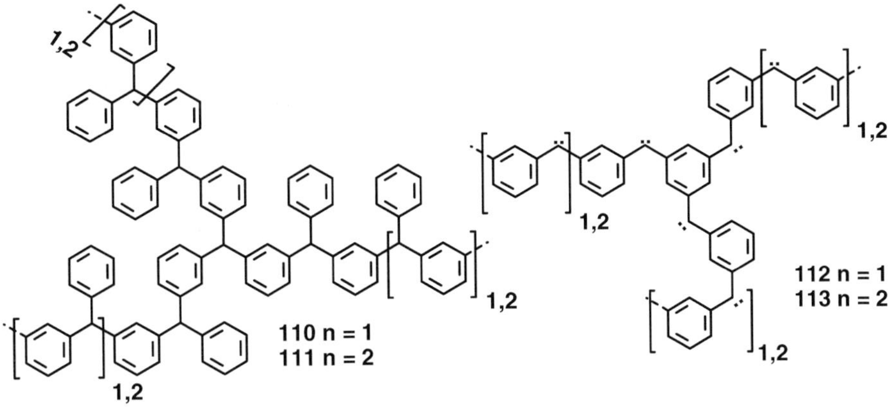

12. High spin molecules directed towards molecular magnets

110 n = 1
111 n = 2

112 n = 1
113 n = 2

The comparison of different hexacarbenes was made including cyclic and further branched derivatives **114-116**.[171,172] Different fine structures were observed and shown to follow Curie behaviour but no complete analysis of the complex spectra was given. A critical behaviour was found for the further extended branched nonacarbenes **114**, which did not yield the anticipated S=18/2 spin state, but S = 7. Subsequent sample analysis indicated a loss of two active centers most probably due to chemical bond formation between the outer carbenes which should be spin allowed.[172] Since

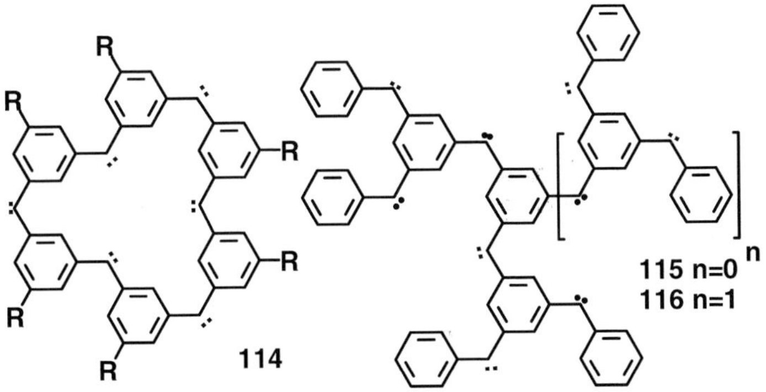

114

115 n=0
116 n=1

such unsubstituted carbenes generally suffer from high reactivity and destruction at ambient temperatures new halide substituted derivatives, as mentioned for the radical sites, are very promising. An example for such an extended stable hexaradical **117** was given by Tomioka et al.[11]

Upon the search of stable building blocks with very high spin, Rajca extended manifold the structures of polyarylmethyl radicals.[157,173] One major goal was to establish substructures where even a defect does not hinder the interaction between all remaining spins. He achieved that perfectly using Calix[4]arene rings like **118**, and many more. For such well defined molecules full cw-EPR spectral analysis is possible. For **118** the ZFS components were determined to be D' = 0.00127 cm^{-1}, E' ~0.

Some remarkable resolutions of spin states in polymers by 2D-ESTN were achieved, where standard cw-EPR just showed a single broad line.[174] High spin components in the polymer **119** were identified as having S = 1/2-4 from their different nutation frequencies. Many more polyradicals like **120**, **121** with phenoxy, nitroxy and other radical sources attached under topological control to conjugated linear polymer chains or to star shaped, dendritic or hyperbranched structures were still published in the 90's.[175-181] The overall problem as in many other polymeric materials is the complete conversion of all active precursors to the radical sites without defects or a "soft" polymerisation with the radicals as shown nicely by Miura et al.[179] Therefore usually only relatively low spin concentrations of 0.2-0.9 per repeat unit are found, displayed by an average of discrete S states, and no ferromagnetic domains are formed.

12. High spin molecules directed towards molecular magnets

So far the only way in this field to very high spin organic molecules and polymers is based on Rajca's extension[182,183] of ladder type and star-branched polymers with multiple pathways of spin interacting moieties. Intriguing example are the ladder type structures **122** (S = 12.4/2 measured instead of S=14) and the cyclic branched molecule **123** (S = 10 measured instead of S = 12) which have further been extended to polymers accessing high spin states of S=50, 100 or now even several 1000's.[184] However EPR does no longer represent a useful technique for the identification of these spin states.

6. CONCLUSION AND OUTLOOK

The highest spin states in organic molecules determined by EPR spectroscopy so far do not exceed the S = 6 state and it remains an open question whether even with new developed and extended pulsed techniques they will become identifiable. Therefore, the inherent usefulness of the EPR methods seems to be based on identification of oligoradicals and their ground state multiplicities. Manifold applications of magnetic susceptibility measurements have shown to be useful and superior for identification of real high spin states (S>10/2) or magnetic probes and can be used in addition.

From the outlines above it should be clear that application of EPR to high spin molecules is very fruitful and an important characterization. As mentioned in the introduction most of the work is directed towards organic magnets. Although it is already an intellectual task to add new building blocks of high spin ground state by design, magnetic ordering desires bulk spin alignment. In order to use the molecules further as building blocks for organic magnets, the molecules in their high spin state should be well defined and very clean, not contaminated partially by lower spin states. Since 1991 a large progress has been made and many more examples of organic ferromagnets became available. Earlier reported ones were shown to be unreproducible and most probably due to impurities, especially those from ill defined polymers. Some of the newly developed organic ferromagnets are based on stable organic mono- or biradicals and their 3-dimensional ordering. For instance after the nitrophenyl-NIT **124** ordering at 0.7 K,[2] Chiarelli and Rassat[185] showed an organic ferromagnet **125** with a ferromagnetic transition temperature of Tc = 1.48 K and Banister et al.[186] succeeded in the alignment of a dithiadiazolyl derivative **126** with a much higher Tc = 36 K.

Also pure organic [60]fullerene charge transfer compounds have been found to exhibit ferromagnetic ordering besides high superconductivity. Already in 1991 Wudl et al.[187] realized a C60 /TDAE complex with Tc = 16

K and prompted further studies for understanding this phenomenon from pure S = ½ spin systems.[188]

Inorganic organic hybrid solids have shown to become very promising and resulted in magnetic ordering at much higher temperatures than for pure organic molecules. For instance Gatteschi mixed hexafluoroacetylacetonato-Mn(II) with NIT and found magnetic transition at Tc < 8K.[189] This approach was further used extensively with NO and NIT radicals as mentioned earlier.[122,123]

Certainly the inorganic complexes prepared by Verdaguer[190] based on iron exchange in Prussian blue by V and Cr (V[Cr(CN)6], Tc = 315 K) and those of Miller[191] complexing vanadocene with tetracyanoethylene under loss of the benzenes (V(TCNE)x(CH2Cl2)y, Tc ~ 400 K) result so far in much higher Tc's. However the organic and inorganic/organic hybrid solids based on organic high spin molecules are still a very promising field of material science research and many more interesting results are anticipated from them in the near future.

Acknowledgement *The author thanks his PhD students and postdoc Anela Ivanova, Giorgio Zoppellaro, Chandrasekar Rajadurai, Ahmed Geies, and colleague Dr. Stoyan Karabunarliev (Houston) for proof reading and the DFG and MPG for continuous financial support. He is indebted to Prof. Hiizu Iwamura for suggesting and supporting the presentation of part of this topic on a workshop during the first Gordon Conference "Organic Structures and Porperties" in Japan, Fukuoka in 1996.*

7. REFERENCES

1. S. Chittapeddi, K. R. Cromack, J. S. Miller, A. Epstein, *Phys. Rev. Letter.* **58** (1987) 2695.
2. M. Tamura, Y. Nakazawa, D. Shiomi, K. Nozawa, Y. Hoosokoshi, M. Ishikawa, M. Takahashi, M. Kinoshita, *Chem. Phys. Lett.* **186** (1991) 401.
3. W. Schlenk, M. Brauns, *Ber. Dtsch. Chem.Ges.* **48** (1915) 661, *ibid.* 669, 716.
4. Tschitschibabin, *Ber. Dtsch. Chem. Ges.* **40** (1907) 1810.
5. N. C. Yang, A.J. Castro, *J. Am. Chem. Soc.* **82** (1960) 6208.
6. G. M. Coppinger, *Tetrahedron*, **18** (1962) 61; *J. Am. Chem. Soc.* **86** (1964) 4385.
7. Y. Teki, S. Miyamoto, K.Iimura, M. Nakatsuji, *J. Am. Chem. Soc.* **122** (2000) 984.
8. Y. Teki, M. Nakatsuji, Y. Miura, *Mol. Phys.* **100** (2002) 1385.
9. T. Takui in *"Molecular Magnetism"* K. Itoh, M. Kinoshita (eds.) Kodansha and Gordon and Breach, Tokyo 2000.

10. A. Schweiger, G. Jeschke *"Principles of pulse electron paramagnetic resonance"*, Oxford University Press, 2001.
11. H. Tomioka , M. Hattori, K. Hirai, K. Sato , D. Shiomi, T. Takui, K.Itoh *J. Am. Chem. Soc.***120** (1998) 1106.
12. J. A. Weil, J. R. Bolton, J. E. Wertz (eds.) *"Electron Paramagnetic Resonance"* John Wiley & Sons, New York 1994.
13. A. M. Trozzolo, R. W. Murray, G. Smolinsky, W. A. Yager, E. Wasserman *J. Am. Chem. Soc.* **85** (1963) 2526.
14. E. Wassermann, R.W. Murray, W.A. Yager, A.M. Trozzolo, G. Smolinsky, *J. Am. Chem. Soc.* **89** (1967) 5076.
15. K. Itoh, *Chem. Phys. Lett.* **1** (1967) 235.
16. Y. Teki, T. Takui, K.Itoh, *J. Chem. Phys.* **88** (1988) 613.
17. W. Adam, M. Baumgarten, W. Maas, *J. Am. Chem. Soc.* **122** (2000) 6735.
18. M. M. Wienk , R. A.J. Janssen, ***Chem. Commun.*** **16** (1996) 1919.
19. B. Kirste, H. van Willigen, H. kurreck, K. Möbius, M. Plato, R. Biehl, *J. Am Chem. Soc.* **100** (1978) 7505.
20. H. Kurreck, B. Kirste, W. Lubitz (eds.) *"Electron Nuclear Double Resonance Spectroscopy of Radicals in Solution"*, Chapter 8, Weinheim 1988.
21. T. Takui, K. Sato, D. Shiomi, K. Itoh in *"Magnetic Properties of Organic Materials"* P.M. Lahti (ed) chapter 11, Marcel Decker, New York,1999, 197 ff.
22. H. Iwamura, *Adv. in Phys. Org. Chem.* **26** (1990) 179.
23. O. Kahn „*Molecular magnets*" VCH, Weinheim, 1993.
24. J. S. Miller, A. J. Epstein, *Angew. Chem.* **106** (1994) 399.
25. P. M. Lahti (ed.) *"Magnetic Properties of Organic Materials"* Marcel Decker, New York,1999.
26. K. Itoh, M. Kinoshita (eds) *"Molecular Magnetism"* Kodansha (Gordon and Breach), Tokyo 2000.
27. J. S. Miller, M. Drillon (eds) *"Magnetism, Molecules to Materials"* Vol I-III Wiley-VCH, 2001.
28. J. Veciana, (ed) *"π-Electron Magnetism"*, Structure and Bonding Vol. **100**, Springer Verlag, 2001.
29. A. Rajca, *Chem. Rev.* **94** (1994) 871
30. W. M. Nau, *Angew. Chem. Int. Ed. Engl.* **36** (1997) 2445.
31. J. A. Crayston, J. N. Devine, J. C. Walton, *Tetrahedron* **56** (2000) 7829
32. H.C. Longuet-Higgins, *J. Chem. Phys.* **18** (1950) 265,275,283
33. A.A. Ovchinnikov, *Theor. Chim Acta.* **47** (1978) 297
34. Tomoika, H.; Harai, K.; Fujii, C. *Acta. Chem. Scand.* **46** (1992) 680.
35. Tomoika, H.; Hattori, M.; Harai, K. *J. Am. Chem. Soc.* **118** (1996) 8723.
36. W. T. Borden, H. Iwamura, J. A. Berson, *Acc. Chem. Res.* **27** (1994) 109
37. W.T. Borden, E.R. Davidson, *J. Am. Chem. Soc.* **99** (1977) 4587
38. W.T. Borden (ed) *"Diradicals"*, Wiley, New York, 1982
39. P. Dowd, K. Sachdev, *J. Am Chem. Soc.* **89** (1967) 715
40. P. Dowd, W. Chang, Y.H. Paik, *J. Am Chem. Soc.* **108** (1986) 7416
41. P. Dowd, W. Chang, Y.H. Paik, *J. Am Chem. Soc.* **109** (1987) 5284
42. P. Du, W.T. Borden, *J. Am Chem. Soc.* **109** (1987) 930
43. B.L.V. Prasad, T.P. Rhadhakrishnan, *J. Phys. Chem.* **96** (1992) 9232
44. P. Nachtigall, K. D. Jordan, *J. Am Chem. Soc.* **114** (1992) 4743.
45. J. Pranata, *J. Am Chem. Soc.* **114** (1992) 10537.
46. P. Nachtigall, K. D. Jordan, *J. Am Chem. Soc.* **115** (1993) 270.
47. K. Matsuda, H. Iwamura, *J. Am. Chem. Soc.* **119** (1997) 7412

48. K. Matsuda, H. Iwamura, *J. Chem. Soc. Perkin II* (1998) 1023
49. M. S. Platz, J. M. McBride, R. D. Little, J.J. Harrison, A. Shaw, S.E. Potter, J.A. Berson, *J. Am. Chem. Soc.* **98** (1976) 5725
50. J. A. Berson in *"Diradicals"* W.T. Borden (ed) chapter 4, Wiley 1982.
51. D. A. Shultz in *"Magnetic Properties of Organic Materials"* P.M. Lahti (ed) chapter 6, Marcel Decker, New York,1999, 103 ff.
52. D.A. Shultz, A.K. Boal, G.T. Farmer, *J. Am. Chem. Soc.* **119** (1997) 3846.
53. M. Rule, A.R. Matlin, D.E. Seeger, E. F. Hilinski, D. A. Dougherty, J. A. Berson, *Tetrahedron* **38** (1982) 787.
54. W.R. Roth, R. Langer, M. Bartmann, B. Stevermann, G. Maier, H.P. Reisenauer, R. Sustmann, W. Müller, *Angew. Chem. Int. Ed.* **26** (1987) 256.
55. W.R. Roth, U. Kowalzik, G. Meier, H.P. Reisenauer, R. Sustmann, W. Müller, *Angew. Chem. Int. Ed.* **26** (1987) 1285.
56. J.H. Reynolds, J. A. Berson, K.K. Kumashiro, J.C. Duchamp, K. Zilm, A. Rubello, P. Vogel, *J. Am. Chem. Soc.* **114** (1992) 763.
57. S.L. Buchwalter, G.L. Closs, *J. Am. Chem. Soc.* **101** (1979) 4688.
58. J.A. Berson, *Acc. Chem. Res.* **30** (1997) 238.
59. L.C. Bush, R.B. Heath, X.W. Feng , P.A. Wang, L. Maksimovic, A.I. Song, W.S. Chung, A.B. Berinstain, J.C. Scaiano, J.A. Berson, *J. Am. Chem. Soc.* **119** (1997) 1406.
60. A. Calder, A.R. Forrester, P.G. James, G.R. Luckhurst *J. Am. Chem. Soc.* **91** (1969) 3724.
61. M. Dvolaitzki, R. Chiarelli, A. Rassat *Angew. Chem. Int. Ed. Engl.* **31** (1992) 180.
62. R. Chiarelli, S. Gambarelli, A. Rassat, *Mol. Cryst. Liq. Cryst.* **305** (1997) 455.
63. F. Kanno, K. Inoue, N. Koga, H. Iwamura, *J. Am. Chem. Soc.* **115** (1993) 847.
64. S. Fang, M.-S. Lee, D. A. Hrovat, W.T. Borden, *J. Am. Chem. Soc.* **117** (1995) 6727.
65. J. Fujita, M. Tanaka, H. Suemune, N. Koga, K. Matsuda, H. Iwamura, *J. Am Chem. Soc.* **118** (1996) 9347.
66. A. Rajca, K. Lu, S. Rajca, C. R. Ross, *Chem. Commun.* 1999, 1249.
67. K. Okada, T. Imakura, M. Oda, M. Baumgarten, *J. Am. Chem. Soc.* **118** (1996) 3047.
68. J. Friedrich, Phd thesis Mainz, 1997.
69. J. Veciana, J. Vidal, N. Jullian, *Mol. Cryst. Liq. Cryst.* **176** (1989) 443.
70. U. Müller, M. Baumgarten, *J. Am. Chem. Soc.* **117** (1995) 5840.
71. M. Baumgarten, L. Gherghel, J. Friedrich, M. Jurczok, W. Rettig, *J. Phys. Chem .* **104** (1990) 1130.
72. K. Inoue, H. Iwamura, *Angew. Chem. Int. Ed.* **34** (1995) 927.
73. H. Kumagai, Y. Hosokoshi, A.S. Markosyan, K. Inoue, *Polyhedron* **20** (2001) 1329.
74. S. Karabunarliev, M. Baumgarten, *Chem. Phys.* **244** (1999) 35.
75. S. Karabunarliev, M. Baumgarten, *Chem. Phys.* **254** (2000)239.
76. Harfmann, PhD thesis, Ludwig Maximilian Univ., Munich 1996.
77. M. Saunders, R. Berger, A. Jaffe, J.M. McBride, J. O'Neill, R. Breslow, J.M. Hoffmann, C. Perchonock, E. Wassermann, R.S. Hutton, V.J. Kuck, *J. Am. Chem. Soc.* **95** (1973) 3017 .
78. R. Breslow, R. Hill, E. Wassermann, *J. Am. Chem. Soc.* **86** (1964) 5349.
79. R. Breslow, H.W. Chang, R. Hill, E. Wassermann, *J. Am. Chem. Soc.* **89** (1967) 1112.
80. E. Wassermann, R.S. Hutton, V.J. Kuck, E.A. Chandross, *J. Am. Chem. Soc.* **96** (1974) 1965.
81. R. Breslow, H.W. Chang, W.A. Yager, *J. Am. Chem. Soc.* **85** (1963) 2033.
82. W. Broser, H. Kurreck, P. Siegele, *Chem. Ber.* **100** (1967) 788.
83. A. Halleux, R. H. Martin, G.S.D. King, *Helv. Chim. Acta* **41** (1958) 1177.

84. R.E. Jesse, P. Biloen, R. Prins, J.D.W. van Voorst, G.J. Hoijtink, *J. Chem.Phys.* **6** (1963) 633.
85. H. van Willigen, J.A.M. van Broekhaven, E. de Boer, *Mol. Phys.* **12** (1967) 533.
86. K. Bechgaard, V.D. Parker, *J. Am. Chem. Soc.* **94** (1972) 4749.
87. R. Breslow, B. Jaun, R.Q. Kluttz, C.-Z. Xia, *Tetrahedron* **38** (1981) 863.
88. T. G. LePage, R. Breslow, *J. Am. Chem. Soc.* **109** (1987) 6412.
89. R. Breslow, P. Maslak, J. S. Thomaides,. *J. Am. Chem. Soc.* **106** (1984) 6453.
90. R. Breslow, *Mol. Cryst. Liq. Cryst.* **125** (1985) 261; *ibid* **176** (1989) 199.
91. R. Breslow, in *"Magnetic Properties of Organic Materials"* P.M. Lahti (ed) chapter 3, Marcel Decker, New York,1999, 27 ff.
92. J. S. Thomaides, P. Maslak, R. Breslow , *J. Am. Chem. Soc.* **113** (1991) 3970.
93. H.M. McConell, Proc. Robert A. Welch Foundation, *Conf. Chem. Res.* **11** (1967) 144.
94. J. S. Miller, D. A. Dixon, J. C. Calabrese, *Sience* **240** (1988) 1185.
95. D.A. Dixon, J.C. Calabrese, R.L. Harlow, J.S. Miller, *Angew. Chem.* **101** (1989) 81.
96. M. Glasbeek, J.D.W. van Voorst, G.J. Hoijtink, *J. Chem. Phys.* **45** (1966) 1852.
97. P. J. Krusic, E.J. Wassermann, *J. Am. Chem. Soc.* **113** (1991) 2322.
98. M. Baumgarten, L. Gherghel, M. Wagner, A. Weitz, M. Rabinovitz, Cheng, L.T. Scott, *J. Am. Chem. Soc.* **117** (1995) 52.
99. L. Gherghel, J.-D. Brandt, M. Baumgarten, K. Müllen *J. Am. Chem. Soc.* **121** (1999) 8104.
100. M. Baumgarten, A. Gügel, L. Gherghel, *Adv. Mater.* **5** (1993) 458.
101. J. Friedrich, M. Baumgarten, *Appl. Magn. Res.* **13** (1997) 393.
102. M.C.B.L. Shohoji, M.L.T.M.B Franco, M.C.R.L.R Lazana, S. Nakazawa,K. Sato, D. Shiomi, T. Takui, *J. Am. Chem. Soc.* **122** (2000) 2962.
103. T.F. Fässler, R. Hoffmann, S. Hoffmann, M. Wörle, *Angew. Chem. Int. Ed.* **39** (2000) 2091.
104. G. Kothe, E. Ohmes, J. Brickmann, H. Zimmermann, *Angew. Chem. Int. Ed.* **10** (1971) 938.
105. J. Brickmann, G. Kothe, *J. Chem, Phys.* **59** (1973): 2807.
106. A. Rajca, S. Utampanya, *J. Am. Chem. Soc.* **115** (1993) 2396.
107. K. Yoshizawa, A. Chano, A. Ito, K. Tanaka, T. Yamabe, H. Fujita, J. Yamauchi, M. Shiro, *J. Am. Chem. Soc.* **114** (1992) 5994.
108. K.R. Stickley, S.C. Blackstock, *J. Am. Chem. Soc.* **116** (1994) 11576.
109. F.R. Dollish, W.K. Dall, *J. Phys. Chem.* **69** (1965) 2127.
110. W. Wilker, G. Kothe, H. Zimmermann, *Chem. Berichte* **108** (1975) 2124 .
111. K. Sato, M. Yano, M. Furuichi, D. Shiomi, T. Takui, K. Abe, K. Itoh, A. Higuchi, K. Katsuma, Y. Shirota *J. Am. Chem. Soc.* **119** (1997) 6607.
112. M. M. Wienk, R.A. J. Janssen, *J. Am. Chem. Soc.* **119** (1997) 4492.
113. M. M. Wienk, R.J.A. Janssen, *Synth. Met.* **85** (1997) 1725.
114. M.P. Struik PhD thesis *"High Spin through Bond and Space"* Eindhoven 1991.
115. T. D. Selby, K. R. Stickley, S. C. Blackstock, *Org. Lett.* **2** (2000) 171.
116. J. P. Zhang, M. Baumgarten, *Chem. Phys.* **214** (1997) 291.
117. J.P. Zhang, R.S. Wang, L. X. Wang, M. Baumgarten, *Chem. Phys.* **246** (1999) 209.
118. T. Nakai, K. Sato, D. Shiomi, T. Takui, K. Itoh, M. Kazaki, K. Okada, *Synth. Met.* **102** (1999) 2265.
119. T. Ishida, H. Iwamura, *J. Am. Chem. Soc.* **113** (1991) 4238.
120. F. Kanno, K. Inoue, N. Koga, H. Iwamura, *J. Phys. Chem.* **97** (1993) 13267.
121. K. Inoue, H. Iwamura, *J. Am. Chem. Soc.* **116** (1994) 3173.
122. K. Inoue, H. Iwamura, *Adv. Mater.* **8** (1996) 73.

12.High spin molecules directed towards molecular magnets 527

123. K. Inoue, "*Metal Aminoxyl-Based Molecular Magnets*" *Structure& Bonding* **100** (2001) 61ff.
124. T. Itoh, K. Matsuda, H. Iwamura, *Angew. Chem.* **111** (1999) 1886.
125. T. Itoh, K. Matsuda, H. Iwamura, K. Hori, *J. Am. Chem. Soc.* **122** (2000) 2567.
126. S. Nakatsuji, H. Anzai, *J. Mater. Chem.* **7** (1997) 2161.
127. D. Shiomi, M. Tamura, H. Sawa, R. Kato, M. Kinoshita, *Synt. Met.* **56** (1993) 3279.
128. L. Catala, P. Turek, J. LeMoigne, A. De Cian, N. Kyritsakas, *Tetrahedron Lett.* **41** (2000) 1015.
129. M. Matsuchita, T. Momose, T. Shida, Y. Teki, T. Takui, K. Itoh, *J. Am. Chem. Soc.* **112** (1990) 4700.
130. M. Matsuchita, T. Nakamura, T. Momose, T. Shida, Y. Teki, T. Takui, T. Kinoshita, K. Itoh, *J. Am. Chem. Soc.* **114** (1992) 7470.
131. M. Matsuchita, T. Nakamura, T. Momose, T. Shida, Y. Teki, T. Takui, T. Kinoshita, K. Itoh, *Bull. Chem. Soc. Jpn.* **1** (1993) 1333.
132. P.M. Lahti, B. Esat, R. Walton , *J. Am. Chem. Soc.* **120** (1998) 5122.
133. P.M. Lahti, B. Esat, Y. Liao, P. Serwinski, J. Lan, R. Walton, *Polyhedron* **20** (2001) 1647.
134. K. Matsuda, H. Iwamura, *Chem. Commun.* 1996, 1131.
135. K. Matsuda, H. Iwamura, *Mol. Cryst. Liq. Cryst.* **306** (1997) 89.
136. M. Tanaka, K. Matsuda, T. Itoh, H. Iwamura, *J. Am. Chem. Soc.* **120** (1998) 7168.
137. A. Izuoka, M. Hiraishi, T. Abe, T. Sugawara, K. Sato, T. Takui, *J. Am. Chem. Soc.* **122** (2000) 3234.
138. R. Kumai, M. Matsushita, A. Izuoka, T. Sugawara, *J. Am. Chem. Soc.* **116** (1994) 4523.
139. T. Sugawara, A. Izuoka, *Mol. Cryst. Liq. Crst.* **305** (1997) 41,1001.
140. D. A. Seeger, J. A. Berson, *J. Am. Chem. Soc.* **105** (1983) 5144.
141. S. Murata, H. Iwamura, *J. Am. Chem. Soc.* **113** (1991) 5547.
142. Y. Teki, T. Takui, K. Itoh, *Chem. Phys. Lett.* **142** (1987) 181.
143. Y. Teki, I. Fujita, T. Takui, T. Kinoshita, K. Itoh, *J. Am. Chem. Soc.* **116** (1994) 11499
144. S. Murata, T. Sugawara, H. Iwamura, , *J. Am. Chem. Soc.* **109** (1987) 1266.
145. T. Matsumoto, T. Ishida, N. Koga, H. Iwamura , *J. Am. Chem. Soc.* **114** (1992) 9952.
146. C. Ling, M. Minato, P. M. Lahti, H. van Willigen, *J. Am. Chem. Soc.* **114** (1992) 9959.
147. K. Itoh, *Pure Applied Chem.* **50** (1978) 1251.
148. K. S. Kalgutkar, P. M. Lahti, *J. Am. Chem. Soc.* **119** (1997) 4771.
149. P.M. Lahti (ed) "*Magnetic Properties of Organic Materials*" chapter **31**, Marcel Dekker, New York 1999.
150. S. J. Jacobs, D.A. Shultz, R. Jain, J. Novak, D.A. Dougherty, *J. Am. Chem. Soc.* **115** (1993) 1744.
151. A. P. West, S. K. Silvermann, D. A. Dougherty, *J. Am. Chem. Soc.* **118** (1996) 1452.
152. J. Y. Bae, M. Yano, K. Sato, D. Shiomi, T. Takui, T. Kinoshita, K. Abe, K. Itoh, D. Hong, *Synth. Met.* **103** (1999) 2261.
153. D. Shiomi, K. Ito, M. Nishizawa, S. Hase, K. Sato, T. Takui, K. Itoh, *Mol. Cryst. Liq. Cryst.* **334** (1999) 99.
154. R. Ziessel, G. Ulrich, R. C. Lawson, L. Echegoyen, *J. Mater. Chem.* **9** (1999) 1435.
155. S. V. Chapyshev, R. Walton, J. A. Sanborn, P.M. Lahti, *J. Am. Chem. Soc.* **122** (2000) 1580.
156. A.Rajca, S. Utampanya, *J. Am. Chem. Soc.* **115** (1993) 2396.
157. A. Rajca, S. Rajca, S. R. Desai, *J. Am. Chem. Soc.* **117** (1995) 806.
158. M. Baumgarten, *Mol. Cryst. Liqu. Cryst.* **272** (1995) 109.
159. M. Baumgarten, L. Gherghel, T. Wehrmeister, *Chem. Phys. Lett.* **267** (1997) 175.
160. N. Hirota, S. I. Weissman, *Mol. Phys.* **5** (1962) 537.

161. J. Fujita, Y. Matsuoka, K. Matsuo, M. Tanaka, T. Akita, N. Koga, H. Iwamura, *Chem. Commun.* 1997, 2393.
162. T. Teki, K. Itoh, *Chem. Phys. Lett.* **19** (1973) 120.
163. Y. Teki, T. Takui, K. Itoh, H. Iwamura, K. Kobayashi, *J. Am. Chem. Soc.* **105** (1983) 3722.
164. I. Fujita, Y. Teki, T. Takui, T. Kinoshita, K. Itoh, F. Miko, Y. Sawaki, H. Iwamura, A. Izuoka, T. Sugawara, *J. Am. Chem. Soc.* **112** (1990) 4074.
165. W. Adam, M. Baumgarten, W. Maas, *J. Am. Chem. Soc.* **122** (2000) 6735.
166. W. Adam, C. van Barneveld, S.E. Bottle, H. Engert, G. Hanson, H.M. Harrer, C. Heim, W.M. Nau, D. Wang, *J. Am. Chem. Soc.* **118** (1996) 3974.
167. W. Adam, W. Maas, *J. Org. Chem.* **65** (2000) 7650.
168. A. Rajca, S. Utampanya, S. Thayumanavan, *J. Am. Chem. Soc.* **114** (1992) 1884.
169. N.Nakamura, K. Inoue, H. Iwamura, T. Fujioka, Y. Sawaki, *J. Am. Chem. Soc.* **114** (1992) 1484.
170. N. Nakamura, K. Inoue, H. Iwamura, *Angew. Chem. Int. Ed.* **32** (1993) 872.
171. K. Matsuda, N. Nakamura, K. Takahashi, K. Inoue, N. Koga, H. Iwamura, *J. Am. Chem. Soc.* **117** (1995) 5550.
172. K. Matsuda, N. Nakamura, K. Inoue, N. Koga, H. Iwamura, *Chem. Eur. J.* **2** (1996) 259.
173. A. Rajca, S. Rajca, S. Desai, *J. Am. Chem. Soc.* **117** (1995) 806.
174. T. Takui, K. Sato, D. Shiomi, K. Itoh, T. Kaneko, E. Tsuchida, H. Nishide, *Mol. Cryst. Liq. Cryst.* **279** (1996) 155.
175. H. Nishide, M. Miyasaka, E. Tsuchida, *J. Org. Chem.* **63** (1998) 7399.
176. H. Nishide, M. Miyasaka, E. Tsuchida, *Angew. Chem. Int. Ed.* **37** (1998) 2400 .
177. H. Nishide in "*Magnetic Properties of Organic Materials*" P.M. Lahti (ed) chapter **14**, 285ff, Marcel Decker, New York,1999.
178. H. Nishide, T. Maeda, K. Oyaizu, E. Tsuchida, *J. Org. Chem.* **64** (1999) 7129.
179. Y. Miura, in "*Magnetic Properties of Organic Materials*" P.M. Lahti (ed) chapter 13, 267ff Marcel Decker, New York,1999.
180. Y. Miura, T. Issiki, Y. Ushitani, Y. Teki, K. Itoh, *J. Mater. Chem.* **6** (1996) 1745.
181. H. Oka, T. Tamura, Y. Miura, Y. Teki, *J. Mater. Chem.* **9** (1999) 1227.
182. A. Rajca, K. Lu, S. Rajca, *J. Am. Chem. Soc.* **119** (1997) 10335.
183. A. Rajca, S. Rajca, J. Wongsriratanakul, *J. Am. Chem. Soc.* **121** (1999) 6308.
184. A. Rajca, J. Wongsriratanakul, S. Rajca, *Science* **294** (2001) 1503.
185. R. Chiarelli, M. A. Novak, A. Rassat, J. L. Tholance, *Nature* **363** (1993) 147.
186. A.J. Banister, N. Bricklebank, I. Lavender, J. M. Rawson, C.I. Gregory, B.K. Tannner, W. Clegg, M. R. J. Egglewood, F. Palacio, *Angew. Chem. In.t Ed.* **35** (1996) 2533.
187. Allemand, K.C. Khemani, A. Koch, F. Wudl, K. Holczer, S. Donovan, G. Gruner, J.D. Thompson, *Science* **253** (1991) 301.
188. B. Narymbetov, A. Omerzu , V. Kabanov, M. Tokumoto , H. Kobayashi, D. Mihailovic, *Phys. Sol. State.* **44** (2002) 437.
189. A. Caneschi, Gatteschi, J. Laugier, P. Rey, R. Sessoli, C. Zanchini, *J. Am. Chem. Soc.* **110** (1988) 2795.
190. S. Ferlay, T. Mallah, R. Ouahes, P. Veillet, M. Verdaguer, *Nature* **378** (1995) 701.
191. J. M. Manriquez, G.T. Yee, R.S. McLean, A.J. Epstein, J.S. Miller, *Science* **252** (1991) 1415.

Chapter 13

ELECTRON TRANSFER AND STRUCTURE OF PLANT PHOTOSYSTEM II

Asako Kawamori
School of Science and Technology, Kwansei Gakuin University, Sanda, Japan

Key words: Photosystem II, Electron Transfer, Structure, S-state, PELDOR, ESEEM, Mn-cluster, Tyrosine, P680

Abstract: Electron transfer and structure of plant photosystem II were studied by advanced EPR techniques. Pulsed EPR, pulsed electron electron double resonance and spin polarized radical pair ESEEM to observe dipole interaction between radical pairs of electron transfer cpmponents were applied to determine the distance between them. The determined distances and their orientations were compared with recently observed X-ray data. EPR of the manganese cluster in water oxidizing complexes in photosystem II were discussed with respect to their functions.

1. INTRODUCTION

1.1 Primary photochemistry in photosynthesis

EPR of a plant photosystem was first observed in the 1950s, when the pigment P700 was bleached during light illumination of chloroplast suspension by Commoner et al.[1] A second overlapping signal was observed even if the chloroplasts were kept in the dark. At that time these two signals were denoted as Signal I and II, respectively. Signal I was recognized as oxidized P700, the oxidized form of chlorophyll *a*. The origin of Signal II was found to be tyrosine radical by site specific deuteration[2] in 1987, thirty years after the first discovery of EPR signals in the plant.

The primary event of photosynthesis starts with light absorption followed by electron transfer from the donor to the acceptor where NADPH stores chemical reductants to synthesize carbohydrate. Various photosystems from bacteria to higher plants have developed during the long four billion year history of life on earth. The ancestors of the first bacterial systems are rather simple and were studied in the early 1970s. The reaction center is bacteriochlorophyll *a* that was found to be a dimer as demonstrated using the ENDOR technique by Feher *et al*.[3] The crystal structure of *Rps. viridis* was analyzed by Deisenhofer *et al*.[4] in 1985.

In higher plants and cyanobacteria there are two photosystems; photosystem I (PS I) and photosystem II (PS II). The donor is water, which is different from that in photosynthetic sulfur bacteria in that their donor is hydrogen sulfide. To oxidize water, a strong oxidant is necessary. Therefore, two photosystems co-operate to oxidize the water by absorbing two light quanta with wavelength 700 and 680 nm, respectively, at one time and to evolve one oxygen four absorptions are necessary. The overall reaction is usually expressed by a Z-scheme as shown in Fig. 1. X-ray analysis of the crystal structure of PS I in a cyanobacterium *Synechococcus elongatus* was first reported in 1992 by Witt *el al*.[5] with 6 Å resolution. The electron transfer path from P700 to the final iron sulfur center A and B has been analyzed at 2.5 Å resolution.[6] Finally, in 2000,[7-9] the PS II single crystals capable of water oxidation were grown and the structure was analyzed at 3.8 Å resolution in 2001.[10]

EPR techniques have been used as microscopic monitors for function of photosynthesis such as oxygen evolving activity and for structural studies to measure dipolar interaction between radicals. In this chapter the structure and function of PS II as determined by EPR methods are described, especially the unique characteristics related to water oxidations. To study photosynthesis, not only both cw EPR and pulsed techniques, but also ENDOR, ODMR etc, have been used. Other spectroscopic techniques employed are FTIR, optical absorption, fluorescence, EXAFS, and XANES, which will be referenced where appropriate.

Sample preparation is an important factor when studying photosynthesis. Chloroplasts from various plants were the objects of the original investigations. Isolation of PS I and PS II particles was implemented later in order to discriminate various functions. Biochemical treatments were found to change functions. Physical preparations such as control of illumination, temperature and trapping are also important for obtaining appropriate sample conditions.

Photosynthesis is a complicated phenomenon which depends on the concerted action of a number of enzymes and proteins. A number of studies on PS II have described the techniques required to obtain information about the important functions. Some recent advances, such as single crystal growth, X-ray analysis, and EPR of PS II during the last two years, are summarized in the chapter.

1.2 Charge separation and electron transfer in plant photosystem

Light illumination induces excitation of the P680 reaction center of chlorophyll*a* followed by electron transfer to pheophytin within hundreds of picoseconds and then to the first electron acceptor quinone, plastoquinone, within microseconds.[11] Oxidized P680$^+$ is reduced by tyrosine Z (Y_Z) in 10 – 500 nanoseconds at physiological temperature. Oxidized $Y_Z^{\bullet}$ is reduced by water oxidizing complexes containing four manganese atoms. The electron on the primary quinone Q_A is transferred to the secondary quinone Q_B and then to the plastoquinone pool. The photosystem I accepts the electron from photosystem II after excitation (see Figure 1).

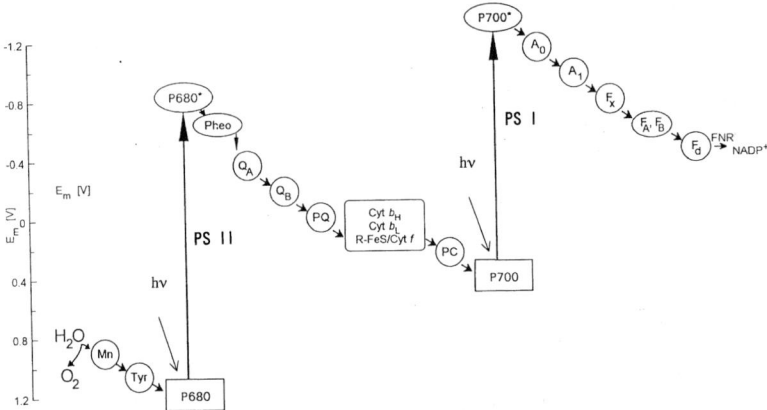

Figure 1. Electron transport pathways of plants, so-called Z-scheme. Excitation by two quanta makes an electron to flow from water to NADP$^+$, nicotinamide adenine nucleotide phosphate. Membrane bound electron transfer protein cofactors Mn to Q_B belong to PSII, while P700, to F_A and F_B, iron sulphur centers, do to PS I. PQ, plastoquinone pool, Cytochrome *bf* complexes and PC, plastocyanin are the mobile proteins between PS II and PS I. E_m is the potential of the transport electron.

A purified PS II reaction center is composed of D1, D2 and Cytochrome $b559$ protein subunits. However, photosynthetic function cannot be realized without water in these systems. The function of water oxidase was discovered by flash illumination during oxygen evolution measurements in which the oxygen yield oscillated with a period of four. The corresponding states were named as S-states or oxygen clocks by Kok et al.[12] The S-state cycles from S_0 to S_4 by absorption of four light quanta.

Figure 2 shows a model of the PS II reaction center with its electron transfer chain. The acceptor side of the structure was considered to be similar to that of the purple bacterial reaction center. The donor side was quite different from it. The water oxidase is a unique enzyme of higher plants for which crystallization has been difficult until recently. Single crystals of active PS II include water oxidizing complexes (WOC) with some extrinsic proteins and several other subunits.

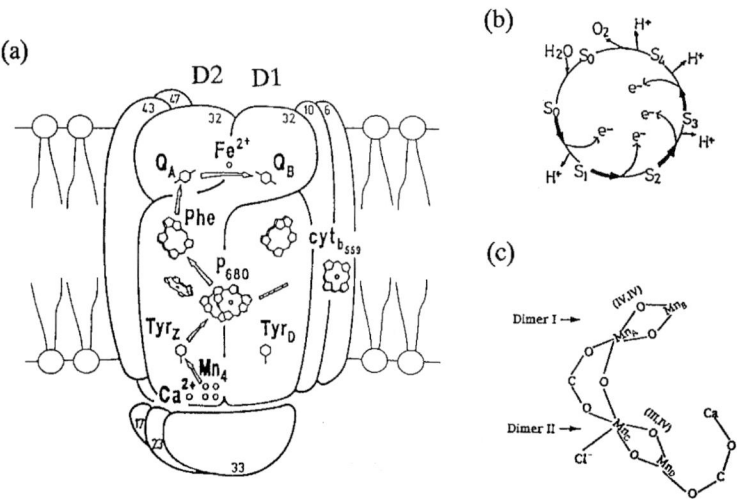

Figure 2. A proposed model structure of photosystem II. (a) Reaction center P680 is supposed to be a chlorophyll*a* dimer located at the symmetric center of two D1 and D2 proteins. Electron transfer occurs on the D1 side from water to quinone Q_A. (b) In Kok's oxygen clock each absorption of light quanta by P680 (bold arrows) draws an electron from the Mn cluster. (c) The structure of the manganese cluster is suggested from EXAFS.

In 2001, X-ray analysis of single crystals of *Synechococcus elongatus* was reported. The resolution was 3.8 Å and a number of the chlorophylls and the Mn-cluster were identified.[10] However, the positions of tyrosine and quinone were not accurately determined because of their small molecular sizes. The distance between radical species has been determined with the accuracy of 1 Å by spin polarized ESEEM and pulsed ELDOR.

The electron transfer rate in biological systems has been thoroughly studied by Marcus and Sutin.[13] The knowledge of the distance and its direction between electron transfer components are essential for understanding the electron transfer mechanism. EPR techniques to determine distances are expected to yield more detailed structural information.

1.3 Electron transfer components in photosystem II

Figure 2 shows the electron transfer pathway and its components studied by EPR and optical absorption. The chlorophyll function is important because a number of chlorophylls serve as antennas for energy transfer and a specific pair act as the reaction center for electron transfer.

1.3.1 Chlorophylls

a. P680
There are six chlorophyll molecules in the purified reaction center complex, two of them are the reaction center chlorophyll dimer called P680, two others are monomer chlorophylls situated between P680 and pheophytin. The last two are rather separated from P680 and one of them is believed to facilitate the side electron transfer pathway from Cytochrome $b559$ (Cyt b_{559}) to P680. The re-reduction kinetics of oxidized P680 has been studied optically by observation of the decay-rate at the 820 nm absorption which has a short life-time. The life time was found to be dependent on pH and temperature. At low temperatures below 200 K donation of electron from tyrosine Z (Y_Z) was inhibited. The life time was determined by charge recombination with the primary acceptor quinone to be about 3 ms.[14]

When the primary acceptor quinone has been reduced, P680 triplet state is formed. The triplet state EPR was observed during illumination at low temperature by Rutherford *et al.*[15] in 1986. This triplet state configuration was found to be different from that in a bacterial system.[16] The triplet state EPR signal had a large zerofield splitting comparable to that in a monomer chlorophyll. FTIR study[17] also proved that the triplet electrons reside on one of the accessory chlorophylls near by.

The reaction center in a bacterial system is composed of dimer bacteriochlorophyll molecules, each one belonging to L and M proteins, separated by 8 Å between Mg atoms in the porphyrin rings.[18] The Mg atoms in P680 are separated by 10 Å. The EPR signal of the oxidized P680 are difficult to obtain because of its short life time. However, the signal of oxidized P680 has been observed transiently by several workers.[19-21] at low temperatures.

b. Chlorophyll Z

This radical signal appears usually in an inhibited system such as during illumination at temperatures below 200 K. The line width is about 1.1 mT peak to peak, characteristic of a monomer chlorophyll radical. A side electron transfer path has been proposed from Cyt b_{559} through P680 to Q_A by Thomson and Brudvig.[22] As D1 and D2 proteins are located in C_2 symmetry, there should be two chlorophylls with each belonging to D1 and D2. Only one is active in donating an electron to the oxidized P680$^+$ when Cyt b_{559} has been oxidized. Chlorophyll Z is named by an analogy with tyrosine Z and is suggested to be located on the D1 protein.[23] Recent X-ray analysis has revealed two chlorophylls corresponding to Chlorophyll Z and Chlorophyll D that belong to D1 and D2 polypeptides, respectively. However, X-ray analysis cannot discriminate their functions.

1.3.2 Tyrosines Y_D and Y_Z

In photosystem II (PS II) particles, deuterated tyrosine was introduced, resulting in a narrowing of the line shape of Sig. II.[2] The stable radical signal in PS II called Sig. II$_s$, was recognized as tyrosine D (Y_D), in which s means slow or stable. This radical is located on the D2 protein, and has been well studied during the past 30 years.

The other radicals induced by illumination could be detected transiently and called Sig. II$_f$ (fast)[24] and Sig. II$_{vf}$ (very fast) according to their life times, as the signals had the same line shape as Sig. II$_s$. They have been ascribed to another tyrosine Z (Y_Z) on the D1 protein, since the proof by isotopic labeling of cyanobacterial PS II was performed.[25] The WOC supplies an electron to $Y_Z^{\bullet}$ in microsecond to hundred nanosecond (Sig II$_{vf}$). When the four manganese are depleted, the life time of $Y_Z^{\bullet}$ becomes longer because of slow electron donation by back reaction from the acceptor side and the EPR signal (Sig. II$_f$) is visible during illumination.

13. Electron transfer and structure of plant photosystem II

1.3.3 Plastoquinone Q_A

In photosystems, bacterial and plant PS II, the acceptor side includes a non-heme iron between the primary and the secondary quinones. EPR signals of Q_A^- cannot be observed except for low temperature below 15 K due to broadening by exchange coupling with the ferrous iron. When this iron is depleted or substituted by zinc, a sharp typical radical signal with the line width of 0.9 mT is observed in reduced condition. The signal has been studied by ENDOR by MacMillan et al.[26] These studies provided the knowledge of hyperfine coupling and hydrogen bonding with surroundings.

1.3.4 Cytochrome b559 (Cyt b_{559})

Oxidized cytochrome b559 can be observed at low temperatures over a wide range of applied magnetic field from $g_z = 2.9$ to $g_x = 1.5$.[27] These g-values manifest low potential and high potential forms depending on biological systems. In oriented membranes the signal position and intensity depend on magnetic field direction. The g_y direction is along the membrane normal as observed in the oriented membranes.[28] The role of this cofactor has not yet been clarified because it works only at an inhibited state. The axis of the heme plane of Cyt b_{559} connects α and β subunits and is suggested to be on the acceptor side by Stewart and Brudvig[29] which the recent crystal structure analysis has also shown.[10]

1.3.5 Mn-cluster in water oxidizing complex

The water oxidizing complex (WOC) includes four manganese atoms that accumulate oxidized equivalents by four light quanta and each state has been named S_0 to S_4 by Kok's S-state or oxygen clock as shown in Figure 2 (b).[12] The first manganese signal was discovered by Dismukes and Siderer[30] in 1980 at 5 K. The signal was very similar to that in a model Mn(III)-Mn(IV) compound[31] in which the EPR signal consists of 16 lines separated by about 8 mT over the field range 180 mT centered at $g = 2$. The spectrum was interpreted by antiferromagnetic coupled dimer with the resultant spin $S = S_1 + S_2 = 1/2$ ($S_1=2$ and $S_2=3/2$). The signal was induced by one flash illumination and its intensity oscillated with Kok's S-state by further flashes. The first maximum was induced by 1 flash and then decreased by 2, 3 and 4 flashes. The second maximum appeared on the fifth flash. Therefore, the signal was assigned to the S_2 state, because the S_1 state has been known to be most stable in the dark. This signal can be observed by illumination at 200

K and has been used as a monitor for S_2 formation after biochemical treatment such as depletion of extrinsic proteins and Ca^{2+}.

A $g = 4.1$ signal was observed after illumination at 140 K.[32] Elimination of cofactors Ca^{2+} and Cl^- in WOC resulted in inactivation of oxygen evolution. Similar $g = 4.1$ signals were observed for PS II samples after Cl^- depletion or acetate treatment by illumination at a physiological temperature as reviewed by Britt.[33]

An S_1-state signal observed by parallel polarization EPR below 5 K in a dark adapted PS II sample was reported by Dexheimer & Klein[34] in 1992. Its resonance field was at $g = 4.9$ with the width about 60 mT. By simulation they assigned the signal to $S = 1$ and zero field splitting D of 0.14 cm^{-1}. However, nobody could reproduce this signal during the following several years. In 1997 Yamauchi et al.[35] reproduced this signal and assigned it to excited triplet states with $S = 1$ separated from the singlet ground state by 2.5 K based on its temperature dependence.

A similar multiline signal was reported in 1997 on the biochemically reduced S_0-state by Messinger et al.[36] or after 3 flashes of illumination by Åhrling et al.[37] The signal was very similar to that observed in the S_2-state. The difference was the number of lines and the overall width was also larger. Saturation behavior in S_0 and S_2-states was also different. The S_0-state signal was observable only with addition of 2-3 % of methanol. This signal has been ascribed to antiferromagnetic coupling between Mn(III) and Mn(II) with the resultant spin $S = 1/2$.

S_3-state signals were observed with a dual mode cavity by Matsukawa et al.[38] This signal was also suggested to arise from one of excited states based on the temperature dependence of the signal intensity and assigned tentatively to $S = 1$ state separated by about 3.6 K from the ground singlet state. Compared to the S_2 and S_0-states that have spin 1/2, the S_1 and S_3 states are difficult to observe by EPR, because the spin numbers are integer and zero field splitting is usually comparable or larger than the microwave frequency.

At present all the Kok's S-state signals except for the S_4-state have been observed in the oxygen evolving photosystem II. The S_4-state is only transient state and cannot be trapped. Instead, the S_0-state could be observed as a ground state. In Figure 3 typical manganese signals are shown for the S_0- to S_3-states together with the stable radical signal of $Y_D^{\bullet}$.

13. *Electron transfer and structure of plant photosystem II* 537

From the EPR and EXAFS studies a model structure called a dimer of dimers has been presented for the Mn-cluster as shown in Figure 2 (c) by Yachandra, *et al.*[39] Most of the electron transport components in Figure 2 (a) have been studied by EPR. Some typical experiments will be described in later sections.

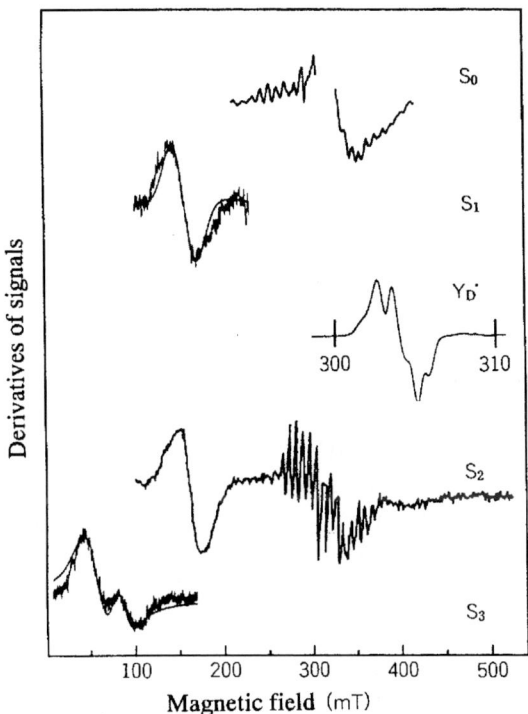

Figure 3. EPR signals of Mn-cluster observed in S_0- to S_3-states and $Y_D^\bullet$ radical in the oxygen evolving PS II particles. The signals for S_1 and S_3 were observed by parallel mode in which smooth lines show simulations. The other signals were observed by ordinary perpendicular mode.

2. METHODS APPLIED TO PHOTOSYNTHESIS

2.1 Distance determination

2.1.1 Spin lattice relaxation measurement

One of the popular methods to derive the distance between paramagnetic species is spin lattice relaxation time measurements. In PS II, we observe usually a radical tyrosine D ($Y_D^\bullet$), while the non-heme iron and manganese are strong relaxers that affect relaxation of the radical called relaxee. The relaxation rate is given by,

$$1/T_1 = \text{const.}/r^6 \left[\tau_c/(1 + \omega_0^2 \tau_c^2) \right] \tag{1}$$

where τ_c is the correlation time of the relaxer.

We usually do not know about the correlation time. Instead, by observing temperature dependence, the maximum point of $1/T_1$ at $\omega_0 \tau_c \sim 1$ can be obtained. Then we derive the distance between the relaxee and relaxer according to the formula $\text{const}/r^6 = 1/T_1/\tau_c$. This method was applied by Hirsh et al.[40] to calculate the distance of Y_D from the non-heme iron.

2.1.2 Selective hole burning

This method was proposed by Dzuba et al.[41] to measure the distribution of radicals in coal and was applied to photosystem II later. The pulse sequence consists of a long period of the first $\pi/2$ or π-pulse to make a hole into a part of a broad line and of ordinary strong detection $\pi/2-\pi$ pulses to observe the electron spin echo (ESE) of the whole line as shown in Figure 4(a).

Let us consider a pair of interacting spins A and B shown in Figure 5(a). The ESE signal of spins A is observed using a pulse sequence shown in Figure 4(b). In this pulsed sequence the second $\pi/2$ and third π pulses separated by the time interval t form the ESE signal with the amplitude determined by the transverse relaxation time T_2:

$$V(2t) \propto \exp(-t/T_2) \tag{2}$$

The spin B produces an extra dipolar field on the spin A in addition to the applied magnetic field H_0 as shown by

$$\Delta H = \mu_B/r^3 (3\cos^2\theta - 1) m_S \tag{3}$$

where m_S is the projection of the spin B on the applied field direction. The extra field usually varies with space and time in a random way contributing to the EPR line width and to the phase memory time $T_M \sim T_2$ in ESE.

By varying the time of detection T after the hole burning we can observe the broadening of the hole as a result of spin relaxation of the relaxer due to change of m_S. We do not have to know the spin relaxation of

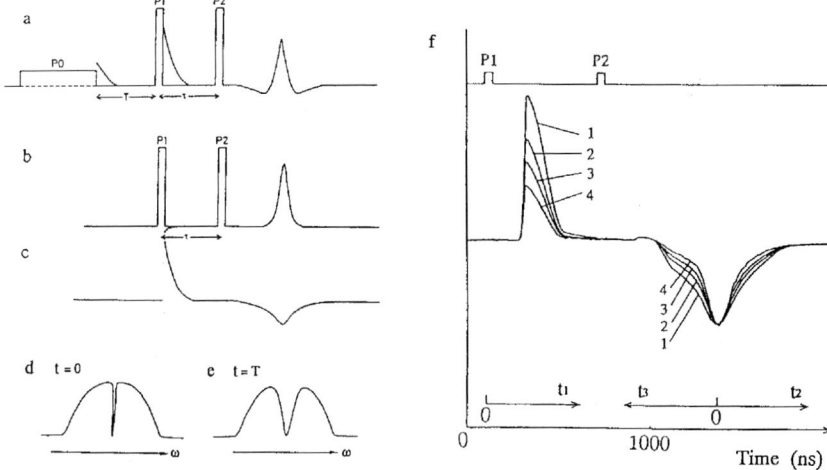

Figure 4. A method of hole burning and detection of the hole broadening. (a) The echo signal after P2 pulse is preceded by a hole-burning pulse P0. (b) The echo signal without P0 pulse. (c) Subtraction of (b) from (a) produces the time domain hole shape at time T after P0. (d) A hole burned and (e) broadened after time T by spectral diffusion and relaxation by relaxers as shown by frequency domain. (f) Variation of hole shapes in time domain with increasing time T as observed in coal sample.

the relaxer. However, the spin relaxation time should be in an appropriate range to make the dipole interaction between the interacting paramagnetic species observable as indicated in Kodera *et al.*[42] and Hara *et al.*[43] After a time T, the echo shape can be given by the following formula for $S = 1/2$ of B spin,

$$f_T(t) = 0.5(1 + \exp(-T/T_1)) + 0.5(1 - \exp(-T/T_1)) \cos(\varepsilon(r,\theta)t) \qquad (4)$$

where $\varepsilon(r,\theta) = \gamma_A \mu_B (3\cos^2\theta - 1)/r^3$ is the dipolar interaction constant from which we can derive the distance r.

2.1.3 Pulsed Electron-Electron Double Resonance (PELDOR)

Two-dimensional ELDOR was introduced by Freed[44] to investigate spin labeled systems. In photosystems, the concentrations of radicals are not enough to apply it and instead two frequencies must be employed.[45]

Let us consider a pair of interacting spins A and B shown in Figure 5a. The electron spin echo (ESE) signal of the spin A is observed using a pulse sequence shown in Figure 5(b). In this pulsed sequence the first $\pi/2$ and third π pulses separated by the time interval τ form the ESE signal with the amplitude determined by the transverse relaxation time T_2 in Eq. (2).

The spin B produces an extra dipolar field on the spin A in addition to the applied magnetic field H_0 as shown by Eq. (3), where m_S is the projection of the spin B on the applied field direction. The extra field usually varies with space and time in a random way contributing to the EPR linewidth and to the phase memory time $T_M \sim T_2$ in ESE. When the second pulse is applied to the spin B at the time τ' to turn the B spins (i. e. to change m_S), the sudden change of the extra field given by Eq. (4) produces a periodic change in the echo height of $V(2\tau)$ depending on τ' as shown in Figure 5(b).

$$V(\tau') \propto -p[1-\cos(\Delta\omega\tau')], \text{ with } \Delta\omega = \varepsilon(r,\theta) \qquad (5)$$

where p is a portion of B spins turned by the second pulse that will not be important.

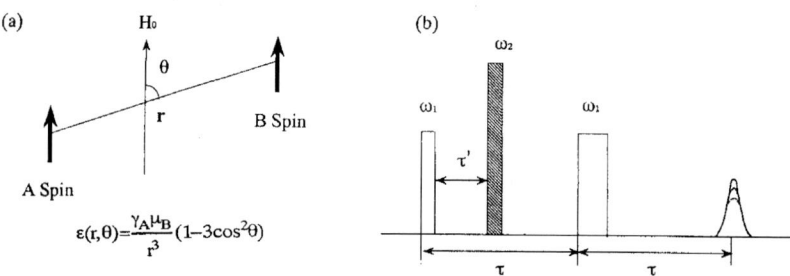

Figure 5. A pulse sequence to detect magnetic dipole interaction by PELDOR (a) Two interacting spins A and B. (b) Three pulse sequence; the first $\pi/2$ and third π pulses form the ESE signal of A spins and the second pulse turns the orientation of spins B.

For a randomly oriented system, $V(\tau')$ should be averaged over the orientations θ.

13. Electron transfer and structure of plant photosystem II

$$V(\tau) \propto <\cos(\Delta\omega\tau)>_\theta \propto \cos(\Delta\omega\tau)\sin\theta\, d\theta \tag{6}$$

In an oriented membrane system the same integral should be multiplied by a Gaussian distribution function $\exp[-\delta\theta^2/2<\Delta\theta^2>]$ in which $\delta\theta$ is deviation of the membrane normal from their average, and $<\Delta\theta^2>$ is a mean square deviation.

When the signals of A and B overlap each other, the same resonance frequency can be applied to excitation and detection which is called a '2 + 1' pulse sequence as described for $Y_Z{}^\bullet$ and $Y_D{}^\bullet$ radical pairs in (**4.1**).

2.1.4 ESEEM of spin polarized radical pairs

The spin polarized radical pair ESEEM for $P700^+A_1^-$ in Photosystem I (PS I) was observed by Moënne-Loccoz et al.[46] in 1994. Dzuba et al.[47] observed the similar ESEEM for $P860^+Q_A^-$ in a bacterial reaction center and derived the distance between the pair from the obtained frequency corresponding to the dipolar interaction. In Figure 6, the principle of spin polarized ESEEM is shown. As the state has 100 % polarized spins, strong temperature independent signal intensity is usually observed. In the later section (**4. 2**), ESEEM of $P700^+A_1^-$ in PS I and $P680^+Q_A^-$ in PS II will be described.

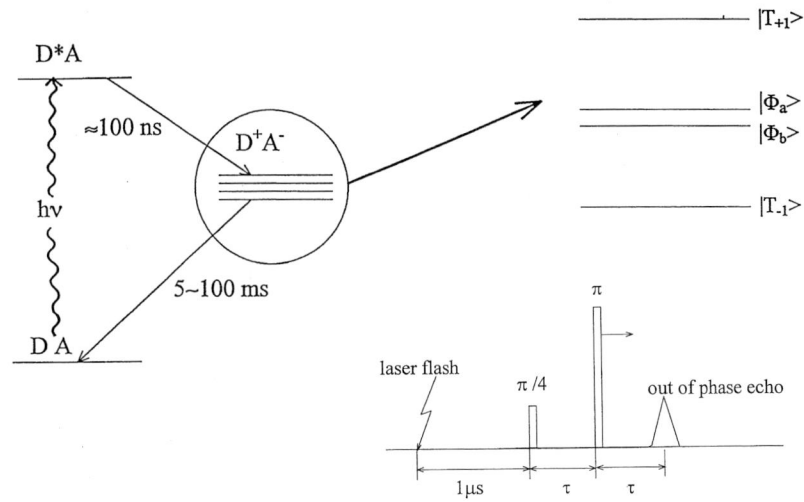

Figure 6. Spin polarized radical pair state induced by laser flash. The radical pair state is a mixed state of T_0 and S with the resultant spin along x direction varying with time τ.

2.2 Other methods

2.2.1 Dual mode EPR

EPR signals of paramagnetic ions with an integer spin have been difficult to observe because of large zero field splitting and large linewidth induced by anisotropic interaction. Forbidden transitions can be observed with *rf* field parallel to the static magnetic field, detecting $\Delta m_S = 0$ transition induced by the matrix element S_z. The transition often appears at about half the resonance field with less line width. The *rf*-field can be applied to a dual mode cavity in TE_{012} mode, while an ordinary perpendicular mode can be observed by tuning to TE_{102} mode with a slightly different frequency. In PS II only a parallel mode signal was observed for the S_1-state of WOC. Both parallel and perpendicular modes were observed for the S_3-state as shown in the later section (**4.5**).

2.2.2 Time resolved EPR

In photochemical reactions, electron transfer occurs rapidly, and time resolved optical absorption has been used to detect intermediate states. Some intermediate states such as triplet states of reaction centers of photosynthetic bacteria[48,44] and PS I[45] have been studied by transient EPR for systems with the acceptor quinone reduced. Dipolar interaction between the electrons was analysed and from the orientation dependence of the signal, the triplet electrons were suggested to reside on the reaction center chlorophylls in these two systems.

The oxidized reaction center chlorophylls were usually observed by cw EPR method for the bacterial and PS I systems during illumination. On the other hand in PS II the signal of oxidized $P680^+$ could be observed only by the time resolved method at low temperatures[20,31] where spin-lattice relaxation times were investigated.

2.2.3 High frequency EPR

High frequency EPR has been developed in the field of solid state physics in the last two decades and applied to photosystems recently. The equipments for W-band (95 GHz) are now available from the Bruker Biospin company.

13. Electron transfer and structure of plant photosystem II　　　　543

A 250 GHz system has been employed to observe tyrosine radicals in PS II and others as referenced in Doret *et al.*[52]

2.2.4　ENDOR

CW and pulsed ENDOR methods have been described in detail by Brustolon in another chapter. The use of these methods is very convenient to study electron spin distribution and to get structural information of the cofactor molecules related to their functions for tyrosines, quinones and manganese clusters.

2.2.5　Optically Detected Magnetic Resonance (ODMR)

Most of the work on ODMR have been developed by Hoff's group on triplet states of bacterial reaction centers.[53] The detection system was optical absorption or fluorescence depending on the relaxation of the relevant levels of the energy state where electron spin resonance occurred. This method has proven useful for investigating intermediate states of primary electron transfer processes, because the life times are generally too short to detect transient microwave absorption or emission. In PS II, the triplet state P680 was studied by absorption detected magnetic resonance (ADMR).[54]

2.2.6　ESEEM applied to study molecular structure of radicals

Electron Spin Echo Envelope Modulation (ESEEM) is a popular methods for studying electron distributions on atoms in a molecule and chemical bonding. It has been described in detail in the book by Dikanov and Tsvetkov.[55] In photosystem II most of the radicals show proton ESEEM by two-pulse but these are not useful for analysing the state. However, by replacing the hydrogen in water by deuteron, deuteron ESEEM has been observed which gave information about coordination numbers around the Mn-cluster.[56] Three pulse ESEEM of nitrogen has been applied to investigate coupling of amino acids with quinone binding on the acceptor side as shown in **4.3.**[57]

3. SAMPLE PREPARATIONS

3.1 Oxygen evolving PS II membranes

Oxygen evolving PS II membranes (about 400 µmol O_2/mg Chl/h) were prepared from market spinach by the method of Kuwabara & Murata[58] or BBY[59] and suspended in a MES buffer (at pH 6.5) with Chlorophyll concentration 3 to 15 mg depending on the experiment with 50 % glycerol as a cryoprotectant added. The membranes were stored in 77 K until use.

3.2 Oriented membranes

PS II membranes were painted on mylar sheets and were dried under 90 % humidity.[28] The distribution of the orientation may be from 15 to 20° as defined by root mean square deviation. The sample sheets were cut into the strips of 3×20 mm^2, and a bundle of piled five or six sheets was inserted into a quartz tube with the inner diameter of 4 mm.

3.3 Single crystals

Single crystals PS II from cyanobacteria were prepared at Max Volmer Institute and Ruhr University for *Synechococcus elongates*,[7,8] and Institute of Physical and Chemical Research in Harima (RIKEN) for *Synechococcus vulcanus*.[9] The space group belongs to orthorhombic $P2_12_12_1$ with unit cell dimension $130 \times 227 \times 308$ Å^3.[10] The local C_2-rotation axis is along to the membrane normal. The sizes of crystals were too small to be studied by EPR. Only tyrosine D was investigated with W-band spectrometer as described in (**4.1**).

3.4 Biochemical treatment

Of the various treatments to inhibit oxygen evolution, two methods will be described. All handling was carried out under dim green light except for Tris-treatment.

a. Tris-treatment

PS II membranes were suspended in Tris (tris(hydroxylmethyl) aminomethane) buffer at pH 8.7 and incubated under room light with gentle stirring for 30 min at 4 °C.[60] The suspension was centrifuged and the solids were washed with an ordinary MES buffer at pH 6.5. This treatment eliminates all manganese and three extrinsic proteins on the donor side, making the $Y_Z^{\bullet}$ radical visible by CW EPR because quick donation of electrons is inhibited.

b. Ca^{2+}-depletion

Oxygen evolving PS II membranes were treated with citric acid at pH 3.0 as described in Ono and Inoue.[61] The treated membranes were suspended in MES buffer at pH 6.5 without Ca^{2+} and exhibited no appreciable oxygen evolution. However, addition of 10-20 mM $CaCl_2$ restored the oxygen evolution activity reversibly.[62]

3.5 Site-directed isotope labelling and mutagenesis

In 1988 a wild type of *Synechocystis* 6803 with Y_D was deuterated. EPR of the deuterated PSII show no resolvable hyperfine structure as shown by Barry and Babcock.[2] Until then the entity of Signal II_s had been considered to be plastoquinone. In the 1990s site-directed mutagenesis has often been carried out and EPR for mutants of various alga has become important for discriminating physiological functions. In this chapter a mutant of *chlamydomonas reinhardtii* lacking Y_D 160 will be investigated to show the distance from the $Y_Z^{\bullet}$ radical without interference from $Y_D^{\bullet}$ signal.[63]

3.6 Physical treatment

a. Illumination

To induce charge separation, illumination of PS II samples with appropriate intensity and wavelength is essential. About 500 W tungsten halogen lamp is recquired for continuous illumination, while the second harmonics of pulse Nd-YAG laser with 532 nm wavelength is used for pulse irradiation for time resolved experiments. Illumination of oxygen evolving PS II samples at 200 K produces the S_2-state of WOC at high yield. On the other hand, all S-states except for S_4 can be produced by flash illumination by laser or Xenon pulse light.

b. Trapping

Below 243 K, electron transfer from the primary Q_A to the secondary acceptor quinone Q_B is inhibited. Trapping below 200 K is necessary to produce radicals such as $Y_Z^{\bullet}$ and Q_A^{-} immediately after illumination above 253 K or to stabilize generally a charge separated state, $D^{+}A^{-}$.

c. Cryoprotectant

To assure complete illumination or to protect proteins from freezing by water, 20 to 70 % glycerol is often used.

4. STUDIED COMPONENTS

4.1 Tyrosines Y_D and Y_Z

a. Molecular environments of Y_D and Y_Z studied by cw and pulsed ENDOR

All of photosynthetic reaction centers have C2 symmetrical arrangements of polypeptides, whereas their functions are asymmetric. D1 and D2 hetero-dimers form the photosynthetic reaction center. Two redox active tyrosine residues, Y_Z 161 in D1 and Y_D 160 in D2 polypeptides are

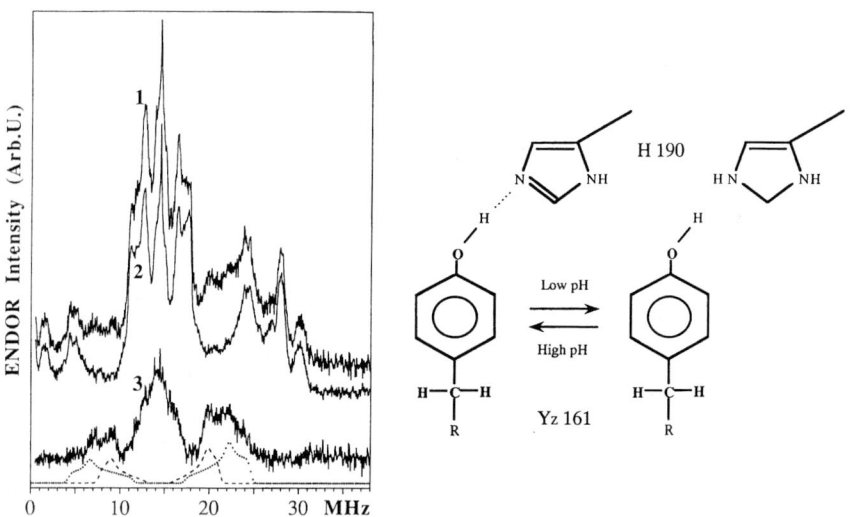

Figure 7. Pulsed ENDOR spectra of tyrosine radicals. (Left) Pulsed ENDOR spectra observed at pH 5.5 in the Mn-depleted PS II. The sample was illuminated at 253 K and rapidly frozen to 77K (trace 1), dark adapted for 30 min at 273 K (trace 2). Subtraction spectrum of trace 2 from trace 1 (trace 3). Traces 2 and 3 show the pulsed ENDOR spectra of Y_D and Y_Z radicals, respectively. (Right) A proposed model for proton coupling to explain the EPR and ENDOR line shapes of Y_Z radical at low pH.

identical. However, they have quite different functional roles. Y_Z is the secondary donor and donates an electron to P680$^+$, while Y_D is an auxiliary donor, when it has been reduced. Above pH 6.5, these radicals show similar EPR line shapes with the hyperfine couplings due to two equivalent ring protons and one of the β-methylene protons. Matrix ENDOR shows the different relaxation rates of both radicals in the proteins, i.e. Y_Z located in the hydrophilic site compared to Y_D.[60] However, below pH 6.5, the EPR line shape of the $Y_Z^{\bullet}$ radical changes dramatically, while that of the $Y_D^{\bullet}$ radical doesn't. The $Y_Z^{\bullet}$ radical at low pH shows EPR/ENDOR line shapes characteristic for a cation radical.[64] In the isolated solution, the dissociation of the proton in the tyrosine molecule is considerably low (pK$_a$ = −2). Therefore, the relationship between the function of the local structure in the protein and the alternation of Y_Z radical depending on pH remains as an unresolved problem.

b. The distance between Y_D and Y_Z studied by '2+1' pulse method

Astashkin et al.[63] applied '2 + 1' pulse sequence for the first time to photosystem II to determine the distance between Y_D and Y_Z. $Y_Z^{\bullet}$ radical was trapped at 200 K immediately after illumination for 20 s at 253 K. The distance was determined to be 29.7 ± 0.3 Å from the observed time profiles for the radical pair $Y_D^{\bullet}$-$Y_Z^{\bullet}$ in non-oriented membranes, which was a similar

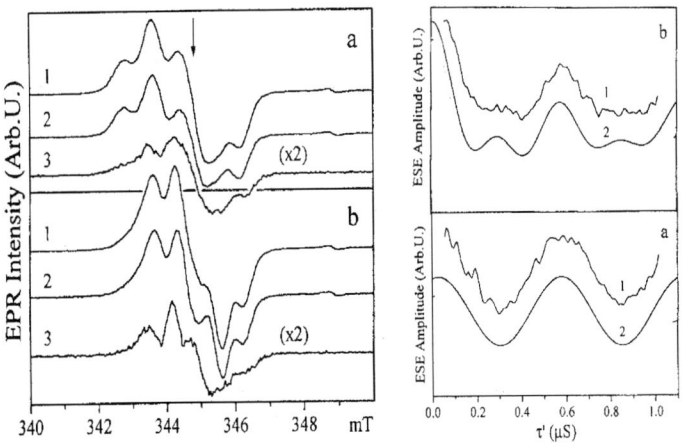

Figure 8. '2+1' Pulse sequence applied to $Y_D^{\bullet}$ - $Y_Z^{\bullet}$ pair. (Left) CW EPR observed in oriented membranes. The membrane normal parallel (a) and perpendicular (b) to the magnetic field. Trace 1 shows an overlapping $Y_D^{\bullet}$ and $Y_Z^{\bullet}$ signal, and trace 2, the $Y_D^{\bullet}$ signal after dark adaptation at 0 °C. Trace 3 shows $Y_Z^{\bullet}$ signal obtained by subtraction of trace 2 from 1. (Right) '2+1' time profiles for 0° (a) and 90° (b) orientation of the membrane normal relative to the field direction. Lines show simulation for the vector Y_Z-Y_D orientation of 80° from the membrane normal ***n***.

Figure 8 (left) shows the cw EPR spectra observed for trapped $Y_Z^\bullet$-$Y_D^\bullet$ radical pair for 0° and 90° orientations, respectively, in the oriented membranes. As the line shapes are different for these angles, $Y_D^\bullet$ EPR is used for setting orientation conveniently. By applying '2 + 1' pulse a doubled frequency was observed for 90° in Figure 8b on the right. This suggests the distance vector is approximately along the membrane plane. By simulation the best fit orientation was determined to be 80° or 100°.[66]

c. High frequency EPR applied to determine molecular orientation

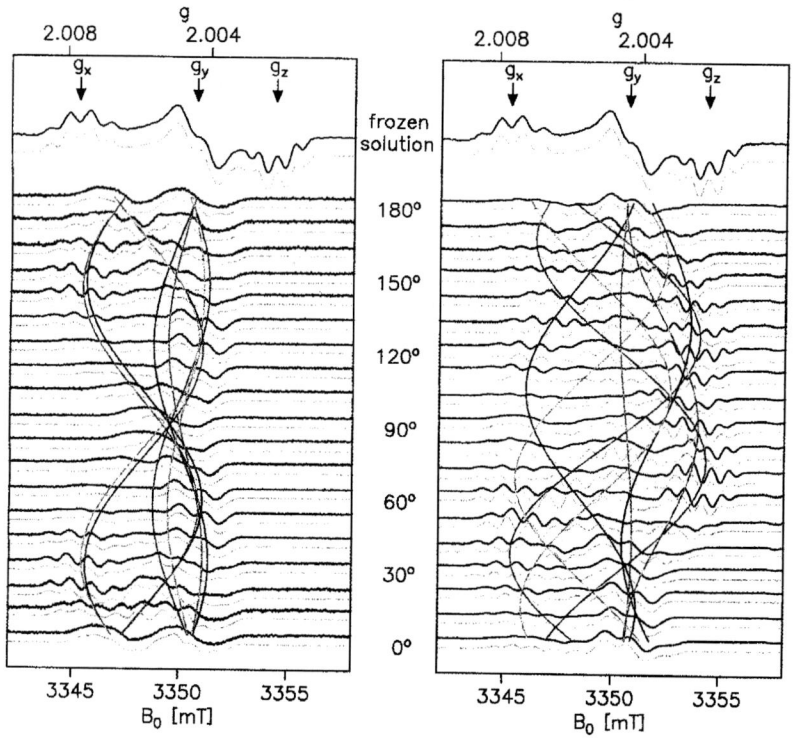

Figure 9. W-band EPR spectra of $Y_D^\bullet$ in frozen solution and single crystals of PS II from *S. elongatus*. (Left) Crystal rotated approximately about the crystallographic *a* axis. (Right) Arbitrary rotation axis. Dim shaded lines show simulations and curved lines indicate the calculated angular dependence of the effective *g* value of $Y_D^\bullet$ residue in the unit cell. Simulation parameters: $g_{x/y/z}$ = 2.00767/2.00438/2.00219, and $A_{x/y/z}(3/5)$ = −26.2/−8/−19.5 MHz. (Taken and modified from Ref. Hofbauer *et al.*[65])

A high-frequency (94 GHz) study using the single crystals of PS II from *Synechococcus elongatus* gave accurate *g*-values deduced from the angular variation of $Y_D^\bullet$ signals shown in Figure 9.[67] Molecular axes and *g*-

13. Electron transfer and structure of plant photosystem II

axes are shown in Figure 10a, The derived orientation of the Y_D moleule is shown in Figure 10b. For spinach, high-frequency spectra were observed at 250 GHz in oriented membranes by Doret et al.[52] The angle α in the crystal is 10° different from that in spinach PS II membranes, which suggests that the crystal c-axis might be inclined relative to the membrane normal. It should be noted that the Y_D axis is directed differently in cyanobacteria and higher plants.

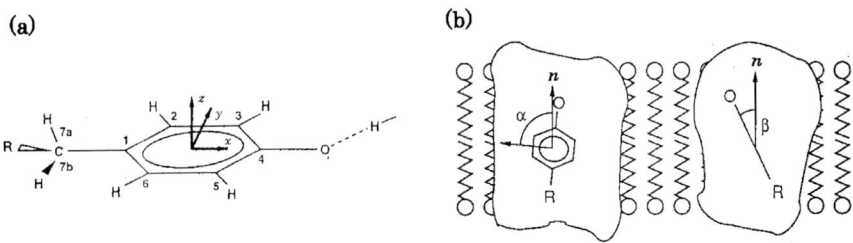

Figure 10. (a) Tyrosine radical with numbering scheme and orientation of *g* tensor principal axes. (b) Orientation of the phenoxl group of Y_D in PS II from *S. elongatus* with respect to the membrane normal (parallel to the C2 symmetry axis) as derived from the single crystal EPR spectra. The angle $\alpha = 84°$ is between the g_y direction and the membrane normal **n**; $\beta = 26°$ is the phenoxyl ring plane with respect to **n**.

d. Distances of Y_D and P860 from the non-heme iron

Measurement of a bacterial reaction centre has been carried out to investigate the applicability of the selective hole burning method for a system where the distance from X-ray data and the magnetic susceptibility data of the non-heme iron on the acceptor are available.[68] According to Eq. (4) in **2.2** the echo shape depends on T and varies with temperature because of variation of T_1. Furthermore, the echo shape is a function of spin number, and Eq. (4) should be that expressed by a general value of S.

Temperature variation of the echo shape after the hole burning of oxidized $P860^+$ in *Rb. sphaeroides R 26* is shown in Figure 11a. From the curve for 40 K, where the value of S was 2 and T/T_1 was considered to be approximately infinite, the distance was determined to be 27 Å. This value is the same as that obtained by X-ray diffraction.[69] In Photosystem II, where the acceptor side was considered to be analogous with that of the purple bacterial system, the effect of hole burning could not be observed at temperatures higher than 10 K because of the short spin lattice relaxation time. At 9 K the higher sublevels of $S = 2$ spin state are not populated enough and the assumption of $S = 2$ is not applicable for this temperature.

The effective spin number $S_{\text{eff}} = 0.8$ and the same zero-field splitting as that for the bacterial reaction centre were assumed. The obtained value of $r = 42 \pm 2$ Å is close to the value of 39 Å for Y_D-Q_A obtained by the '2+1' method.[70] This value can be considered to be more reliable than the value of 38 ± 5 Å obtained based on the spin lattice relaxation of $Y_D^{\bullet}$ because of no assumption for the T_1 value.[40]

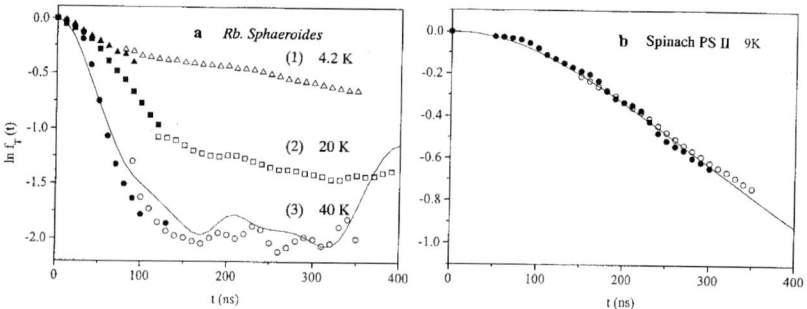

Figure 11. Temperature dependence of the burned hole shapes of $P860^+$ and $Y_D^{\bullet}$ after $T = 260$ μs. The hole shapes observed by ESE and FID (Left) for bacterial (a) and (Right) for PS II (b). Experimental values obtained from ESE are shown by closed circles and those from FID by open circles.

4.2   Chlorophylls

a. Chlorophyll triplet state

A triplet state signal similar to that in a bacterial reaction center was first observed in PS II during illumination at a reduced condition of Q_A by Rutherford *et al.*[15] as shown in Figure 12. The triplet state was produced by a radical pair mechanism from $P680^+Pheo^-$. The pattern was interpreted as *aeeaae* (*a*; absorption and *e*; emission). The deduced *D* value was 0.290 cm^{-1} that is similar to the value obtained for monomeric chlorophyll. The angular dependence of the EPR spectra in oriented membranes showed that the ring plane of the triplet of $^3P680$ was tilted by 30° from the membrane plane.[71] This result suggests that the $^3P680$ may be assigned not to P680 RC but to the monomer chlorophyll nearby.

13. Electron transfer and structure of plant photosystem II

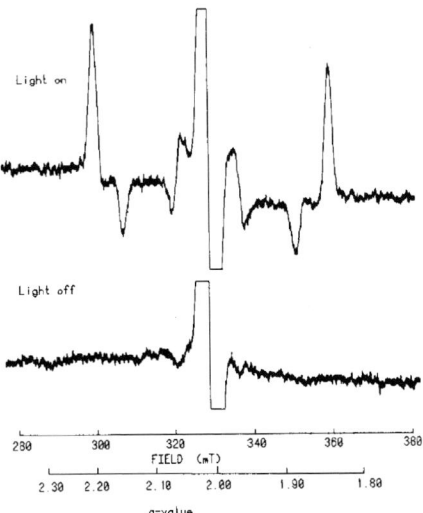

Figure 12. Spectra of ^{3}P680 state. The effect of illumination at 3-5 K on the EPR spectrum of PS II particles. The sample was poised at an Eh of -7mV.

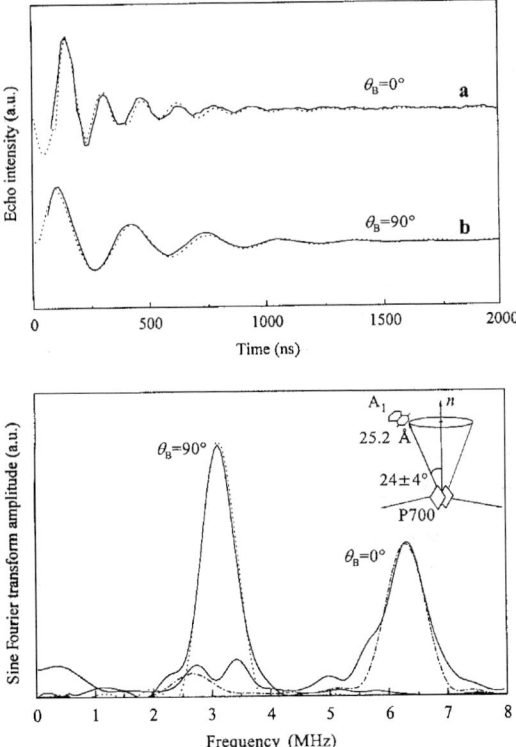

Figure 13. Spin polarized radical pair ESEEM of P700$^+$A$_1^-$ in oriented PS I. Time profiles (upper) and Fourier transforms (lower) for orientations 0° and 90° respectively.

b. ESEEM of spin polarized radical pairs: $P700^+A_1^-$ and $P680^+Q_A^-$

Figure 13 shows the 2 pulse ESEEM observed by out of phase detection in PS I oriented core particles 1 ms after laser excitation.[72] The period of oscillation for the magnetic field direction parallel to the membrane normal n is approximately twice that for the perpendicular direction. The distance was determined by analysis of the ESEEM for non-oriented membranes to be 25.6 Å.[73,74] The angle of the distance vector from the n-axis was determined to be $24 \pm 4°$. In the single crystal of *Synechococcus eleongatus*, the angle was determined to be $27 \pm 5°$.[75] Both values are identical within the experimental error. The accuracy in distance of 0.3 Å is much higher than that by X-ray analysis.

The same method has been applied to $P680^+Q_A^-$ radical pairs in PS II by Zech *et al.*[76] and Hara *et al.*[77] and the distance of 27.4 Å was obtained. The angle of $21° \pm 5°$ was determined in the oriented membranes.[72]

c. Temperature dependence of microwave saturation of oxidized $P680^+$

The oxidized $P680^+$ has only a short life time in oxygen evolving PS II particles at physiological temperature and cannot be observed even by time resolved EPR. However, at temperatures below 200 K, $P680^+$ can be re-reduced by the primary acceptor quinone on a millisecond time scale and EPR signals have been observed. The line width was about 0.9 mT, which was a little narrower than that of monomer chlorophyll. The temperature dependence of microwave saturation of the peak height of the $P680^+$ signal in PS II in the S_2-state was investigated between 77 and 200 K.[21] The Mn-cluster in the S_2-state was the relaxer to $P680^+$ that was found to be most effective at about 90 K. By applying the Carr-Purcell method, the T_2 value was found to be 0.7 μs. The distance from the Mn-cluster was determined to be 21-25 Å.

d. Chlorophyll Z (Chl_Z)

At temperatures below 220 K, the normal electron transfer pathway is inhibited. A side path from Cyt b_{559} through Chl_Z to $P680^+$ was suggested by Thomson *et al.*[22], though the position of Chl_Z was not clarified. For the case where Cyt b_{559} was pre-oxidized, the $Chl_Z^+Q_A^-$ radical pairs were stabilized. Since $P680^+Q_A^-$ recombination on this side path was inhibited, the transient $P680^+$ signal decreased. This is another reason why the $P680^+$ EPR signal is not easily observed.

Stewart *et al.*[23] proposed that Chl_Z is situated on the D1 protein. X-ray analysis identified two accessory chlorophylls corresponding to this function. However, X-ray data cannot yield information about the function.

13. Electron transfer and structure of plant photosystem II 553

The distance between Y_D and Chl_Z was determined to be 29.4 Å.[70] Furthermore, the orientation of the Y_D-Chl_Z vector was determined to be 50 ± 5° relative to the membrane normal based on the time profiles of '2+1' pulse sequence in the oriented membrane.[78] However, PELDOR was not observed for the $Y_Z{}^\bullet$-Chl_Z pair in the Y_D-less mutant of *Chlamydomonas reinhardtii*, suggesting a distance greater than 50 Å between them.[63]

4.3   ESEEM and two-dimensional study of $Q_A^{\overline{=}}$

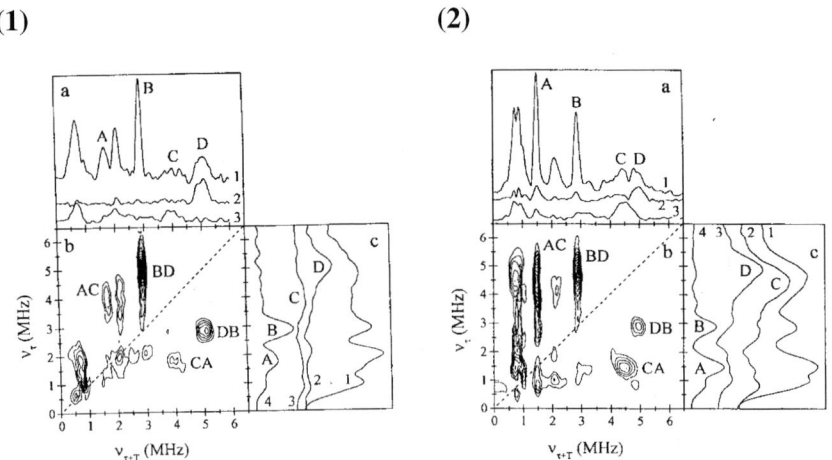

Figure 14. 2D spectrum of stimulated ESEEM of Q_A^- in CN^--treated (1) and Zn^{2+} substituted (2) PS II (b), and its projections on $v_{\tau+T}$ (a) and v_τ (c). Traces 1,2, and 3 in panel (a) are the projections of the whole 2D spectra and of cross peaks DB and CA, respectively. Traces 1,2,3 and 4 in panel (c) are the projections of the whole 2D spectra of BD and AC, and the region containing cross peaks DB and CA, respectively.

Non-heme iron connecting Q_A and Q_B can be eliminated or replaced by zinc. In such PS II, we can observe Q_A^- signals at the g = 2.0046 position with a peak to peak width 0.9 mT. The magnetic exchange coupling of the ferrous iron can be eliminated by cyanide treatment[79] or in an alkaline buffer of pH 11[80] However, these treatments produce different couplings of the iron with surrounding amino acids. Three cyanides coordinate with the iron instead of three histidine molecules and also result in the diamagnetic state of the ferrous ion that decouples the exchange with Q_A^-.

The chemical environment of Q_A^- was studied by Astashkin *et al*[81] by 2D ESEEM as shown in Figure 14 where the authors compared the two states for zinc-substituted and cyanide-treated PS II. Little change was observed for the Q_A site in both preparations. Two nitrogen nuclei were found to contribute to the spectra in both. One of these nitrogens is, probably, an amino nitrogen in the imidazole ring of histidine 214 of the D2 protein. The other nitrogen has been assigned to the peptide group of alanine 261 of the D2 protein.[81, 82]

4.4 Cytochrome b559 (Cyt b_{559})

The oxidized Cytochrome b559 can be observed at low temperatures over a wide field range from $g = 2.9$ to $g = 1.5$. g-values depend on, so-called low potential and high potential forms.[27] In oriented membranes the signal position and intensity depend on the magnetic field direction. PELDOR was detected at the $g = 1.988$ position with the partner of primary acceptor quinone Q_A^- ($g = 2.005$) by Kuroiwa et al.[83] As PELDOR was observed only for the contribution from some selective orientations of Cyt b_{559}, angular selection was taken into the simulation. Assuming that the g_y-axis is directed along the membrane normal, the distance between Cyt b_{559} and Q_A was determined to be 40 ± 5 Å and its orientation relative to the n-axis, to be $80 \pm 5°$. The distance was determined by X-ray to be more than 45 Å.

4.5 Mn-cluster in water oxidizing complex (WOC)

Since the discovery of the S_2 state with its manganese multiline signal, a number of works have been carried out to elucidate the mechanism of water oxidation and the structure of the Mn cluster.[84,33] It has been known since 1980s that the water oxidizing system includes four manganese, calcium, chloride and three extrinsic proteins in higher plants, because PS II particles deficient of any of these cofactors lose oxygen evolving activity. Among the electron transfer components in PS II, the structure and function of the Mn cluster are most poorly understood at present. EPR signals for all the Kok's S_0 to S_3-states have been observed. However, the mechanism to produce these signals has not yet been clarified. Some theoretical consideration for the signal source has been done and the hyperfine structure has been ascribed to magnetic coupling between four manganeses.[85, 86] Several studies of manganese model compounds and theoretical analysis to elucidate exchange coupling schemes of the Mn-cluster have also been reported as reviewed by V. L. Pecoraro and W-U. Hsieh.[87]

a. The S_3-state signal observed by a dual mode EPR
 S_3-state signals were first observed at 5 K using dual modes (parallel and perpendicular) by Matsukawa et al.[38] in the 240 K illuminated and/or

13. Electron transfer and structure of plant photosystem II

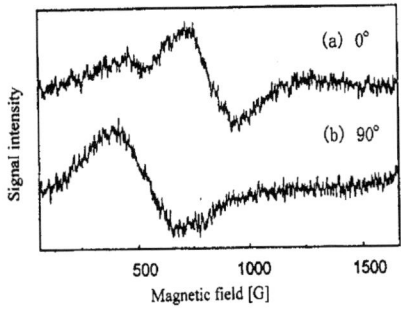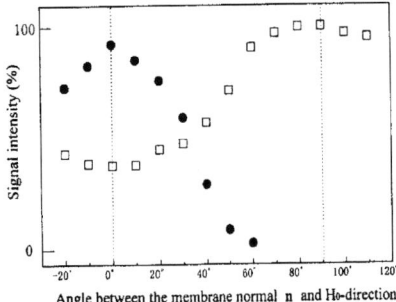

Figure 15. The S_3-state signal observed in oriented membranes. (Left) (a) ***n*** along the magnetic field and (b) perpendicular to it. (Right) Angular dependence of intensity.

Neglecting the intra-dipolar interaction, the value of D may be related to the dipole interaction between the spins on the two manganese atoms as suggested by the Mn-Mn direction between the central two atoms as presented in Figure 2(c). In the S_1-state the orientation of the D-axis was suggested to be perpendicular to the ***n*** axis.[88] The change in the D-axis with S-state advancement may be due to the change in the spin distribution induced by oxidation of manganese.

b. Abnormal signal in Ca^{2+}-depleted PS II and its origin

A modified form of the multiline signal was observed by illumination at 273K of Ca^{2+}-depleted PS II.[61] The multiline in the Ca^{2+}-depleted PS II is stable having a life time of a few hours after illumination at 273K. The multiline signal is almost similar to that of the normal S_2-state differing only in the increased number of lines and slightly decreased hyperfine separation. By immediate trapping of the illuminated sample, a split signal at the $g = 2$ position having a separation of about 16 mT and another overlapped broad signal were formed.[90] There have been several contradictory views expressed concerning the origin of the split signal of the overlapped abnormal S-state. One is that the signal arises from the dipole interaction between two organic radicals,[91] and the other is that it originates from the exchange interaction between an organic radical and the manganese cluster.[91,92] Both views are in agreement that the $Y_Z^\bullet$ radical is involved in the split signal based on the pulsed ENDOR spectra. Figure 16(a) shows the pulsed ENDOR spectrum of the split signal and (b) shows the pulsed ENDOR induced ESE that has provided proof that one partner of the split signal is Y_Z. For the other overlapped signal Y_Z ENDOR signal was not observed, showing that Ca^{2+}-depletion might have produced centres with different states of WOC.

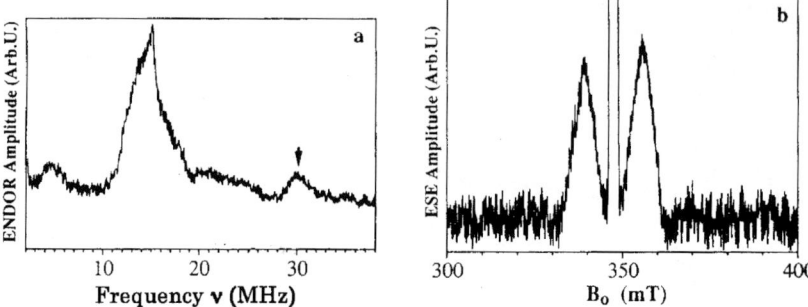

Figure 16. Pulsed ENDOR induced ESE of the split signal. Ca^{2+}-depleted PS II was illuminated at 273 K for 3 min. (a) Pulsed ENDOR signal of the split signal that is coincident with ENDOR spectra of $Y_Z{}^{\bullet}$. (b) Pulsed ENDOR induced ESE spectrum observed at the ENDOR frequency shown by the arrow in (a).

c. Distances of spin centers of Mn-cluster in S_0 and S_2 states

The PELDOR method has been used to measure the distance between the S_2-state manganese cluster and tyrosine D ($Y_D{}^{\bullet}$) and its relative orientation with respect to the membranes. The distance was found to be 27.3 Å[93] and the orientation of the distance vector was 70° from the membrane normal.[66] Recently, we applied the method to the system of the Mn-cluster in the S_0-state and Y_D radical[94] that was prepared by reduction of the PS II membranes with NH_2OH.[36] The yield of about 30 % was much less compared to the S_2 state's 100 %. However, it was more difficult to get the S_0 state by flash illumination,[37] because the sample was dilute and it was contaminated with S_2-state that gave a very similar multiline signal. The PELDOR signal is shown in Figure 17 in comparison with that for the S_2-state.

This result supports the conclusion that the change of position in the spin center of the manganese cluster is accompanied by oxidation of WOC due to the change in the exchange coupling scheme of the four manganese complex. As there are magnetic exchange coupling between these manganese, the spin projection on the four manganese should be taken into consideration to get more precise distances as shown in the work on a three 3Fe-S center by Elssasser *et al.*[95]

13. Electron transfer and structure of plant photosystem II

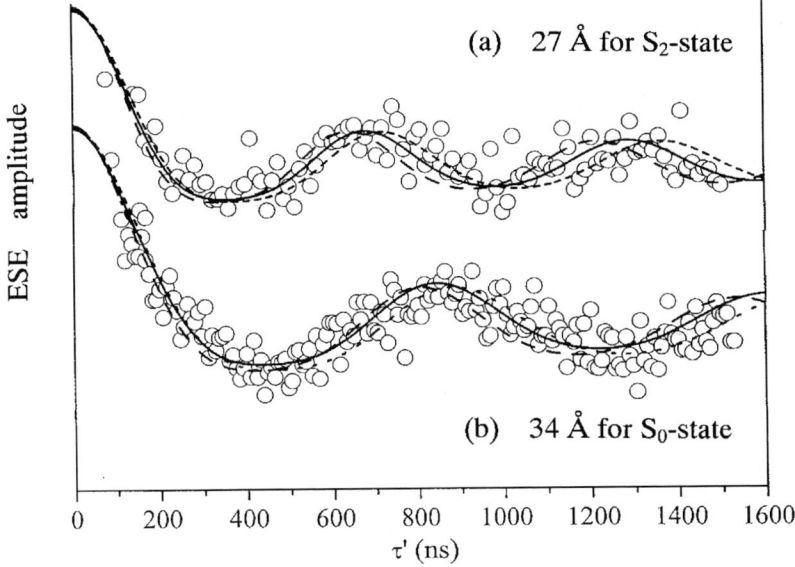

Figure 17. PELDOR signals for the Mn-cluster in the S_2-state (a) and S_0-state (b). The partner is Y_D radical. The full lines show the simulation for the distance 27 and 34 Å respectively. The broken lines show simulations with ± 0.5 Å.

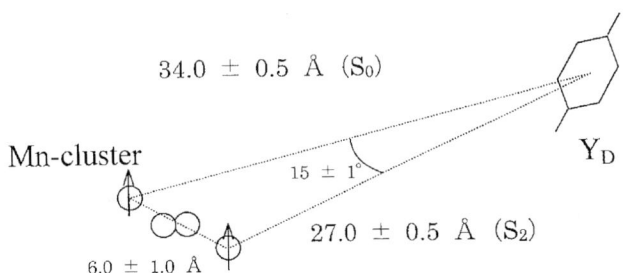

Figure 18. The positions of spin centers of the S_2 and S_0-state Mn-cluster. The configuration of the four manganese atoms are shown based on the distance between the centers, 6 Å, and the angles, 15°, derived from the X-ray data. The arrows show the spin center of S_0 and S_2, respectively.

5. STRUCTURE OF PS II; COMPARISON WITH X-RAY ANALYSIS

So far we have obtained the distance and its orientation independently for each pair of paramagnetic species and radicals. PELDOR on three spin systems is presently being undertaken to determine relative positions of electron transfer components more accurately. Results for the Y_Z-Y_D-Q_A system has been reported.[97] The recorded PELDOR profile fits well with the simulation for the known distances, 30 Å[65] of Y_Z-Y_D, and 39 Å[68] of Y_D-Q_A

and by assuming 34 Å for Y_Z-Q_A that is coincident with the obtained distances from Y_D-less mutant of *C. reinhardtii*[63] and *S. elongatus*.[98]

Table 1 shows distances and their orientations relative to the membrane normal (crystal *c*-axis) as obtained by EPR. Accurate values for distances were obtained by ESEEM and PELDOR. Other EPR methods give only approximate values comparable to the resolution from X-ray analysis.

Table 1. The derived distances (Å) and angles of electron transfer components in PS II studied by EPR

Paramagnetic Pairs	Distances EPR	X-ray	Angles (°) from n axis	METHODS
$P680-Q_A$	$27.4\pm0.3^{74,77}$	26.9	21 ± 5^{72}	Spin polarized ESEEM
Y_D-Q_A	38.5 ± 0.8^{70}	40*	28 ± 5^{96}**	'2+1' pulse
Y_Z-Q_A	$34\pm1^{63,98}$	34.2*		ESEEM, PELDOR
Y_D-Y_Z	29.5 ± 0.5^{65}	29.2*	80 ± 2^{66}	'2+1' pulse
Y_D-Chl_Z	29.4 ± 0.5^{70}	26.8_{D2}*	50 ± 5^{78}	'2+1' pulse
$Y_D-Mn_4(S_2)$	$27.3\pm0.2^{42,94}$	30.3*	70 ± 2^{66}	PELDOR
$\tilde{Y}_D Mn_4(S_0)$	34 ± 0.5^{94}	30.3*		PELDOR
$Q_A-Cyt\ b_{559}$	40 ± 3^{83}	47.8	78 ± 5^{83}	PELDOR
Y_D-non-heme Fe	$38-42^{40,43,99}$	37.6*		Selective hole burning T_1, and saturation
$P680-Mn_4(S_2)$	$21-25^{21}$	18.2		Time resolv. saturation
$\tilde{Y}_Z Mn_4(S_2)$	$6-9^{91}$			P-ENDOR
$\tilde{Y}_Z Mn_4(S_1)$	$>15^{42,100}$	8.4*		T_1 measurement

* The distance was measured from C_δ of the tyrosine molecule, other distances without * were measured from average positions.
** The corrected value is shown after re-calculation.

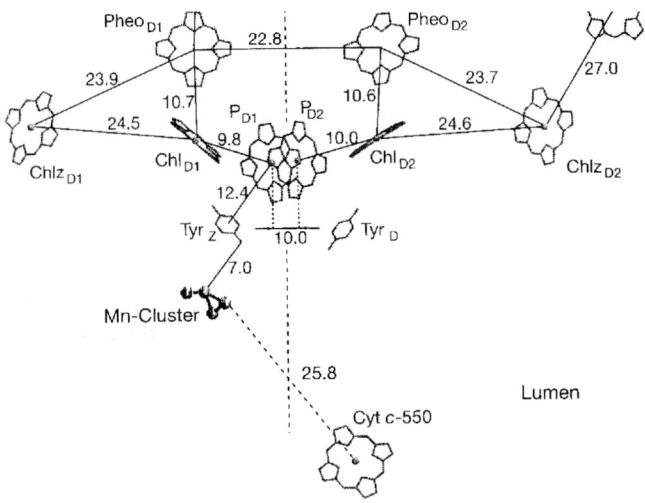

Figure 19. The side view of the configuration of cofactors in PS II determined by X-ray analysis. The unit of distance is in Å. P_{D1} and P_{D2} are the reaction center chlorophyll molecules with the separation between Mg atoms 10 Å. The crystal c-axis is along the line from the center of P680 to Fe atoms. Cyt c-550 belongs to an extrinsic protein specific to cyanobacteria in the place of 23 kD protein in spinach. (Taken from Ref. 10).

Figure 19 shows the side view of the structure of PS II deduced from X-ray analysis by Zouni et al.[10] In Table 1 the distances derived from the Protein Data Bank are also shown for comparison. The distances from P680 to Q_A and the Mn cluster seem to be consistent with EPR results. The distance from Cyt b_{559} to Q_A obtained by EPR is much shorter than that by X-ray data. However, the value depends on the assumed EPR line width in simulation because angular selection has been taken into account. The values of distances for Y_D and Y_Z to P680 and to the Mn-cluster by X-ray seem to be coincident within the resolution of 3.8 Å for X-ray.

As the three spin system fixes relative positions of electron transfer components, the configurations of electron transfer components could be derived using the triangles and the orientation of distance vector relative to the membrane normal. However, the errors in orientation in EPR data were too large to give the accurate accurate positions. Only an approximate configuration of electron transfer components was derived. The coincidence in the distances were fairly good except for those related to Cyt b_{559}. The distances related to small molecules were also coincident with those derived from EPR data. These distances obtained by EPR provide information for further refinement of X-ray determination of the electron transfer components.

6. REFERENCES

1. B. Commoner, J.J. Heise .and J. Townsend, *Proc. Natl. Acad. Sci. USA*, **426** (1956) 710.
2. B.A. Barry and G.T. Babcock, *Proc. Natl Acad. Sci. USA*, **84** (1987) 7099.
3. G. Feher, A.J. Hoff, R.A. Isaacson and L.C. Ackerson, *Ann. New York Acad. Sci.* **244** (1975) 239.
4. J. Deisenhofer, O. Epp, K. Micher, K.Miki, R. Huber and H. Micher, *Nature* **318** (1985) 618.
5. H.T. Witt, N. Krauss, W. Hindrichs, I. Witt, P. Fromme, W. Saenger, In: *Research in Photosynthesis, Proc. Int. 9th Cong. Photosynthesis*, Ed. N. Murata, Kluwer Academic, (Dordrecht), 1992, 521.
6. P. Jordan, P. Fromme, O. Klukas, H.T. Witt, W. Saenger, N. Krauss, *Nature* **411** (2001) 909.
7. A. Zouni, R. Jordan, E. Schlodder, P. Fromme, H.T. Witt, *Biochim. Biophys. Acta,* **1457** (2000) 103.
8. H. Kuhl, J. Kruip, A. Seidler, A. Krieger-Lidszkay, M. Bünker, D. Bald, A.J. Scheidig and M. Rögner, *J. Biol. Chem.,* **275** (2000) 20652.
9. J.-R. Shen and N. Kamiya, *Biochemistry* **39** (2000) 14739.
10. A. Zouni, H.T. Witt, J. Kern, P. Fromme, N. Krauß, W. Saenger and P. Orth, *Nature* **409** (2001) 739.
11. B.A. Diner and G.T. Babcock, In *Oxygenic Photosynthesis: The Light Reactions*, Ed. D.R. Ort and C.F. Yocum, Kluweer Academic (Dordrecht), 1996, p213.
12. B. Kok, B. Forbush and M. McGloin, *Photochem. Photobiol.* **11** (1970) 457.
13. R.A. Marcus and N. Sutin, *Biochim. Biophys. Acta.* **811** (1985) 265.
14. J.H.A. Nugent, M.CW. Evans, B.A. Diner, *Biochim. Biophys. Acta.* **682** (1982) 106.
15. A.W. Rutherford, D.R. Paterson and J.E. Mullet, *Biochim. Biophys. Acta.* **635** (1981) 205.
16. M.C. Thurnauer, J.J. Katz and J. R. Norris, *Proc. Natl. Acad. Sci. USA* **72** (1975) 3279.
17. T. Noguchi, T. Tomo and Y. Inoue, *Biochemistry* **40** (2001) 2176.
18. J.P. Allen, G. Feher, T.O. Yeates, H. Komiya and D.C. Rees, *Proc. Natl. Acad. Sci. USA* **84** (1987) 5730.
19. C.H. Bock, S. Gerken D. Stehlik and H.T. Witt, *FEBS Lett.* **227** (1988) 53.
20. C.W. Hognason and G.T. Babcock, *Biochemistry* **28** (1989) 1448.
21. Y. Kodera, K. Takura and A. Kawamori, *Biochim. Biophys. Acta.* **1101** (1992) 23.
22. L.K. Thomson and G.T. Brudvig, *Biochemistry* **27** (1988) 6653.
23. D.H. Stewart, A. Cua, D.A. Chisolm, B.A. Diner, D.F. Pocian and G.W. Brudvig, *Biochemiostry* **37** (1998) 10040.
24. G.T. Babcock and K. Sauer, *Biochim. Biophys. Acta.* **350** (1975) 315.
25. R.J. Boerner and B.A. Barry, *J. Biol. Chem.* **268** (1993) 17151.
26. M. MacMillan, F.H. Lendzian, G. Renger and W. Lubitz, *Biochemistry* **34** (1996) 8144.
27. A-F. Miller and G.W. Brudvig, *Biochim. Biophys. Acta* **1056** (1990) 1.
28. A.W. Rutherford, *Biochim Biophys Acta,* **807** (1985) 189.
29. D.H. Stewart and G.W. Brudvig, *Biochim. Biophys. Acta* **1367** (1998) 63.
30. G.C. Dismukes and Y. Siderer, *FEBS Lett.* **121** (1980) 78.
31. S.R. Cooper, G.C. Dismukes, M.P. Klein and M. Calvin, *J. Am. Chem. Soc.* **100** (1978) 7248.
32. J.C. De Paula, J.B. Innes and G.W. Brudvig, *Biochemistry* **24** (19985) 8114.
33. R.D. Britt, In *Oxygenic Photosynthesis: The Light Reactions*, Ed. D.R. Ort and C.F. Yocum, Kluwer Academic (Dordrecht) 1996, p137.
34. S.L. Dexheimer and M.P. Klein, *J. Am. Chem. Soc.* **114** (1992) 2821.

35. T. Yamauchi, H. Mino, T. Matsukawa, A. Kawamori and T. Ono, *Biochemistry* **36** (1997) 7520.
36. J. Messinger, J.H.A. Nugent, M.CW. Evans, *Biochemistry* **36** (1997) 11055.
37. K.A. Åhrling, S. Peterson and S. Styring, *Biochemistry* **36** (1997) 13148.
38. T. Matsukawa, H. Mino, D. Yoneda and A. Kawamori, *Biochemistry* **38** (1999) 4072.
39. V.K. Yachandra, V.J. DeRose, M.J. Latimer, I. Mukkerji, K. Sauer and M.P. Klein, *Science* **260** (1993) 675.
40. D.J. Hirsh, W. F. Beck, J.B. Innes and G.W. Brudvig, *Biochemistry* **31** (1992) 532.
41. S.A. Dzuba and A. Kawamori, *Concepts in Mag Reson.* **8** (1996) 49.
42. Y. Kodera, S.A. Dzuba, H. Hara and A. Kawamori, *Biochim. Biophys. Acta* **1186** (1994) 91.
43. H. Hara and A. Kawamori, *Appl. Magn. Reson.* **13** (1997) 241.
44. J.H. Freed, *Annu. Rev. Phys. Chem.* **51** (2000) 655.
45. A.D. Milov, A.B. Ponomalrev and Y.D. Tsvetkov, *Chem. Phys. Lett.*, **110** (1984) 67.
46. T-P. Moënne-Loccoz, P. Heathcote, D.J. Maclachlan, M.C. Berry, I.H. Davis and M.CW. Evans, *Biochemistry* **33** (1994) 10037.
47. S.A. Dzuba, P. Gast and A.J. Hoff, *Chem. Phys. Lett.* **236** (1995) 595.
48. A.J. Hoff, P. Gast and J.C. Romijn, *FEBS Lett.* **73** (1977) 185.
49. A. van der Est, R. Bittl. E.C. Abresch, W. Lubitz and D. Stehlik, *Chem. Phys. Lett.* **212** (1993) 561.
50. G. Kothe, S. Weber, E. Ohmes, M.C. Thurnauer and J.R. Norris, *Chem. Phys. Lett.* **186** (1991) 474.
51. M.Bosch, I.I. Proskuryakov, P. Gast and A.J. Hoff, *J Phys Chem* **100** (1996) 2384.
52. P. Dorlet, A.W. Rutherford and S. Un, *Biochemistry* **39** (2000) 7826.
53. A.J. Hoff, in *Biophysical Techniques in Photosynthesis*, Ed. J. Amesz and A.J. Hoff, Kluwer Academic (Dordrecht), 1996, p277.
54. R. van der Vos, P.J. van Leeuwen, P. Brown and A.J. Hoff, *Biochim. Biophys. Acta* **1140** (1992) 184.
55. S.A. Dikanov and Y.D. Tsvetkov, *Electron Spin Echo Envelope Modulation (ESEEM) Spectroscopy*, CRC Press (Boca Raton USA) 1992.
56. D.A. Force, D.W. Randall, G.A. Lorigan, K.L. Clemens and R.D. Britt, *J. Am. Chem. Soc.* **120** (1998) 13321.
57. A.V. Astashkin, A. Kawamori, Y. Kodera, S. Kuroiwa and K. Akabori, *J. Chem. Phys.* **102** (1995) 5583.
58. T. Kuwabara and N. Murata, *Plant Cell Physiol.* **23** (1982) 533.
59. D.A.Berthold, G.T. Babcock and C.F. Yocum, *FEBS Lett.* **134** (1981) 231.
60. H. Mino and A. Kawamori, *Biochim. Biophys. Acta* **1185** (1994) 213.
61. T-A. Ono and Y. Inoue, *Biochemistry* **31** (1992) 7648.
62. T-A. Ono and Y. Inoue, *Biochim. Biophys. Acta* **973** (1989) 443.
63. A. Kawamori, N. Katsuta, H. Mino A. Ishii, J. Minagawa and T-A. Ono, *J. Biological Phys.* (2002), In print.
64. H. Mino, A.V. Astashkin and A. Kawamori, *Specrochimica Acta* **A53** (1997) 1565.
65. A.V. Astashkin, Y. Kodera and A. Kawamori, *Biochim. Biophys. Acta* **1187** (1994) 89.
66. A.V. Astashkin, H. Hara and A. Kawamori, *J. Chem. Phys.* **108** (1998) 3805.
67. W. Hofbauer, A. Zouni, R. Bittl R, J. Kern, P. Orth, F. Lendzian, P. Fromme, H.T. Witt and W. Lubitz, *Proc. Natl Acad Sci USA* **98** (2001) 6623.
68. W.F. Butler, R. Calvo, D.R. Fredkin, R.A. Isaacson, M.Y. Okamura and G. Feher, *Biophys. J.* **45** (1984) 947.
69. T.O.Yeates, H. Komiya, D.C. Rees, J. P. Allens and G. Feher, *Proc. Natl. Acad. Sci. USA* **84** (1987) 6438.

70. K. Shigermori, H. Hara, A.Kawamori and K. Akabori, *Biochim. Biophys. Acta.* **1363** (1998) 187.
71. F.J.E. van Mieghem, K. Satoh and A.W. Rutherford, *Biochim. Biophys. Acta* **1058** (1991) 379.
72. T. Yoshii, H. Hara, A. Kawamori, K. Akabori, M. Iwaki and S.Itoh, *Appl Magn. Reson.* **16** (1999) 565.
73. S.A. Dzuba, H. Hara, A. Kawamori, M. Iwaki, S. Itoh and Y.D. Tsvetkov, *Chem. Phys. Lett.* **264** (1997) 238.
74. R. Bittl, S.G. Zech, *J. Phys. Chem.* **B101** (1997) 1429.
75. R. Bittl, S.G. Zech, P. Fromme, H.T. Witt and W. Lubitz, *Biochemistry* **36** (1997) 12001.
76. S.G. Zech, J. Kurreck, H-J. Eckert, G. Renger, W. Lubitz and R. Bittl, *FEBS Lett* **414** (1997) 454.
77. H. Hara[b], S.A. Dzuba, A. Kawamori, K. Akabori, T. Tomo, K. Satoh, M. Iwaki and S. Itoh, *Biochim. Biophys. Acta* **1332** (1997) 77.
78. M. Tonaka, A. Kawamori, H. Hara, and A.V. Astashkin, *Appl. Magm. Reson.* **19** (2000) 141.
79. Y. Sanakis,V. Petrouleas and B. Diner, *Biochemistry* **33** (1994) 9922.
80. Y. Deligiannakis, A. Boussac A and AW Rutherford, *Biochemistry* **34** (1995) 16030.
81. A.V. Astashkin[b], H. Hara, S. Kuroiwa, A. Kawamori and K. Akabori, *J. Chem. Phys.* **108** (1998a) 10143.
82. A.V. Astashkin, A. Kawamori, Y. Kodera, S. Kuroiwa and K. Akabori, *J. Chem. Phys.* **102** (1995) 5583.
83. S. Kuroiwa, M. Toaka, A. Kawamori, K. Akabori, *Biochim. Biophys. Acta,* **1460** (2000) 330.
84. R.J. Debus, *Biochim. Biophys. Acta* **1102** (1992) 269.
85. M. Zheng and G. C. Dismukes, *Inorg. Chem.* **35** (1996) 3307.
86. K. Hasegawa, M. Kusunoki, Y. Inoue and A. Ono, *Biochemistry* **33** (1998) 9457.
87. V.L. Pecorao and W-U. Hsieh; *Manganese and its role in biological processes: Metal Ions in Biological systems* Ed. A. Sigel and H. Sigel, **vol 37** (2000) p429. Marcell Decker (NY).
88. T. Matsukawa, A. Kawamori and H. Mino, *Spectrochim. Acta* **A55** (1999) 895.
89. A.V. Astashkin, H. Mino, A. Kawamori, T-A. Ono, *Chem. Phys. Letts.* **272** (1997) 506.
90. H. Mino and A. Kawamori, *Biochim. Biophys. Acta* **1503** (2001) 112.
91. J.M. Pelloquin, K.A.Cambell, R.D.Britt, *J. Am. Chem. Soc.* **120** (1998) 6840.
92. J.M. Pelloquin and R.D. Britt, *Biochim. Biophys. Acta* **1503** (2001) 96.
93. H. Hara, A. Kawamori, A.V. Astashkin and T-A. Ono, *Biochim. Biophys. Acta,* **1276** (1996) 140.
94. S. Arao, S. Yamada, A. Kawamori, J. -R. Shen, N. Ionnidis and V. Petrouleas, The *Proceedings of the 3rd Asia Pacific EPR/ESR Symposium,* Ed. A. Kawamori, J. Yamauchi and H. Ohta, Elsevier Science (Amsterdam) 2002, p 455.
95. C. Elsasser, M. Brecht and R Bittl, *J. Am Chem. Soc.* **124** (2002) 12606.
96. T. Yoshii, A. Kawamori, M. Tonaka and K. Akabori, *Biochim. Biophys. Acta.* **1313** (1999) 43.
97. H Hara, A Kawamori and N. Katsuta, *The proceedings of the 3rd Asia Pacific EPR/ESR Symposium* Ed. A. Kawamori, J. Yamauchi and H. Ohta, Elsevier Science (Amsterdam) 2002, p658.
98. S.G. Zech, J. Kurreck, H-J. Eckert, G. Renger, W. Lubitz and R. Bittl, *FEBS Lett* **442** (1999) 79.
99. S. Un, L-C Brunel, T.M Brill, J-L Zimmermann and A.W. Rutherford. *Proc. Natl. Acad. Sci. USA,* **91** (1994) 5262.

100. Y. Kodera, H. Hara, A.V. Astashkin, A. Kawamori A and T. Ono T, *Biochim. Biophys. Acta* **1232** (1995) 43.

Chapter 14

RECENT DEVELOPMENT OF EPR DOSIMETRY

Nicola D. Yordanov and Veselka Gancheva
PR Laboratory, Institute of Catalysis, Bulgarian Academy of Sciences, 1113 Sofia Bulgaria

Key words: Applied EPR, dosimetry, identification of irradiated foodstuffs, post-radiation, emergency and *in vivo* dosimetry.

Abstract: In the last two decades EPR spectrometry has expanded significantly in the field of practical applications. This is due mainly to the high selectivity and sensitivity of the method. Moreover, it is a non-destructive technique and in some cases samples may be kept as documents for future inspection. The aim of the present chapter is to highlight the current status, as well as the near future trends in development of Solid State/EPR dosimetry. The main subjects considered are the recent developments in SS/EPR dosimetry in standard and emergency accidental cases, identification of irradiated foodstuffs as well as new approaches to the extension of the period of identification by EPR.

1. INTRODUCTION

Despite the fact that EPR was discovered in 1944[1] and that very intensive studies and development of the method have been done since then with regard to theoretical, methodical and instrumentation aspects, its technique has not been widely used for practical purposes. One of the main reasons for not exploiting such advantages as high selectivity, sensitivity and non-destructivity is the expensive, heavy, and sophisticated equipment required. Up to now, there are only three internationally recognised applications, all connected with the effects induced in the solid state by high-energy radiation. One of them is alanine/EPR dosimetry, which gives the possibility to estimate the absorbed dose of high-energy radiation by the alanine dosimeter.[2-7] The other two provide procedures for identification of irradiated foodstuffs containing cellulose[8] and bone material[9]. There are many books and specialised meeting proceedings devoted to EPR. But the difficulty of its practical application are not, or very rarely, included in them although many authors are working to increase the applicability of these methods. Such problems are discussed only in a few specialised reviews, books and meetings.[10-16]

The aim of the present chapter is to highlight the present state, as well as the near future trends in the development of Solid State/EPR dosimetry. Three main subjects are considered – recent developments in SS/EPR dosimetry in standard applications, in emergency situations after an accident, as well as new approaches for the extension of the identification period of irradiated foodstuffs by EPR.

2. PRINCIPLES OF EPR DOSIMETRY AS A METHOD FOR ESTIMATION OF THE ABSORBED DOSE AND FOR POST RADIATION PROCESSING DETECTION

There has been an increased practical application of radiation processing in the last decades. It is used for example, for sterilization of medical and pharmaceutical products because of the ability of ionizing radiation to kill pathogenic microorganisms. The radiation sterilization is accepted and applied in many countries for improving the hygienic quality of foodstuffs in order to extend their shelf life. However, the manipulation needs a dosimetric system to control the radiation process and to estimate the absorbed dose by the material. Nowadays, there are many different dosimetric methods for this purpose and among them EPR is the only method characterized by a simple and time saving procedure of dose estimation. It is also a method that does not destroy the radiation sensitive material under study, making it possible to preserve it for future inspection. However, there are some problems with dose evaluation. Most of them are due to the fact that quantitative aspects of EPR spectrometry are not well developed[17,18] and because the EPR spectrometry is not a calibrated method up to now. Some new approaches were recently developed to overcome these problems in order to improve the reproducibility of the results obtained with SS/EPR dosimeters by different laboratories and to extend the period of reading the dose as well.

2.1 Formation of Paramagnetic Centres upon Irradiation

The action of ionising radiation on matter primary yields charged particles (molecular ions and electrons) and excited molecules. In solid inorganic materials the liberated free electrons may be trapped by the crystal lattice whereas in solutions the solvent is also under irradiation and thus plays an important role in the overall process. In solid organic compounds, to which the present chapter is devoted, the primary active particles are transformed into free radicals and some final products of radiolysis. As an example we will consider alanine, an amino acid which has an important application in EPR dosimetry and will be extensively referred to in this chapter.

The first investigations by EPR of the processes taking place in alanine irradiated with γ-rays were done long ago. In order to find the primary, unstable particles the studies were carried out on single crystals of α-alanine in the temperature interval 77-300 K. The results led to the following scheme about the effect of radiation on alanine molecules.[19-22]

T = 77 K

$$CH_3CH(N^+H_3)COO^- \xrightarrow{h\nu} [CH_3CH(N^+H_3)COO^-]^+ + e^-$$

$$CH_3CH(N^+H_3)COO^- + e^- \longrightarrow CH_3CH(N^+H_3)\overset{\bullet}{C}O^-O^-$$

$$CH_3CH(N^+H_3)\overset{\bullet}{C}O^-O^- + H^+ \longrightarrow CH_3CH(N^+H_3)\overset{\bullet}{C}O^-OH$$

T > 77 K

$$CH_3CH(N^+H_3)\overset{\bullet}{C}O^-OH \longrightarrow NH_3 + CH_3\overset{\bullet}{C}HCOOH$$

Immediately after irradiation such dynamic changes are common and it takes time before the sample under investigation reaches the steady-state conditions necessary for EPR dosimetry. The period is specific for every substance and it is ca. 72 h. for alanine.[23]

It is now accepted that the radical, which is stable at room temperature corresponds to the deaminated alanine. The EPR spectrum of this radical, obtained in irradiated powder of α-alanine is stable, quasi-isotropic and consists of five lines with approximate intensity distribution of 1:4:6:4:1 (Fig. 1). However, recent ENDOR studies[24,25] have shown that there are at least two major and one minor paramagnetic species in the irradiated alanine sample.

Fig. 1 EPR spectrum of irradiated powder L-α-alanine.

2.2 Lifetime and Yield of Paramagnetic Centres ("G-value") of a Given Irradiated Material. Minimal Detectable Absorbed Dose of Radiation by EPR

There are two main characteristics of the free radicals generated by ionising radiation of matter, which are important for dosimetric purposes:

- the life time of the free radicals;
- the radiation sensitivity of the material *i.e.* the number of stable free radicals produced by the absorbed radiation per 100 eV which is so called "G-value" of the matter.

The lifetime of the obtained free radicals of different materials varies over a very broad time scale - some of them recombine within a few milliseconds whereas others are stable for periods of several hundreds or thousands of years. The lifetime depends on many factors such as the structure of the final free radicals, their ability to recombine in the solid state as well as on the conditions of their storage (temperature, humidity, exposure to UV light, presence of O_2 etc.). Radiation sensitive materials for dosimeters should have the longest possible lifetime. For example, free radicals of α-alanine are extremely stable in the solid state whereas β-alanine free radicals are not.[26]

The other parameter, the G-value of the matter, gives information about the number of detected free radicals obtained by absorption of 100 eV ionising energy. G is equal to 1 if the number of free radicals produced by the absorbed dose of 1 Gy is 6.3×10^{16}/kg.[11] The G-value is specific for the material under investigation and its magnitude is usually determined for γ-radiation. The G-value data for alanine determined by different authors, varies from 2 to 7.7.[27-31] There is no explanation about the differences in the literature and it is assumed that radiation interacts directly with alanine molecules only. However, some factors not considered up to now may affect this interaction as, for example, the influence of the nature of binding materials and/or ingredients on the sensitivity of alanine/EPR dosimeters. These materials could be assumed to play a sensitising or an inhibiting role on the yield of paramagnetic centres by other processes. Another reason for the high dispersion of alanine G-value may be connected with the EPR technique. The problem is that until recently the quantitative aspects of EPR spectrometry were not well developed.[17,18] Nowadays, there is some progress in this direction but it is not certain that quantitative EPR measurements give comparable results. At best, semi-quantitative or comparative EPR studies using reference standards are commonly accepted.

For heavy particles the G-value depends on particle identity, on their energy (LET) and typically is lower in comparison with γ-rays. This is because whereas paramagnetic centers created by gamma-rays are randomly spread in the volume of the solid matter, the heavy particles make tracks in them with a very high number of closely situated paramagnetic species. This high density of paramagnetic species provides spin-spin interaction, connected with line broadening and also facilitates recombination processes. For example the G-value of alanine estimated for fast neutrons is between 0.4 and 0.65.[32,33] In order to increase the EPR response (G-value) some authors add boric acid to the alanine pellet.[34-37] The method is based on the production of additional alanine free radicals caused by α-particles. The α particles are coming from the neutron capture $^{10}_{5}B(n, \alpha)\ ^{7}_{3}Li$ reaction. It is reported[34,35] that in this case the G-value is strongly dependent on the alanine/boric acid ratio. Increase in the EPR response between 20 and 40 times is observed.

The minimum detectable number of spins by the EPR spectrometer depends on its sensitivity, which for a commercial apparatus is ca. 10^{11} spins/mT. This value is valid for a point sample with dielectric constant $\varepsilon = 1$, situated in the middle of the EPR cavity with respect to the sample axis. The sensitivity profile of the EPR cavity is known[38] to have a bell shape along the sample axis with a maximum in the centre of the cavity, falling off to zero at the upper and bottom walls. However, real samples have finite dimensions and the spectrometer sensitivity depends on many other factors. These cannot be calculated, but recording of 5×10^{11} spins/mT in one scan at a signal/noise ratio 2:1 can be considered as realistic. If we assume a 10 mm length and ca. 0.2 g weight of a dosimeter positioned in the centre of the cavity, irradiated with 1 Gy and G = 1, the expected number of spins is calculated to be ca. $0.2 \times 6.3 \times 10^{13}$ spins/mT = 1.26×10^{13} spins/mT. However, the spectral response, as for any spectral method is given by the area under the absorption line. Since the first derivative of the absorption line is used in EPR, the response (R) is given as the product of peak-to-peak intensity (I_R) and the square of the line width (ΔH) of the first derivative spectrum, *i.e.* $R \sim I_R (\Delta H)^2$. The line width of most of the radiation sensitive materials known up to now is at least 1 mT. Having this in mind, we can expect to obtain with G = 1 an EPR response of ca. 0.5 Gy radiation dose (at $\varepsilon = 1$). However, because for many materials G > 1 we can record even lower doses with one scan. Using spectra accumulation, the spectrometer sensitivity could be improved with increasing number of accumulated scans (n), whereas in the same time the intensity of the noise decreases with square root ($\sqrt{n}$). At present, it is assumed that for a radiation sensitive material with low dielectric constant an absorbed dose of 0.05 Gy could be detected without any difficulty.

3. SOLID STATE/EPR DOSIMETERS

3.1 Earlier Studies

In 1962 Bradshow *et al.*[39] used for the first time α-alanine as a radiation detector. Thus, they opened the possibility to use alanine as a dosimetric material. Regulla and Defner[40] and other groups carried out intensive studies on this property of alanine. Alanine can be considered to be the best studied material in the field of Solid State/EPR dosimetry and at present it is formally accepted by IAEA[2-5], NIST[6] and NPL[7] (the abbreviations IAEA, NIST and NPL stand for International Atomic Energy Agency (Vienna, Austria), National Institute for Standards and Technology (USA) and National Physical Laboratory (UK)) as a secondary reference and transfer dosimeter for high (industrial) dose irradiation. The main advantages of the alanine dosimeter are the following:
- linear (within ±3%) EPR response in a wide range of doses - from 10 up to 5×10^4 Gy;[41]

14. Recent development of EPR dosimetry

- high stability of the radiation induced free radicals under normal conditions;
- similarity to biological systems;
- simple and rapid data interpretation;
- no sample treatment before EPR measurement of the signal;
- cheap radiation detector that could be kept as a document.

Alanine dosimeter is currently produced in the form of pellets, rods, films and cables with different binding materials.

However, there are various sources of uncertainties in the evaluation of absorbed dose, the main one being the necessity to calibrate each EPR spectrometer and each batch of dosimeters before use. There is currently one method for estimation of absorbed dose in alanine/EPR dosimetry. According to this procedure each laboratory should have available for everyday calibration purposes a set of standards prepared from the same alanine dosimeter batches as those under study. These standards have to be irradiated in advance with known doses. Using these standards the EPR spectrometer is calibrated by preparing a calibration graph of the EPR intensity versus absorbed dose ($I_{alanine}/D$) before measurement of an unknown sample. During the calibration and during all following measurements the EPR instrument's settings (P_{MW}, H_{mod}) must remain unchanged. (This calibration graph may be used only for the batch of dosimeters, EPR spectrometer and cavity used in the calibration. A different calibration graph will be valid for another spectrometer, cavity and/or other dosimeter batch.) Finally, the intensity of the EPR signal of the unknown sample is compared with the graph.

This procedure has several obvious disadvantages connected with the general problems of quantitative EPR measurements:
- the consecutive procedure which is used for dose estimation is not as precise as the simultaneous;[17]
- every laboratory should have for everyday use a set of standards for every type (or batch) of dosimeters in use. Even with the use of similar standards, the results obtained in a given laboratory are typically not comparable with the results of other laboratories or other spectrometers in the same laboratory.

In order to overcome these disadvantages some authors have recently inserted a reference EPR standard permanently in the EPR spectrometer cavity and have simultaneously recorded the signals of alanine and standard.[42-45] The following parts of the procedure described above remain unchanged. In principle, simultaneous recording of EPR spectra of reference and unknown sample gives much better results[17] and the reproducibility and accuracy of the reading are increased. However, it is only valid for a given cavity and spectrometer. The results obtained with other cavities, spectrometers and settings, or laboratories remain still different. Having in mind these problems IAEA and NIST have offered dosimeters by mail-transfer service called "International Dose Assurance Service (IDAS)".[4,7,46,47] The overall procedure includes distribution of dosimeters by these organizations to the customers, which are sent back after irradiation for dose estimations and users receive the results. In this case the problem of calibration of EPR spectrometers is avoided. However, the procedure is concentrated in only a few laboratories and takes considerable time after the irradiation. The IDAS procedure is not for everyday use. It is only used for periodical calibrations or re-calibration of the irradiation units and after some reconstruction or changes in them.

3.2 Recent Developments

A new generation of dosimeters, called "self-calibrated", was recently proposed.[48-50] Each pellet of these dosimeters incorporates radiation sensitive material, a quantity of EPR active substance and binding material. The incorporated EPR active substance acts as an internal standard permanently present in the dosimeter pellet and its signal is recorded together with the signal of the radiation sensitive material. In this way, the procedure for dose evaluation includes irradiation and simultaneous recording of the EPR signals for both EPR active substances (standard and radiation sensitive material) in one pellet, the results do not depend on:
- the type of EPR cavity;
- the positioning of dosimeter pellet in it;
- the spectrometer used.

This is because the reading of this type of dosimeter is not a measurement of the intensity of the EPR line of the radiation sensitive material (I_{RS}), in our case the central line of alanine, but is given as the ratio of I_{RS} versus the EPR intensity of the internal standard (I_{ST}). The calibration of the ratio (I_{RS}/I_{ST}) for dose estimation may be directly obtained by one of the following two ways chosen by the operator, which may:
- irradiate in advance one or two self-calibrated dosimeters with a known dose (D, in Gy), estimate (I_{RS}/I_{ST})/D and find the ratio (I_{RS}/I_{ST}) per Gy, called "calibration coefficient" of the used dosimeter.
- ask the producer of alanine self-calibrated dosimeters to supply him with the dosimeter calibration coefficient.

14. Recent development of EPR dosimetry

Since the self-calibrated dosimeter may be kept as a document and its calibration coefficient $(I_{RS}/I_{ST})/D$ as well as its response (I_{RS}/I_{ST}) are stored, it can be checked at any time without constructing a calibration graph in advance.

The materials used as the internal standard in the self-calibrated dosimeters have to satisfy some important requirements such as:
- to be EPR active before and after irradiation;
- to have radiation independent EPR response;
- to have easily and unambiguously distinguished EPR lines from alanine lines.

Mn^{2+} magnetically diluted in MgO has been used for this purpose because studies have shown[51] that after γ-irradiation (at room temperature) with doses up to 10^5 Gy the host lattice (MgO) is EPR silent. In addition no changes have been found in the intensity and other EPR parameters of the Mn^{2+} spectrum after γ-irradiation of an Mn^{2+}/MgO sample with doses up to 100 kGy. Mn^{2+} lines are narrow, easily and unambiguously distinguishable from those of the radiation sensitive material. The response of the described dosimeters for γ-rays in the range of absorbed dose from 100-50 000 Gy exhibits excellent linearity and reproducibility.[48,49]

Fig. 2 shows the full EPR spectrum of the new self-calibrated alanine dosimeter. As seen from Fig. 2 it contains the six EPR lines of Mn^{2+} and those of alanine appear in the central part of the Mn^{2+} spectrum. Each of the following four Mn^{2+} lines (1, 2 or 5, 6) may be used for dose estimation.

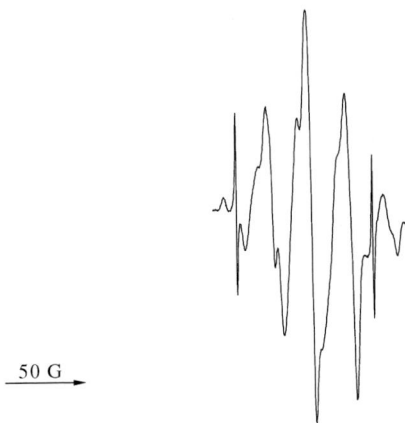

Fig. 2 Full EPR spectrum of irradiated self-calibrated alanine dosimeter.

Recently, small portable fully computer controlled and cheap EPR spectrometers were commercially offered.[41,52-59] The use of these spectrometers as dosimeter readers is a very attractive idea and the self-calibrated dosimeters were especially developed for them. These spectrometers are operating with a permanent magnet and magnetic field sweep is limited to 15-20 mT. The EPR spectrum of the self-calibrated alanine dosimeter measured with them exhibits only part of its full spectrum including the alanine spectrum and two of the Mn^{2+} lines (3rd and 4th lines of Mn^{2+}) situated on both sides of the alanine spectrum (Fig. 3). The central alanine line, which is used for measurements, is not disturbed, but the 3rd and 4th lines of Mn^{2+} overlap with alanine wing lines. Recent attempts show that this effect can be overcome and very precise results can be obtained by considering the ratio ($I_{alanine}/I_{Standard}$),[48] where $I_{Standard} = (I^3+I^4)/2$, and I^3, I^4 are the 3rd and 4th lines of Mn^{2+}.

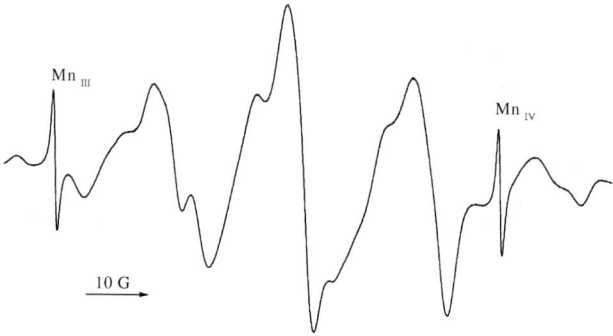

Fig. 3 Typical EPR spectrum for the new generation of alanine/EPR dosimeter.

Because after irradiation self-calibrated dosimeters contain two EPR active substances, the EPR instrumental setting parameters and conditions for simultaneous recording of undistorted spectra of both substances become important. The studies in this direction[49] have shown that an undistorted spectrum of both substances could be obtained if microwave power is less than 1 mW and modulation amplitude is less than 0.5 mT. Using these parameters it is possible to perform intercomparison with data obtained by other instruments or laboratories. Higher sensitivity can be achieved by consecutive recording of two separate spectra in the same pellet - that of Mn^{2+} using the above parameters in order to reach a basis for the calibration and after that the alanine free radical using increased spectrometer gain, modulation amplitude and microwave power. All these parameters have to be calibrated in advance.

The advantages of the self-calibrated dosimeters were demonstrated during an International intercomparison trial[49] in which six laboratories from Europe were supplied with these dosimeters and asked to:
- irradiate them with a known dose;
- record the EPR response using the above EPR instrumental settings (P < 1 mW, H_{mod} < 0.5 mT) on their spectrometers (without any restriction in respect to the used cavities);
- estimate the ratio ($I_{alanine}/I_{Standard}$);
- report the obtained results.

The results of eight estimations obtained by several different spectrometers equipped with different EPR cavities were of extremely high precision for such international comparison trial - statistical processing of all independent measurements showed a standard deviation of 0.3%.[49] All this confirms the advantages and flexibility of the self-calibrated SS/EPR dosimeters because they may be used without any limitations with respect to spectrometer and operator experience.

Dosimeters of this type are usable for calibration of the source of irradiation itself as well as for everyday routine use and it is not necessary to calibrate the spectrometer before every reading. Moreover, they can be used several times because each new absorbed dose is added to the previous and can be recorded with only one scan. They may be kept as documents and re-examined at any time with only one recording of the spectrum. And finally, they work with equal accuracy for all kinds of EPR spectrometers.

Very recently another attempt to improve the accuracy of absorbed dose estimations has been proposed by Internet calibration service for the radiation processing industry.[61] The Internet-based transfer calibration service is developed by NIST, in collaboration with Bruker Biospin. It is designed especially for alanine pellets and film shaped dosimeters. The special feature of the EPR reader is the permanently installed reference standard (ruby) in its cavity. The absorbed dose calibration is performed by

two different sets of pre-irradiated dosimeters - one for pellets and the other for film shaped dosimeters (*vide supra*). Thus all calibrations and measurements are performed using the reference standard. The server, with which the readers are connected, keeps in memory the data from all of them and in this way the estimations become comparable. This is similar to the IDAS procedure. The difference is in the simultaneous recording of the reference standard with pre-irradiated dosimeters or with unknown sample and the presentation of results not as $I_{alanine}/D$ but as a ratio $I_{alanine}/I_{Standard})/D$ (*vide supra*). In this way the accuracy of the estimations of the different readers is increased. This sub-procedure has some disadvantages because the customers have to:
- subscribe to the NIST service;
- use only certain types of dosimeters and pre-irradiated standards recommended by NIST;
- obtain new EPR readers from a given producer;
- pay for each dose point.

Another problem is that this procedure will be usable in only some countries but not yet all over the world. And finally, it could result in some kind of monopoly in this area.

4. TRENDS IN THE FUTURE STUDIES ON SS/EPR DOSIMETERS

Up to now SS/EPR dosimeters have received wide application for high dose radiation. Alanine/EPR dosimeters are very successfully used for estimation of γ-ray doses in the region of 50-100 000 Gy. The recently published data, reviewed above, suggest increased accuracy of dose measurement and acceptable intercomparison of the results. The following directions could be expected to develop in the near future connected with:
- increasing the sensitivity of the method and especially for the low doses (0.5-10 Gy) used in human radiation therapy;
- searching for alternatives to alanine as a radiation sensitive material;
- increasing the EPR response of the radiation sensitive material with respect to high-LET radiation such as protons, neutrons, α- and β-rays as well as different heavy particles.

4.1 Estimations of Low Absorbed Dose Used in the Human Radiation Therapy by SS/EPR Dosimetry

Use of the alanine/EPR dosimeter was reported for the first time in 1984 for estimations of absorbed doses below 10 Gy[62] and there are several papers on this topic after that time.[63-66] Because the irradiation of humans must be carefully controlled, measurements must be very precise and applicability of alanine is under question.

Although alanine is a very suitable radiation sensitive material it has some disadvantages particularly for low dose dosimetry:
- Only its central line is used for dosimetric purposes. As mentioned before, the alanine EPR spectrum consists of five quasi-isotropic lines with intensity distribution 1:4:6:4:1. Using only the central line for dosimetric purposes we employ the fraction 6/16 of its radiation induced EPR response. On the other hand integration of the full EPR spectrum in order to get full EPR response is connected with some uncertainties because the spectrometer gain is high at low doses thus giving rise to noise and base line drift;
- Very often a background signal appears from alanine itself and from the binding material, as a result of high temperature and/or pressure treatment during the preparation of the dosimeters or from the spectrometer. Because the induced EPR response of alanine dosimeters is weak at low doses and overlaps with this background signal of the dosimeter,[67,68] computer-aided procedures for subtraction of the background signal from alanine response are used[40] to find the neat EPR response. Nevertheless, all stated problems still remain. Subtracting of the background signal is also connected with some uncertainties, especially at low doses.

4.2 Searching for New Radiation Sensitive Materials

There are several very promising studies on new radiation sensitive materials, for example on sugar,[69-72] acetates, phosphates and lactates,[73-75] ammonium tartrate,[76,77] anhydrous $MgSO_4$,[78,79] alkaline-earth metal dithionates, $MeS_2O_6 \cdot xH_2O$,[80-82] Li_2CO_3 and $CaSO_4$,[83] etc. The recent investigations have shown that some of them are 2 - 3 times more sensitive than alanine.[84] At equal EPR spectrometer settings the increased sensitivity is attributed mainly to more suitable EPR spectrum with narrow lines and small (or absent) hyperfine splitting.[84] It may be expected that screening of other materials will improve the sensitivity of SS/EPR dosimetry towards low ($D \leq 10$ Gy) absorbed doses.

Studies on some additives assuming the role of sensitizers for the radiation sensitive materials can also be considered as a promising direction for future studies.

Principally, dosimeters devoted to industrial purposes have to be cheap. However, this is not the case for dosimeters used for estimation of low doses (0.5-10 Gy). Because of the application of some expensive radiation sensitive materials for measurement of absorbed doses used in radiation therapy, approaches for their regeneration after a given cycle of irradiation can be considered as a promising direction for future work.

4.3 Recent Development of the Self-calibrated SS/EPR Dosimeters

Very recently a new self-calibrated dosimeter was constructed containing sugar (Fig. 4)[85] as radiation sensitive material. The characteristic feature is that the EPR spectrum of this radiation sensitive material has narrower spectral width thus avoiding overlapping with the Mn^{2+} lines. It can be expected that its detailed examination for low dose estimations will be realised in the near future.

Fig. 4. EPR spectrum of irradiated self-calibrated sugar dosimeter.

4.4 Recent Developments in the Independent Calibration of SS/EPR Dosimeters Using UV Spectroscopy

Irradiated sugar, or sucrose, which turns to brown color after irradiation in the solid state is reported[85] to keep its colour after dissolution in water with an absorption maximum at 37 400 cm^{-1}. The magnitude of this absorption is proportional to the absorbed dose of high-energy radiation. This feature opens a new way for independent calibration of the radiation source, the irradiated material and/or dosimeter itself.[85]

4.5 Instrumentation

Up to now EPR spectrometers or readers working in X-band were used for dosimetric purposes. These spectrometers have high sensitivity, depending on the dielectric constant of the sample, and are relatively expensive. The diameter of the dosimeter species used in these spectrometers is limited to maximum 5 mm. If they are in the form of pellets their length is between 3 and 10 mm, or if they have a rod shape the length is maximally 40 mm. The third option, film shaped dosimeters, have a width of maximum 5 - 6 mm and thickness less than 0.5 mm. Keeping in mind these limitations it may be assumed that other EPR spectrometers (or readers) working at lower frequency for example S- or L-band (3 or 1 GHz) will be more useful. Such spectrometers are cheaper and not so heavy because the magnetic field (obtained by iron magnet and/or with only Helmholtz coils) at which they work is lower, and this makes them portable. The loss of sensitivity, compared to the X-band, may be compensated by the increased sample dimensions, and especially of the dosimeter diameter.

5. IDENTIFICATION OF RADIATION PROCESSING OF FOODSTUFFS BY EPR

Radiation processing was advocated for sterilising of arterial[86] and bone grafts[87] in the beginning of 1950. In the next 20 - 30 years its application was extended to foodstuffs and now it is considered as a clean, inexpensive and effective method for sterilisation. However, recombination processes of the free radicals formed in the food during irradiation generate new, unknown chemical substances with unknown effects with regard to human health. For this reason, radiation processing, import and export of irradiated foodstuffs is forbidden in some countries. It is worthwhile noting that currently there is no evidence of any hazard for living organisms consuming irradiated foodstuffs. However, the simple fact of the presence of unknown substances in the irradiated foodstuffs justifies the necessity for control of radiation processing. Reviews,[88-90] as well as books[15,16] describing different methods used for identification of irradiated foodstuffs have been published. Finally, after many studies, the European Community Bureau of Reference concerted action[91] led to protocols, which were adopted in 1996 by the European Committee of Normalisation and published in the beginning of 1997.[8,9,92-94] Two of these Protocols[8,9] use EPR, the first one for cellulose and the second for bone containing foodstuffs. Following these Protocols the problem which has to be solved by EPR is on a qualitative level - whether an appropriate food sample has been irradiated or not. In the case of samples containing hard tissues qualitative estimation of the absorbed dose of γ-rays by EPR is unambiguous because the radiation generated signals are extremely stable with time. Moreover, it is possible to estimate the magnitude of the absorbed dose by the methods applied in post-irradiation dosimetry (*vide infra*). However this is not the case with cellulose containing foodstuffs because radiation generated EPR signals can be recorded only during a limited period of time. In view of this, some new approaches were recently developed for extension of the identification period of irradiated cellulose containing foodstuffs by EPR.

5.1 Detection of Irradiated Foods Containing Hard Tissues

The hard tissues are bio-minerals consisting of inorganic and organic part. The inorganic part of bones and teeth contains hydroxyapatite. The structure of hydroxyapatite, $Ca_{10}(PO_4)_6(OH)_2$, may be changed by replacing its cation and/or anion by other ions. For example Ca^{2+} can be replaced by Mg^{2+}, Fe^{2+} Mn^{2+}, Sr^{2+}, etc., PO_4^{3-} by divalent CO_3^{2-} or SO_4^{2-} and OH^- by CO_2^{2-}, monovalent anions or even neutral H_2O. Shells contain $CaCO_3$ as a mineral part in which Ca^{2+} can be replaced by other ions (*vide supra*) and CO_3^{2-} by SO_3^{2-}. The organic part of bones is the collagene and that of shells is conchiolin. Theeth consist of enamel, dentine and cement. Enamel contains mainly hydroxyapatite and a few percents of organic material.

Thus, impurities sometimes appearing in hard tissues may have an effect on the formation of paramagnetic species upon irradiation and increasee or decreasee of the EPR response due to their presence cannot be excluded. In most cases before irradiation there are no observable EPR signals in hard tissues such as meat bones[95-97] and sea[98-101], egg[101-104] and snail[105] shells. However, there are some exceptions, like the EPR spectrum in Fig. 5 of a sea

14. Recent development of EPR dosimetry

shell from the family *Ostrea SP* recorded before irradiation with significant contamination of Mn^{2+} ions[101] which does not interfere with the radiation induced EPR signal. (It is worth noting that there is no internationally accepted protocol for identification of previous radiation treatment of molluscs and shell containing foods.)

After bone irradiation an anisotropic EPR signal with $g_\parallel = 1.996$ and $g_\perp = 2.002$ appears due to radiation induced CO_2^- free radicals.[9,95-97] The signal is stable if the bone is kept cold, dried or boiled.[97] The bone EPR signal exhibits some angular dependence with respect to the positioning of the sample towards the magnetic field (Fig. 6)[97] but its presence in the bone piece is unambiguous evidence of previous radiation treatment of the meat.

The EPR signals appearing after irradiation of fish bones are not stable. They disappear shortly after the radiation processing.[106]

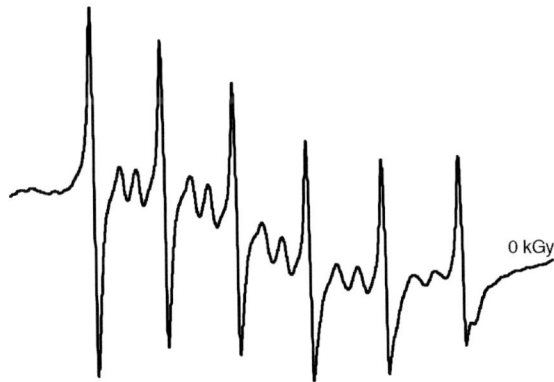

Fig. 5. EPR spectrum of sea shell of the family of *Ostrea SP* recorded before irradiation.

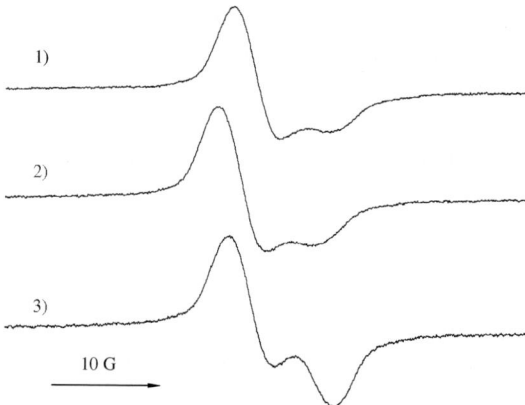

Fig. 6. EPR spectra of irradiated bone: 1) without any treatment, 2) dried and 3) boiled and dried (D = 8 kGy).

In calcified tissues of molluscs, snails (Fig. 7) and eggshells (Fig. 8), the radiation induced EPR signal is complex, suggesting three different EPR active species. They are characterised with the following parameters: $g_1 = 2.0055$, $g_2 = 2.003$, $g_3 = 2.002$, $g_z = 1.996$.[101-105] These EPR signals are attributed to SO_2^- (isotropic signal with $g = 2.0055$[107]) and SO_3^- (isotropic with $g = 2.003$[107,108] and anisotropic with $g_z = 1.996$, $g_1 = 2.002$ and $g_2 = 2.000$[109,110]) free radicals[111] and have long lifetimes. The presence of such EPR signals is considered as unambiguous evidence for previous radiation treatment.

14. Recent development of EPR dosimetry

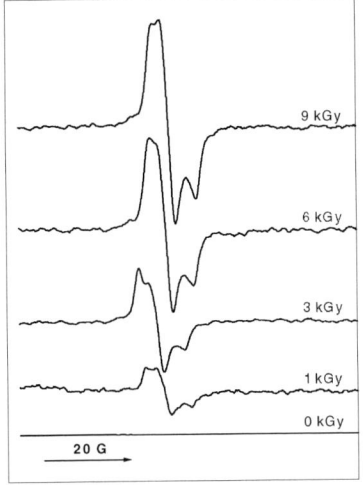

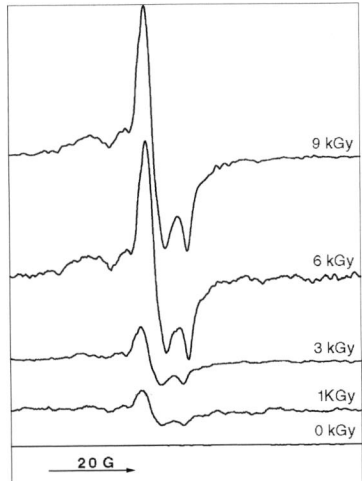

Fig. 7. A typical set of EPR spectra of sea snails before and after irradiation.

Fig. 8. A set of EPR spectra of egg shells before and after irradiation with doses between 1 and 9 kGy.

It is worth noting that EPR has a big advantage in identification of previously irradiated foods containing hard tissues since there is no need of special procedures for sample preparation, one measurement takes only 30 min and it is unambiguous. The recommended alternative method, gas chromatography,[92-93] is not useful in routine practice because it is complicated, time consuming (it needs c.a. 72 h) and expensive.[97]

5.2 Detection of Irradiated Cellulose Containing Foods

Before irradiation, cellulose containing food samples exhibit only a weak EPR signal with g = 2.0050 ± 0.0005 and line width of c.a. 0.6 mT (Fig. 9.1.).[112,113] After irradiation there is a significant increase in the amplitude of this signal with simultaneous appearance of a pair of weak lines on both sides (Fig. 9.2.).

These two satellite lines,[8] attributed to cellulose free radicals generated by radiation,[8,112] are considered to be unambiguous evidence of previous radiation processing. However, the applicability of this procedure is strongly limited by the lifetime of the radiation-induced free radicals. The main problem is that the satellite lines are relatively weak and disappear within a few weeks or months depending on storage conditions.[113,114] Therefore, after this period EPR cannot give any information about the radiation history of the sample under investigation. In such case thermoluminescence analysis[94] has to be used but it is not suitable for routine use because it is a time-consuming procedure - one measurement needs 72 h including one additional irradiation of the sample,[115] which complicates the problem.

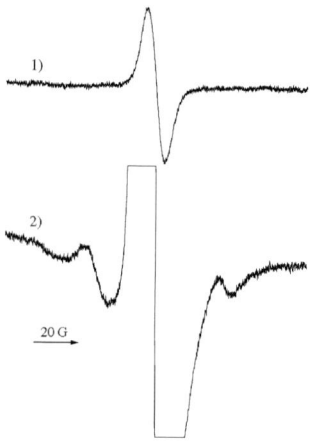

Fig. 9. EPR spectra of cellulose containing food: 1) before and 2) after irradiation.

5.3 New Approach for Extension the Period for Identification of Irradiated Cellulose Containing Foodstuffs

To overcome the time limit problem a new approach for extending the period of identification of irradiated cellulose containing foodstuffs using EPR was recently published.[113] It considers the intensity of the central line (g = 2.0050) which strongly increases after irradiation and remains enormously high for much longer periods of time than both radiation induced satellite lines. According to this approach, after the disappearance of the satellite lines, the sample is inserted into a special EPR quartz sample tube containing in the bottom a finger filled with Mn^{2+}/MgO[51,113] used as a reference standard (Fig. 10) and the EPR spectrum is recorded.

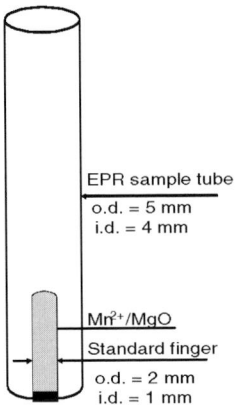

Fig. 10. Sketch of the EPR tube used for the simultaneous recording and processing of herb samples.

The intensity of the central line (I_{SA}) of the foodstuff is compared with that of a given Mn^{2+} line (I_{ST}) thus obtaining the ratio (I_{SA}/I_{ST}). Then the sample tube with the material under investigation is transferred to a standard laboratory drying unit equipped with a high precision ($\pm$ 0.1°) thermometer and is heated for one hour at 60°C. After the sample has reached room temperature the EPR spectrum is recorded again and the ratio of the intensities of the same lines (I_{SA}/I_{ST}) are compared. For previously irradiated samples there is ca. 50-70 % decrease of the ratio I_{SA}/I_{ST} and for non-irradiated samples the decrease of the ratio I_{SA}/I_{ST} is ca. 10 - 20% (Fig. 11). Therefore, more than 40-50% decrease of the ratio I_{SA}/I_{ST} after the described procedure is unambiguous evidence of previous radiation treatment of the sample under investigation. In this way the period of identification can be increased ca. 2 - 3 times.

The described method[113,114] is very easily performed but it is worth noting that it is strongly dependent on the conditions under which the samples of the foodstuff under study are kept after the radiation processing. The most critical factors are the humidity and temperature of the room in which they are stored and also the packing material - paper, polyethylene, etc. In spite of this disadvantage the recommendation is first to use the EPR procedure and only if the results are not clear to use thermoluminescence.

In the case of fruits having a shell or containing stones the characteristic radiation induced EPR signal (Fig. 12) remains unchanged for more than one year.[114]

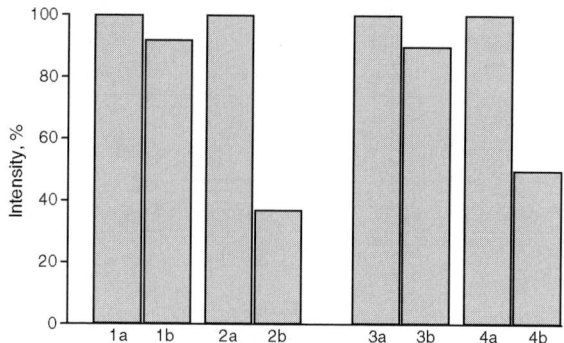

Fig. 11. Ratio of intensity of the sample vs. the intensity of Mn^{2+} (1) non-irradiated white pepper, (2) irradiated white pepper, (3) non-irradiated hot paprika, (4) irradiated hot paprika, (a) before and (b) after heating.

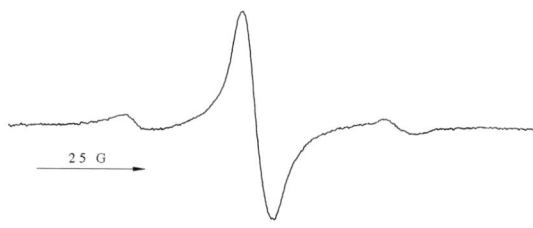

Fig. 12. EPR spectra of shelled fruit one year after irradiation.

6. DETECTION OF PHARMACEUTICALS STERILIZED BY HIGH ENERGY RADIATION

In the last years ionizing radiation has attracted increasing interest as a method for sterilization of medical devices as well as for improvement of the hygienic quality of pharmaceuticals.[116-119] The major advantage of radio sterilization is that it is carried out in their final packing and keeps them sterile until their use. It is a preferable method, especially in the case of heat-sensitive drugs. On the other hand, apart from the microbiological

14. Recent development of EPR dosimetry

aspect of sterilization by irradiation the problems with the radiolytic products and their pharmacological and toxicological action should be considered. According to the regulations of the European Community the radiation treatment of drugs is accepted only in cases when no toxicological hazard is possible.[120] At the moment the maximal allowable dose in the case of medical devices and drug sterilization is 25 kGy. To prevent the unauthorized and uncontrolled use of radiation processing a suitable method for distinguishing between irradiated and non-irradiated drugs is needed. The studies performed on some irradiated drugs have shown the applicability of EPR spectroscopy as a method for detection.

Intensive studies on the ionizing radiation effect on cephalosporins have been done, because of their susceptibility to degradation and their sensitivity to irradiation.[121-124] On the other hand cephalosporins are thermosensitive and can be regarded as potential candidates for non-authorized radiation treatment.

Investigation on cephradine has shown the non-feasibility of radiation sterilisation of this substance because of the formation of foreign products and long-lived free radicals.[124] In the case of radiation treated powder samples of cefuroxime and cefotaxime a composite EPR signal which is a superposition of a singlet and a weak multiplet, mostly hidden by the singlet, has been detected.[123] The different dependence of the singlet and multiplet lines on the microwave power suggests the presence of two different paramagnetic centres. The studies on the fading of the radiation induced free radicals have shown that they are more stable in cefotaxime than in cefuroxime. After 150 days of storage at low temperature and darkness the fading of the intensity of the singlet line is 30 % for cefotaxime and 70% for cefuroxime.

Extended studies on the effect of radiosterilization on ceftazidime carried out at 4.2-295 K have been done using EPR. Three types of free radicals were detected. The first one giving a septet signal in the EPR spectrum and decaying at 230 K has been attributed to the $\bullet C(CH_3)_2COOH$ radical. The second one with a triplet spectrum and decay at 293 K has been assigned to the iminoxyl radical ($>C=N-O\bullet$). The third one presents a broad singlet line and remains unchanged at 295 K.[121] It was also found that the yields of free radicals in irradiated ceftazidime increase linearly with dose of irradiation up to 10 kGy both at 77 K and 295 K. Half of the free radicals giving the broad singlet line decay upon storage at 277 K after 50 days. Nevertheless, after 160 days of storage at the same conditions it is still possible to observe the broad singlet EPR spectrum of irradiated ceftazidime.[121]

The EPR spectrum of irradiated ampicilin consists of a broad doublet signal.[122] The shape of the spectra of ampicilin irradiated with 12.5 kGy does not change upon storage at 277 K for 140 days but the intensity of the lines decreases to ca. 60 % of the initial intensity recorded immediately after the irradiation. The radical yield measured just after the irradiation at 295 K increases linearly with dose up to 12.5 kGy and slightly deviates from linearity above 12.5 kGy.

The initial results started an extensive study on a large number of irradiated antibiotics belonging to the group of cephalosporins.[125] Radiation induced EPR signals have been detected for 12 out of 13 cephalosporines irradiated with a 25 kGy dose. Only in the case of non-irradiated cefaclor a weak signal of the same shape as in the irradiated one has been recorded. The studies on the fading of the EPR signal have shown that the stability of the created free radicals in all investigated samples is comparable to the shelf life of the antibiotics when stored under the proper conditions.

Recently an EPR study on five antibiotics belonging to the groups of cephalosporins and penicillins has been carried out.[126] The influence of irradiation and storage conditions on the concentration of the radiation induced free radicals has been investigated. The samples irradiated at 77 K as well as at room temperature have shown complex EPR spectra. The influence of different factors such as: radiation dose, microwave power, temperature and storage time vary from one EPR signal to another indicating the presence of mixture of radicals in every irradiated antibiotic.

EPR spectroscopy has also been applied to study the effect of gamma rays on three nitronimidazoles.[127] The nonirradiated samples did not show any signal. After irradiation with gamma rays at a dose of 25 kGy stable paramagnetic centers were detected only in ornidazole and metronidazole samples. The intensity and the shape for both substances were similar. It was found that a significant portion of the created free radicals decayed in several days after irradiation, whereas 35 and 10 % of the initial radicals respectively in metronidazole and ornidazole could be detected after 135 days of storage.

High performance liquid chromatography (HPLC) and EPR spectrometry have been used to study the degradation of theodrenaline after gamma radiation treatment.[128] The effect of storage on the free radical concentration was studied.

Free radicals induced in drugs and excipients by radiation and mechanical treatments have been studied by EPR.[129] Special attention is focused on the use of this method as well as other methods such as thermoluminescence and gas chromatography as proof of radiation treatment of drugs.

Different substances belonging to the groups of cytostatic, anti-carcinogenic, blood circulation regulating and other drugs have been investigated before and after irradiation with high-energy irradiation. Weak EPR signals were detected in 3 of 15 non-irradiated polycrystalline samples. The EPR measurements have shown that the radiation induced radicals in all examined drugs can be detected after 4 and 8 weeks of storage. The authors also made some suggestions about the structure of the observed radical species.[130]

7. EMERGENCY DOSIMETRY

Nuclear technologies were developed during the World War II but the risk of their use was recognised many years before. Now they are considered safe and secure operations. Many different dosimetric techniques have been developed and used for control purposes during normal work as well as in the cases of accidents. The characteristic feature of an accident is that the dose of ionising radiation absorbed by the victims is unknown and as a rule it is over the limits of the personal monitors used every day at normal conditions. In these cases the method of EPR dosimetry may be, and is, helpful for reconstruction of the absorbed dose.

The commonly accepted procedure for estimation of the absorbed dose by EPR in the cases of post-radiation dosimetry is the method of the "additional" dose. It is based on the assumption that the EPR response is zero before irradiation.[11,131] The procedure is as follows: The EPR response (R_x) due to the unknown, accidental, radiation dose (D_x) absorbed by the sample, is measured at the beginning. After that, without any changes or treatments, the sample is additionally irradiated several (j) times with some well known doses (ΔD_j, $x \leq j \leq i$) and after every irradiation the new EPR response ($R_j = R_x + \Sigma \Delta R_j$) is estimated. Then, the obtained data points, D_x and all ΔD_j, are plotted on the abscissa axis in Gy, starting with D_x, in such a way that every incremental difference ($\Delta D_{j+1} - \Delta D_j$) may be used for calibration of the abcissa. The corresponding magnitudes of the EPR response, $R_j = R_x + \Sigma \Delta R_j$ at each data point D_x and ΔD_j, are plotted on the ordinate in arbitrary units (Fig. 13).

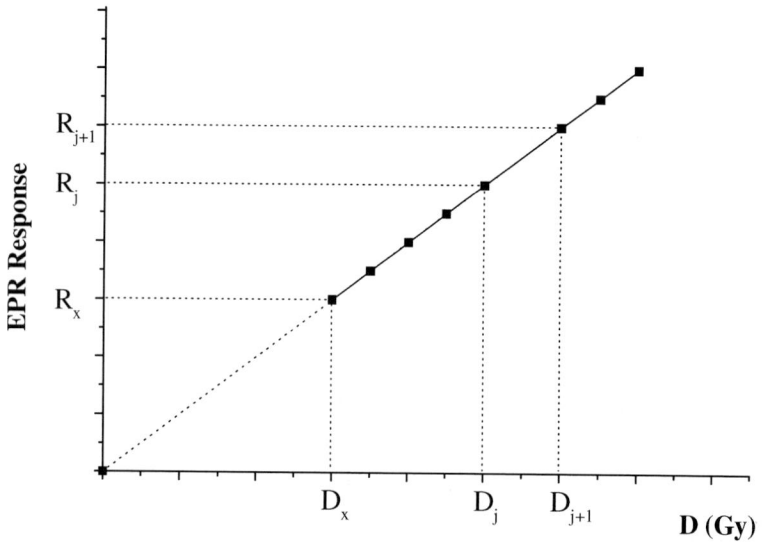

Fig. 13. Graph representing the method of "additional" dose, used in post-radiation dosimetry.

Initially it is not known where the origin of the co-ordinate system $R = f(D)$ is positioned. The extrapolation of the data points of R to the zero ordinate (using the least-square fitting method) gives the origin of the co-ordinate system $R = f(D)$. Finally, the difference between it and the point marked with D_x gives the magnitude of accidental radiation dose D_x. In general, there are two types of such reconstruction graphs – linear or approximately linear at the beginning and with saturation at high doses. This is the reason that the number of data points for the additional irradiations (j) must be between 5 and 10.

The application of EPR for accidental irradiation, especially for X- or gamma rays, was proposed for the first time in 1965.[95,96] After that many studies were reported mainly on the survivors of the A-bomb radiation at Hiroshima and Nagasaki by using the EPR response of shell buttons, tooth enamel and quartz grains.[11] Other accidents, with a local character, were also investigated using EPR.[132,133]

In 1986 the Chernobyl disaster appeared and after that other local accidents were also described in the literature (for example [134-138]). It was a turning point for the acceleration of emergency dosimetry. Thus after Chernobyl the application of EPR in the field of emergency dosimetry became extremely useful. Now there are numerous investigations devoted to the emergency EPR dosimetry. Many different materials as tooth enamel,[11,139-142] bones[143] shell buttons,[11,144,145] clothing,[145-148] some fabrics,[145-147] fingernails and hair,[145,146,149-152] pharmaceuticals,[11,132,146] sugar,[69-72,146] quartz grains in the rocks or house building materials,[11,153-155] etc. have been tested as EPR post-radiation dosimeters. All these materials have potential use in emergency cases but among them shell buttons, sugar and tooth enamel exhibit long-lived radiation induced EPR signals. For clothing the radiation induced EPR signal disappears very shortly after irradiation, whereas the EPR response of fingernails in addition to being short lived is dependent on the particle size.[151,152] Recently most of the studies are concerned mainly with tooth enamel. This is because the studies have shown that tooth enamel is the only living tissue which retains indefinitely its radiation history, and EPR can record a radiation signal down to 10 cGy. Moreover, the EPR active defects in tooth enamel from the extracted teeth exhibit extremely long lifetimes, calculated to be up to 10^7 years[158] and may be used as personal cumulative radiation dosimeter for the staff working at nuclear facilities. Thus, tooth enamel has been used for estimation of the radiation doses from individuals exposed to A-bomb radiation at Hiroshima and Nagasaki,[11] in the Chernobyl accident,[11,159-162] at the uranium processing plant Mayak in the USSR[163] and in the Ural region[164]. Extracted and irradiated *in vitro* bone (grafts) has a long lifetime of the radiation induced EPR signal, but after implantation in living organisms the signal disappears within a few months.[156] Usually the investigations of the accidental cases are performed immediately after it happened. So, the bone EPR dosimetry is fully usable (the problem is only that it is invasive). Recently such an effect was also reported for the tooth EPR response.[157] In addition, sugar samples collected from different places around the accident site have been used for dose examination at Chernobyl[165] and the JCO uranium facility in Japan[135].

The problem of dose estimation by using tooth enamel appears to be complicated.[166] There are several reasons for this - the EPR signal of an irradiated tooth consists of two main overlapping components, that of enamel (CO_2^- radical with $g_z = 1.9975$ and $g_\perp = 2.0018$) and a background signal ($g = 2.0045$) which is due to the organic part of the tooth.[159] The enamel exhibits crystalline structure[167] and for EPR studies it is necessary that it be crushed in very small particles in order to avoid an angular dependence of the signal. A complicating factor is that the EPR response depends on the mechanical treatment before investigation.[168,169] Several different procedures have been tested to obtain the true radiation induced signal - computer subtraction,[170,171] microwave saturation of the background signal,[172] and chemical treatment.[173,174] Recently, review articles have appeared[175-177] and results from an organised intercomparison trial were reported.[178,179] They both indicate some progress but some problems still remain. And finally all studies performed on bone and tooth are *in vitro* i.e. the samples have to be extracted from the living human body.

7.1 *In Vivo* Emergency EPR Dosimetry

Recently, estimations of the absorbed dose *in vivo*, *i.e.* without extracting the teeth from the patient, was carried out using the recently developed portable EPR reader operating at X band (10 GHz.).[11,180-182] Its specific features are:
- special, new type of cavity permitting contact of the front tooth with it;
- special configuration of the magnet.

   This method is very attractive but has some disadvantages:
- Due to the fact that the measurements are *in vivo* additional re-irradiation cannot be used in order to obtain the real EPR dose response;
- A calibration graph cannot be prepared;
- The tooth anisotropy must be taken into consideration.

Despite all these considerations this is currently the most promising method for estimation of the absorbed dose *in vivo*. Instead of the "method of additional dose" it is possible to use a calibration graph prepared on the basis of wide collection of data *in vitro*. In this case dose measurements over 0.5 Gy can be performed with sufficient accuracy. Thus, we look forward to future developments in this field.

   More recently[183] a low frequency EPR spectrometer operating in the L-band (1.2 GHz) was reported to be useful for *in vivo* measurement of the absorbed dose by a tooth. It is estimated that the current low limit for detecting radiation doses in human teeth *in vivo* is 0.5-1.0 Gy. This is sufficient for identification of a person who has been exposed to life threatening dose of ionising radiation.

14. Recent development of EPR dosimetry   593

8. CONCLUSIONS

The present paper describes some of the recent developments in the field of EPR dosimetry. Such studies are very promising and can be expected to extend to many other subjects in the future. It is worth noting that prospects for SS/EPR dosimetry in the future are good even if the dream of "Green" movements in the world to close all facilities operating with radioactive materials (nuclear power stations, irradiation facilities, *etc.*) is realised since the waste radioactive materials will remain along with the need of EPR dosimetry for control purposes. In addition, there are still no alternatives to radiation therapy which will continue to serve the health needs of a large number of people.

9. REFERENCES

1. E. Zavoisky, *J. Phys.(USSR)* **9** (1945) 211.
2. K. Mehta, "*Report of the second research co-ordination meeting (RCM) for the co-ordination research project (CRP E2 40 06) on Characterisation and Evaluation of high-dose dosimetry techniques for quality assurance in radiation proseccing*", SSDL Newsletter No. 39, IAEA, Vienna (1998) 19.
3. K. Mehta, High-dose dosimetry programme of the IAEA, in "*Techniques for high dose dosimetry in industry, agriculture and medicine*", Extended Synopses of Symposium held in Vienna, November 2-5 (1998) paper # IAEA-SM-356/R2
4. K. Mehta, R. Girzikowski, Alanine-ESR dosimetry for radiotherapy. IAEA Experience, *Appl. Radiat. Isot.* **47** (1996) 1189.
5. K. Mehta, R. Girzikowsky, IAEA reference dosimeter: Alanine-ESR, in "*Techniques for high dose dosimetry in industry, agriculture and medicine*", Proc. Symp., IAEA-TECDOC-1070, 1999, 299.
6. M. Desrosiers at al., Alanine dosimetry at the NIST, *Intl. Conf. Biodosimetry and 5th Intl. Sym. ESR Dosimetry and Applications*, Moscow/Obninsk, Russia, June 22 1998. Final Programme and Book of Abstracts, 149.
7. P. H. G. Sharpe, J. P. Sephton, Alanine dosimetry at NPL - the development of a mailed reference dosimetry service at radiotherapy dose level, in *Techniques for high dose dosimetry in industry, agriculture and medicine*, Proc. Symp., IAEA-TECDOC-1070, 1999, p. 299.
8. CEN Protocol EN 1787(1997), *Determination of irradiated food containing cellulose: analysis by EPR*.
9. CEN Protocol EN 1786 (1997), *Detection of irradiated food containing bone: analysis by electron paramagnetic resonance*.
10. M. Ikeya (Ed.) "*EPR Dating and Dosimetry*" Ionics Publ., Tokyo, Japan, 1985.
11. M. Ikeya, "*New Applications of Electron Paramagnetic Resonance, Dating, Dosimetry and Microscopy*", World Sci., Singapore 1993.
12. D. F. Regulla, EPR dosimetry - present and future, in *Techniques for high dose dosimetry in industry, agriculture and medicine*, Proc. Symp., IAEA-TECDOC-1070, 1999, p. 171.

13. International Conference on "EPR Dosimetry", organized regulary in different countries. Usually their Proceedings appear in *Appl. Radiat. Isot.*
14. International Workshops on "*Electron Magnetic Resonance of Disordered Systems*" *(EMARDIS)*. Meetings organised biannually by the Bulgarian EPR Society.
15. D. E. Johnson, M. H. Stevenson (Eds.) *Food Irradiation and the Chemist*, Royal Soc. Chem., London, 1990.
16. C. H. McMurray, E. M. Stewart, R. Gray, J. Pearce (Eds.) *Detection Methods for Irradiated Foods*. Current Status., Royal Soc. Chemistry, London, 1996.
17. N. D. Yordanov, *Appl. Mag. Res.* **6** (1994) 241.
18. N. D. Yordanov, M. Ivanova, *Appl. Mag. Res.* **6** (1994) 333.
19. H. Shields, P. J. Hamrick Jr., C. Smith and Y. Haven, *J. Chem. Phys.* **58** (1973) 3420.
20. I. Miyagawa, N.Tamura, J. W. Cook, *J. Chem. Phys.* **51** (1969) 3520.
21. E. A. Friday, I. Miyagawa, *J. Chem. Phys.* **55** (1971) 3589.
22. P.-O. Samskog, G. Nilson, A. Lund and T. Gillbro, *J. Chem. Phys.* **84** (1980) 2819.
23. V. Yu. Nagy, M. F. Desrosiers, *Appl. Radiat. Isot.* **47** (1996) 789.
24. E. Sagstuen, E. O. Hole, S. R. Haugedal, A. Lund, O. I. Eid, R. Erickson, *Nucleonica* **42** (1997) 85.
25. E. Sagstuen, E. O. Hole, S. R. Haugedal, W. H. Nelson, *J. Phys. Chem.* **101** (1997) 9763.
26. S. Prydz and T. Henriksen, *Acta Chem. Scand.* **17** (1961) 791.
27. T. Henrikson, T. Sanner, A. Pihl, *Radiat. Res.* **18** (1963) 147.
28. P. P. Panta, G. Strzelczak-Burlinska, Z. Tomasinski, *Appl. Radiat. Isot.* **40** (1989) 971.
29. D. F. Regulla, U. Defner, *Appl. Radiat. Isot.* **33** (1982) 1101.
30. K. Nakagawa, S. S. Eaton, G. R. Eaton, *Appl. Radiat. Isot.* **44** (1993) 73.
31. Z. Stuglik, J. Sadlo, *Appl. Radiat. Isot.* **47** (1996) 1219.
32. K. Katsamura, Y. Tabata, in *EPR Dating and Dosimetry (M. Ikeya, Ed.)* Ionics, Tokyo, (1985) p. 415.
33. Y. Katsumura, Y. Tabata, T. Srguchi, N. Morishita, T. Kolima, *Radiat. Phys. Chem.* **28** (1986) 337.
34. S. Galindo, F. Urena-Nunez, *Radiat. Res.* **133** (1993) 387.
35. B. Ciesielski, L. Wielopolski, *Radiat. Res.* **144** (1995) 59.
36. F. Uena-Nunez, S. Galindo, J. Azorin, *Appl. Radiat. Isot.* **49** (1998) 1657.
37. F. Uena-Nunez, S. Galindo, J. Azorin, *Appl. Radiat. Isot.* **50** (1999) 763.
38. N. D. Yordanov, B. Genova, *Analyt. Chim. Acta* **353** (1997) 99.
39. W. W. Bradshaw, D. G. Cadena, G. W. Crawford, H. A. Spetzler, *Radiat. Res.* **17** (1962) 11.
40. D. F. Regulla, U. Deffner, *Intl. J. Appl. Radiat. Isot.* **33** (1982) 1101.
41. T. Kojima, Y. Haruyama, H. Tachibana, R. Tanaka, J. Okamoto, H. Hara, Y. Yamamoto, *Appl. Radiat. Isot.* **44** (1993) 361.
42. E. H. Hasskell, R. B. Hayes, G. H. Kenner, *Radiat. Prot. Dosim.* **77** (1998) 43.
43. R. B. Hayes, E. H. Haskell, A. Wieser, A. A. Romanyukha, B. L. Hardy, F. K. Barrus, *Nucl. Instr. Methods Phys. Res.* A. (2000) 453.
44. V. Nagy, *Appl. Radiat. Isot.* **52** (2000) 1039.
45. V. Nagy, O. F. Sleptchonok, M. F. Desrosiers, R. T. Weber, A. H. Heiss, *Radiat. Phys. Chem.* **59** (2000) 429.
46. W. J. Nam, D. F. Regulla, *Appl. Radiat. Isot.* **40** (1989) 953.
47. K. Mehta, High dose dosimetry programme of the Agency, in *Techniques for high dose dosimetry in industry, agriculture and medicine*, Proc. Symp., IAEA-TECDOC-1070, 1999, p. 11.
48. N. D. Yordanov, V. Gancheva, *J. Radioanalyt. Nuclear Chem.* **240** (1999) 619.

14. Recent development of EPR dosimetry

49. N. D. Yordanov, V. Gancheva, *J. Radioanalyt. Nuclear Chem.* **245** (2000) 323.
50. BG Pat. (1997). #344Y1.
51. N. D. Yordanov, V. Gancheva, V. A. Pelova, *J. Radioanalyt. Nuclear Chem.* **240** (1999) 215.
52. V. N. Lyniov, in *Electron Magnetic Resonance of Disordered Systems*, N. D. Yordanov (Ed.),, World Scientific, Singapore, 1991, p.53.
53. D. Maier, D. Schmalbein, *Appl. Radiat. Isot.* 44 (1993) 345.
54. D. Maier, *BRUKER Report* **140** (1994) 22.
55. M. Ikeya, M. Furusawa, *Appl. Radiat. Isot.* 40 (1989) 845.
56. T. Kojima, R. Tanaka, *Apll. Radiat. Isot.* 40 (1989) 851.
57. A. Nakanishi, N. Sagawara, A. Fuse, *Appl. Radiat. Isot.* **44** (1993) 357.
58. T. Herrling, N. Groth, F. Klein, J. Rehberg, *Spectrochim. Acta* (A) **56** (2000) 417.
59. C. Yamanaka, M. Ikeya, K. Meguro, A. Nakanashi, *Radiat. Meas.* **18** (1991) 279.
60. Sumitomo Special Metals Co. Ltd., Japan, *EPR Newsletter* **4** (1992) 10.
61. V. Nagy ,M. F. Desrosoers (2001) Private communication.
62. A. Bartolotta, F. L. Indovina, S. Onori, A. Rosati, *Radiat. Prot. Dosim.* **9** (1984) 277.
63. O. Bugay, V. Bartchuk, S. Kolesnik, M. Mazin, H. Gaponenko, in *"Techniques for high dose dosimetry in industry, agriculture and medicine"*, Proc. Symposium held in Vienna, November 2-5 (1998), p. 191.
64. P. H. G. Sharpe, J. P. Sephton, *Appl. Radiat. Isot.* **47** (1996) 1171.
65. P. H. G. Sharpe, J. P. Sephton, in *"Techniques for high dose dosimetry in industry, agriculture and medicine"*, Proc. Symposium held in Vienna, November 2-5 (1998), p. 183.
66. F. Chen, D. T. Covas, O. Baffa, *Appl. Radiat. Isot.* **55** (2001) 13.
67. O. Bugai, V. Bartchuk, S. Kolesnik, M. Mazin, H. Gaponenko, Alanine EPR dosimetry of therapeutic irradiators, in *Techniques for high dose dosimetry in industry, agriculture and medicine*, Proc. Symp., IAEA-TECDOC-1070, 1999, p. 191.
68. P. H. G. Sharpe, J. P. Sephton, Alanine dosimetry at NPL - the development of a mailed reference dosimetry service at radiotherapy dose level, in *Techniques for high dose dosimetry in industry, agriculture and medicine*, Proc. Symp., IAEA-TECDOC-1070, 1999, p. 299.
69. T. Nakajima, *Health. Phys.* **55** (1988) 951.
70. T. Nakajima, *Brit. J. Radiology* **62** (1989) 148.
71. T. Nakajima, T. Ohtsuki, *Appl. Radiat. Isot.* **41** (1990) 359.
72. T. Nakajima, *Appl. Radiat. Isot.* **46** (1995) 819.
73. G. M. Hassan, M. Ikeya, S. Toyoda. *Appl. Radiat. Isot.* **49** (1998) 823.
74. G. M. Hassan, M. Ikeya, *Appl. Radiat. Isot.* **52** (2000) 1247.
75. M. Ikeya, G. M. Hassan, H. Sasaoka, Y. Kinoshita, S. Takaki, C. Yamanaka, *Appl. Radiat. Isot.* **52** (2000) 1209.
76. S. K. Olson, S. Bagherian, E. Lund, G. Alm Carlsson, A. Lund, *Appl. Radiat. Isot.* **50** (1999) 955.
77. S. Olson, E. Lund, A. Lund, *Appl. Radiat. Isot.* **52** (2000) 1235.
78. P. N. Keizer, J. R. Morton, K. F. Preston, *J. Chem. Soc. Faraday Soc.* **87** (1991) 3147.
79. J. R. Morton, F. J. Ahlers, C. C. J. Schneider, *Radiat. Prot. Dosim.* **47** (1993) 263.
80. S. E. Bogushevich, V. Makatin, A. K. Potapovich, I. Ugolev, *Zh. Prikl. Spektrosk.* **55** (1991) 613.
81. S. E. Bogushevich, I. I. Ugolev, A. K. Potapovich, *Zh. Prikl. Spektrosk.* **63** (1996) 258.
82. S. E. Bogushevich, I. I. Ugolev, *Appl. Radiat. Isot.* **52** (2000) 1217.
83. S. Murali, V. Natarajan, R. Venkataramani, Pusharja, M. D. Sastry, *Appl. Radiat. Isot.* **55** (2001) 253.

84. A. Lund, S. Olson, M. Bonora, E. Lund, *Spectrochim. Acta* (A) **58** (2002) 1301.
85. N. D. Yordanov, V. Gancheva, E. Georgieva, *Radiat. Phys. Chem.* **65** (2002) 269.
86. I. A. Meeker, R. E. Gross, *Surgery* **30** (1951) 19.
87. T. C. Turner, C. A. L. Bassett, J. W. Pate, P. N. Sawyer, *J. Bone Joint Surg.* **38A** (1956) 862.
88. N. J. F. Dood, A. J. Swallow, F. J. Ley, *Radiat. Phys. Chem.* **26** (1985) 451.
89. H. Delincee, D. A. E. Ehermann, *Radiat. Phys. Chem.* **34** (1989) 877.
90. K. W. Boegl, *Appl. Radiat. Isot.* **40** (1989) 1203.
91. J. Raffi, H. Delinecee, E. Marchioni C. Hasselmann, A.-M. Sjoberg, M. Leonardi, M. Kent, K.-W. Bogl, G. Schreiber, M. H. Stevenson, W. Meier, CEC, BCR, Luxemburg, 1994 (EUR 15261 EN).
92. CEN Protocol EN 1784. *Detection of irradiated food containing fat: gas chromatography (GC) analysis of hydrocarbons.*
93. CEN Protocol EN 1785. *Detection of irradiated food containing fat: gas chromatography (GC) analysis of 2-alkylcyclobutanones.*
94. CEN Protocol EN 1788. *Detection of irradiated food containing silicate materials: analyzis by thermoluminescence, TL.*
95. H. M. Swartz, *Radiat. Res.* **24** (1965) 579.
96. J. M. Brady, N. O. Aarestad, H. M. Swartz, *Health Phys.* **15** (1968) 43.
97. N. D. Yordanov, V. Gancheva, R. Tarandjiiska, R. Velkova, L Kulieva, B. Damyanova, S. Popov, *Spectrochim. Acta* (A) **54** (1998) 2421.
98. J. Raffi, P. Stocker, *Appl. Magn. Res.* **10** (1996) 357.
99. E. M. Stewart, *Appl. Magn. Res.* **10** (1996) 375.
100. B. Ziegelmann, K. W. Boegel, N. D. Schreiber, *Radiat. Phys. Chem.* **54** (1999) 413.
101. N. D. Yordanov, B. Mladenova, *Bull. Chem. Technol. Macedonia* **19** (2000) 171.
102. M. F. Desrosiers, *Appl. Radiat. Isot.* **42** (1991) 617.
103. D. F. Regulla, H. Y. Goekus, A. Vogenauer, A. Wisser, *Appl. Radiat. Isot.* **45** (1994) 371.
104. C. Corredos, J. Diaz, J. M. Diaz, H. A. Farach, C. P: Poole, *Jr., Appl. Magn. Res.* **9** (1995) 613.
105. N. D. Yordanov, B. Mladenova, *Radiat. Phys. Chem.* **60** (2001) 191.
106. F. Abdel-Rehim, A. A. Bester, H. A. Al-Kahtan, H. M. Abu-Tarboush, *Appl. Radiat. Isot.* **48** (1997) 241.
107. M. Barabas, *Nuclear Tracks* **20** (1992) 453.
108. A. Kai, T. Miki, M. Ikeya, *Radiat. Phys. Chem.* **40** (1992) 469.
109. O. Katzenger, R, Debuyst, P. Dwcanniere, F, Bejehet, D. Apes, M, Barabas, *Appl. Radiat. Isot.* **40** (1989) 1113.
110. Ref. 11, p. 185.
111. P. W.Atkins, A. Horsefield, M. C. R. Symons, *J. Chem. Soc.* (1964) 5220.
112. J. Raffi, J. P. L. Angel, *Radiat. Phys. Chem.* **34** (1989) 891.
113. N. D. Yordanov, V. Gancheva, *Appl. Radiat. Isot.* **52** (2000) 195.
114. J. Raffi, N. D. Yordanov, S. Chabane, L. Douifi, V. Gancheva, S. Ivanova , *Spectrochim. Acta* (A) **56** (2000) 409.
115. N. D. Yordanov, V. Gancheva, M. Radicheva, B. Hristova, M. Guelev, O. Penchev, *Spectrochim. Acta* (A) **54** (1998) 2413.
116. W. Bögl, *Radiat. Phys. Chem.* **25** (1985) 425.
117. G. P. Jacobs, *Radiat. Phys Chem.* **26** (1985) 133.
118. N. G. S. Gopal, K. M. Patel, G. Sharma, H. L. Bhalla, A. Wills., N. Hilmy, *Radiat. Phys. Chem.* **32** (1988) 619.
119. F. Zeegers, A. S. Crucq, M. Gibella, B. Tilquin, *J. Chim. Phys.* **90** (1993) 1029.

120. Commission of the European Communities, 1992. CPMP working party on quality of medicinal products. Ionizing irradiation in the manufacture of medicinal products. III Suppl. N.2.
121. T. Miyazaki, T. Kaneko, T. Yoshimura, A.-S. Crucq, B. Tilquin, *J. Pharm. Sci.* **83** (1994) 68.
122. T. Miyazaki, J. Arai, T. Kaneko, K. Yamamoto, M. Gibella, B. Tilquin,. *J. Pharm. Sci.* **83** (1994) 1643.
123. E. Ciranni Signoretti, L. Valvo, P. Fattibene, S. Onori., M. Pantaloni, *Drug Dev. Ind. Pharm.* **20** (1994) 2493.
124. E. C. Signoretti., S. Onori, L. Valvo, P. Fattibene, A. L. Savella, C. DeSena, S. Alimonti, *Drug Dev. Ind. Pharm.* **19** (1993) 1693.
125. S. Onori, M. Pantaloni, P. Fattibene, E. C. Signoretti, L. Valvo, M. Santucci, *Appl. Radiat. Isot.* **47** (1996) 1569.
126. M. Gibella, A.-S. Crucq, B. Tilquin, P. Stocker, G. Lesgards, J. Raffi, *Radiat. Phys. Chem.* **58** (2000) 69.
127. J. L. Duroux, J. P. Basly, B. Penicaut, M. Bernard, *Appl. Radiat. Isot.* **47** (1996) 1565.
128. J. P. Basly, I. Longy, M. Bernard, *Int. J. Pharm.* **152** (1997) 201.
129. J. Raffi, S. Gelly, L. Barral, F. Burger, P. Piccerelle, P. Prinderre, M. Baron, A. Chamayou, *Spectrochim. Acta (A)* **58** (2002) 1313.
130. H. B. Ambroz, E. M. Kornacka, B. Marciniec, M. Ogrodowczyk, G. K. Przybytniak, *Radiat. Phys. Chem.* **58** (2000) 357.
131. R. Gruen, Die ESR-Altersbestimmungs-methode, Springer, Berlin, 1989.
132. E. Sagstuen, H. Theisen, T. Henriksen, *Health Phys.* **45** (1983) 961.
133. D. F. Regulla, U. Deffner, *Appl. Radiat. Isot.* **40** (1989) 1039.
134. M. F. Desrosiers, *Health Phys.* **61** (1991) 859.
135. K. Shirashi, S. Kimura, H. Yonehara, J. Takada, M. Ishikawa, Y. Igarashi, M. Aoyama, K. Komura, T. Nakajima, *Adv. EPR Appl.* **16** (2000) 9.
136. D. A. Schauer, B. M. Coursey, C. E. Dick, W. L. Mclaughlin, J. M. Puhl, M. F. Desrosiers, A. D. Jacobson, *Health Phys.* **65** (1993) 131.
137. A. M. Rossi, C.C. Wafcheck, E. E. de Jesus, F. Pelegrini, *App. Rad. Isot.* **52** (2000) 1297.
138. G. Hütt, L. Brodski, V. Polyakov, *Appl. Radiat. Isot.* **47** (1996) 1329.
139. B. Pass, J. E. Aldrich, *Med. Phys.* **12** (1985) 305.
140. J. E. Aldrich, B. Pass, *Radiat. Prot. Dosim.* **17** (1986) 175.
141. M. Ikeya, J. Miyajima, S. Okajima, *Jpn. J. Appl. Phys.* **23** (1984) L697.
142. M. Hoshi, S. Sawada, M, Ikeya, in *ESR Dating and Dosimetry, Ionica*, Tokyo, 1985, p. 407.
143. W. Stachowich, J. Michalik, A. Dziedzia-Goclawska, R. Ostrowski, *Nucleonika* **19** (1974) 843.
144. T. Nakajima, *Health Phys.* **53** (1987) 405.
145. T. Nakajima, *Int. J. Appl. Radiat. Isot.* **33** (1982) 1077.
146. G. Delgano, J. D. McClymond, *Appl. Radiat. Isot.* **40** (1989) 1013.
147. V. Kamenopoulou, J. Barthe, C. Hickman, G. Portal, *Radiat. Prot. Dosim.* **17** (1986) 185.
148. V. Chumak, S. Sholom, L. Pasalskaya, *Radiat. Prot Dosim.* **84** (1999) 515.
149. J. Barthe, V. Kamenopoulou, B. Cattoire, G. Portal, *Appl. Radiat. Isot.* **40** (1989) 1029.
150. H. Chandra, M. C. R. Symons, *Nature* **328** (1987) 833.
151. R. Kudinski, J. Kudinska, H. A. Buckmaster, *Appl. Radiat. Isot.* **45** (1994) 645.
152. M. C. R. Symons H. Chandra, J. L. Wyatt, *Radad. Prot. Dosim.* **58** (1995) 11.

153. B. C. Cope, L. Hopegood, R. J. Latham, R. G. Linford, J. D. Reilly, M. C. R. Symons, F. A. Taiwo, *J. Mater. Sci.* **8** (1998) 43.
154. A. F. Usatyi, N. V. Verein, *Appl. Radiat. Isot.* **47** (1996) 1351.
155. H. Y. Goeksu, A. Wieser, D. Stoneham, I. K. Bailiff, M. Figel, *Appl. Radiat. Isot.* **47** (1996) 1369.
156. S. F. Ginsbourg, T. A. Babushkina, L. B. Basova, T. P. Klimova, *Appl. Radiat. Isot.* **47** (1996) 1381.
157. T. Slager, M. J. Zucker, E. B. Reilly, *Radiat. Res.* **22** (1964) 556.
158. A. Brik, V. Baraboy, O. Atamanenko, Yu. Shevchenko, V. Brik, *Appl. Rad. Isot.* **52** (2000) 1305.
159. G. J. Hennig, W. Herr, E. Weber, N. I. Xirotiris, *Nature* **292** (1981) 533.
160. H. Ishii, M. Ikeya, S. Okano, *J. Nucl. Sci. Tech.* **27** (1990) 1153.
161. G. Gualtieri, S. Colacicchi, R. Sgattoni, M. Giannoni, *Appl. Radiat. Isot.* **55** (2001) 71.
162. A. Serezhenkov, E. V. Domracheva, G. A. Klavezal, S. M. Kulikov, A. Kuznetsov, P. I. Mordvintcev, L. I. Sukhovskaya, N. E. Schklovsky-Kordi, A. F. Vanin, N. V. Voevodskaya, A. I. Vorobev, *Radiad. Prot. Dosim.* **42** (1992) 33.
163. A. Romanyukha, D. Regulla, E. Vasilenko A. Wieser, *Appl. Radiat. Isot.*, **45** (1994) 1195.
164. A. Romanyukha, E. A. Ignatev, M. O. Degteva, V. P. Kozheurov, Wieser, P. Jacob, *Nature* **381** (1996) 199.
165. T. Nakajima, T. Ohtsuki, I. Likhariov, *J. Nucl. Sci. Technol.* **28** (1991) 71.
166. K. Ostriwski, A. Dziedzic-Gostawska, W. Stachowich W., in ``*Free Radicals in Biology*``, vol. 4 (W. Pryor, Ed.) Academic Press NY (1980) p. 321.
167. P. Cevch, M. Schara, C. Ravnic, *Radiat. Res.* **51** (1972) 581.
168. M. F. Desrosiers, M. G. Simic, F. C. Eichmiller, A. D. Johnston, R. L. Bowen, *Appl. Radiat. Isot.* **40** (1989) 1195.
169. D. Aragno, P. Fattibene, *Appl. Radiat. Isot.* **55** (2001) 375.
170. M. Ikeya, J. Miyajima, S. Okajima, *Jap. J. Appl. Phys.* **23** (1984) L699.
171. E. H. Haskell, R. B. Haves, G. H. Kenner, A. Wieser, D. Aragno, P. Fattibene, Onori, *Radiat. Protect. Dosim.* **84** (1999) 527.
172. E. A. Ignatiev, A. A. Romanyucha, A. A. Koshta, A. Wieser, *Appl. Radiat. Isot.* **47** (1996) 333.
173. A. Wieser, E. Haskell, G. Kenner, F. Bruenger, *Appl. Radiat. Isot.* **45** (1994) 525.
174. A. L. Ivannikov, D. D. Tikunov, V. G. Skovortsov, V. L. Stepanenko, V. V. Khomichyonok, L. G. Khamidova, D. D. Skiripnik, L. L. Bozadjiev, M. Hoshi, *Appl. Radiat. Isot.* **55** (2001) 701.
175. A. Wieser, S. Onori, D. Aragano, P. Fattibene, A. Romanyukha, E. Ignatiev, A. Koshta, V. Skvortzov, A. Ivannikov, V. Stepenko, V. Chumak, S. Shalom, R. Hayes, G. Kenner, *Appl. Rad. Isot.* **52** (2000) 1059.
176. A. Romanyukha, M. F. Desrosiers, D. F. Regulla, *Appl. Rad. Isot.* **52** (2000) 1265.
177. M. Desrosiers, D. A. Schauer, *Nucl. Instrum. Meth. Phys. Res.* (B) **184** (2001) 219.
178. A. Wieser, K. Mehta, S. Amira, D. Agano, S. Bercea, A. Brik, A. Bugai, F. Callens, V. Chumak, B. Ciesielski, R. Debuyst, S. Dubrovsky, O.O. Duliu, P. Fattibene, H. Haskell, R. B. Hayes, E. A. Ignatiov, A. Ivannikov, V. Kirilov, N. Nakamura, Nather, J. Nowak, S. Onori, B. Pass, S. Pivovarov, A. Romanyukha, O. Scherbina, A. I. Shames, S. Sholom, V. Skvortsov, V. Stepanenko, D.D. Tikhonov. S. Toyoda, *Radiat. Meas.* **32** (2000) 549.
179. G. Vanhaelewyn, S. Amira, R. Debuyst, F. Callens, Th. Glorieux, G. Leloup, H. Thierens, *Radiat. Meas.* **33** (2001) 417.
180. M. Ikeya, H. Ishii, *Appl. Radiat. Isot.* **40** (1989) 1027.

181. H. Ishii, M. Ikeya, *Jap. J. Appl. Phys.* **29** (1990) 871.
182. M. Ikeya, M. Yamamoto, H. Ishii, *Rev. Sci. Instrum.* **65** (1994) 3670.
183. M. Miyake, K. J. Liu, T. Walczak, H. M. Swartz, *Appl. Radiat. Isot.* **52** (2000) 1031.

Chapter 15

OPTICALLY DETECTED MAGNETIC RESONANCE OF DEFECTS IN SEMICONDUCTORS

Weimin M. Chen
Department of Physics and Measurement Technology, Linkoping University, S-581 83 Linkoping, Sweden

Key words: cw-ODMR, time-resolved ODMR, photoluminescence, ESR, defects, semiconductors, radiative, nonradiative, Si, SiGe.

Abstract: This chapter provides a review of the optically detected magnetic resonance (ODMR) techniques. The principle and methods of a variety of the ODMR technique, namely ODMR by different ways of optical detection, ODMR in zero field or in the presence of an external magnetic field, cw- and time-resolved ODMR, will be described. The ability of the ODMR technique to provide important information on physical properties of defects in semiconductors, such as chemical identification, electronic and geometric structure, related carrier recombination, etc., will be demonstrated. Recent progress, trends and prospects in achieving high spectral, time and spatial resolution of the ODMR techniques will also be outlined.

1. INTRODUCTION

The optically detected magnetic resonance (ODMR) technique is a combination of electron spin resonance (ESR) and photoluminescence (PL). Since its first application for the excited state of mercury atoms in 1952,[1] ODMR has been extensively employed to investigate excited states of solids in particular in retrieving detailed microscopic information on defects as well as recombination processes in semiconductors.[2-6] With recent rapid developments in semiconductor thin films and quantum structures, ODMR has attracted increasing attention as the traditional ESR technique has failed to meet the challenge of a significantly decreasing number of spins in the newly emerging materials.

In this chapter we shall provide an overview of the physics, methods, capabilities and limitations of the ODMR technique, which will hopefully be found helpful for the readers who are interested to employ ODMR in studies of various materials. It is therefore not intended to provide a complete survey of the literature and the research field. The chapter will be organized in the following way. A background and methods of the ODMR techniques will be given in Sections 2 and 3. Examples of successful applications of the ODMR technique will be presented in Section 4, to demonstrate the ability of the ODMR technique in retrieving detailed electronic and microscopic information on the defects. The chosen examples are from our own work, i.e. the S-Cu complex in silicon and the non-radiative defects in Si/SiGe epitaxial films and quantum structures. Recent developments and trends for high spectral, time and spatial resolution of the magnetic resonance techniques will be outlined in Section 5.

2. BACKGROUND OF THE ODMR TECHNIQUE

2.1 What is ODMR

The spin resonance part of the ODMR technique, such as the Zeeman splitting and the spin resonance selection rule, is identical to ESR. The essential difference between the conventional ESR and the ODMR technique is the detection scheme (Fig. 1).

In the presence of an external magnetic field **B**, the angular momentum of the electron spin (assuming an effective spin S=1/2) will be quantized along the magnetic field giving rise to two magnetic sublevels (M_S=-1/2, +1/2). The energy separation between these two levels scales with the field strength, described by an electronic Zeeman term in the spin Hamiltonian treatment. When this separation coincides with the energy of the applied microwave photons, a spin resonance transition can occur by either absorbing or emitting the microwave photons. In the traditional ESR experiments, the resulting change of the microwave power is detected.[7,8] In ODMR, a change induced by spin resonance in optical response, either in emission or in absorption, is monitored instead of the microwave power change (Fig. 1). (Only the ODMR technique via optical emission will be discussed in this chapter, but not via optical absorption. For the latter the reader is referred to Ref.9).

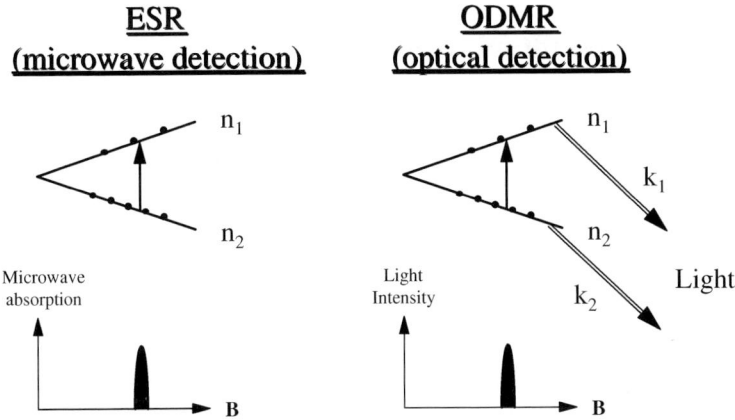

Figure 1. Difference between ESR and ODMR. k_i and n_i denote the radiative decay rate and population of the spin sublevel i.

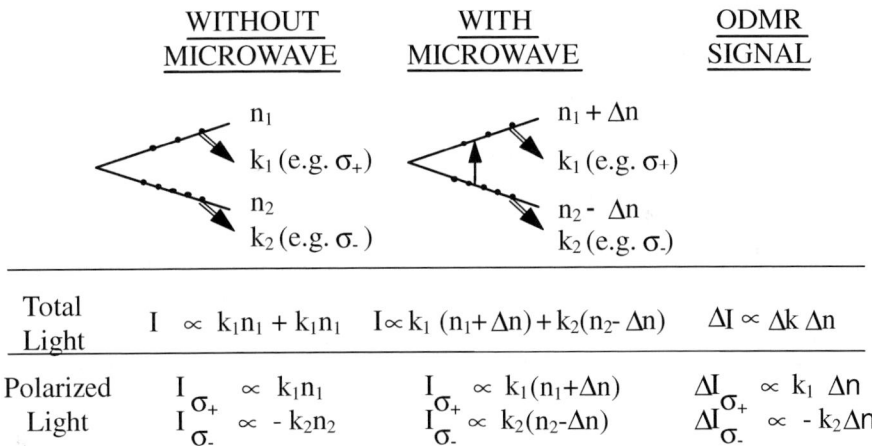

Figure 2. Principle of the ODMR signals obtained by monitoring total intensity or polarization of emitting light. Δn corresponds to the population change induced by the spin resonance.

In Fig. 2 schematic pictures are displayed to describe the principle of the ODMR signal. As it can be seen, the microwave induced spin transitions can now be sensitively detected by their resulting changes in the total PL intensity ΔI or polarizations ($\Delta I_{\sigma+}$ or $\Delta I_{\sigma-}$ for examples) of the electric-dipole allowed optical transitions that give rise to the light emission, provided that the radiative decay rates or the senses of light polarization of the involved sublevels are different.

2.2 Spin Hamiltonian analysis

Energies of magnetic sublevels revealed from ODMR studies can be analyzed by the spin Hamiltonian description,[7-10] which is identical to that for analysis of ESR data,

$$H = \mu_B \mathbf{S} \cdot \mathbf{g} \cdot \mathbf{B} + \mathbf{S} \cdot \mathbf{D} \cdot \mathbf{S} + \Sigma \, \mathbf{S} \cdot \mathbf{A}_i \cdot \mathbf{I}_i \qquad (1)$$

Here μ_B is the Bohr magneton. **S** and $\mathbf{I}_i$ represent the effective electronic spin and the nuclear spin of the defect or ligand atom *i*. **g** is the g-tensor. **D** and $\mathbf{A}_i$ are the fine-structure and hyperfine-structure (HF) tensors. The first term describes the usual linear term of electronic Zeeman splitting. The second term in Eq.(1) introduces a fine structure (*i.e.* zero-field splitting) only for $S>1/2$, but not for $S=1/2$ except an identical shift in energy of both spin sublevels. The third term in Eq. (1) for the hyperfine structure has no effect if $I_i=0$. The nuclear Zeeman term and other higher order terms have not been included in Eq.(1) due to their negligible effects in most cases of ESR and ODMR investigations.

The **g**-tensor should be isotropic and has the principal value of 2.0023 for a free electron in free space. For electrons (either free or bound at a defect) in a semiconductor crystal, **g** should be a tensor reflecting anisotropy of the crystal lattice and a defect. In other words, it reflects the local symmetry of the defect. The deviation of the **g**-tensor from the free electron g-value in the material is largely contributed by orbital angular momentum, *e.g.* via spin-orbit interaction if orbital angular momentum is initially quenched.[8]

On the other hand, anisotropy of the **D**-tensor can be attributed to both spin-orbit interaction and magnetic dipole-dipole interaction,[8,10] of which both are a function of defect symmetry. When the contribution from the spin-orbit interaction is negligible, the magnetic dipole-dipole interaction can sometimes provide useful insight on the separation of the localized magnetic dipoles (spins). In a favorable case of a paired defect, the separation between the defect atoms forming the pair can be estimated.[8]

The hyperfine structure is extremely valuable for a positive chemical identification of a defect, since the nuclear spin number I_i and the natural

15. ODMR of defects in semiconductors

abundance of the isotope *i* are unique for each element in the periodic table. The value and the anisotropy of the A_i-tensor provide detailed information on the localization and the character of the electronic wavefunction of the defect.

The information on a defect gained from the spin Hamiltonian analysis of ODMR data is therefore the same as that from ESR and is summarized below:
- symmetry
- chemical identification
- local geometric arrangement
- electronic structure

2.3 Why ODMR

Though the ESR technique has in the last half century played an indispensable role in identification of defects in bulk semiconductors and in determining their electronic properties,[9,11] the importance of the traditional ESR technique in characterization of modern epitaxial layers and layered structures has so far been undermined by its limited sensitivity as the volume of the materials and thus the total number of spins decreases significantly.

The main limitations of the otherwise powerful ESR technique are:
- paramagnetic ground state required.
- no information on carrier recombination.
- low sensitivity due to microwave detection.

The first limitation can be overcome if another charge state of the same defect can be reached by changing the Fermi level position, *e.g.* by doping with shallow impurities. Fortunately, the most important defects in semiconductors are either electrically or optically active by introducing energy levels within the forbidden bandgap. The presence of a defect level within the bandgap implies that a charge transition can occur when the Fermi level moves across the defect level. The first limitation can also be bypassed by performing ESR under non-equilibrium conditions such as under optical excitation, so-called photo-ESR. Non-equilibrium population of spin-active states is usually very low, however, as the spin system tends to return to its thermal equilibrium ground state, which has largely limited the success of the photo-ESR technique.

To remove the second limitation requires a direct probe of carrier recombination processes. This can also be fulfilled for example via optical excitation and detection by monitoring photoluminescence. The last limitation of a low sensitivity (thus volume demanding) has been the main reason preventing ESR from a wide application in semiconductor epitaxial layers, layered and quantum structures. Optical detection apparently provides a solution to the problem due to (a) the fact that PL transitions are electric-dipole allowed thus much more intense and easier to detect as compared to the magnetic-dipole allowed microwave transitions, and (b) the availability of sensitive optical detectors in the visible and near-infrared spectral region as compared to the microwave range.

All these limitations imposed on the ESR technique call for a solution by the optical detection method, namely the optically detected magnetic resonance technique.

ODMR has in this context emerged to be the technique of choice, which combines highly sensitive optical spectroscopies with the microscopically informative ESR technique. Apart from the capabilities shared with the ESR technique as to providing the electronic and microscopic information of defects, the ODMR technique offers new features much needed for applications in modern semiconductor thin films and layered structures. These include
- High sensitivity due to
  - electric-dipole optical transition
  - sensitive optical detectors
  suitable for thin, layered and quantum structures
- High selectivity due to
  - spectral resolution and direct probe of corresponding optical transitions
  suitable for studies of carrier recombination processes
- High energy resolution (~ 10^{-7} eV) as compared to that of photoluminescence (typically 10^{-4} eV)
- Excited states and non-equilibrium conditions
- New possibilities for complementary optical spectroscopies.

The main limitations of the ODMR techniques can be summarized as follows:
- Light emission required
- Paramagnetic excited states
- Spectral broadening due to lifetime effect, *etc.*
- Not quantitative (*e.g.* in defect concentration)

3. VARIETIES OF THE ODMR TECHNIQUE

In the past a variety of the ODMR technique has been explored in terms of various ways of optical detection, with or without applying an external magnetic field, cw or time-resolved spectroscopy. Below we shall give a brief account of these variations of the ODMR technique.

3.1 ODMR by monitoring total PL intensity and polarization

If two spin states involved in a spin transition have different radiative decay rates, a spin resonance transition leads to a net change of their total PL intensity. The so-obtained ODMR signal is proportional to the product of the difference in their radiative decay rates and the population change induced by the spin resonance transition (see Fig. 2 and Fig. 3). The vast majority of the ODMR studies reported to date have employed this type of the ODMR technique.

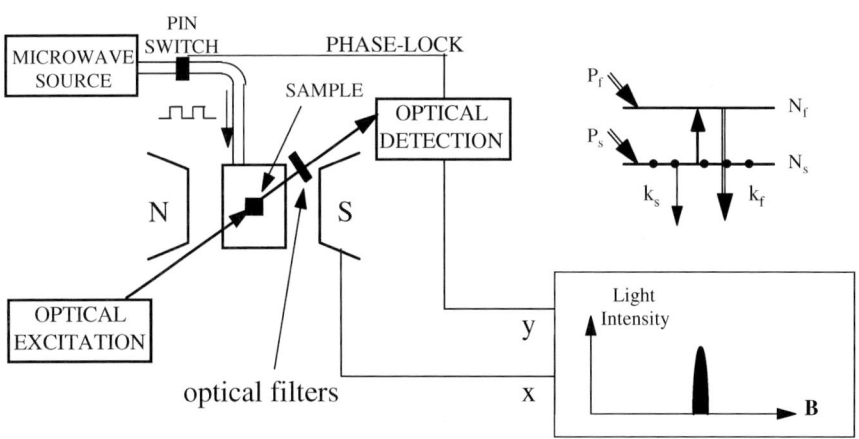

Figure 3. Schematic picture of the ODMR technique by monitoring total PL intensity. P, N, and k denote the populating, population and radiative decay rate of the corresponding spin level. The subscript f and s refer to the spin levels with a faster and slower decay rate, respectively. S and N represent the poles of an external magnet.

When the radiative decay rates are the same for the involved spin levels, the ODMR signal by monitoring total PL intensity vanishes. It is, however, still possible to observe an ODMR signal by monitoring a specific polarization or the difference of the two polarizations if the two spin levels emit light of different polarizations (Fig. 2 and Fig. 4). This is usually a more sensitive way of optical detection as compared to the above method of detecting total PL intensity.

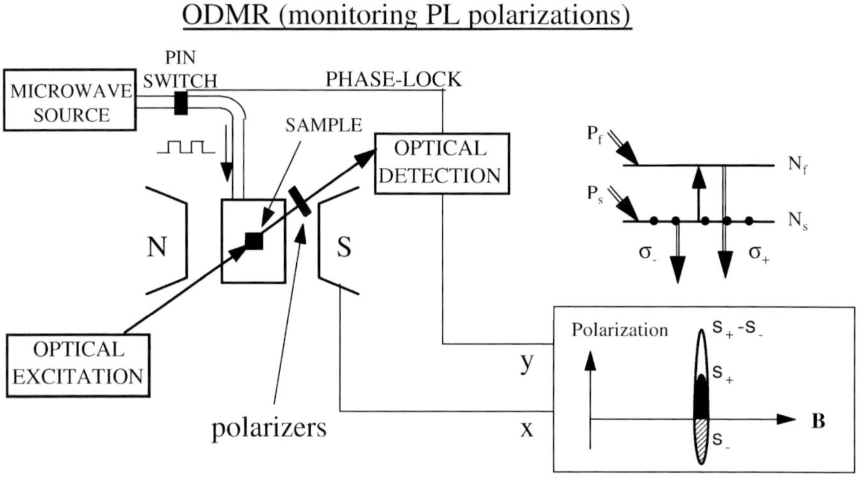

Figure 4. Schematic picture of the ODMR technique, by monitoring a specific light polarization or the difference of the two polarizations.

3.2 ODMR in zero field and in applied magnetic fields

In most of ODMR studies reported to date, a sweeping external magnetic field is applied to tune Zeeman splitting of spin sublevels into resonance with the energy of the microwave photons of a fixed frequency, in the same way as a conventional ESR experiment does. In a zero field ODMR experiment,[12] on the other hand, the energy of the microwave photons is tuned by sweeping the frequency of the microwave field to match zero-field splitting of the spin multiplet (*e.g.* a spin triplet as shown in Fig. 5).

The main advantages of the zero field ODMR are as follows:
- The hyperfine interaction is quenched up to the second order since $<T_u|H_{HF}|T_u>=0$, where T_u denotes the eigenstate of the sublevel u of the spin triplet (Fig. 5). This leads to a narrower ODMR linewidth as compared to that in the presence of a magnetic field[13] and gives a better chance to resolve the quadrupole interaction, which contains important information about the nuclear spin of a defect atom (thus toward a positive identification)[14] and the electric field gradient near the defect.

15. ODMR of defects in semiconductors

- A stronger ODMR signal is expected due to the orientational degeneracy.
- No external magnetic field is required.
- Relatively simple states due to the absence of mixing of states by an external magnetic field.

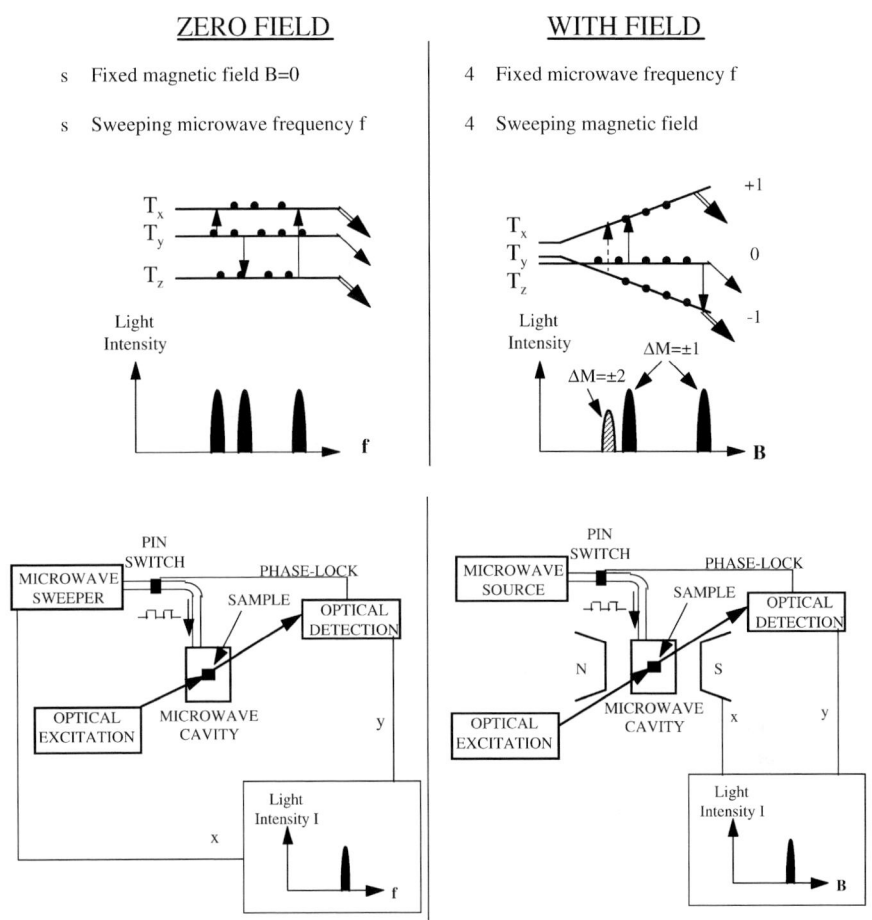

Figure 5. Principle and setup of the ODMR technique in zero field and in an applied magnetic field.

Though the zero-field ODMR can infer the highest symmetry allowed for the defect from the resolved level degeneracy, it is not possible to definitely determine the symmetry of a defect. For that one has to resort to the ODMR technique in an applied magnetic field. It is also rather difficult to obtain a microwave resonant cavity that is tunable in frequency and can be synchronized with the sweeping frequency of the microwave source.

3.3 CW and time-resolved ODMR

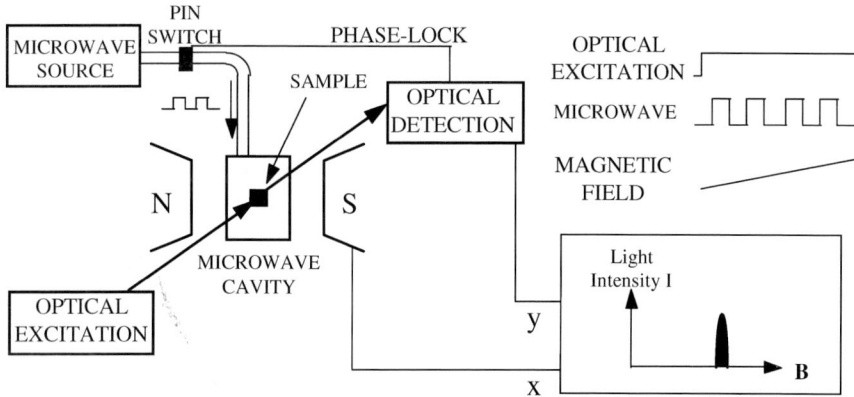

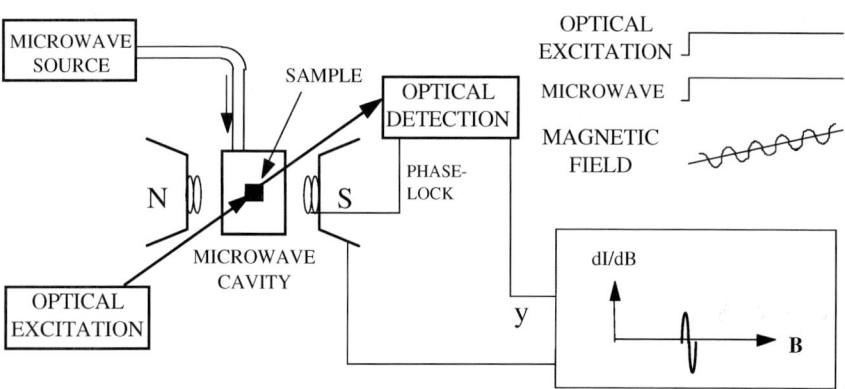

Figure 6. Simple pictures of the cw ODMR experimental setup, (a) with an on-and-off microwave modulation, and (b) with a magnetic field modulation.

In a cw ODMR experiment, an ODMR signal is obtained in a time-integrated manner. Typically, a cw optical excitation is employed to elevate a defect to its excited state from which spin dependent recombination can be studied by ODMR. To enhance the ODMR signal, an on-and-off microwave modulation or a magnetic field modulation has often been applied to allow sensitive lock-in detection of the microwave-induced change in the optical emission intensity, see Fig. 6. The so-obtained ODMR signal intensity can

15. ODMR of defects in semiconductors

be estimated from an analysis of rather simple rate equations under steady-state conditions.[15]

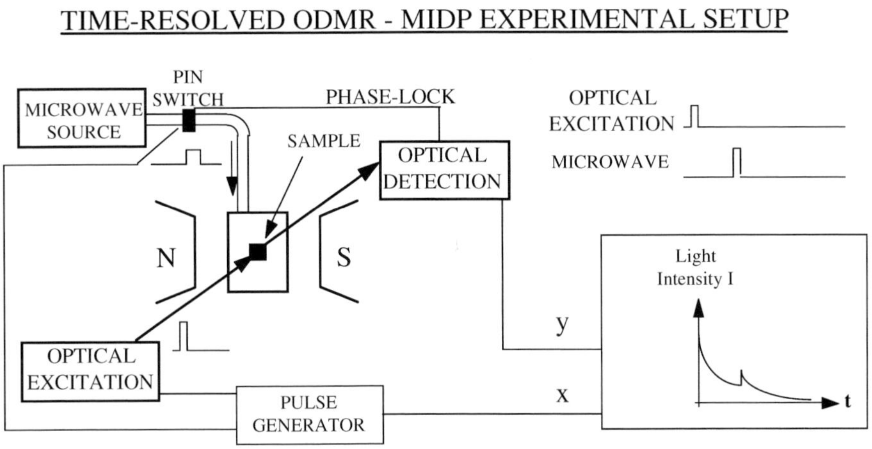

TIME-RESOLVED ODMR - MIDP EXPERIMENTAL SETUP

Figure 7. Principle and setup of the time-resolved ODMR technique - MIDP.

In a time-resolved ODMR experiment, on the other hand, a pulsed optical excitation (usually by a laser pulse) is used and the transient decay of the excited spin states is monitored. A microwave pulse applied some time after the laser pulse will change the population of the sublevels brought into resonance by the microwave field. If the radiative decay rates are different between these two sublevels, this will lead to a change in PL intensity – often called by microwave induced delayed phosphorescence (MIDP)[15] (see Fig. 7).

Useful dynamical properties of the defect excited states can be obtained by MIDP such as populating rates, total decay rates, relative radiative decay rates, *etc.*, provided that spin-lattice relaxation rates are slow compared to the decay rates. To ensure that this condition applies, MIDP experiments are usually carried out at the lowest temperature possible. The method to obtain the dynamical properties of a defect excited state will be presented below in Section 4.1.2 in connection with the example case of ODMR applications.

4. APPLICATIONS OF THE ODMR TECHNIQUE

Below we shall present some examples to demonstrate the capability of the ODMR technique in retrieving useful information on physical properties of defects in semiconductors. We will show that ODMR can not only be

612   Chapter 15

directly applied to studies of radiative defects but also be employed to indirectly study nonradiative defects by monitoring radiative recombination processes of other competing defects.

4.1 Radiative defects: example by the S-Cu complex in silicon

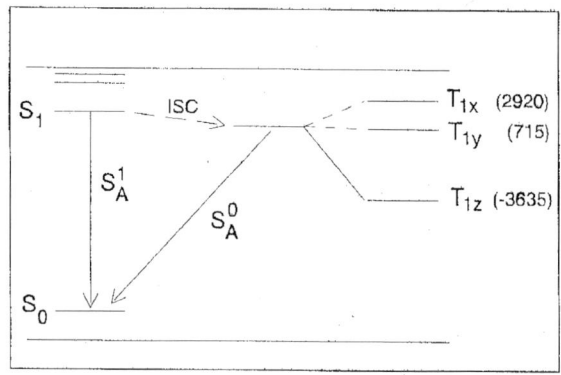

Figure 8. The electronic level scheme of the S-Cu defect in silicon. Solid and broken arrows represent the radiative and nonradiative transitions, respectively. The zero-field splitting parameters are indicated in MHz. (From Ref.17).

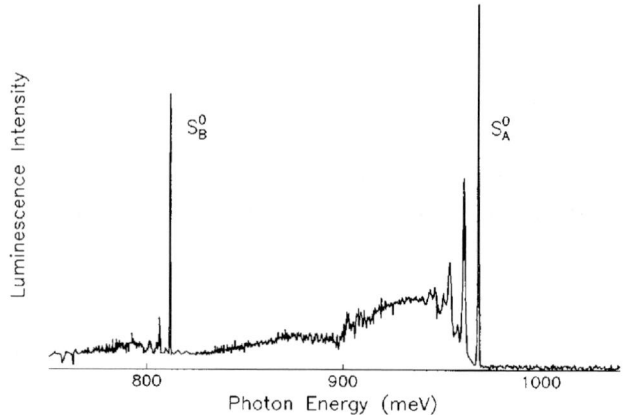

Figure 9. PL spectrum of the S-Cu defect in silicon obtained at 4.2 K.

15. ODMR of defects in semiconductors 613

As an example of successful applications of the ODMR technique for radiative defects in semiconductors, we present below ODMR results from the S-Cu complex defect in silicon.[14,17,18] The stable configuration of this defect gives rise to the zero-phonon PL line at 0.968 eV (S_A^0) at low temperature,[19-23] originating from the photo-excited spin triplet state of the bound exciton at the defect, Figs. 8 and 9. With increasing temperature the PL line at 0.9771 eV (S_A^1) arising from the singlet excited state becomes dominating. The zero-phonon PL line at 0.812 eV (S_B^0) arises from the triplet excited state of the same defect in its metastable configuration,[14,17-23] but will not be discussed further here.

4.1.1 CW zero-field ODMR

Due to the presence of a Cu atom in the defect with a nuclear spin I=3/2 for both isotopes (^{63}Cu, I=3/2, g_n=1.484, Q=-0.356x10^{-43} C cm^2, natural abundance 69.1% and ^{65}Cu, I=3/2, g_n=1.588, Q=-0.312x10^{-43} C cm^2, natural abundance 30.9%), the electronic structure of the spin triplet (S=1) excited state of the S-Cu defect including interactions with the nuclear spins is rather complicated. By neglecting the difference in the nuclear g_n values and quadrupole moments of the two Cu isotopes, the following spin Hamiltonian was used to describe the interaction between the electron spin S=1 and a single nuclear spin I=3/2:

$$H = H_{ss} + H_Q + H_{HF} = \mathbf{S}\cdot\mathbf{D}\cdot\mathbf{S} + \mathbf{I}\cdot\mathbf{P}\cdot\mathbf{I} + \mathbf{S}\cdot\mathbf{A}\cdot\mathbf{I} \qquad (2)$$

Here the first term describes the zero-field splitting of the electron spin induced by the spin-orbit interaction and spin-spin interactions. In its principal axes system, denoted by x,y,z, H_{ss} can be written as

$$H_{ss} = D_{zz}S_z^2 + D_{yy}S_y^2 + D_{xx}S_x^2 \qquad (3)$$

H_{ss} is diagonal on the basis of the triplet function T_z, T_y, and T_x. These functions are linear combinations of the eigenfunctions of the S_z operator:

$$T_x = \frac{1}{\sqrt{2}}(|+1\rangle - |-1\rangle)$$
$$T_y = \frac{i}{\sqrt{2}}(|+1\rangle + |-1\rangle) \qquad (4)$$
$$T_z = |0\rangle$$

The functions have the property

$$S_x T_y = -S_y T_x = iT_z$$
$$S_u T_u = 0, u = x, y, z \qquad (5)$$

The result is that matrix elements of the type $<T_u|H_{HF}|T_u>=0$ and the hyperfine interaction reduces to a second-order effect.

The second term H_Q of Eq.(2) represents the quadrupole splitting of the spin states of the I=3/2 nucleus. Its principal axis system x', y', z' is determined by the electric field gradient at the nucleus and H_Q can be written as

$$H_Q = P_{z'z'}I_{z'}^2 + P_{y'y'}I_{y'}^2 + P_{x'x'}I_{x'}^2$$
$$= \frac{1}{2}P_{z'z'}\left[3I_{z'}^2 - I(I+1) + \eta(I_{x'}^2 - I_{y'}^2)\right] \qquad (6)$$

with

$$\eta = \frac{P_{x'x'} - P_{y'y'}}{P_{z'z'}} \quad (0 < h < 1)$$

The last term H_{HF} describes the central hyperfine interaction between the electron spin and the nuclear spin. Since no indication was found of an anisotropy of **A** we assume that H_{HF} can be written as

$$H_{HF} = A(S_x I_x + S_y I_y + S_z I_z) \qquad (7)$$

The calculated positions of the triplet sublevels with the aid of the spin Hamiltonian are shown in Fig. 10.[14] The allowed and "forbidden" spin transitions are indicated by the solid and broken vertical arrows, respectively. Due to the mixing of states induced by the hyperfine interaction, the formerly forbidden electron spin transitions become partially allowed and can be observed at high microwave power. The experimentally obtained ODMR signals arising from the transitions T_y-T_z (4.3 GHz) and T_x-T_z (6.5 GHz) are shown in Fig. 11. The effect of microwave power on the ODMR spectra can be understood as being due to the relative increase of the intensities of the forbidden transitions due to saturation of the allowed ones with increasing microwave power.

15. ODMR of defects in semiconductors

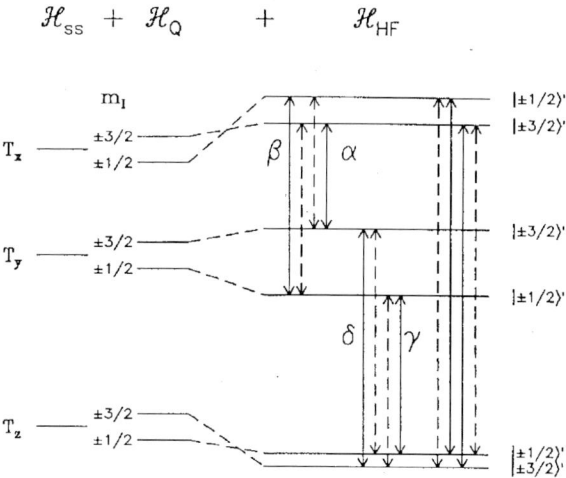

Figure 10. The calculated positions of the spin triplet sublevels of the S-Cu defect in Si from Eq.(2). The allowed and "forbidden" spin transitions are indicated by the solid and broken vertical arrows, respectively. α, β, γ, and δ transitions give rise to the ODMR signals shown in Fig. 12. (From Ref.14).

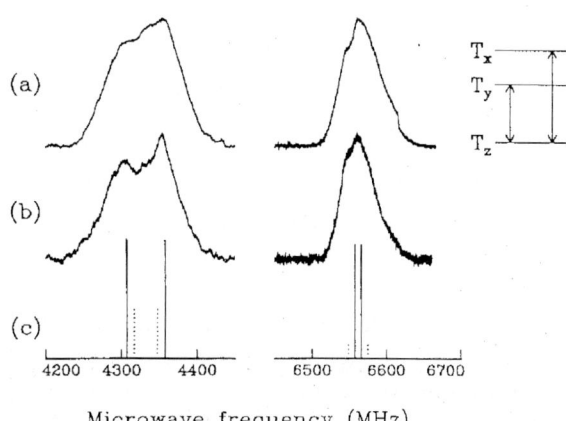

Figure 11. (a) and (b) The T_y-T_z and T_x-T_z ODMR transitions of the S_A spin triplet of the S-Cu defect in Si at high (200 mW) and low (10 mW) microwave power. (c) The calculated positions for the forbidden (dashed lines) and allowed (solid lines) transitions as obtained from the spin Hamiltonian Eq.(2). The length of the sticks indicates the relative values of the transition probabilities. (From Ref.14).

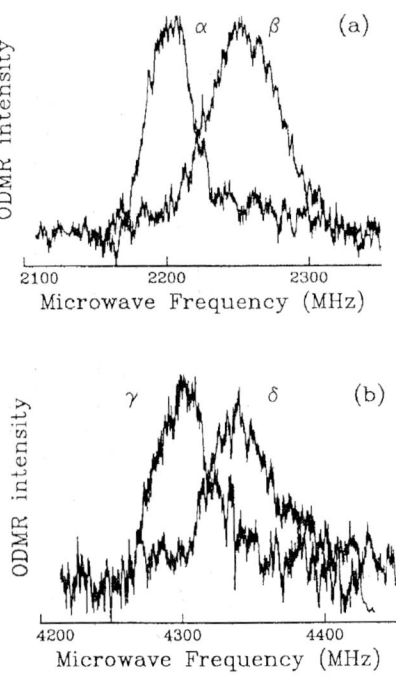

Figure 12. The double-resonance ODMR transitions of the S_A spin triplet of the S-Cu defect in Si. (a) Transitions α and β are obtained by fixing the frequency of the first cw microwave field at 4350- and 4300-MHz, respectively, while scanning amplitude modulated second microwaves in the 2.2-GHz region. (b) Transitions γ and δ are obtained by fixing the frequency of the first cw microwave field at 2260- and 2200-MHz, respectively, while scanning amplitude modulated second microwaves in the 4.3-GHz region. (From Ref.14).

The cw zero-field ODMR performed in a similar way has failed to detect the T_x-T_y (2.2 GHz) transition, however, because the two sublevels involved have the same radiative decay rate (supported by the time-resolved ODMR results of the same defect to be discussed below). This difficulty was successfully overcome by employing a double resonance technique. Here we first pump the T_y-T_z transition with cw microwaves at a fixed frequency. When using a saturating microwave power the populations of the two levels are equalized and become $(N_y+N_z)/2$. To make the T_x-T_y transition visible we subsequently scan a saturating amplitude modulated microwave field through the T_x-T_y resonance region and detect again synchronously in the PL intensity. This second microwave field will change the population of T_x, T_y and T_z to $(N_x+N_y+N_z)/3$. Since the radiative decay rate of T_z is different from T_x and T_y, this will lead to a change in PL intensity. The results are shown in Fig. 12. The double-resonance ODMR allowed us to prove that the T_x-T_y and the T_y-T_z transitions consist of two components as expected for the

15. ODMR of defects in semiconductors

involvement of a Cu atom with a nuclear spin I=3/2. The double-resonance ODMR also allowed us to determine the linewidth of the transitions to be about 50 MHz, considerable smaller than the linewidth of 700 MHz (25 mT) observed in the X-band ODMR experiments in applied magnetic fields. [18] This reduction of the ODMR linewidth is caused by the quenching of the hyperfine interaction in first order in zero magnetic field. The obtained spin Hamiltonian parameters are given in Table 1.

Table 1. The set of parameters from the best fit of the spin Hamiltonian Eq.(2) to the experimental data. (From Ref.14).

D_{xx} (MHz)	D_{yy} (MHz)	D_{zz} (MHz)	A (MHz)	$P_{z'z'}$ (MHz)	η
-2905	-705	3610	175	10-20	0-1

4.1.2 Time-resolved zero-field ODMR

The MIDP method allows us to determine the total decay rates k_i (i=x, y, and z), the relative radiative decay rates k_i^r, and the relative populating rates P_i of the spin triplet excited state. [17] In the MIDP experiment a laser flash, with a short pulse duration compared to the decay times, is applied to excite the defect from the ground state S_0 to the singlet excited state S_1. The resulting populations of the triplet sublevels N_x, N_y and N_z are proportional to their populating rates P_x, P_y and P_z. The decay of the S_A^0 PL intensity from the triplet to the ground state S_0 as a function of time is given by

$$I_{PL}(t) = c \sum_{i=x,y,z} k_i^r N_i e^{-k_i t} \tag{8}$$

where c is a constant depending on the experimental conditions. At a time t_d after the laser flash a microwave pulse, resonant with, for example, the T_x-T_z transition at 6555 MHz is applied. The pulse length is chosen short compared to the lifetimes k_x^{-1} and k_z^{-1}, so that the total population of T_x and T_z is approximately constant during the microwave pulse. A typical result of this experiment is shown in Fig. 13. The evolution of the PL intensity following the microwave pulse is then

$$I_{PL}(t) = cN_y(t_d)k_y^r e^{-k_y(t-t_d)} + c[N_x(t_d) - f\Delta N_{x-z}(t_d)]k_x^r e^{-k_x(t-t_d)} \\ + c[N_z(t_d) + f\Delta N_{x-z}(t_d)]k_z^r e^{-k_z(t-t_d)} \tag{9}$$

where $\Delta N_{x-z}(t_d) = N_x(t_d) - N_z(t_d)$

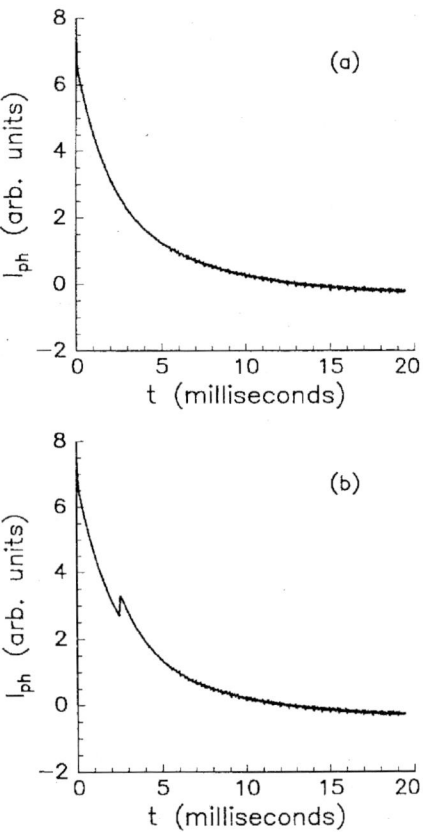

Figure 13. The decay of the S_A^0 PL of the S-Cu defect in Si (a) without and (b) with the application of a microwave pulse at 6555 MHz, resonant with the T_x-T_z transition at t_d=2.5 ms. The PL is selectively excited by a pulse laser at 977.1 meV into the excited singlet state S_1. T=1.2 K. (From Ref.17).

The factor f in Eq.(9) represents the effect of the transfer of population between T_x and T_z induced by the resonant microwaves. The MIDP signal is obtained from the difference of the PL decay with and without microwaves, respectively. The result of this subtraction for t_d=2.5 ms is shown in Fig. 14. The evolution of this signal is described by the difference of Eqs.(9) and (8) for $t \geq t_d$.

$$I_{MIDP}(t) = cf[N_x(t_d) - N_z(t_d)] \times [k_z^r e^{-k_z(t-t_d)} - k_x^r e^{-k_x(t-t_d)}] \quad (10)$$

From this equation it follows that the MIDP signal can be fitted to a biexponential function with time constants that represent the decay rates of the two sublevels connected by the microwave field. The ratio of the

15. ODMR of defects in semiconductors

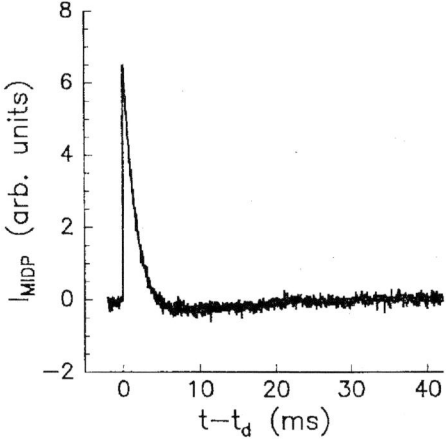

Figure 14. The MIDP signal - the difference of the PL decay curves with and without the application of a microwave pulse resonant with the T_x-T_z transition at 6555 MHz. T=1.2 K. (From Ref.17).

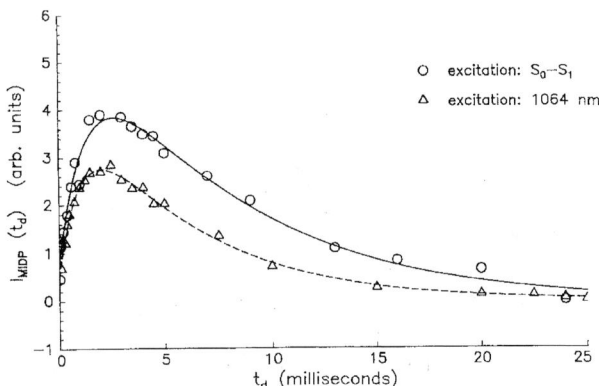

Figure 15. The maximum amplitude of the MIDP signal, upon a microwave pulse at 6555 MHz saturating the T_x-T_z transition, as a function of t_d. The signals have been obtained either following selective excitation at 977.1 meV into the S_1 excited state or following excitation at 1164 meV (1064 nm) over the band gap of Si. The drawn lines represent a fit to Eq.(11). T=1.2K. (From Ref.17).

prefactors of the two exponentials yields the relative radiative decay rates k_x^r and k_z^r. The maximum amplitude of the MIDP signal at $t=t_d$ as a function of t_d is described by the following relation:

$$I_{MIDP}(t_d) = cf(k_z^r - k_x^r)\left[N_x(0)e^{-k_x t_d} - N_z(0)e^{-k_z t_d}\right] \tag{11}$$

In this way one again finds a decay curve which is described by a biexponential function with k_x and k_z as the time constants (Fig. 15). Here the ratio of the prefactors yields the relative populations at $t_d=0$ and thus the relative populating rates of the two sublevels involved in the microwave transition. A similar set of experiments on the T_y-T_z transition at 4350 MHz yields k_y^r, k_z^r, k_y, k_z, P_y and P_z. They are summarized in Table 2.

Table 2. The decay rates k_i, the relative radiative decay rates k_i^r, and the relative populating rates P_i of the first excited triplet state T_1 of the S-Cu defect in Si in zero field. T=1.2 K. (From Ref.17).

	k_i	k_i^r	P_i
T_x	$(0.11\pm0.02)\times10^3$	1.0 ± 0.2	1.0 ± 0.15
T_y	$(0.11\pm0.02)\times10^3$	1.0 ± 0.2	1.0 ± 0.15
T_z	$(0.56\pm0.08)\times10^3$	4.0 ± 1.0	1.0 ± 0.15

4.1.3 CW ODMR in magnetic fields

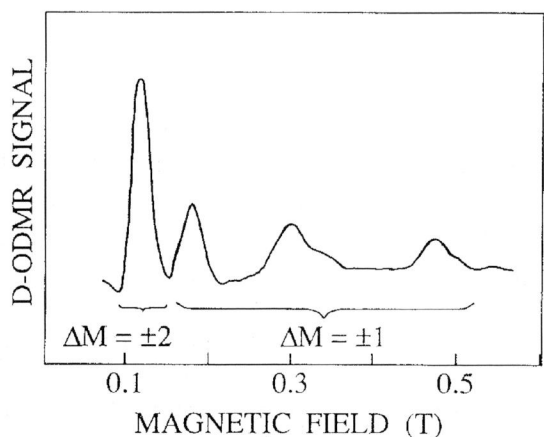

Figure 16. A typical ODMR spectrum from the S_A spin triplet of the S-Cu defect in Si, at 4 K and 9.24 GHz. (From Ref.18).

In the X-band ODMR experiments, the microwave frequency is fixed at 9.24 GHz while sweeping magnetic field. At certain field positions, the Zeeman splitting of the triplet sublevels can be brought into resonance with the microwave photon energy causing a spin resonance transition between the involved sublevels. If these sublevels have different radiative decay rates, such a spin transition can be detected optically leading to an ODMR signal. A typical ODMR spectrum obtained by the quasi-cw delayed ODMR technique[19] is shown in Fig. 16, where both the allowed $\Delta M=\pm1$ and the forbidden ΔM=±2 transitions are observed. The observation of the forbidden transitions is attributed to large mixing of states by the strong fine-structure and hyperfine interactions of the defect. From the angular dependence study of the ODMR signals, the principal defect axes x, y, z can be determined and the defect symmetry is concluded to be monoclinic-I.[18]

4.2 Nonradiative defects: examples by the vacancy-oxygen complex and the P-donor in MBE-Si/SiGe

The scope of the ODMR applications actually reaches beyond radiative defects and recombination channels. The ODMR technique is also sensitive to strong nonradiative recombination channels and defects, via their competition in carrier recombination. This provides the framework for the optical detection of nonradiative defects,[20] as also illustrated in Fig. 17.

Fig. 18 shows an ODMR spectrum observed via deep PL bands in a wide spectral range 0.7 eV-1.05 eV occurring in Si epitaxial layers and Si/SiGe quantum well structures grown by molecular beam epitaxy (MBE) at low temperatures.[21-25] Even though the origins of various PL bands are known to be different, the same ODMR spectrum could be obtained since the nonradiative defects undergoing the spin resonance transitions compete with all the radiative defects in carrier recombination. From detailed analysis of the ODMR data, two of the nonradiative defects were identified to be the shallow P donor and the deep vacancy-oxygen (V-O) complex.[23] While a deep center like the V-O complex is largely expected to act as a recombination center, the similar observation for the P donor is rather surprising as its traditional role should be an electron trap. The physical mechanism has in the past been shown in bulk Si as being due to efficient charge transfer from the P donor to other efficient recombination centers (Fig. 17c).[26] The introduction of the V-O defect in the as-grown material was facilitated by a low surface adatom mobility and a substrate bias during potential-enhanced growth in combination with a low oxygen desorption from the surface at the low growth temperatures. Fortunately, the V-O defect is known from earlier studies in bulk Si to be readily annealed out at 500 °C. A similar postgrowth annealing treatment can thus be expected to remove this nonradiative defect, leading to an improvement in the optical properties of the structures. This has indeed been experimentally demonstrated where a significant improvement has been achieved in thermal quenching behavior of the PL emissions by annealing.[24,25] This gives a good example on the feedback from the knowledge gained from the ODMR investigations to the defect engineering.

Competing carrier recombination

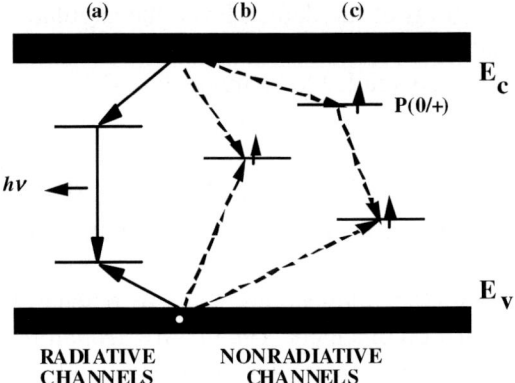

Figure 17. Principle for ODMR studies of nonradiative defects. Spin-resonance enhanced carrier recombination via non-radiative channels [(b) by a single nonradiative defect and (c) by nonradiative inter-center charge transfer] reduces the number of carriers available for radiative recombination channel (a), leading to a corresponding decrease in PL intensity.

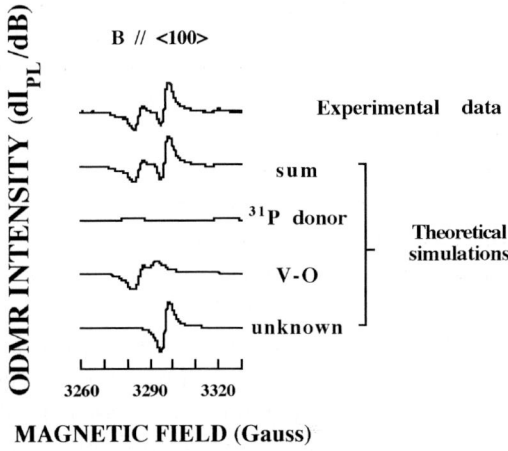

Figure 18. The top curve shows an ODMR spectrum from MBE-Si grown at 420°C, taken at 9.23 GHz. From the analysis, three nonradiative defects (the P donor, the V-O complex and an unknown defect) were revealed and their corresponding ODMR spectra could be deconvoluted.[23] The derivative lineshape is due to the field modulation in the ODMR experiments.

5. RECENT DEVELOPMENTS AND TRENDS

Substantial effort has been made over the past years in instrumentation developments of spin resonance techniques. It has mainly been directed to the following three fronts: high spectral resolution, better time resolution and high spatial resolution. The first two fronts have now been met with the commercialization of the W-band (95-GHz) ESR spectrometer and the pulsed ESR spectrometer in both X-band (9-GHz) and W-band. The W-band system offers ten times better in spectral resolution as compared to that of the commonly used X-band system.[4] The best time resolution available in the commercial spectrometers is about 20 ns at present. High frequency (95-140 GHz) ODMR has been developed in several research labs worldwide.

The quest for unprecedented spatial resolution has been made through various approaches. The first approach utilizes selective optical excitation by single-frequency laser to single out a single spin system from the inhomogeneously broadened spectral line of an ensemble under only a slight fluctuation in local potential. This approach has been shown to be rather successful for molecular systems, where ODMR from a single molecule has been reported.[27,28] Though the single-spin ODMR was achieved by improved spectral resolution rather than by spatial resolution, it has demonstrated the ability of the ODMR technique for single-spin detection. The second approach applied submicron optical apertures made by a metal mask patterned directly on samples with electron-beam lithographic technique, so that only few spins are monitored. By this approach, single quantum dots in GaAs/AlGaAs semiconductor quantum structures were studied by optically detected nuclear magnetic resonance (OD-NMR) reaching a lateral spatial resolution of about 10 nm.[29] The third approach for high spatial resolution is based on local probe microscopies. The most remarkable progress in this aspect has been the magnetic resonance force microscopy (MRFM),[30] where the force sensitivity has been shown to be feasible for single spin detection. It also offers possibilities for three-dimensional imaging. The best spatial resolution achieved so far is still far from atomic scale, though. No local probe ODMR has been reported so far. The most recent years have witnessed a sharp increase of the interest towards this issue among the scientific community of various disciplines, and surely this trend will continue to grow.

ACKNOWLEDGEMENTS *I am grateful to many colleagues for their valuable collaboration and contributions in connection to the work presented in this chapter, in particular J. Schmidt, A.M. Frens, M.T.*

Bennebroek, M.E. Braat, I.A. Buyanova, B. Monemar, A. Henry, E. Janzén, W.X. Ni, and G.V. Hansson. The financial support by the Swedish Research Council is also greatly appreciated.

6. REFERENCES

1. J. Brossel and F. Bitter, *Phys. Rev.* **86** (1952) 308.
2. B.C. Cavenett, *Adv. Physics* **30** (1981) 475.
3. J.J. Davies, *J. Cryst. Growth* **86** (1988) 599.
4. W.M. Chen, B. Monemar, A.M. Frens, M.T. Bennebroek, J. Schmidt, *Mat. Sci. Forum* **143-147** (1994) 1345.
5. T.A. Kennedy and E.R. Glaser, *Semicond. And Semimetals* **51**A (1998) 93.
6. W.M. Chen, *Thin Solid Films* **364** (2000) 45.
7. C.P. Slichter: *"Principles of Magnatic resonance"*, Springer Ser. in Solid State Sci. Vol. **1**, Springer, Berlin, 1990.
8. A.Abragam, B. Bleaney: *"Electron Paramagnetic Resonance of Transition Ions"*, Clarendon, Oxford, 1970.
9. J.-M. Spaeth, J.R. Niklas, and R.H. Bartram, *"Structural Analysis of Point Defects in Solids"*, Springer-Verlag, New York, 1992.
10. W.M. Chen, B. Monemar and M. Godlewski, *Defects and Diffusuion Forum* **62/63** (1989) 133.
11. G.D. Watkins, *Semicond. And Semimetals* **51**A (1998) 1.
12. E.H. Salib and B.C. Cavenett, *J. Phys.* C**17** (1984) L251.
13. J. Schmidt and J.H. van der Waals, *Chem. Phys. Lett.* **3** (1969) 546.
14. A.M. Frens, M.T. Bennebroek, J. Schmidt, W.M. Chen and B. Monemar, *Phys. Rev.* B**46** (1992) 12316.
15. K. Morigaki, *Japanese J. Appl. Phys.* **22** (1983) 375.
16. J. Schmidt, D.A. Antheunis and J.H. van der Waals, *Molecular Physics* **22** (1971) 1.
17. A.M. Frens, M.E. Braat, J. Schmidt, W.M. Chen and B. Monemar, *Phys. Rev.* B**52** (1995) 8848.
18. W.M. Chen, M. Singh, B. Monemar, A. Henry, E. Janzén, A.M. Frens, M.T. Bennebroek and J. Schmidt, *Phys. Rev.* B**50** (1994) 7365.
19. W.M. Chen and B. Monemar, *J. Appl. Phys.* **68** (1990) 2506.
20. W.M. Chen and B. Monemar, *Appl. Phys.* A**53** (1991) 130.
21. I.A. Buyanova, W.M. Chen, A. Henry, W.X. Ni, G.V. Hansson and B. Monemar, *Appl. Phys. Lett.* **67** (1995) 1642.
22. W.M. Chen, I.A. Buyanova, A. Henry, W.X. Ni, G.V. Hansson and B. Monemar, *Appl. Phys. Lett.* **68** (1996) 1256.
23. W.M. Chen, I.A. Buyanova, W.X. Ni, G.V. Hansson and B. Monemar, *Phys. Rev. Lett.* **77** (1996) 4214.
24. I.A. Buyanova, W.M. Chen, G. Pozina, B. Monemar, W.X. Ni and G.V. Hansson, *Appl. Phys. Lett.* **71** (1997) 3676.
25. W.M. Chen, I.A. Buyanova, W.X. Ni, G.V. Hansson and B. Monemar, *Appl. Phys. Lett.* **70** (1997) 369.
26. W.M. Chen, B.Monemar, E. Janzén and J.L. Lindström, *Phys. Rev. Lett.* **67** (1991) 1914.
27. J. Köhler, J.A.J.M. Disselhorst, M.C.J.M. Donckers, E.J.J. Groenen, J. Schmidt, W.E. Moerner, *Nature* **363** (1993) 242.

15. ODMR of defects in semiconductors

28. J. Wrachtrup, C. von Borczyskowski, J. Bernard, M. Orrit, R. Brown, *Nature* **363** (1993) 244.
29. D. Gammon, S.W. Brown, E.S. Snow, T.A. Kennedy, D.S. Katzer, D. Park, *Science* **277** (1997) 85.
30. D. Rugar, C.S. Yannoni, J.A. Sidles, *Nature* **360** (1992) 563.

Index

β-proton, 31
(-(interactions, 292
(biogenic) hydrocarbons, 327
(MuCH2CCl2() sorbed in silica-gel and in kaolin, 309
(-radicals, 295
(TMS+), 172
[60]fullerene, 507
'2 + 1' pulse sequence, 547
"light" molecular fragments, 129
"modified Heinzer" method, 5
"spin-other-orbit" (SOO) terms, 275
"stretched-exponential" decay, 105
"surface science", 330
"unrelaxed" radical cations, 378
"zero-order regular approximation" (ZORA) method, 269
1,1,2-trimethylallyl, 308
1,1-dichloroethene, 305
1,1-dichloroethyl radicals, 305, 309
^{14}N quadrupole coupling, 11
1-Me-cSiC5-2,2-d2+, 187
2,5-dimethylcyclohexadienyl radicals, 320
2+1 ESE, 73
^{23}Na hyperfine and quadrupolar interactions, 12
2D effect, 189
2D electron spin nutation spectroscopy, 469
2D isotope effects, 154, 185
2D mass effects, 178
2-mm-band CW-EPR, 105
2p- and 3p-ESE experiments, 49
2p echo decays, 87

2p-ESE, 44
2p-ESE decay, 66
2p-ESEEM, 75
3p-ESE, 51
3p-ESEEM, 78
90o jump motion, 325
A1-lines, 156
ab initio, 241
ab initio and density functional methods, 268
ab initio calculations, 272
ab initio MO calculations, 172, 175, 183, 186
acetals, 395
activation energies, 27, 107
activation energy, 28
activation entropies, 315
activation entropy, 323
adiabatic, 117
adiabatic contribution, 142
adsorption, 33
alanine, 565, 567, 569, 570, 572, 574, 575
Alanine, 571, 576
alfa-proton, 127
A-lines, 159
alkane radical cations, 187, 371, 393
alkyl-substituted cyclohexanes, 182
allowed and "forbidden" spin transitions, 614
allowed |M=±1, 620
AMFI SO operator, 280
amine molecules, 28
amino acid residues, 296
amino acids, 240
aminoacid residues, 145
amplitude of libration, 110

analytical exact expressions, 411
analytical expressions, 408, 415, 426, 431, 432, 433, 435
Analytical expressions, 431
anisole, 324
anisotropic (dipolar) muon hyperfine coupling, 307
anisotropic ESR spectroscopy, 412, 417
anisotropic hyperfine coupling constants, 243
annealing, 621
antiaromatic annulenes, 506
antisymmetric tensor, 420
apical ("a") site, 173
apical "a" site, 170
approximation, 272
Ar, 156, 159
Ar matrix, 162, 163
argon, 377, 385, 392
aromatic hydrocarbons, 372
asymmetrically distorted structures, 174
atmosphere, 303
atmospheric aerosols, 308
atomic-meanfield approximation, 273
barrier height., 106
basis set, 244
benzaldehyde, 320, 323
benzene, 320, 329
Benzene, 322
benzene radical cation, 385
Benzoylenebenzene, 471
Berry phase, 464, 466
beta protons, 104
beta-protons, 103
biological electron transfer processes, 288
biological radicals, 240

biradical, 419, 420, 421, 422, 437, 441, 442, 443, 444, 445, 446, 447, 448
Biradical, 439
blind spots, 79, 83
Bloch, 42
Boltzman distribution, 179
Boltzmann distribution, 184
Boltzmann factor, 440
Bonding at surfaces, 33
bone, 565, 579, 581, 582
bones, 591
boson, 157
Breit-Pauli (BP) Hamiltonian, 269
$C(CH_3)_4+$, 169
C_2 symmetry, 534
C_{2h} distorted structure, 183
C_2H_5 radical, 167
C_{2v} structure, 171, 176
C_{2v} symmetry, 172
C_{60}, 507
C_{60} fullerene, 471
Ca^{2+}-depletion, 545
cages, 320
calibration graph, 571, 592
calix[4]arene-based, 421
calix[4]arene-based biradical, 419, 422
canonical orientations, 418, 421, 433, 435
canonical peaks, 422, 432
carbenes, 501
carbonaceous aerosol, 322
carrier recombination, 621
Carr-Purcell-Meiboom-Gill, 64
catalytic processes, 331
cation-exchanged zeolite X, 315, 317, 320
cavity-type hosts, 374
CD_2H_2+, 170
CD_3 radical, 159
CD_3 Radical, 157

Index 629

CD3H+, 170
CD3OCHD2+, 178
CDH3+, 170
cellulose, 565, 580, 583
Cellulose, 584
CH2D, 158
CH2DF+, 180
CH3 radical, 190
CH3 Radical, 156
CH3 radicals, 166
CH3OCH2D+, 177
CH3OCH3+, 177
CH3OCHD2+, 177
CH3SiD2, 190
CH4+, 170
channels, 320
charge transfer, 621
CHCl=CCl2, 329
CHD2, 158
Chlorophyll triplet state, 550
Chlorophyll Z (ChlZ), 552
classical free rotator, 314
Clay particles, 308
coherences, 51
collective excitation, 409
columnar motif, 471
component tensor, 414, 415, 427
computational scheme, 242, 251
computational time, 210, 212, 213, 215, 219
computational times, 205, 212, 219
conducter-like polarized continuum method, 250
conformation change, 98
conformational changes, 135
conformational defects, 397, 398
contact coupling, 103
contact term, 138
convergence of DFT-based methods, 247
Coppinger's6 biradicals, 491

corannulene, 507
Corba, 199, 201
coronene, 507
correlation functions, 117
correlation time, 100, 114
coupled cluster, 352
coupled-perturbed Hartree-Fock treatments, 274
CPMG, 64, 67
crystal field, 348
cSiC5-2,2,6,6-d4+, 187
Curie, 502
Curl's equation, 279
CW, 610
CW ODMR, 620
CW zero-field ODMR, 613
CW-EPR, 4
Cyclic radicals, 25
Cyclohexadienyl, 320
cyclohexadienyl radical, 322
cyclohexadienyl radicals, 315, 317
cyclopentanediyl-based hydrocarbons, 410
Cytochrome b559, 535
D···D2 pair, 165
D3d, 181
D3h symmetry, 176
Davies ENDOR, 83
D-E strain, 218
decacyclene, 471
decay rates, 611, 617
defect symmetry, 621
defects, 601
degenerate HOMO, 181
delayed ODMR, 620
delocalized defects, 298
denaturation, 145
density matrix, 47, 443
density-functional theory, 271
dephasing, 146

deprotonation of radical cations, 388
desaturation, 132
desorption, 33
deuterated cyclohexane (cC6) radical cations, 182
deuterated methane radical cations, 170
deuterium (2D) isotope effects, 177
deuterium isotope effects, 172
Deuterium isotope effects, 162
deuterium labelled radicals, 153
Deuterium labelling, 154
dicarbenes, 495, 513
Diffuse functions, 245
DIFFUSION, 31
diffusion limit, 125
diffusional motion, 104
dimensionality, 412, 424, 472, 473
dimerization, 509
Dimethylether cations, 177
dinitrenes, 495, 513
diphenothiazine, 504
dipolar broadening, 472
dipolar spin wave, 409
dipole dipole, 202
dipole interaction, 539
dipole-dipole, 198, 229, 233
Dirac equation, 269
direct process, 57
disjoint, 501
Distance determination, 537
distant nuclei, 131
distribution, 204, 210, 211, 214, 215, 217, 218, 227, 230, 234
distributions, 218, 234
Distributions, 234
dosimeters, 575
Dosimetry, 592

Double Electron-Electron Resonance, 73
double resonance, 616
doubly-disjoint, 514
Douglas-Kroll-Hess (DKH) transformation, 269
drift diffusion model, 96
drug sterilization, 587
drugs, 587, 589
D-tensor, 604
Dual mode EPR, 542
dynamical properties, 611
dynamics, 96
echo, 41
echo decay, 53
echo detected EPR, 64
EchoEPR, 64, 68
Eenergy level, 224
effective, 410
effective fine-structure tensor, 420
effective spin Hamiltonian, 268, 420, 421, 427, 452
Effective spin Hamiltonian, 412
effective viscosity, 326
effects, 249
eigenenergy method, 426
eigenfield, 207, 208, 217
Eigenfield, 207, 213, 214
eigenfield approach, 426
eigenfield equation, 429, 432, 433, 434
eigenfield matrix, 426, 433
eigenfield method, 426, 428, 430, 431, 434, 435
Eigenfield method, 428
EIGENFIELD METHOD, 425
eigenvalues, 207, 208, 210, 211, 213, 214, 215
eigenvectors, 207, 208, 213, 215
electron correlation, 245, 284
electron paramagnetic resonance, 337

electron scavenger, 377
electron spin echo (ESE), 540
electron spin resonance (ESR), 601
electron transfer, 532
electron transfer chain, 294
electron Zeeman, 198, 233
Electron Zeeman, 202, 228, 232, 233
electronic Zeeman, 602
Electron-nuclear multiple resonance spectroscopy, 410
E-lines, 156, 158
Emergency, 592
emergency dosimetry, 591
enamel, 591
Enamel, 580
ENDOR, 74, 423, 475, 543
ENDOR enhancement, 131
ENDOR frequencies, 16
ENDOR intensity, 16
ENDOR spectra, 15
ENDOR transition, 9
energy level, 206, 214
Energy level diagram, 213
Energy level Diagram, 224
energy level diagrams, 225
Energy level diagrams, 198, 201, 224, 235
Energy level Diagrams, 224
enthalpies ((H‡) and entropies ((S‡) of activation, 317
environmental, 249
environmental effects, 279
EPR cryostat, 368
EPR dosimetry, 565, 567, 570, 571, 589, 593
EPR Dosimetry, 577
EPR measurables, 96
equatorial ("e") site, 173
equatorial "e" site, 170
error analysis, 14

ESEEM, 74, 543
ESR, 308
ethyl benzene, 383
exact analytical formulae, 431
Exact analytical formulae, 434
excess energy, 366
exchange, 198, 231, 232
Exchange, 202
exchange correlation times, 122
exchange interaction, 413, 437, 438, 441, 442, 445, 448
EXCHANGE INTERACTION, 436
exchange interactions, 409, 443, 455, 473
Exchange interactions, 436
exchange narrowing Lorentzian line shapes, 472
exchange-correlation energy functional, 246
exchange-correlation functional, 274
exchange-coupled magnetic systems, 407, 408
exchange-coupled transition metal ion clusters, 409
exchanges, 171
excited high-spin multiplet states, 479
excited state, 610, 613
Excited states, 606
exclusion of EPR transitions, 128
Extended Time Excitation, 97
extra lines, 412, 417, 418
Extra lines, 417
extreme narrowing, 115
faujasites, 315
Fermi contact analysis, 351
fermion, 158
Ferrimagnetic, 443
ferrimagnetics, 408, 443, 444
ferrimagnets, 444, 445

Ferrimagnets, 443
ferromagnetic microstructures, 409
ferromagnetism, 491
field segmentation, 197, 201, 206
Field-Sweep Pulse-Detected EPR, 108
fine structure constants, 410, 415, 416, 417
fine structure ESR, 407, 415, 432
fine structure tensors, 416, 471
fine-structure, 604
fine-structure constants, 408, 412, 418, 461, 471, 479
fine-structure ESR, 407, 408, 409, 411, 412, 416, 417, 418, 421, 422, 435, 462, 468, 469, 479
Fine-structure ESR, 415, 468
fine-structure tensors, 468
First order analysis, 11
flip-flop mechanism, 139
foodstuffs, 584
forbidden transitions, 230, 412, 417, 418, 421, 425, 435, 495
forbidden $\Box M=\pm 2$ transitions, 621
Fortran 77 programme, 17
Fortran 77/90, 25
free energies ((G‡), 317
Free Induction Decay, 43
free radicals, 305
free rotor, 326
Freon matrices, 371
Freon-11, 371
Freon-113, 371
frequency swept, 204
frozen biological samples, 141
FTIR spectroscopy, 380
full Breit-Pauli SO Hamiltonian, 280
g-A strain, 218, 234
galactose oxidase, 285

gas phase, 33
gas-phase microwave data, 278
gauge-including-atomic-orbitals (GIAO56), 278
gauge-invariant theory, 271
Gaussian, 203, 210, 211, 212, 218, 472
Gaussian distribution, 22
Gaussian envelope, 25
Gaussian-type functions, 244
generalized gradient approximation, 276
generalized gradient approximations, 246
geometry optimizations, 251
GGA-corrected exchange functionals, 246
glassy materials, 141
goodness of fit, 219
g-shift, 472
g-shift tensors, 292
g-tensor, 604
g-tensor anisotropy, 268
G-value, 568, 569
H↑ ↑CH3 radical pairs, 160
H···D2, 164
H···H2 pair, 163
H···HD, 164
Hahn Echo, 44
halocarbon matrix, 25
halogenated VOC, 309
Hartree-Fock approach, 272
Hartree-Fock calculations, 244
Hartree-Fock limit, 245
Heinzer model, 26
Heinzer program, 29
Heisenberg antiferromagnets, 472
Heisenberg exchange, 437, 445
Heisenberg-Dirac types exchange, 410
Heterocyclic radicals, 27
heterospin molecules, 511

hexaazahydrocoronene, 507
hexabenzocoronene, 507
hexaradical, 518
H-H2 pair, 165
high field approximation, 496
High frequency (95-140 GHz) ODMR, 623
High frequency EPR, 542
high resolution ESR, 160
HIGH RESOLUTION ESR, 155
High resolution ESR spectra, 156, 166
high spin chemistry, 407, 408, 435, 458, 469
HIGH SPIN CHEMISTRY, 451
higher order terms, 462, 465
higher-order fine-structure terms, 413, 417, 435, 464
higher-order perturbation, 425, 428
higher-order terms, 416
high-field high-frequency EPR spectroscopy, 268
high-field/high-frequency ESR, 407, 408, 423, 424, 435, 453, 454, 456, 457, 458, 462, 463, 468, 474
High-field/high-frequency ESR, 416, 418, 463
HIGH-FIELD/HIGH-FREQUENCY ESR, 451
high-LET radiation, 576
high-resolution ESR, 154
High-resolution ESR spectra, 167
high-resolution ESR spectroscopy, 165
high-spin ESR spectroscopy, 410, 411
high-spin molecular clusters, 408, 411
high-spin nitrene chemistry, 412
hindered methyl, 130

hindered methyl fragments, 135
hole transfer, 382
homogeneous broadening, 472
Hooke and Jeeves, 202, 220, 223
Hot fragmentation, 389
hybrid, 435
HYBRID, 425
hybrid eigenfield approach, 426, 428
hybrid eigenfield method, 430, 436
Hybrid eigenfield method, 407
hybrid functionals, 246, 277
hydrocarbon-based high-spin clusters, 471
hydrogen abstraction reactions, 190
HYDROGEN ATOM - HYDROGEN MOLECULE PAIR, 162
HYDROGEN ATOM ABSTRACTION VIA TUNNELLING, 189
hydrogen bonding, 288, 291, 422, 468
hydrogen-bonded molecule-based high-spin clusters, 468
Hydrogen-bonded molecule-based high-spin clusters, 468
hydroxyl radicals, 303
hydroxyproline-derived radicals, 254
hyperconjugation, 285
hyperfine, 198, 218, 232
Hyperfine, 202, 228, 232
hyperfine coupling tensors, 8
hyperfine interaction, 242, 608, 614
hyperfine interactions, 621
hyperfine structure, 411, 421, 475, 476, 604
hyperfine-structure, 417, 604

hyperfine-structure ESR, 478
HYSCORE, 79
hystereses, 409, 458
Ice surfaces, 329
IDAS, 572, 576
identification, 604
immobilized nitroxide, 146
individual-gauges-for-localized-orbitals (IGLO57), 278
inert matrices, 278
inertial effects, 125
inhomogeneous broadening, 118
instantaneous diffusion, 65, 88, 140, 145
instrumentation, 623
intensity, 603
interchange, 184
intermediate exchange coupling, 415
intermediate exchange-coupling, 415
Internal motion, 25
internal standard, 572
Internet calibration service, 575
intramolecular exchange, 123, 188
intrinsic broadening, 121
Inversion recovery, 59
inversion recovery, saturation recovery, 56
irradiated, 584, 587
irradiated alanine, 17
irradiated foods, 583
irradiated foodstuffs, 565, 579
irreducible tensor, 414
isochromats, 43
isoelectronic pairs, 128
isosteric enthalpies of adsorption of benzene, 317
isotope effects, 190
isotope labelling, 545
isotopic substitution, 125

isotropic hyperfine coupling constant, 243
iterative fitting procedure, 13
Jahn-Teller (J-T), 154
Jahn-Teller (J-T) distortion, 168
Jahn-Teller (J-T) effects, 182
Jahn-Teller distorted structures of cyclohexane cation, 181
Jahn-Teller distortions, 507
Jahn-Teller species, 385
J-T effects, 179
jump diffusion model, 133
kaolin, 305, 328
ketone-based dianion, 469
Kivelson, 217
Kohn-Sham calculations, 275
Kohn-Sham orbitals, 276
Kok's S-state, 535
krypton, 392
L-(-alanine, 253
Landau-Zener model, 466
large-amplitude motions, 102
lattice temperature, 113
least squares, 20
least squares fitting, 14
libration, 98
libration amplitude, 111
lifetime broadening, 117, 121, 133, 134
lifetimes, 617
limitations of the ODMR, 606
line shape, 22
linear response theory, 273, 472
lineshape, 203, 204, 212, 214, 223, 230
Lineshape, 200
lineshape function, 24
linewidth, 617
Liouville representation, 443
Liouville space formalism, 448
Liouville-von Neumann equation, 49

Index

local concentration, 145
local field, 138
local spin density approximation, 246
lock-in detection, 610
long time tail, 424
longitudinal relaxation, 114
LONGITUDINAL RELAXATION, 56
long-range order, 108
long-time tailing, 473
looping transitions, 210, 227, 232, 233
Lorentzian, 25, 203
Lorentzian absorption, 22
Lorentzian function, 427
Lorentzian line shapes, 440, 472
low dimensional magnetic systems, 410
low temperature, 153, 189
low-dimensional magnetic materials, 407
low-dimensional magnetic systems, 473
low-dimensional spin assemblages, 424
Low-temperature techniques, 367
LTT, 424, 473
macromolecules, 396
magnetic dipole-dipole interaction, 412, 419, 604
magnetic field modulation, 610
magnetic fields, 608
magnetic properties, 339
magnetic resonance force microscopy (MRFM), 623
magnetic susceptibility measurements, 410, 436, 453, 457
magnetic-dipole interactions, 243
magnetization, 41
malonic acid radical, 24

Marquardt, 360
matrix, 156
matrix catalysis, 389
matrix diagonalisation, 15
matrix isolation, 127, 338, 368, 376
Matrix isolation, 369
Matrix-assisted deprotonation, 386
matrix-isolation EPR, 279
McConnell model II, 507
$Me_2C=O$, 329
mean total 1H hf splitting, 184
mean total hf splitting, 178
Methyl fluoride cations, 179
methyl group conformation, 177
methyl- substituted nitroxides, 129
methyl tert-butyl ether, 372
methylal, 392, 394
Methylsilane (CH_3SiH_3), 190
methylsilyl radical (CH_3SiH_2), 190
methyl-type rotors, 126
microcrystallite, 108
microheterogeneous systems, 398
micropores, 323
microwave induced delayed phosphorescence (MIDP), 611
microwave modulation, 610
Microwave power effects, 23
microwave pulse, 611
microwave saturation, 552
MIDP, 618
Mims ENDOR, 83
mixed CH_2D and CHD_2 rotors, 128
mixing of states, 614, 621
models for approximating surrounding molecules, 250
modified Heinzer method, 28
modulation depth, 76

modulation frequency, 22
molecular beam epitaxy (MBE), 621
molecular dynamics, 249, 356
molecular dynamics approaches, 299
molecular high-spin clusters, 407, 409, 411, 412, 419, 453, 469, 471
molecular magnetism, 468
molecular orientation, 548
molecule-based exchange-coupled systems, 468, 478
molecule-based high-spin clusters, 451, 469
molecule-based magnetics, 407, 408, 409, 410, 439
molecule-based magnetism, 443, 458, 478
Møller-Plesset, 352
moment, 471
moment of inertia, 126
Monte Carlo, 360
Monte-Carlo, 222
montmorillonite, 308, 328
mordenite, 320
mosaic misorientation, 197, 201, 202, 210, 212
Mosaic misorientation, 210
Mosaic Misorientation, 210, 211
motif of herringbone types, 471
motional correlation time, 327
motional effects on the g-tensors, 299
motional model, 112
Motions and spin-lattice relaxation:, 58
m-phenylene, 501
multibilayer, 143
multifrequency, 202, 220, 227
Multifrequency, 198, 229
multiline ESR, 475

multiline signal, 475, 476
Multiline signals, 475
multiple resonance, 411
multiplets, 446, 450, 468
multi-reference configuration-interaction (MRCI), 273
muon, 304
muon spin rotation, 304
muonium, 304
Muonium, 305
MuSR, 304
mutagenesis, 545
m-Xylene, 500
Na-mordenite, 320
Na-ZSM5, 320
NBMO's, 500
NBMOs, 501
neon, 376, 385
neon matrix, 170
new radiation sensitive materials, 577
nitronylnitroxide, 493
nitroxide spin labels, 283
Nitroxide Spin Labels, 295
nitroxides, 510
NO2, 167
non selective, 44
non-disjoint, 501
non-heme iron, 549
non-Kekulé, 500, 501
non-linear Arrhenius plot, 191
nonlinear least squares, 219
non-linear least squares, 15
Non-linear least squares method, 11
nonperturbative, 124
nonradiative defect, 621
nonradiative defects, 621
Nonradiative defects, 621
nonvanishing orbital angular momenta, 468, 478

Index

non-vanishing orbital angular momenta, 468
nuclear spin I=3/2, 617
nuclear spin operators, 446
nuclear spin-rotation couplings, 154, 158
nuclear Zeeman, 198
Nuclear Zeeman, 202
nutation angles, 140
nutation or Rabi frequency, 42
nutation spectra, 469, 470, 479
nutation spectroscopy, 453, 469, 479
octahedral symmetry, 413
ODMR, 602, 607, 608
ODMR linewidth, 608
off-axis extra lines, 412, 417
off-principal-axis, 412, 417
off-principal-axis absorption, 417
Off-principal-axis absorption, 412
Off-principal-axis extra lines, 407
off-principal-axis lines, 412, 417
o-H2, 165
oligoarylmethanes, 516
oligopolycarbenes, 410
oligopolynitrenes, 410
one-center approximation, 285
one-component approach, 282
Onsager model, 250
optical axis, 143
optical detection, 608
optical emission, 602
optical excitation and detection, 606
optically detected magnetic resonance (ODMR), 601
Optically Detected Magnetic Resonance (ODMR), 543
optically detected nuclear magnetic resonance (OD-NMR), 623

optimisation, 197, 201, 219, 220, 223, 224, 234
Optimisation, 200, 202, 219, 223, 234
Orbach-Aminov process, 57
orbital angular momenta, 408, 410, 468
order parameter, 109, 112
organic ferrimagnets, 444, 451
Organic glasses, 370
orientation dependent relaxation, 142
Oriented membranes, 544
ortho-D2 (o-D2), 164
out-of–phase, 426
Out-of-phase ESEEM, 74
oxidation, 303
oxophenalenyl, 478
oxo-verdazyl, 493
Oxygen evolving PS II membranes, 544
oxygen-evolving complex (OEC), 457, 474
Oxygen-evolving high-spin complexes, 474
P donor, 621
P680, 533
P680+, 552
P860, 549
Para-D2 (p-D2), 164
para-hydrogen (p-H2), 165
parallel microwave polarization, 435
parallelisation, 197, 201
Parallelisation, 218, 219
paramagnetic, 197, 198, 203, 217, 235
paramagnetic atoms or small ions encapsulated in cages, 298
paramagnetic impurities, 119
paramagnetic susceptibility measurements, 447

Partial orientation, 167
Pauli principle, 154, 158
Pauli repulsion, 350
P-donor, 621
PELDOR, 540
pentadienyl, 506
pentaradicals, 517
peroxyl radicals, 241
perturbation, 202, 204, 207, 208, 210, 217
perturbation approach, 425, 426
perturbations, 208
p-H2, 166, 168
phase memory decay, 133
phase memory time, 63
phenalenyl radicals, 471
phenoxyl radical, 241
Phenoxyl Radicals, 283
photochemistry, 373
photo-ESR, 605
photoluminescence (PL), 601
photosynthesis, 136
photosynthetic reaction centers, 288
photosystem I, 298, 531
photosystem II, 457, 474, 476, 531
Photosystem II, 475, 476
PHOTOSYSTEM II, 474
Photosystems II., 477
phylloquinone, 288
pinene, 327, 329
pinenes, 327
PL intensity, 607, 617
Plastoquinone QA, 535
plastoquinone-9, 288
polarisability, 392
polarization, 607
polarization functions, 245
polarization transfer, 83
polarizations, 603
polyarylmethylradicals, 518

polyethylene, 396
polyoriented sample, 107
polyradicals, 436
Pople diagram, 245
populating rates, 611, 620
population, 616, 617
population change, 607
populations, 617
porous carbon, 322
post-irradiation dosimetry, 580
post-radiation dosimetry, 589
post-radiation dosimetry., 590
potential barrier, 107
power dependence, 19
power parameter, 25
principal components, 242
principal g-values, 7
product operators, 54
Program packages, 425
program softwares, 425
progressive saturation, 56
protein environments, 291
proton transfer, 388
pseudo J-T distortion, 154
pseudo J-T effect, 188
Pseudorotation, 102
pulse, 44
pulse delays, 141
pulsed ELDOR, 73
Pulsed ENDOR, 82
pulsed ESR, 424, 469, 473, 479, 623
PULSED ESR, 451
Pulsed experiments, 41
pulsed optical excitation, 611
p-xylene, 320
pyramidal inversion, 102
QA^-, 565
QTM, 461, 462, 463, 464
quadratic, 215, 221, 223
Quadratic, 202, 220, 221
quadrupole, 198, 614

Index

Quadrupole, 202, 232
quadrupole interaction, 608
quantitative g-tensor calculations, 298
quantum chemical programs, 268
quantum effects, 127
Quantum effects, 153
quantum solid, 154
quantum spin mixing, 415, 416, 457
quantum tunneling, 458, 461, 464, 465, 466
quantum tunneling of spin magnetization, 409
quartet, 493
quenching of the hyperfine interaction, 617
quinone radicals, 241
quintet, 493
quintet state, 513
radiative decay, 607
radiative decay rate, 616
radiative decay rates, 611
radiative defects, 613
Radiative defects, 612
radical cations, 365
radical concentration, 139
Radical pair, 159
radical pair mechanism, 550
radical pairs, 33
Radical structure, 6
radicals on surfaces, 28
Raman process, 57
randomly orientated, 197, 198, 199, 201, 202, 203, 204, 206, 211, 223, 227, 235
rare gas matrices, 376, 385
rare-gas-matrix, 127
rate equations, 611
rate of reaction, 122
ray irradiation, 471

Rayleigh-Schrödinger perturbation theory, 270
Reaction selectivity, 190
recombination, 601
recombination center, 621
reconstituted reaction centers, 294
redox switchable, 513
relative radiative decay rates, 620
relativistic mass correction to the spin-Zeeman term, 271
relaxation function, 423, 424, 472, 473
relaxation functions, 473
relaxation matrix, 442, 448
relaxation of excess energy, 390
relaxation pathways, 131
reorientation time, 314, 326
residual anisotropy, 111
resolution, 606
resonance frequency, 244
resonant field, 204, 206, 207, 208, 209, 211, 213, 214, 215, 216, 217
resonant fields, 214, 215, 216
restricted open-shell HF (ROHF), 245
resultant tensors, 414
ribonucleotide reductase, 297
role of clouds in ozone depletion, 331
rotational angular momentum, 130
rotational constant B, 128
rotational diffusion, 168
Rotational diffusion, 31
S0-state, 536
S1-state, 536
S2 state, 535
S3-state, 536
saturation, 5
saturation curve, 20
Saturation properties, 17

Saturation recovery, 59
Schlenk hydrocarbon3, 491
Schonland method, 6
Schonland procedure, 4
Schonland type analysis, 14
S-Cu complex, 612, 613
S-Cu defect, 615, 618, 620
second-order effect, 614
second-order spin-orbit/orbital Zeeman cross term, 270
secular equation, 431, 433
secular equations, 433
selective, 44
Selective hole burning, 538
selectivity, 606
self-calibrated, 572, 574, 575
self-calibrated alanine, 573
self-calibrated dosimeter, 578
self-calibrated sugar dosimeter, 578
self-consistent reaction field, 250
semiconductors, 601
semiempirical approaches, 272
semi-empirical MO calculations, 284
semiquinone radical anions, 283, 290
semisecular, 147
sensitivity, 606
short range, 108
Si, 620, 621
SI unit for hyperfine coupling constants, 244
Si(CH3)2(CD3)2+ (TMS-d6+), 172
Si(CH3)3CD3+ (TMS-d3+), 172
Si(CH3)4+, 172
Si/SiGe, 621
silacyclohexane (cSiC5) radical cation, 187
silacyclohexane radical cations, 189

silicalite, 320
silicon, 612, 613
simplex, 221, 222
Simplex, 202, 220, 221, 223, 360
simulated annealing, 220, 222, 223
Simulated annealing, 222
Simulated Annealing, 202, 222
single crystal, 197, 198, 199, 201, 202, 203, 235
Single Crystal, 200, 233
single crystal measurements, 6
Single crystals, 544
single molecule, 623
single-component, 444
single-component ferrimagnetics, 444, 445
single-molecule magnets, 409
Single-molecule magnets, 458
Single-Molecule Magnets, 458
single-spin detection, 623
single-spin ODMR, 623
singlet-quintet complete mixing, 416
site-selective reactivity, 393
skeleton bond fragmentation, 389
slow exchange, 133
slow passage, 22
slow tumbling, 123
slow-motional spectra, 32
SO operator, 270
SO pseudopotentials, 276
solid-state spectra, 268
solution ESR spectroscopy, 410, 439
SOLUTION ESR SPECTROSCOPY, 436
solvent interactions, 291
sophe, 199, 428
Sophe, 197, 199, 200, 201, 202, 207, 208, 210, 212, 217, 219, 220, 223, 225, 234, 235

Index 641

SOPHE, 201, 205, 206, 208, 209, 210, 211
Sophe Interpolation, 202
SOPHE interpolation, 197, 201, 206, 209, 212
SOPHE Interpolation, 211
SOPHE partition scheme, 197, 201, 205
spatial resolution, 623
spectral density function, 116
spectral diffusion, 53, 56
spectral resolution, 623
spectral simulation, 411, 412, 415, 425, 437, 448, 452, 479
Spectral simulation, 440
SPIN HAMILTONIAN, 227
spin alignment, 492
spin contamination, 245
spin density, 243, 296
spin dependent recombination, 610
spin dephasing, 64
spin diffusion, 120, 147
spin dynamics, 408, 409, 410, 423, 424, 473
spin flips, 120
spin Hamiltonian, 198, 201, 203, 206, 211, 219, 220, 223, 224, 227, 231, 232, 234, 235, 604, 613
Spin Hamiltonian, 200, 201, 229, 231
spin Hamiltonian parameters, 617
spin Hamiltonians, 198
Spin lattice relaxation, 537
spin packet, 107
spin packets, 44, 137
spin polarisation, 500
spin polarization, 501
spin polarization model, 248
spin polarized radical pairs, 541, 552
spin population, 243
spin probe, 167
spin quantum mixing, 415, 416
spin transition, 607
spin triplet, 608, 613, 615
spin waves, 409
spin-excitation transfer, 139
spin-Hamiltonian, 410
spinics, 408
spin-lattice interaction, 313
Spin-lattice relaxation processes, 57
spin-lattice relaxation rates, 611
spin-orbit (SO) coupling, 269
spin-orbit coupling, 350, 408, 416, 420, 479
spin-orbit interaction, 604
spin-other-orbit (SOO) term, 271
spin-phonon processes, 113
spin-polarization effects, 294
spin-restricted formalism, 275
spin-spin distance determination, 73
spin-spin interaction, 412, 418, 420
spin-spin interactions, 65
spin-spin interactions (spectral diffusion), 88
spin-spin relaxation, 114
spin-Zeeman gauge correction terms, 271
state mixing, 227, 232
State mixing, 232, 233
State Mixing, 232
states, 493
steady-state conditions, 611
stimulated echo, 51
Stimulated or 3p-ESE, 60
Stone's, 272
Stone's perturbation model7,8, 284
stopped methyl rotor, 107

stopped quantum rotors, 130
stretched exponential, 87, 147
strong exchange-coupling limit, 413
strong narrowing, 114, 116
substituent effects, 288
sugar, 577, 591
sulfur-containing radicals, 296
sum-over-states (SOS) approach, 273
superexchange interactions, 455, 459
super-hf coupling, 163
superhyperfine, 198, 202, 234
superhyperfine interaction, 353
superlattices, 409
supermolecular models, 250
superparamagnets, 408, 409, 461
suspended particles, 303, 331
T1, 313
T1 minima, 313
T2, 313
Td symmetry, 168, 172
techniques, 331
temperature, 502
temperature dependence, 29
Temperature dependent 1H hf splittings, 178
temperature dependent ESR spectra, 184
Temperature Dependent ESR Spectra, 188
templates, 28
tensor-based analyses, 412, 471
tensor-based expressions, 414
tensor-based spectral simulation, 423
terpenes, 322
Terpenes, 327
tetrahydrofuran, 387
tetramethylenebenzene, 502
Tetramethyleneethane, 501

Tetramethylsilane radical cations, 172
the local fields, 139
The performance of various levels of conventional theory, 248
theoretical calculations, 241
theoretical spectral simulation, 412, 419, 421, 436
Theoretical spectral simulation, 419
thermal activation, 493
thermal quenching, 621
thermally activated, 116
thianthrene, 513
three-dimensional imaging, 623
time correlation function, 423, 424, 472
time resolution, 623
Time resolved EPR, 542
time-correlation function, 424, 473
time-dependent interactions, 442
time-dependent perturbations, 442
Time-dependent perturbations, 437
time-resolved ODMR, 610, 611
Time-resolved zero-field ODMR, 617
times T2 and T1, 42
tissue-factor/factor-VIIa protein complex, 296
TME, 502
TMM, 502
toluene, 323
Toluene, 323
topological symmetry, 410, 412, 478
torsional motion, 143
torsional oscillations, 129
torsional states, 135
total electron spin, 439, 440, 441, 448

total spin, 441, 455, 461
transient nutation, 494
transition, 469
transition moment, 430, 436, 479
transition moment spectroscopy, 436, 469
transition probabilities, 408, 413, 416, 417, 427, 430, 431, 434, 435, 615
transition probability, 203, 204, 209, 215, 234
transition roadmaps, 198, 201, 225, 235
Transition roadmaps, 225
Transition Roadmaps, 224
transition surfaces, 198, 201, 214, 217, 225, 235
Transition surfaces, 214, 225
Transition Surfaces, 224
Transition-state theory, 317
translational diffusion, 31
transverse and a longitudinal relaxation, 42
transverse relaxation, 412, 423, 424, 473
TRANSVERSE RELAXATION, 63
triazine, 509
trication, 508, 509
Trimethylenemethane, 501
Trimethylenemethane radical cation, 175
triphenylamines, 509
triphenylbenzene, 507
triphenylene, 507
triphenylmethylenes, 509
triplet, 617, 620
triradical, 443, 444, 445, 446, 447, 448, 451
Triradical, 446
triradicals, 508
Trisphenylaminobenzene, 508

Tris-treatment, 545
tropospheric oxidation of pinenes, 322
tryptophan (Trp) residues, 292
tryptophan radicals, 297
Tschitschibabin's, 491
T-stacked arrangement, 292
T-stacked hydrogen bonding, 294
Tunneling quantum rotor, 130
tunneling splitting, 465, 466
tunnelling, 127, 154
tunnelling frequency, 135
tunnelling reaction, 163
tunnelling-rotor, 130
twist angle, 103
two-component method, 281
two-dimensional electron spin transient nutation spectroscopy, 407, 436
two-site exchange, 122
two-site model, 27
Two-trap model, 381
tyrosine D, 534
tyrosine Z, 534
Tyrosines YD, 546
tyrosyl radicals, 283
ubiquinone-10, 288
uncoupled DFT approach, UDFT, 276
unfolding, 145
unrestricted HF (UHF), 245
unrestricted reorientation, 133
vacancy-oxygen (V-O) complex, 621
vacancy-oxygen complex, 621
van der Waals attraction, 350
variable temperature, 230, 231
Variable temperature, 202, 231, 232
vector model, 41
vibrational averaging, 249
vibronic coupling, 249

vinyl cyclopropane, 372
vinyl monomers, 372
viscous friction, 326
viscous medium, 123
volatile organic compounds, 303
Walker mode, 409
Walker's mode, 409
water oxidizing complex (WOC), 535
W-band (95-GHz) ESR, 623
weak exchange-coupling limit, 415
Wigner-Eckart theorem, 414
X Windows graphical user interface, 197
X-band ODMR, 617, 620
Xemr, 360
xenon, 377, 385, 392
Xepr, 198, 201, 224, 235
XeprView, 197, 198, 199, 200, 201, 220, 234, 235
X-ray radiolysis, 162
Xsophe, 220
XSophe, 197, 199, 200, 201, 209, 217, 220, 224, 225, 226, 234, 235
XSOPHE, 199, 234

X-Windows graphical user interface, 199
Yang's, 491
Zeeman interaction, 268
zeolites, 28, 308
Zeolites, 374
zero field, 608
zero field ODMR, 608
zero field splitting, 494
zero point vibrational energy (ZPVE), 154, 172
zero-field ODMR, 616
zero-field splitting, 604, 613
zero-field splittings, 160
zero-order regular approximation, 274
Zero-point vibrational energy (ZPVE), 183
ZFS, 494
ZPVE, 178, 181
Z-scheme, 530
ZSM5, silicalite, 320
zwitterionic amino acid radicals, 259
zwitterionic isomers of the amino acid radicals, 257
$\Delta m_s = 1$ transition, 160

Progress in Theoretical Chemistry and Physics

1. S. Durand-Vidal, J.-P. Simonin and P. Turq: *Electrolytes at Interfaces.* 2000
   ISBN 0-7923-5922-4
2. A. Hernandez-Laguna, J. Maruani, R. McWeeny and S. Wilson (eds.): *Quantum Systems in Chemistry and Physics.* Volume 1: Basic Problems and Model Systems, Granada, Spain, 1997. 2000   ISBN 0-7923-5969-0; Set 0-7923-5971-2
3. A. Hernandez-Laguna, J. Maruani, R. McWeeny and S. Wilson (eds.): *Quantum Systems in Chemistry and Physics.* Volume 2: Advanced Problems and Complex Systems, Granada, Spain, 1998. 2000   ISBN 0-7923-5970-4; Set 0-7923-5971-2
4. J.S. Avery: *Hyperspherical Harmonics and Generalized Sturmians.* 1999
   ISBN 0-7923-6087-7
5. S.D. Schwartz (ed.): *Theoretical Methods in Condensed Phase Chemistry.* 2000
   ISBN 0-7923-6687-5
6. J. Maruani, C. Minot, R. McWeeny, Y.G. Smeyers and S. Wilson (eds.): *New Trends in Quantum Systems in Chemistry and Physics.* Volume 1: Basic Problems and Model Systems. 2001   ISBN 0-7923-6708-1; Set: 0-7923-6710-3
7. J. Maruani, C. Minot, R. McWeeny, Y.G. Smeyers and S. Wilson (eds.): *New Trends in Quantum Systems in Chemistry and Physics.* Volume 2: Advanced Problems and Complex Systems. 2001   ISBN 0-7923-6709-X; Set: 0-7923-6710-3
8. M.A. Chaer Nascimento: *Theoretical Aspects of Heterogeneous Catalysis.* 2001
   ISBN 1-4020-0127-4
9. W. Schweizer: *Numerical Quantum Dynamics.* 2001   ISBN 1-4020-0215-7

KLUWER ACADEMIC PUBLISHERS – DORDRECHT / LONDON / BOSTON